MATHEMATICS
for Elementary Teachers
with Activity Manual

MATHEMATICS
for Elementary Teachers
with Activity Manual

THIRD EDITION

SYBILLA BECKMANN
UNIVERSITY OF GEORGIA

Addison-Wesley

Boston   Columbus   Indianapolis   New York   San Francisco   Upper Saddle River
Amsterdam   Cape Town   Dubai   London   Madrid   Milan   Munich   Paris   Montreal   Toronto
Delhi   Mexico City   Sao Paulo   Sydney   Hong Kong   Seoul   Singapore   Taipei   Tokyo

Acquisitions Editor: Marnie Greenhut
Executive Project Manager: Christine O'Brien
Editorial Assistant: Elle Driska
Senior Managing Editor: Karen Wernholm
Associate Managing Editor: Tamela Ambush
Senior Production Project Manager: Sheila Spinney
Designer Supervisor: Andrea Nix
Senior Design Specialist: Heather Scott
Digital Assets Manager: Marianne Groth
Media Producer: Christine Stavrou
Executive Marketing Manager: Roxanne McCarley
Marketing Assistant: Kendra Bassi
Senior Author Support/Technology: Joe Vetere
Senior Prepress Supervisor: Caroline Fell
Rights and Permissions Advisor: Michael Joyce
Manufacturing Manager: Evelyn Beaton
Senior Manufacturing Buyer: Carol Melville
Senior Media Buyer: Ginny Michaud
Text Design Studio: Montage
Production Coordination, Composition, and Illustrations: Aptara Corporation

Cover photo: Butterflies, Shutterstock, stylized by Tamara Newman.

For permission to use copyrighted material, grateful acknowledgment is made to the copyright holders on page C-1, which is hereby made part of this copyright page.

Many of the designations used by manufacturers and sellers to distinguish their products are claimed as trademarks. Where those designations appear in this book, and Pearson was aware of a trademark claim, the designations have been printed in initial caps or all caps.

Library of Congress Cataloging-in-Publication Data
Beckmann, Sybilla.
Mathematics for elementary teachers with activity manual / Sybilla
Beckmann—3rd ed.
    p. cm.
Includes bibliographical references and index.
ISBN 0-321-65427-7
    1. Mathematics—Study and teaching (Elementary) I. Title.
QA135.6.B43 2010
372.7'044—dc22
                              2009039944

3 4 5—CRW—14 13 12 11 10

Addison-Wesley
is an imprint of

www.pearsonhighered.com

ISBN-10:   0-321-64694-0
ISBN-13: 978-0-321-64694-1

To Will, Joey, and Arianna

BRIEF CONTENTS

CONTENTS

Throughout the text, each section concludes with Practice Problems, Answers to Practice Problems, and Problems.

FOREWORD

by Roger Howe, Ph.D., Yale University

We owe a debt of gratitude to Sybilla Beckmann for this book.

Mathematics educators commonly hear that teachers need a "deep understanding" of the mathematics they teach. In this text, this pronouncement is not mere piety, it is the guiding spirit.

With the 1989 publication of its *Curriculum and Evaluation Standards for School Mathematics*, the National Council of Teachers of Mathematics (NCTM) initiated a new era of ferment and debate about mathematics education. The NCTM *Standards* achieved widespread acceptance in the mathematics education community. Many states created or rewrote their standards for mathematics education to conform to the NCTM *Standards*, and the National Science Foundation funded large-scale curriculum development projects to create mathematics programs consistent with the *Standards*' vision.

But this rush of activity largely ignored a major lesson from the 1960s' "New Math" era of mathematics education reform: in order to enable curricular reform, it is vital to raise the level of teachers' capabilities in the classroom. In 1999, the publication of *Knowing and Teaching Elementary Mathematics* by Liping Ma finally focused attention on teachers' understanding of mathematics principles they were teaching. Ma adapted interview questions (originally developed by Deborah Ball) to compare the basic mathematics understanding of American teachers and Chinese teachers. The differences were dramatic. Where American teachers' understanding was foggy, the Chinese teachers' comprehension was crystal clear. This vivid evidence showed that the difference in Asian and American students' achievement, revealed in many international comparisons, correlated to a difference in the mathematical knowledge of the teaching corps.

The Mathematical Education of Teachers, published by the Conference Board on the Mathematical Sciences (CBMS), was one response to Ma's work. Its first recommendation gave official voice to the dictum: "Prospective teachers need mathematics courses that develop a deep understanding of the mathematics they will teach." This report provided welcome focus on the problem, but the daunting task of creating courses to fulfill this recommendation remained.

Sybilla Beckmann has risen admirably to that challenge. Keeping mathematical principles firmly in mind while listening attentively to her students and addressing the needs of the classroom, she has written a text that links mathematical principles to their day-to-day uses. For example, in Chapter 4, Multiplication, the first section is devoted to the meaning of multiplication. First, it is defined through grouping: $A \times B$ means "the number of objects in A groups of B objects each." Beckmann then analyzes other common situations where multiplication arises to show that the definition applies to each. The section problems, then, do not simply provide practice in multiplication; they require students to show how the definition applies.

Subsequent sections continue to connect applications of multiplication (e.g., finding areas, finding volumes) to the definition. This both extends students' understanding of the definition and unites varied applications under a common roof. Reciprocally, the applications are used to develop the key properties of multiplication, strengthening the link between principle and practice. In the next chapter, the definition of multiplication is revisited and adapted to include fraction multiplication as well as whole numbers. Rather than emphasizing the procedure for multiplying fractions, this text focuses on how the procedure follows from the definitions.

Here and throughout the book, students are taught not merely specific mathematics, but the coherence of mathematics and the need for careful definitions as a basis for reasoning. By inculcating the point of view that mathematics makes sense and is based on precise language and careful reasoning, this book conveys far more than knowledge of specific mathematical topics: it can transmit some of the spirit of doing mathematics and create teachers who can share that spirit with their students. I hope the book will be used widely, with that goal in mind.

PREFACE

Introduction

I wrote this book to help future elementary school teachers develop a deep understanding of the mathematics they will teach. People commonly think that elementary school mathematics is simple and that it shouldn't require college-level study to teach it. But to teach mathematics well **teachers must know more than just *how* to carry out basic mathematical procedures; they must be able to explain *why*** mathematics works the way it does. Knowing why requires a much deeper understanding than knowing how. For example, it is easy to multiply fractions—multiply the tops and multiply the bottoms—but why do we multiply fractions that way? When we *add* fractions, we can't just add the tops and add the bottoms. The reasons we can in one case and can't in the other are not obvious; they require study. By learning to explain why mathematics works the way it does, teachers will learn to make sense of mathematics. I hope they will carry this "sense of making sense" into their own future classrooms.

Because I believe in deep understanding, this book focuses on *explaining why*. Prospective elementary school teachers will learn to explain why the standard procedures and formulas of elementary mathematics are valid, why nonstandard methods can also be valid, and why other seemingly plausible ways of reasoning are not correct. The book emphasizes key concepts and principles, and it guides prospective teachers in giving explanations that draw on these key concepts and principles. In this way, teachers will come to organize their knowledge around the key concepts and principles of mathematics so they will be able to help their students do likewise.

A number of activities and problems **examine common misconceptions**. Since most misconceptions have a certain plausibility about them, it is important to understand what makes them mathematically incorrect. By examining what makes misconceptions incorrect, teachers deepen their understanding of key concepts and principles, and they develop their sense of valid mathematical reasoning. I also hope that, by studying and analyzing these misconceptions, teachers will be able to explain to their students why an erroneous method is wrong, instead of just saying "You can't do it that way."

Other activities and problems **examine calculation methods that are nonstandard but nevertheless correct**. When explaining why nonstandard methods are correct, teachers have further opportunities to draw on key concepts and principles and to see how these concepts and principles underlie calculation methods. By examining nonstandard methods, teachers also learn there can be more than one correct way to solve a problem. They see how valid logical reasoning, not convention or authority, determines whether a method is correct. I hope that by having studied and analyzed a variety of valid solution methods, teachers will be prepared to value their students' creative mathematical activity. Surely a student who has found an unusual but correct solution method will be discourage if told the method is incorrect. Such a judgment also conveys to the student entirely the wrong message about what mathematics is.

Studying nonstandard methods of calculation provides valuable opportunities, but the common methods deserve to be studied and appreciated. These methods are remarkably clever and make highly efficient use of underlying principles. Because of these methods, we know that a wide range of problems can always be solved straightforwardly. The common methods are major human achievements and part of the world's heritage; like all mathematics, they are especially wonderful because they cross boundaries of culture and language.

In addition to knowing how to explain mathematics, prospective teachers should also know **how mathematics is used**. Therefore, this book emphasizes knowing which kinds of problems can be solved by the four basic operations of addition, subtraction, multiplication, and division. Practicing teachers common complaint is that their students simply guess which operations to use to solve a problem. To help student learn, teachers must know and be able to convey what the four operations mean, and they must be able to write a variety of story problems for the operations. It is also helpful for teachers to know how mathematics is applied in science and daily life. To this end, I have included various applications such as calculating discounts and tips, using visualization skills to explain the phases of the moon, using spheres to explain how the Global Positioning System works, and using random samples to estimate populations with the capture–recapture method.

Because of this unique approach. I believe *Mathematics for Elementary Teachers with Activity Manual* is an excellent fit for the recommendations of the Conference Board of the Mathematical Sciences regarding the mathematical preparation of teachers. I also believe that the book helps prepare teachers to teach in accordance with the principles and standards of the National Council of Teachers of Mathematics (NCTM).

Activity Manual

Class Activities, which were developed alongside the text, are central and integral to the text. All good teachers of mathematics know mathematics is not a spectator sport: We can't learn mathematical ideas solely by watching someone present them. Instead, we must actively think through mathematical ideas to make sense of them for ourselves. When students work on problems in the class activities—first on their own, then in a pair or a small group, and then within a whole class discussion—they have a chance to think through the mathematical ideas several times. By discussing mathematical ideas and explaining their solution methods to each other, students can deepen and extend their thinking. As every mathematics teacher knows, students really learn mathematics when they have to explain it to someone else. The activities are presented in a manual correlated with the textbook.

To get a feel for the class activities, look at the following:

- Class Activity 2S: Can We Reason This Way? (page 34)
- Class Activity 3P: A Third-Grader's Method of Subtraction (page 54)
- Class Activity 4X: Showing the Algebra in Mental Math (page 91)
- Class Activity 5H: Explaining Why We Place the Decimal Point Where We Do When We Multiply Decimals (page 103)
- Class Activity 6K: Interpreting the Standard Division Algorithm as Dividing Bundled Toothpicks (page 124)
- Class Activity 8H: Spirograph Flower Designs (page 164)
- Class Activity 9O: Solving Story Problems with Strip Diagrams and with Equations (page 200)
- Class Activity 10O: How Big is the Reflection of Your Face in a Mirror? (page 249)
- Class Activity 12D: Determining Areas of Triangles in Progressively Sophisticated Ways (page 284)
- Class Activity 15D: Using Random Samples to Estimate Population Size by Marking (Capture–Recapture) (pages 358)

Content and Organization

The book is **organized around the operations** instead of around the different types of numbers. In my view, there are two key advantages to focusing on the operations. The first advantage is a more advanced, unified perspective, which emphasizes that a given operation (addition, subtraction, multiplication, or division) retains its meaning across all the different types of numbers. Prospective teachers who have already studied numbers and operations in the traditional way for years will find that method enables them to take a broader view and to consider a different perspective. A second advantage is that fractions, decimals, and percents—traditional weak spots—can be studied repeatedly throughout a course, rather than only at the end. The repeated coverage of fractions, decimals, and percents allows students to gradually become used to reasoning with these numbers, so they aren't overwhelmed when they get to multiplication and division of fractions and decimals.

Students will use **visualization** not only in traditional mathematical contexts but also to understand basic astronomical phenomena, such as the phases of the moon, the reason for the seasons, and the rotation of the earth around its axis every day.

There is ample coverage of **volume and surface area**. Both the text and the activity manual promote a deeper understanding of these concepts with the aid of hands-on exploration.

Unique content in Chapter 9 on algebra introduces U.S. teachers to the impressive diagrammatic method presented in the math texts for grades 3–6 used in Singapore, whose children get the top math scores in the world. This method helps students make sense of and solve a variety of **algebra and other story problems** without using variables. The text helps students see the relationship between the Singaporean diagrammatic method and standard algebraic problem-solving methods.

A section on **solving problems and explaining solutions** appears near the beginning of Chapter 2 on fractions. Fractions provide an especially rich domain for explanations and for challenging problems. The section helps students think about how explanations in math can be different from explanations in other fields of knowledge.

Arithmetic in bases other than 10 is not explicitly discussed, but other bases are implied in several activities and problems. For example, see Class Activity 3R: Regrouping with Dozens and Dozens of Dozens, which involves base 12, and Class Activity 3S: Regrouping with Seconds, Minutes, and Hours, which involves base 60. Problems 8–12 on pages 119 and 120 of Section 3.3 also involve the idea of other bases. These activities and problems allow students to grapple with the significance of the base in place value without getting bogged down in the mechanics of arithmetic in other bases.

The **counting methods of probability** are uniquely explained. When determining probabilities, we must frequently multiply to determine all possible outcomes of an experiment. Rather than have students simply accept a rule stating that we should multiply to determine a total number, we encourage students to understand where this rule comes from: It follows directly from the meaning of multiplication.

Pedagogical Features

Practice Exercises give students the opportunity to try out problems. Solutions appear in the text immediately after the Practice Exercises. These solutions provide students with many examples of the kinds of good explanations they should learn to write. By attempting the Practice Exercises themselves and checking their solutions against the solutions provided, students will be better prepared to provide good explanations in their homework.

Problems are opportunities for students to explain the mathematics they have learned, without being given an answer at the end of the text. Problems are typically assigned as homework. Solutions appear in the *Instructor's Solutions Manual* and can be provided online for students at the discretion of the instructor.

To get a feel for the problems, take a look at the following:

- Problem 13 on page 25 of Section 1.2 (Decimals and Negative Numbers)
- Problem 18 on page 47 of Section 2.1 (The Meaning of Fractions)
- Problem 7 on page 119 of Section 3.3 (Why the Common Algorithms for Adding and Subtracting Numbers in the Decimal System Work)
- Problem 8 on page 180 of Section 4.5 (Properties of Arithmetic, Mental Math, and Single-Digit Multiplication Facts)
- Problem 5 on page 200 of Section 5.1 (Multiplying Fractions)

- Problem 13 on page 294 of Section 7.2 (Solving Proportion Problems by Reasoning with Multiplication and Division)

- Problem 15 on page 386 of Section 9.4 (Solving Algebra Story Problem with Strip Diagrams and with Algebra)

- Problem 4 on page 626 of Section 14.5 (Areas, Volumes, and Scaling)

A **chapter summary** and study items are provided at the end of each chapter to help students organize their thinking and focus on key ideas as they study for exams.

 The **core icon** denotes central material. These problems and activities (or similar ones) are highly recommended for mastery of the material.

The **focal points icon** indicates that the section addresses one of the NCTM focal points at the given grade level. The treatment of the topic goes beyond what students at that grade level are expected to do, because teachers need to know how mathematical ideas develop and progress.

This **extension icon** denotes material that is beneficial to cover but not critical. These problems and activities are either more challenging, involve an extended investigation, or are designed to extend students' thinking beyond the central areas of study.

Content Changes for the Third Edition

This edition has been enhanced extensively:

- The text draws on the findings of recent reports, including the NCTM Focal Points for Prekindergarten through Grade 8 Mathematics and the report of the National Research Council Committee on Early Childhood Mathematics.

- The book now begins with a chapter on numbers and the decimal system. The chapter includes an enhanced treatment of early number ideas that teachers of very young children need to know.

- The introduction to fractions in Chapter 2 is a more gradual presentation to guide students to mathematically valid ways of reasoning about fractions. The second section of the chapter is an interlude on solving problems and explaining solutions. The final section on percent takes the perspective of equivalent fractions, thus building on the work of previous sections.

- In Chapter 3 on addition and subtraction, material on negative numbers has been moved to the end of the chapter. Material on properties of arithmetic, mental math, and single-digit facts now occurs earlier in the chapter. The chapter includes enhanced material on children's learning paths for single-digit addition and subtraction.

- Chapter 4 on multiplication begins with a discussion of different types of multiplication story problems.

- In Chapter 6 on division, the section on fraction division was separated into two sections. The first takes the easier "how many groups?" perspective, and the second takes the "how many in one group?" perspective, which can also be considered a proportion perspective.

- A new chapter on proportional reasoning, Chapter 7, was created from material on ratio and proportion and features a section on percent increase.

- The chapters on number theory and algebra were moved to follow the chapters on numbers and operations.

- A new chapter on solid shapes and their volume and surface area, Chapter 13, was created to unify the discussion on solid shapes and to highlight the distinction between surface area and volume.

- In Chapter 15, Statistics, there is additional material on errors with data displays and errors with the mean and median. The section on data distribution has improved coverage of types of distributions and of comparing distributions.

- Material on counting that had previously been in the multiplication chapter was moved to the probability chapter.

- Throughout the book, there is a closer alignment among Class Activities, Practice Exercises, Problems, and the sample test problems that are available to instructors.

- The *Instructor's Resource Manual* recommends IMAP (Integrating Mathematics and Pedagogy) video clips from the CD ISBN 0-13-119854-8; as well as other videos available on the Internet, to show in conjunction with a number of the Class Activities.

I am enthusiastic about these changes and am sure students and professors will be as well.

Supplements

Instructor's Resource Manual
0-321-64695-9/978-0-321-64695-8
This manual, written by the author, includes teaching ideas for each chapter, tips and solutions for the activities, and course organization and grading suggestions.

Instructor's Solutions Manual
0-321-64697-5/978-0-321-64697-2
Michael Edward Matthews, University of Nebraska at Omaha, *Christopher Danielson*, Normandale Community College

The solutions manual contains worked-out solutions to all problems in the text.

Instructor's Testing Manual
This manual, written by the author, contains sample test problems and can be downloaded from the Instructor's Resource Center at www.pearsonhighered.com/irc.

NEW! ### Insider's Guide
0-321-65350-5/978-0-321-65350-5
By instructors, for instructors, this supplement provides teaching tips, suggestions, classroom ideas, activities, and other helpful support from instructors currently using the text and the inquiry-based approach to teaching the course.

NEW! ### Active Teachers, Active Learners DVD
0-321-65347-5/978-0-321-65347-5
This new DVD was created to help instructors enrich their classroom by expanding their knowledge of teaching using an inquiry-based approach. The DVD features classroom video of Beckmann and her students discovering concepts using activities and small group discussion. Optional voice-over commentary has Beckmann explaining what is happening during her class, and why, providing further insight into the inquiry-based approach. Also included is classroom footage from other instructors using this book and inquiry-based approach in their courses. The DVD is the ideal resource for instructors who are considering switching to a more inquiry-based approach from a more traditional lecture style, teaching an inquiry-based course for the first time, or for instructors who seek new ideas to integrate into their existing inquiry-based classes.

Companion Web site
www.pearsonhighered.com/beckmann
This Web site contains links to relevant Web sites and resources the author references in the text.

ACKNOWLEDGMENTS

In writing this book, I have benefited from much help and advice, for which I am deeply grateful. I would like to thank Deborah Ball, Hyman Bass, Jennifer Belton, Andrew Chen, Michael Ching, Herb Clemens, Doug Clements, Christopher Danielson, Sarah Donaldson, Greg Dresden, Bree Ettinger, Victor Ferdinand, Chris Franklin, Sol Friedberg, Karen Fuson, Whitney George, Wei Shen Hsia, Will Kazez, Cathy Kessel, Harvey Keynes, Hulya Kilic, Steffen Lempp, Matt Mastin, Michael Matthews, Betsy McNeal, Denise Mewborn, Eileen Murray, Ira Papick, Tom Parker, Ana Rusch, Bill Sargeant, Akihiko Takahashi, Mark Hoover Thames, Tad Watanabe, Emilie Wiesner, and Carrie Wright for helpful comments and conversations. I thank my students for generously sharing their thinking with me and for teaching me many interesting and surprising solution methods.

I would also like to express my appreciation to the following reviewers and class testers for many helpful comments on this edition and earlier drafts:

Gerald Beer, California State University, Los Angeles
Barbara Britton, Eastern Michigan University
*Christopher Danielson, Normandale Community College
Maria Diamantis, Southern Connecticut State University
Ron English, Niagara County Community College
Larry Feldman, Indiana University of Pennsylvania
Richard Francis, Southeast Missouri State University
Judy Kidd, James Madison University
*Michael Edward Matthews, University of Nebraska at Omaha
*Doris J. Mohr, University of Southern Indiana
Michelle Moravec, McLennan Community College
Shirley Pereira, Grossmont College
Kevin Peterson, Columbus State University
*Cheryl Roddick, San Jose State University
*David Ruszkiewicz, Milwaukee Area Technical College
*Ayse A. Sahin, DePaul University
*William A. Sargeant, University of Wisconsin – Whitewater
*Dan Schultz-Ela, Mesa State College
*Julia Shew, Columbus State Community College
Darla Shields, Slippery Rock University
*Wendy Hagerman Smith, Radford University
Ginger Warfield, University of Washington
David C. Wilson, Buffalo State College
Carol Yin, LaGrange College

The Active Teachers, Active Learners DVD was a huge undertaking that could not have been possible without the contributions of my colleagues. Few people feel comfortable being videotaped in their classrooms, so I would like to thank the following contributors and their students for taking that chance and helping make the DVD a reality:

Sarah Donaldson, University of Georgia
Michael Edward Matthews, University of Nebraska at Omaha

Janice Rech, University of Nebraska at Omaha

Bridgette Stevens, University of Northern Iowa

I am very grateful for the support and excellent work of everyone at Pearson Addison-Wesley who has contributed to this project. I especially want to thank my editor, Marnie Greenhut, for her enthusiastic support, as well as Joe Vetere, Sheila Spinney, Christine O'Brien, Elle Driska, Christine Stavrou, Roxanne McCarley, Kendra Bassi, Heather Scott, and Christina Gleason for invaluable assistance. It is only through their dedicated hard work that this book has become a reality. I thank my family—Will, Joey, and Arianna Kazez, and my parents Martin and Gloria Beckmann—for their patience, support, and encouragement. My children long ago gave up asking, "Mom, are you still working on that book?" and are now convinced that books take forever to write.

Sybilla Beckmann
Athens, Georgia

Numbers and the Decimal System

What are numbers and where do they come from? What humans consider to be numbers has evolved over the course of history, and the way children learn about numbers parallels this development. When did humans first become aware of numbers? The answer is uncertain, but it is at least many tens of thousands of years ago. Some scholars believe that numbers date back to the beginning of human existence, citing as a basis for their views the primitive understanding of numbers observed in some animals. (See [20] for a fascinating account of this, as well as of the human mind's capacity to comprehend numbers.)

Today numbers are everywhere, and the ability to work fluently with numbers is essential for effective functioning in society. In this chapter (and the next), we discuss beginning ideas about numbers, how to write and represent numbers, how to interpret the meaning of written numbers, and how to compare numbers. Even the most basic numbers—the counting numbers—reveal surprising intricacies that we are scarcely aware of as adults. We will study the decimal system, which is a remarkably powerful and efficient system for writing numbers and is a major achievement in human history. Not only does the decimal system allow us to express arbitrarily large numbers and arbitrarily small numbers—as well as everything in between—but it also enables us to quickly compare numbers and assess the ballpark size of a number. The decimal system is familiar to adults, but its slick compactness hides its inner workings. We will examine with care those inner workings of the decimal system that children must grasp to make sense of numbers.

The National Council of Teachers of Mathematics (NCTM) has set standards for the mathematics that students from grades prekindergarten through 12 should know and be able to do [64].

Instructional programs concerning numbers and operations should enable all students from prekindergarten through grade 12 to

- *understand numbers, ways of representing numbers, relationships among numbers, and number systems;*
- *understand meanings of operations and how they relate to one another;*
- *compute fluently and make reasonable estimates.*

For the NCTM's specific number and operation recommendations for grades PreK–2, grades 3–5, grades 6–8, and grades 9–12, go to www.pearsonhighered.com/beckmann.

This chapter and the next several chapters will deepen your own understanding of numbers and operations, so that you will be prepared to teach these topics. In addition, these chapters will help you acquire a vision for the way topics develop beyond the grade level you may teach.

1.1 The Counting Numbers

 Focal Points
PreK, K, Grade 1, and Grade 2

What are numbers and why do we have them? Numbers provide us with a precise means for describing, representing, and reasoning about quantities. We use numbers to tell us "how many" or "how much," so that we can communicate specific, detailed information about collections of things and about quantities of stuff. Although there are many different kinds of numbers (e.g., fractions, decimals, and negative numbers), the most basic numbers, and the starting point for young children, are the **counting numbers**, namely the numbers 1, 2, 3, 4, 5, 6, ...

counting numbers

There are two distinctly different ways to think about the counting numbers. Connecting these two views of counting numbers is a major mathematical idea that very young children must grasp before they can do school arithmetic. What are these two different ways of thinking about the counting numbers and how are they related?

The Counting Numbers as a List and for Describing Set Size

> **Class Activity**   *Now Turn to Class Activities Manual*
>
> **1A** The Counting Numbers as a List, p. 1

One way to think about the counting numbers is as a list. The list of counting numbers starts with 1; every number in the list has a unique successor, and except for the number 1, every number in the list has a unique predecessor. So the list of counting numbers is an ordered list. Every counting number appears exactly once in this ordered list. The ordering of the list of counting numbers is important because of the second way of thinking about the counting numbers.

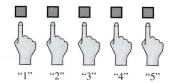

FIGURE 1.1 Counting 5 blocks makes a one-to-one correspondence between the list 1, 2, 3, 4, 5 and the blocks.

The second way of thinking about the counting numbers is as "telling how many." In other words, a counting number describes how many things are in a set[1] of things. The number of things in a set is called the **cardinality** of a set. Think for a moment about how surprisingly abstract the notion of cardinality is. We use the number 3 to quantify a limitless variety of collections—3 cats, 3 toy dinosaurs, 3 jumps, 3 claps, and so on. The number 3 is the abstract, common aspect that all these limitless examples of sets of 3 things share.

cardinality of a set

For sets of up to about 3, 4, or 5 objects, we can usually recognize the number of objects in the set immediately, without counting the objects one by one. The process of immediate recognition of the exact number of objects in a set is called *subitizing* and is discussed further in [68]. But in general, we must count the objects in a set to determine how many there are. The process of counting the objects in a set connects the "list view" of the counting numbers with the "cardinality view." As adults, this connection is so familiar that we are usually not even aware of it. But for young children, this connection is not obvious, and grasping it is an important milestone (see [68]).

Class Activity   *Now Turn to Class Activities Manual*

1B 🏛 Connecting Counting Numbers as a List with Cardinality, p. 2

one-to-one correspondence

When we count a set of objects one by one we make a **one-to-one correspondence** between an initial portion of the list of counting numbers and the set. For example, when a child counts a set of 5 blocks, the child makes a one-to-one correspondence between the list 1, 2, 3, 4, 5 and the set of blocks. This means that each block is paired with exactly one number and each number is paired with exactly one block, as indicated in Figure 1.1. Such a one-to-one correspondence connects the "list" view of the counting numbers with cardinality. However, there is another critical piece of understanding that this connection relies on, a piece of understanding that adults typically take for granted but is not obvious to young children: *The last number we say when we count a set of objects tells us the total number of objects in the set*, as indicated in Figure 1.2. It is for this reason that the order of the counting numbers is so important, unlike the order of the letters of the alphabet, for example.

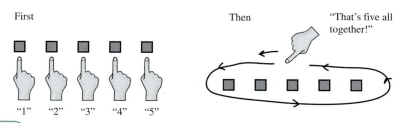

FIGURE 1.2 When we count objects, the last number we say tells us the total number of objects.

---

[1] A **set** is a collection of distinct "things." These things can include concepts and ideas, such as the concept of an infinitely long straight line, and imaginary things such as heffalumps.

FIGURE 1.3

A shepherd's
tally of sheep

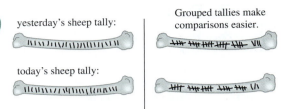

yesterday's sheep tally:

today's sheep tally:

Grouped tallies make
comparisons easier.

Is it the same number of sheep?

The connection between the list and cardinality views of the counting numbers is especially important in understanding that numbers that are *later* in the list correspond with *larger* quantities and that numbers that are *earlier* in the list correspond with *smaller* quantities. In particular, starting at any counting number, the next number in the list describes the size of a set that has one more object in it, and the previous number in the list describes the size of a set that has one less object in it.

Representing Counting Numbers and the Origins of the Decimal System

Let's think about the list of counting numbers and the symbols we use to represent these numbers. The symbols for the first nine counting numbers 1, 2, 3, 4, 5, 6, 7, 8, 9 have been passed along to us by tradition and could have been different. Instead of the symbol 4, we could be using a completely different symbol. In fact, one way to represent 4 is simply with 4 tally marks. So why don't we just use tally marks to represent counting numbers?

Think back to a shepherd living thousands of years ago who might have used tally marks to keep track of his sheep. If his tally marks are not organized, he will have a hard time comparing the number of sheep he has on different days, as shown in Figure 1.3. But if he groups his tally marks, it becomes easier to compare the number of sheep.

As people began to live in cities and engage in trade, they needed to work with larger numbers. But tally marks are cumbersome to write in large quantities. Instead of writing tally marks, it's more efficient to write a single symbol that represents a group of tally marks, such as the Roman numeral V, which represents 5. To record 50 sheep, a person long ago might have written this:

VVVVVVVVVV

However, it's hard to read all those Vs, so the Romans devised new symbols: X for 10, L for 50, C for 100, and M for 1000. These symbols are fine for representing numbers up to a few thousand, but what about representing 10,000? Once again,

MMMMMMMMMM

is difficult to read. What about 100,000 or 1,000,000? To represent these, one might want to create yet more new symbols.

But there is a problem with creating ever more new symbols; namely, the list of counting numbers is infinitely long. So how can each counting number in this infinitely long list be represented in its own unique way? Starting with our modern day symbols—1, 2, 3, 4, 5, 6, 7, 8, 9—how can one continue the list without resorting to creating an endless string of new symbols? The solution to this problem was not obvious and was a significant achievement in the history of human thought. The **decimal system**, or **base-ten** **system**, is the ingenious system we use today to write (and say) counting numbers without resorting to creating more and more new symbols. The decimal system requires using only ten distinct symbols, namely, the **digits** 0, 1, 2, 3, 4, 5, 6, 7, 8, 9. The key innovation of the decimal system is that rather than using new symbols to represent larger and larger numbers, the decimal system uses place value.

decimal system
base-ten system

digit

Place Value in the Decimal System

place value   **Place value** means that the quantity that a digit in a number represents depends—in a very specific way—on the position of the digit in the number.

| | | | | | | | | | | | | |
1 2 3 4 5 6 7 8 9 10 11 12 13 14

FIGURE 1.4 The important role that ten plays is not obvious even when counting more than 10 items.

Do the next Class Activity before you read on.

Class Activity   *Now Turn to Class Activities Manual*

1C How Many Are There? p. 4

How did you organize the toothpicks in the Class Activity? If you made bundles of 10 toothpicks and then bundled 10 bundles of 10 to make bundles of 100, then you began to reinvent place value and the decimal system. Place value works by creating larger and larger units by *repeatedly bundling in groups of ten.*

Ten plays a special role in the decimal system, but its importance is not obvious to children. Teachers must repeatedly draw children's attention to the role that ten plays. For example, a young child might be able to count that there are 14 beads in a collection, such as the collection shown in Figure 1.4, but the child may not realize that 14 actually stands for 1 ten and 4 ones. With the perspective presented in Figure 1.4, the numbers 10, 11, 12, 13, 14 are just the counting numbers that follow 9. But when young children learn to organize collections of between 10 and 19 small objects into one group of ten and some ones, as in Figure 1.5, and when they understand that the digit 1 in 14 does not stand for "one" but instead stands for *1 group of ten*, they have begun to learn about place value and the decimal system. Some teachers like to help young students learn the meaning of the digit 1 in the numbers 10 to 19 with the aid of cards, such as the ones shown in Figure 1.6. These cards show how a number such as 17 is made up of 1 ten and 7 ones.

Children extend their understanding of the decimal system in two ways: when they view a group of ten as a unit in its own right, and when they understand that a two-digit number such as 37 stands for 3 tens and 7 ones and can be represented with bundled objects and simple drawings like those in Figure 1.7

Just as 10 ones are grouped to make a new unit of ten, 10 tens are grouped to make a new unit of one hundred and 10 hundreds are grouped to make a new unit of one-thousand, as indicated in Figure 1.8. Continuing in this way, 10 thousands are grouped to make a new unit of ten-thousand, and so on, so that arbitrarily large units can be made by grouping 10 of the previously made units. These increasingly large units are represented in successive places to the left in a number, so that *the value of each place in a*

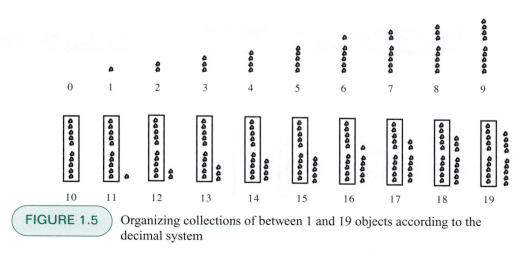

FIGURE 1.5 Organizing collections of between 1 and 19 objects according to the decimal system

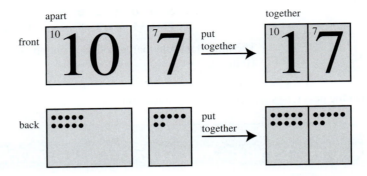

FIGURE 1.6 Cards that can be put together and taken apart to show teen numbers as tens and ones.

number is ten times the value of the place to its immediate right, as indicated in Figure 1.9. Figure 1.9 shows the standard names (used in the United States) of the place values up to the trillions place.

In general, a string of digits, such as 1234, stands for the total amount that all its places taken together represent. Within the string, each digit stands for that many of that place's value. So in 1234, the 1 stands for 1 thousand, the 2 stands for 2 hundreds, the 3 stands for 3 tens, and the 4 stands for 4 ones. Overall then, 1234 stands for

1 thousand and 2 hundreds and 3 tens and 4 (ones),

which is the total number of toothpicks pictured in Figure 1.10.

decimal representation A string of digits that represents a number, such as 1234, is called the **decimal representation** of the number. To clarify the meaning of the decimal representation of a number, we sometimes write it in **expanded form**, namely, in one of the following forms:

1 thousand + 2 hundreds + 3 tens + 4 ones

$$1 \cdot 1000 + 2 \cdot 100 + 3 \cdot 10 + 4$$

$$1000 + 200 + 30 + 4$$

Notice that the way the toothpicks in Figure 1.10 are organized corresponds with the decimal representation and the expanded form for the total number of toothpicks that are depicted. Notice also how compactly this rather large number of toothpicks is represented by the short string 1234. Now imagine showing ten times as many toothpicks as are in the bundle of 1000. This would be a lot of toothpicks; yet to write this number, 10,000, only takes 5 digits! To appreciate how quickly the value of the places in the decimal system grow, do the next Class Activity.

Class Activity *Now Turn to Class Activities Manual*

1D Showing the Values of Places in the Decimal System, p. 5

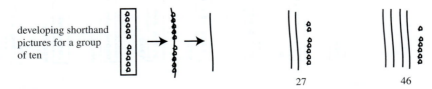

FIGURE 1.7 Simple pictures for representing two-digit numbers as tens and ones.

value of the thousands place

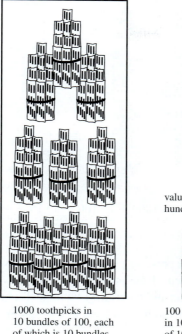

value of the
hundreds place

value of the
tens place

value of the
ones place

—

1000 toothpicks in
10 bundles of 100, each
of which is 10 bundles
of 10

100 toothpicks
in 10 bundles
of 10

10 toothpicks
in a bundle

thousands place *hundreds place* *tens place* *ones place*

FIGURE 1.8 Bundled toothpicks representing values of places in the decimal system.

Although the decimal system is highly efficient and practical, children have difficulty learning what written numbers represent because they must keep the values of the places in mind. Because the values of the places are not shown explicitly, even interpreting written numbers requires a certain level of abstract thinking.

In summary, the decimal system allows us to write any counting number, no matter how large, using only the ten digits 0 through 9, by the use of place value. The key idea of place value is to create larger and larger units, which are the values of places farther and farther to the left, by taking the value of

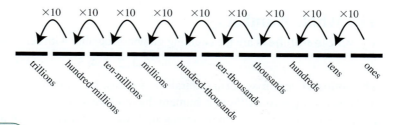

FIGURE 1.9 Each place's value is ten times the value of the place to its right.

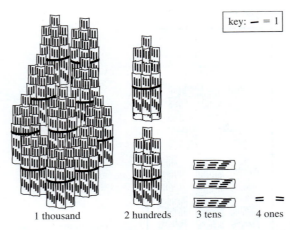

key: — = 1

1 thousand   2 hundreds   3 tens   4 ones

FIGURE 1.10   Representing $1234 = 1 \cdot 1000 + 2 \cdot 100 + 3 \cdot 10 + 4 \cdot 1$ with bundled toothpicks.

each place to be ten times the value of the previous place to the right. We can think of the value of each place as obtained by bundling together ten of the previous place's value, and we can even show this process of repeated bundling (up to about the thousands) with small objects such as toothpicks. By using place value, every counting number can be expressed in a unique way as a string of digits.

Saying the Counting Numbers

Before young children learn to write the symbols for the counting numbers, they learn to say the number words. Unfortunately, some of the words used to say the counting numbers in English do not correspond well with the decimal representations of these numbers. This makes the early learning of numbers more difficult for children who speak English than for children who speak some other languages.

In English, as in other languages, the words we use for the first ten counting numbers are arbitrary and could have been different. For example, instead of the word *four* we could be using a completely different word, such as the word for four in other languages. Thereafter, matters become more difficult for English speakers than for speakers of some other languages. The difficulty arises because the way we say the counting numbers from 11 to 19 in English does not fit well with their decimal representations. Notice that "eleven" does not sound like "one ten and one," which is what 11 stands for; nor does "twelve" sound very much like "one ten and two," which is what 12 stands for. To add to the confusion, notice that "thirteen, fourteen, . . . , nineteen" sound like the reverses of "one ten and three, one ten and four, . . . , one ten and nine," which is what 13, 14, . . . , 19 stand for. From 20 onward, most of the English spoken words for counting numbers do fit fairly well with their decimal representations. For example, "twenty" sounds roughly like "two tens," "sixty-three" sounds very much like "six tens and three," and "two-hundred eighty-four" sounds very much like "2 hundreds and eight tens and four," which is what 284 stands for. Note, however, that it's easy for children to confuse decade numbers with teen numbers because their pronunciation is so similar. For example, "sixty" and "sixteen" sound similar.

The Whole Numbers

whole numbers   The **whole numbers** are the counting numbers together with zero, as follows:

$$0, 1, 2, 3, 4, 5, \ldots$$

The notion of zero may seem natural to us today, but our early ancestors struggled to discover and make sense of zero. Although humans have always been acquainted with the notion of "having none," as in having no sheep or having no food to eat, the concept of 0 as a number was introduced far later than the counting numbers—not until sometime before 800 A.D. (See [10].) Even today, the notion of 0 is difficult for many children to grasp. This difficulty is not surprising: Although the

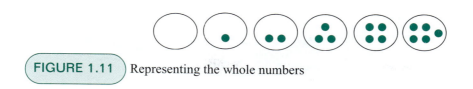

FIGURE 1.11 Representing the whole numbers

counting numbers can be represented nicely by sets of objects, you have to show *no* objects in order to represent the number 0 in a similar fashion. But how does one *show* no objects? We might use a picture as in Figure 1.11.

Notice that place value requires the use of 0. To write three-hundred, for example, we must show that the 3 is in the hundreds place and we must show that there are no tens and no ones, which, of course, we do by writing 300.

The Counting Numbers on Number Paths and The Whole Numbers on Number Lines

The counting numbers can be used to count the number of physical things in a collection, but they can also be used to count events. For example, children might count how many times they have hopped or jumped or how many steps they have taken. Many children's games involve moving a game piece along a path. If the path is labeled with successive counting numbers, like the one in Figure 1.12, we can call **number path** the path a **number path**. Number paths are informal precursors to the concepts of distance and length, and to the mathematical concept of a number line.

number line A **number line** is a line on which one location has been chosen as 0, and another location, to the right of 0, has been chosen as 1. Number lines stretch infinitely far in both directions, although in practice only a small portion of a number line can be shown (and that portion may or may not include 0 and 1). **unit** The distance from 0 to 1 is called a **unit**, and the choice of a unit is called the **scale** of the number line. **scale** Once choices for the locations of 0 and 1 have been made, each counting number is represented by the point on the number line that is that many units to the right of 0.

Number lines are an important way to represent numbers because they allow the concept of number to be expanded to decimals, fractions, and negative numbers, and they unify all these different kinds of numbers and present them as a coherent whole.

Although number lines are important, and number paths and number lines are similar, there is a critical distinction between them, and this distinction makes number paths better for use with the youngest children than number lines. Number paths are suitable for use with young children because they clearly show distinct "steps" along the path that children can count, just as they might count their own steps, hops, or jumps. In contrast, to interpret a number line correctly, we must rely on the ideas of length and distance from 0, as indicated in Figure 1.13. Instead of focusing on length, young children tend to count "tick marks" along a number line. The habit of counting tick marks instead of attending to length can lead to omitting 0 and to misinterpretations about locations of fractions on number lines. We will examine some of these errors when we study fractions in Chapter 2.

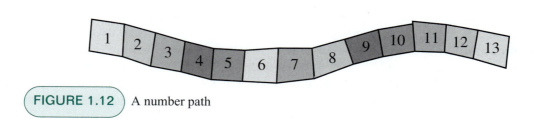

FIGURE 1.12 A number path

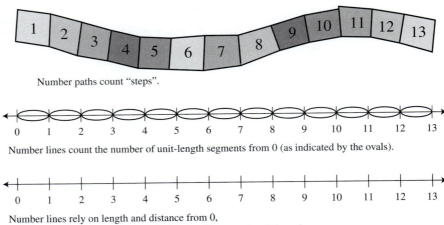

Number paths count "steps".

Number lines count the number of unit-length segments from 0 (as indicated by the ovals).

Number lines rely on length and distance from 0, but number lines are sometimes misinterpreted by counting tick marks.

FIGURE 1.13   Number paths versus number lines: counting steps versus length

---

Practice Exercises for Section 1.1

Answers to these practice exercises begin on page 00.

1. If a young child can correctly say the number word list to five—"one, two, three, four, five"—will the child necessarily be able to determine how many bears are in a collection of 5 toy bears that are lined up in a row? Discuss why or why not.

2. If a young child can correctly say the number word list to five— "one, two, three, four, five" and point one by one to each bear in a collection of 5 toy bears while saying the number words, does the child necessarily understand that there are 5 bears in the collection? Discuss why or why not.

3. What problem in the history of mathematics did the development of the decimal system solve?

4. How are the values of adjacent places in decimal representations related?

5. Draw a picture showing how to organize 19 objects in a way that fits with the structure of the decimal system.

6. Describe how to organize 100 toothpicks in a way that fits with the structure of the decimal system. Explain how your organization reflects the structure of the decimal system and how it fits with the way we write the number 100.

7. What difficulty do English speakers face in learning to say some of the counting numbers?

8. Describe the difference between a number path and a number line.

Answers to Practice Exercises for Section 1.1

1. No, the child might not be able to determine that there are 5 bears in the collection because the child might not be able to make a one-to-one correspondence between the numbers 1, 2, 3, 4, 5 and the bears. For example, the child might point twice to one of the bears and count two numbers for that bear, or the child might skip over a bear while counting. See also the next practice exercise and its answer.

2. No, the child might not understand that the last number word that is said while counting the bears tells how many bears there are in all in the collection.

3. The decimal system solved the problem of having to invent more and more new symbols to stand for larger and larger numbers. By using the decimal system, and place value, every counting number can be written using only the ten digits 0, 1, . . . 9.

4. See text.

5. See Figure 1.5.

6. First, bundle all the toothpicks into bundles of 10. Then gather those 10 bundles of 10 into a single bundle. This repeated bundling in groups of 10 is the basis of the decimal system. The 1 in 100 stands for this 1 large bundle of 10 bundles of 10.

7. See text.

8. See text.

Problems for Section 1.1

1. In your own words, discuss the connection between the counting numbers as a list and the counting numbers as they are used to describe how many objects are in sets. Include a discussion of what you will need to attend to if you are teaching young children who are learning to count.

2. If you give a child in kindergarten or first grade a bunch of beads or other small objects and ask the child to show you what the 3 in 35 stands for, the child might show you 3 of the beads. You might be tempted to respond that the 3 really stands for "thirty" and not 3. Of course it's true that the 3 does stand for thirty, but is there a better way you could respond, so as to draw attention to the decimal system? How could you organize the beads to make your point?

3. For each of the following collections of small objects, draw a simple picture and write a brief description for how to organize the objects in a way that corresponds to the way we use the decimal system to write the number for that many objects.

  a. 47 beads

  b. 328 toothpicks

  c. 1000 toothpicks

4. In your own words, describe how you can use collections of objects (such as toothpicks or Popsicle sticks) to show how the values of adjacent places in decimal representations of numbers are related.

5. In your own words, discuss the beginning ideas of place value and the decimal system that young children who can count beyond ten must begin to learn. Include a discussion of some of the hurdles faced by English speakers.

6. Children sometimes mistakenly read the number 1001 as "one hundred one." Why do you think a child would make such a mistake? Draw pictures that indicate how to represent 1001 with bundled objects.

7. Explain why the bagged and loose toothpicks pictured in Figure 1.14 are not organized in a way that fits with the structure of the decimal system. Describe how to alter the appearance of these bagged and loose toothpicks so that the same total number of toothpicks are organized in a way that is compatible with the decimal system.

FIGURE 1.14 Why are these not organized according to the decimal system?

8. The students in Ms. Caven's class have a large poster showing a million dots. Now, the students would really like to see a billion of something. Think of at least two different ways that you might attempt to show a billion of something and discuss whether your methods would be feasible. Be specific and back up your explanations with calculations.

1.2 Decimals and Negative Numbers

Focal Points
Grades 4, 5

The counting numbers are the most basic kinds of numbers, followed by the whole numbers. However, many situations, both practical and theoretical, require other numbers, such as fractions, decimals, and negative numbers. As different as fractions, decimals, and negative numbers may appear to be, they become unified when they are placed together on a number line.

Origins of Decimals and Negative Numbers

Why do we have decimals (and fractions) and negative numbers? How did these numbers arise?

In ancient times, a farmer filling bags with grain might have had only enough grain to fill the last bag half full. When trading goods, the farmer needed a way to express partial quantities. In modern times, we buy gasoline by the gallon (or liter), but we don't always buy a whole number of gallons. So we need a precise way to describe numbers that are in between whole numbers. Actually, we have two ways of describing numbers in between whole numbers: as decimals and as fractions. Both fractions and decimals arise by creating new units that are less than 1, but the decimals are created by extending the decimal system.

The introduction of negative numbers was relatively late in human development. Although the ancient Babylonians may have had the concept of negative numbers around 2000 B.C., negative numbers were not always accepted by mathematicians even as late as the sixteenth century A.D. ([10]). The difficulty lies in interpreting the meaning of negative numbers. How can negative numbers be represented? This problem may seem perplexing at first, but in fact there is a nice interpretation of negative numbers, namely, as amounts owed. For example, −7 can represent *owing* 7 marbles. In comparison, 7 can represent *having* 7 marbles. (See Figure 1.15.)

A common way that negative numbers are used nowadays is for temperatures below zero degrees.

Decimals Extend the Decimal System

The essential structure of the decimal system is that the value of each place is ten times the value of the place to its right. So, moving to the *left* across the places in the decimal system, the value of the places are successively *multiplied* by 10. But then moving to the *right* across the places in the decimal system, the values of the places are successively *divided* by 10, as indicated in Figure 1.16.

The decimals are created by creating places to the right of the ones place while retaining the essential structure of the decimal system. So, starting at the ones place, dividing the unit 1 into 10 equal pieces creates a new unit, a tenth, and tenths are recorded in the place to the right of the ones place. We indicate **decimal point** the location of the ones place by placing a **decimal point** (.) to its right. At the tenths place, dividing a tenth into 10 equal pieces creates a new unit, a hundredth, and hundredths are recorded in the place to the right of the tenths place. Figure 1.17 depicts the process of dividing by 10 to create values of places to the right of the ones place.

The process of dividing a unit at a given place by 10 to create a new, smaller unit that is recorded in the place to the right continues without end. When we use the decimal system to represent a number as a string of digits, possibly including a decimal point, and possibly having infinitely many nonzero digits to

FIGURE 1.15

Representing negative integers

Having 7 marbles represents 7 🔵 🔵 🔵 🔵 🔵 🔵 🔵

Owing 7 marbles represents −7 ⚪ ⚪ ⚪ ⚪ ⚪ ⚪ ⚪

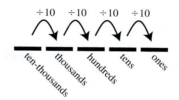

FIGURE 1.16   Moving to the right, the values of places in the decimal system are divided by 10.

decimal notation

decimal

decimal number

decimal representation

decimal expansion

the right of the decimal point, we say the number is in **decimal notation** and we call the number a **decimal** or a **decimal number.** We may also refer to the string of digits representing the number as a **decimal representation** or **decimal expansion** of the number.

Just as 2345 stands for the combined amount of 2 thousands, 3 hundreds, 4 tens, and 5 ones, so too the decimal 2.345 stands for the combined amount of 2 ones, 3 tenths, 4 hundredths, and 5 thousandths and the decimal 23.45 stands for the combined amount of 2 tens, 3 ones, 4 tenths, and 5 hundredths.

The decimal system has the same structure to the left and right of the decimal point, so we can represent (some) decimals with bundled objects in the same way that we represent whole numbers with bundled

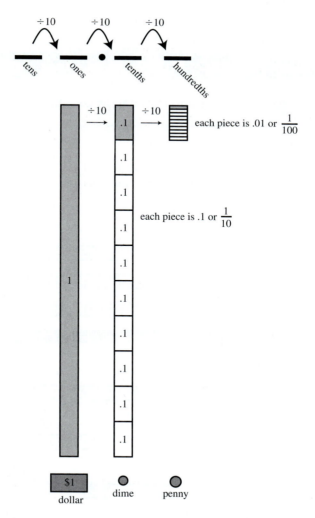

FIGURE 1.17   Tenths are created by dividing 1 into 10 equal pieces; hundredths are created by dividing a tenth into 10 equal pieces.

objects, as the next Class Activity shows. The only difference is that a single object must be allowed to represent the value of a decimal place that is less than 1. Although this may seem surprising at first, it is a common idea. After all, a penny represents $0.01.

Class Activity   *Now Turn to Class Activities Manual*

1E   Representing Decimals with Bundled Objects, p. 7

Decimals as Lengths and on Number Lines

A good way to represent positive decimals is as lengths; this way of representing decimals fits in a natural way with the way decimals are placed on number lines. Figure 1.18 indicates how to represent the decimals 1.2, 1.23, and 1.234 as lengths. The meter, which is the main unit of length used in the metric system, is a natural unit to use when representing decimals as lengths because the metric system was designed to be compatible with the decimal system.

Class Activity   *Now Turn to Class Activities Manual*

1F   Representing Decimals as Lengths, p. 8

If you did the Class Activity you used strips of paper, which are not very durable. For instruction, you might want to use a more durable material, such as lengths of plastic tubing (see [84]).

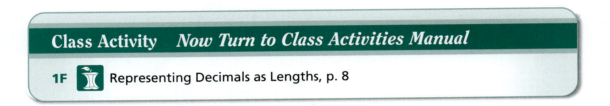

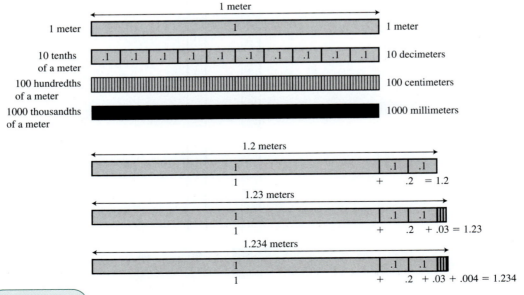

FIGURE 1.18   Representing decimals as lengths using metric length units

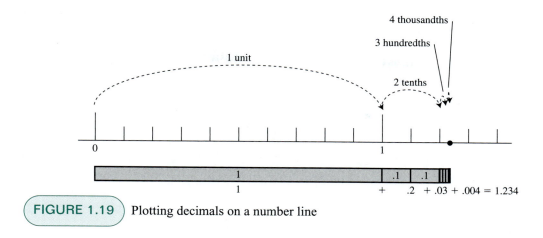

FIGURE 1.19   Plotting decimals on a number line

To connect lengths with number lines, imagine representing a positive decimal as a length by using strips of paper (or pieces of plastic tubing) as in the Class Activity. Now imagine placing the left end of the length of paper at 0. Then the right end of the length of paper lands on the decimal that the length represents, as in Figure 1.19. In this way, we can view number lines as related to lengths, and we can view points on number lines in terms of their distances from 0. More generally, a positive number N is located to the right of 0, at a distance of N units away from 0.

One way to think about decimals is as "filling in" the locations on the number line between the whole numbers. You can think of plotting decimals on the number line in successive stages in a way that fits with the structure of the decimal system. At the first stage, the whole numbers are placed on a number line so that consecutive whole numbers are one unit apart. (See Figure 1.20).

At the second stage, the decimals that have entries in the tenths place, but no smaller place, are spaced equally between the whole numbers, breaking each interval between consecutive whole numbers into 10 smaller intervals each one tenth unit long. See the stage-two number line in Figure 1.20. Notice that, although the interval between consecutive whole numbers is broken into 10 intervals,

FIGURE 1.20

Decimal numbers "fill in" number lines

Stage 1: representing whole numbers on a number line

Stage 2: Each unit is divided into tenths.

Stage 3: Each tenth is divided into hundredths.

Stage 4: Each hundredth is divided into thousandths.

Stage 5: Each thousandth is divided into ten-thousandths.

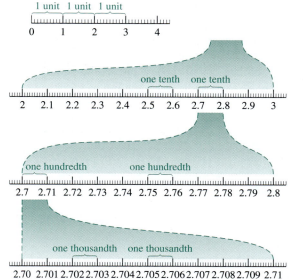

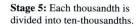

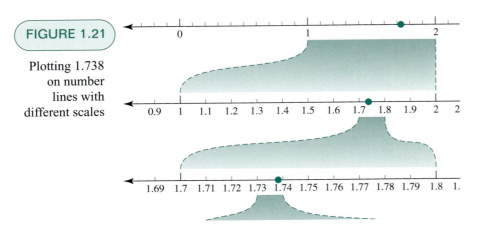

FIGURE 1.21

Plotting 1.738
on number
lines with
different scales

there are only 9 tick marks for decimal numbers in the interval, one for each of the 9 nonzero entries, 1 through 9, that go in the tenths places.

At the third stage, the decimals that have entries in the hundredths place, but no smaller place, are spaced equally between the previously plotted decimal numbers, breaking each interval between previously plotted consecutive decimals into 10 smaller intervals, each one hundredth unit long. See the stage-three number line in the middle of Figure 1.20. Notice that the number lines shown in Figure 1.20 all have different scales, so that they can show clearly the locations of the decimals. These locations also exist in the first number line at the top, but they are not shown due to space limitations.

At each stage in the process of filling in the number line, we plot new decimals in between previously plotted decimal numbers. The tick marks for these new decimals should be smaller than the tick marks of the decimal numbers plotted at the previous stage, as shown in Figure 1.20. We use smaller tick marks to distinguish among the stages and to show the structure of the decimal system.

Notice that where you plot a decimal on a number line depends on the scale of the number line. For example, Figure 1.21 shows the location of 1.738 on number lines of different scales.

The idea of filling in locations on the number line, starting from the whole numbers, is like starting with the notion of dollars as a denomination, deciding that a smaller denomination is needed, and therefore creating dimes. Ten dimes make a dollar, and a dime is worth one tenth of a dollar, just as the intervals on a number line between consecutive integers are broken into 10 intervals, each one tenth of a unit long. If dimes don't make a small enough denomination, pennies are created. Ten pennies make a dime, and a penny is worth one hundredth of a dollar, just as the intervals between consecutive decimals with no entries to the right of the tenths place are broken into 10 smaller intervals, each one hundredth of a unit long. If we felt that we needed a smaller denomination than the penny, we would create a new coin. In this case, 10 of the new coins would make a penny. In theory, we could keep going, creating smaller and smaller denominations of money, just as the values of places in the decimal system get smaller and smaller as one moves to the right, and just as we can repeatedly break intervals on a number line into 10 smaller intervals.

Class Activity   *Now Turn to Class Activities Manual*

1G   Zooming In on Number Lines, p. 9

See the Web site *www.pearsonhighered.com/beckmann* for an amazing experience of zooming in through powers of ten.

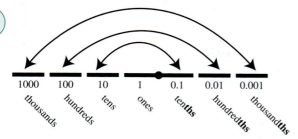

FIGURE 1.22

Symmetry in the place value names is around the ones place

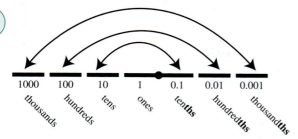

Class Activity   *Now Turn to Class Activities Manual*

1H Numbers Plotted on Number Lines, p. 10

Saying and Writing Decimals

The names for the values of the places to the right of the decimal point are symmetrically related to the names of the values of the places to the left of the decimal point, as shown in Figure 1.22.

There are several common errors associated with the place value names for decimals. One error is not distinguishing clearly between the values of places to the left and right of the decimal point. For example, students sometimes confuse tens with ten**ths** or hundreds with hundred**ths** or thousands with thousand**ths**. The pronunciation is similar, so it's easy to see how this confusion can occur! Teachers must take special care to pronounce the place value names clearly and to make sure students understand the difference. Another error occurs because students expect the symmetry in the place value names to be around the decimal point, not around the ones place. Some students expect there to be a "oneths place" immediately to the right of the decimal point, and they may mistakenly call the hundredths place the tenths place because of this misunderstanding.

When plotting decimals on number lines or when comparing, adding, or subtracting decimals, it is often useful to append zeros to the right-most nonzero digit to express explicitly that the values in these smaller places are zero. For example, 1.78, 1.780, 1.7800, 1.78000, and so on all stand for the same number. These representations show explicitly that the number 1.78 has 0 thousandths, 0 ten-thousandths, and 0 hundred-thousandths. Similarly, we may append zeros to the left of the left-most nonzero digit in a number to express explicitly that the values in these larger places are zero. For example, instead of writing .58, we may write 0.58, which perhaps makes the decimal point more clearly visible.

A cultural convention is to say decimals according to the value of the right-most nonzero decimal place and to say "and" for the decimal point. For example, we usually say 3.84 as "three and eighty-four hundredths" because the right-most digit is in the hundredths place, and we say 1.592 as "one and five-hundred ninety-two thousandths" because the right-most digit is in the thousandths place. From a mathematical perspective, however, it is perfectly acceptable to say 3.84 as "3 and 8 tenths and 4 hundredths" or "three point eight four." In fact, we can't use the usual cultural conventions when saying decimals that have infinitely many digits to the right of the decimal point. For example, the number pi, which is 3.1415 . . . must be read as "three point one four one five . . ." because there is no right-most nonzero digit in this number! Furthermore, there is reasoning that explains why the conventional way of saying decimals is logical, but this reasoning will not be immediately obvious to students who are just learning about decimals and place value. We will discuss this reasoning in the section on adding and subtracting fractions in Chapter 3.

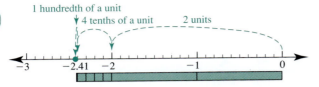

FIGURE 1.23

Plotting −2.41

Negative Numbers

negative
minus sign

integers

For any number, *N*, its **negative** is also a number and is denoted −*N*. The symbol − is called a **minus sign**. For example, the negative of 4 is −4, which can be read *negative four* or *minus four*. The set of numbers consisting of 0, the counting numbers, and the negatives of the counting numbers, is called the **integers**.

$$\ldots -5, -4, -3, -2, -1, 0, 1, 2, 3, 4, 5, \ldots$$

Negative numbers are commonly used to denote amounts owed, temperatures below zero, and even for locations below ground or below sea level. For example, we could use −100 to represent owing 100 dollars. The temperature −4° Celsius stands for 4 degrees below 0 degrees Celsius. An altitude of −50 feet stands for 50 feet below sea level. In some places, negative numbers are even used on floor levels. The photo to the left shows a floor directory in a French department store. Floor 0 is ground level and Floor −1 is one flight below ground level.

Number lines allow us to display the negative numbers on an equal footing with 0 and the other numbers on the number line. To the right of 0 on the number line are the positive numbers. To the left of 0 on the number line are the negative numbers. The number 0 is considered neither positive nor negative.

Given a positive number *N* that is located to the right of zero, *N* units from 0 on the number line, its negative, −*N* is located to the *left* of 0 on the number line, *N* units from 0. For example, the negative number −2.41 is located to the left of 0 at a distance of 2 units plus 4 tenths of a unit plus 1 hundredth of a unit away from 0. So to locate this number on a number line, start at 0, move 2 units to the left, then move another 4 tenths of a unit to the left, and finally move yet another 1 hundredth of a unit to the left. (See Figure 1.23) In this way, the number −2.41 is plotted a full 2.41 units away from 0.

> **Class Activity**   *Now Turn to Class Activities Manual*
>
> **1I** Negative Numbers on Number Lines, p. 11

Decimals with Infinitely Many Nonzero Entries

Some decimals extend infinitely far to the right. For example, it turns out that the decimal representation of $\sqrt{2}$ (the positive number that when multiplied with itself is 2) is

$$\sqrt{2} = 1.41421356237\ldots$$

which goes on forever, never ending in zeros. *Every* decimal, even a decimal that extends infinitely to the right, has a definite location on a number line. However, such a decimal will never fall *exactly* on a tick mark, no matter what the scale of the number line. Figure 1.24 shows where $\sqrt{2}$ is located on number lines of various scales. The fact that even a decimal with infinitely many nonzero entries to the right of the decimal point has a location on the number line may seem like a murky idea; in fact, it is a very subtle point, whose details were not fully worked out by mathematicians until the late nineteenth century.

FIGURE 1.24

Plotting $\sqrt{2} =$ 1.41421356237 . . . on a number line

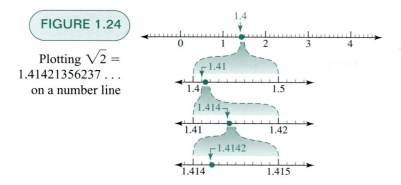

Practice Exercises for Section 1.2

Answers to these practice exercises begin on page 20.

1. Describe three ways discussed in the text to represent a decimal such as 1.234. For each way, show how to represent 1.234.

2. Describe how to represent 0.0278 with bundles of small objects in a way that fits with and shows the structure of the decimal system. In this case, what does one small object represent?

3. Suppose you are teaching fourth-graders about decimals and how the structure of the decimal system remains the same to the left and right of the decimal point. You have bundles of toothpicks like the ones shown in Figure 1.25 and you want to use these bundled toothpicks to represent a decimal. List at least three decimals that you could use these bundles to represent and explain your answer in each case.

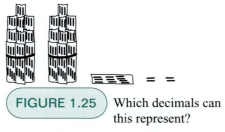

FIGURE 1.25   Which decimals can this represent?

4. Describe and draw pictures showing how to represent 1.369 and 1.07 as lengths by using strips of paper in a way that fits with and shows the structure of the decimal system.

5. Draw a number line on which long tick marks are whole numbers and on which 0.003 can be plotted (in its approximate location). Now show how to "zoom in" on smaller and smaller portions of the number line (as in Figure 1.21 and Class Activity 1G, part 1) until

you have zoomed in to a portion of the number line in which the long tick marks are thousandths and so that 0.003 can be plotted on each number line. Label the long tick marks on each number line, and plot 0.003 on each number line (in its approximate location).

6. Draw number lines like the ones in Figure 1.26. Label the tick marks on your three number lines in three different ways. In each case, your labeling should fit with the structure of the decimal system and the fact that the tick marks at the ends of the number lines are longer than the other tick marks.

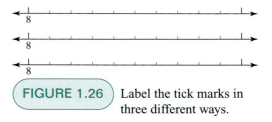

FIGURE 1.26   Label the tick marks in three different ways.

7. Draw number lines like the ones in Figure 1.27. Label the tick marks on your three number lines in three different ways. In each case, your labeling should fit with the structure of the decimal system and the fact that the tick marks at the ends of the number lines are longer than the other tick marks.

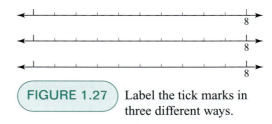

FIGURE 1.27   Label the tick marks in three different ways.

8. Draw a number line like the one in Figure 1.28 for the exercises that follow. For each exercise, label all the tick marks on the number line. The number to be plotted need not land on a tick mark.

   a. Plot 28.369 on a number line on which the long tick marks are tenths.

   b. Plot 1.0601 on a number line on which the long tick marks are thousandths.

   c. Plot 14.8577 on a number line on which the long tick marks are whole numbers.

   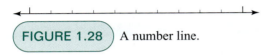

   FIGURE 1.28   A number line.

9. Draw number lines like the ones in Figure 1.29. Label the tick marks on your number lines with appropriate decimals.

   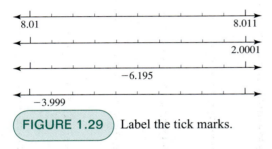

   8.01   8.011

   2.0001

   −6.195

   −3.999

   FIGURE 1.29   Label the tick marks.

10. Describe where the negative numbers are located on the number line. Then describe how to plot the numbers −2 and −2.41 on a number line on which 0 and 1 have been plotted.

11. Draw a number line like the one in Figure 1.28 for the exercises that follow. For each exercise, label all the tick marks on the number line. The number to be plotted need not land on a tick mark.

    a. Plot −7.65 on a number line on which the long tick marks are whole numbers.

    b. Plot −0.0118 on a number line on which the long tick marks are hundredths.

    c. Plot −1.584 on a number line on which the long tick marks are tenths.

12. Give examples of decimal numbers that cannot be represented with bundles of toothpicks—even if you had as many toothpicks as you wanted.

Answers to Practice Exercises for Section 1.2

1. The decimal 1.234 can be represented with bundled objects, as a length, or on a number line. To represent 1.234 with bundled toothpicks, let 1 toothpick represent one thousandth. Then 1.234 is represented with 4 single toothpicks, 3 bundles of ten, 2 bundles of one hundred (each of which is 10 bundles of 10) and 1 bundle of a thousand (which is 10 bundles of 100). See the text for representing 1.234 as a length.

2. Represent 0.0278 as 2 bundles of 100 objects (each of which is 10 bundles of 10), 7 bundles of 10 objects, and 8 individual objects. In this case, each individual object must represent one ten-thousandth, since 0.0278 is 2 hundredths and 7 thousandths and 8 ten-thousandths.

3. If one toothpick represents 1, then Figure 1.25 represents 214. The following table shows several other possibilities:

If 1 Toothpick Represents	Then Figure 1.25 Represents
100	21,400
10	2140
1	214
0.1	21.4
0.01	2.14
0.001	0.214

4. Let a long strip of paper (e.g., 1 meter long) represent 1 unit of length, as at the top of Figure 1.30. Make another 1-unit-long strip, and subdivide it into 10 strips of equal length. Each of these strips are then 0.1 unit long. Make another 0.1-unit-long strip, and subdivide this strip into 10 strips of equal length. Each of these smaller strips are then 0.01 unit long. Make another 0.01-unit-long strip, and subdivide this strip into 10 strips of equal length. Each of these tiny strips are then 0.001 unit long.

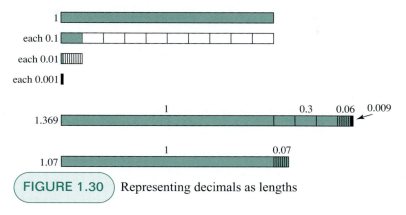

FIGURE 1.30 Representing decimals as lengths

To represent 1.369 as a length, make a long strip by laying the 1-unit-long strip next to 3 of the 0.1-unit-long strips, 6 of the 0.01-unit-long strips, and 9 of the 0.001-unit-long strips, as in the middle of Figure 2.29.

To represent 1.07 as a length, make a long strip by laying the 1-unit-long strip next to 7 of the 0.01-unit-long strips (don't use any 0.1-unit-long strips or 0.001-unit-long strips), as at the bottom of Figure 1.30.

5. See Figure 1.31.

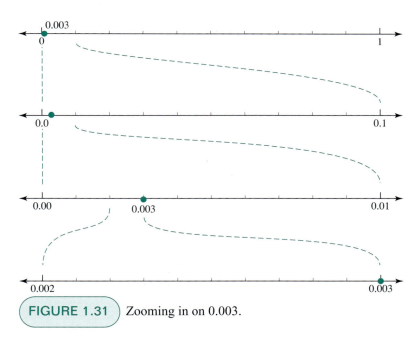

FIGURE 1.31 Zooming in on 0.003.

6. See Figure 1.32 for one way to label the tick marks. We can think of the second number line as "zoomed in" on the left portion of the first, and we can think of the third number line as "zoomed in" on the left portion of the second.

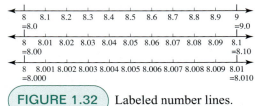

FIGURE 1.32 Labeled number lines.

7. See Figure 1.33 for one way to label the tick marks. We can think of the second number line as "zoomed in" on the right portion of the first and the third number line as "zoomed in" on the second.

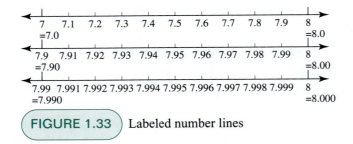

FIGURE 1.33   Labeled number lines

8. See Figure 1.34.

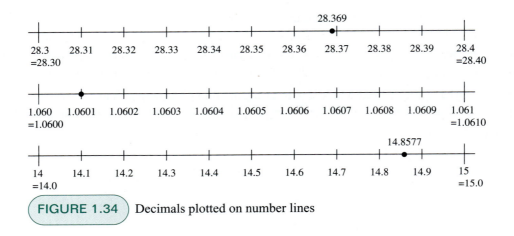

FIGURE 1.34   Decimals plotted on number lines

9. See Figure 1.35.

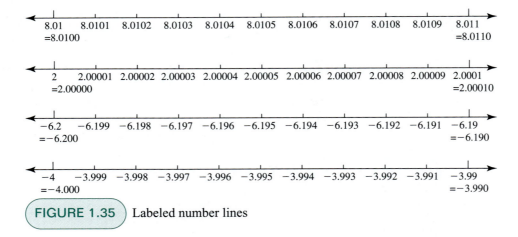

FIGURE 1.35   Labeled number lines

10. The negative numbers are located to the left of 0 on the number line. The number -2 is located 2 units from 0 to the left of 0 (and recall that the distance of 1 unit is the distance between 0 and 1). See the text for how to locate -2.41.

11. See Figure 1.36.

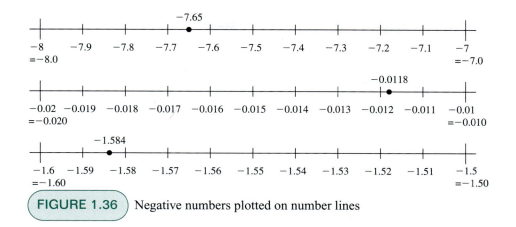

FIGURE 1.36   Negative numbers plotted on number lines

12. Realistically, we would be hard pressed to represent numbers with more than 4 nonzero digits with toothpicks. Even 999 would be difficult to represent. However, there are some numbers whose expanded form *can't* be represented by toothpicks (in the manner described in the text)—even if you had as many toothpicks as you wanted. For example, consider

0.333333333 . . .

where the 3s go on forever. Which place's value would you pick to be represented by 1 toothpick? If you did pick such a place, you would have to represent the values of the places to the right by tenths of a toothpick, hundredths of a toothpick, thousandths of a toothpick, and so on, forever, in order to represent this number.

As an aside, here is a surprising fact: We *can* represent 0.333333333 . . . with toothpicks, but in a different way (not by bundling so as to show the places in the decimal number). Namely, we can represent 0.333333333 . . . by one third of a toothpick, because it so happens that one third = 0.333333333 . . . (which you can see by dividing 1 by 3).

Problems for Section 1.2

1. Suppose you are teaching fourth-graders about decimals and how the structure of the decimal system remains the same to the left and right of the decimal point. You have bundles of toothpicks like the ones shown in Figure 1.37 and you want to use these bundled toothpicks to represent a decimal. List at least three decimals that you could use these bundles to represent and explain your answer in each case.

FIGURE 1.37   Which decimals can these bundled toothpicks represent?

2. Draw rough pictures of small bundled objects to show how to represent the accompanying decimals. Your pictures should correspond to the decimal representation of the numbers. In each case, list two other decimals that your picture could represent. Explain how to interpret your picture as representing these decimals.

a. 0.26   **c.** 1.28

b. 13.4   **d.** 0.000032

3. Describe and draw pictures showing how to represent 1.438 and 0.804 as lengths by using strips of paper in a way that fits with and shows the structure of the decimal system.

4. Jerome says that the unlabeled tick mark on the number line in Figure 1.38 should be 7.10. Why might Jerome think this? Explain to Jerome why he is not correct. Describe how you could help Jerome understand the correct answer.

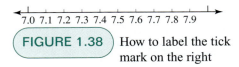

FIGURE 1.38   How to label the tick mark on the right

5. Students are sometimes uncertain about which zeros in decimals can be dropped and which can't (without changing the meaning of the number). Give examples of zeros in decimals that can be dropped and zeros that can't be dropped. Consider whole numbers as well as decimals that require a decimal point. Choose a small set of examples that nevertheless covers all the types of cases where zeros can be dropped and where zeros can't be dropped.

6. Draw a number line on which the tick marks are whole numbers. Now show how to zoom in on smaller and smaller portions of the number line (as in Figure 1.21 and Class Activity 1G, part 1) until you have zoomed in to a portion of the number line in which the long tick marks are thousandths. (You may choose which portions of the number line to zoom in on, but of course, your example should be different from Figure 1.21 and Class Activity 1G.)

7. ![icon] Draw a number line on which the long tick marks are whole numbers and on which 7.0028 can be plotted (in its approximate location). Show how to zoom in on smaller and smaller portions of the number line (as in Figure 1.21 and Class Activity 1G, part 1) until you have zoomed in to a portion of the number line in which the long tick marks are thousandths and so that 7.0028 can be plotted on each number line. Label the long tick marks on each number line, and plot 7.0028 on each number line (in its approximate location).

8. ![icon] Use a number line like the one in Figure 1.39 for the problems that follow. For each problem, label all the tick marks on the number line. The number to be plotted need not land on a tick mark.

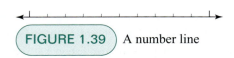

FIGURE 1.39   A number line

a. Plot 13.58 on a number line on which the long tick marks are whole numbers.

b. Plot 0.193 on a number line on which the long tick marks are tenths.

c. Plot 26.9999 on a number line on which the long tick marks are thousandths.

d. Plot 2.379 on a number line on which the long tick marks are tenths.

e. Plot 7.148 on a number line on which the long tick marks are whole numbers.

f. Plot 9.075132 on a number line on which the long tick marks are thousandths.

9. Cierral plots the decimal number 7.001 in the location shown on the number line in Figure 1.40. Is the label legitimate? If so, how should Cierral label the other tick marks?

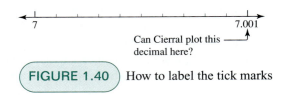

Can Cierral plot this decimal here?

FIGURE 1.40   How to label the tick marks

10. Juan plots the decimal number 9.999 in the location shown on the number line in Figure 1.41. Is this label legitimate? If so, how should Juan label the other tick marks?

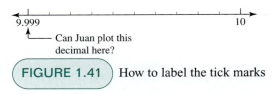

Can Juan plot this decimal here?

FIGURE 1.41   How to label the tick marks

11. For each number line in parts (a)–(d), draw three copies of the line. Use your number lines to show three different ways to label the tick marks in the original line. In each case, your labeling should fit with the structure of the decimal system and the fact that the tick marks at the ends of the number lines are longer than the other tick marks.

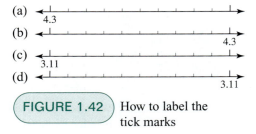

FIGURE 1.42   How to label the tick marks

12. Using the examples -1, -2, and -1.68, describe in your own words where negative numbers are located on a number line based on the locations of 0 and 1.

13. Some students confuse decimals and negative numbers and mix up the locations of 0 and 1 when comparing decimals (see [83]).

   a. Plot 0, 1, 0.6, -0.7, and -0.06 on a carefully drawn number line. You may "zoom in" on portions of the number line in order to show them more clearly.

   b. Put 0, 1, 0.6, -0.7, and -0.06 in order from least to greatest.

14. Students sometimes get confused about the locations of decimals and negative numbers on number lines. Draw a number line on which you have plotted 0, 1, and -1. Then give at least three examples of numbers that are between 0 and 1 and at least three examples of numbers that are between 0 and

-1, and plot all your examples on your number line. Pick examples that give students some sense of the variety of numbers that are between 0 and 1 and between 0 and -1. Why do you think your choice of examples will be good for students to see?

15. Draw a number line like the one in Figure 1.39 for the problems that follow. For each problem, label all the tick marks on the number line. The number to be plotted need not land on a tick mark.

   a. Plot -4.3 on a number line on which the long tick marks are whole numbers.

   b. Plot -0.28 on a number line on which the long tick marks are whole numbers.

   c. Plot -0.28 on a number line on which the long tick marks are tenths.

   d. Plot -6.193 on a number line on which the long tick marks are tenths.

   e. Plot -6.193 on a number line on which the long tick marks are hundredths.

1.3 Comparing Numbers in the Decimal System

**F.P Focal Points
Grades 2, 4**

If Timothy Elementary School raised $1023 for a fundraiser and Barrow Elementary School raised $789, which school raised more money? Of course you know right away that Timothy Elementary raised more, but why is the method we use to determine which of 1023 and 789 is greater legitimate? How do we know that we can compare numbers that way? The rationale for why we can compare numbers the way we do relies on the nature of the decimal system, and we will investigate the rationale in this section. We will also discuss the concepts of "greater than" and "less than" and the idea of comparing numbers by viewing them as representing quantities and by viewing numbers as located on number lines.

Comparing Numbers by Viewing Them as Amounts

greater than   If A and B are numbers that are not negative, then we say that A is **greater than** B, and we write

$$A > B$$

less than   if A represents a larger quantity than B. Similarly, we say that A is **less than** B, and we write

$$A < B$$

if A represents a smaller quantity than B. A good way to remember which symbol to use is to notice that the "wide side" faces the larger number. Some teachers have their students draw alligator teeth on the symbols, as in Figure 1.43, to help them: Alligators are hungry and always want to eat the larger amount.

$2 \triangleleft 5$

FIGURE 1.43

Alligator teeth on a "less than" symbol

Understanding Why We Can Compare Numbers the Way We Do

Think for a moment about how you compare two numbers to decide which one is greater. For example, how do you know that 584,397 is less than 600,214? As adults, we are used to comparing numbers, so we usually don't stop to think about why it works. But how do we know that we can compare numbers the way we do? The answer lies in the nature of the decimal system.

Class Activity   *Now Turn to Class Activities Manual*

1J Places of Larger Value Count More Than Lower Places Combined, p. 12

If you did the Class Activity, you saw that even if you make the largest possible number that uses only the hundreds, tens, and ones places, this number—999—is still smaller than the smallest number you can make with a nonzero entry in the thousands place, namely, 1000. So the thousands place counts more than the lower hundreds place, tens place, and ones place combined.

Similarly, as we see in Figure 1.44, the tenths place counts more than the lower hundredths place and thousandths place combined. Even if we make the largest possible number that uses only the hundredths and thousandths places, namely 0.099, this number is still less than the smallest number we can make with a nonzero entry in the tenths place, namely 0.1. Notice that in Figure 1.44, starting with the

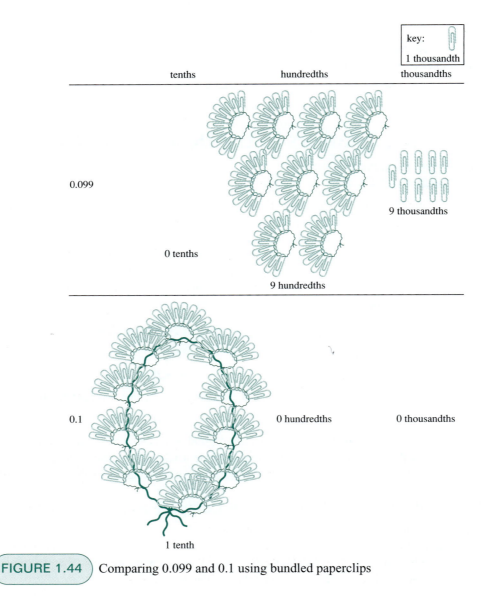

FIGURE 1.44   Comparing 0.099 and 0.1 using bundled paperclips

9 bundles of 10 paperclips and 9 loose paperclips representing 0.099, we would have to add one more paperclip to bundle with the 9 loose paperclips to make another bundle of 10 that could then be bundled with the 9 bundles to make 10 bundles of 10 representing 0.1. Therefore,

$$0.099 < 0.1$$

In general, as illustrated by the previous paragraphs, *decimal places of larger value count more than the largest combined value made with lower places.*

Therefore, to compare the sizes of decimal numbers, we should compare numbers by comparing like places and by starting at the place of largest value in which at least one of the numbers has a nonzero entry.

For example,

$$1234 > 789$$

because 1234 has a 1 in the thousands place and 789 has 0 in the thousands place (even though this zero is usually not written).

$$1234$$
$$\updownarrow$$
$$0789$$

Similarly,

$$1.2378 < 1.24$$

because both numbers have a 1 in the ones place and a 2 in the tenths place, but 1.24 has a larger digit, namely, 4, in the hundredths place than does 1.2378, which only has a 3.

$$1.2378$$
$$\updownarrow\updownarrow\updownarrow$$
$$1.24$$

Notice that determining the greater number is similar to putting words in alphabetical order, except that for decimal numbers, we compare values in *like places*, whereas for words, we compare letters starting from the *beginning* of each word. Maybe this is why a decimal such as 1.01 might at first glance look as if it is less than 0.998.

A place counts more than all the places of lower value combined; therefore, use the following general rule for comparing two nonnegative numbers.

How to Compare Numbers in the Decimal System *Starting from the place of largest value* represented in the numbers (the left-most place), compare the digits of both numbers in that place.

- If the digits in that place are not equal, then the decimal number with the larger digit is greater than the other number.
- If the digits in that place are equal, keep moving one place to the right, comparing the digits of the numbers in like places, until one of the numbers has a larger digit than the other. The number with this larger digit is greater than the other number.

Class Activity *Now Turn to Class Activities Manual*

1K Misconceptions in Comparing Decimals, p. 12

1L Finding Smaller and Smaller Decimals, p. 14

1M Finding Decimals Between Decimals, p. 15

A Technical Exception to the Rule for Comparing Numbers in the Decimal System There is a rare exception to the preceding method for determining which of two decimals is greater. The exception occurs when one of the decimals has an infinitely repeating 9, such as

$$37.569999999999\ldots$$

In this case, the preceding method would lead us to say that

$$37.57 > 37.569999999999\ldots$$

However, this is *not true*. In fact, as we will see in Section 8.6,

$$37.57 = 37.569999999999\ldots$$

Using Number Lines to Compare Numbers

It is especially easy to compare numbers when they are plotted on a number line. When we use a number line, we can compare negative numbers as well as positive ones.

If A and B are numbers plotted on a number line, then the following statements are true:

$A < B$ provided that A is to the left of B on the number line.

$A > B$ provided that A is to the right of B on the number line.

These interpretations of "greater than" and "less than" are consistent with the interpretation we developed previously for positive numbers, where we viewed positive numbers as representing quantities. Why? If a positive number A represents a larger quantity than another positive number B, then A will be plotted farther from 0 than B. Since both are plotted to the right of 0, A will be to the right of B. Likewise, if A represents a smaller quantity than B, then we will plot A closer to 0 than B. Since both are to the right of 0, A will be to the left of B.

With the number line interpretation of $>$ and $<$, it is easy to compare negative numbers. For example, we plot -2.7 to the left of -1.5, as in Figure 1.45; therefore,

$$-2.7 < -1.5$$

Comparing Negative Numbers by Viewing Them as Owed Amounts

In addition to explaining why $-2.7 < -1.5$ by using a number line, we can explain why $-2.7 < -1.5$ by considering these negative numbers as amounts owed. Think of -2.7 as owing \$2.70 and -1.5 as owing \$1.50; in other words, to have $-\$2.70$ is to owe \$2.70, and to have $-\$1.50$ is to owe \$1.50. If you owe \$2.70, you have less than if you owe \$1.50; therefore,

$$-2.7 < -1.5$$

In general, the more you owe, the less you have, and the less you owe, the more you have. In other words,

If $A > B$, then $-A < -B$.

If $A < B$, then $-A > -B$.

For example,

$$7 > 4$$

Therefore,

$$-7 < -4$$

Notice that the negative signs cause the $>$ symbol to reverse.

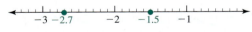

(**FIGURE 1.45**) Showing that -2.7 is less than -1.5

Class Activity   *Now Turn to Class Activities Manual*

1N Decimals Between Decimals on Number Lines, p. 15

1O "Greater Than" and "Less Than" with Negative Numbers, p. 16

Practice Exercises for Section 1.3

1. Explain in your own words why we compare numbers in the decimal system by starting at the leftmost place at which there is a nonzero digit and then proceeding to the right if necessary. For example, we compare 341 and 87 by starting at the hundreds place and we compare 1097 and 1233 by starting at the thousands place and then moving to the hundreds place. What is it about the decimal system that allows us to compare numbers this way?

2. Which is greater, 0.01 or 0.0099999999999?

3. Draw a picture that shows bundled objects representing 1.2 and 0.89 and demonstrates that $1.2 > 0.89$.

4. Describe and draw pictures showing how to represent 1.2 and 0.89 as lengths by using strips of paper that are 1 unit long, 0.1 unit long, 0.01 unit long, and 0.001 unit long as in Class Activity 1F. Use these representations to explain why $1.2 > 0.89$.

5. Use a number line to show that $1.2 > 0.89$.

6. Use a number line to show that $-1.2 < -0.89$.

7. Explain why $-1.2 < -0.89$, without using a number line.

8. Find a number between 1.4142133 and 1.41422, and plot all three numbers visibly and distinctly on a number line in which the tick marks fit with the structure of the decimal system. The numbers need not land on tick marks.

9. Give two different decimals that are between 3.456 and 3.457.

10. Children who have heard of a googol, which is the number that is written as a 1 followed by one hundred zeros, will often think it's the largest number. Is it? Is there a largest number?

11. The smallest whole number that is greater than zero is 1. Is there a smallest decimal that is greater than zero?

Answers to Practice Exercises for Section 1.3

1. See text.

2. The number 0.01 is greater because is has a 1 in the hundredths place, whereas 0.0099999999999 has a 0 in the hundredths place and all higher places.

3. See Figure 1.46.

4. Let a long strip of paper (e.g., 1 meter long) represent 1 unit of length, as at the top of Figure 1.47. Make another 1-unit-long strip, and subdivide it into 10 strips of equal length. Each of these strips is then

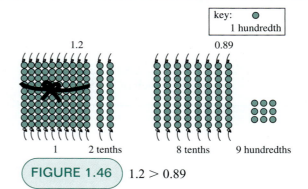

FIGURE 1.46 $1.2 > 0.89$

0.1 unit long. Make another 0.1-unit-long strip, and subdivide this strip into 10 strips of equal length. Each of these smaller strips is then 0.01 unit long. We won't need it, but we could make another 0.01-unit-long strip, and subdivide this strip into 10 strips of equal length. Each of these tiny strips would then be 0.001 unit long.

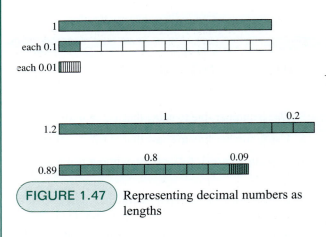

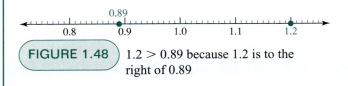

FIGURE 1.47   Representing decimal numbers as lengths

To represent 1.2 as a length, make a long strip by laying the 1-unit-long strip next to 2 of the 0.1-unit-long strips, as in the middle of Figure 1.47. To represent 0.89 as a length, make a long strip by laying 8 of the 0.1-unit-long strips next to 9 of the 0.01-unit-long strips (don't use a 1-unit-long strip), as at the bottom of Figure 1.47. Since the 1.2-unit-long strip is longer than the 0.89-unit-long strip, we conclude that $1.2 > 0.89$.

5. See Figure 1.48.

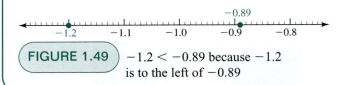

FIGURE 1.48 $1.2 > 0.89$ because 1.2 is to the right of 0.89

6. See Figure 1.49.

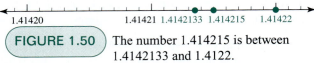

FIGURE 1.49 $-1.2 < -0.89$ because -1.2 is to the left of -0.89

7. To explain why $-1.2 < -0.89$, consider having -1.2 as owing $1.20 and having -0.89 as owing $0.89. If you owe $1.20, then you have less than if you owe only $0.89. Therefore, having -1.2 means you have less than if you have -0.89. So, $-1.2 < -0.89$.

8. The number 1.414215 is one example of number that is in between 1.4142133 and 1.41422. There are many other examples. See Figure 1.50.

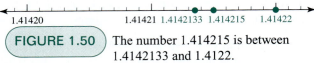

1.41420   1.41421 1.4142133 1.414215   1.41422

FIGURE 1.50   The number 1.414215 is between 1.4142133 and 1.4122.

9. The numbers 3.4563 and 3.4567 both lie in between 3.456 and 3.457.

10. The number a googol plus one, for example, is greater than a googol. There is no largest number because, no matter what number you choose as candidate for the largest number, that number plus one is a larger number.

11. No, there is no smallest decimal that is greater than 0. Consider the following list of numbers:

$$0.1$$
$$0.01$$
$$0.001$$
$$0.0001$$
$$0.00001$$
$$0.000001$$

Imagine the list continuing forever. The numbers in this list get smaller and smaller, getting ever closer to 0 without ever reaching 0. No matter what decimal you choose that is greater than 0, it will have a nonzero entry somewhere. Based on where the first nonzero entry in your chosen number is, you can pick a decimal from the preceding list that is even smaller than your chosen number. Therefore, there can be no smallest decimal that is greater than zero.

Problems for Section 1.3

1. Explain in your own words why we compare numbers in the decimal system by starting at the leftmost place at which there is a nonzero digit and then proceeding to the right if necessary. For example, we compare 234 and 219 by starting at the hundreds place and then moving to the tens place. We compare 1122 and 987 by starting at the thousands place. What is it about the decimal system that allows us to compare numbers this way?

2. Draw a picture that shows bundled objects representing 1.1 and 0.99 and that demonstrates that 1.1 > 0.99.

3. Explain how to show which of 1.1 and 0.999 is greater in the following three ways:

   a. Describe and draw (approximate) pictures showing how to represent 1.1 and 0.999 as lengths by using strips of paper that are 1 unit long, 0.1 unit long, 0.01 unit long, and 0.001 unit long as in Class Activity 1F. Use these representations to show which of 1.1 and 0.999 is greater.

   b. Use a carefully drawn number line to show which of 1.1 and 0.999 is greater. You may wish to "zoom in" on portions of the number line.

   c. Draw pictures showing how to represent 1.1 and 0.999 with bundled objects in a way that fits with the structure of the decimal system. Use these representations to show which of 1.1 and 0.999 is greater.

4. Explain how to show which of 0.1 and 0.095 is greater in the following three ways:

   a. Describe and draw (approximate) pictures showing how to represent 0.1 and 0.095 as lengths by using strips of paper that are 1 unit long, 0.1 unit long, 0.01 unit long, and 0.001 unit long as in Class Activity 1F. Use these representations to show which of 0.1 and 0.095 is greater.

   b. Use a carefully drawn number line to show which of 0.1 and 0.095 is greater. You may wish to "zoom in" on portions of the number line.

   c. Draw pictures showing how to represent 0.1 and 0.095 with bundled objects in a way that fits with the structure of the decimal system. Use these representations to show which of 0.1 and 0.095 is greater.

5. Some students have difficulty comparing decimals with 0 or comparing decimals that have zeros in some places (see [88] and [83]). Plot each of the given pairs or triples of decimal numbers on a carefully drawn number line to show which of the pair or triple is greatest. You may "zoom in" on portions of your number lines in order to show locations more clearly.

   a. Compare 0.6 and 0.

   b. Compare 0.00 and 0.7.

   c. Compare 3.00, 3.0, and 3.

   d. Compare 3.7777 and 3.77.

6. Mary is labeling tick marks on a number line. Starting at a tick mark labeled 4.90, she labels the next tick marks to the right with 4.91, 4.92, 4.93, . . . , 4.99. Then she labels the next tick mark to the right of 4.99 with 4.100. Mary then says that 4.100 is greater than 4.99.

   a. Discuss: What error is Mary making? Why might Mary make this error?

   b. Describe how you might help Mary understand her error and how to correct it.

7. Mark says that 0.178 is greater than 0.25. Explain in at least two different ways why Mark's statement is not correct.

8. Find a number between 3.24 and 3.241, if there is one. If there is no number between them, explain why not.

9. Is there more than one decimal between 8.45 and 8.47? If so, find two such decimals.

10. a. Find a number between 3.8 and 3.9, and plot all three numbers visibly and distinctly on a number line like the one in Figure 1.51, which has a set of longer tick marks and a set of shorter tick marks. Your labeling of the tick marks should fit with the structure of the decimal system.

FIGURE 1.51   A number line

   b. Describe how to use money to find a number between 3.8 and 3.9.

11. For each of the following pairs of numbers, find a decimal between the two numbers, and plot all three numbers visibly and distinctly on a number line like the one in Figure 1.51, which has a set of longer tick marks and a set of shorter tick marks. Label all the longer tick marks. Your labeling of the tick marks should fit with the structure of the decimal system.

   a. The numbers 2.981 and 2.982
   b. The numbers 13 and 12.9999
   c. The numbers 13 and 13.0001

12. Explain in two different ways why $-8 < -5$.

13. Explain in two different ways why $-3.25 < -1.4$.

14. For each of the following pairs of numbers, find a decimal between the two numbers, and plot all three numbers visibly and distinctly on a number line like the one in Figure 1.51, which has a set of longer tick marks and a set of shorter tick marks. Label all the longer tick marks. Your labeling of the tick marks should fit with the structure of the decimal system.

   a. The numbers -1.5 and -1.6
   b. The numbers -34.9714 and -34.9835
   c. The numbers -7.834 and -7.83561

15. Describe an infinite list of decimals, all of which are greater than 3.514, but get closer and closer to 3.514.

16. The smallest integer that is greater than 2 is 3. Is there a smallest decimal that is greater than 2? Explain why or why not in your own words.

1.4 Rounding Numbers

Before you read on, do the Class Activity.

Class Activity   *Now Turn to Class Activities Manual*

1P Why Do We Round? p. 16

We often use numbers to describe the size of actual quantities. The distance between two locations might be described as 2.7 miles; the population of a city might be given as 87,000; a tax rebate might be described as $1.3 trillion. These numbers are not the *exact* amounts: There aren't exactly 87,000 people living in the city, and the tax rebate isn't $1.3 trillion on the nose. Instead, we understand that these numbers have been *rounded.* Sometimes we round because we don't know a quantity exactly, and sometimes we round because we want to convey the approximate size of a quantity. For example, a company might know that it spent $174,586.74 on some equipment, but in a report to management, an employee might describe the expenditure as $175,000. This round number is more quickly and easily grasped and is close enough to the actual expenditure for the discussion at hand. We also round numbers to estimate the result of a calculation. For example, to estimate 5.9×8.2 we could round 5.9 to 6 and 8.2 to 8, estimating 5.9×8.2 as $6 \times 8 = 48$.

How to Round

round   To **round** a number means to find a nearby number that has fewer (or no more) nonzero digits. We can round to the nearest 100, to the nearest 10, to the nearest 1, to the nearest tenth, to the nearest hundredth, and so on. For any place value, we can round to that place's value. To round a given number to the nearest 100, we must find the number that is closest to our given number and has only zeros in places smaller

FIGURE 1.52

Rounding to the
nearest hundred

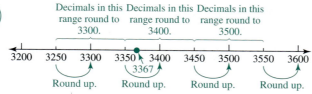

than the hundreds place. To round a given number to the nearest tenth, we must find the number that is closest to our given number and has only zeros in places smaller than the tenths place. In general, to round a given number to a given place's value, we must find the number that is closest to our given number and that has only zeros in places smaller than that place.

To understand rounding, think about zooming in or zooming out on a number line so that you can see the tick marks that fit with the desired type of rounding. When rounding to the nearest hundred, zoom out so that you see large tick marks labeled with numbers in the hundreds, as in Figure 1.52. The numbers at the large tick marks are exactly those numbers that have only zeros in places smaller than the hundreds place. To round to the nearest hundred, find the large tick mark that is closest to the location of the number you are rounding. Thus, to round 3367 to the nearest hundred, notice that 3367 lies between the hundreds 3300 and 3400. But 3367 is closer to 3400 than to 3300 because 3367 is greater than 3350, which is halfway between 3300 and 3400. Therefore, round 3367 to 3400.

When rounding to the nearest tenth, zoom in on the number line so that you see large tick marks labeled with tenths, as in Figure 1.53. The numbers at these large tick marks are exactly those numbers that have only zeros in places smaller than the tenths place. To round to the nearest tenth, find the large tick mark that is closest to the location of the number you are rounding. Thus, to round 27.839 to the nearest tenth, notice that 27.839 lies between the tenths 27.8 and 27.9. But 27.839 is closer to 27.8 than to 27.9 because 27.839 is less than 27.85, which is halfway between 27.8 and 27.9. Therefore, round 27.839 to 27.8.

What do we do about numbers that are exactly halfway in between two tick marks? For example, how do we round 3450 when we are rounding to the nearest hundred? How do we round 27.45 when we are rounding to the nearest tenth? We need a convention to break ties like these. The most common convention, and the one you should use unless otherwise specified, is to *round up* when there is a tie. So when rounding 3450 to the nearest hundred, round to 3500, as shown in Figure 1.52. When rounding 27.85 to the nearest tenth, round to 27.9, as shown in Figure 1.53. Another convention for rounding—one that is often used by engineers—is to *round to the nearest even* when there is a tie. With this convention, when rounding to the nearest hundred, 3450 would be rounded to 3400, rather than 3500, because 4 is even. Similarly, when rounding to the nearest tenth, 27.85 would be rounded to 27.8, rather than 27.9, because 8 is even.

Class Activity   *Now Turn to Class Activities Manual*

1Q Explaining Rounding by Zooming Out, p. 16

1R Rounding with Number Lines, p. 18

1S Can We Round This Way? p. 19

FIGURE 1.53

Rounding to the
nearest tenth

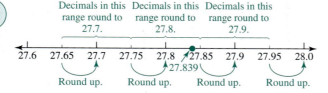

Working with Numbers That Represent Actual Quantities

When we use a number to describe the size of an actual quantity, such as a distance or a population, we generally assume that this number has been rounded. Furthermore, we assume that *the way a number is written indicates the rounding that has taken place.* For example, when the distance between two locations is described as 6.2 miles, we assume from the presence of a digit in the tenths place, but no smaller place, that the actual distance has been rounded to the nearest tenth of a mile. Therefore, the actual distance could be anywhere between 6.15 and 6.25 miles (but less than 6.25 miles). Similarly, if the population of a city is described as 93,000, we assume from the nonzero entry in the thousands place and the zeros in all lower places that the actual population has been rounded to the nearest thousand. Therefore, the actual population could be anywhere between 92,500 and 93,500 (but less than 93,500).

Because of rounding, there is a slight difference in the meaning of numbers in the abstract and decimal numbers when they are used to describe the size of an actual quantity. In the abstract,

$$8.00 = 8$$

However, when we write 8 for the weight of an object in kilograms, it has a different meaning than when we write 8.00 for the weight of an object in kilograms. Writing 8 indicates that the actual weight has been rounded to the nearest one, whereas writing 8.00 means that the actual weight has been rounded to the nearest hundredth. If the weight is reported as 8.00 kilograms, then the actual weight could be anywhere between 7.995 and 8.005 kilograms. If, however, the weight is reported as 8 kilograms, then the actual weight could be anywhere between 7.5 and 8.5 kilograms. Therefore, writing 8.00 conveys that the weight is known much more accurately than if the weight is reported as 8 kilograms.

When you solve a problem that involves real or realistic quantities, round your answer so that it does not appear to be more accurate than it actually is. Your answer cannot be any more accurate than the numbers you started with, so round your answer to fit the rounding of your initial numbers. For example, suppose that the population of a city is given as 1.6 million people. After some calculations, you project that the city will have 1.95039107199 million people in 10 years. Although the decimal number 1.95039107199 may be the exact answer to your calculations, do not report your answer this way, because it makes your answer appear to be more accurate than it actually is. You must assume that the initial number 1.6 is rounded to the nearest tenth. Therefore, you should also round the answer, 1.95039107199, to the nearest tenth and report the projected population in 10 years as 2.0 million.

Practice Exercises for Section 1.4

1. Round 6.248 to the nearest tenth. Explain in words why you round the number the way you do. Use a number line to support your explanation.

2. Round 39,995 to the nearest ten. Explain in words why you round the number the way you do. Use a number line to support your explanation.

3. Round 173.465 to the nearest hundred, to the nearest ten, to the nearest one, to the nearest tenth, and to the nearest hundredth.

4. The distance between two cities is described as 1500 miles. Should you assume that this is the exact distance between the cities? If not, what can you say about the exact distance between the cities?

Answers to Practice Exercises for Section 1.4

1. You are rounding to the nearest tenth, so you must first find the "tenths" that 6.248 lies between. In other words, on a number line on which the tick marks are tenths, you must find which tick marks 6.248 lies between. The number 6.248 is between the tenths 6.2 and 6.3. But because of the 4 in the hundredths place, 6.248 is less than 6.25, which is halfway between 6.2 and 6.3. Therefore, 6.248 is closer to 6.2 than to 6.3, so 6.248 rounded to the nearest tenth is 6.2.

   The number line in Figure 1.54 illustrates what was just stated.

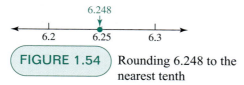

   FIGURE 1.54 Rounding 6.248 to the nearest tenth

2. Since you are rounding to the nearest ten, you must first find the "tens" that 39,995 lies between. In other words, on a number line on which the tick marks are tens, you must find which tick marks 39,995 lies between. The number 39,995 lies exactly halfway between 39,990 and 40,000, as shown in Figure 1.55. By the "round a 5 up" convention, round up to 40,000.

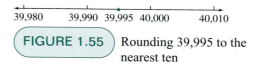

   FIGURE 1.55 Rounding 39,995 to the nearest ten

3. To the nearest hundred: 200

   To the nearest ten: 170

   To the nearest one: 173

   To the nearest tenth: 173.5

   To the nearest hundredth: 173.47

4. Because the reported distance has zeros in the tens and ones places, assume that the distance has been rounded to the nearest hundred. Therefore, the exact distance between the two cities is probably not 1500 miles; it could be anywhere between 1450 miles and 1550 (but less than 1550).

Problems for Section 1.4

1. Round 2.1349 to the nearest hundredth. Explain in your own words why you round the number the way you do. Use a number line to support your explanation.

2. Round 27,003 to the nearest hundred. Explain in your own words why you round the number the way you do. Use a number line to support your explanation.

3. Round 9995.2 to the nearest ten. Explain in your own words why you round the number the way you do. Use a number line to support your explanation.

4. Adam has made up his own method of rounding. Starting at the right-most place in a number, he keeps rounding to the value of the next place to the left until he reaches the place to which the number was to be rounded. For example, Adam would use the following steps to round 11.3524 to the nearest tenth:

   $$11.3524 \rightarrow 11.352 \rightarrow 11.35 \rightarrow 11.4$$

   Is Adam's method a valid way to round? Explain why or why not.

5. The label on a snack food package says that one serving of the snack food contains 0 grams of trans fat. Does this mean that there is no trans fat in one serving of the snack food? If not, what can you say about the amount of trans fat in a serving of the snack food? Discuss these questions in light of the information presented in this section.

6. The weight of an object is reported as 12,000 pounds. Should you assume that this is the exact weight of the object? If not, what can you say about the exact weight of the object? Explain.

7. In a report, a population is given as 2700. Should you assume that this is the exact population? If not, what can you say about the exact population? Explain.

Chapter Summary and Study Items

Section 1.1 The Counting Numbers

The counting numbers are the numbers 1, 2, 3, 4, ... There are two distinct ways to think about the counting numbers: on the one hand, the counting numbers form an ordered list; on the other hand, counting numbers tell how many objects are in a set (i.e., the cardinality of a set). When we count the number of objects in a set, we connect the two views of counting numbers by making a one-to-one correspondence between an initial portion of the list of counting numbers and the objects in the set. The last number word that we say when we count a number of objects tells us how many objects there are.

In the decimal system, every counting number can be written using only the ten symbols 0, 1, 2, . . . 9. The decimal system uses place value, which means that the value that a digit in a number represents depends on the location of the digit in the number. The value of each place is ten times the value of the place to its immediate right. We can represent numbers in the decimal system by repeated bundling of small objects in groups of ten. If one object represents 1, then a bundle of ten objects represents 1 ten. Ten bundles of ten objects represent 1 hundred. Ten bundles of one hundred (each of which is ten bundles of ten) represents 1 thousand, and so on. This repeated bundling corresponds to the value of places in the decimal system: The value of a place is ten times the value of the place to its immediate right.

Unfortunately, the way we say the English names of the counting numbers 11 through 19 does not fit well with the way we write these numbers.

The counting numbers can be displayed on number paths, which are informal precursors to number lines. The whole numbers, 0, 1, 2, 3, . . . , can be displayed on number lines.

Key skills and understandings:

- Describe the two views of the counting numbers—as a list and as used for cardinality. Discuss the connections between the list and cardinality views of the counting numbers.

- Explain what it means for the decimal system to use place value, and explain why the decimal system is so useful.

- Describe the relationship between values of decimal places.

- Describe and draw rough pictures to represent a given counting number in terms of bundled objects in a way that fits with the decimal representation for that number of objects.

- Discuss the difference between a number path (as described in the text) and a number line.

Section 1.2 Decimals and Negative Numbers

The decimals extend the decimal system and are formed by creating decimal places to the right of the ones place while preserving the essential structure of the decimal system: the value of each place is ten times the value of the place to its immediate right. Just as the counting numbers can be represented with bundled objects, finite (positive) decimals can also be represented with bundled objects, provided that a single object stands for the value of the lowest place one needs to represent.

The positive decimals can be represented as lengths in a way that fits with the structure of the decimal system. This way of representing decimals fits nicely with the metric system (meters, decimeters, centimeters, and millimeters.) Decimals can also be represented on number lines in a way that is compatible with viewing positive decimals as lengths: the location of a positive decimal is its distance from 0. We can also think of the decimals as filling in the number line: First, we plot whole numbers, then we plot tenths, then hundredths, and so on.

The negative numbers are the numbers to the left of zero on the number line.

Key skills and understandings:

- Describe and draw rough pictures to represent a given positive decimal in terms of bundled objects in a way that fits with and shows the structure of the decimal system. (Take care to state the meaning of one of the objects.)

- Describe and draw rough pictures to represent a given positive decimal as a length in a way that fits with and shows the structure of the decimal system.

- Discuss how decimals fill in a number line, label tick marks on number lines, and plot numbers on number lines.

- Show how to zoom in on portions of number lines to see the portions in greater detail.

- View negative numbers as owed amounts, and plot a given negative number on a number line.

Section 1.3 Comparing Numbers in the Decimal System

The value of a decimal place is greater than the largest number made from places of lower value. For this reason, we compare numbers in the decimal system by looking first at the place of greatest value, then moving to places of lower value if the numbers agree in the place of greatest value. On number lines, numbers become greater as one moves to the right. Given any two distinct numbers, there is always another number in between. (In fact, there are infinitely many numbers in between.) Taking the negative of two numbers reverses the comparison between them.

Key skills and understandings:

- Given any two numbers in the decimal system (including negative numbers), determine which is greater, and put any collection of numbers in order from least to greatest (or vice versa).

- Explain the rationale for comparing numbers in the decimal system by lining up the decimal points and looking first at the place of greatest value (and then moving to places of lower value).

- Draw rough pictures of bundled objects to show that one number is greater than another.

- Describe and draw rough pictures representing numbers as lengths and showing that one number is greater than another.

- Use a number line to demonstrate that one number is greater than another (including negative numbers).

- Describe negative numbers as owed amounts to explain why one negative number is greater than another.

Section 1.4 Rounding Numbers

To round a number to a given place's value means to find the closest number that has zeros in all smaller places. When we work with actual quantities, the way the number is written generally indicates the rounding that has taken place and, therefore, the precision with which the quantity is known.

Key skills and understandings:

- Given any number, round it to a given place.

- Use an appropriate number line to explain how to round a given number to a given place.

- Recognize that numbers representing actual quantities have generally been rounded.

Fractions

Much of the information we receive in daily life is presented in terms of fractions or percent. A news report might begin, "Nearly $\frac{3}{4}$ of the people surveyed . . . ," or a sign in a store might report, "35% off." To function effectively in society, we must understand fractions and percents.

Throughout the chapter, we will study the meaning of fractions and percents, how to represent fractions and percents, and how to compare fractions. Fractions and percents are defined in relation to a whole—or unit amount—by dividing the whole into equal parts. The notion of dividing into equal parts may seem simple, but it can be problematic. Although we use pairs of numbers to represent fractions, a fraction stands for a single number, and as such, has a location on the number line. Number lines provide an excellent way to represent improper fractions, which represent an amount that is more than the related whole.

We will see why different fractions—in fact, infinitely many different fractions—can be used to represent the same number. We must be able to represent fractions with different denominators in order to work simultaneously with several fractions, for example, to compare fractions. Percents are special fractions, namely, ones with denominator 100. We will study several different ways to reason about and calculate with percents. We will also consider some of the difficulties in understanding fractions and percents and learn key aspects that will help us overcome these difficulties.

The study of fractions and percents offers many delightful and challenging opportunities to practice mathematical reasoning. Logical reasoning lies at the heart of mathematics. Instruction in fractions and percents that focuses only on the mechanics of procedures and not on reasoning misses valuable opportunities to guide students in developing this core mathematical skill.

2.1 The Meaning of Fractions

 **Focal Points**
Grades 3, 4

This section explains the meaning of fractions, reviews some of the common difficulties in understanding the meaning of fractions, and describes how to use simple pictures to represent fractions.

Fractions arise naturally whenever we want to consider one or more parts of an object or quantity that is divided into pieces. Consider how fractions are used in the following ordinary situations:

$$\frac{1}{8} \quad \text{of a pizza}$$

$$\frac{2}{3} \quad \text{of the houses in the neighborhood}$$

$$\frac{9}{10} \quad \text{of the profit}$$

$$\frac{5}{4} \quad \text{of a cup of water}$$

$$\frac{3}{4} \quad \text{of the distance from Anklescratch to Buniontown}$$

$$\frac{4}{100} \quad \text{of } \$20,000$$

All these examples use the word *of*, and all the fractions represent part *of* some object, collection of objects, or quantity.

How are fractions defined? In this section, we define what we mean by a fraction $\frac{A}{B}$ of an object, collection or quantity. This will be the basis for describing in the next section what we mean by a fraction $\frac{A}{B}$ as a number. So suppose there is an object (such as a sandwich), a collection (such as a group of people or a bunch of bagels), or a quantity (such as a quantity of water or of money). Let's call this object, collection, or quantity of our *whole* (or *unit amount*). Suppose also that A and B are whole numbers, but B is not zero. What do we mean by $\frac{A}{B}$ of the whole we are considering?

unit fraction $\frac{1}{B}$

fraction $\frac{A}{B}$ of an object, collection or quantity

In order to define $\frac{A}{B}$, define the **unit fraction $\frac{1}{B}$** first. If the whole we are considering can be divided into B equal parts, then the amount formed by 1 of those parts is what we mean by $\frac{1}{B}$ of our whole. In other words, an amount is $\frac{1}{B}$ of our whole, if B copies of it joined together make the whole.

Then the **fraction $\frac{A}{B}$** of our whole is the amount formed by A parts (or copies of parts), each of which is $\frac{1}{B}$ of the whole.

numerator denominator

If $\frac{A}{B}$ is a fraction, then A is called the **numerator** and B is called the **denominator**. These names make sense because the word *numerator* comes from *number*, and the numerator tells you the number of parts. The word *denominator* comes from *name* and is related to *denomination*, which tells what type something is. For example, in money, the denomination of a bill tells you what type of bill it is—in other words, what value it has. In religion, a denomination is a group with the same type of religious belief and practice. The denominator of a fraction tells you what type of parts are being created. These parts can be halves, thirds, fourths, fifths, and so on. Hence, a fraction tells you how many of what type of parts.

$$\frac{A}{B} \quad \begin{array}{l} \leftarrow \quad \text{numerator: the number of parts} \\ \leftarrow \quad \text{denominator: the type or name of the parts} \end{array}$$

Figure 2.1 illustrates fractions of objects or collections of objects. In each case, the fraction $\frac{A}{B}$ is represented by shading A parts, each of which is $\frac{1}{B}$ of the whole object, collection, or quantity. In the pizza

FIGURE 2.1

Fractions
of objects,
collections,
and quantities

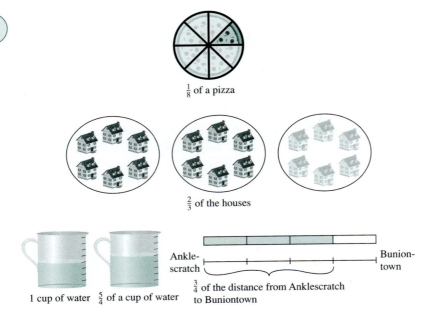

$\frac{1}{8}$ of a pizza

$\frac{2}{3}$ of the houses

1 cup of water $\frac{5}{4}$ of a cup of water

Ankle-
scratch

Bunion-
town

$\frac{3}{4}$ of the distance from Anklescratch
to Buniontown

example, the pizza was divided into 8 equal parts. One of those parts is shaded, so $\frac{1}{8}$ of the pizza is shaded. In the houses example, the collection of houses was divided into 3 equal parts. Two of the parts are shaded, so the shaded part is $\frac{2}{3}$ of the houses. In the water example, 1 cup of water was divided into 4 equal parts. Five copies of those parts are shown shaded, so the shaded portion is $\frac{5}{4}$ of a cup of water. In the distance example, the distance between Anklescratch and Buniontown was divided into 4 equal parts. Three of those parts are indicated by shading, so these 3 parts represent $\frac{3}{4}$ of the distance from Anklescratch to Buniontown.

Class Activity   *Now Turn to Class Activities Manual*

2A   Fractions of Objects, p. 20

A Fraction is Associated with a Whole

Notice the crucial word *of* in the examples of fractions of objects, fractions of collections of objects, and fractions of quantities described in the following:

$$\frac{2}{5} \text{ of the land}$$

$$\frac{2}{3} \text{ of the cars on the road}$$

$$\frac{1}{10} \text{ of the water in a lake}$$

Fractions are defined *in relation to a whole*, and this whole can be just one object or it can be a collection of objects, such as the cars on the road or 18 houses. The whole can also be a quantity, such as a quantity of water, or it can simply be the number 1.

When working with fractions, always keep in mind that there is an associated whole—whether or not that whole has been made explicit or drawn to your attention. Emphasize the relationship of a fraction to its whole by using language that draws attention to the whole. For example, you can describe the shaded part of the pizza in Figure 2.1 as "$\frac{1}{8}$ of the pizza" rather than as "1 out of 8 parts of pizza." Students from elementary school through college can correct many mistakes in their work with fractions if they can identify the whole associated with a fraction. That is, they need to understand what the fraction is "*of*."

Class Activity   *Now Turn to Class Activities Manual*

2B The Whole Associated with a Fraction, p. 21

2C Relating a Fraction to Its Whole, p. 21

2D Comparing Quantities with Fractions, p. 22

2E Fractions of Non-Contiguous Wholes, p. 22

Equal Parts

Before you read on, do the next Class Activity.

Class Activity   *Now Turn to Class Activities Manual*

2F Is the Meaning of Equal Parts Always Clear? p. 24

If you did the Class Activity, you saw that the meaning of equal parts is not always completely clear. If you want to divide a collection of toys into 4 equal parts, you might not simply put the same number of toys in each part: All the toys may not be considered equal; some may be more desirable than others. If you are asked to divide a shape, such as the one in Figure 2.2, into 3 equal parts, you might not know whether the parts should be equal in size and shape or whether they only need to be the same size. Problems that ask for a fraction of a shape to be shaded should be interpreted as asking for that fraction of the *area* of the shape to be shaded, even though this may not be stated explicitly.

In realistic situations, it is seldom a problem to determine what constitutes equal parts. In most cases, equal parts will be defined by dollar value, length, area, volume, or number. To divide the items in an estate into equal parts, you would probably make the parts of equal dollar value. For rope, parts of the same length constitute equal parts. For land, parts of the same area would generally constitute equal parts. For a pile of gravel, parts of the same volume constitute equal parts. On the other hand, when talking about $\frac{2}{3}$ of the stamps in a collection, $\frac{2}{3}$ of the cars on the road, or $\frac{2}{3}$ of the people in the room, we treat each stamp, car, or person as equal, even though some stamps may be more valuable than others, some cars may be more valuable or bigger than others, and some people are bigger than others. Notice, however, that the *dollar value* of $\frac{2}{3}$ of the stamps in a collection might not be $\frac{2}{3}$ of the value of the collection.

FIGURE 2.2

What does "equal parts" mean?

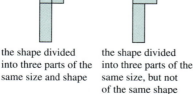

a shape   the shape divided   the shape divided
           into three parts of the   into three parts of the
           same size and shape   same size, but not
                                    of the same shape

Improper Fractions

proper fraction
improper fraction

A fraction $\frac{A}{B}$ (where A and B are whole numbers and B is not zero) is called a **proper fraction** if the numerator A is smaller than the denominator B. Otherwise, it's called an **improper fraction**. For example,

$$\frac{2}{5}, \quad \frac{3}{8}, \quad \frac{15}{16}$$

are proper fractions, whereas

$$\frac{6}{5}, \quad \frac{11}{8}, \quad \frac{27}{16}$$

are improper fractions.

Students sometimes think that improper fractions don't make sense because "we don't have enough pieces." For example, what is $\frac{3}{2}$ of a candy bar? According to the definition, to determine $\frac{3}{2}$ of a candy bar we must first think of the candy bar as divided into 2 equal pieces. Then $\frac{3}{2}$ of the candy bar is the amount formed by 3 copies of a piece. So, in order to conceptualize $\frac{3}{2}$ of a candy bar, we must be able to conceive of a third part that is a copy of the two equal parts that make up the whole candy bar. We must therefore be able to think of the candy bar and its parts as replicable. For example, we must be able to see 5 candy bars as 5 copies of 1 candy bar. So, two ideas are needed to interpret $\frac{3}{2}$ of a candy bar: the idea that half of a candy bar is an entity in its own right, and the idea that half a candy bar can be replicated 3 times, even though only 2 of those copies are present in one full candy bar.

Section 2.3 describes an excellent way to make sense of improper fractions with a number line.

Class Activity *Now Turn to Class Activities Manual*

2G Improper Fractions, p. 25

Practice Exercises for Section 2.1

1. This rectangle of x's is $\frac{3}{4}$ of another (original) rectangle of x's.

$$
\begin{array}{cccc}
x & x & x & x \\
x & x & x & x \\
x & x & x & x \\
x & x & x & x \\
x & x & x & x \\
x & x & x & x \\
\end{array}
$$

Show the original rectangle.

2. This rectangle of x's is $\frac{8}{3}$ of another (original) rectangle of x's.

x x x x x x x x x x x x x x x
x x x x x x x x x x x x x x x
x x x x x x x x x x x x x x x

Show the original rectangle.

3. Simone got $\frac{3}{4}$ of a full bar of chocolate. Hank was supposed to get $\frac{1}{4}$ of the full chocolate bar, but he has to get his share from Simone. What fraction of Simone's chocolate should Hank get? Draw a picture that helps you solve this problem. Use your picture to help you explain your solution. For each fraction in this problem, and in your solution, describe the whole associated with this fraction. In other words, describe what each fraction is *of*.

4. If one serving of juice gives you $\frac{3}{2}$ of your daily value of vitamin C, how much of your daily value of vitamin C will you get in $\frac{2}{3}$ of a serving of juice? Draw a picture that helps you solve this problem. Use your picture to help you explain your solution. For each fraction in this problem, and in your solution, describe the whole associated with this fraction. In other words, describe what each fraction is *of*.

5. The strips in Figure 2.3 show the relative amounts of money that Annie and Aria have saved. Write two sentences, one in which you use a fraction to describe how Aria's savings compare with Annie's and another in which you use a fraction to describe how Annie's savings compare with Aria's. For each fraction, state what its whole is.

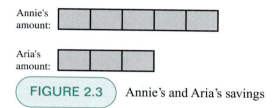

FIGURE 2.3 Annie's and Aria's savings

6. Show $\frac{1}{8}$ of the combined amount in the 3 pies in Figure 2.4 and explain why your answer is correct.

FIGURE 2.4 Three pies

7. Hermione has a potion recipe that calls for 4 drams of snake liver oil. Hermione wants to make $\frac{2}{3}$ of the potion recipe. Rather than calculate $\frac{2}{3}$ of the number 4, Hermione measures $\frac{2}{3}$ of a dram of snake liver oil 4 times and uses that amount of snake liver oil in her potion. Use pictures and the meaning of fractions to explain why Hermione's method is valid.

8. Divide the shaded shape in Figure 2.5 into 4 equal parts in two different ways: one where all parts have the same area and shape, and one where all parts have the same area, but some parts do not have the same shape.

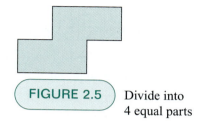

FIGURE 2.5 Divide into 4 equal parts

9. Explain why we can use the fractions in (a), and (b) to describe the shaded region in Figure 2.6. How can two different numbers describe the same shaded region? How must we interpret the shaded region in each case?

a. $\dfrac{5}{6}$

b. $\dfrac{5}{3}$

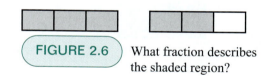

FIGURE 2.6 What fraction describes the shaded region?

Answers to Practice Exercises for Section 2.1

1. Because the rectangle shown in the practice problem is $\frac{3}{4}$ of another rectangle, it must consist of 3 parts of the other rectangle, when the other rectangle is divided into 4 equal parts. So the rectangle must consist of 3 equal parts, as shown in the following diagram:

x	x	x	x
x	x	x	x
x	x	x	x
x	x	x	x
x	x	x	x
x	x	x	x

The original rectangle that we are looking for must consist of 4 of these parts:

x	x	x	x
x	x	x	x
x	x	x	x
x	x	x	x
x	x	x	x
x	x	x	x
x	x	x	x
x	x	x	x

2. Because the rectangle shown in the practice problem is $\frac{8}{3}$ of another rectangle, it must consist of 8 parts of the other rectangle, when the other rectangle is divided into 3 equal parts. So the rectangle shown in the practice problem must consist of 8 equal parts, as shown in the following diagram:

x x	x x	x x	x x	x x	x x	x x	x x
x x	x x	x x	x x	x x	x x	x x	x x
x x	x x	x x	x x	x x	x x	x x	x x

The original rectangle must consist of 3 of these parts, as shown in the following diagram:

x x x x x x
x x x x x x
x x x x x x

3. As Figure 2.7 shows, we can divide the full chocolate bar into 4 equal parts, and Simone has 3 of those 4 parts. One of Simone's 3 parts should go to Hank. Therefore, Hank should receive $\frac{1}{3}$ of Simone's chocolate.

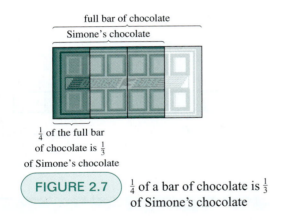

full bar of chocolate
Simone's chocolate

$\frac{1}{4}$ of the full bar
of chocolate is $\frac{1}{3}$
of Simone's chocolate

FIGURE 2.7 $\frac{1}{4}$ of a bar of chocolate is $\frac{1}{3}$ of Simone's chocolate

Notice that the 1 part that is to go to Hank is $\frac{1}{4}$ of the full bar of chocolate, but it is also $\frac{1}{3}$ of Simone's chocolate. In the problem, the full bar of chocolate is the whole associated with the fractions $\frac{3}{4}$ and $\frac{1}{4}$; in other words, both fractions are *of* the full bar of chocolate. But Simone's chocolate is the whole associated with the answer, $\frac{1}{3}$; in other words, the fraction $\frac{1}{3}$ is *of* Simone's chocolate.

4. Because one serving of juice provides $\frac{3}{2}$ of the daily value of vitamin C, one serving of juice represents 3 parts, when the daily value of vitamin C is divided into 2 equal parts. The daily value of vitamin C is represented in 2 of those parts, as shown

in Figure 2.8. If you drink $\frac{2}{3}$ of a serving of juice, you will get those 2 parts of the daily value of vitamin C, so you will get the full daily value of vitamin C in $\frac{2}{3}$ of a serving of juice.

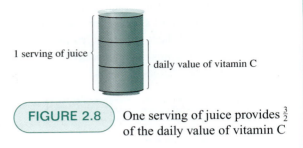

1 serving of juice

daily value of vitamin C

FIGURE 2.8   One serving of juice provides $\frac{3}{2}$ of the daily value of vitamin C

There are two different wholes associated with the fractions in this problem: a serving of juice and the daily value of vitamin C. The whole associated with the $\frac{3}{2}$ in the practice problem is the "daily value of vitamin C." The whole associated with the $\frac{2}{3}$ in the practice problem is "a serving of juice."

5. Aria's savings are $\frac{3}{5}$ as much Annie's savings. Annie's savings are $\frac{5}{3}$ as much as Aria's savings. You can also say that Annie's savings are $1\frac{2}{3}$ times as much as Aria's savings. The whole for $\frac{3}{5}$ is Annie's savings. The whole for $\frac{5}{3}$ (or $1\frac{2}{3}$) is Aria's savings.

6. If you shade $\frac{1}{8}$ of each pie individually, as in Figure 2.9, then collectively, the 3 shaded parts, which are labeled 1, are $\frac{1}{8}$ of the 3 pies combined. Why? Because when 8 copies of the 3 shaded parts are joined, they form the whole three pies.

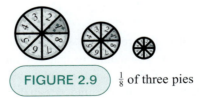

FIGURE 2.9 $\frac{1}{8}$ of three pies

7. If we represent the 4 drams of snake liver oil with drawings of 4 jars, as at the top of Figure 2.10, then the shaded regions in the middle of Figure 2.10 represent $\frac{1}{3}$ of the 4 drams of snake liver oil, because 3 copies of these shaded regions make the full 4 drams. Therefore, 2 copies of the shaded

regions in the middle of Figure 2.10 form $\frac{2}{3}$ of the 4 drams of snake liver oil, as shown at the bottom of Figure 2.10.

4 drams of snake liver oil

4 drams of snake liver oil divided into 3 equal pieces

$\frac{1}{3}$ of the 4 drams of snake liver oil

$\frac{2}{3}$ of the 4 drams of snake liver oil

FIGURE 2.10 Snake liver oil

8. See Figure 2.11.

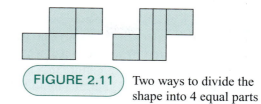

FIGURE 2.11 Two ways to divide the shape into 4 equal parts

9. We can use different numbers to describe the shaded part in Figure 2.6 by choosing different wholes. If you take the whole to be the full two bars in Figure 2.6, then because the whole consists of 6 equal parts and because 5 of those parts are shaded, the shaded part is $\frac{5}{6}$ of the chosen whole.

If you take the whole to be 1 bar (of 3 parts), then because the whole consists of 3 equal parts, each part is $\frac{1}{3}$ of a bar. So 5 parts are $\frac{5}{3}$ of a bar.

Problems for Section 2.1

1. Michael says that the dark marbles in Figure 2.12 can't represent $\frac{1}{3}$ because "there are 5 marbles and 5 is more than 1, but $\frac{1}{3}$ is supposed to be less than 1." In a short paragraph, discuss the source of Michael's confusion, and discuss what Michael must learn about fractions to overcome his confusion.

FIGURE 2.12 Dark and light marbles

2. When Jean was asked to say what the 3 in the fraction $\frac{2}{3}$ means, Jean said that the 3 is the whole. Explain why it is not completely correct to say that "3 is the whole." What is a better way to say what the 3 in the fraction $\frac{2}{3}$ means?

3. This rectangle of x's is $\frac{4}{5}$ of another (original) rectangle of x's.

$$x \quad x \quad x \quad x$$
$$x \quad x \quad x \quad x$$
$$x \quad x \quad x \quad x$$
$$x \quad x \quad x \quad x$$
$$x \quad x \quad x \quad x$$

Show the original rectangle. Explain why your solution is correct.

4. This rectangle of x's is $\frac{6}{5}$ of another (original) rectangle of x's.

$$x \quad x \quad x \quad x \quad x$$
$$x \quad x \quad x \quad x \quad x$$
$$x \quad x \quad x \quad x \quad x$$
$$x \quad x \quad x \quad x \quad x$$
$$x \quad x \quad x \quad x \quad x$$
$$x \quad x \quad x \quad x \quad x$$

Show the original rectangle. Explain why your solution is correct.

5. 🏺 Kaitlyn gave $\frac{1}{2}$ of her candy bar to Arianna. Arianna gave $\frac{1}{3}$ of the candy she got from Kaitlyn to Cameron. What fraction of a candy bar did Cameron get? Draw a picture that helps you solve this problem. Use your picture to help you explain your solution. For each fraction in this problem, and in your solution, describe the whole associated with this fraction. In other words, describe what each fraction is *of*.

6. One fourth of the beads in Maya's bead collection are blue. Of the blue beads, $\frac{1}{2}$ are small. What fraction of Maya's bead collection consists of small, blue beads? Draw a picture that helps you solve this problem. Use your picture to help you explain your solution. For each fraction in this problem, and in your solution, describe the whole associated with this fraction. In other words, describe what each fraction is *of*.

7. You were supposed to use $\frac{2}{3}$ of a cup of cocoa to make a batch of cookies, but you only have $\frac{1}{3}$ of a cup of cocoa. What fraction of the cookie recipe can you make with your $\frac{1}{3}$ cup of cocoa (assuming you have enough of the other ingredients)? Draw a picture that helps you solve this problem. Use your picture to help explain your solution. For each fraction in this problem and in your solution, describe the whole associated with this fraction. In other words, describe what each fraction is *of*.

8. 🏺 Susan was supposed to use $\frac{5}{4}$ of a cup of butter in her recipe, but she only used $\frac{3}{4}$ of a cup of butter. What fraction of the butter that she should have used did Susan actually use? Draw a picture that helps you solve this problem. Use your picture to help you explain your solution. For each fraction in this problem, and in your solution, describe the whole associated with this fraction. In other words, describe what each fraction is *of*.

9. If $\frac{3}{4}$ of a cup of a snack food gives you your daily value of calcium, then what fraction of your daily value of calcium is in 1 cup of the snack food? Draw a picture that helps you solve this problem. Use your picture to help you explain your solution. For each fraction in this problem, and in your solution, describe the whole associated with this fraction. In other words, describe what each fraction is *of*.

10. A container is filled with $\frac{5}{2}$ of a cup of cottage cheese. What fraction of the cottage cheese in the container should you eat if you want to eat 1 cup of cottage cheese? Draw a picture that helps you solve this problem. Use your picture to help you explain your solution. For each fraction in this problem, and in your solution, describe the whole associated with this fraction. In other words, describe what each fraction is *of*.

11. Make up a story problem or situation where *one* object (or collection, or quantity) is *both* $\frac{1}{2}$ of something and $\frac{1}{3}$ of something else.

12. Give two different fractions that you can legitimately use to describe the shaded region in Figure 2.13. For each fraction, explain why you can use that fraction to describe the shaded region. Write an unambiguous question about the shaded region in Figure 2.13 that can be answered by naming a fraction. Explain why your question is not ambiguous.

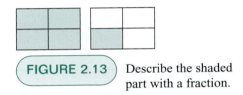

FIGURE 2.13   Describe the shaded part with a fraction.

13. Give two different fractions that you can legitimately use to describe the shaded region in Figure 2.14. For each fraction, explain why you can use that fraction to describe the shaded region. Write an unambiguous question about the shaded region in Figure 2.14 that can be answered by naming a fraction. Explain why your question is not ambiguous.

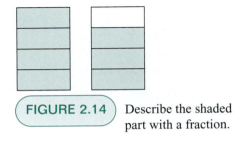

FIGURE 2.14   Describe the shaded part with a fraction.

14. 🏺 Marquez says that the shaded region in Figure 2.15 represents the fraction $\frac{9}{12}$. Carmina says the shaded region in Figure 2.15 represents the fraction $\frac{9}{4}$. Explain why each of the

two student's answers can be considered correct. Then write an unambiguous question about the shaded region in Figure 2.15 that can be answered by naming a fraction. Explain why your question is not ambiguous.

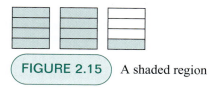

FIGURE 2.15   A shaded region

15. The strips in Figure 2.16 show the relative amounts of books that Rachel and Leah have read so far this school year. (Note that each rectangle represents some fixed number of books, but this number may be greater than 1).

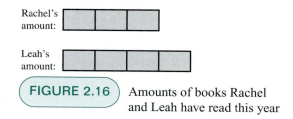

FIGURE 2.16   Amounts of books Rachel and Leah have read this year

   a. Write a sentence in which you use a fraction to describe how the amount of books that Rachel has read compares with the amount of books Leah has read. Explain briefly.

   b. Write a sentence in which you use a fraction to describe how the amount of books that Leah has read compares with the amount of books Rachel has read. Explain briefly.

16. a. Draw a "strip diagram" like the one in Figure 2.16 to show that Adam has read $\frac{5}{6}$ as many books as Joseph this year. Explain briefly.

   b. Based on your strip diagram in part (a), write a sentence in which you use a fraction to describe how the amount of books that Joseph has read compares with the amount of books that Adam has read.

17. *Cake problem:* Figure 2.17 shows a diagram representing two cakes of different sizes. Each cake was divided into 12 equal pieces. Marla ate one piece from each cake. What fraction of the total amount of cake (in the two cakes combined) did Marla eat?

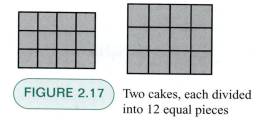

FIGURE 2.17   Two cakes, each divided into 12 equal pieces

   a. Ben says: "Marla ate $\frac{2}{24}$ of the cake because she ate 2 pieces out of a total of 24 pieces." Discuss Ben's reasoning. Is it correct? Explain why or why not. What if the two cakes were the same size instead of different sizes, would Ben's reasoning be correct in that case? Explain.

   b. Seyong says: "Marla ate 2 pieces, each of which was $\frac{1}{12}$ of cake, so according to the definition of fraction, Marla ate $\frac{2}{12}$ of the cake." Discuss Seyong's reasoning. Is it correct? Explain why or why not. What if the two cakes each weighed 1 pound. In that case, could Seyong's reasoning be used to make a correct statement? Explain.

   c. Explain how to solve the cake problem with valid reasoning that uses the definition of fraction (and that does not involve further subdividing the cake pieces).

18. Harry wants to make $\frac{3}{4}$ of a recipe of a potion. The full recipe calls for 2 vials full of newt blood. Instead of calculating $\frac{3}{4}$ of 2, Harry measures $\frac{3}{4}$ of a vial of newt blood 2 times and uses this amount to make $\frac{3}{4}$ of his potion recipe. Is Harry's method valid? Using the meaning of fractions, explain why or why not.

19. Figure 2.18 shows a picture of three pies. Draw a copy of the three pies. Shade $\frac{3}{8}$ of the combined amount in the three pies, and use the meaning of fractions to explain why your answer is correct.

FIGURE 2.18   Three pies

20. Discuss which of the items on the following list are good, and which are not as good, for showing improper fractions. Explain your choices.

string

a cake

a box of cereal and some cup measures

apples
something else—your choice

21. NanHe made a design that used hexagons, rhombuses, and triangles like the ones shown in Figure 2.19. NanHe counted how many of each shape she used in her design and determined that $\frac{4}{11}$ of the shapes she used were hexagons, $\frac{5}{11}$ were rhombuses, and $\frac{2}{11}$ were triangles. You may combine your answers to parts (a) and (b).

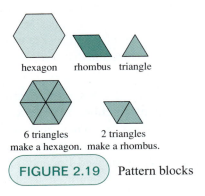

hexagon   rhombus   triangle

6 triangles   2 triangles
make a hexagon. make a rhombus.

 **FIGURE 2.19**   Pattern blocks

a. Even though $\frac{4}{11}$ of the shapes NanHe used were hexagons, does this mean that the hexagons in NanHe's design take up $\frac{4}{11}$ of the area of her design? If not, what fraction of the area of NanHe's design do the hexagons take up?

b. Write a short paragraph describing how the notion of "equal parts" relates to your answer in part (a) of this problem.

22. A county has two elementary schools, both of which have an after-school program for the convenience of working parents. In school A, $\frac{1}{3}$ of the children attend the after-school program; in school B, $\frac{2}{3}$ of the children attend the after-school program. There are a total of 1500 children in the two schools combined.

a. If school A has 600 students and school B has 900 students, then what is the fraction of elementary school students in the county who attend the after-school program? What if it's the other way around, and school A has 900 students and school B has 600 students?

b. Using the data in the problem statement, make up two more examples of numbers of students in each school. (These don't have to be entirely realistic numbers.) For each of your examples, determine the fraction of elementary school children in the county who attend the after-school program.

c. Jamie says that if $\frac{1}{3}$ of the children in school A attend the after-school program, and if $\frac{2}{3}$ of the children in the school B attend the after-school program, then $\frac{1}{2}$ of the children from the two schools combined attend the after-school program. Jamie arrives at this answer because she says $\frac{1}{2}$ is halfway between $\frac{1}{3}$ and $\frac{2}{3}$. Is Jamie's answer of $\frac{1}{2}$ *always* correct? Is Jamie's answer of $\frac{1}{2}$ *ever* correct, and if so, under what circumstances?

2.2   Interlude: Solving Problems and Explaining Solutions

F·P **Focal Points**
      All Grades

Now we briefly interrupt our study of fractions to examine some simple but sensible guidelines for solving problems and to think about what qualifies as a good explanation *in mathematics*. A main theme of this book is explaining why: Why are the familiar procedures and formulas of elementary mathematics valid? Why is a student's response incorrect? Why is different way of carrying out a calculation often perfectly correct? We will be seeking mathematical answers to these questions.

Solving Problems

The main reason for learning mathematics is to be able to solve problems. Mathematics is a powerful tool that can be used to solve a vast variety of problems in technology, science, business and finance, medicine, and daily life. The potential uses of mathematics are limited only by human ingenuity. Solving problems is not only the most important *end* of mathematics; it is also a *means* for learning mathematics.

Mathematicians have long known that good problems can deepen our thinking about mathematics, guide us to new ways of using mathematical techniques, help us recognize connections between topics in mathematics, and force us to confront mathematical misconceptions we may hold. By working on good problems, we learn mathematics better. This is why the National Council of Teachers of Mathematics (NCTM) advocates that instructional programs for prekindergarten through grade 12 focus on problem solving.

The NCTM, which is over 75 years old and has more than 100,000 members, advocates the highest quality mathematics education for all students. (See *www.pearsonhighered.com/beckmann* for links to NCTM.) To this end, the NCTM has prepared *Principals and Standards for School Mathematics* [64], which includes following standard on problem solving.

NCTM Standard: Problem Solving Instructional programs from prekindergarten through grade 12 should enable all students to

- build new mathematical knowledge through problem solving;
- solve problems that arise in mathematics and in other contexts;
- apply and adapt a variety of appropriate strategies to solve problems;
- monitor and reflect on the process of mathematical problem solving.

Polya's Four Problem-Solving Steps In 1945, the mathematician George Polya presented a four-step guideline for solving problems in his book *How to Solve It* [72]. These four steps are simple and sensible, and they have helped many students improve their problem-solving abilities.

Polya's Steps
1. Understand the problem.
2. Devise a plan.
3. Carry out the plan.
4. Look back.

The first step, *understand the problem*, is the most important. It may seem obvious that if you don't understand a problem, you won't be able to solve it, but it is easy to rush into a problem and try to do "something like we did in class" before you think about what the problem is asking. So *slow down* and read problems carefully. In some cases, drawing a diagram or a picture can help you understand the problem.

Try to be creative and flexible in the second step, when you *devise a plan*. There are many different types of plans for solving problems. In devising a plan, think about what information you know, what information you are looking for, and how to relate these pieces of information.

The third step, *carry out the plan*, is often the hardest part. If you get stuck, modify your plan or try a new plan. It often helps to cycle back through steps 1 and 2. Monitor your progress: If you are stuck, is it because you haven't tried hard enough to make your plan work, or is it time to try a new plan? Don't give up too easily. Students sometimes think they can solve a problem only if they've seen one just like it before, but this is not true. Your common sense and natural thinking abilities are powerful tools that will serve you well if you use them.

The fourth step, *look back*, gives you an opportunity to catch mistakes. Check to see if your answer is plausible. For example, if the problem was to find the height of a telephone pole, then answers such as 2.3 feet or 513 yards are unlikely—look for a mistake somewhere. Looking back also gives you an opportunity to make connections: Have you seen this type of answer before? What did you learn from this problem? Could you use these ideas in some other way? Is there another way to solve the problem? When you look back, you have an opportunity to learn from your own work.

Solving Problems for Yourself Students sometimes wonder why they need to solve problems themselves: Why can't the teacher just *show* us how to solve the problem? Of course, teachers do show solutions to many problems. However, sometimes teachers should step back and *guide* their students, helping them to use fundamental concepts and principles to *figure out* how to solve a problem. Why? Because the process of grappling with a problem can help students understand the underlying concepts and principles. Teachers who are too quick to tell students how to solve problems may actually rob them of valuable learning experiences.

When you solve problems yourself, you will inevitably get stuck some of the time, and you will try things that don't work. This may seem unproductive, but it is not. Even if you are not able to solve a problem, the process of getting stuck and trying to get unstuck primes you to better understand a solution presented by someone else.

Stop for a moment and think about something that you know well, for example, an academic topic or a topic from sports, music, religion, or any aspect of your life. Chances are that the topic you know well is something about which you have thought deeply, something with which you have grappled and maybe even struggled. You probably learned a lot about the topic from others, but you also likely put effort into making sense of it for yourself. The process of making sense of things for yourself is the essence of education—use it in mathematics and do not underestimate its power.

Explaining Solutions

After you solved a problem, the natural next step is to explain why your solution is valid. Why does your method work? Why does it give the correct answer to the problem? As a teacher, you will need to explain why the mathematics you are teaching works the way it does. However, there is an even more compelling reason for providing explanations. When you try to explain something to someone else, you clarify your own thinking and you learn more yourself. When you try to explain a solution, you may find that you don't understand it as well as you thought. The exercise of explaining is valuable because it provides an opportunity to learn more, to uncover an error, or to clear up a misconception. Even if you understood the solution well, you will understand it better after explaining it. Therefore, the NCTM includes the following standard, communication [64].

NCTM Standard: Communication Instructional programs from prekindergarten through grade 12 should enable all students to

- organize and consolidate their mathematical thinking through communication;
- communicate their mathematical thinking coherently and clearly to peers, teachers, and others;
- analyze and evaluate the mathematical thinking and strategies of others;
- use the language of mathematics to express mathematical ideas precisely.

Communicating about mathematics gives both children and adults an opportunity to make sense of mathematics. According to the NCTM [64, p. 56],

> *From children's earliest experiences with mathematics*, it is important to help them understand that assertions should always have reasons. Question such as "Why do you think it is true?" and "Does anyone think the answer is different, and why do you think so?" help students see that statements need to be supported or refuted by evidence.

When we communicate about mathematics in order to explain and convince, we must use *reasoning*. Logical reasoning is the essence of mathematics. In mathematics, everything but the fundamental starting assumptions ahs a reason, and the whole structure of mathematics is built up by reasoning. Therefore, the NCTM includes the following standard on reasoning and proof [64].

NCTM Standard: Reasoning and Proof Instructional programs from prekindergarten through grade 12 should enable all students to

- recognize reasoning and proof as fundamental aspects of mathematics;
- make and investigate mathematical conjectures;

- develop and evaluate mathematical arguments and proofs;
- select and use various types of reasoning and methods of proof.

Explaining *why* is so fundamental to mathematics that it is emphasized in every topic in this book.

What Is an Explanation in Mathematics?
What qualifies as an explanation? The answer depends on the context. In Mathematics, we seek particular kinds of explanations: those using logical reasoning which are based on initial assumptions that are either explicitly stated or assumed to be understood by the reader or listener.

Explanations can vary according to different areas of knowledge. There are many different kinds of explanations—even of the same phenomenon. For example, consider this question: Why are there seasons? (See Figure 2.20.)

The simplest answer is "because that's just the way it is." Every year, we observe the passing of the seasons, and we expect to see the cycle of spring, summer, fall, and winter continue indefinitely. The cycle of seasons is an observed fact that has been documented since humans began to keep records. We could stop here, but when we ask why there are seasons, we are searching for a deeper explanation.

A poetic explanation for the seasons might refer to the cycles of birth, death, and rebirth around us. In our experiences, nothing remains unchanged forever, and many things are parts of a cycle. The cycle of the seasons is one of the many cycles that we observe.

FIGURE 2.20

Why are there seasons?

 Spring

 Summer

Winter

Autumn

Most cultures have stories that explain why we have seasons. The ancient Greeks, for example, explained the seasons with the story of Persephone and her mother, Demeter, who tends the earth. When Pluto, god of the underworld, stole Persephone to become his bride, Demeter was heartbroken. Pluto and Demeter arranged a compromise, and Persephone could stay with her mother for half a year and return to Pluto in the underworld for the remaining half of every year. When Persephone is in the underworld, Demeter is sad and does not tend the earth. Leaves fall from the trees, flowers die, and it is fall and winter. When Persephone returns, Demeter is happy again and tends the earth. Leaves grow on the trees, flowers bloom, and it is spring and summer. This is a beautiful story, but we can still ask for another kind of explanation.

Modern scientists explain the reason for the seasons by the tilt of the earth's axis relative to the plane in which the earth travels around the sun. When the northern hemisphere is tilted toward the sun, it is summer there; when it is tilted away from the sun, it is winter. Perhaps this settles the matter, but a seeker could still ask for more. Why are the earth and sun positioned the way they are? Why does the earth revolve around the sun and not fly off alone into space? These questions can lead again to poetry, or to the spiritual, or to further physical theories. Maybe they lead to an endless cycle of questions.

How Are Good Explanations Written?   While an oral explanation helps you develop your solution to a problem, written explanations push you to polish, refine, and clarify your ideas. This is as true in mathematical writing as in any other kind of writing, and it is true at all levels. You should write explanations of your solutions to problems, and your students should write explanations of their solutions, too. Some elementary school teachers have successfully integrated mathematics and writing in their classrooms and use writing to help their students develop their understanding of mathematics.

Like any kind of writing, it takes work and practice to write good mathematical explanations. When you solve a problem, do not attempt to write the final draft of your solution right from the start. Use scratch paper to work on the problem and collect your ideas. Then, write your solution as part of the *looking back* stage of problem solving. Think of your explanations as an essay. As with any essay that aims to convince,

Characteristics of Good Explanations in Mathematics

1. The explanations is factually correct, or nearly so, with only minor, inconsequential flaws.

2. The explanation addresses the specific question or problem that was posed. It is focused, detailed, and precise. Key points are emphasized. There are no irrelevant or distracting points.

3. The explanation is clear, convincing, and logical. A clear and convincing explanation is characterized by the following:

    a. The explanation could be used to teach another (college) student, possibly even one who is not in the class.

    b. The explanation could be used to convince a skeptic.

    c. The explanation does not require the reader to make a leap of faith.

    d. If applicable, supporting pictures, diagrams, and equations are used appropriately and as needed.

    e. The explanation is coherent.

    f. Clear, complete sentences are used.

what counts is not only factual correctness but also persuasiveness, explanatory power, and clarity of expression. In mathematics, we persuade by giving a thorough, logical argument, in which chains of logical deductions are strung together connecting the starting assumptions to the desired conclusion.

Good mathematical explanations are thorough. They should not have gaps that require leaps of faith. On the other hand, a good explanation should not belabor points that are well-known to the audience or not central to the explanation. For example, if your solution contains the calculation $356 \div 7$, a college-level explanation need not describe how the calculation is carried out, except when it is necessary for the solution. Unless your instructor tells you otherwise, assume that you are writing your explanations for your classmates.

The box on the previous page lists characteristics of good mathematical explanations. When you write an explanation, check whether it has these characteristics. The more you work at writing explanations, and the more you ponder and analyze what makes good explanations, the better you will write explanations, and the better you will understand the mathematics involved. Note that the solutions to the practice exercises in each section provide you with many examples of the kinds of explanations you should learn to write.

2.3 Fractions as Numbers

 **Focal Points**
Grades 3, 4

So far we have discussed fractions of objects, collections, or quantities. Now let's look at fractions abstractly, as numbers in their own right.

We create the notion of the whole numbers by abstracting from our experiences with objects. For example,

2 apples,   7 balls,   25 people, …

is really like saying (somewhat awkwardly)

2 of apple,   7 of ball,   25 of person, …

which abstracts to the following notion of number:

2,   7,   25, …

fractions as numbers   In the same way, we create the following notion of **fractions as numbers** by abstracting from fractions of objects:

$\frac{2}{3}$ of a pie, $\frac{11}{10}$ of an acre of land, $\frac{7}{8}$ of the population of the United States, …

These abstract to

$$\frac{2}{3}, \quad \frac{11}{10}, \quad \frac{7}{8}, \dots$$

But even when fractions are viewed abstractly as numbers, they are still "of a whole." Just as 5 is "five ones," so, too, $\frac{3}{4}$ is "$\frac{3}{4}$ of 1."

Viewing fractions as locations on number lines, shows even more clearly how fractions are numbers.

Fractions as Numbers on Number Lines

Fractions can be plotted on a number line in a natural way; therefore, fractions are numbers in the same way that whole numbers, decimals, and negative numbers are.

If A and B are whole numbers and B is not zero, then the **fraction $\frac{A}{B}$ is the number** that is located on a number line according to the rule for locating decimal numbers on number lines given in Section 1.2. Thus, $\frac{A}{B}$ is located to the right of zero at a distance of $\frac{A}{B}$ units from 0. Therefore, $\frac{A}{B}$ is at the right end of a line segment whose left end is at 0 and whose length is $\frac{A}{B}$ of one unit.

FIGURE 2.21

Plotting $\frac{5}{3}$ on a number line

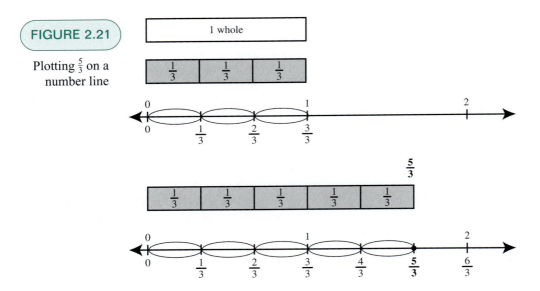

For example, where is $\frac{5}{3}$ on a number line? The fraction $\frac{5}{3}$ is at the right end of a line segment whose left end is at 0 and whose length is $\frac{5}{3}$ of a unit. To construct such a line segment, use the meaning of $\frac{5}{3}$: Divide the line segment from 0 to 1 into 3 equal parts, and make a line segment consisting of 5 copies of those parts. Place this five-part line segment on the number line so that its left endpoint is at 0, as shown in Figure 2.21. Then the right endpoint of the five-part line segment is at the fraction $\frac{5}{3}$.

To emphasize the important point that a number is plotted on a number line according to its *distance* from 0, some teachers find it helpful to circle segments on the number line as shown in Figure 2.21. Circling the segments helps students focus on their length, and circling highlights that the segment between 0 and $\frac{5}{3}$ consists of 5 pieces, each of which is $\frac{1}{3}$ of a unit long. When students pay attention to the lengths of the segments, they avoid some of the common number line errors that are due to counting tick marks without attending to length and distance. The next Class Activity examines some of these errors.

Class Activity   *Now Turn to Class Activities Manual*

2H Number Line Errors with Fractions, p. 26

Number lines are especially good for showing improper fractions. When we show improper fractions with pieces of pie or pieces of some other object, it is often unclear which whole the fraction refers to. For example, does the shaded region in Figure 2.22 represent $\frac{5}{3}$ or $\frac{5}{6}$? We can't say unless we know what the fraction is supposed to be *of.* The shaded region is $\frac{5}{3}$ of the three-piece rectangle, but it is also $\frac{5}{6}$ of the two rectangles. In contrast, on number lines, the whole is always the line segment between 0 and 1, so no ambiguity arises.

A good number line activity for children is to label fractions on a number line (attending to length as previously noted) like the one shown in Figure 2.23. Since the denominator stays the same throughout the labeling process, and since the children essentially count by the numerator, this activity helps develop the understanding that the numerator represents the *number* of pieces and the denominator represents the *type* of pieces.

FIGURE 2.22

Showing $\frac{5}{3}$
with pieces of
rectangles

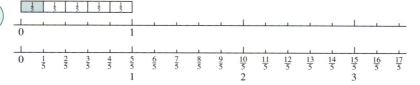

Does the shaded region represent $\frac{5}{3}$ or $\frac{5}{6}$?
We can't say unless the whole is specified.

FIGURE 2.23

Counting by
fractions on a
number line

Class Activity   *Now Turn to Class Activities Manual*

21   Fractions on Number Lines, Part 1, p. 26

Decimal Representations of Fractions

Because fractions can be plotted on number lines, and because every location on a number line corresponds to a decimal number, it must be possible to represent fractions as decimal numbers. In fact, it is easy to write a fraction $\frac{A}{B}$ as a decimal number. Just divide A by B as follows:

$$\frac{A}{B} = A \div B$$

We will see why it makes sense to write a fraction in decimal notation by dividing in Chapter 6.

So,

$$\frac{5}{16} = 5 \div 16 = 0.3125$$

$$\frac{1}{12} = 1 \div 12 = 0.083333\ldots$$

$$\frac{2}{7} = 2 \div 7 = 0.285714285714\ldots$$

Practice Exercises for Section 2.3

1. Starting with a number line on which 0 and 1 have already been plotted, explain where to plot $\frac{5}{8}$ and explain why that location fits with the definition of fraction.

2. According to the text, why are number lines especially good for showing improper fractions?

3. Plot 0, 1, and $\frac{8}{7}$ on a number line like the one in Figure 2.24 in such a way that each number falls on a tick mark. Lengthen the tick marks of whole numbers.

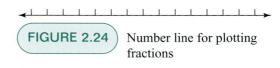

FIGURE 2.24   Number line for plotting fractions

4. Place equally spaced tick marks on the number line in Figure 2.25 so that you can plot $\frac{1}{7}$ on a tick mark. Then plot $\frac{1}{7}$. Explain your reasoning.

FIGURE 2.25   Plot $\frac{1}{7}$

5. Place equally spaced tick marks on the number line in Figure 2.26 so that you can plot 1 on a tick mark. Then plot 1. Explain your reasoning.

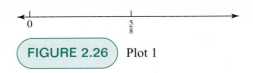

FIGURE 2.26   Plot 1

6. Plot $\frac{11}{8}$ on a number line like the one in Figure 2.27.

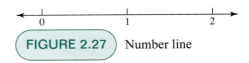

FIGURE 2.27   Number line

7. Plot $\frac{29}{3}$ on a number line like the one in Figure 2.28.

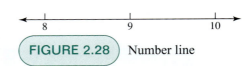

FIGURE 2.28   Number line

Answers to Practice Exercises for Section 2.3

1. First, divide the line segment between 0 and 1 into 8 pieces of equal length. According to the definition of unit fractions, each of these segments is $\frac{1}{8}$ of a unit long. Use tick marks to show where these segments begin and end, but note that it turns out only 7 tick marks will be inserted between 0 and 1. Plot $\frac{5}{8}$ at the end of the fifth segment from 0. By the definition of fraction, the length of an interval made from 5 segments, each of which is $\frac{1}{8}$ of a unit long, is $\frac{5}{8}$ of a unit. So the location of $\frac{5}{8}$ fits with the definition of fraction.

2. See page 53.

3. To plot $\frac{8}{7}$, we need to work with sevenths. Therefore, we must divide the 1-unit segment from 0 to 1 into 7 equal parts, as shown in Figure 2.29.

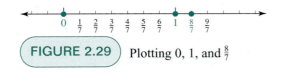

FIGURE 2.29   Plotting 0, 1, and $\frac{8}{7}$

4. On a number line on which adjacent tick marks are $\frac{1}{7}$ apart, the point $\frac{3}{7}$ will be the third tick mark to the right of 0, and $\frac{1}{7}$ is the first tick mark, as shown in Figure 2.30.

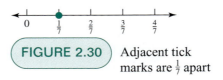

FIGURE 2.30   Adjacent tick marks are $\frac{1}{7}$ apart

5. On a number line on which adjacent tick marks are $\frac{1}{8}$ apart, $\frac{5}{8}$ will be at the fifth tick mark to the right of 0 and $1 = \frac{8}{8}$ will be at the eighth tick mark to the right of 0, as shown in Figure 2.31.

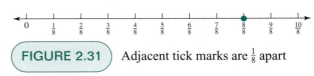

FIGURE 2.31   Adjacent tick marks are $\frac{1}{8}$ apart

6. To plot $\frac{11}{8}$ on a number line, divide the 1-unit segment from 0 to 1 into 8 equal parts, and measure off 11 of those parts, as shown in Figure 2.32.

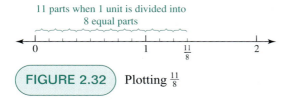

11 parts when 1 unit is divided into 8 equal parts

FIGURE 2.32 Plotting $\frac{11}{8}$

7. The fraction $\frac{29}{3}$ is located 29 parts away from 0, where each part is $\frac{1}{3}$ of a unit long (i.e., 3 parts are 1 unit long). Nine sets of 3 parts will use up 27 parts, and these 27 parts will end at 9 on the number line because $\frac{27}{3} = 27 \div 3 = 9$. Two more parts will get to the location of $\frac{29}{3}$ on the number line, as shown in Figure 2.33.

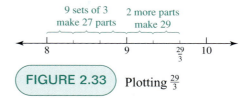

9 sets of 3 make 27 parts   2 more parts make 29

FIGURE 2.33 Plotting $\frac{29}{3}$

Problems for Section 2.3

1. 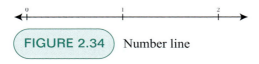 Draw a number line like the one in Figure 2.34 and plot $\frac{1}{3}$ and $\frac{2}{3}$ on your number line. In your own words, explain why those locations for these fractions fit with the definition of fraction.

FIGURE 2.34 Number line

2. Draw a number line like the one in Figure 2.34 and plot $\frac{5}{3}$ on your number line. In your own words, explain why this location fits with the definition of fraction.

3. 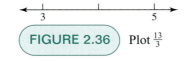 Discuss why it can be confusing to show an improper fraction such as $\frac{7}{3}$ with pieces of pie or pieces of some other object. What is another way to show the fraction $\frac{7}{3}$?

4. Draw a number line like the one in Figure 2.35. Then plot 0, 1, $\frac{4}{3}$, and $\frac{11}{3}$ on your number line in such a way that each number falls on a tick mark. Lengthen the tick marks of whole numbers.

FIGURE 2.35 Number line

5. Draw a number line like the one in Figure 2.36. Plot $\frac{13}{3}$ on your number line.

3   5

FIGURE 2.36 Plot $\frac{13}{3}$

6. Draw a number line like the one in Figure 2.37. Plot $\frac{47}{6}$ on your number line.

6   7   8

FIGURE 2.37 Plot $\frac{47}{6}$

7. Erin says the tick mark indicated in Figure 2.38 should be labeled 2.2. Is Erin right or not? If not, why not, and how can she label the tick mark properly?

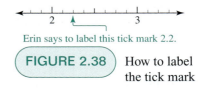

2   3

Erin says to label this tick mark 2.2.

FIGURE 2.38 How to label the tick mark

8. Liam says the tick mark indicated in Figure 2.39 should be labeled 1.7. Is Liam right or not? If not, why not, and how can the tick mark be labeled properly?

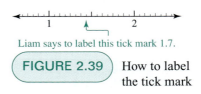

1   2

Liam says to label this tick mark 1.7.

FIGURE 2.39 How to label the tick mark

2.4 Equivalent Fractions

 **Focal Points**
Grades 4, 5

When we write a whole number in decimal notation, there is only one way to do so without the use of a decimal point. The only way to write 1234 in decimal notation without a decimal point is 1234. But in the case of fractions, every fraction is equal to infinitely many other fractions. For example,

$$\frac{1}{2} = \frac{2}{4} = \frac{3}{6} = \frac{4}{8} = \frac{5}{10} = \cdots$$

In general,

$$\frac{A}{B} = \frac{A \cdot 2}{B \cdot 2} = \frac{A \cdot 3}{B \cdot 3} = \frac{A \cdot 4}{B \cdot 4} = \frac{A \cdot 5}{B \cdot 5} = \cdots$$

In this section, we will see why every fraction is equal to infinitely many other fractions, and we will study some consequences of this fact.

Class Activity   *Now Turn to Class Activities Manual*

2J Equivalent Fractions, p. 27

2K Misconceptions about Fraction Equivalence, p. 28

Why Every Fraction Is Equal to Infinitely Many Other Fractions

As we consider why every fraction is equal to infinitely many other fractions, let's start with an example. Why is

$$\frac{3}{4}$$

of a rod the same amount of the rod as

$$\frac{3 \cdot 5}{4 \cdot 5} = \frac{15}{20}$$

of the same rod? We can use the meaning of fractions of objects to explain this equivalence. To form $\frac{3}{4}$ of the rod, first divide the rod into 4 equal parts. Then $\frac{3}{4}$ of the rod consists of 3 parts. Divide each of those 4 equal parts into 5 small, equal parts, as shown in Figure 2.40. Now the rod consists of a total of $4 \cdot 5 = 20$ small parts. The 3 parts representing $\frac{3}{4}$ of the rod have each been subdivided into 5 equal parts; therefore, these 3 parts have become $3 \cdot 5 = 15$ smaller parts, as shown in the shaded portions of Figure 2.40. So, 3 out of the original 4 parts of the rod is the same amount of the rod as $3 \cdot 5$ smaller parts out of the total $4 \cdot 5$ smaller parts. Therefore,

$$\frac{3}{4} = \frac{3 \cdot 5}{4 \cdot 5} = \frac{15}{20}$$

FIGURE 2.40

Subdivide each part into 5 parts.

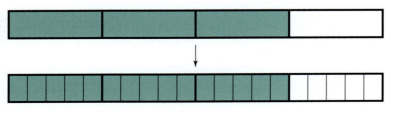

When showing this equation to children, you might prefer to condense it as such:

$$\times 5$$

$$\frac{3}{4} = \frac{15}{20}$$

$$\times 5$$

There wasn't anything special about the numbers 3, 4, and 5 used here—the same reasoning will apply to other counting numbers substituted for these numbers. Therefore, in general, if N is any counting number, $\frac{A}{B}$ of a rod (or any other object or collection of objects) is the same amount of the rod (or object or collection) as $\frac{A \cdot N}{B \cdot N}$ of the rod (or object or collection). In other words,

$$\frac{A}{B} = \frac{A \cdot N}{B \cdot N}$$

Another way to explain why $\frac{A}{B} = \frac{A \cdot N}{B \cdot N}$ is to multiply by 1 in the form of $\frac{N}{N}$:

$$\frac{A}{B} = \frac{A}{B} \cdot 1$$

$$= \frac{A}{B} \cdot \frac{N}{N}$$

$$= \frac{A \cdot N}{B \cdot N}$$

This explanation for why $\frac{A}{B} = \frac{A \cdot N}{B \cdot N}$ relies on the more advanced notion of fraction multiplication, whereas the previous explanation relies only on the meaning of fractions.

Common Denominators

Because every fraction is equal to infinitely many other fractions, you have a lot of flexibility when you work with fractions. When you start with a fraction, say, $\frac{3}{4}$, you know that it is always equal to many other fractions with larger denominators, for example,

$$\frac{3}{4} = \frac{3 \cdot 7}{4 \cdot 7} = \frac{21}{28}$$

$$\frac{3}{4} = \frac{3 \cdot 25}{4 \cdot 25} = \frac{75}{100}$$

The fractions $\frac{21}{28}$ and $\frac{75}{100}$ are just two of infinitely many fractions equal to $\frac{3}{4}$, such as

$$\frac{3}{4} = \frac{6}{8} = \frac{9}{12} = \frac{12}{16} = \frac{15}{20} = \cdots$$

common denominators When working with two fractions simultaneously, it is often desirable to give them **common denominators**, which just means the *same* denominators. Before you read on, do the next Class Activity.

Class Activity   *Now Turn to Class Activities Manual*

2L Common Denominators, p. 29

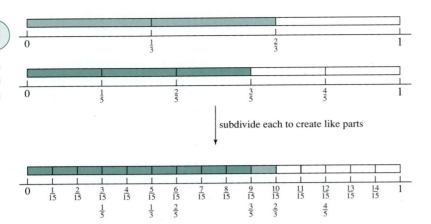

FIGURE 2.41

Common denominators give fractions like parts.

As you saw in the activity, when we give fractions common denominators, we describe the fractions in terms of like parts. The fractions $\frac{2}{3}$ and $\frac{3}{5}$ are in terms of thirds and fifths, respectively. We can write these two fractions with the common denominator $3 \cdot 5 = 15$:

$$\frac{2}{3} = \frac{2 \cdot 5}{3 \cdot 5} = \frac{10}{15}$$

$$\frac{3}{5} = \frac{3 \cdot 3}{5 \cdot 3} = \frac{9}{15}$$

As you see in Figure 2.41, when we give $\frac{2}{3}$ and $\frac{3}{5}$ common denominators, we subdivide the thirds of $\frac{2}{3}$ and the fifths of $\frac{3}{5}$ into fifteenths, so that both fractions are described with like parts.

The fractions $\frac{2}{3}$ and $\frac{3}{5}$ have many other common denominators as well, such as 30 and 45, but 15 is the smallest one.

For any two fractions, multiplying the denominators always produces a common denominator; however, it may not be the least common denominator. The number 24 is a common denominator for the fractions $\frac{3}{8}$ and $\frac{5}{6}$, whereas $8 \cdot 6 = 48$. We will discuss least common denominators again in Section 8.2 when we discuss least common multiples.

Class Activity *Now Turn to Class Activities Manual*

2M Solving Problems by Changing Denominators, p. 30

2N Fractions on Number Lines, Part 2, p. 31

The Simplest Form of a Fraction

Every fraction is equal to infinitely many other fractions, but in a collection of fractions that are equal to each other, there is one that is the simplest.

simplest form   A fraction of whole numbers $\frac{A}{B}$ (where B is not zero) is said to be in **simplest form** (or in **lowest terms**) if there is no whole number other than 1 that divides both A and B evenly.

The fraction

$$\frac{3}{4}$$

is in simplest form because no whole number other than 1 divides both 3 and 4 evenly. The fraction

$$\frac{30}{35}$$

is not in simplest form because the number 5 divides both 30 and 35 evenly. Notice, however, that we can put the fraction $\frac{30}{35}$ in simplest form as follows:

$$\frac{30}{35} = \frac{6 \cdot 5}{7 \cdot 5} = \frac{6}{7} \qquad (2.1)$$

This equation is the reverse of the equation that gives $\frac{6}{7}$ the denominator 35:

$$\frac{6}{7} = \frac{6 \cdot 5}{7 \cdot 5} = \frac{30}{35}$$

When showing Equation 2.1 to children, you might prefer to condense it as

Before you read on, do the next Class Activity.

Class Activity *Now Turn to Class Activities Manual*

20 Simplifying Fractions, p. 32

In general, if

$$\frac{A}{B}$$

is a fraction of whole numbers that is not in simplest form, then we can put it in simplest form as follows: First, we find the largest whole number N that divides both A and B evenly. We write A and B in the form $C \cdot N$ and $D \cdot N$, respectively, where C and D are whole numbers. Then

$$\frac{A}{B} = \frac{C \cdot N}{D \cdot N} = \frac{C}{D}$$

and $\frac{C}{D}$ is the simplest form of $\frac{A}{B}$.

To write $\frac{18}{24}$ in simplest form, notice that 6 is the largest whole number that divides both 18 and 24 evenly. Because

$$18 = 3 \cdot 6$$

and

$$24 = 4 \cdot 6$$

we have

$$\frac{18}{24} = \frac{3 \cdot 6}{4 \cdot 6} = \frac{3}{4}$$

FIGURE 2.42

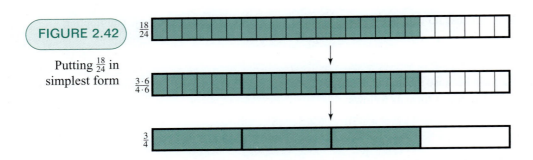

Putting $\frac{18}{24}$ in simplest form

Or in other words,

Thus, $\frac{3}{4}$ is the simplest form of $\frac{18}{24}$.

If you think in terms of rods, putting a fraction in simplest form is like joining together small rod pieces to make larger rod pieces, as shown in Figure 2.42. In other words, putting a fraction in simplest form reverses the procedure of subdividing rod pieces that was demonstrated earlier in Figure 2.40.

We can also use several steps to put a fraction in simplest form. For example, to put $\frac{18}{24}$ in simplest form, we don't have to know right away that 6 is the largest whole number that divides both 18 and 24 evenly. Instead, we can take several steps, as follows:

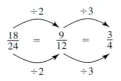

We will discuss putting fractions in simplest form again in Section 8.2 when we discuss greatest common factors.

Practice Exercises for Section 2.4

1. Use the meaning of fractions to explain why

$$\frac{4}{5} = \frac{4 \cdot 3}{5 \cdot 3} = \frac{12}{15}$$

2. Write $\frac{3}{8}$ and $\frac{5}{6}$ with common denominators in three different ways.

3. Plot $1, \frac{5}{3}$, and $\frac{7}{4}$ on a number line like the one in Figure 2.43 in such a way that each number falls on a tick mark. Lengthen the tick marks of whole numbers.

FIGURE 2.43   Number line for plotting fractions

4. Plot 1, 0.9, and $\frac{5}{4}$ on a number line like the one in Figure 2.43 in such a way that each number falls on a tick mark. Lengthen the tick marks of whole numbers.

5. The line segment in Figure 2.44 has length $\frac{9}{8}$ unit. Explain how to subdivide the line segment and possibly add pieces to the line segment to create a line segment of length $\frac{3}{4}$ unit. Do this without first creating a segment of length 1 unit. Explain how you know the resulting line segment is the correct length.

$\frac{9}{8}$ units

FIGURE 2.44 Create a line segment of length $\frac{3}{4}$ units given that this line segment has length $\frac{9}{8}$ units.

6. Place equally spaced tick marks on the number line in Figure 2.45 so that you can plot $\frac{2}{3}$ on a tick mark. Then plot $\frac{2}{3}$. Explain your reasoning.

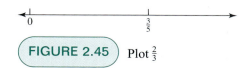

FIGURE 2.45 Plot $\frac{2}{3}$

7. Kelsey has $\frac{3}{5}$ of a chocolate bar. Kelsey wants to give some of her chocolate bar to Janelle. What fraction of her chocolate bar should Kelsey give Janelle so that Kelsey will be left with $\frac{1}{2}$ of the original chocolate bar?

   a. Draw pictures to help you solve this problem. Explain why your answer is correct.

   b. In solving this problem, how do $\frac{3}{5}$ and $\frac{1}{2}$ appear in different forms?

   c. What are the different wholes associated with the fractions in this problem? In other words, for each fraction in this problem and its solution, what is the fraction *of*?

8. Ken ordered $\frac{3}{4}$ of a ton of gravel. Ken wants $\frac{1}{4}$ of his order of gravel delivered now and $\frac{3}{4}$ delivered later. What fraction of a ton of gravel should Ken get delivered now?

   a. Draw pictures to help you solve this problem. Explain why your answer is correct.

   b. In solving the problem, how does $\frac{3}{4}$ appear in a different form?

   c. What are the different wholes associated with the fractions in this problem? In other words, for each fraction in this problem, and its solution, what is the fraction *of*?

9. Use the diagram in Figure 2.46 to help you explain why the following equations that put $\frac{10}{15}$ in simplest form make sense:

$$\frac{10}{15} = \frac{2 \cdot 5}{3 \cdot 5} = \frac{2}{3}$$

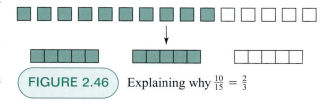

FIGURE 2.46 Explaining why $\frac{10}{15} = \frac{2}{3}$

10. Put the following fractions in simplest form:

$$\frac{45}{72}, \frac{24}{36}, \frac{56}{88}$$

Answers to Practice Exercises for Section 2.4

1. If you have a pie that is divided into 5 equal parts, and 4 are shown shaded, then if you divide each of the 5 parts into 3 smaller parts, the shaded amount will now consist of $4 \cdot 3$ small parts, and the whole pie will consist of $5 \cdot 3$ small parts. Thus, the shaded part of the pie can be described both as $\frac{4}{5}$ of the pie and as

$$\frac{4 \cdot 3}{5 \cdot 3} = \frac{12}{15}$$

   of the pie.

2. There are infinitely many ways to write $\frac{3}{8}$ and $\frac{5}{6}$ with common denominators. Here are three:

$$\frac{3}{8} = \frac{3 \cdot 6}{8 \cdot 6} = \frac{18}{48}$$

$$\frac{5}{6} = \frac{5 \cdot 8}{6 \cdot 8} = \frac{40}{48}$$

$$\frac{3}{8} = \frac{3 \cdot 3}{8 \cdot 3} = \frac{9}{24}$$

$$\frac{5}{6} = \frac{5 \cdot 4}{6 \cdot 4} = \frac{20}{24}$$

$$\frac{3}{8} = \frac{3 \cdot 9}{8 \cdot 9} = \frac{27}{72}$$

$$\frac{5}{6} = \frac{5 \cdot 12}{6 \cdot 12} = \frac{60}{72}$$

3. Because

$$\frac{5}{3} = \frac{5 \cdot 4}{3 \cdot 4} = \frac{20}{12}$$

$$\frac{7}{4} = \frac{7 \cdot 3}{4 \cdot 3} = \frac{21}{12}$$

and

$$1 = \frac{12}{12}$$

the numbers $\frac{5}{3}$, $\frac{7}{4}$ and 1 can be plotted as shown in Figure 2.47.

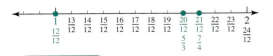

FIGURE 2.47   Plotting 1, $\frac{5}{3}$, and $\frac{7}{4}$

4. Because

$$0.9 = \frac{9}{10} = \frac{9 \cdot 2}{10 \cdot 2} = \frac{18}{20}$$

$$\frac{5}{4} = \frac{5 \cdot 5}{4 \cdot 5} = \frac{25}{20}$$

and

$$1 = \frac{20}{20}$$

the numbers 1, 0.9, and $\frac{5}{4}$ can be plotted as shown in Figure 2.48.

FIGURE 2.48   Plotting 1, 0.9, and $\frac{5}{4}$

5. Because the original line segment is $\frac{9}{8}$ units long, it consists of 9 equal parts, each of which is $\frac{1}{8}$ of a unit long.

Since

$$\frac{3}{4} = \frac{3 \cdot 2}{4 \cdot 2} = \frac{6}{8}$$

we should create a line segment consisting of 6 of the $\frac{1}{8}$-unit-long parts, as shown in Figure 2.49.

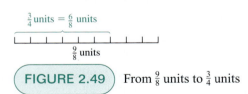

FIGURE 2.49   From $\frac{9}{8}$ units to $\frac{3}{4}$ units

6. $\frac{3}{5}$ and $\frac{2}{3}$ can both be expressed in terms of fifteenths. So, if we plot these fractions on a number line on which adjacent tick marks are $\frac{1}{15}$ apart, they will land on tick marks. The fraction $\frac{3}{5}$ will be at the ninth tick mark to the right of 0, and $\frac{2}{3}$ will be at the tenth tick mark to the right of 0, as shown in Figure 2.50.

FIGURE 2.50   When tick marks are $\frac{1}{15}$ apart, $\frac{3}{5} = \frac{9}{15}$ and $\frac{2}{3} = \frac{10}{15}$ land on tick marks.

7. a. See Figure 2.51. If the original chocolate bar is divided into 10 equal parts, then the amount of chocolate that Kelsey has is $\frac{6}{10}$ of the original chocolate. The amount that Kelsey wants to keep is $\frac{5}{10}$ of the original chocolate; so Kelsey should give 1 part out of her 6 parts to Janelle. Therefore, Janelle should get $\frac{1}{6}$ of Kelsey's chocolate.

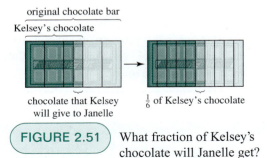

FIGURE 2.51   What fraction of Kelsey's chocolate will Janelle get?

b. In solving this problem, $\frac{3}{5}$ becomes

$$\frac{3 \cdot 2}{5 \cdot 2} = \frac{6}{10}$$

and $\frac{1}{2}$ becomes

$$\frac{1 \cdot 5}{2 \cdot 5} = \frac{5}{10}$$

c. The chocolate bar is one of the wholes in this problem—Kelsey's chocolate is $\frac{3}{5}$ of this whole, as is the $\frac{1}{2}$ that Kelsey wants to keep. Kelsey's chocolate is also a whole in this problem because we describe the amount of chocolate that Janelle will get as a fraction of this whole. Janelle will get $\frac{1}{6}$ of Kelsey's chocolate.

8. a. In Figure 2.52, 1 ton of gravel is represented by a rectangle, and Ken's order is represented by $\frac{3}{4}$ of the rectangle. Then Ken's order is divided into 4 equal parts and of those parts one is darkly shaded. This darkly shaded amount is $\frac{3}{16}$ of the original rectangle representing 1 ton. Therefore, $\frac{1}{4}$ of Ken's order is $\frac{3}{16}$ of a ton.

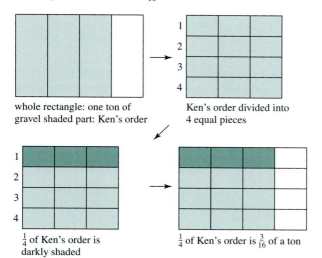

whole rectangle: one ton of gravel shaded part: Ken's order

Ken's order divided into 4 equal pieces

$\frac{1}{4}$ of Ken's order is darkly shaded

$\frac{1}{4}$ of Ken's order is $\frac{3}{16}$ of a ton

FIGURE 2.52   What fraction of a ton should Ken get now?

b. In solving this problem, the $\frac{3}{4}$ of a ton of gravel appears as

$$\frac{3 \cdot 4}{4 \cdot 4} = \frac{12}{16}$$

of a ton of gravel.

c. A full ton of gravel is one of the wholes in this problem. Ken's order of gravel ($\frac{3}{4}$ of a ton) is another whole, because in this problem we determine what $\frac{1}{4}$ of Ken's order of gravel is as a fraction of 1 ton of gravel.

9. If we consider the collection of squares at the top of Figure 2.46, we see that 10 out of 15 of these squares are shaded. Therefore, $\frac{10}{15}$ of the squares are shaded. The squares can be put in groups of 5. Viewing the squares in groups, we see that $2 \cdot 5$ out of $3 \cdot 5$ squares are shaded, so we can also say that $\frac{2 \cdot 5}{3 \cdot 5}$ of the squares are shaded. The squares form 3 groups, and 2 of those groups are shaded. So $\frac{2}{3}$ of the squares are shaded. Therefore, we have three ways of writing the fraction of squares that are shaded, and these three ways must all be equal:

$$\frac{10}{15} = \frac{2 \cdot 5}{3 \cdot 5} = \frac{2}{3}$$

10.

$$\frac{45}{72} = \frac{5 \cdot 9}{8 \cdot 9} = \frac{5}{8}$$

$$\frac{24}{36} = \frac{2 \cdot 12}{3 \cdot 12} = \frac{2}{3}$$

$$\frac{56}{88} = \frac{7 \cdot 8}{11 \cdot 8} = \frac{7}{11}$$

Notice that we could also simplify each of these fractions in several steps. For example,

$$\frac{45}{72} = \frac{15 \cdot 3}{24 \cdot 3} = \frac{15}{24} = \frac{5 \cdot 3}{8 \cdot 3} = \frac{5}{8}$$

Problems for Section 2.4

1. Use the meaning of fractions to explain in your own words why

$$\frac{3}{4} = \frac{3 \cdot 3}{4 \cdot 3}$$

Do not use multiplication by 1. Draw a picture to support your explanation.

2. Use the meaning of fractions to explain in your own words why

$$\frac{6}{5} = \frac{6 \cdot 2}{5 \cdot 2}$$

Do not use multiplication by 1. Draw a picture to support your explanation.

3. Explain in two different ways why

$$\frac{2}{3} = \frac{2 \cdot 4}{3 \cdot 4}$$

4. Explain in two different ways why

$$\frac{5}{6} = \frac{5 \cdot 2}{6 \cdot 2}$$

5. Using the fractions $\frac{2}{3}$ and $\frac{3}{4}$, describe how to give two fractions common denominators. In terms of pictures, what are you doing when you give fractions common denominators?

6. Using the fractions $\frac{5}{6}$ and $\frac{3}{8}$, describe how to give two fractions common denominators. In terms of pictures, what are you doing when you give fractions common denominators?

7. Using the example

$$\frac{6}{9}$$

describe how to put a fraction in simplest form, and explain why this procedure makes sense. In terms of pictures, what are you doing when you put a fraction in simplest form?

8. Using the example

$$\frac{12}{16}$$

describe how to put a fraction in simplest form, and explain why this procedure makes sense. In terms of pictures, what are you doing when you put a fraction in simplest form?

9. Draw a number line like the one in Figure 2.53. Then plot 0, $\frac{3}{8}$, and $\frac{1}{3}$ on your number line in such a way that each number falls on a tick mark. Lengthen the tick marks of whole numbers.

FIGURE 2.53   Number line

10. Draw a number line like the one in Figure 2.53. Then plot $\frac{1}{2}$, $\frac{3}{5}$, and $\frac{5}{8}$ on your number line in such a way that each number falls on a tick mark. Lengthen the tick marks of whole numbers (if there are any).

11. Draw a number line like the one in Figure 2.53. Then plot 0, 0.3, and $\frac{3}{5}$ on your number line in such a way that each number falls on a tick mark. Lengthen the tick marks of whole numbers.

12. Draw a number line like the one in Figure 2.53. Then plot 1, $\frac{5}{6}$, and $\frac{8}{7}$ on your number line in such a way that each number falls on a tick mark. Lengthen the tick marks of whole numbers.

13. Draw a number line like the one in Figure 2.54. Place equally spaced tick marks on your number line so that you can plot $\frac{1}{3}$ on a tick mark. Then plot $\frac{1}{3}$. Explain your reasoning.

FIGURE 2.54   Plot $\frac{1}{3}$

14. Draw a number line like the one in Figure 2.55. Place equally spaced tick marks on your number line so that you can plot $\frac{1}{2}$ on a tick mark. Then plot $\frac{1}{2}$. Explain your reasoning.

FIGURE 2.55   Plot $\frac{1}{2}$

15. Draw a number line like the one in Figure 2.56. Place equally spaced tick marks on your number line so that you can plot $\frac{5}{12}$ on a tick mark. Then plot $\frac{5}{12}$. Explain your reasoning.

FIGURE 2.56   Plot $\frac{5}{12}$

16. The line segment shown next has length $\frac{3}{5}$ unit. Draw a line segment like this one. Explain how to subdivide, and possibly add pieces to, your line segment to create a line segment of length $\frac{1}{2}$ unit. Do this without first creating a segment of length 1 unit. Explain how you know your line segment is the correct length.

$$\overline{}$$
$$\frac{3}{5}\text{unit}$$

17. Ted put $\frac{3}{4}$ of a cup of chicken stock in his soup. Later, Ted added another $\frac{2}{3}$ of a cup of chicken stock to his soup. All together, what fraction of a cup of chicken stock did Ted put into his soup?

   a. Draw pictures to help you solve this problem. Explain why your answer is correct.

   b. In solving the problem, how do $\frac{3}{4}$ and $\frac{2}{3}$ appear in different forms?

   c. What are the different wholes associated with the fractions in this problem? In other words, for each fraction in this problem, and its solution, what is the fraction *of*?

18. First you poured $\frac{3}{4}$ of a cup of water into an empty bowl. Then you scooped out $\frac{1}{3}$ of a cup of water. How much water is left in the bowl?

   a. Draw pictures to help you solve this problem. Explain why your answer is correct.

b. In solving the problem, how do $\frac{3}{4}$ and $\frac{1}{3}$ appear in different forms?

c. What are the different wholes associated with the fractions in this problem? In other words, for each fraction in this problem, and its solution, what is the fraction *of*?

19. First you poured $\frac{3}{4}$ of a cup of water into an empty bowl. Then you scooped out $\frac{1}{2}$ of the water in the bowl. How much water is left in the bowl?

a. Draw pictures to help you solve this problem. Explain why your answer is correct.

b. In solving the problem, how does $\frac{3}{4}$ appear in a different form?

c. What are the different wholes associated with the fractions in this problem? In other words, for each fraction in this problem, and its solution, what is the fraction *of*?

20. You want to make a recipe that calls for $\frac{2}{3}$ of a cup of flour, but you only have $\frac{1}{2}$ of a cup of flour left. Assuming you have enough of the other ingredients, what fraction of the recipe can you make?

a. Draw pictures to help you solve this problem. Explain why your answer is correct.

b. In solving the problem, how do $\frac{2}{3}$ and $\frac{1}{2}$ appear in different forms?

c. What are the different wholes associated with the fractions in this problem? In other words, for each fraction in this problem, and its solution, what is the fraction *of*?

21. Ken got $\frac{2}{3}$ of a ton of gravel even though he ordered $\frac{3}{4}$ of a ton of gravel. What fraction of his order did Ken get?

a. Draw pictures to help you solve this problem. Explain why your answer is correct.

b. In solving this problem, how do $\frac{2}{3}$ and $\frac{3}{4}$ appear in different forms?

c. What are the different wholes associated with the fractions in this problem? In other words, for each fraction in this problem, and its solution, what is the fraction *of*?

22. Two thirds of a cup of Healthy SnackOs provides your full daily value of vitamin B. You ate $\frac{1}{2}$ of a cup of Healthy SnackOs. What fraction of your daily value of vitamin B did you get in the Healthy SnackOs you ate?

a. Draw pictures to help you solve this problem. Explain why your answer is correct.

b. In solving this problem, how do $\frac{2}{3}$ and $\frac{1}{2}$ appear in different forms?

c. What are the different wholes associated with the fractions in this problem? In other words, for each fraction in this problem, and its solution, what is the fraction *of*?

23. Your cookie recipe calls for $\frac{2}{3}$ of a cup of butter for a batch of cookies. You decide that you only want to make $\frac{1}{3}$ of a batch of cookies. How much butter will you need?

a. Draw pictures to help you solve this problem. Explain why your answer is correct.

b. In solving the problem, how does $\frac{2}{3}$ appear in a different form?

c. What are the different wholes associated with the fractions in this problem? In other words, for each fraction in this problem, and its solution, what is the fraction *of*?

24. One serving of DietMuck is $\frac{2}{3}$ of a cup. Jane wants to eat $\frac{2}{3}$ of a serving of DietMuck. What fraction of a cup of DietMuck should Jane eat?

a. Draw pictures to help you solve this problem. Explain why your answer is correct.

b. In solving the problem, how does $\frac{2}{3}$ appear in a different form?

c. What are the different wholes associated with the fractions in this problem? In other words, for each fraction in this problem, and its solution, what is the fraction *of*?

25. If one serving of DietDelights provides $\frac{1}{3}$ of your daily value of vitamin C, then what fraction of your daily value of vitamin C is in $\frac{3}{4}$ of a serving of DietDelights?

a. Draw pictures to help you solve this problem. Explain why your answer is correct.

b. In solving the problem, how does $\frac{1}{3}$ appear in a different form?

c. What are the different wholes associated with the fractions in this problem? In other words, for each fraction in this problem, and its solution, what is the fraction *of*?

26. Becky moves into an apartment with two friends on August 1. Becky's friends have been in the apartment since July 1. The electric bill

comes every two months, and the next one will be for the electricity used in July and August. The bill is not broken down by month. What fraction of the July/August electric bill should Becky pay, and what fraction should her two friends pay, if they want to divide the bill fairly?

a. Solve Becky's electric bill problem if the apartment has a large communal area that is frequently used by all the friends, the friends typically eat meals together, and not much electricity is used in the separate sleeping areas. Explain your reasoning.

b. Solve Becky's electric bill problem if the friends spend most of their time in their separate rooms, don't eat meals together, and don't spend much time together in their communal area. Explain your reasoning.

2.5 Comparing Fractions

 Focal Points
Grade 5

Given two numbers in decimal representation, we can determine which one is greater by comparing the digits of the decimal numbers. However, comparing fractions is more complicated, because every fraction is equal to infinitely many other fractions. Unlike whole numbers, we can't always tell just by comparing digits of fractions whether they are equal or whether one is greater than the other. You will probably recognize that a familiar fraction such as $\frac{1}{2}$ is equal to $\frac{3}{6}$ or $\frac{50}{100}$, but can you tell right away, just by looking, that

$$\frac{411}{885} \quad \text{and} \quad \frac{548}{1180}$$

are equal? It is not obvious.

There are four standard methods for determining whether two fractions are equal, or if not, which one is greater. The four methods for comparing fractions are as follows:

1. Converting to decimals
2. Using common denominators
3. Using common numerators
4. Cross-multiplication

Comparing Fractions by Converting to Decimals

We can convert every fraction to a decimal number by dividing the denominator into the numerator. Therefore, we can compare two fractions simply by converting both fractions to decimals and comparing the decimals.

Is $\frac{17}{35}$ equal to $\frac{43}{87}$, or if not, which is greater?

$$\frac{17}{35} = 17 \div 35 = 0.4857\ldots$$

$$\frac{43}{87} = 43 \div 87 = 0.4942\ldots$$

Since

$$0.4942\ldots > 0.4857\ldots$$

it follows that

$$\frac{43}{87} > \frac{17}{35}$$

Comparing Fractions by Using Common Denominators

When two fractions have the same denominator, we can determine whether they are equal, or if not, which is greater, just by looking at them. The fractions

$$\frac{784}{953} \quad \text{and} \quad \frac{621}{953}$$

have the same denominator. Since

$$784 > 621$$

it follows that

$$\frac{784}{953} > \frac{621}{953}$$

In general, if two fractions have the same denominator, then the one with the greater numerator is greater; if the numerators and denominators are equal, then the fractions are equal.

Why does this rule for comparing fractions with like denominators make sense? Say we have two fractions that have the same denominator. If we think of the two fractions as fractions of two identical pies, then each pie is divided into the same number of pieces. Therefore, each piece of pie in the two pies is the same size. The numerator describes the number of pieces we have of the pie. Since the pie pieces are all the same size, if we have more pieces of one pie, then we have more of that pie. See Figure 2.57. If we have the same number of pieces of each pie, then we have the same amount of each pie. Therefore, when two fractions have the same denominators, the fraction with the greater numerator is greater; and when two fractions have the same denominator and the same numerator, the fractions are equal.

But what if you have two fractions that don't have the same denominator? How can you find out if they are equal; or if they are not, how can you find out which fraction is greater? You can give the two fractions a common denominator. It can be any common denominator—it does not have to be the least one. A common denominator that always works is obtained by multiplying the two denominators.

Consider the following two fractions:

$$\frac{8}{12} \quad \text{and} \quad \frac{10}{15}$$

We can give these fractions the common denominator

$$12 \cdot 15$$

FIGURE 2.57

Comparing fractions with equal denominators $\frac{4}{10} > \frac{3}{10}$

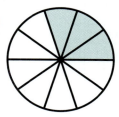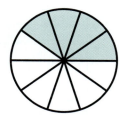

Then,

$$\frac{8}{12} = \frac{8 \cdot 15}{12 \cdot 15} = \frac{120}{180}$$

and

$$\frac{10}{15} = \frac{10 \cdot 12}{15 \cdot 12} = \frac{120}{180}$$

The two fractions $\frac{8}{12}$ and $\frac{10}{15}$ are equal because when both are written with the denominator $12 \cdot 15 = 180$, they have the same numerator. We could also have given these two fractions the common denominator of 60 in order to compare them (because $60 = 12 \cdot 5$ and $60 = 15 \cdot 4$). Or we could have noticed that $\frac{8}{12}$ and $\frac{10}{15}$ are both equal to $\frac{2}{3}$.

What are we really doing when we give two fractions common denominators in order to compare the sizes of the fractions? As we saw in Section 2.4, when we give fractions common denominators, we create *like parts*. For example, to compare the fractions

$$\frac{4}{9} \quad \text{and} \quad \frac{3}{5}$$

we can give both fractions the common denominator of $9 \cdot 5 = 45$, as follows:

$$\frac{4}{9} = \frac{4 \cdot 5}{9 \cdot 5} = \frac{20}{45}$$

$$\frac{3}{5} = \frac{3 \cdot 9}{5 \cdot 9} = \frac{27}{45}$$

As shown in Figure 2.58, the process of giving the fractions common denominators corresponds to subdividing ninths and fifths into pieces of the same size, namely, forty-fifths. The denominators are now the same, and the numerator 27 is greater than the numerator 20. Therefore,

$$\frac{27}{45} > \frac{20}{45}$$

so that

$$\frac{3}{5} > \frac{4}{9}$$

FIGURE 2.58

Comparing $\frac{4}{9}$ and $\frac{3}{5}$

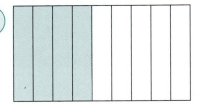 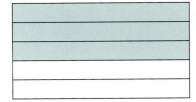

When we give fractions common denominators, we subdivide to create like parts.

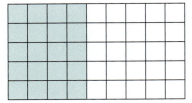 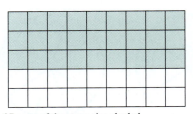

$4 \cdot 5 = 20$ parts shaded versus $3 \cdot 9 = 27$ parts of the same size shaded

FIGURE 2.59

Comparing $\frac{5}{8}$ and $\frac{5}{9}$

Both have 5 parts shaded, but 8ths are bigger than 9ths.

Before you read on, do the next Class Activity.

Class Activity   *Now Turn to Class Activities Manual*

2Q What Is Another Way to Compare These Fractions? p. 33

Comparing Fractions by Using Common Numerators

Which fraction is greater,

$$\frac{5}{8} \quad \text{or} \quad \frac{5}{9}?$$

Both fractions represent 5 parts, but $\frac{5}{8}$ is 5 parts when the whole is divided into 8 parts, whereas $\frac{5}{9}$ is 5 parts when the whole is divided into 9 parts. But if an object is divided into 8 equal parts, then each part is larger than if the object is divided into 9 equal parts—fewer parts making up the same whole means that each part has to be larger. So 5 of 8 equal parts must be greater than 5 of 9 equal parts. Thus,

$$\frac{5}{8} > \frac{5}{9}$$

as shown in Figure 2.59.

The reasoning we just used shows us that when two fractions have the same numerator, the fraction with the greater denominator is *less than* the fraction with the smaller denominator. We can see this not only by thinking about subdividing pies, as in Figure 2.59, but also by considering number lines, as in Figure 2.60. As we subdivide the interval between 0 and 1 into more pieces, each piece becomes smaller, and fractions with the same numerator, but increasing denominators, become smaller.

FIGURE 2.60

Fractions with the same numerator and increasing denominators become smaller

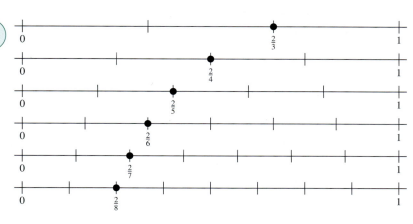

If we want to compare two fractions and the fractions don't already have a common numerator, then we can give the fractions a common numerator to compare them. For example, which fraction is greater,

$$\frac{4}{235} \quad \text{or} \quad \frac{6}{301}?$$

We can give the two fractions the common numerator of 12 to compare them:

$$\frac{4}{235} = \frac{4 \cdot 3}{235 \cdot 3} = \frac{12}{705}$$

$$\frac{6}{301} = \frac{6 \cdot 2}{301 \cdot 2} = \frac{12}{602}$$

Since 705 is greater than 602,

$$\frac{12}{705} < \frac{12}{602}$$

because 12 pieces of an object that has been divided into 705 equal pieces is a smaller amount than 12 pieces of the object if it has been divided into 602 equal pieces.

Comparing Fractions by Cross-Multiplying

When we compared fractions by giving them common denominators, we showed that $\frac{8}{12}$ and $\frac{10}{15}$ are equal by showing that

$$\frac{8}{12} = \frac{8 \cdot 15}{12 \cdot 15} = \frac{120}{180}$$

and

$$\frac{10}{15} = \frac{10 \cdot 12}{15 \cdot 12} = \frac{120}{180}$$

Notice that all we needed to check was that

$$8 \cdot 15 = 10 \cdot 12$$

because $8 \cdot 15$ and $10 \cdot 12$ were the numerators when we gave $\frac{8}{12}$ and $\frac{10}{15}$ a common denominator by multiplying the two denominators. So, all we really had to do was to multiply the numerator of each fraction $\frac{8}{12}$ and $\frac{10}{15}$ by the denominator of the other fraction, and check if those two resulting numbers were equal.

Similarly, to show that

$$\frac{3}{5} > \frac{4}{9}$$

all we need to check is that

$$3 \cdot 9 > 4 \cdot 5$$

because $3 \cdot 9$ and $4 \cdot 5$ are the numerators when we give $\frac{4}{9}$ and $\frac{3}{5}$ the common denominator of $9 \cdot 5$.

Now let's generalize. If $\frac{A}{B}$ and $\frac{C}{D}$ are two fractions of whole numbers (where neither B nor D is zero), then we can give these fractions the common denominator $B \cdot D$:

$$\frac{A}{B} = \frac{A \cdot D}{B \cdot D}$$

$$\frac{C}{D} = \frac{C \cdot B}{D \cdot B}$$

To compare $\frac{A}{B}$ and $\frac{C}{D}$, we can compare

$$\frac{A \cdot D}{B \cdot D} \quad \text{and} \quad \frac{C \cdot B}{D \cdot B}$$

Because the denominators $D \cdot B$ and $B \cdot D$ are equal, we need to compare only the numerators

$$A \cdot D \quad \text{and} \quad C \cdot B$$

We can make these conclusions:

- The fractions $\frac{A}{B}$ and $\frac{C}{D}$ are equal exactly when $A \cdot D$ and $C \cdot B$ are equal.

- $\frac{A}{B}$ is greater than $\frac{C}{D}$ exactly when $A \cdot D$ is greater than $C \cdot B$.

- $\frac{A}{B}$ is less than $\frac{C}{D}$ exactly when $A \cdot D$ is less than $C \cdot B$.

Stated symbolically,

$$\frac{A}{B} = \frac{C}{D} \qquad \text{exactly when} \qquad A \cdot D = C \cdot B$$

$$\frac{A}{B} > \frac{C}{D} \qquad \text{exactly when} \qquad A \cdot D > C \cdot B$$

$$\frac{A}{B} < \frac{C}{D} \qquad \text{exactly when} \qquad A \cdot D < C \cdot B$$

cross-multiplying This method for checking whether two fractions are equal—or if not, which one is greater—is often called the **cross-multiplying** method. Notice that it is just a way to check if the numerators are equal when the two fractions are given the common denominator obtained by multiplying the two original denominators.

Using Other Reasoning to Compare Fractions

The standard methods just described for comparing fractions are useful because they are efficient and they always work, but sometimes it is possible to use other reasoning to compare fractions. Why would we want to use other reasoning when we already have several good methods? In some cases, other reasoning can be more efficient. Even more importantly, however, reasoning in other ways can help us think about the meaning of fractions and can help us understand fractions better.

Before you read on, do the next Class Activity.

> **Class Activity   Now Turn to Class Activities Manual**
>
> **2R** Comparing Fractions by Reasoning, p. 34

Consider the case of determining whether $\frac{3}{5}$ and $\frac{4}{9}$ are equal, and if not, which is larger. We can reason that $\frac{4}{9}$ is less than $\frac{1}{2}$,

$$\frac{4}{9} < \frac{1}{2}$$

because if we divide an object into 9 equal pieces, then it would take four and a half pieces to make half of the object, and 4 pieces is certainly less than four and a half pieces. Similarly, we can say that

$$\frac{1}{2} < \frac{3}{5}$$

because if we divide an object into 5 equal pieces, then two and a half of them would make half the object, and three pieces is more than two and a half pieces. Since $\frac{4}{9}$ is less than a half, and $\frac{3}{5}$ is more than a half, $\frac{3}{5}$ is the larger fraction.

benchmark (or landmark) When we compared $\frac{3}{5}$ and $\frac{4}{9}$ by comparing both fractions with $\frac{1}{2}$, we used $\frac{1}{2}$ as a **benchmark (or landmark)**. The fractions $\frac{1}{2}, \frac{1}{4}, \frac{3}{4}, \frac{1}{3}$, and 1 are good to use as benchmarks.

> **Class Activity   Now Turn to Class Activities Manual**
>
> **2S** Can We Reason This Way? p. 34

Practice Exercises for Section 2.5

1. Compare the sizes of the following pairs of fractions in two ways: by giving the fractions common denominators and by cross-multiplying. How are these two methods related?

$$\frac{2}{3} \quad \text{and} \quad \frac{3}{5}$$

$$\frac{8}{13} \quad \text{and} \quad \frac{13}{21}$$

$$\frac{15}{20} \quad \text{and} \quad \frac{6}{8}$$

$$\frac{1}{4} \quad \text{and} \quad \frac{3}{8}$$

$$\frac{4}{14} \quad \text{and} \quad \frac{5}{15}$$

2. Complete the following to make true statements about comparing fractions:

   a. If two fractions have the same denominators, then the one with _____ is greater.

   b. If two fractions have the same numerator, then the one with _____ is greater.

   Use the meaning of fractions to explain why your answers make sense.

3. For each of the following pairs of fractions, determine which is larger. Use reasoning other than finding common denominators, cross-multiplying, or converting to decimal representation to explain your answers.

   a. $\frac{6}{11}$ versus $\frac{6}{13}$

   b. $\frac{5}{8}$ versus $\frac{7}{12}$

   c. $\frac{5}{21}$ versus $\frac{7}{24}$

   d. $\frac{21}{22}$ versus $\frac{56}{57}$

   e. $\frac{97}{100}$ versus $\frac{35}{38}$

4. Explain why

$$\frac{A}{B} > \frac{C}{D} \quad \text{exactly when} \quad A \cdot D > C \cdot B$$

5. Without using a calculator, order the following numbers from largest to smallest: $\frac{41}{20}$, 2, $\frac{11}{5}$, -2, -2.3, $-\frac{19}{10}$. Explain your reasoning.

6. Find a fraction between $\frac{2}{11}$ and $\frac{3}{11}$ whose numerator and denominator are whole numbers.

7. Find a decimal number between $\frac{21}{34}$ and $\frac{34}{55}$, and plot all three numbers visibly and distinctly on a number line. Be sure the labeling fits with the structure of the decimal system. The numbers need not land on tick marks.

Answers to Practice Exercises for Section 2.5

1.

Fractions		Common Denominator	Cross-Multiplying	Conclusion
$\frac{2}{3}$	$\frac{3}{5}$	$\frac{2 \cdot 5}{3 \cdot 5} > \frac{3 \cdot 3}{5 \cdot 3}$	$2 \cdot 5 > 3 \cdot 3$	$\frac{2}{3} > \frac{3}{5}$
$\frac{8}{13}$	$\frac{13}{21}$	$\frac{8 \cdot 21}{13 \cdot 21} < \frac{13 \cdot 13}{21 \cdot 13}$	$8 \cdot 21 < 13 \cdot 13$	$\frac{8}{13} < \frac{13}{21}$
$\frac{15}{20}$	$\frac{6}{8}$	$\frac{15 \cdot 8}{20 \cdot 8} = \frac{6 \cdot 20}{8 \cdot 20}$	$15 \cdot 8 = 6 \cdot 20$	$\frac{15}{20} = \frac{6}{8}$
$\frac{1}{4}$	$\frac{3}{8}$	$\frac{1 \cdot 8}{4 \cdot 8} < \frac{3 \cdot 4}{8 \cdot 4}$	$1 \cdot 8 < 3 \cdot 4$	$\frac{1}{4} < \frac{3}{8}$
$\frac{4}{14}$	$\frac{5}{15}$	$\frac{4 \cdot 15}{14 \cdot 15} < \frac{5 \cdot 14}{15 \cdot 14}$	$4 \cdot 15 < 5 \cdot 14$	$\frac{4}{14} < \frac{5}{15}$

Notice that to compare $\frac{15}{20}$ and $\frac{6}{8}$ with the common denominator method, you could choose a smaller common denominator than $20 \cdot 8$. For instance, you could choose 40 as the common denominator. In this case, you compare the numerators $2 \cdot 15$ and $5 \cdot 6$ rather than comparing $8 \cdot 15$ and $20 \cdot 6$, which is what you compare with the cross-multiplying method. You can use a similar procedure with $\frac{1}{4}$ and $\frac{3}{8}$. Of course, it's easiest just to use the common denominator 8 in that case.

When we give two fractions the common denominator obtained by multiplying the two denominators, the numerators are exactly the quantities we compare when we cross-multiply. Therefore, the cross-multiplying method is really just a shortcut for comparing two fractions by giving them the common denominator that is the product of the two denominators.

2. **a.** If two fractions have the same denominators, then the one with the greater numerator is greater. Think of the two fractions as fractions you have eaten of two identical pies. Since the two fractions have the same denominator, the number of pieces that both pies were divided into is the same. Because the pie pieces are the same size, you have eaten more of the pie from which you ate the largest number of pieces. Therefore, you have eaten the greater fractions of pie for the fractions that has the greater numerator.

b. If two fractions have the same numerator, then the one with the smaller denominator is greater. Think of the two fractions as portions you have eaten of two identical pies. The fractions with the smaller denominator corresponds to the pie that was divided into fewer pieces. In the pie that was divided into fewer pieces, each piece is bigger than in the other pie. For example, the pieces in a 6-piece pie are bigger than the pieces in an 8-piece pie—fewer pieces making up the same size pie means that each piece must be bigger, as seen in Figure 2.61. Because the numerators of the two fractions are the same, you have eaten the same number of pieces from each pie. So, you have eaten the most pie from the pie that was divided into fewer pieces. Therefore, you have eaten the greater fractions of pie for the fraction that has the smaller denominator.

 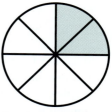

FIGURE 2.61 Two pieces are eaten in each, but the pies have different size pieces.

3. **a.** $\frac{6}{11} > \frac{6}{13}$. If you divide a pie into 13 pieces, then each piece will be smaller than if you divide that same pie into 11 pieces. So, 6 pieces of a pie that is divided into 11 pieces will be more than 6 pieces of a pie that is divided into 13 pieces.

b. $\frac{5}{8}$ is one piece more than half $(\frac{4}{8})$, and $\frac{7}{12}$ is one piece more than half $(\frac{6}{12})$. However, when a pie is divided into 8 pieces, each piece is larger than when an identical pie is divided into 12 pieces—fewer pieces making up the same amount means each piece must be larger. The eighths pieces are larger than the twelfths pieces, and since each fraction, $\frac{5}{8}$ and $\frac{7}{12}$, is one piece more than one half, it follows that $\frac{5}{8} > \frac{7}{12}$.

c. Notice that both fractions are close to $\frac{1}{4}$. The fractions $\frac{1}{4}$ and $\frac{5}{20}$ are equal; therefore, $\frac{5}{21}$ is a little less than $\frac{1}{4}$. (See the reasoning in the previous answer.) The fractions $\frac{1}{4}$ and $\frac{6}{24}$ are equal; therefore, $\frac{7}{24}$ is a little bigger than $\frac{1}{4}$. (See the reasoning in the previous answer.) So, $\frac{5}{21} < \frac{7}{24}$.

d. Notice that each fraction is "one piece less than 1 whole." The fraction $\frac{21}{22}$ is $\frac{1}{22}$ less than a whole, and $\frac{56}{57}$ is $\frac{1}{57}$ less than a whole. But $\frac{1}{22}$ is bigger than $\frac{1}{57}$ because, if you divide a pie into 22 pieces, each piece will be bigger than if you divide an identical pie into 57 pieces. Therefore, $\frac{21}{22} < \frac{56}{57}$ because $\frac{21}{22}$ is a bigger piece away from a whole than is $\frac{56}{57}$.

e. $\frac{97}{100} > \frac{35}{38}$. The same reasoning used in the answer to the previous part applies here, too. This time, each fraction is 3 pieces away from 1 whole.

4. To compare two fractions $\frac{A}{B}$ and $\frac{C}{D}$, we can give them the common denominator $B \cdot D = D \cdot B$. Then

$$\frac{A}{B} = \frac{A \cdot D}{B \cdot D}$$

and

$$\frac{C}{D} = \frac{C \cdot B}{D \cdot B}$$

Since the fractions $\frac{A \cdot D}{B \cdot D}$ and $\frac{C \cdot B}{D \cdot B}$ have the same denominator, the fraction $\frac{A \cdot D}{B \cdot D}$ is greater than $\frac{C \cdot B}{D \cdot B}$ exactly when the numerator $A \cdot D$ is greater than the numerator $C \cdot B$. This in turn means that the original fraction $\frac{A}{B}$ is greater than $\frac{C}{D}$ exactly when $A \cdot D$ is greater than $C \cdot B$. In other words,

$$\frac{A}{B} > \frac{C}{D}$$

exactly when

$$A \cdot D > C \cdot B$$

5. Every positive number is greater than every negative number, so we know that $\frac{41}{20}$, 2, and $\frac{11}{5}$ are greater than -2, -2.3, and $-\frac{19}{10}$. We can give $\frac{41}{20}$, 2, and $\frac{11}{5}$ the common denominator of 20 by letting

$$2 = \frac{2}{1} = \frac{2 \cdot 20}{1 \cdot 20} = \frac{40}{20}$$

and

$$\frac{11}{5} = \frac{11 \cdot 4}{5 \cdot 4} = \frac{44}{20}$$

Comparing the numerators, we see that

$$\frac{11}{5} > \frac{41}{20} > 2$$

Since $\frac{19}{10} = 1.9$, and since

$$2.3 > 2 > 1.9$$

it follows that

$$-1.9 > -2 > -2.3$$

Therefore,

$$\frac{11}{5} > \frac{41}{20} > 2 > -1.9 > -2 > -2.3$$

6. Since $\frac{2}{11} = \frac{4}{22}$ and $\frac{3}{11} = \frac{6}{22}$, the fraction $\frac{5}{22}$ is between $\frac{2}{11}$ and $\frac{3}{11}$.

7. The decimal representation of $\frac{21}{34}$ is 0.6176... and the decimal representation of $\frac{34}{55}$ is 0.6181...; so, 0.618 is one example of a decimal number that is in between $\frac{21}{34}$ and $\frac{34}{55}$. There are many other examples. See Figure 2.62.

FIGURE 2.62   The number 0.618 is between $\frac{21}{34}$ and $\frac{34}{55}$.

Problems for Section 2.5

1. Julie says that the picture in Figure 2.63 shows that

$$\frac{1}{4} > \frac{1}{2}$$

by comparing the areas. Explain carefully the error in Julie's reasoning.

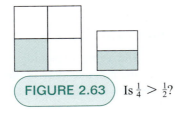

FIGURE 2.63   Is $\frac{1}{4} > \frac{1}{2}$?

2. In your own words, explain in detail why we can determine which of two fractions is greater by giving the two fractions common denominators. What is the rationale behind this method? What are we really doing when we give the fractions common denominators?

3. In your own words, explain in detail why we can determine which of two fractions is greater by using the cross-multiplying method. What is the rationale behind this method? What are we really doing when we cross-multiply in order to compare fractions?

4. In your own words, explain in detail why it is the case that if two fractions have the same numerator, the fraction with the smaller denominator is greater.

5. Which fraction is greater, $\frac{35}{109}$ or $\frac{36}{104}$? To explain your answer, use the meaning of fractions along with reasoning other than finding common denominators, common numerators, cross-multiplying, or converting to decimal numbers.

6. ⬛ Which fraction is greater, $\frac{15}{31}$ or $\frac{23}{47}$? To explain your answer, use the meaning of fractions along with reasoning other than finding common denominators, common numerators, cross-multiplying, or converting to decimal numbers.

7. ⬛ Which fraction is greater, $\frac{19}{20}$ or $\frac{27}{28}$? To explain your answer, use the meaning of fractions along with reasoning other than finding common denominators, common numerators, cross-multiplying, or converting to decimal numbers.

8. Without using a calculator, order the following numbers from smallest to largest: $\frac{11}{10}$, 1.2, $\frac{9}{8}$, -1.1, -0.98, -1. Explain your reasoning.

9. Find a number between $\frac{23}{84}$ and $\frac{29}{98}$ and plot all three numbers visibly and distinctly on a number line like the one in Figure 2.64, which has a set of longer tick marks and a set of shorter tick marks. The labeling of the number line should fit with the structure of the decimal system. Label all the longer tick marks. The numbers need not land on tick marks.

FIGURE 2.64   Number line

10. Find a number between $\frac{78}{134}$ and $\frac{124}{213}$ and plot all three numbers visibly and distinctly on a number line like the one in Figure 2.64, which has a set of longer tick marks and a set of shorter tick marks. The labeling of the number line should fit with the structure of the decimal system. Label all the longer tick marks. The numbers need not land on tick marks.

11. Find two different fractions between $\frac{2}{5}$ and $\frac{3}{5}$ whose numerators and denominators are whole numbers, and plot all four fractions on a number line.

12. Give two different methods for solving the following problem: Find two different fractions in between $\frac{3}{5}$ and $\frac{2}{3}$ whose numerators and denominators are whole numbers.

13. Give two different methods for solving the following problem: Find two different fractions between $\frac{5}{7}$ and $\frac{6}{7}$ whose numerators and denominators are whole numbers.

14. A student says that $\frac{1}{5}$ is halfway between $\frac{1}{4}$ and $\frac{1}{6}$. Use a carefully drawn number line to show that this is not correct. What fraction is halfway between $\frac{1}{4}$ and $\frac{1}{6}$? Explain your reasoning.

15. Sam has a method for comparing fractions: He just looks at the denominator. Sam says the fraction with the larger denominator is smaller because, if there are more pieces, each piece is smaller. Discuss Sam's ideas.

16. Minju says that fractions that use bigger numbers are greater than fractions that use smaller numbers. Make up two problems for Minju to help her reconsider her ideas. For each problem, explain how to solve it, and explain why you chose that problem for Minju.

17. ⬛ Malcolm says that

$$\frac{8}{11} > \frac{7}{10}$$

because

$$8 > 7 \text{ and } 11 > 10$$

Discuss Malcolm's reasoning. Even though it is true that $\frac{8}{11} > \frac{7}{10}$, is Malcolm's reasoning correct? If Malcolm's reasoning is correct, clearly explain why. If Malcolm's reasoning is not correct, give Malcolm two examples that show why not.

18. You may combine your answers to all three parts of this problem.

   a. Is it valid to compare

   $$\frac{30}{70} \quad \text{and} \quad \frac{20}{50}$$

   by "cancelling" the 0s and comparing

   $$\frac{3}{7} \quad \text{and} \quad \frac{2}{5}$$

   instead? Explain your answer.

   b. Is it valid to compare

   $$\frac{15}{25} \quad \text{and} \quad \frac{105}{205}$$

   by "cancelling" the 5s and comparing

   $$\frac{1}{2} \quad \text{and} \quad \frac{10}{20}$$

   instead? Explain your answer.

   c. Write a paragraph discussing the distinction between your answer in (a) and your answer in (b).

19. ↻ Consider the following list of fractions:

$$\frac{1}{1}, \frac{2}{1}, \frac{3}{2}, \frac{5}{3}, \frac{8}{5}, \dots$$

You do not have to explain your answers to the following parts:

a. Describe a pattern in the list of fractions. Use your description to find the next 5 entries in the list after $\frac{8}{5}$. You will now have the first 10 entries in the list of fractions.

b. Use either the cross-multiplying method or the common denominator method to compare the sizes of the 1st, 3rd, 5th, 7th, and 9th fractions in the list. Describe a pattern in the sizes of these fractions. Describe a pattern that occurs when you compare the fractions.

c. Use either the cross-multiplying method or the common denominator method to compare the sizes of the 2nd, 4th, 6th, 8th, and 10th fractions in the list. Describe a pattern in the sizes of these fractions. Describe a pattern that occurs when you compare the fractions.

d. Convert the 10 fractions on your list to decimals, and plot them on a number line. Zoom in on portions of your number line so that you can show clearly where each decimal number is plotted relative to the others.

e. If you could find more and more entries in the list of fractions, and plot them on a number line, in what region of the number line would they be located? Do you think these numbers would get closer and closer to a particular number?

20. Suppose you start with a proper fraction and you add 1 to both the numerator and the denominator. For example, if you started with $\frac{2}{3}$, then you'd get a new fraction $\frac{2+1}{3+1} = \frac{3}{4}$.

a. Give at least 5 examples of proper fractions $\frac{A}{B}$. In each example, compare the sizes of $\frac{A}{B}$ and $\frac{A+1}{B+1}$. What do you notice? Make sure you are working with proper fractions (where the numerator is less than the denominator).

b. Frank says that if $\frac{A}{B}$ is a proper fraction, then $\frac{A+1}{B+1}$ is always greater than $\frac{A}{B}$ because $\frac{A+1}{B+1}$ has more parts. Regardless of whether Frank's conclusion is correct, discuss whether Frank's reasoning is valid. Did Frank give a convincing explanation that $\frac{A+1}{B+1}$ is greater than $\frac{A}{B}$? If not, what objections could you make to Frank's reasoning?

c. Explain the phenomenon you discovered in part (a). Compare the sizes of $\frac{A}{B}$ and $\frac{A+1}{B+1}$. Explain why the fraction you say is larger really *is* larger.

2.6 Percent

F·P **Focal Points**
        Grade 7

Almost any time we open a newspaper, walk into a store, or listen to a sports broadcast, we encounter percentages. A percentage is a number that expresses a relationship between two other numbers. In this way, percentages are like fractions: Just as a fraction is *of* an associated whole, a percentage is *of* some number. In fact, percentages can be viewed as special kinds of fractions, namely, those with denominator 100. Every fraction has a decimal representation; therefore, every percentage also has a decimal representation. So our study of percent extends our study of decimals and fractions. We will also study the basic types of percent problems and a variety of methods for solving these problems.

The Meaning of Percent

percent   The word **percent**, which is usually represented by the symbol %, means "of each hundred" or "out of a hundred" (per = "of each" or "for each" or "out of" and cent = "hundred"). So 35% means 35 out of 100, or $\frac{35}{100}$, and in general,

$$P\% = \frac{P}{100}$$

When we work with percentages, we can apply the definition of fraction. For example, if 35% of the trees at an arboretum are evergreen, then this means that the fraction of evergreen trees at the arboretum is $\frac{35}{100}$. According to the definition of $\frac{35}{100}$, the trees can be divided into 100 equal parts so that 35 of those parts consist of evergreen trees, as shown in Figure 2.65.

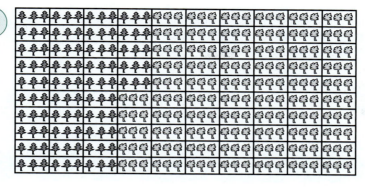

FIGURE 2.65

35% of the trees at an arboretum are evergreen.

On the other hand, the fraction of evergreen trees at the arboretum can also be expressed as $\frac{\text{\# evergreens}}{\text{\# trees}}$. So we have two ways of expressing the fraction of trees at the arboretum that are evergreen; therefore, these two fractions must be equivalent, so

$$\frac{35}{100} = \frac{\text{\# evergreens}}{\text{\# trees}}$$

In general, when we write a given fraction as a percent, we express that fraction as an equivalent fraction with denominator 100. Why use percents when we could use ordinary fractions? By using the denominator 100, it becomes easy to compare fractional amounts of different quantities. For example, if the fraction of evergreen trees at one arboretum is $\frac{160}{500}$ and the fraction of evergreen trees at another arboretum is $\frac{200}{800}$, we have to do some calculating to tell which arboretum has the greater fraction of evergreen trees. On the other hand, if we are told that 32% of the trees at one arboretum are evergreens whereas 25% of the trees at another arboretum are evergreens, then we know immediately that the first arboretum has a greater fraction of evergreen trees.

Percents, Fractions, and Pictures

Percents that we encounter frequently are highlighted in Table 2.1, along with their equivalent fractions expressed in simplest form. These percentages are also shown in Figure 2.66 as the shaded portion of each diagram.

If you are familiar with the fraction representations of the percentages in Table 2.1, you can easily extend them to nearby percentages and fractions, as follows:

55% is a little more than $\frac{1}{2}$ and is halfway between 50% and 60%,

or

15% is $\frac{1}{4}$ minus 10%.

Some of the class activities, exercises, and problems ask you to use pictures to solve percent problems. Although drawing pictures is usually not an efficient way of solving percent problems, pictures can help you develop a better understanding of percents. Just as it would not be reasonable to pull out bundled toothpicks whenever we work with decimal numbers, it would not be reasonable to do all percent calculations with pictures. However, both physical objects and pictures are valuable learning aids.

FIGURE 2.66

Pictures of percentages

    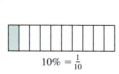

$25\% = \frac{1}{4}$ $50\% = \frac{1}{2}$ $75\% = \frac{3}{4}$ $10\% = \frac{1}{10}$ $20\% = \frac{1}{5}$ $5\% = \frac{1}{20}$

TABLE 2.1 Common percentages expressed as fractions

$25\% = \dfrac{25}{100} = \dfrac{1}{4},$	$50\% = \dfrac{50}{100} = \dfrac{1}{2},$	$75\% = \dfrac{75}{100} = \dfrac{3}{4}$
$10\% = \dfrac{10}{100} = \dfrac{1}{10},$	$20\% = \dfrac{20}{100} = \dfrac{1}{5},$	$5\% = \dfrac{5}{100} = \dfrac{1}{20}$

Class Activity   *Now Turn to Class Activities Manual*

2T Pictures, Percentages, and Fractions, p. 35

Solving Percent Problems

A percent expresses a fraction as an equivalent fraction that has denominator 100. So a basic percent problem involves a statement that there are two equivalent ways to express what fraction a portion is of a whole amount, namely that

$$P\% = \frac{\text{portion}}{\text{whole amount}}$$

or in other words that

$$\frac{P}{100} = \frac{\text{portion}}{\text{whole amount}} \tag{2.2}$$

Another formulation of this equation is

$$P\% \text{ of the whole amount is the portion}$$

or

$$\text{the portion is } P\% \text{ of the whole amount}$$

A helpful way to organize the information in a basic percent problem is with a *percent table,* namely, a table of this form:

$$P\% \quad \rightarrow \quad \text{portion}$$
$$100\% \quad \rightarrow \quad \text{whole amount}$$

There are three basic kinds of percent problems: one for each of the cases where one of P (the percent), the "portion," or the "whole amount" is unknown and to be determined, and the other two amounts are known. To solve such a problem, we must solve Equation 2.2 for the unknown amount. In this section, we consider three different (but interrelated) ways to solve Equation 2.2:

1. Using standard techniques of algebra, such as cross-multiplying Equation 2.2 to obtain the equation

$$P \cdot (\text{whole amount}) = 100 \cdot (\text{portion})$$

   and then solving this equation;

2. using a percent table and either reasoning about benchmark fractions, "going through 1%," or "going through 1;"

3. making equivalent fractions (without cross-multiplying), which can involve "going through 1."

We focus now on the second and third methods, because these methods can help us reason about percentages while developing our number sense and connecting equivalent fractions with percents.

Finding the Portion When the Percent and the Whole Amount Are Known In some basic percent problems the percent, P, and the whole amount are known and the portion is to be calculated:

Susie must pay 6% tax on her purchase of $44. How much tax must Susie pay?

To solve this problem, we must solve the equation

$$\frac{6}{100} = \frac{\text{tax}}{\$44}$$

Solving algebraically, we find that

$$\text{tax} = 0.06 \cdot \$44 = \$2.64$$

The next Class Activity will help you use percent tables and and benchmark fractions (or percentages) to calculate percents of quantities.

Class Activity   *Now Turn to Class Activities Manual*

2U Calculating Percents of Quantities by Using Benchmark Fractions and Percent Tables, p. 35

Part 3 of Class Activity 2U demonstrated a general method for calculating a percentage of a quantity. We can call this method "going through 1%." To calculate 6% of $44 by going through 1%, we can reason as follows. Since 100% is $44, we can find 1% by dividing by 100. So, 1% is $0.44. Then, 6% is 6 times as much, which is $6 \cdot \$0.44 = \$2.64$. We can summarize this reasoning in the percent table in Table 2.2. Notice that to find 6% by going through 1%, we did not have to know how to set up a proportion or how to multiply two decimals. We had to know only how to divide by 100 and how to multiply a decimal by the whole number 6. The method of going through 1% is nicely demonstrated in the sixth-grade texts used in Singapore (where children do very well in mathematics). See [79], volume 6A.

Sometimes it is easy to find a percentage of a quantity by thinking in terms of equivalent fractions. For example, how much is a 6% tax on a purchase of $50? We can solve this problem mentally by thinking that a 6% tax means $6 in tax for every $100. Since $50 is half as much as $100, we pay half as much tax, namely, $3. Formulated in terms of equivalent fractions, this reasoning corresponds to solving

$$\frac{6}{100} = \frac{\text{tax}}{50}$$

TABLE 2.2 Using a percent table to calculate 6% of $43.95

100%	→	$44
1%	→	$44 ÷ 100 = $0.44
6%	→	6 · $0.44 = $2.64

by dividing the numerator and denominator of $\frac{6}{100}$ by 2 to create an equivalent fraction with denominator 50:

$$\frac{6}{100} = \frac{\text{tax}}{50}$$

with $\div 2$ above and $\div 2$ below

Thus, tax $= 6 \div 2 = 3$, and so the tax is $3.

Finding the Percent When the Whole Amount and the Portion Are Known
In some problems, the whole amount and the portion of some quantity are known, and the percent that the portion is of the whole amount is to be calculated:

Nellie leaves a $9 tip for a meal that cost $50. What percent was Nellie's tip?

If $P\%$ is this percentage, then we have the equivalent fractions

$$\frac{P}{100} = \frac{9}{50}$$

Solving algebraically, we find that

$$P = 100 \cdot (9 \div 50) = 18$$

So Nellie left an 18% tip. Notice that to solve this type of problem, you just divide the portion by the whole amount and then multiply by 100 to find the percentage.

Another way to solve the problem about Nellie's tip is to work with equivalent fractions and to change $\frac{9}{50}$ into an equivalent fraction that has denominator 100:

$$\frac{9}{50} = \frac{18}{100}$$

with $\times 2$ above and $\times 2$ below

Class Activity *Now Turn to Class Activities Manual*

2V Calculating Percentages with Equivalent Fractions, p. 37

If you did part 7 of the Class Activity 2V, you may have used the method of going through 1. Here is another example of going through 1. What percent is 3 of 8? To solve this problem, we must solve

$$\frac{P}{100} = \frac{3}{8}$$

for P. In other words, we must find a fraction that is equivalent to $\frac{3}{8}$ and has denominator 100. To do so, we can go through 1 as in the next equations:

$$\frac{3}{8} = \frac{0.375}{1} = \frac{37.5}{100}$$

with $\div 8$ and $\times 100$ above and $\div 8$ and $\times 100$ below

So we conclude that 3 is 37.5% of 8.

Another way to determine what percent 3 is of 8 is with a percent table, as shown in Table 2.3.

TABLE 2.3 Using a percent table to calculate what percent 3 is of 8 by going through 1

100%	→	8
100% ÷ 8 = 12.5%	→	1
3 · 12.5% = 37.5%	→	3

Class Activity   *Now Turn to Class Activities Manual*

2W Calculating Percentages with Pictures and Percent Tables, p. 37

Finding the Whole Amount when the Percent *P* and the Portion Are Known In another basic type of percent problem, we know the percent, *P*, and the portion, and we must calculate the whole amount:

A store gave $15,000 to schools in the community. This $15,000 represents 3% of the store's annual profit. What was the store's annual profit?

To solve this problem, we must solve

$$\frac{3}{100} = \frac{15,000}{\text{annual profit}}$$

Solving algebraically, we find that

$$\text{annual profit} = (15,000 \cdot 100) \div 3 = 500,000$$

So the store's annual profit was $500,000. We can reach the same conclusion by making equivalent fractions:

$$\overset{\times 5,000}{\frac{3}{100} = \frac{15,000}{\text{annual profit}}}\underset{\times 5,000}{}$$

Yet another way to reach the same conclusion is with the percent table in Table 2.4. We can reason that if 3% of the store's annual profit is $15,000, then to find 1% we should divide by 3. So 1% of the store's annual profit is $15,000 ÷ 3 = $5,000. Then 100% of the store's annual profit is 100 times as much, which is 100 · $5,000 = $500,000.

TABLE 2.4 Calculating 100% of an amount if 3% of the amount is $15,000

3%	→	$15,000
1%	→	$15,000 ÷ 3 = $5,000
100%	→	100 × $5,000 = $500,000

Class Activity   *Now Turn to Class Activities Manual*

2X Calculating a Quantity from a Percentage of It, p. 38

2Y Percent Problem Solving, p. 38

Practice Exercises for Section 2.6

1. For each of the shapes shown in Figure 2.67, determine what percent of the shape is shaded. Give your answer rounded to the nearest multiple of 5 (i.e., 5, 10, 15, 20, …).

FIGURE 2.67   What percent of each shape is shaded?

2. A restaurant server received a $7.00 tip on a meal he served. If this tip represents 20% of the cost of the meal, then how much did the meal cost? Solve this problem by using a common fraction and a picture. Then use a percent table to solve this problem.

3. If $12.3 million is 75% of the budget, then how much is the full budget? Solve the problem with the aid of a picture and a percent table. Explain your reasoning.

4. There were 4800 gallons of water in a tank. Some of the water was drained out, leaving 65% of the original amount of water in the tank. How many gallons of water are in the tank? Solve the problem with the aid of a picture and a percent table. Explain your reasoning.

5. If your daily value of vitamin C is 60 milligrams, then how many milligrams is 95% of your daily value of vitamin C? Solve the problem with the aid of a picture and a percent table. Explaining your reasoning.

6. George was given 9 grams of medicine, but the full dose that he was supposed to receive is 20 grams. What percent of his full dose did George receive? First, solve the problem with the aid of a picture and a percent table. Explaining your reasoning. Then show how to solve the problem by making equivalent fractions (without cross-multiplying).

7. a. 1260 is 150% of what number?

   b. What percent of 760 is 950?

8. Show how to use a percent table and either the method of going through 1% or the method of going through 1 to solve the following problems:

   a. What percent of 200 is 15?

   b. 210 is 84% of what number?

   c. What percent of 80 is 3?

   d. 375 is 125% of what number?

9. Show how to solve the following problems by working with equivalent fractions (without cross-multiplying):

   a. A dress cost $75. Now the dress is being sold at a $6 discount. What percent is the discount?

   b. Of the rabbits on a rabbit farm, 35% are white. If there are 70 white rabbits on the rabbit farm, how many rabbits in all are on the farm?

   c. There are 25 children in a class. If 16% of the children in the class get free lunch, how many children get free lunch?

10. Billy has 20% more marbles than Sammy. If Billy has 12 more marbles than Sammy, then how many marbles does Billy have? Explain your solution.

11. There are 200 marbles in a bucket. Of the 200 marbles, 80% have swirled colors and 20% are solid colored. How many swirled marbles must be removed so that 75% of the remaining marbles are swirled?

Answers to Practice Exercises for Section 2.6

1. See Figure 2.68.

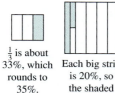

$\frac{1}{3}$ is about 33%, which rounds to 35%.

Each big strip is 20%, so the shaded part is 15%.

Each big strip is 25%. The small strips are $\frac{1}{5}$ of that, so 5%. So 80% is shaded.

The large shaded portion is 75%. The small shaded piece must be about 5% because 10% would be close to half of the remaining 25% of the pie. So, about 80% is shaded.

FIGURE 2.68   Percent shaded

2. Twenty percent is equal to $\frac{1}{5}$. If $7.00 represents one-fifth of the cost of the meal, then the meal must cost 5 times $7.00, which is $35.00. See Figure 2.69 for the corresponding picture. This reasoning is also summarized in a percent table (Table 2.5).

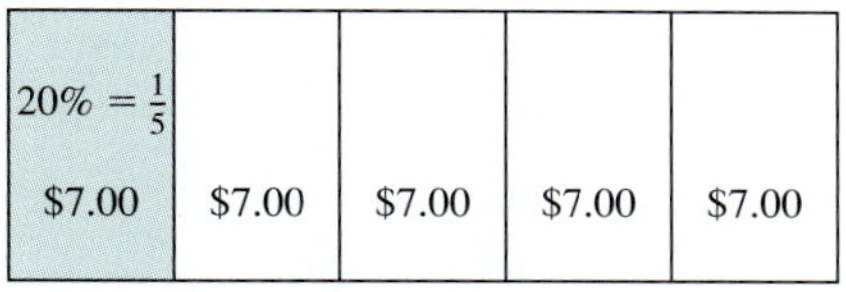

20% = $\frac{1}{5}$				
$7.00	$7.00	$7.00	$7.00	$7.00

FIGURE 2.69   Calculating 100% from 20%

TABLE 2.5 Using a percent table

20%	→	$7.00
100%	→	5 × $7.00 = $35.00

3. Because

$$75\% = \frac{3}{4}$$

we can think of the $12.3 million that make up 75% of the budget as distributed equally among the

$4.1 million	$4.1 million
$4.1 million	$4.1 million

$12.3 million is distributed equally among 3 parts.

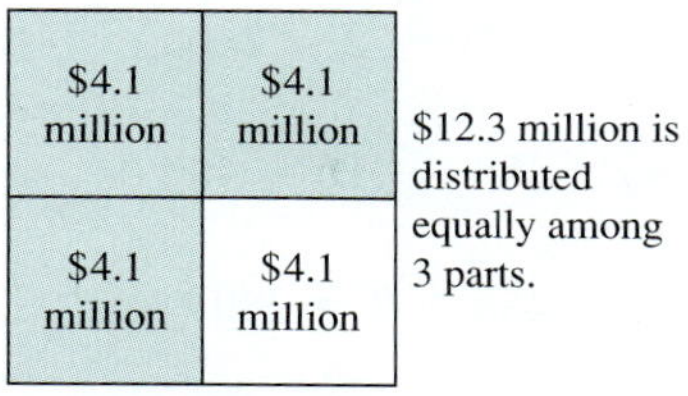

FIGURE 2.70   Calculating the full budget

TABLE 2.6 Using a percent table

75%	→	$12.3 million
25%	→	$12.3 ÷ 3 = $4.1 million
100%	→	4 × $4.1 = $16.4 million

3 parts of $\frac{3}{4}$, as shown in Figure 2.70. Each of those 3 parts must, therefore, contain $4.1 million. The full budget is made of 4 of those parts. Thus, the full budget must be 4 × $4.1 = $16.4 million. This reasoning is also summarized in a percent table (Table 2.6).

4. As Figure 2.71 shows, we can think of 65% as 50% + 10% + 5%. Now, 50% of 4800 is $\frac{1}{2}$ of 4800, which is 2400. Since 10% of 4800 is $\frac{1}{10}$ of 4800, which is 480, it follows that 5% is half of 480, which is 240. Therefore, 65% of 4800 is 2400 + 480 + 240, which is 3120, so 3120 gallons of water are left in the tank. This reasoning is also summarized in a percent table (Table 2.7).

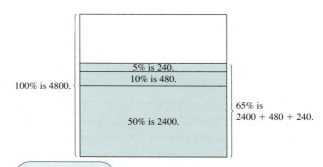

100% is 4800.

5% is 240.
10% is 480.

50% is 2400.

65% is 2400 + 480 + 240.

FIGURE 2.71   Calculating 65% of 4800

TABLE 2.7 Using a percent table

100%	→	4800
50%	→	4800 ÷ 2 = 2400
10%	→	4800 ÷ 10 = 480
5%	→	480 ÷ 2 = 240
65%	→	2400 + 480 + 240 = 3120

5. As Figure 2.72 shows, 95% is 5% less than 100%. The 10 vertical strips in Figure 2.72 are each 10%, and half of one of these vertical strips is 5%. Because 10% of 60 milligrams is $\frac{1}{10}$ of 60 milligrams, which is 6 milligrams, 5% of 60 milligrams is half of 6 milligrams, which is 3 milligrams. So, 95% of 60 milligrams is 3 milligrams less than 60 milligrams, which is 57 milligrams. This reasoning is also summarized in a percent table (Table 2.8).

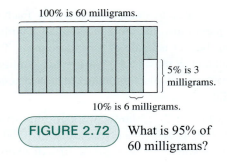

100% is 60 milligrams.

5% is 3 milligrams.

10% is 6 milligrams.

FIGURE 2.72   What is 95% of 60 milligrams?

TABLE 2.8 Using a percent table

100%	→	60 mg
10%	→	6 mg
5%	→	3 mg
95%	→	60 − 3 = 57 mg

6. One way to determine what percent 9 grams is of 20 grams is shown in Figure 2.73. If a rectangle representing a full dose of 20 grams of medicine is divided into 10 equal parts, then each of those parts represents 10% of a full dose of medicine. Each part also must represent 2 grams of medicine. So 9 grams

of medicine is represented by 4 full parts and half of a fifth part. Therefore, 9 grams of medicine is 45%. This reasoning is also summarized in a percent table (Table 2.9).

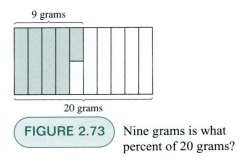

9 grams

20 grams

FIGURE 2.73   Nine grams is what percent of 20 grams?

TABLE 2.9 Using a percent table

100%	→	20 grams
10%	→	2 grams
5%	→	1 gram
45%	→	4 × 2 + 1 = 9 grams

We can also solve the problem by working with equivalent fractions:

$$\underset{\times 5}{\overset{\times 5}{\frac{9}{20} = \frac{45}{100}}} = 45\%$$

7. a. If 150% of a number is 1260, then to find 50% of the number, first divide by 3. So 50% of the number is 420. Therefore, 100% of the number is twice as much, which is 840. So the number is 840. The next percent table records this line of reasoning:

150%	→	1260
50%	→	420
100%	→	840

b. Using equivalent fractions:

$$\frac{950}{760} = \frac{95}{76} = \frac{5 \cdot 19}{4 \cdot 19} = \frac{5}{4} = \frac{125}{100} = 125\%$$

So 950 is 125% of 760.

8. a.

$$100\% \rightarrow 200$$
$$100\% \div 200 = \frac{1}{2}\% \rightarrow 1$$
$$15 \cdot \frac{1}{2}\% = 7.5\% \rightarrow 15$$

So 15 is 7.5% of 200.

b.
$$84\% \rightarrow 210$$
$$1\% \rightarrow 210 \div 84 = 2.5$$
$$100\% \rightarrow 100 \cdot 2.5 = 250$$

So 84% of 250 is 210.

c.
$$100\% \rightarrow 80$$
$$100\% \div 80 = 1.25\% \rightarrow 1$$
$$3 \cdot 1.25\% = 3.75\% \rightarrow 3$$

So 3 is 3.75% of 80.

d. $125\% \rightarrow 375$
$$1\% \rightarrow 375 \div 125 = 3$$
$$100\% \rightarrow 100 \cdot 3 = 300$$

So 125% of 300 is 375.

9. a.

$$\frac{6}{75} = \frac{2}{25} = \frac{8}{100} = 8\%$$

($\div 3$, $\times 4$)

So 6 is 8% of 75, and the discount on the dress is 8%.

b.

$$35\% = \frac{35}{100} = \frac{70}{200}$$

($\times 2$)

So 70 is 35% of 200; therefore, there are 200 rabbits on the farm.

c.

$$16\% = \frac{16}{100} = \frac{4}{25}$$

($\div 4$)

So 16% of 25 is 4, and 4 children get free lunch.

10. Billy has 20% more marbles than Sammy and this is 12 marbles; this means that 20% of Sammy's marbles is 12 marbles. Therefore, 100% of Sammy's marbles is 5 times as much, which is $5 \times 12 = 60$

marbles, as summarized in the next percent table and as shown pictorially in Figure 2.74.

$$20\% \text{ of Sammy's marbles} \rightarrow 12$$
$$100\% \text{ of Sammy's marbles} \rightarrow 5 \times 12 = 60$$

Since Sammy has 60 marbles and Billy has 12 more, Billy has $60 + 12 = 72$ marbles.

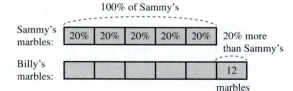

FIGURE 2.74 Billy has 20% more marbles than Sammy and this is 12 marbles

11. Because 80% of the 200 marbles are swirled and 20% are solid colored, 160 marbles are swirled and 40 are solid colored. We want these 40 solid colored marbles to be 25% of the remaining marbles. So, if 25% of the remaining marbles is 40, then 100% of them is 4 times as much, namely, 160.

Since $200 - 160 = 40$, we must remove 40 swirled marbles.

Another way to solve the problem is with a picture, as in Figure 2.75.

The 200 marbles are shown in 10 sections, with the swirled marbles in 8 sections (80%) and the solid marbles in 2 sections (20%). Those 2 sections of solid marbles must become 25% of the marbles. Since 25% is $\frac{1}{4}$, the figure shows that 2 sections of swirled marbles must be removed. Since each section has 20 marbles ($200 \div 10 = 20$), the 2 sections to be removed consist of 40 marbles. So, 40 swirled marbles must be removed.

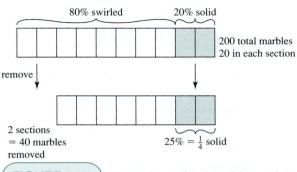

FIGURE 2.75 Removing swirled marbles so that the solid marbles become 25%

Problems for Section 2.6

1. If your daily value of carbohydrates is 300 grams, then how many grams is 95% of your daily value of carbohydrates? Solve the problem with the aid of either a picture, or a percent table, or both, explaining your reasoning.

2. A road crew ordered $\frac{5}{2}$ tons of gravel, but they only received $\frac{3}{2}$ tons of gravel. What percent of their order did the road crew receive? Solve the problem with the aid of a picture or a percent table, or both, explaining your reasoning.

3. If your daily value of dietary fiber is 25 grams and if you only ate 80% of your daily value of dietary fiber, then how many grams of dietary fiber did you eat? Solve the problem with the aid of either a picture or a percent table, or both, explaining your reasoning.

4. If 36,000 people make up 15% of a population, then what is the total population? Solve the problem with the aid of either a picture or a percent table, or both, explaining your reasoning.

5. If 180 milligrams of potassium constitute 5% of your daily value of potassium, then how many milligrams is your full daily value of potassium? Solve the problem with the aid of either a picture or a percent table, or both, explaining your reasoning.

6. In Happy Valley, the average rainfall in July is 5 inches, but this year only 3.5 inches of rain fell in July. What percent of the average July rainfall did Happy Valley receive this year? Solve the problem with the aid of either a picture or a percent table, or both, explaining your reasoning.

7. If a $\frac{3}{4}$-cup serving of cereal provides your full daily value of vitamin B6, then what percentage of your daily value of vitamin B6 will you receive in $\frac{1}{2}$ of a cup of the cereal? Solve the problem with the aid of either a picture or a percent table, or both, explaining your reasoning.

8. a. Mentally determine what percent 225 is of 250. Use a percent table to help you explain why your method makes sense.

    b. Show how to determine what percent 225 is of 250 by finding equivalent fractions (without cross-multiplying).

    c. Mentally determine what percent 960 is of 1600. Use a percent table to help you explain why your method makes sense.

    d. Show how to determine what percent 960 is of 1600 by finding equivalent fractions (without cross-multiplying).

9. Show how to use a percent table and either the method of going through 1% or the method of going through 1 to solve the following problems:

    a. 690 is 23% of what number?

    b. What percent of 40 is 3?

    c. 210 is 14% of what number?

    d. What percent of 800 is 28?

10. If a $\frac{2}{3}$-cup serving of cereal provides your full daily value of folic acid, then what percentage of your daily value of folic acid will you receive in $\frac{1}{2}$ of a cup of the cereal? Solve the problem with the aid of either a picture or a percent table, or both, explaining your reasoning.

11. The Biggo Corporation hopes that 95% of its 4600 employees will participate in the charity fund drive. How many employees does the Biggo Corporation hope will participate in the fund drive? Solve the problem with the aid of either a picture or a percent diagram, or both, explaining your reasoning.

12. A company has bought 3.4 acres of land out of the 4 acres of land that it plans to buy. What percent of the land has the company already bought? Solve the problem with either a picture or a percent table, or both, explaining your reasoning.

13. The mayor says that $3.6 million has been spent and that this represents 75% of the money allocated for a project. What was the total amount of money that was allocated for the project? Solve the problem with either a picture or a percent table, or both, explaining your reasoning.

14. Sixty percent of a city's population of 84,000 live within 5 miles of the library. How many people live within 5 miles of the library? Solve the problem with either a picture or a percent table, or both, explaining your reasoning.

15. Frank ran 80% as far as Denise. How far did Denise run as a percentage of Frank's running distance?

Draw a picture or diagram to help you solve the problem. Use your picture to help explain your answer.

16. GrandMart sells 115% as much soda as BigMart. How much soda does BigMart sell, calculated as a percentage of GrandMart's soda sales? Explain your answer.

17. Connie and Benton paid for identical plane tickets, but Benton spent more than Connie (and Connie's ticket was not free).

    a. If Connie spent 75% as much as Benton, then did Benton spend 125% as much as Connie? If not, then what percentage of Connie's ticket price did Benton spend? Explain your answer.

    b. If Benton spent 125% as much as Connie, then did Connie spend 75% as much as Benton? If not, then what percentage of Benton's ticket price did Connie spend? Explain your answer.

18. At a newsstand, 75% of the items sold are newspapers. Does this mean that 75% of the newsstand's income comes from selling newspapers? Why or why not? Write a paragraph to explain. Include examples to illustrate your points.

19. Brand A soup is 20% solids and 80% water. Use this information about brand A soup in all parts of this problem.

    a. If a can of brand A soup is mixed with a full can of water, what percent of the mixture is water? Explain your answer.

    b. The manufacturer of brand A soup has found a way to remove half of the water in the soup to make a more concentrated soup. What percent of this concentrated soup is water? Explain your answer.

    c. What percent of the water in the soup would the manufacturer of brand A soup have to remove so that the remaining, concentrated soup becomes 50% solids? Explain your answer.

20. Suppose that 200 pounds of freshly picked cucumbers are 99% water by weight. After several days, some of the water from the cucumbers has evaporated, and the cucumbers are now 98% water. How much do the cucumbers weigh now? Explain your solution.

21. At a store, there is a display of 300 cans of beans. Of the 300 cans, 60% are brand A and 40% are brand B. How many cans of brand B beans must be removed so that 75% of the remaining cans are brand A? Explain your solution.

22. A company produces two types of handbags and sells a total of 500 handbags per day. Of the 500 bags sold per day, 30% are style A and 70% are style B. Suppose the company begins to sell additional style A bags every day, but does not sell any additional style B bags. How many more style A bags would the company have to sell such that 50% of their handbag sales are style A bags? Explain your solution.

23. A county has two elementary schools, school A and school B. At school A, 30% of the children speak Spanish at home. At school B, 20% of the children speak Spanish at home. Use this information about the schools in all parts of this problem.

    a. Tom says that 50% of the elementary school children in the county speak Spanish at home. Tom found 50% by adding 30% and 20%. Is Tom correct? Explain why or why not.

    b. DeShun says that 25% of the elementary school children in the county speak Spanish at home. DeShun found 25% by averaging 30% and 20%.

       i. If there are 400 children in school A and 100 children in school B, is DeShun correct that 25% of the elementary school children in the county speak Spanish at home? If not, what percent is it?

       ii. If there are 100 children in school A and 400 children in school B, is DeShun correct that 25% of the elementary school children in the county speak Spanish at home? If not, what percent is it?

       iii. Find circumstances under which DeShun is correct that 25% of the elementary school children in the county speak Spanish at home.

Chapter Summary and Study Items

Section 2.1 The Meaning of Fractions

Fractions of objects are defined as follows.

If A and B are whole numbers, and B is not zero, and if an object can be divided into B equal parts, then $\frac{1}{B}$ of the object is the amount formed by 1 part, and $\frac{A}{B}$ of the object is the amount formed by A parts (or copies of parts). When working with fractions, pay close attention to the *whole,* namely, the object, collection, or quantity that the fraction is *of.*

Key skills and understandings:

1. Find fractional amounts of an object, collection, or quantity, and justify your reasoning.

2. In problems, determine the whole associated with a fraction appearing in the problem.

3. Use fractions to compare quantities.

Section 2.2 Interlude: Solving Problems and Explaining Solutions

Problem solving is central to mathematics. The mathematician George Polya developed a four-step guideline for solving problems. The first step is to understand the problem; the second step is to devise a plan for solving the problem: the third step is to carry out the plan; and the fourth step is to look back to see if your answer makes sense and to see what you can learn from having solved the problem.

As a teacher, it will be especially important for you to give mathematically correct explanations and to evaluate your students' explanations for mathematical correctness. Good mathematical explanations are factually correct. They address the specific question or problem that was posed and emphasize key points; they do not include irrelevant or distracting points. Good mathematical explanations are clear, convincing, and logical. They could be used to teach a student and to convince a skeptic, and as such, they do not require the reader to make a leap of faith. Good mathematical explanations use supporting pictures, diagrams, and equations appropriately and where needed. Explanations are coherent and use clear, complete sentences.

Key skills and understandings:

1. Make good attempts at solving problems by applying Polya's four-step method.

2. Make good attempts at giving good mathematical explanations for why a solution to a problem is correct.

Section 2.3 Fractions as Numbers

Even though fractions are expressed in terms of a pair of numbers, a fraction represents a single number, and as such, it can be plotted on a number line.

Key skills and understandings:

• Plot fractions, including improper fractions, on number lines and explain why the location fits with the definition of fraction.

Section 2.4 Equivalent Fractions

Any fraction $\frac{A}{B}$ is equivalent to infinitely many other fractions via this relationship:

$$\frac{A}{B} = \frac{A \cdot N}{B \cdot N}$$

We use equivalent fractions to give two fractions common denominators. When we give fractions common denominators, we express the fractions in terms of like parts. The simplest form of a fraction is an equivalent fraction that has the smallest possible numerator and denominator.

Key skills and understandings:

1. Given a fraction, use the meaning of fractions, pictures, and number lines to explain why multiplying the numerator and denominator by the same counting number produces an equivalent fraction.

2. Explain that giving fractions common denominators expresses the fractions in terms of like parts.

3. Solve problems in which you will need to give fractions common denominators, and justify your solutions.

4. Put fractions in simplest form, and explain the process in terms of pictures and the meaning of fractions.

5. Given two fractions on a number line, place equally spaced tick marks to plot other fractions on the number line as well.

Section 2.5 Comparing Fractions

We can compare fractions by converting them to decimals, by giving the fractions common denominators, by giving the fractions common numerators, or by cross-multiplying the fractions. Other reasoning can also be used to compare fractions: for example, comparing each fraction to a benchmark, such as $\frac{1}{2}$ or 1.

Key skills and understandings:

1. Compare fractions by giving them common denominators, and use pictures and the meaning of fractions to explain the rationale for this method of comparison.

2. Compare fractions that have the same numerator, and use pictures and the meaning of fractions to explain the rationale for this method of comparison.

3. Compare fractions using cross-multiplication, and explain that this method can be viewed as a shortcut for comparing fractions by giving them a common denominator.

4. Compare fractions by comparing them to benchmarks such as $\frac{1}{2}$ and 1. Compare fractions by reasoning about the number of parts and the sizes of the parts.

Section 2.6 Percent

Percents are special fractions, namely, ones that have denominator 100. A basic percent problem states that a *portion* is *P*% of a *whole amount,* which states that two fractions are equivalent:

$$\frac{P}{100} = \frac{\text{portion}}{\text{whole amount}}$$

Key skills and understandings:

1. Use several methods to solve basic percent problems in which two of the percent, P, the portion, and the whole amount are known, and one is unknown.

2. Know the following methods for solving basic percent problems: working with equivalent fractions; using a percent table; going through 1%; going through 1; and using pictures, benchmark fractions, and mental calculation if appropriate.

Addition and Subtraction

*I*n this chapter, our focus is on addition and subtraction. We begin by considering what addition and subtraction mean and what kinds of problems addition and subtraction solve. The types of problems we can solve with addition and subtraction are surprisingly rich and varied. We'll relate story problems with equations and with simple diagrams that can help children understand and solve problems. We'll also view addition and subtraction as movement along number lines.

Next, we'll see how the commutative and associative properties of addition are building blocks for addition and subtraction. These properties allow us to flexibly take apart and rearrange addition and subtraction problems to make them easier to solve. In particular, the properties of addition underlie methods young children learn to become fluent with the facts of single-digit addition and subtraction.

We'll then study the efficient general methods for adding and subtracting in the decimal system and with fractions and analyze how these methods work. What's the actual meaning of that "little 1" we write when we "carry" in an addition problem? Why do we add and subtract fractions in the way we do rather than by just adding the numerators and adding the denominators? We'll study the logical explanations for why the standard processes work. Finally, we'll see why we add and subtract negative numbers the way we do.

3.1 Interpretations of Addition and Subtraction

F·P **Focal Points**
Grades 1, 2

The most familiar ways to think of addition and subtraction are as *combining* and *taking away*. But as we'll see, there are other kinds of situations in which we add or subtract. In particular, we often use subtraction when comparing quantities. We can also view addition and subtraction on number lines, which will allow us to understand addition and subtraction of negative numbers.

Addition and Subtraction as Combining and Taking Away

sum   Typically, if A and B are two numbers, then the **sum**

$$A + B$$

basic meaning of addition   represents the total number of objects we will have if we start with A objects and then get B more objects. The numbers A and B in a sum are called **terms**, **addends**, or **summands**.

We can represent the sum

$$149 + 85$$

as the total number of toothpicks we will have if we start with 149 toothpicks and we get 85 more toothpicks.

We can represent the sum

$$34.5 + 7.89$$

as the total number of dollars we will have if we start with $34.5 = 34.50$ dollars and then get 7.89 more dollars.

We can represent the sum

$$\frac{3}{4} + \frac{2}{3}$$

as the total number of cups of flour in a batch of dough if we start with $\frac{3}{4}$ cup of flour in the dough and add another $\frac{2}{3}$ cup of flour.

difference   Typically, if A and B are two numbers, the **difference**

$$A - B$$

basic meaning of subtraction   represents the total number of objects we will have if we start with A objects and take away B of those objects. The numbers A and B in a difference can be called **terms**. The number A is sometimes called the **minuend**, and the number B is sometimes called the **subtrahend**.

We can represent

$$142 - 83$$

as the number of toothpicks we have left if we start with 142 toothpicks and give away 83 toothpicks.

We can represent

$$319 - 148.2$$

as the number of dollars we have left if we start with $319.00 and spend $148.20.

We can represent

$$\frac{3}{4} - \frac{1}{8}$$

as the number of acres of land we have left if we start with $\frac{3}{4}$ acre and sell $\frac{1}{8}$ acre.

Relating Addition and Subtraction

Class Activity   *Now Turn to Class Activities Manual*

3A Relating Addition and Subtraction—The Shopkeeper's Method of Making Change, p. 40

Every statement about subtraction corresponds to a statement about addition. Namely, to say that

$$A - B = C$$

is equivalent to saying that

$$C + B = A$$

Why? We can use the equation $A - B = C$ to represent the situation where we have A apples, we give away B apples, and we are left with C apples. Reversing the process, we can start with the C apples we have left. If we now get back the B apples, we will then have the A apples we originally started with. This latter scenario corresponds to the equation

$$C + B = A$$

Notice that the strip diagram in Figure 3.1 fits naturally with the equation $A - B = C$, as well as with the equation $C + B = A$ (or $B + C = A$), showing us visually that the two equations are equivalent.

We will see that the relationship between addition and subtraction gives rise to problems that can be solved by addition, but are not "add to" problems, and problems that can be solved by subtraction, but are not "take away" problems.

Other Addition and Subtraction Scenarios

We defined addition in terms of "adding to" or combining, and we defined subtraction in terms of "taking away." These interpretations of addition and subtraction are the most basic and simple interpretations, but there are other kinds of situations in which we add or subtract. Before you read on, do these Class Activities: 3B, 3C, 3D.

Class Activity   *Now Turn to Class Activities Manual*

3B Types of Addition and Subtraction Story Problems, p. 40

3C Why Can't We Rely on Keywords to Solve Story Problems?, p. 42

3D Using Simple Diagrams to Decide Whether to Add or Subtract to Solve a Story Problem, p. 43

FIGURE 3.1

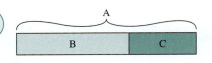

Representing $A - B = C$ and $C + B = A$ with a strip diagram

The basic addition equation, $A + B = C$, gives rise to a problem when one of A, B, or C is unknown, as in the following equations:

$$12 + 9 = ? \tag{3.1}$$

$$12 + ? = 21 \tag{3.2}$$

$$? + 12 = 21 \tag{3.3}$$

The basic subtraction equation, $D - E = F$, gives rise to a problem when one of D, E, or F is unknown, as in the following equations:

$$21 - 12 = ? \tag{3.4}$$

$$21 - ? = 12 \tag{3.5}$$

$$? - 9 = 12 \tag{3.6}$$

Several different types of story problems can be created to fit with these equations. The problems are often called add to (or join), take away (or separate)—which are two types of change problems—part-part-whole, and compare problems.

Change Problems Change problems are addition and subtraction story problems that involve a change over time. The change problems consist of the most basic types of addition and subtraction story problems, namely the **add to** (or **join**) and the **take away** (or **separate**) that we used to define addition and subtraction. For example,

> *Add to problem* for Equation 3.2: Rachel had 12 CDs. After Rachel got some more CDs, she had 21 CDs. How many CDs did Rachel get? (Fits with Figure 3.3, D.)

> *Take away problem* for Equation 3.6: Rachel had some CDs. After she gave away 9 CDs, she had 12 CDs left. How many CDs did Rachel have at first? (Fits with Figure 3.2, A.)

> *Take away problem* for Equation 3.5: Rachel had 21 CDs. After she gave some CDs away, she had 12 CDs left. How many CDs did Rachel give away? (Fits with Figure 3.3, D.)

Part-Part-Whole Problems Part-part-whole problems are addition and subtraction story problems that involve two distinct parts that make a whole but that don't involve change over time. For example,

> *Part-part-whole problem* for Equation 3.1: Rachel has 12 hip-hop CDs and 9 rap CDs (but no other CDs). How many CDs does Rachel have in all? (Fits with Figure 3.2, A.)

> *Part-part-whole problem* for Equation 3.2 or 3.3: Rachel has 21 CDs. Some of the CDs are hip-hop and the rest are rap. Rachel has 12 hip-hop CDs. How many rap CDs does Rachel have? (Fits with Figure 3.3, D.)

Notice that Equations 3.4, 3.5, and 3.6 don't fit naturally with the way part-part-whole problems are phrased.

FIGURE 3.2

Strip diagrams for Equations 3.1 and 3.6.

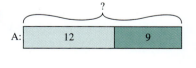

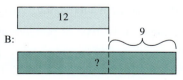

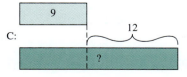

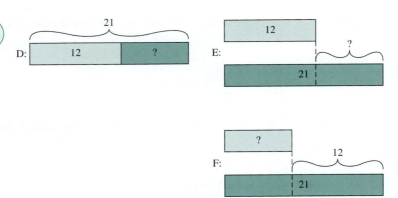

FIGURE 3.3

Strip diagrams
for Equations
3.2, 3.3, 3.4,
and 3.5.

Compare Problems Compare problems are addition and subtraction story problems that involve the comparison of two quantities. These problems involve a larger quantity, a smaller quantity, and the difference between the two quantities. For example,

> *Compare problem* for Equation 3.1: Rachel has 12 CDs. Benny has 9 more CDs than Rachel. How many CDs does Benny have? (Fits with Figure 3.2, B.)

> *Compare problem* for Equation 3.4: Rachel has 21 CDs. Benny has 12 fewer CDs than Rachel. How many CDs does Benny have? (Fits with Figure 3.3, F.)

> *Compare problem* for Equation 3.2 or 3.4: Rachel has 12 CDs. Benny has 21 CDs. How many more CDs does Benny have than Rachel? (Fits with Figure 3.3, E.)

> *Compare problem* for Equation 3.5 or 3.4: Rachel has 12 CDs. Benny has 21 CDs. How many fewer CDs does Rachel have than Benny? (Fits with Figure 3.3, E.)

The complex wording of compare problems can be difficult for children to grasp. Children will first need experience with comparing quantities without having to decide specifically how much more, or how much less, one quantity is than another, as in the next problems:

> Anna has 7 butterflies. Katie has 9 butterflies. Who has more butterflies?

> Anna has 7 butterflies. Katie has 9 butterflies. Who has fewer butterflies? (or: Who has less?)

Why We Can't Rely on Keywords to Solve Story Problems It's important to realize that the use of keywords is not reliable for solving story problems. There simply isn't any substitute for reading and understanding a story problem! For example, consider this problem:

> Tanya has 12 ladybugs. How many *more* ladybugs does she need to have 21 ladybugs *altogether*?

A child who relies on the keywords *more* and *altogether* might attempt to solve this problem incorrectly by adding 12 and 21 instead of subtracting 12 from 21.

Notice that, regardless of the type of problem used, story problems for Equations 3.1 and 3.6 can be solved by adding

$$12 + 9$$

even if the problems aren't add to problems. These problems fit naturally with parts A, B, or C in Figure 3.2.

Similarly, story problems for Equations 3.2, 3.3, 3.4, and 3.5 can be solved by subtracting

$$21 - 12$$

even if the problems aren't take away problems. These problems fit naturally with parts D, E, or F in Figure 3.3.

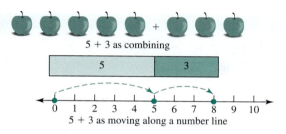

FIGURE 3.4

From addition as combining to addition as moving along a number line

5 + 3 as combining

5 + 3 as moving along a number line

Using Simple Diagrams to Help Understand and Solve Problems

Keywords aren't reliable, so what can children do to help them understand and determine how to solve a problem? Children from about third or second grade on can draw and use simple strip diagrams, such as those in Figures 3.2 and 3.3, to help them determine how the quantities in a problem are related and to figure out whether to add or subtract to solve the problem. Strip diagrams are quick and easy to draw. Drawing a long strip to represent 21 ladybugs is a lot quicker and easier than drawing 21 ladybugs! By drawing strip diagrams, a child can spend more time focusing on mathematical relationships than on drawing details that are not relevant to the mathematics. In contrast, a child who spends time drawing spots and legs on ladybugs when this is not relevant to the math problem may lose track of the mathematical ideas. In the next section you will study some methods that very young children—from prekindergarten through about first grade—can do to help them solve problems.

Addition and Subtraction on Number Lines

The interpretations of addition and subtraction that we have discussed so far encompass most ordinary addition and subtraction situations. But what about

$$\sqrt{2} + \sqrt{3}?$$

It doesn't really make sense to have $\sqrt{2}$ of an apple and get another $\sqrt{3}$ of an apple. Addition and subtraction of all numbers, whether positive or negative, and whether represented by decimals or fractions, can be interpreted easily by the use of number lines. By interpreting addition and subtraction in terms of number lines, we can extend the definition of addition and subtraction to all numbers.

Our interpretation of addition on number lines must fit with our definition of addition in terms of combining. So consider the situation of starting with 5 apples and getting 3 more. Then we have 5 + 3 apples, as at the top of Figure [3.4.] We can represent this situation more abstractly with the strip diagram in the middle of Figure [3.4.] The strip diagram fits naturally with the view of addition as combining lengths. Once we view addition as combining lengths, it is natural to proceed to viewing addition as moving along a number line, as at the bottom of Figure [3.4.]

Similarly, our interpretation of subtraction on number lines must fit with our definition of subtraction in terms of taking away. So consider the situation of starting with 5 apples and taking away 3. Then we have 5 − 3 apples, as at the top of Figure [3.5.] We can represent the situation more abstractly with the strip diagram in the middle of Figure [3.5.] The strip diagram fits naturally with the view of subtraction as decreasing a length. Once we view subtraction as decreasing a length, it is natural to proceed to viewing subtraction as movement along a number line, as at the bottom of Figure [3.5.]

FIGURE 3.5

From subtraction as taking away to subtraction as moving along a number line

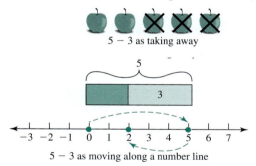

5 − 3 as taking away

5

3

5 − 3 as moving along a number line

Practice Exercises for Section 3.1

1. Write a part-part-whole problem that fits naturally with the equation

$$8 + ? = 14$$

2. Write an add to problem that fits naturally with the equation

$$? + 7 = 15$$

3. Write a compare problem that fits naturally with the equation

$$23 - 9 = ?$$

Draw a strip diagram, similar to one in Figure 4.2 or Figure 4.3, that could help students determine how to solve the story problem.

4. Write a take away problem that fits naturally with the equation

$$23 - ? = 9$$

Draw a strip diagram, similar to one in Figure 3.2 or Figure 3.3, that could help students determine how to solve the story problem.

5. Write a story problem that a child who relies only on keywords might solve incorrectly by subtracting $31 - 28$, but that is solved correctly by adding $31 + 28$. Discuss why the child might solve the problem incorrectly.

Answers to Practice Exercises for Section 3.1

1. Pavel has 14 blocks in all. Eight of Pavel's blocks are red and the rest are blue. How many blue blocks does Pavel have?

2. Pavel had some blocks. After he got 7 more blocks, he had 15 blocks in all. How many blocks did Pavel have at first?

3. Problem: Becky has 23 rocks. Sam has 9 fewer rocks than Becky. How many rocks does Sam have?

A student could draw the strip diagram on the left in Figure 3.6 to determine how to solve the problem.

We can also write the following different type of problem: Becky has 23 rocks. Sam has 9 rocks. How many more rocks does Becky have than Sam?

A student could draw the strip diagram on the right in Figure 3.6 to determine how to solve the problem.

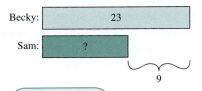

FIGURE 3.6　Strip diagrams for compare story problems

4. Problem: Nico has 23 snap-together building blocks. After Nico used some of his blocks to build a robot, he had 9 blocks left. How many blocks did Nico use to build his robot?

A student could draw the strip diagram in Figure 3.7 to determine how to solve the problem.

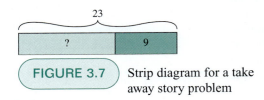

FIGURE 3.7　Strip diagram for a take away story problem

5. Story problem: Katie had some pencils. After she gave away 28 pencils, she had 31 pencils left. How many pencils did Katie have at first?

The words "gave away" and "left" in the problem indicate subtraction and may lead a child who is not thinking carefully about the problem statement to simply subtract 28 from 31. In fact, the subtraction equation, $? - 28 = 31$, fits with the problem. But this does not mean that 28 should be subtracted from 31 to solve the problem.

Problems for Section 3.1

1. **a.** Write an add to story problem that fits naturally with the equation

$$? + 17 = 42$$

Draw a strip diagram, similar to one in Figure 3.2 or Figure 3.3, that could help students determine how to solve the story problem.

   b. Write an add to story problem that fits naturally with the equation

$$17 + ? = 42$$

Draw a strip diagram, similar to one in Figure 3.2 or Figure 3.3, that could help students determine how to solve the story problem.

   c. Write a part-part-whole story problem that fits naturally with the equation

$$? + 17 = 42 \quad \text{or} \quad 17 + ? = 42$$

Draw a strip diagram, similar to one in Figure 3.2 or Figure 3.3, that could help students determine how to solve the story problem.

2. **a.** Write a compare story problem that fits naturally with the equation

$$17 + ? = 42$$

Draw a strip diagram, similar to one in Figure 3.2 or Figure 3.3, that could help students determine how to solve the story problem.

   b. Write a compare story problem that fits naturally with the equation

$$42 - ? = 17$$

Draw a strip diagram, similar to one in Figure 3.2 or Figure 3.3, that could help students determine how to solve the story problem.

3. **a.** Write a compare story problem that fits naturally with the equation

$$? - 17 = 25$$

Draw a strip diagram, similar to one in Figure 3.2 or Figure 3.3, that could help students determine how to solve the story problem.

   b. Write a compare story problem that fits naturally with the equation

$$25 + 17 = ?$$

Draw a strip diagram, similar to one in Figure 3.2 or Figure 3.3, that could help students determine how to solve the story problem.

4. **a.** Write a take away story problem that fits naturally with the equation

$$42 - 17 = ?$$

Draw a strip diagram, similar to one in Figure 3.2 or Figure 3.3, that could help students determine how to solve the story problem.

   b. Write a compare story problem that fits naturally with the equation

$$42 - 17 = ?$$

Draw a strip diagram, similar to one in Figure 3.2 or Figure 3.3, that could help students determine how to solve the story problem.

5. **a.** Write a take away story problem that fits naturally with the equation

$$? - 17 = 25$$

Draw a strip diagram, similar to one in Figure 3.2 or Figure 3.3, that could help students determine how to solve the story problem.

   b. Write a take away story problem that fits naturally with the equation

$$42 - ? = 25$$

Draw a strip diagram, similar to one in Figure 3.2 or Figure 3.3, that could help students determine how to solve the story problem.

6. a. Write an add to story problem that can be solved by subtracting numbers in the problem. Show how a child could use a strip diagram to see why the problem can be solved by subtracting.

b. Write a take away story problem that can be solved by adding numbers in the problem. Show how a child could use a strip diagram to see why the problem can be solved by adding.

3.2 The Commutative and Associative Properties of Addition, Mental Math, and Single-Digit Facts

 **Focal Points**
Grades 1, 2

In this section we study the commutative and associative properties of addition. We'll see how these properties underlie mental methods of addition that both children and adults can use to solve problems quickly and to enhance the flexibility of their thinking. The commutative and associative properties of addition underlie the increasingly advanced methods that young children learn as they travel along the path to fluency with single-digit addition and subtraction facts.

Why emphasize the commutative and associative properties of addition? These properties, together with the other properties of arithmetic (which include the commutative and associative properties of multiplication and the distributive property) form the building blocks of all of arithmetic. Ultimately, every calculation strategy, whether a mental method of calculation or a standard algorithm, relies on these properties. These properties allow us to take numbers apart, to break arithmetic problems into pieces that are easier to solve, and to put the pieces back together. The strategy of decomposing into simpler pieces, analyzing the pieces, and then putting them back together is important at every level of mathematics and in all branches of mathematics.

In our study of the properties of addition, and later in our study of other properties of arithmetic, we will consider mathematical expressions involving three or more numbers. Sometimes, we will want to group the numbers in an expression in certain ways. To do this, we will need to use parentheses.

Parentheses in Expressions with Three or More Terms

A sum

$$A + B + C$$

with 3 terms means the sum of $A + B$ and C. In other words, according to the meaning of $A + B + C$, this sum is to be calculated by first adding A and B, and then adding C. Likewise, if there are 4 or more terms in a sum, the sum stands for the result obtained by adding from left to right.

But what if we want to indicate the sum of A with $B + C$? We can show this with parentheses. In ordinary writing, parentheses are used for asides, but in mathematical expressions, parentheses are used to group numbers and operations ($+, -, \times, \div$). To indicate the sum of A with $B + C$, simply write

$$A + (B + C)$$

So,

$$17 + (18 + 2) = 17 + 20$$
$$= 37$$

Notice that we can express the meaning of the sum

$$A + B + C$$

by using parentheses:

$$(A + B) + C$$

Class Activity   *Now Turn to Class Activities Manual*

3E Mental Math, p. 44

The Associative Property of Addition

How can you make the problem

$$7384 + 999 + 1$$

easy to solve mentally? Rather than adding from left to right, it is easier to first add 999 and 1 to make 1000, and then add 7384 and 1000 to get 8384. In other words, rather than adding from left to right, calculating

$$(7384 + 999) + 1$$

according to the meaning of the sum $7384 + 999 + 1$, it is easier to group the 999 with the 1 and calculate

$$7384 + (999 + 1)$$

instead. Why is it legitimate to switch the way the numbers in the sum are grouped? Because, according to the associative property of addition, both ways of calculating the sum give equal results; in other words,

$$(7384 + 999) + 1 = 7384 + (999 + 1)$$

associative property of addition The **associative property of addition** tells us that when we add any three numbers, it doesn't matter whether we add the first two and then add the third, or whether we add the first number to the sum of the second and the third—either way we will always get the same answer. In other words, the associative property of addition says that if A, B, and C are any three numbers, then

$$(A + B) + C = A + (B + C)$$

That is, the sum of $A + B$ and C is equal to the sum of A and $B + C$.

It might help to understand the terminology *associative property* by remembering that to *associate* with someone means to keep company with them. In

$$(7384 + 999) + 1 = 7384 + (999 + 1)$$

the number 999 can either *associate* with 7384, or it can *associate* with 1.

We assume that the associative property of arithmetic is true for all numbers, but we can use an example to see why this property of addition makes sense. Suppose there is a group of 2 marbles, a second group of 3 marbles, and a third group of 4 marbles, as at the top of Figure 3.8. If we put the first and second groups together, we can describe the total number of marbles as

$$(2 + 3) + 4$$

But if we put the second and third groups of marbles together, we can describe the total number of marbles as

$$2 + (3 + 4)$$

Since the total number of marbles is the same no matter how we describe it, these two expressions for the total must be equal. In other words,

$$(2 + 3) + 4 = 2 + (3 + 4)$$

FIGURE 3.8

Demonstrating the associative property of addition

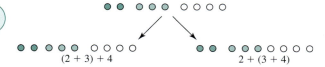

Of course, the same kind of relationship would hold if other numbers of marbles replaced the numbers 2, 3, and 4. The associative property of addition states that this kind of relationship always holds, no matter what numbers are involved, even if these numbers are fractions, decimals, or negative numbers.

The associative property of addition allows us to group sums of numbers any way we like. For example, to add

$$0.01 + 2.47 + 3.97 + 0.02 + 0.01$$

we can group the first two terms and the last three terms to make the addition easy:

$$(0.01 + 2.47) + (3.97 + 0.02 + 0.01) = 2.48 + 4.00 = 6.48$$

The Commutative Property of Addition

How can we make the addition problem

$$2997 + 569 + 3$$

easy to solve mentally? Adding 2997 and 569 is hard, but if we combine the 2997 with the 3 to make 3000, then it's easy to add the 569 to make 3569. In other words, rather than adding the numbers from left to right in the order they appear, it's easier to switch the order of the 569 and the 3, and calculate

$$2997 + 3 + 569$$

instead. Why is it legitimate to switch the order of 569 and the 3 in the original sum? Because

$$569 + 3 = 3 + 569$$

and this is true by virtue of the commutative property of addition.

commutative property of addition

The **commutative property of addition** states that for all real numbers A and B,

$$A + B = B + A$$

To illustrate the commutative property, assume that you start with 3 apples and you get 2 more. Then you have

$$3 + 2$$

apples. But what if you had started instead with 2 apples and then got 3 more? In this case you'd have

$$2 + 3$$

apples. Common sense tells us it doesn't matter in which order you got the apples; either way you'll have the same number of apples in all, as shown in Figure 3.9. Another way to say that you have the same amount of apples either way is to say

$$3 + 2 = 2 + 3$$

The commutative property of addition says that no matter what the numbers A and B are—even if they are fractions, decimals, or negative numbers—$A + B$ is equal to $B + A$. In other words,

$$A + B = B + A$$

for all numbers A and B.

In this setting, the word *commute* means *to change places with*. In going from $3 + 2$ to $2 + 3$, the 2 and the 3 change places. The term **turnarounds** is sometimes used with children instead of *commutative property*.

Students sometimes get confused between the commutative and associative properties. Notice that the commutative property involves changing the order of terms; in other words, the commutative property

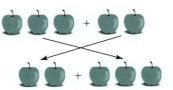

FIGURE 3.9

The commutative property:
$3 + 2 = 2 + 3$

involves *rearranging* terms. The associative property involves *regrouping* terms by changing the location of parentheses.

Combining the Commutative and Associative Properties of Addition

Combined, the commutative and associative properties of addition allow for great flexibility in calculating sums: You can rearrange the terms any way you like, and you can combine the terms with each other any way you like. This flexibility is valuable not only for mental calculations, but it is also an essential skill in algebra.

When we combine like terms in the expression

$$3x + 5 + 2x^2 - 4 + 8x - 5x^2$$

in order to rewrite it as

$$-3x^2 + 11x + 1$$

we have used the commutative and associative properties of addition.

Children's Learning Paths for Single-Digit Facts and the Commutative and Associative Properties of Addition

According to the NCTM's Focal Points [59], by about second grade or so, children should be fluent with the single-digit addition facts

$$1 + 1 = 2 \quad 1 + 2 = 3 \quad \ldots \quad 1 + 9 = 10$$
$$\vdots \qquad\qquad \vdots \qquad\qquad \vdots$$
$$9 + 1 = 10 \quad 9 + 2 = 11 \quad \ldots \quad 9 + 9 = 18$$

and the associated single-digit subtraction facts from $2 - 1 = 1$ to $18 - 9 = 9$. Accomplished teachers and strong programs help children develop this fluency in several ways: by helping them make sense of addition and subtraction through story problems, by helping them progress with understanding through increasingly sophisticated solution methods, and by helping them understand the connection between subtraction and addition.

Before you read on, please do the next Class Activities, which introduce you to the levels of increasingly sophisticated solution methods that young children gradually learn. Some of these solution methods rely on the commutative and associative properties of addition, even though children may not be aware that they are using these properties.

Class Activity   *Now Turn to Class Activities Manual*

3F Children's Learning Paths for Single-Digit Addition, p. 44

3G Children's Learning Paths for Single-Digit Subtraction, p. 46

3H Solving Addition and Subtraction Story Problems: Easier and Harder Subtypes, p. 47

The Commutative Property and Single-Digit Addition As you saw in Class Activity 3F, the commutative property of addition is important for fluency with single-digit addition problems because it allows children to replace addition problems of the form

smaller number + larger number

with the problem

larger number + smaller number

FIGURE 3.10

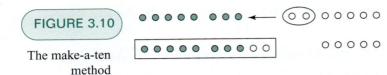

The make-a-ten
method

For example, instead of solving $2 + 7 = ?$, a child can solve $7 + 2 = ?$ if the child understands the commutative property of addition (even if she doesn't know the term). If the child doesn't yet have quick recall of either $2 + 7$ or $7 + 2$, then the child can "count on" from 7 as described in the Class Activity. Notice how important the commutative property of addition is for lightening the load of learning all the single-digit addition facts: It reduces these facts to knowing the "doubles," $1 + 1, 2 + 2, 3 + 3, \ldots, 9 + 9$, and the facts of the form (larger number) + (smaller number).

The Associative Property and Derived Fact Methods Children use the associative property of addition (whether or not they know this term) when they use the more advanced *derived fact methods* described in Class Activity 3F to solve single-digit addition problems. Especially important is the **make-a-ten method**. To use the make-a-ten method to calculate $8 + 7$, a child breaks 7 into $2 + 5$, joins the 2 with the 8 to make a 10, then adds the remaining 5 to the 10 to get the solution, 15. The dot picture in Figure 3.10 illustrates this method. The next equations show that this method uses the associative property of addition.

make-a-ten
method

$$8 + 7 = 8 + (2 + 5)$$
$$= (8 + 2) + 5$$
$$= 10 + 5 = 15$$

The associative property was used in going from the first line of the equation to the second line: the 2 goes from associating with the 5 to make 7 to associating with the 8 to make a 10. So a child who knows that 8 needs 2 more to make a 10 and who knows that 7 breaks into $2 + 5$ can derive that $8 + 7 = 15$.

Why is the make-a-ten method particularly important? The decimal system is based on grouping by tens. The make-a-ten method draws children's attention to ten and helps reinforce the understanding of the teen numbers 11 through 19 as a ten and some ones. For example, when children use the *subtract from ten method* described in Class Activity 3G to calculate $13 - 8$, they view 13 as $10 + 3$ and subtract the 8 from the 10, leaving 2 to add to 3 to make 5. The understanding of teen numbers as a ten and some ones is critically important to making sense of multi-digit addition and subtraction. When children solve a multi-digit addition problem such as $28 + 47$, the first step in the common method is

$$\overset{1}{28}$$
$$+\ 47$$
$$\overline{5}$$

To understand this regrouping method, a child must know not only that $8 + 7 = 15$ but also that this 15 is 1 ten and 5 ones.

In general, the derived fact methods are important because they can help children connect and relate addition facts and derive facts they don't yet know from ones they have already learned. Derived fact methods give children the opportunity to engage in a powerful technique that is used across arithmetic, geometry, and algebra and at all levels of mathematics: the technique of taking apart, analyzing, and putting back together.

Viewing Subtraction Problems as Unknown Addend Problems Class Activity 3G described two *unknown addend methods* children can use for subtraction. To solve $12 - 7 = ?$ a child can first view this problem as the unknown addend problem $7 + ? = 12$. Then one way children can solve this unknown addend problem is to count on from 7 to 12. Another way children can solve the unknown addend problem is to add 3 to make 10, then add another 2 to make 12, and then recognize the unknown addend as $3 + 2 = 5$. Of course children could solve $12 - 7 = ?$ by counting down 7 from 12;

however, teachers consider this method slow and error prone. In contrast, teachers have found that using an unknown addend method to subtract makes subtraction as easy for children as addition. Learning unknown addend methods to subtract helps children understand the link between subtraction and addition.

Proper Use of the Equal Sign and Writing Equations

We have seen that the associative and commutative properties of addition can sometimes be used to make mental addition problems easier. In order to show a line of reasoning for a calculation, it is useful to write sequences of equations that correspond to the line of reasoning.

equation An **equation** is a mathematical statement saying that two numbers or expressions are equal. The familiar equal sign, =, shows this equivalence. Remember that the equal sign means one and only one thing: = means *equals* or *is equal to*.

An equation is actually just a sentence, and you should read it and interpret it as a sentence. The equation

$$7 + (6 + 4) = 17$$

is equivalent to the following sentence:

7 plus the sum of 6 and 4 is equal to 17.

To show a method of calculation, we often string several equations together. Rather than writing

$$198 + 357 + 2 = 198 + 2 + 357$$
$$198 + 2 + 357 = 200 + 357$$
$$200 + 357 = 557$$

it is neater and easier to follow this sequence of steps:

$$198 + 357 + 2 = 198 + 2 + 357$$
$$= 200 + 357$$
$$= 557$$

When several equations are strung together, it is for the purpose of concluding that the *first* expression written is equal to the *last* expression written. The purpose of the previous equations is to conclude that $198 + 357 + 2$ is equal to 557. This string of equations used the following general property of equality, which we assume holds true:

If A is equal to B and if B is equal to C, then A is also equal to C.

It is common for students at all levels (including college) to make careless, incorrect use of the equal sign. Because the proper use of the equal sign is essential in algebra, and because elementary school mathematics lays the foundation for learning algebra, it is especially important for you to use the equal sign correctly.

When you use an equal sign, *make sure that the quantities before and after the equal sign really are equal to each other*. It is easy to make a careless error with the equal sign when solving a problem in several steps. Suppose you calculate

$$499 + 165$$

by taking 1 from 165, adding this 1 to 499 to make 500, and then adding on the remaining 164 to get 664. How can you show this method of calculation with equations? It is easy to make the following mistake:

$$\boxed{\text{Warning: incorrect}} \quad 499 + 1 = 500 + 164$$
$$= 664$$

These equations are incorrect because $499 + 1$ *is not equal to* $500 + 164$. Instead, you can write the following correct equations:

$$499 + 165 = 499 + 1 + 164$$
$$= 500 + 164$$
$$= 664$$

Mental Methods for Multi-Digit Addition and Subtraction

Some of the methods that are part of children's learning paths for single-digit addition and subtraction generalize to multi-digit situations, thereby providing children and adults with flexible, quick ways to solve addition and subtraction problems mentally.

Make-a-Round-Number Method The make-a-round-number method is just like the make-a-ten method, except that instead of just making a ten, we make other round numbers. The make-a-round-number method uses the associative property of addition to shift one piece of an addend and join the piece with the other addend. To use this method to calculate a sum, look for a nice round number that is close to one of the addends. For example, to add $376 + 199$ mentally, break 376 into $375 + 1$ and join the 1 with 199 to make 200. Therefore, the sum is $375 + 200 = 575$. The next equations show that we used the associative property of addition in applying this method:

$$376 + 199 = (375 + 1) + 199$$
$$= 375 + (1 + 199)$$
$$= 375 + 200 = 575$$

Rounding and Compensating Another way to calculate $376 + 199$ is to round and compensate: Suppose we add 200 to 376 instead of adding 199 to 376. This makes 576. But we added 1 too many, so we must take 1 away from 576. Therefore, $376 + 199 = 575$.

We can write corresponding equations as follows:

$$376 + 199 = 376 + 200 - 1$$
$$= 576 - 1$$
$$= 575$$

We can also round and compensate in subtraction problems. Consider the problem $684 - 295$. The number 295 is close to 300. So, to calculate

$$684 - 295$$

we can reason as follows:

Taking 300 objects away from 684 objects leaves 384 objects. But when we took 300 away, we took away 5 more than we should have (because 300 is 5 more than 295). This means that we must *add* 5 to 384 to get the answer, 389.

The following equations correspond to this line of reasoning:

$$684 - 295 = 684 - 300 + 5$$
$$= 384 + 5$$
$$= 389$$

Subtraction Problems as Unknown Addend Problems We saw previously that it can be helpful for children to view a subtraction problem such as $14 - 9 = ?$ as an unknown addend problem, $9 + ? = 14$. Viewing a subtraction problem as an unknown addend problem can also be helpful in solving multi-digit subtraction problems mentally. Consider once again the example $684 - 295 = ?$. We can view this subtraction problem as the unknown addend problem $295 + ? = 684$.

With this point of view, we start with 295 and keep adding numbers until we reach 684:

$$295 + 5 = 300$$
$$300 + 300 = 600$$
$$600 + 84 = 684$$

or

$$295 + 5 + 300 + 84 = 684$$

All together, starting with 295, we added

$$5 + 300 + 84 = 389$$

to reach 684. Therefore, $684 - 295 = 389$. This is the method that shopkeepers frequently use to make change.

Class Activity   *Now Turn to Class Activities Manual*

3I Using Properties of Addition in Mental Math, p. 47

3J Writing Correct Equations, p. 48

3K Writing Equations That Correspond to a Method of Calculation, p. 49

3L Other Ways to Add and Subtract, p. 50

Practice Exercises for Section 3.2

1. State the associative property of addition and describe how to use a collection of small objects to illustrate why this property makes sense.

2. State the commutative property of addition and describe how to use a collection of small objects to illustrate why this property makes sense.

3. Describe how a young child who is learning the single-digit addition facts could use the commutative property of addition (even if the child doesn't know the term *commutative property*).

4. Describe how a child who is learning the single-digit addition facts can use the make-a-ten method to add $7 + 4$. Write equations that show this method. Which property of addition does this method use?

5. Describe two ways children who are learning the single-digit addition and subtraction facts can solve subtraction problems by viewing subtraction problems as unknown addend problems.

6. Give an example that demonstrates how you can use the associative property of addition to make a multi-digit addition problem easier to do mentally. Describe in words how you can solve the problem mentally. Write equations that show where you used the associative property.

7. The sequence of equations that follows shows a way of using properties of addition to calculate a sum. Say specifically which properties of addition were used, and where.

$$
\begin{aligned}
27 + 89 + 13 &= 27 + (89 + 13) \\
&= 27 + (13 + 89) \\
&= (27 + 13) + 89 \\
&= 40 + 89 \\
&= 129
\end{aligned}
$$

8. For each of the addition problems that follow, write equations which correspond to a mental method for calculating the sum that uses the associative and/or the commutative properties of addition. Say specifically which properties of addition were used, and where.

   a. $993 + 2389$

   b. $398 + (76 + 2)$

9. Each arithmetic problem in this exercise has a description for how to solve the problem. In each case, write a sequence of equations that correspond to the given description.

   a. Problem: $23 + 45$

   Solution: 23 plus 40 is 63, and then 5 more makes 68.

   b. Problem: $800 - 297$

   Solution: $800 - 300 = 500$, but subtracting 300 subtracts 3 too many, so we must add 3 back, making 503.

10. Nancy writes the following equations to solve $37 + 14$:

$$30 + 10 = 40 + 7 = 47 + 4 = 51$$

Write correct equations that solve $37 + 14$ and that incorporate Nancy's solution strategy.

11. Jim writes the following equations to solve $85 - 15$:

$$85 - 10 = 75 - 5 = 70$$

Write correct equations that solve $85 - 15$ and that incorporate Jim's solution strategy.

12. Find ways to solve the next set of addition and subtraction problems *other than* by using the common addition or subtraction algorithms. In each case, explain your reasoning, and also write equations that incorporate your thinking.

a. $786 - 47$

b. $427 + 28$

c. $999 + 999$

d. $1002 - 986$

e. $237 - 40$

Answers to Practice Exercises for Section 3.2

1. See text.

2. See text.

3. A young child uses the commutative property of addition when she "counts on from larger." For example, if the child is asked to add $2 + 9$, the child can replace this sum with $9 + 2$ and count on 2 from 9, by saying "10, 11" instead of counting on 9 from 2, which would be much more work. The child uses the commutative property of addition when she replaces $2 + 9$ with $9 + 2$.

4. To add $7 + 4$ with the make-a-ten method, the child breaks 4 into $3 + 1$, joins the 3 with the 7 to make a 10, then joins this 10 with the remaining 1 to make 11. In equations:

$$7 + 4 = 7 + (3 + 1)$$
$$= (7 + 3) + 1$$
$$= 10 + 1 = 11$$

The associative property of addition was used at the second equal sign.

5. A child can view a subtraction problem such as $13 - 9 = ?$ as the unknown addend problem $9 + ? = 13$. If the child is still counting on, the child can count up from 9 to 13, saying "10, 11, 12, 13" and using his fingers to keep track of how many numbers he counted on. Since he counted on 4 numbers, he concludes that $9 + 4 = 13$, so $13 - 9 = 4$. If the child is ready for derived fact methods and no longer needs to count on, then the child can think: "Starting at 9, one more is 10, and then 3 more is

13, so that's 1 and 3 which is 4," again concluding that $9 + 4 = 13$, so that $13 - 9 = 4$.

6. To calculate $49 + 37$ mentally, think of moving 1 from the 37 to the 49, so that the sum becomes $50 + 36$, which we can easily see is 86. The corresponding equations show that the associative property of addition was used:

$$49 + 37 = 49 + (1 + 36)$$
$$= (49 + 1) + 36$$
$$= 50 + 36$$
$$= 86$$

The associative property of addition is used at the second equals sign to switch the placement of parentheses from grouping 1 and 36 together, to grouping 49 and 1 together.

7. The associative property of addition was used at the first equals sign to say that $(27 + 89) + 13 = 27 + (89 + 13)$. The commutative property of addition was used at the second equals sign to change $89 + 13$ to $13 + 89$. The associative property of addition was used at the third equals sign to say that $27 + (13 + 89) = (27 + 13) + 89$.

8. a. $993 + 2389 = 993 + (7 + 2382) = (993 + 7) + 2382 = 1000 + 2382 = 3382$. The associative property of addition was used in rewriting $993 + (7 + 2382)$ as $(993 + 7) + 2382$.

b. $398 + (76 + 2) = 398 + (2 + 76) = (398 + 2) + 76 = 400 + 76 = 476$. The commutative property of addition was used to rewrite $76 + 2$ as $2 + 76$. The associative property of addition was used to rewrite $398 + (2 + 76)$ as $(398 + 2) + 76$

9. a. $23 + 45 = 23 + (40 + 5)$
$= (23 + 40) + 5$
$= 63 + 5$
$= 68$

Note that the associative property of addition was used at the second equals sign to switch the placement parentheses. This had the effect of grouping the 40 with the 23 instead of with the 5 (where it was part of 45).

b. $800 - 297 = 800 - 300 + 3$
$= 500 + 3$
$= 503$

10. $37 + 14 = 30 + 10 + 7 + 4$
$= 40 + 7 + 4$
$= 47 + 4$
$= 51$

11. $85 - 15 = 85 - 10 - 5$
$= 75 - 5$
$= 70$

12. Here are equations you could write:

a. $786 - 47 = 786 - 50 + 3$
$= 736 + 3$
$= 739$

b. $427 + 28 = 427 + 3 + 25$
$= 430 + 25$
$= 455$

c. $999 + 999 = 999 + 1000 - 1$
$= 1999 - 1$
$= 1998$

d. $986 + 4 = 990$
$990 + 10 = 1000$
$1000 + 2 = 1002$

so,

$$986 + 4 + 10 + 2 = 1002$$

so,

$$1002 - 986 = 4 + 10 + 2$$
$$= 16$$

e. $237 - 40 = 240 - 40 - 3$
$= 200 - 3$
$= 197$

Problems for Section 3.2

1. Many teachers have a collection of small cubes that can be snapped end-to-end to make "trains" of cubes. The cubes come in various colors.

a. Describe in detail how you could use snap-to-gether cubes in different colors to demonstrate the commutative property of addition.

b. Describe in detail how you could use snap-to-gether cubes in different colors to demonstrate the associative property of addition.

2. Discuss the difference between the commutative property of addition and the associative property of addition.

3. Figure 3.11 indicates a make-a-10 method for adding $6 + 8$.

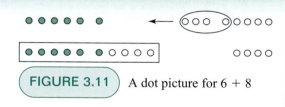

FIGURE 3.11 A dot picture for $6 + 8$

a. Write equations that correspond to the make-a-10 method for adding $6 + 8$ depicted in Figure 3.11. Your equations should make careful and appropriate use of parentheses. Which property of arithmetic do your equations and the picture of Figure 3.11 illustrate?

b. Draw a picture for $7 + 5$ that illustrates a make-a-10 method. Write equations that correspond to the strategy indicated in your picture, making careful and appropriate use of parentheses.

4. Tomaslav has learned the following facts well:

- All the sums of whole numbers that add to 10 or less—Tomaslav knows these facts "forwards and backwards." For example, he knows not only that 5 + 2 is 7, but also that 7 breaks down into 5 + 2 or 2 + 5.
- 10 + 1, 10 + 2, 10 + 3, . . . , 10 + 10.
- the *doubles* 1 + 1, 2 + 2, 3 + 3, . . . , 10 + 10.

For each sum in this problem, describe at least three different ways that Tomaslav could use reasoning, together with the facts he knows well, to determine the sum. In each case, write equations that correspond to the strategies you describe. Take care to use parentheses appropriately and as needed.

a. 8 + 7

b. 6 + 7

c. 8 + 9 (Try to find four or five different ways to solve this.)

5. Give an example of an arithmetic problem that can be made easy to solve mentally by using the associative property of addition. Write equations that show your use of the associative property of addition. Your use of the associative property must genuinely make the problem easier to solve.

6. Give an example of an arithmetic problem that can be made easy to solve mentally by using the commutative property of addition. Write equations that show your use of the commutative property of addition. Your use of the commutative property must genuinely make the problem easier to solve.

7. Describe a way to calculate 304 − 81 mentally, by using reasoning other than the common subtraction algorithm. Then write a coherent sequence of equations that correspond to your reasoning.

8. To calculate 159 − 73, a student writes the following equations:

$$160 - 70 = 90 - 3 = 87 - 1 = 86$$

Although the student has a good idea for solving the problem, his equations are not correct. In words, describe the student's solution strategy and discuss why the strategy makes sense; then write a correct sequence of equations that correspond to this solution strategy. Write your equations in the following form:

$$159 - 73 = \text{some expression}$$
$$= \text{some expression}$$
$$\vdots$$
$$= 86$$

9. To calculate 201 − 88, a student writes the following equations:

$$88 + 2 = 90 + 10 = 100 + 100 = 200 + 1 = 201$$
$$2 + 10 = 12 + 100 = 112 + 1 = 113$$

Although the student has a good idea for solving the problem, his equations are not correct. In words, describe the student's solution strategy and explain why it makes sense; then write correct equations that correspond to this solution strategy.

10. David and Ashley want to calculate $8.27 − $2.98 by first calculating $8.27 − $3 = $5.27. David says that they must *subtract* $0.02 from $5.27, but Ashley says that they must *add* $0.02 to $5.27.

a. Draw a number line (which need not be perfectly to scale) to help you explain who is right and why. Do not just say which answer is numerically correct; use the number line to help you explain why the answer must be correct.

b. Explain in another way who is right and why.

11. Tylishia says that she can calculate 324 − 197 by adding 3 to both numbers and calculating 327 − 200 instead.

a. Draw a number line (which need not be perfectly to scale) to help you explain why Tylishia's method is valid.

b. Explain in another way why Tylishia's method is valid.

c. Could you adapt Tylishia's method to other subtraction problems, such as to the problem 183 − 49? If so, give at least two more examples and show how to apply Tylishia's method in each case.

12. Is there an *associative property of subtraction*? In other words, is

$$A - (B - C) = (A - B) - C$$

true for all real numbers A, B, C? Explain your answer. If the equation is always true, explain why; if the equation is not always true, find another expression that *is* equal to the expression

$$A - (B - C)$$

and explain why the two expressions are equal.

13. Is there a *commutative property of subtraction*? Explain why or why not. In your answer, include a statement of what a commutative property of subtraction would be if there were such a property.

3.3 Why the Common Algorithms for Adding and Subtracting Numbers in the Decimal System Work

F·P **Focal Points**
Grade 2

In school, we learned the common paper-and-pencil methods for adding, subtracting, multiplying, and dividing numbers. But have you ever wondered why they work? All these common methods were developed by people, independently, in many different places throughout the world, and at many different times. How did mathematicians know that their clever methods would yield correct results? Just as you can't bake a cake by randomly combining flour, sugar, butter, and eggs, you also can't come up with the correct answer to an addition problem by randomly combining the numbers involved. The originators of the methods we use in arithmetic today had to think deeply about place value and about decomposing and recomposing numbers according to place value in order to derive those methods. The common methods of arithmetic can provide fertile ground for developing a sense of place value and a sense of how to take numbers apart and recombine them suitably. For this reason, even though calculators are ubiquitous, children should learn not just how to apply the common methods of arithmetic, but also to make sense of these methods and to use them with an understanding of why they work.

What Is an Algorithm?

algorithm

An **algorithm** is a method or a procedure for carrying out a calculation.

When you bake a cake, you probably follow a recipe. A cake recipe is a kind of algorithm: a step-by-step procedure taking various ingredients and turning them into a cake. In the same way, there are step-by-step procedures for adding, subtracting, multiplying, and dividing numbers. Just as there is mystery in how flour, eggs, butter, and sugar can combine to make cake, there is mystery in how the standard algorithms result in correct answers to arithmetic problems. Somehow, we throw numbers in, mix them up in a specific way, and out comes the correct answer to the addition problem. In this book, we won't examine the mysteries of cake baking, but we will uncover the mysteries of the algorithms of arithmetic. In fact, they aren't mysteries at all, but clever, efficient ways of calculating that make complete sense once you know how to look at them.

The Addition Algorithm

When we use the common algorithm to add decimals, such as $34.5 + 7.89$ or $149 + 85$, we put the numbers one under the other, lining up the decimal points if necessary. Then we add column by column, *regrouping* as needed. When adding numbers this way, if the digits in a column add to 10 (or more), we write down the ones digit of the sum and shift the remaining 10 to become a 1 in the next place to the left. Traditionally we write this 1 at the top of the next column to the left, although some educators have found it is better for children to write the 1 at the *bottom* of the next column to the left. This process of shifting a 10 from the sum of the digits in one place to a 1 in the next place to the left is

regrouping

called **regrouping**. Regrouping is also called **trading** or **carrying**.
Regrouping has taken place in the following examples:

$$
\begin{array}{r}
\overset{11}{34.50} \\
+\ \ 7.89 \\
\hline
42.39
\end{array}
\qquad
\begin{array}{r}
\overset{11}{149} \\
+\ \ 85 \\
\hline
234
\end{array}
\qquad
\begin{array}{r}
149 \\
+\ \ 85 \\
\hline
\overset{11}{} \\
234
\end{array}
$$

Notice that even though you don't see any decimal points in $149 + 85$, they are still lined up; we could also write this addition problem as $149.0 + 85.0$. Why do we line up the decimal points like that? Why do we add column by column? Why do we regroup? The answers to these questions lie in the interpretation of addition as combining and in the idea of place value.

Class Activity   *Now Turn to Class Activities Manual*

3M   Adding and Subtracting with Ten-Structured Pictures, p. 52

3N   Understanding the Common Addition Algorithm, p. 52

3O   Understanding the Common Subtraction Algorithm, p. 53

Working with Physical Objects to Understand the Addition Algorithm for Whole Numbers

Consider the sum $149 + 85$. We can represent this sum as the total number of toothpicks when we combine 149 toothpicks with 85 toothpicks. In order to analyze the addition algorithm, represent the numbers 149 and 85 with bundles, as described in Section 1.1. Represent 149 toothpicks as 1 bundle of one hundred (which is ten bundles of ten), 4 bundles of ten, and 9 individual toothpicks, as shown in Figure 3.12. Similarly, represent 85 toothpicks as 8 bundles of ten and 5 individual toothpicks. When all these toothpicks are combined, how many are there? Adding like bundles, we have the following:

14 individual toothpicks (9 plus 5 more)

12 bundles of 10 toothpicks (4 bundles of ten plus 8 more bundles of ten)

1 bundle of 100 toothpicks

Symbolically, we can write this as follows:

$$
\begin{aligned}
1(100) + \ &4(10) + \ 9(1) \\
+ \ &8(10) + \ 5(1) \\
\hline
= 1(100) + \ &12(10) + 14(1)
\end{aligned}
$$

So there are

$$1(100) + 12(10) + 14(1)$$

toothpicks, but this expression is not the expanded form of a number in ordinary decimal notation because there are 12 tens and 14 ones. This is where we regroup our bundles. *The regrouping of bundles of toothpicks is the physical representation of the regrouping process in the addition algorithm.*

To regroup the bundled toothpicks, convert the 14 individual toothpicks into 1 bundle of ten toothpicks and 4 individual toothpicks. This 1 bundle of ten is the small "carried" 1 we write above the 4 in the standard procedure. The small 1 above the tens in the following example really stands for 1 ten:

$$
\begin{array}{r}
{\scriptstyle 11} \\
149 \\
+ \ \ 85 \\
\hline
234
\end{array}
$$

Likewise, regroup the 12 bundles of 10 toothpicks into 1 bundle of 100 (namely 10 bundles of ten) and 2 bundles of ten. This 1 bundle of 100 is the small "carried" 1 we write above the 1 in the standard procedure. That small 1 really stands for 1 hundred. When you collect like bundles, you will have 2 bundles of a hundred, 3 bundles of ten, and 4 individual toothpicks.

FIGURE 3.12

Regrouping
149 + 85
toothpicks

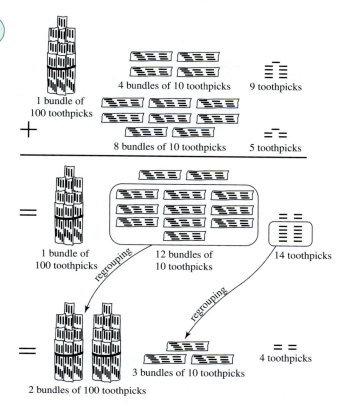

Here is the same example written symbolically:

$$1(100) + 4(10) + 9(1)$$
$$+ 8(10) + 5(1)$$
$$= 1(100) + 12(10) + 14(1)$$
$$+ 1(10) + 4(1) \quad \text{14 ones become 1 ten and 4 ones.}$$
$$+ 1(100) + 2(10) \quad\quad\quad \text{12 tens become 1 hundred and 2 tens.}$$
$$= 2(100) + 3(10) + 4(1)$$

We can also write this regrouping in equation form:

$$1(100) + 12(10) + 14(1) = \quad 1(100) \quad + 10(10) + 2(10) + 10(1) + 4(1)$$
$$= 1(100) + 1(100) + 2(10) + 1(10) + 4(1)$$
$$= \quad 2(100) \quad + \quad 3(10) \quad + \quad 4(1)$$
$$= \quad 234$$

(The diagonal arrows show how the 10 tens become 1 hundred and the 10 ones become 1 ten.) Notice that these equations represent symbolically the *physical actions* of regrouping the toothpicks.

The key point is that the common addition algorithm is just a way to condense the information in equations like the ones shown; therefore, it is a way to quickly and efficiently record the physical action of adding and regrouping actual objects, such as toothpicks. This example demonstrates why the common addition algorithm gives us correct answers to addition problems.

The Addition Algorithm for Decimals When we use the common addition algorithm to add decimals, we follow the same procedure as for whole numbers. The first step in the algorithm is to line up the decimal points. Although the decimal points are usually not shown in whole numbers, this step also occurs when adding whole numbers because the places where the decimal points would be are lined up. Why do we line up the decimal points? We can see why this makes sense by working with bundled toothpicks. The key lies in place value.

As we saw in Section 1.2, bundled objects can be used to represent (finite) decimals, as long as the meaning of 1 object is interpreted suitably. To represent the sum $0.834 + 6.7$ with paperclips, let 1 paperclip represent 0.001. Then a bundle of 10 paperclips represents 0.01, a bundle of 100 paperclips (a bundle of 10 tens) represents 0.1, and a bundle of 1000 paperclips (a bundle of 10 hundreds) represents 1. So 0.834 is then represented by

> 8 bundles of 100 paperclips,
>
> 3 bundles of 10 paperclips, and
>
> 4 individual paperclips,

while 6.7 is represented by

> 6 bundles of 1000 paperclips and
>
> 7 bundles of 100 paperclips.

Notice that we chose 1 paperclip to represent 0.001 in *both* 0.834 and 6.7. In this way, when we use the bundled paperclips to represent the sum $0.834 + 6.7$, thousandths will be added to thousandths, hundredths to hundredths, tenths to tenths, and ones to ones. If 1 paperclip were to represent 0.001 in 0.834, but 0.1 in 6.7, then we would add thousandths to tenths and hundredths to ones, which wouldn't make any sense. This would be like treating a penny as 0.01 of a dollar in one setting and 0.1 of a dollar in another setting. *The consistent choice for the meaning of 1 paperclip when representing both decimals has the same effect as lining up the decimal points of the two decimals. Decimal points must be lined up so that like terms will be added*—tens to tens, ones to ones, tenths to tenths, hundredths to hundredths, and so on.

If you are adding decimals that have no digits in places lower than the hundredths place, then you can think of your addition problem in terms of money. For example, think of the addition problem

$$3.7 + 0.45$$

as

$$\$3.70 + \$0.45$$

When we add these numbers in a column, we must line up the decimal points so that we add pennies to pennies, dimes to dimes, and dollars to dollars. If we didn't line up the decimal points, then we would add 7 dimes to 5 pennies and erroneously conclude that the result is 12 pennies.

Once the decimal points have been lined up, the addition algorithm proceeds in the same way as if the decimal points weren't there, and the explanation for why the algorithm works is the same as for whole numbers.

$$
\begin{array}{r}
1 \\
0.834 \\
+\ 6.700 \\
\hline
7.534
\end{array}
$$

The Subtraction Algorithm

Now let's turn to subtraction. When we subtract numbers in the decimal system, such as $319 - 148.2$ or $142 - 83$, using the standard paper-and-pencil algorithm, we first put the numbers one under the other, lining up the decimal points. Then we subtract column by column, *regrouping* as needed so that we can subtract the numbers in a column without a negative number resulting. When subtracting numbers this way, if the digit at the top of a column is less than the digit below it, we cross out the digit at the top of the next column to the left, changing this digit to one that is 1 less, and we replace the digit at the top of our original column with 10 plus the original digit. (For example, we replace a 2 with 12.) If we can't carry out these steps because there is a 0 in the next column to the left, then we keep moving to the left, crossing out 0s until we come to a digit that is greater than 0. We cross this nonzero digit out and replace it with the digit that is 1 less, we replace all the intervening 0s with 9s, and, as before, we replace the digit at the top of the original column with 10 plus this digit. This process of **regrouping** changing digits so as to be able to subtract in a column is called **regrouping**. Regrouping is also called **trading** or **borrowing**.

Regrouping is used in the following examples:

$$
\begin{array}{r}
319.0 \\
-\,148.2 \\
\end{array}
\quad \rightarrow \quad
\begin{array}{r}
{\scriptstyle 211\ \ 8\ 10} \\
\cancel{3}\cancel{1}9.\cancel{0} \\
-\,1\,4\,8.\,2 \\
\hline
1\,7\,0.\,8
\end{array}
$$

$$
\begin{array}{r}
142 \\
-\ \ 83 \\
\end{array}
\quad \rightarrow \quad
\begin{array}{r}
{\scriptstyle 0\,1\,3\,12} \\
\cancel{1}\cancel{4}\cancel{2} \\
-\ \ \ 83 \\
\hline
59
\end{array}
$$

$$
\begin{array}{r}
100.2 \\
-\ \ \ 5.3 \\
\end{array}
\quad \rightarrow \quad
\begin{array}{r}
{\scriptstyle 0\ 9\ 9\ \ 12} \\
\cancel{1}\cancel{0}\cancel{0}.\cancel{2} \\
-\ \ \ \ 5.\,3 \\
\hline
9\,4.\,9
\end{array}
$$

As with the addition algorithm, we want to understand why this procedure makes sense. We will learn why by interpreting subtraction as taking away and by considering place value.

Working with Physical Objects to Understand the Subtraction Algorithm   As with addition, we can model the subtraction algorithm, particularly the regrouping process, with bundles of toothpicks. Consider the difference $142 - 83$. To represent this with bundled toothpicks, start with 142 toothpicks in 1 bundle of a hundred, 4 bundles of ten, and 2 individual toothpicks. How many toothpicks will be left when we take 83 toothpicks away? When we try to take 3 individual toothpicks away from the 142 toothpicks, we first need to do some unbundling. *This unbundling is a physical representation of the regrouping process.* This process is illustrated in Figure 3.13. One of the 4 bundles of ten can be unbundled and added to the individual toothpicks. The result is 1 bundle of a hundred, 3 bundles of ten, and 12 individual toothpicks. Using equations, we have

$$
\begin{aligned}
142 &= 1(100) + \quad 4(10) \quad + \quad\quad 2(1) \\
&= 1(100) + 3(10) + 1(10) + \quad\quad 2(1) \\
&\hspace{6cm}\searrow \\
&= 1(100) + \quad 3(10) \quad + 10(1) + 2(1) \\
&= 1(100) + \quad 3(10) \quad + \quad\quad 12(1)
\end{aligned}
$$

Now there won't be any problem taking 3 individual toothpicks away, but what about taking away 80, namely 8 bundles of ten? There are only 3 bundles of ten. So we must unbundle the 1 bundle of a hundred as 10 bundles of 10 and combine these 10 bundles of ten with the 3 bundles of ten to make 13 bundles of ten. We continue from the previous equations to get

$$
\begin{aligned}
142 &= 1(100) + \quad\quad 3(10) \quad + 12(1) \\
&\hspace{2.2cm}\searrow \\
&= \quad\quad 10(10) + 3(10) + 12(1) \\
&= \quad\quad 13(10) \quad + 12(1)
\end{aligned}
$$

FIGURE 3.13

Regrouping 142
toothpicks

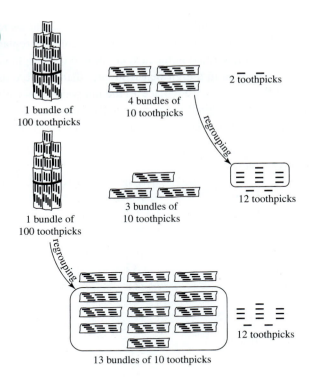

1 bundle of
100 toothpicks

4 bundles of
10 toothpicks

2 toothpicks

regrouping

12 toothpicks

1 bundle of
100 toothpicks

3 bundles of
10 toothpicks

regrouping

13 bundles of 10 toothpicks

12 toothpicks

After regrouping, there is no difficulty taking 3 individual toothpicks away from the 12 individual tooth-picks and 8 bundles of ten away from the 13 bundles of ten. Notice that the physical actions of moving the toothpicks correspond exactly to the regrouping process that takes place in the common subtraction algorithm. In expanded form, the subtraction problem $142 - 83$ can now be rewritten as

$$
\begin{array}{r}
13(10) + 12(1) \\
-\quad 8(10) + 3(1) \\
\hline
=\quad 5(10) + 9(1) = 59
\end{array}
$$

The key point is that the common subtraction algorithm is just a way to condense the information in equations like those given previously. Therefore, it is a quick and efficient way to record the physical actions of regrouping and then taking away actual objects, such as toothpicks. This is why the common subtraction algorithm gives us correct answers to subtraction problems.

Understanding the Subtraction Algorithm for Decimals As with the addition algorithm for decimals, we can generally still represent decimal subtraction problems with bundles of toothpicks by an appropriate interpretation of 1 toothpick. Decimal points are lined up so that like terms are subtracted—tens from tens, ones from ones, tenths from tenths, hundredths from hudredths, and so on.

Class Activity   *Now Turn to Class Activities Manual*

3P A Third-Grader's Method of Subtraction, p. 54

3Q Subtracting Across Zeros, p. 54

3R Regrouping with Dozens and Dozens of Dozens, p. 56

3S Regrouping with Seconds, Minutes, and Hours, p. 56

Practice Exercises for Section 3.3

1. Define these two terms: algorithm and regroup (in addition and in subtraction).

2. Draw a ten-structured picture for 37 + 26, in which tens are represented as lines and ones are represented as small dots (as in Class Activity 3M). Show how a child could use the picture side by side with the common algorithm for adding 37 + 26 to help make sense of the common algorithm. Briefly discuss how the child could use the picture to understand the regrouping process.

3. Draw a ten-structured picture for 41 − 28, in which tens are represented as lines and ones are represented as small dots (as in Class Activity 3M). Show how a child could use the picture side by side with the common algorithm for subtracting 41 − 28 to help make sense of the common algorithm. Briefly discuss how the child could use the picture to understand the regrouping process.

4. Write equations with numbers in expanded form showing how to regroup the number 104 so that 69 can be subtracted from it.

5. Why do we line up decimal points before adding or subtracting decimals?

6. Ellie solves the subtraction problem 2.5 − 0.13 with toothpicks. She represents 2.5 with 2 bundles of 10 toothpicks and 5 individual toothpicks, and she represents 0.13 with 1 bundle of 10 toothpicks and 3 individual toothpicks. Ellie gets the answer 1.2. Is she right? If not, explain why not and discuss how she could use the toothpicks correctly.

7. A store buys action figures in boxes. Each box contains 50 bags, and each bag contains 6 action figures. At the beginning of the month, the store has

   7 unopened boxes, 15 unopened bags, and 3 individual action figures.

   At the end of the month, the store has

   2 unopened boxes, 37 unopened bags, and 5 individual action figures.

   How many action figures did the store sell during the month (assuming they got no additional shipments of action figures)? Write your answer in terms of boxes, bags, and individual action figures. *Work with boxes, bags, and individuals in a sort of expanded form and use regrouping to solve this problem.*

Answers to Practice Exercises for Section 3.3

1. See text.

2. See Figure 3.14 which shows how a child could draw a ten-structured picture for 37 + 26 side by side with the usual way we record the common algorithm. To understand the regrouping process, the child could use the make-a-ten method to join 3 from the 6 in 26 with the 7 in 37 to make a ten. This ten is put with the other tens, which is done by writing the "small 1" over the 3 in 37 in the numerical version of the problem. (The remaining 3 from the 6 becomes the ones digit in the sum.)

3. See Figure 3.15, which shows how a child could draw a ten-structured picture for 41 − 28 side by side with the usual way we record the common algorithm. To

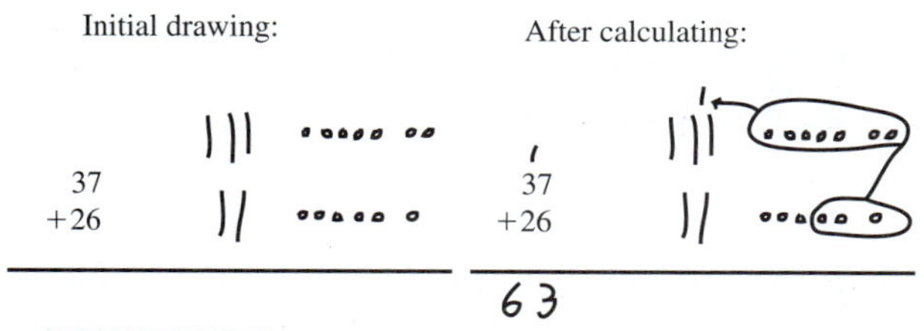

FIGURE 3.14   A ten-structured picture for 37 + 26

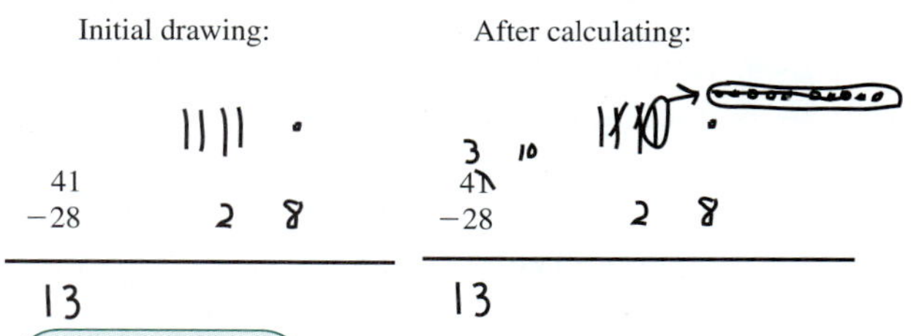

FIGURE 3.15   A ten-structured picture for 41 − 28

understand the regrouping process, the child could unbundle a ten to make 10 ones and then take the 8 in 28 away from 10, leaving 2 ones to be combined with the other 1 in 41. (The 2 ones and the other 1 combined become the 3 in the ones place of the difference.)

4. Here's the regrouping process, with equations in expanded form:

$$
\begin{aligned}
104 &= 1(100) + &0(10) &+ &4(1) \\
&= &10(10) + 0(10) + &&4(1) \\
&= &10(10) &+ &4(1) \\
&= &9(10) + 1(10) + &&4(1) \\
&= &9(10) &+ 10(1) + &4(1) \\
&= &9(10) &+ &14(1)
\end{aligned}
$$

5. See text.

6. No, Ellie's answer is not correct—1 toothpick must represent the same amount when representing both 2.5 and 0.13. Ellie should think of 1 toothpick as representing $\frac{1}{100}$ in both cases. Then 2.5 is represented by 2 bundles of 100 toothpicks and 5 bundles of 10 toothpicks, while 0.13 is represented by 1 bundle of 10 toothpicks and 3 individual toothpicks. Now Ellie should be able to see that she'll need to regroup in order to subtract. It might help Ellie to think in terms of money: 2.5 and 0.13 can be represented by $2.50 and $0.13. Ellie's way of using the toothpicks would be like saying that a dime is equal to a penny.

7. We must solve the following problem:

$$
\begin{aligned}
&7 \text{ boxes} + 15 \text{ bags} + 3 \text{ individual} \\
-\ &(2 \text{ boxes} + 37 \text{ bags} + 5 \text{ individual})
\end{aligned}
$$

We can solve this by first regrouping the 7 boxes, 15 bags, and 3 individual action figures. If we open one of the bags, then there is one less bag, but 6 more individual figures, so there are 7 boxes, 14 bags, and 9 individual action figures.

If we open one of the boxes, then there is one less box, but 50 more bags of action figures, so there are 6 boxes, 64 bags, and 9 individual action figures.

It's still the same number of action figures—they are just arranged in a different way. In equation form we can write this as

$$
\begin{aligned}
&7 \text{ boxes} + 15 \text{ bags} + 3 \text{ individual} \\
= &7 \text{ boxes} + \quad 14 \text{ bags} \quad + (6+3) \text{ individual} \\
= &6 \text{ boxes} + (50+14) \text{ bags} + 9 \text{ individual} \\
= &6 \text{ boxes} + \quad 64 \text{ bags} \quad + 9 \text{ individual}
\end{aligned}
$$

Now we are ready to subtract the 2 boxes, 37 bags, and 5 individual action figures:

$$
\begin{aligned}
&6 \text{ boxes} + 64 \text{ bags} + 9 \text{ individual} \\
-\ &(2 \text{ boxes} + 37 \text{ bags} + 5 \text{ individual}) \\
\hline
&4 \text{ boxes} + 27 \text{ bags} + 4 \text{ individual}
\end{aligned}
$$

So a total of 4 boxes, 27 bags, and 4 individual action figures were sold during the month.

Problems for Section 3.3

1. Refer to Class Activity 3M, Adding and Subtracting with Ten-Structured Pictures. Show how students 1, 2, and 3 might solve the addition problem 29 + 46 and how student 4 might solve the subtraction problem 54 − 28. In each case explain briefly why you think the student would solve the problem that way.

2. Describe how to use bundled toothpicks (or bundles of other objects) to explain regrouping in the addition problem 167 + 59. Draw (simplified) pictures to aid your explanation.

3. Describe how to use bundled toothpicks (or bundles of other objects) to explain

regrouping in the subtraction problem $231 - 67$. Draw (simplified) pictures to aid your explanation.

4. Allie solves the subtraction problem $304 - 9$ as follows:

$$\begin{array}{r} \overset{2\ \ 14}{3\cancel{0}4} \\ -\quad 9 \\ \hline 205 \end{array}$$

Explain to Allie what is wrong with her method, and explain why the correct method makes sense.

5. Zachary added $3.4 + 2.7$ and got the answer 5.11. How might Zachary have gotten this incorrect answer? Explain to Zachary why his answer is not correct and why a correct method for adding $3.4 + 2.7$ makes sense.

6. To solve $512 - 146$, a student writes the following:

$$\begin{array}{ll} 512 & 400 \\ -146 & -\ \ 30 \\ \hline -4 & 370 \quad 512 - 146 = 366 \\ -30 & -4 \\ \hline 400 & 366 \end{array}$$

Describe the student's solution strategy and discuss why the strategy makes sense. Expanded forms may be helpful to your discussion.

7. Here's how Mo solved the subtraction problem $635 - 813$:

$$\begin{array}{r} 635 \\ -813 \\ \hline -222 \end{array}$$

Mo did this by working from right to left, saying

$$5 - 3 = 2$$
$$3 - 1 = 2$$
$$6 - 8 = -2$$

a. Is Mo's answer right?

b. Solve

$$\begin{array}{r} 6(100) + 3(10) + 5(1) \\ -\left[8(100) + 1(10) + 3(1)\right] \end{array}$$

by working with expanded forms. Discuss how Mo's work compares with your work in expanded forms.

8. Problem: Matteo is 4 feet 3 inches tall. Nico is 3 feet 11 inches tall. How much taller is Matteo than Nico?

Sarah solved this problem as follows:

$$\begin{array}{r} \overset{3}{\cancel{4}}\text{ ft }\ \overset{13}{\cancel{3}}\text{ in} \\ -\ 3\text{ ft } 11\text{ in} \\ \hline 2\text{ in} \end{array}$$

So Sarah gave 2 inches as the answer. Is Sarah right? If not, explain what is wrong with her method, and show how to *modify* her method of regrouping to make it correct. (Do not start from scratch.)

9. Problem: A container holds 2 quarts and 4 fluid ounces. The container is now filled with 6 fluid ounces of liquid. How much liquid must be added to the container to make it full?

John solved this problem as follows:

$$\begin{array}{r} \overset{1}{2}\text{ q }\ \overset{14}{\cancel{4}}\text{ fl oz} \\ -\qquad 6\text{ fl oz} \\ \hline 1\text{ q } 8\text{ fl oz} \end{array}$$

So John gave 1 quart and 8 fluid ounces as the answer. Is John right? If not, explain what is wrong with his method, and show how to *modify* his method of regrouping to make it correct. (Do not start from scratch.)

10. On a space shuttle mission, a certain experiment is started 2 days, 14 hours, and 30 minutes into the mission. The experiment takes 1 day, 21 hours, and 47 minutes to run. When will the experiment be completed? Give your answer in days, hours, and minutes into the mission. Work with a sort of expanded form. In other words, work with

$$2(\text{days}) + 14(\text{hours}) + 30(\text{minutes})$$

and

$$1(\text{day}) + 21(\text{hours}) + 47(\text{minutes})$$

and *regroup among days, hours, and minutes* to solve this problem.

11. We can write dates and times in a sort of expanded form. For example, October 4th, 6:53 P.M. can be written as

$$4(\text{days}) + 18(\text{hours}) + 53(\text{minutes})$$

(In some circumstances you might want to include the month and the year, too.) How long is it from 3:27 P.M. on October 4 to 7:13 A.M. on October 19? Give your answer in days, hours, and minutes. Work in the type of expanded form previously described, and *regroup among days, hours, and minutes* to solve this problem.

12. Erin wants to figure out how much time elapsed between 9:45 A.M. and 11:30 A.M. Erin does the following:

$$\begin{array}{r} {\scriptstyle 0\ \ 12\ 1}\\ 1\,\cancel{1}:\cancel{3}\,0 \\ -\ \ 9:4\,5 \\ \hline 1:8\,5 \end{array}$$

and says the answer is 1 hour and 85 minutes.

a. Is Erin right? If not, explain what is wrong with her method, and show how to *modify* her method of regrouping to make it correct. (Do not start from scratch.)

b. Solve the problem of how much time elapsed between 9:45 A.M. and 11:30 A.M. in another way than your modification of Erin's method. Explain your method.

13. The common subtraction algorithm described in the text is not the only correct subtraction algorithm. Some people use the following algorithm instead: Line up the numbers as in the standard algorithm, and subtract column by column, proceeding from right to left. The only difference between this new algorithm and our standard one is in regrouping. To regroup with the new algorithm, move one column to the left and cross out the digit at the *bottom* of the column, replacing it with that digit *plus 1*. (So replace an 8 with a 9, replace a 9 with a 10, etc.) Then, as in the standard algorithm, replace the digit at the top of the original column with the original number plus 10. The following example shows the steps of this new algorithm:

$$\begin{array}{r} 132 \\ -\ \ 79 \\ \hline \end{array} \rightarrow \begin{array}{r} {\scriptstyle 12}\\ 1\ 3\ \cancel{2} \\ -\ \ \cancel{7}\,9 \\ {\scriptstyle 8}\\ \hline 3 \end{array} \rightarrow \begin{array}{r} {\scriptstyle 13\ 12}\\ 1\ \cancel{3}\ \cancel{2} \\ {\scriptstyle 1\ 8}\\ -\ \cancel{0}\,\cancel{7}\,9 \\ \hline 5\ 3 \end{array}$$

a. Use the new algorithm to solve $524 - 198$ and $1003 - 95$. Verify that the new algorithm gives correct answers.

b. Explain why it makes sense that this new algorithm gives correct answers to subtraction problems. What is the reasoning behind this new algorithm?

c. What are some advantages and disadvantages of this new algorithm compared with our common one?

14. The common subtraction algorithm described in the text is not the only correct subtraction algorithm. The next subtraction algorithm is called *adding the complement*. For a 3-digit whole number, N, the **complement** of N is $999 - N$. For example, the complement of 486 is

$$999 - 486 = 513$$

Notice that regrouping is never needed to calculate the complement of a number. To use the adding-the-complement algorithm to subtract a 3-digit whole number, N, from another 3-digit whole number, start by adding the complement of N rather than subtracting N. For example, to solve

$$723 - 486$$

first add the complement of 486:

$$\begin{array}{r} 723 \\ +\ 513 \\ \hline 1236 \end{array}$$

Then cross out the 1 in the thousands column, and add 1 to the resulting number:

$$\cancel{1}236 \rightarrow 236 + 1 = 237$$

Therefore, according to the *adding-the-complement* algorithm, $723 - 486 = 237$.

a. Use the adding-the-complement algorithm to calculate $301 - 189$ and $295 - 178$. Verify that you get the correct answer.

b. Explain why the adding-the-complement algorithm gives you the correct answer to any 3-digit subtraction problem. Focus on the relationship between the original problem and the addition problem in adding the complement. For example, how are the problems $723 - 486$ and $723 + 513$ related? Work with the *complement* relationship, $513 = 999 - 486$, and notice that $999 = 1000 - 1$.

c. What are some advantages and disadvantages of the adding-the-complement subtraction algorithm compared with the common subtraction algorithm described in the text?

15. After you solve part (b) of this problem, think carefully about your explanation in part (a). You may find that you need to revise it.

a. What is the smallest number of coins you can use to make 43 cents if quarters, dimes, nickels, and pennies are available? Explain why your answer is correct.

b. What is the smallest number of coins you can use to make 43 cents if quarters, dimes, and pennies (but no nickels) are available? Explain why your answer is correct.

3.4 Adding and Subtracting Fractions

 **Focal Points**
Grade 5

In the previous sections, we learned the reasoning behind the common addition and subtraction algorithms for whole numbers and decimals. There are also standard procedures for adding and subtracting fractions, and in this section, we will study why these make sense. Because they are defined in terms of addition, we will also study mixed numbers (numbers such as $2\frac{3}{4}$) and the algorithm for converting a mixed number to an improper fraction. Similarly, because we can always express a finite decimal as a sum of fractions by writing the decimal in expanded form, we will see how to write finite decimals as fractions. Finally, we will consider fraction addition and subtraction story problems and a source of errors in fraction addition and subtraction story problems: the failure to work with like wholes.

Class Activity   *Now Turn to Class Activities Manual*

3T 🏛 Why Do We Add and Subtract Fractions the Way We Do? p. 57

Adding and Subtracting Fractions with Like Denominators

The easiest situation to work with when adding fractions is adding fractions that have the same denominator. If two fractions have the same denominator, then you can add or subtract these fractions by adding or subtracting the numerators and leaving the denominator unchanged. For example,

$$\frac{4}{15} + \frac{7}{15} = \frac{4+7}{15} = \frac{11}{15}$$

Why does this make sense? We can interpret the sum

$$\frac{4}{15} + \frac{7}{15}$$

as the total fraction of a fence you have painted if you first paint $\frac{4}{15}$ of the fence and then paint another $\frac{7}{15}$ of the fence. When the fence is divided into 15 equal pieces, each piece is $\frac{1}{15}$ of the fence; $\frac{4}{15}$ of the fence consists of 4 pieces and $\frac{7}{15}$ of the fence consists of 7 pieces. So if you paint 4 pieces of the fence and then paint another 7 pieces of the fence, then you will have painted $4 + 7 = 11$ pieces altogether, as you can see in Figure 3.16. What kind of pieces are they? All 11 pieces are fifteenths of the fence. Therefore, you have painted $\frac{11}{15}$ of the fence in all.

FIGURE 3.16

Adding fractions with like denominators:
$\frac{4}{15} + \frac{7}{15} = \frac{11}{15}$

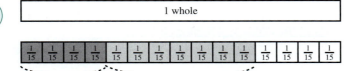

4 fifteenths + 7 fifteenths = 11 fifteenths

Adding and Subtracting Fractions with Unlike Denominators by Finding Common Denominators

How do we add or subtract fractions that have different denominators, such as $\frac{5}{6} + \frac{3}{4}$? We can interpret the sum

$$\frac{5}{6} + \frac{3}{4}$$

as the total distance walked if you first walk $\frac{5}{6}$ of a mile and then walk $\frac{3}{4}$ of a mile. (See Figure 3.17.) In this case, the first distance walked is described in terms of *sixths* of a mile and the second distance walked is described in terms of *fourths* of a mile. Sixths of a mile and fourths of a mile are different-size distances, so we can't just add the 5 in $\frac{5}{6}$ to the 3 in $\frac{3}{4}$. We need a common distance with which to describe both $\frac{5}{6}$ of a mile and $\frac{3}{4}$ of a mile. In other words, we need to break *sixths* and *fourths* into *like parts*.

As in comparing fractions, the process of breaking into like parts is achieved numerically by giving the fractions $\frac{5}{6}$ and $\frac{3}{4}$ a common denominator. *Any* common denominator will do; it does not have to be the least one. You can always produce a common denominator by multiplying the two denominators. So, to add

$$\frac{5}{6} \quad \text{and} \quad \frac{3}{4}$$

we can use the common denominator $6 \cdot 4$, which is 24. If we want to work with smaller numbers, we can use 12, which is a common denominator because $12 = 6 \cdot 2$ and $12 = 4 \cdot 3$. Once we have common denominators, the fractions are described in terms of like parts, so we simply add the numerators to determine the total number of parts:

$$\frac{5}{6} + \frac{3}{4} = \frac{5 \cdot 4}{6 \cdot 4} + \frac{3 \cdot 6}{4 \cdot 6} = \frac{20}{24} + \frac{18}{24} = \frac{38}{24}$$

or

$$\frac{5}{6} + \frac{3}{4} = \frac{5 \cdot 2}{6 \cdot 2} + \frac{3 \cdot 3}{4 \cdot 3} = \frac{10}{12} + \frac{9}{12} = \frac{19}{12}$$

Since

$$\frac{38}{24} = \frac{19 \cdot 2}{12 \cdot 2} = \frac{19}{12}$$

FIGURE 3.17

Using a common denominator to add

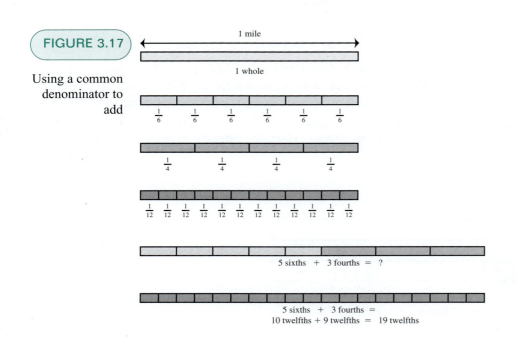

the two ways just shown of adding $\frac{5}{6}$ and $\frac{3}{4}$ produce equal results. When we use the common denominator 12, the resulting sum $\frac{19}{12}$ is in simplest form. If the least common denominator is not used when adding fractions, then the resulting sum will not be in simplest form.

Figure 3.17 shows the addition of $\frac{5}{6}$ and $\frac{3}{4}$ with the common denominator 12, using fraction strips. In this case, each sixth is broken into 2 parts and each fourth is broken into 3 parts to produce twelfths. In general, suppose

$$\frac{A}{B} \quad \text{and} \quad \frac{C}{D}$$

are two fractions (where $B \neq 0$ and $D \neq 0$). Then

$$\frac{A}{B} = \frac{A \cdot D}{B \cdot D}$$

and

$$\frac{C}{D} = \frac{C \cdot B}{D \cdot B}$$

and the latter two fractions

$$\frac{A \cdot D}{B \cdot D} \quad \text{and} \quad \frac{C \cdot B}{D \cdot B}$$

both have denominators equal to $B \cdot D$. Both fractions are expressed in terms of like parts when they have common denominators. Therefore, we can add or subtract these fractions by adding or subtracting the number of parts, (i.e., the numerators). Therefore, in general,

$$\frac{A}{B} + \frac{C}{D} = \frac{A \cdot D + C \cdot B}{B \cdot D}$$

and

$$\frac{A}{B} - \frac{C}{D} = \frac{A \cdot D - C \cdot B}{B \cdot D}$$

> **Class Activity**   *Now Turn to Class Activities Manual*
>
> **3U** How Do We Find a Suitable Common Denominator for Adding and Subtracting Fractions? p. 58

Writing Mixed Numbers as Improper Fractions

mixed number   A **mixed number** or **mixed fraction** is a number that is written in the form

$$A\frac{B}{C}$$

where A, B, and C are whole numbers, and $\frac{B}{C}$ is a proper fraction (i.e., the numerator is less than the denominator). So

$$2\frac{3}{4} \quad \text{and} \quad 5\frac{7}{8}$$

are mixed numbers.

The mixed number

$$A\frac{B}{C}$$

stands for the sum of its whole number part and its fractional part:

$$A + \frac{B}{C}$$

Since a whole number A can also be written as a fraction, namely, $\frac{A}{1}$, we can write the mixed number $A\frac{B}{C}$ as follows:

$$A\frac{B}{C} = A + \frac{B}{C} = \frac{A}{1} + \frac{B}{C}$$

Now we can use the way we add fractions to show that every mixed number can be written as an improper fraction:

$$A\frac{B}{C} = \frac{A \cdot C + B}{C}$$

So,

$$2\frac{3}{4} = \frac{2 \cdot 4 + 3}{4} = \frac{11}{4}$$

We often convert mixed numbers to improper fractions in order to add, subtract, multiply, or divide them.

Class Activity   *Now Turn to Class Activities Manual*

3V   Mixed Numbers and Improper Fractions, p. 59

3W Adding and Subtracting Mixed Numbers, p. 59

3X   Addition with Whole Numbers, Decimals, Fractions, and Mixed Numbers: What Are Common Ideas? p. 60

Writing Finite Decimals as Fractions and Saying Decimals

Every finite decimal stands for a finite sum of fractions. We can see this representation by writing the decimal in its expanded form—for example,

$$2.7 = 2 + \frac{7}{10}$$

$$32.85 = 30 + 2 + \frac{8}{10} + \frac{5}{100}$$

$$0.491 = \frac{4}{10} + \frac{9}{100} + \frac{1}{1000}$$

By giving the fractions in the expanded form of a finite decimal a common denominator, we can write the decimal as a fraction with a denominator that is a value of a decimal place. For example, we have

$$2.7 = 2 + \frac{7}{10}$$

$$= \frac{20}{10} + \frac{7}{10}$$

$$= \frac{27}{10}$$

or

$$0.491 = \frac{4}{10} + \frac{9}{100} + \frac{1}{1000}$$
$$= \frac{400}{1000} + \frac{90}{1000} + \frac{1}{1000}$$
$$= \frac{491}{1000}$$

It is for this reason that we can say 0.491 as "four-hundred ninety-one thousandths."

Observe that this method for writing decimals as fractions applies only to finite decimals. It does not apply to a decimal such as

$$0.4444444\ldots$$

where the 4s continue forever. In fact, it turns out that some numbers, such as $\sqrt{2} = 1.4142\ldots$ cannot be written as fractions of whole numbers.

When Is Combining Not Adding?

The old admonishment "You can't add apples and oranges" is especially important to remember when adding and subtracting fractions, because sometimes fractions of quantities that are to be combined refer to *different wholes*.

Whenever we add two numbers, such as

$$3 + 4 \quad \text{or} \quad \frac{2}{3} + \frac{3}{4}$$

both summands and the answer all refer to the same whole. Similarly, whenever we subtract two numbers, such as

$$4 - 3 \quad \text{or} \quad \frac{3}{4} - \frac{2}{3}$$

the minuend, the subtrahend, and the answer all refer to the same whole.

For example, we can interpret the sum

$$3 + 4$$

as the total number of apples when 3 apples are combined with 4 apples. But if you interpret $3 + 4$ by combining 3 bananas and 4 blocks, as in Figure 3.18, then the best you can say is that you have 7 *objects* in all. Similarly, we can interpret the sum

$$\frac{1}{3} + \frac{1}{4}$$

as the fraction of a pie we will have in all if we have $\frac{1}{3}$ of the pie and then get another $\frac{1}{4}$ *of the same pie or an equivalent pie.* The $\frac{1}{3}$, the $\frac{1}{4}$, and the sum $\frac{1}{3} + \frac{1}{4} = \frac{7}{12}$ *all refer to the same pie,* or to copies of an equivalent pie.

But suppose we have a small pie and a large pie, that the small pie is divided into 3 pieces, and that the large pie is divided into 4 pieces, as shown in Figure 3.19. If we get one piece of each pie, does this represent the sum

$$\frac{1}{3} + \frac{1}{4}?$$

FIGURE 3.18

3 bananas and
4 blocks

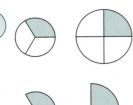

FIGURE 3.19

$\frac{1}{3}$ of a small pie and $\frac{1}{4}$ of a big pie does not make $\frac{1}{3} + \frac{1}{4}$ of a pie.

No, it does not, because the $\frac{1}{3}$ and the $\frac{1}{4}$ refer to *different wholes* that are not equivalent. All we can say is that we have 2 pieces of pie. We can't say that we have $\frac{7}{12}$ of a pie—because what pie would the $\frac{7}{12}$ refer to?

So when working with fraction addition or subtraction story problems, be especially careful that the fractions in question refer to the same underlying wholes. In some cases, fractional amounts that are to be combined or taken away refer to different wholes; therefore, the story problem is not a fraction addition or subtraction problem.

Class Activity   *Now Turn to Class Activities Manual*

3Y Are These Story Problems for $\frac{1}{2} + \frac{1}{3}$? p. 61

3Z Are These Story Problems for $\frac{1}{2} - \frac{1}{3}$? p. 62

3AA What Fraction Is Shaded? p. 63

Practice Exercises for Section 3.4

1. Using the example $\frac{1}{4} + \frac{5}{6}$, explain why we must give fractions a common denominator in order to add them.

2. When we add or subtract fractions, *must* we use the least common denominator? Are there any advantages to using the least common denominator?

3. The usual procedure for converting a mixed number, such as $7\frac{1}{3}$, into an improper fraction is this:

$$7\frac{1}{3} = \frac{7 \cdot 3 + 1}{3} = \frac{22}{3}$$

Explain the logic behind this procedure.

4. You showed Tommy the diagram in Figure 3.20 to explain why $2\frac{1}{4} = \frac{9}{4}$, but Tommy says that it shows $\frac{9}{12}$, not $\frac{9}{4}$. What must you clarify?

FIGURE 3.20   Showing $2\frac{1}{4} = \frac{9}{4}$

5. We usually call the number 0.43 "forty-three hundredths" and not "four tenths and three hundredths." Why do these two mean the same thing?

6. Jessica says that

$$\frac{1}{2} + \frac{2}{3} = \frac{3}{5}$$

and shows the picture in Figure 3.21 to prove it. What is the problem with Jessica's reasoning? Why is it not correct? Don't just explain how to do the problem correctly; explain what is faulty with Jessica's reasoning.

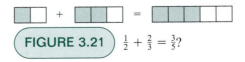

FIGURE 3.21 $\frac{1}{2} + \frac{2}{3} = \frac{3}{5}$?

7. Which of the following problems can be solved by adding $\frac{1}{2} + \frac{1}{3}$? For those problems that can't be solved by adding $\frac{1}{2} + \frac{1}{3}$, solve the problem in

another way if there is enough information to do so, or explain why the problem cannot be solved.

a. In Ms. Dock's class, $\frac{1}{2}$ of the boys want pizza for lunch and $\frac{1}{3}$ of the girls want pizza for lunch. What fraction of the children in Ms. Dock's class want pizza for lunch?

b. $\frac{1}{2}$ of Jane's shirts have the color red on them somewhere. $\frac{1}{3}$ of Jane's shirts have the color pink on them somewhere. What fraction of Jane's shirts have either red or pink on them somewhere?

c. $\frac{1}{2}$ of Jane's shirts have the color red on them somewhere. Of the shirts that belong to Jane and do not have red on them somewhere, $\frac{1}{3}$ have the color pink on them somewhere. What fraction of Jane's shirts have either red or pink on them somewhere?

8. Can the story problem that follows be solved by subtracting $\frac{1}{2} - \frac{1}{3}$? If not, explain why not, and solve the problem in a different way if there is enough information to do so.
Story problem: There is $\frac{1}{2}$ of a pie left over from yesterday. Pratima eats $\frac{1}{3}$ of the leftover pie. How much pie is left?

9. What is wrong with the following story problem? Give two different ways to restate the problem. Explain how to solve your restated problems.
Story problem: Liat has $\frac{3}{4}$ cup of juice. Sumin has $\frac{1}{3}$ less. How much juice does Sumin have?

10. For each square in Figure 3.22, determine the fraction of the square that is shaded. Explain your reasoning. You may assume that all lengths that appear to be equal really are equal.

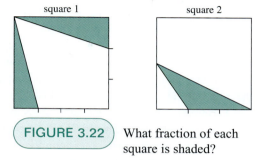

FIGURE 3.22   What fraction of each square is shaded?

11. Each of the pictures in Figure 3.23 shows two adjacent lots of land, lot A and lot B. In each case, 20% of lot A is shown shaded and 40% of lot B is shown shaded. What percent of the *combined amount* of lot A and lot B is shaded in each case?

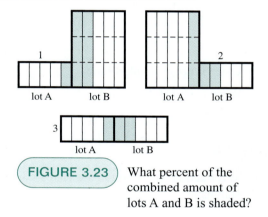

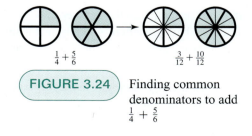

FIGURE 3.23   What percent of the combined amount of lots A and B is shaded?

Answers to Practice Exercises for Section 3.4

1. When we add $\frac{1}{4}$ and $\frac{5}{6}$, we first give the fractions a common denominator so they have *like parts*. We can interpret $\frac{1}{4} + \frac{5}{6}$ as the total amount of pie you have if you start with $\frac{1}{4}$ of a pie and then get $\frac{5}{6}$ of another equivalent (same size) pie. But fourths and sixths are pieces of different size, so we must express both kinds of pie pieces in terms of like pieces. If we divide each of the 4 fourths of a pie into 3 pieces and each of the 6 sixths of the other pie into 2 pieces, then $\frac{1}{4}$ of the first pie becomes $\frac{3}{12}$ and $\frac{5}{6}$ of the second pie becomes $\frac{10}{12}$. Now both fractions of pie are expressed in terms of twelfths of a pie, as

seen in Figure 3.24. Therefore, the total amount of pie you have consists of $3 + 10 = 13$ pieces, and these pieces are twelfths of a pie. Thus, you have $\frac{13}{12}$ of a pie, which is $1\frac{1}{12}$ of a pie.

FIGURE 3.24   Finding common denominators to add $\frac{1}{4} + \frac{5}{6}$

Numerically, this process of subdividing pieces of pie can be expressed succinctly with the following equations:

$$\frac{1}{4} + \frac{5}{6} = \frac{1 \cdot 3}{4 \cdot 3} + \frac{5 \cdot 2}{6 \cdot 2} = \frac{3}{12} + \frac{10}{12} = \frac{13}{12} = 1\frac{1}{12}$$

2. To add fractions, we do not need to use the least common denominator. Any common denominator will do. In practice exercise 1, we could have used the common denominator 24 instead of 12 to add $\frac{1}{4} + \frac{5}{6}$. However, there are some advantages to using the least common denominator. The least common denominator is smaller than other common denominators; therefore, it may be easier to work with. If we use a common denominator that is not the least common denominator, then the resulting sum will not be in simplest form. For example, if we use the common denominator 24 to add $\frac{1}{4} + \frac{5}{6}$, then the resulting sum is

$$\frac{1}{4} + \frac{5}{6} = \frac{6}{24} + \frac{20}{24} = \frac{26}{24}$$

which is not in simplest form.

3. Remember that $7\frac{1}{3}$ stands for $7 + \frac{1}{3}$, and that $7 = \frac{7}{1}$. To calculate

$$7\frac{1}{3} = \frac{7}{1} + \frac{1}{3}$$

as a fraction, we need to first find a common denominator:

$$7\frac{1}{3} = 7 + \frac{1}{3}$$
$$= \frac{7}{1} + \frac{1}{3}$$
$$= \frac{7 \cdot 3}{1 \cdot 3} + \frac{1}{3}$$
$$= \frac{7 \cdot 3 + 1}{3}$$
$$= \frac{22}{3}.$$

Notice that the next-to-last step shows the procedure for turning a mixed number into an improper fraction. This procedure is really just a shorthand way to give 7 and $\frac{1}{3}$ a common denominator and add them.

4. You must identify the whole in the diagram. Tommy is taking the whole to be the full collection of squares, but to interpret the diagram as representing $2\frac{1}{4}$, we must take a strip of 4 squares to be the whole. Without identifying the whole in the diagram, we cannot interpret it unambiguously.

5. Notice that four tenths is the same as forty hundredths, so four tenths and three hundredths is forty hundredths and three hundredths, or forty-three hundredths. We can express this reasoning with the following equations:

$$0.43 = \frac{4}{10} + \frac{3}{100}$$
$$= \frac{40}{100} + \frac{3}{100}$$
$$= \frac{43}{100}$$

6. Jessica's pictures show that $\frac{1}{2}$ of the two-block bar combined with $\frac{2}{3}$ of the three-block bar does indeed make up $\frac{3}{5}$ of a five-block bar. The problem is that Jessica is using three different *wholes*. The fraction $\frac{1}{2}$ refers to a two-block whole, the fraction $\frac{2}{3}$ refers to a 3-block whole, and the fraction $\frac{3}{5}$ refers to a 5-block whole. When we add fractions, such as $\frac{1}{2} + \frac{2}{3}$, both fractions in the sum and the sum itself should all refer to the same whole.

7. None of the problems can be solved by adding $\frac{1}{2} + \frac{1}{3}$.

   a. In this problem, the number of *boys* in Ms. Dock's class is the whole associated with the fraction $\frac{1}{2}$, whereas the number of *girls* in Ms. Dock's class is the whole associated with the fraction $\frac{1}{3}$. These two wholes are different. When we add any two numbers, the numbers must refer to the same wholes. There is not enough information to determine what fraction of the children in Ms. Dock's class want pizza for lunch, because we do not know if there is the same number of girls as boys in Ms. Dock's class.

   b. In this problem, the fractions $\frac{1}{2}$ and $\frac{1}{3}$ refer to the same whole, namely Jane's shirts, but some of the shirts that have pink on them may also have red on them somewhere. So, if we start with the $\frac{1}{2}$ of Jane's shirts that have red on them, and if we then want to add on the remaining shirts that have pink on them, we do not know what fraction of Jane's shirts these remaining shirts are. We can't solve this problem because we don't know what fraction of Jane's shirts have both red and pink on them.

c. In this problem, the wholes that the fraction $\frac{1}{2}$ and $\frac{1}{3}$ refer to are not the same. The fraction $\frac{1}{2}$ refers to all of Jane's shirts, whereas the fraction $\frac{1}{3}$ refers to the half of Jane's shirts that don't have red on them. We can solve this problem, however, because the shirts that have pink but not red on them are $\frac{1}{3}$ of $\frac{1}{2}$ of Jane's shirts, which is $\frac{1}{6}$ of Jane's shirts. (You can see this by drawing a diagram.) Therefore, the fraction of Jane's shirts that have either red or pink on them is

$$\frac{1}{2} + \frac{1}{6} = \frac{4}{6} = \frac{2}{3}$$

8. The problem cannot be solved by subtracting $\frac{1}{2} - \frac{1}{3}$ because the fractions $\frac{1}{2}$ and $\frac{1}{3}$ refer to different wholes. The $\frac{1}{2}$ refers to the whole pie, whereas the $\frac{1}{3}$ refers to the pie that is left over. We can solve this problem because when Pratima eats $\frac{1}{3}$ of the $\frac{1}{2}$ pie that is left over, she is eating $\frac{1}{6}$ of the pie (as you can see by drawing a diagram). So we know the following:

$$\frac{1}{2} + \frac{1}{6} = \frac{4}{6} = \frac{2}{3}$$

of the pie has been eaten. Therefore, $\frac{1}{3}$ of the pie is left.

9. The problem is with the statement "Sumin has $\frac{1}{3}$ less." Does this mean that Sumin has $\frac{1}{3}$ *cup* less juice than Liat, or does it mean that the *amount of juice* Sumin has is $\frac{1}{3}$ less than the amount of juice Liat has? It is not clear which meaning is intended. To correct the problem, replace the statement, "Sumin has $\frac{1}{3}$ less" either with "Sumin has $\frac{1}{3}$ cup less juice than Liat" or with "Sumin has $\frac{1}{3}$ less juice than Liat." In the first case, the problem is solved by calculating

$$\frac{3}{4} - \frac{1}{3} = \frac{9}{12} - \frac{4}{12} = \frac{5}{12}$$

so that Sumin has $\frac{5}{12}$ cup of juice. In the second case, the problem is solved by first calculating $\frac{1}{3}$ of $\frac{3}{4}$ cup of juice, which is $\frac{1}{4}$ cup of juice. (There are 3 equal parts in $\frac{3}{4}$, and each part is $\frac{1}{4}$.) Therefore, Sumin has $\frac{1}{4}$ cup less juice than Liat, so Sumin has $\frac{3}{4} - \frac{1}{4} = \frac{1}{2}$ cup of juice.

10. The shaded region in square 1 consists of two pieces. One piece is $\frac{1}{8}$ of the square, as shown on the left in Figure 3.25. The other piece is $\frac{1}{6}$ of the square, as shown on the right in Figure 3.25. Therefore, the shaded region in square 1 is $\frac{1}{8} + \frac{1}{6} = \frac{7}{24}$ of the square.

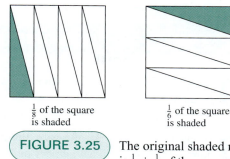

$\frac{1}{8}$ of the square is shaded $\frac{1}{6}$ of the square is shaded

FIGURE 3.25 The original shaded region is $\frac{1}{8} + \frac{1}{6}$ of the square.

We can think of the shaded region in square 2 as the difference between the shaded region on the left in Figure 3.26 and the shaded region on the right in Figure 3.26. Therefore, the shaded region in square 2 is $\frac{1}{4} - \frac{1}{12} = \frac{1}{6}$ of the square.

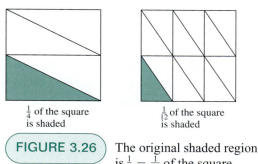

$\frac{1}{4}$ of the square is shaded $\frac{1}{12}$ of the square is shaded

FIGURE 3.26 The original shaded region is $\frac{1}{4} - \frac{1}{12}$ of the square.

11. See Figure 3.27.

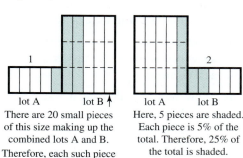

1   2

lot A   lot B   lot A   lot B

There are 20 small pieces of this size making up the combined lots A and B. Therefore, each such piece is $\frac{1}{20}$, which is 5%. Since 7 are shaded, that means 35% of the combined amount is shaded.

Here, 5 pieces are shaded. Each piece is 5% of the total. Therefore, 25% of the total is shaded.

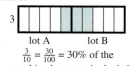

3

lot A   lot B

$\frac{3}{10} = \frac{30}{100} = 30\%$ of the combined amount is shaded.

FIGURE 3.27 Various percents of the combined amount of lots A and B are shaded.

Problems for Section 3.4

1. Using the example $\frac{2}{3} + \frac{3}{4}$, explain why we add fractions the way we do. In other words, what is the logic behind this procedure? Explain your answer.

2. Using the example $3\frac{5}{6}$, describe the procedure for turning a mixed number into an improper fraction, and explain in your own words why this procedure makes sense. What is the logic behind this procedure? Draw pictures to support your explanation.

3. Use two number lines, one labeled with (fractions and) mixed numbers, the other labeled with (proper and) improper fractions, to help you explain why the procedure for turning mixed numbers into improper fractions described on page 126 makes sense.

4. **a.** For each of the following decimals, show how to write the decimal as a fraction by first putting the decimal in expanded form:

   i. 2.34

   ii. 124.5

   iii. 7.938

   b. Based on your results in part (a), describe a quick way to rewrite a finite decimal as a fraction. Illustrate with the example 2748.963.

5. **a.** Show how to calculate the sum $\frac{2}{5} + .25$ and show how to write the answer as a fraction and as a decimal.

   b. Show how to calculate the sum $3.8 + \frac{3}{8}$ and show how to write the answer as a fraction and as a decimal.

   c. Discuss briefly what kinds of errors you think students who are just learning about fraction and decimal addition might make with the problems in parts (a) and (b).

6. Show how to calculate $5\frac{3}{4} + 1\frac{2}{3}$ in two different ways. In each case, express your answer as a mixed number. Explain why both of your methods are legitimate.

7. Show how to calculate $4\frac{2}{3} - 1\frac{3}{4}$ in two different ways. In each case, express your answer as a mixed number. Explain why both of your methods are legitimate.

8. Find two *different*, positive fractions whose sum is $\frac{1}{11}$.

9. John says $\frac{2}{3} + \frac{2}{3} = \frac{4}{6}$ and uses the picture in Figure 3.28 as evidence. Discuss what is wrong with John's reasoning. What underlying misconception does John have? *Don't* just explain how to do the problem correctly; explain why John's reasoning is faulty.

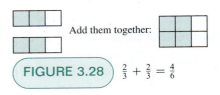

FIGURE 3.28 $\frac{2}{3} + \frac{2}{3} = \frac{4}{6}$

10. Denise says that $\frac{2}{3} - \frac{1}{2} = \frac{1}{3}$ and gives the reasoning indicated in Figure 3.29 to support her answer. Is Denise right? If not, what is wrong with her reasoning and how could you help her understand her mistake and fix it? *Don't* just explain how to solve the problem correctly; explain where Denise's reasoning is flawed.

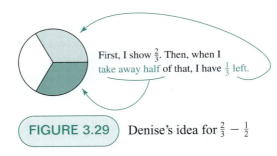

First, I show $\frac{2}{3}$. Then, when I take away half of that, I have $\frac{1}{3}$ left.

FIGURE 3.29   Denise's idea for $\frac{2}{3} - \frac{1}{2}$

11. Arnold says that $2\frac{2}{3} = \frac{4}{5}$, and he uses the picture in Figure 3.30 to support his conclusion. What is wrong with Arnold's reasoning? Do not just state the correct way to convert $2\frac{2}{3}$; explain why Arnold's reasoning is not valid.

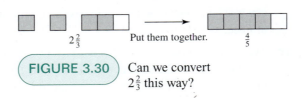

$2\frac{2}{3}$ Put them together. $\frac{4}{5}$

FIGURE 3.30   Can we convert $2\frac{2}{3}$ this way?

12. Can the following story problems about voters be solved by adding $\frac{1}{2} + \frac{1}{3}$? If so, explain why. If not, explain why not. Solve the problems if

they can be solved. Write a different story problem about voters that can be solved by adding $\frac{1}{2} + \frac{1}{3}$.

Story problem 1 about voters: In Kneebend County, $\frac{1}{2}$ of the female voters and $\frac{1}{3}$ of the male voters voted for a certain referendum. Altogether, what fraction of the voters voted for the referendum in Kneebend County?

Story problem 2 about voters: In Kneebend County, $\frac{1}{2}$ of the female voters and $\frac{1}{3}$ of the male voters voted for a certain referendum. There are the same number of women voters as men voters in Kneebend County. Altogether, what fraction of the voters voted for the referendum in Kneebend County?

13. Can the following story problems about a bird feeder be solved by subtracting $\frac{3}{4} - \frac{1}{2}$? If so, explain why. If not, explain why not. Solve the problems if they can be solved. Write a different story problem about a bird feeder that can be solved by subtracting $\frac{3}{4} - \frac{1}{2}$.

Story problem 1 about a bird feeder: A bird feeder was filled with $\frac{3}{4}$ of a full bag of bird seed. The birds ate $\frac{1}{2}$ of what was in the bird feeder. What fraction of a full bag of bird seed did the birds eat?

Story problem 2 about a bird feeder: A bird feeder was filled with $\frac{3}{4}$ of a full bag of bird seed. The birds ate $\frac{1}{2}$ of what was in the bird feeder. What fraction of a full bag of bird seed is left in the bird feeder?

14. Can the following story problems about Sarah's bead collection be solved by adding $\frac{1}{4} + \frac{1}{5}$? If so, explain why. If not, explain why not. Solve the problems if they can be solved. Write a different story problem about Sarah's bead collection that can be solved by adding $\frac{1}{4} + \frac{1}{5}$.

Story problem 1 about Sarah's bead collection: One-fourth of the beads in Sarah's collection are pink. One-fifth of the beads in Sarah's collection are oblong. What fraction of the beads in Sarah's collection are either pink or oblong?

Story problem 2 about Sarah's bead collection: One-fourth of the beads in Sarah's collection are pink. One-fifth of the beads in Sarah's collection that are not pink are oblong. What fraction of the beads in Sarah's collection are either pink or oblong?

15. Can the following story problem about Jim's medicine be solved by subtracting $\frac{1}{3} - \frac{1}{4}$? If so, explain why. If not, explain why not and solve the problem if it can be solved. Write a different story problem that can be solved by subtracting $\frac{1}{3} - \frac{1}{4}$.

Story problem about Jim's medicine: Jim has a small container filled with one tablespoon of medicine. Jim poured out $\frac{1}{3}$ tablespoon of medicine. Then Jim poured out $\frac{1}{4}$ tablespoon of medicine. Now how much medicine is left in the container?

16. **a.** Write and solve a story problem for $\frac{3}{4} + \frac{2}{3}$.

b. Write and solve a story problem for $\frac{3}{4} - \frac{2}{3}$.

17. **a.** Write and solve a story problem for $2\frac{1}{2} + 1\frac{1}{3}$.

b. Write and solve a story problem for $2\frac{1}{2} - 1\frac{1}{3}$.

18. **a.** Write and solve a compare story problem for $\frac{2}{3} + ? = \frac{3}{4}$ (or $\frac{3}{4} - \frac{2}{3} = ?$).

b. Write and solve a compare story problem for $? - 1\frac{1}{2} = \frac{2}{3}$.

19. **a.** Write and solve a part-part-whole story problem for $\frac{1}{2} + ? = \frac{2}{3}$.

b. Write and solve a part-part-whole story problem for $\frac{1}{2} + \frac{2}{3} = ?$.

20. For each square in Figure 3.31, determine the fraction of the square that is shaded. Explain your reasoning. You may assume that all lengths that appear to be equal really are equal. Do not use any area formulas.

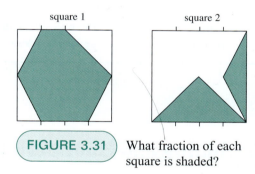

square 1   square 2

FIGURE 3.31   What fraction of each square is shaded?

21. For each square in Figure 3.32, determine the fraction of the square that is shaded. Explain your reasoning. You may assume that all lengths that appear to be equal really are equal. Do not use any area formulas.

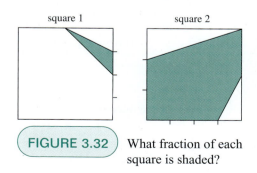

square 1   square 2

FIGURE 3.32   What fraction of each square is shaded?

22. Broad Street divides Popperville into an east side and a west side. On the east side of Popperville, 20% of the children qualify for a reduced-price lunch. On the west side of Popperville, 30% of the children qualify for a reduced-price lunch. Is it correct to calculate the percentage of children in all of Popperville who qualify for reduced-price lunch by adding 20% and 30% to get 50%? If the answer is no, why not? Explain in detail and calculate the correct percentage in at least two examples.

23. Anklescratch County and Kneebend County are two adjacent counties. In Anklescratch County, 30% of the miles of road have bike lanes, whereas in Kneebend County, only 10% of the miles of road have bike lanes.

   a. Elliot says that in the two-county area, 40% of the miles of road have bike lanes. Elliot found 40% by adding 30% and 10%. Is Elliot correct or not? Explain.

   b. What percent of miles of road in the two-county area have bike lanes if Anklescratch County has 200 miles of road and Kneebend County has 800 miles of road? What if it's the other way around and Anklescratch County has 800 miles of road and Kneebend County has 200 miles of road?

   c. Ming says that 20% of the miles of road in the two-county area have bike lanes because 20% is the average of 10% and 30%. Explain why Ming's answer could be either correct or incorrect. Under what circumstances will Ming's answer be correct?

24. There are two elementary schools in the town of South Elbow. In the first school, 20% of the children are Hispanic. In the second school, 30% of the children are Hispanic. Write a brief essay about what you can and cannot tell from these data alone. Include examples to support your points.

25.  Suppose you start with a fraction and you add 1 to both the numerator and the denominator. For example, if you started with $\frac{2}{3}$, then you'd get a new fraction $\frac{2+1}{3+1} = \frac{3}{4}$. Is this procedure of adding 1 to the numerator and the denominator the same as adding the number 1 to the original fraction? (For example, is $\frac{2+1}{3+1}$ equal to $\frac{2}{3} + 1$?) Explain your answer.

26. In the first part of the season, the Bluejays play 18 games and win 7, while the Robins play 3 games and win 1. In the second part of the season, the Bluejays play 2 games and win 1, while the Robins play 17 games and win 8.

   a. Which team won a larger fraction of its games in the first part of the season?

   b. Which team won a larger fraction of its games in the second part of the season?

3.5 Adding and Subtracting Negative Numbers

F·P **Focal Points**
Grade 7

Class Activity   *Now Turn to Class Activities Manual*

3BB Story Problems and Rules for Adding and Subtracting with Negative Numbers, p. 63

How do we interpret the meaning of sums and differences such as

$$2 + (-3)$$

and

$$1 - 5?$$

How can we "take away" 5 objects if we only have 1 object? Many addition and subtraction problems involving negative numbers can be interpreted in terms of amounts owed, temperatures below zero, locations below ground (such as in a mine or a building), or locations below sea level. In general, how should we interpret $A + (-B)$ and $A - (-B)$? We will see why it makes sense to interpret

$$A + (-B) \quad \text{as} \quad A - B$$

and

$$A - (-B) \quad \text{as} \quad A + B$$

To do so, we first consider sums such as $(-3) + 3$ and $(-7) + 7$.

Adding a Number to Its Negative

How can we interpret $(-3) + 3$? A story problem for $(-3) + 3 = ?$ follows:

It was $-3°$ C at dawn. In the meantime, the temperature went up $3°$ C. Now what is the temperature? Since $-3°$ C means 3 degrees C below 0, the new temperature must be $0°$ C, as we see in Figure 3.33, so $(-3) + 3 = 0$. In general,

$$(-N) + N = 0$$

for any number N. This equation states that $-N$ and N are **additive inverses**—in other words, that $-N$ and N add up to 0.

Interpreting $A + (-B)$ as $A - B$

How can we interpret the sum $5 + (-3)$?

You may know the rule that $5 + (-3) = 5 - 3$, but where does this rule come from and why does it make sense? We can deduce this rule from the associative property of addition and the link between subtraction and addition.

To explain why the sum $5 + (-3)$ must be equal to $5 - 3$, we will first evaluate the 3-term sum $5 + (-3) + 3$ in two different ways by applying the associative property of addition. Then we will apply the link between subtraction and addition.

So consider the sum

$$5 + (-3) + 3 \tag{3.7}$$

According to the associative property of addition, we can associate the -3 in Equation 3.7 with either the 3 or with the 5 and we will get the same end result either way. When we associate the -3 with the 3 we get this:

$$5 + \underbrace{(-3) + 3}_{0} = 5 + 0 = 5$$

because $(-3) + 3 = 0$, as previously discussed. On the other hand, when we associate the -3 with 5 and represent the unknown $5 + (-3)$ as ? we get this:

$$\underbrace{5 + (-3)}_{?} + 3 = ? + 3$$

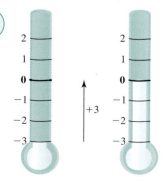

FIGURE 3.33

Using an increase in temperature to explain why $(-3) + 3 = 0$

Whichever way we evaluate $5 + (-3) + 3$ we must get the same answer. Therefore

$$? + 3 = 5 \qquad\qquad (3.8)$$

But according to the link between addition and subtraction, Equation 3.8 is equivalent to the subtraction equation

$$5 - 3 = ?$$

Since we are using ? to stand for $5 + (-3)$, we conclude that

$$5 + (-3) = 5 - 3$$

The same reasoning applies with other numbers; so, in general, it makes sense to interpret

$$A + (-B) \quad \text{as} \quad A - B$$

Interpreting $A - (-B)$ as $A + B$

How do we interpret a difference such as $5 - (-3)$? A compare story problem for $5 - (-3) = ?$ is as follows:

The temperature was 5° C in West Lafayette. At the same time it was −3° C in Indianapolis. How much warmer was it in West Lafayette than in Indianapolis?

Since −3° C means 3 degrees C below 0, it is $3 + 5 = 8$ degrees warmer in West Lafayette than in Indianapolis, as we see in Figure 3.34. So

$$5 - (-3) = 5 + 3$$

The same reasoning applies with other numbers, so, in general, we should interpret

$$A - (-B) \quad \text{as} \quad A + B$$

Extending Addition and Subtraction on Number Lines to Negative Numbers

In Section 3.1 we saw how to represent addition and subtraction on number lines when only positive numbers are involved. How should we interpret addition and subtraction on a number line when negative numbers are involved? For example, where should $5 + (-3)$ be on the number line?

Since

$$5 + (-3) = 5 - 3$$

and in general, since

$$A + (-B) = A - B$$

when we use a number line to add a negative number, we should move to the *left* instead of to the right. Similarly, where should $5 - (-3)$ be on the number line? Since

$$5 - (-3) = 5 + 3$$

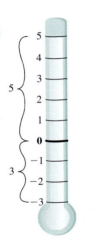

FIGURE 3.34

Using a comparison of temperatures to explain why $5 - (-3) = 5 + 3$

FIGURE 3.35

Using a number
line to show that
$2 + (-3) = -1$

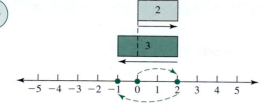

and in general, since

$$A - (-B) = A + B$$

when we use a number line to subtract a negative number, we should move to the *right* instead of to the left.

These discussions indicate that we should interpret addition and subtraction on number lines in the following ways:

If A and B are any two numbers, then they are represented by points on a number line. The sum

$$A + B$$

corresponds to the point on the number line that is located as follows:

**number line
meaning of
addition**

1. Go to A on the number line.
2. Move a distance equal to the number of units that B is away from 0.

   - Move to the right if B is positive.
   - Move to the left if B is negative.

3. The resulting point is the location of $A + B$.

Where is $2 + (-3)$ on a number line? As shown in Figure 3.35, go to 2 and move 3 units to the left. Move left because -3 is negative. You end up at -1; so,

$$2 + (-3) = -1$$

If A and B are any two numbers, the difference

$$A - B$$

corresponds to the point on the number line that is located as follows:

**number line
meaning of
subtraction**

1. Start at A on the number line.
2. Move a distance equal to the number of units that B is away from 0.

   - Move to the *left* if B is positive.
   - Move to the *right* if B is negative.

3. The resulting point is the location of $A - B$.

Where is $(-2) - (-5)$ on a number line? As shown in Figure 3.36, go to -2 and move 5 units to the *right*. Move right because -5 is negative and we are subtracting. You end up at 3; so, $(-2) - (-5) = 3$.

Although the combining and taking away interpretations of addition and subtraction fit with our common sense and intuition, they do not work well with certain numbers. The number line interpretation of addition and subtraction, although more abstract, applies to *all* numbers.

FIGURE 3.36

Using a number
line to show that
$(-2) - (-5) = 3$

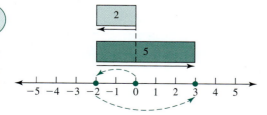

Practice Exercises for Section 3.5

1. Write a compare story problem that fits naturally with the equation

$$(-3) - 7 = ?$$

and in which one quantity is -3 and the other quantity unknown and is 7 less than -3.

2. Write a compare story problem that fits naturally with the equation

$$4 - (-3) = ?$$

and in which one quantity is 4, the other quantity is -3, and the difference between the two quantities

is unknown. Then solve the story problem and use your solution to explain why it makes sense that subtracting -3 from 4 is equivalent to adding 3 to 4, or in other words why it makes sense that

$$4 - (-3) = 4 + 3$$

3. Use a number line to calculate $0 - (-2)$. Briefly explain the method.

4. Use a number line to calculate $(-3) - (-3)$. Briefly explain the method.

Answers to Practice Exercises for Section 3.5

1. Problem: The temperature was $-3°$ C in Buffalo. At the same time, it was $7°$ C colder in Syracuse. What was the temperature in Syracuse?

As we see in Figure 3.37, the temperature that is 7 degrees C colder than $-3°$ C is $-10°$ C.

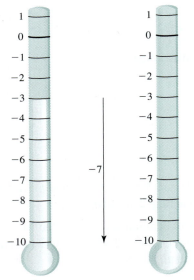

FIGURE 3.37   Seven degrees colder than $-3°$ C is $-10°$ C, so $(-3) - 7 = -10$.

2. Story problem: Mary is on floor 4 of a building while Jonathan is on floor -3 of the same building. How many floors does Jonathan have to go up to get to Mary's floor?

To get from Jonathan's floor to Mary's floor you first neet to go up 3 floors to get to floor 0, and then you need to go up another 4 floors to get to Mary's floor. Altogether, that means going up $4 + 3$ floors. So based on this solution, we see that subtracting -3 from 4 is equivalent to adding 3 to 4, or in other words, that

$$4 - (-3) = 4 + 3$$

3. See Figure 3.38. To find the location of $0 - (-2)$, go to 0 first. Since we are subtracting, we might think we will move to the left, but since -2 is a negative number, we move 2 units to the *right*, ending at 2.

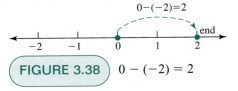

FIGURE 3.38 $0 - (-2) = 2$

4. See Figure 3.39. To find the location of $(-3) - (-3)$, go to -3 first. Since we are subtracting, we might think we will move to the left, but since -3 is a negative number, we move 3 units to the *right*, ending at 0.

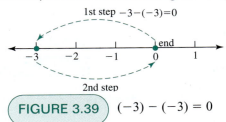

FIGURE 3.39 $(-3) - (-3) = 0$

Problems for Section 3.5

1. a. Write an add to story problem that fits naturally with the equation

$$(-4) + 3 = ?$$

Solve the problem, and explain why the solution makes sense.

b. Write a take away story problem that fits naturally with the equation

$$(-4) - 3 = ?$$

Solve the problem, and explain why the solution makes sense.

2. **a.** Write a compare story problem that fits naturally with the equation

$$(-7) + ? = -1$$

(or $(-1) - (-7) = ?$). Solve the problem, and explain why the solution makes sense.

b. Write a compare story problem that fits naturally with the equation

$$(-7) + 6 = ?$$

(or $? - (-7) = 6$). Solve the problem, and explain why the solution makes sense.

3. **a.** Show how to use a number line to calculate $-1 + (-2)$.

b. Show how to use a number line to calculate $1 - (-2)$.

c. Show how to use a number line to calculate $-1 - 2$.

d. Show how to use a number line to calculate $-1 - (-2)$.

4. a. For each of these four expressions,

$$-2 + 6, \quad -(2 + 6), \quad -2 - 6, \quad -(2 - 6)$$

draw a number line (with a consistent scale) and plot the expression on the number line. In each case, show clearly why the expression is located where it is based on the rules for plotting sums, differences, and negatives of numbers on number lines.

b. Based on your work in part (a), compare and contrast the four expressions in part (a).

Chapter Summary and Study Items

Section 3.1 Interpretations of Addition and Subtraction

The most basic way to interpret addition is as combining or adding to, and the most basic way to interpret subtraction is as taking away. But we also use addition and subtraction to solve problems that arise in part-part-whole situations and in situations where we compare two quantities. If we view addition and subtraction in terms of movement along a number line, we can make sense of addition and subtraction when any numbers are involved.

Key skills and understandings:

1. Write add to, take away, part-part-whole, and compare story problems to go along with an addition or subtraction equation.
2. Explain how to use a number line to add and subtract numbers (for nonnegative numbers only).

Section 3.2 The Commutative and Associative Properties of Addition, Mental Math, and Single-Digit Facts

The associative property of addition says that, for all real numbers A, B, and C,

$$(A + B) + C = A + (B + C)$$

The commutative property of addition says that, for all real numbers A and B,

$$A + B = B + A$$

We can use these properties to help children learn the basic addition facts. The commutative property allows children to cut the number of facts that must be memorized almost in half. The associative property is used in the important make-a-ten strategy. The commutative and associative properties of addition can also help make some mental addition problems easier to carry out.

Key skills and understandings:

1. State and describe the associative property of addition, give examples to show how to use it to make problems easier to do mentally, including the make-a-ten strategy.

2. State and describe the commutative property of addition, give examples to show how to use it to make problems easier to do mentally, including the "count on from the larger addend" methods, and explain how it helps children cut down on the memorization of basic addition facts.

3. Describe how to view subtraction problems as unknown addend problems, explain how this can help young children with basic subtraction facts, and explain how this can be applied to other mental math problems.

4. Identify where the commutative or associative properties of addition have been used in calculations.

5. Write correct equations to go along with a mental method of addition or subtraction.

6. Explain how to add or subtract by using methods other than the common addition and subtraction algorithms.

Section 3.3 Why the Common Algorithms for Adding and Subtracting Decimal Numbers Work

The common addition and subtraction algorithms work in terms of objects grouped in ones, tens, hundreds, and so on. Regrouping occurs when we make or break bundles.

Key skills and understandings:

1. Explain the addition and subtraction algorithms in terms of bundled objects, paying special attention to regrouping.

2. Explain why we must line up decimal points when adding decimals.

Section 3.4 Adding and Subtracting Fractions

To add or subtract fractions, we first give the fractions common denominators. We give fractions common denominators so as to work with like parts. Once we have like parts, we can simply add or subtract the number of parts. A mixed number stands for the sum of a whole number and a fraction; when we add the whole number part to the fraction part, the result is an improper fraction. When we work with fraction addition and subtraction story problems, we must pay close attention to underlying wholes and to the wording of the problem.

Key skills and understandings:

1. Describe how to add or subtract fractions, and explain why the process makes sense. In particular, explain why we give the fractions common denominators.

2. Use pictures and number lines to help explain the logic behind the procedure of turning a mixed number into an improper fraction.

3. Write fraction addition and subtraction story problems.

4. Determine if a given story problem fits a given fraction addition or subtraction problem.

Section 3.5 Adding and Subtracting Negative Numbers

We can use story problems—as well as properties of addition and the link between addition and subtraction—to make sense of why the rules for adding and subtracting with negative numbers make sense. Using these rules, we can extend addition and subtraction on number lines to the case of negative numbers.

Key skills and understandings:

1. Write and solve story problems involving addition and subtraction with negative numbers.
2. Explain how to use a number line to add and subtract numbers, including negative numbers.

Multiplication

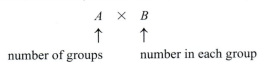

In this chapter, we will study the meaning of multiplication, ways of representing multiplication, the properties of multiplication, and the procedures we use to multiply whole numbers. Because we use multiplication so often, we may take the process for granted, and we may think that it is easy. However, even though it is easy for adults to carry out the procedure of multiplying, the underlying concept of multiplication is much more subtle. As we will see, the procedures we use when we multiply are based on very clever uses of the properties of arithmetic.

4.1 Interpretations of Multiplication

F·P **Focal Points**
Grade 3

How do you think about multiplication? In this section we give a definition of multiplication and we examine some categories of problems that can be solved with multiplication.

The Meaning of Multiplication

basic meaning of multiplication

What does multiplication mean? If A and B are nonnegative numbers, then

$$A \times B \quad \text{or} \quad A \cdot B$$

which we read as "A times B," means the total number of objects in A groups if there are B objects in *each* group. Shorthand: $A \times B$ means the total in A groups of B. The result

product
factors

of $A \times B$ is called the **product** of A and B, and the numbers A and B are called **factors**. Notice that according to this definition, the first factor, A, stands for the number of groups, and the second factor, B stands for the number of objects in each group.

$$A \quad \times \quad B$$
$$\uparrow \qquad \uparrow$$

number of groups   number in each group

FIGURE 4.1

Grouped objects
showing
multiplicative
structure

These interpretations of the first and second factors in a product are conventions (which are sometimes different in other countries). For consistency and clear communication, we will stick to this convention for interpreting the first and second factors in a product.

Given a whole number A, a whole number B that can be written as a whole number times A is called a

multiple   **multiple** of A. For example, 15 is a multiple of 5 because $15 = 3 \times 5$, and the numbers

$$4, 8, 12, 16, 20, 24, \ldots$$

are multiples of 4.

How to Tell If a Problem Is Solved by Multiplication

If you have a story problem, how can you tell if it is solved by multiplication instead of in some other way, such as by addition or division? This question is an important one for teachers, because students just guess at which operation to use to solve a problem if they don't understand what addition, subtraction, multiplication, and division mean. Multiplication applies to situations that involve equal groups. In some cases, the groups are evident. For example, Figure 4.1 clearly shows 3 groups with 4 dots in each group. So according to the meaning of multiplication, there are 3×4 dots in all. Similarly, if there are 237 bags and each bag contains 46 potatoes, then the total number of potatoes is 237×46 (which turns out to be 10,902 potatoes.) Whenever a collection of objects is arranged into A groups, and there are B objects in each group, then we know that, according to the meaning of multiplication, there are $A \times B$ objects in all.

Next, we will examine types of story problems that can be solved with multiplication. In some cases the groups in these story problems are not obvious, and we must first figure out what the groups are in order to determine that the problem can be solved by multiplication.

Class Activity   *Now Turn to Class Activities Manual*

4A Showing Multiplicative Structure, p. 65

Array Problems   In some problems, we want to determine the total number of objects that are in an array.

Array problem: How many soft drink cans are in the case shown in Figure 4.2?

array   A two-dimensional **array** is a rectangular arrangement of things into (horizontal) rows and (vertical) columns, such that each row has the same number of things and each column has the same number of things.

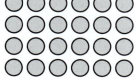

FIGURE 4.2

An array of soft
drink cans

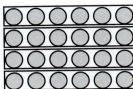

FIGURE 4.3

Subdividing an
array of soft
drink cans into
groups

View each row as a group.
4 groups of 6
4 × 6 total

View each column as a group.
6 groups of 4
6 × 4 total

We can multiply to determine how many cans are depicted in the array in Figure 4.2. Why? View each *row* in the array as a group, as on the left in Figure 4.3. With this view, the array consists of 4 groups, and there are 6 cans in each group. Therefore, according to the meaning of multiplication, the total number of cans in the array is 4 × 6.

We can also view each *column* in the array of cans as a group, as shown on the right in Figure 4.3. With this perspective, the array consists of 6 groups, and there are 4 cans in each group. Therefore, according to the meaning of multiplication, the total number of cans in the array is 6 × 4 (which is the reverse of the previous 4 × 6).

Arrays are especially important because they can be divided into natural groups in two different ways. The ability to see multiplication problems in two different ways is very helpful to children as they learn single-digit multiplication facts.

Ordered Pair Problems   In some problems, we want to determine how many ordered pairs of things can **ordered pair**   be made. A pair of things (two things) is an **ordered pair** if one of the things is designated as first and the other is designated second.

> *Ordered pair problem:* A restaurant serves cheese sandwiches that are made from a piece of bread and a piece of cheese. There are 3 types of bread to choose from: wheat, white, and rye, and there are 4 types of cheese to choose from: cheddar, provolone, Swiss, and American. How many types of cheese sandwiches can the restaurant make with these choices?

We can multiply to determine how many types of cheese sandwiches the restaurant can make. Why? Each type of cheese sandwich can be considered an ordered pair consisting of a type of bread and a type of cheese. (Note that the type of bread was designated first and the type of cheese second, although it could just as well be the other way around as long as consistency is maintained.) These pairs are organized in an array in a natural way, as in Figure 4.4, where each row of the array shows all the different types of cheese sandwiches that can be made with one particular kind of bread. Viewing each row as a group, there are 3 groups, and there are 4 pairs in each group, so according to the meaning of multiplication, there are 3 × 4 pairs in all. Therefore, there are 3 × 4 different types of cheese sandwich that the restaurant can make.

Instead of organizing the types of cheese sandwiches into an array, they can also be organized into a list, as on the right in Figure 4.5, or into a tree diagram, as on the left in Figure 4.5.

FIGURE 4.4

Types of cheese
sandwiches
organized in an
array

(white, cheddar)	(white, provolone)	(white, Swiss)	(white, American)
(wheat, cheddar)	(wheat, provolone)	(wheat, Swiss)	(wheat, American)
(rye, cheddar)	(rye, provolone)	(rye, Swiss)	(rye, American)

FIGURE 4.5

Types of cheese
sandwiches
organized in a
tree diagram
and an
organized
list

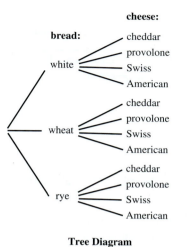

bread:

cheese:

white — cheddar, provolone, Swiss, American

wheat — cheddar, provolone, Swiss, American

rye — cheddar, provolone, Swiss, American

Tree Diagram

bread, cheese:

white, cheddar
white, provolone
white, Swiss
white, American

wheat, cheddar
wheat, provolone
wheat, Swiss
wheat, American

rye, cheddar
rye, provolone
rye, Swiss
rye, American

Organized List

tree diagram   A **tree diagram** is a diagram consisting of line segments, called branches, that connect pieces of information. To read a tree diagram, start at the far left and follow branches all the way across to the right.

The list and the tree diagram in Figure 4.5 each have a structure consisting of 3 groups of 4, as indicated in Figure 4.6. Therefore, as before, we conclude that the total number of types of cheese sandwiches is 3 × 4.

Class Activity   *Now Turn to Class Activities Manual*

4B Problems about Pairs, p. 66

FIGURE 4.6

Showing equal
groups in a
tree diagram
and an
organized list

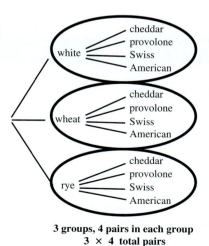

white — cheddar, provolone, Swiss, American

wheat — cheddar, provolone, Swiss, American

rye — cheddar, provolone, Swiss, American

white, cheddar
white, provolone
white, Swiss
white, American

wheat, cheddar
wheat, provolone
wheat, Swiss
wheat, American

rye, cheddar
rye, provolone
rye, Swiss
rye, American

3 groups, 4 pairs in each group
3 × 4 total pairs

3 groups, 4 pairs in each group
3 × 4 total pairs

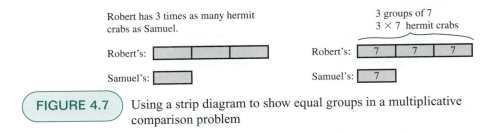

FIGURE 4.7   Using a strip diagram to show equal groups in a multiplicative comparison problem

Multiplicative Comparison Problems Some problems describe a multiplicative relationship between two quantities.

> *Multiplicative comparison problem:* Robert and Samuel each have a hermit crab collection. Robert has 3 times as many hermit crabs as Samuel. Samuel has 7 hermit crabs. How many hermit crabs does Robert have?

The wording of this problem indicates that it involves multiplication, but can we use equal groups to describe the situation in the problem? Yes. In Figure 4.7 there is a simple strip diagram depicting Robert's hermit crab collection as 3 equal groups, each of the same size as Samuel's collection of 7 hermit crabs. Therefore, according to the meaning of multiplication, Robert has 3×7 hermit crabs.

Strip diagrams can be especially helpful for multiplicative comparison problems because the "N times as many as" language of these problems is often difficult for students to grasp. By drawing a strip diagram, students think about size comparisons and the meaning of the phrase "N times as many as." Importantly, strip diagrams are simple and easy to draw, and because of their simplicity, they highlight the relationship between quantities.

Strip diagrams are math drawings that are used to aid mathematical reasoning. In contrast, if a student were to make an artistic drawing of individual hermit crabs, the drawing would take longer to do, the student might lose track of the relationship between the quantities, and the student might become distracted by drawing features that aren't related to solving the math problem, such as claws, eyes, or shell details on the hermit crabs.

Notice that we can view multiplicative comparison situations in terms of fractions. In the problem about Robert's and Samuel's hermit crabs, we can say that Samuel has $\frac{1}{3}$ as many hermit crabs as Robert, which we can see from the strip diagram Figure 4.7.

Class Activity   *Now Turn to Class Activities Manual*

4C Writing Multiplication Story Problems, p. 67

In this section, we examined some types of story problems that can be solved with multiplication. Problems about areas of rectangles and volumes of boxes (rectangular prisms) are also important types of problems that can be solved with multiplication; we will study these types of problems in Section 4.3. But first, we study the special role that multiplication by 10 plays in the decimal system.

Practice Exercises for Section 4.1

1. Write your own array, ordered pair, and multiplicative comparison multiplication story problems. In each case, explain why the problem can be solved with multiplication by applying the meaning of multiplication given in this section.

2. For each of the following problems, explain why the problem can be solved by multiplication according to the meaning of multiplication given in this section.

   a. If apples cost $2 per pound, how much will 5 pounds of apples cost?

   b. There are 1000 milliliters in a liter. How many milliliters are in 4 liters?

   c. If you drive a steady 120 kilometers per hour for 2 hours, how far will you have driven?

3. Explain the difference among the next three problems and the way they are solved.

   a. Your laundry basket contains 4 plain socks: a red one, a blue one, a yellow one, and a green one. The basket also contains 4 striped socks: a red striped one, a blue striped one, a yellow striped one, and a green striped one. If you want to wear a plain sock on your left foot and a striped sock on your right foot, how many options do you have?

   b. Your laundry basket contains 4 plain socks: a red one, a blue one, a yellow one, and a green one. The basket also contains 4 striped socks: a red striped one, a blue striped one, a yellow striped one, and a green striped one. If you want to pick out a pair of socks consisting of one plain sock and one striped sock, and then wear that pair of socks how many options do you have?

   c. Your laundry basket contains 4 plain socks: a red one, a blue one, a yellow one, and a green one. The basket also contains 4 striped socks: a red striped one, a blue striped one, a yellow striped one, and a green striped one. If you reach into the laundry basket, pick out a sock, and put it on your left foot and then reach in again, pick out another sock, and put it on your right foot, how many different possible outcomes could there be?

Answers to Practice Exercises for Section 4.1

1. See the text for examples of story problems and for explanations for why the problems can be solved with multiplication. Be sure to use correct language in your multiplicative comparison problems. Problems should include a phrase of the form "*N* times as many as," although the word *many* can be replaced with *much, long, tall, wide,* or similar words. Be sure also in your explanation for why your problem can be solved by multiplying $A \times B$, that you explain how to see A groups with B objects in each group.

2. a. View each pound as a group. Then 5 pounds is 5 groups. Each group contains $2 worth of apples. So the total cost of the apples is the total dollar value of 5 groups of $2, which is 5×2 dollars, according to the meaning of multiplication as described in this section.

   b. View each of the 4 liters as a group. Each 1-liter group contains 1000 milliliters. Four liters can therefore be viewed as 4 groups, each of which contains 1000 milliliters. So according to the meaning of the multiplication as described in this section, the total number of milliliters in 4 liters is 4×1000.

   c. View each hour as a group. During each 1-hour time period you travel 120 kilometers. So each 1-hour group "consists of" 120 kilometers. So during 2 hours, you travel a total of "2 groups of 120 kilometers," which is 2×120 kilometers.

3. The discussion here is only a brief outline. Problem 6 asks you to solve the three problems in detail.

Part (a) is an example of an ordered pair problem. The information can be organized into an array, a tree diagram, or an ordered list, each of which has the structure of 4 groups of 4. So there are $4 \times 4 = 16$ options.

To solve part (b), multiply the solution for part (a) by 2. So there are $2 \times 16 = 32$ options.

For part (c), the possible outcomes can be organized into 8 groups of 7, so there are $8 \times 7 = 56$ possible outcomes.

Problems for Section 4.1

1. Use the meaning of multiplication to explain why each of the following problems can be solved by multiplying:

  a. There are 3 feet in a yard. If a rug is 5 yards long, how long is it in feet?

  b. There are 5280 feet in a mile. How long in feet is a 4-mile-long stretch of road?

  c. Will is driving 65 miles per hour. If he continues driving at that speed, how far will he drive in 3 hours?

2. Write an array story problem that can be solved by multiplying 6×8. Explain clearly why the problem can be solved by multiplying 6×8 by using the meaning of multiplication as described in this section. (You may replace the numbers 6 and 8 with different numbers if the context of your story problem fits better with different numbers.)

3. Write an ordered pair story problem that can be solved by multiplying 6×8. Explain clearly why the problem can be solved by multiplying 6×8 by using the meaning of multiplication as described in this section.

4. Write a multiplicative comparison story problem that can be solved by multiplying 6×8. Explain clearly why the problem can be solved by multiplying 6×8 by using the meaning of multiplication as described in this section.

5. a. Write a multiplicative comparison story problem that can be solved by multiplying 3×5.

  b. Draw a strip diagram for your problem in part (a) and explain how the meaning of multiplication applies to solve the problem.

  c. Reword your problem in part (a) so that it is the same problem but you use a fraction in the statement of the problem.

6. Solve the three problems that are given in Practice Exercise 3: parts (a), (b), (c). Explain the solutions in detail.

7. John, Trey, and Miles want to know how many two-letter secret codes there are that don't have a repeated letter. For example, they want to count BA and AB, but they don't want to count doubles such as ZZ or XX. John says there are $26 + 25$ because you don't want to use the same letter twice; that's why the second number is 25. Trey says he thinks it should be times, not plus: 26×25. Miles says the number is $26 \times 26 - 26$ because you need to take away the double letters. Discuss the boys' ideas. Which answers are correct, which are not, and why? Explain your answers clearly and thoroughly, drawing on the meaning of multiplication.

8. a. A 40-member club will elect a president and then elect a vice president. How many possible outcomes are there?

  b. A 40-member club will elect a pair of copresidents. How many possible outcomes are there?

  c. Are the answers to parts (a) and (b) the same or different? Explain why they are the same or why they are different.

4.2 Why Multiplying Numbers by 10 Is Easy in the Decimal System

Focal Points
Grade 4

Why is multiplying by 10, by 100, by 1000, and so on, so easy in the decimal system? The answer lies in how the values of the places in the decimal system are related.

Class Activity   *Now Turn to Class Activities Manual*

4D   Multiplying by 10, p. 68

Multiplying by 10, 100, 1000, and so on, is easy because of the special structure of place value in the decimal system: In the decimal system, the value of each place is 10 times the value of the place to its immediate right. Consider what happens when we multiply the number 34 by 10. The number 34 stands for 3 tens and 4 ones and can be represented by 3 bundles of 10 toothpicks and 4 individual toothpicks, as shown in Figure 4.8. Then 10×34 stands for the total number of toothpicks in 10 groups of 34 toothpicks. As Figure 4.9 shows, when we form 10 groups of 34 toothpicks, each of the 3 original tens becomes bundled into 1 group of 100 and each of the 4 original individual toothpicks is bundled into 1 group of 10. Therefore, when we multiply 34 by 10, the 3 in the tens place moves one place over to the hundreds place and the 4 in the ones place moves one place over to the tens place. *Notice that this shifting occurs precisely because the value of the hundreds place is 10 times the value of the tens place and the value of the tens place is 10 times the value of the ones place.*

The situation is similar for other numbers.

We can think of

$$10 \times 0.034$$

as the total in 10 groups of 0.034. Each group of 0.034 consists of 3 hundredths and 4 thousandths, which can be represented with 4 small objects (each representing one thousandth) and 3 bundles of ten of the objects. In fact, if we let 1 toothpick in Figure 4.9 stand for 1 thousandth, then Figure 4.9 shows this:

The 4 thousandths move one place to the left to become 4 hundredths because the value of the hundredths place is 10 times the value of the thousandths place.

The 3 hundredths move one place to the left to become 3 tenths because the value of the tenths place is 10 times the value of the hundredths place.

$$10 \times 0.034: \quad \underline{.0\ 0\ 3\ 4}$$

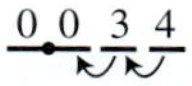

In general, when we multiply a number by 10, all the digits move one place to the left.

FIGURE 4.8

The number 34 represented by 3 bundles of 10 and 4 individual toothpicks

FIGURE 4.9

Ten groups of 34 is bundled into 3 hundreds and 4 tens so the 3 tens become 3 hundreds and the 4 ones become 4 tens.

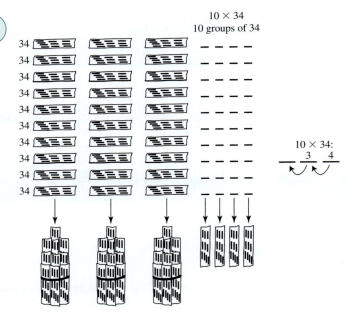

What about multiplying by 100, or 1000, or 10000, and so on? Because

$$100 = 10 \times 10$$

multiplying by 100 has the same effect as multiplying by 10 twice. Therefore, multiplying by 100 moves each digit in the decimal representation of a number *two* places to the left. Similarly, because

$$1000 = 10 \times 10 \times 10$$

multiplying by 1000 has the same effect as multiplying by 10 three times. Therefore, multiplying by 1000 moves each digit in the decimal representation of a number *three* places to the left.

Class Activity   *Now Turn to Class Activities Manual*

4E   Multiplying by Powers of 10 Explains the Cycling of Decimal Representations of Fractions, p. 69

Practice Exercise for Section 4.2

1. a. What are the limitations of the following two statements?

- To multiply a number by 10, put a 0 at the end of the number.

- To multiply a number by 10, move the decimal point one place to the right.

b. What is a better way to describe what multiplying by 10 does to the decimal representation of a number than either of the statements in part (a)?

2. Using the example 10×3.4 to illustrate, explain why we move the digits in the decimal representation of a number one place to the left when we multiply by 10.

1. a. Although the first statement is valid for whole numbers, it is not correct for decimals. For example, 10×2.8 is not equal to 2.80. The second statement is correct, but you wouldn't want to use it with students who are only studying whole number multiplication.

   b. A better way to describe what multiplying by 10 does to a number is to say that each digit moves one place to the left.

2. Using bundled toothpicks, we find that the number 3.4 can be represented as shown in Figure 4.8, as long as 1 toothpick represents $\frac{1}{10}$ and a bundle of 10 toothpicks represents 1. Now use the explanation that is given in the text and that accompanies Figure 4.9.

Problems for Section 4.2

1. Using the example 10×47 to illustrate, explain why we move the digits in a decimal number one place to the left when we multiply by 10.

2. Mary says that $10 \times 3.7 = 3.70$. Why might Mary think this? Explain to Mary why her answer is not correct and why the correct answer is right. If you tell Mary a procedure, be sure to tell Mary why the procedure makes sense.

3. Now that you understand why multiplying a number by 10 shifts the digits one place to the left, explain how we can deduce that multiplying by 10,000 shifts the digits 4 places to the left and multiplying by 100,000 shifts the digits 5 places to the left. How should we think about the numbers 10,000 and 100,000 to make these deductions?

4. a. Find the decimal representation of $\frac{1}{37}$ to at least 6 places (or as many as your calculator shows). Notice the repeating pattern.

   b. Now find the decimal representations of $\frac{10}{37}$ and of $\frac{26}{37}$ to at least 6 places. Compare the repeating patterns with each other and to the decimal representation of $\frac{1}{37}$. What do you notice?

   c. Write $10 \times \frac{1}{37} = \frac{10}{37}$ and $100 \times \frac{1}{37} = \frac{100}{37}$ as mixed numbers (with a whole number part and a fractional part).

   d. What happens to the decimal representation of a number when it is multiplied by 10? By 100? Use your answer, and part (c), to explain the relationships you noticed in part (b).

5. a. Find the decimal representation of $\frac{1}{41}$ to at least 10 decimal places. Notice the repeating pattern.

   b. Use your answer in part (a) to find the decimal representations of the numbers

   $$10 \times \frac{1}{41}, \quad 100 \times \frac{1}{41}, \quad 1000 \times \frac{1}{41},$$

   $$10,000 \times \frac{1}{41}, \quad 100,000 \times \frac{1}{41}$$

   without a calculator.

   c. Write the numbers

   $$10 \times \frac{1}{41} = \frac{10}{41},$$
   $$100 \times \frac{1}{41} = \frac{100}{41},$$
   $$1000 \times \frac{1}{41} = \frac{1000}{41},$$
   $$10,000 \times \frac{1}{41} = \frac{10,000}{41},$$
   $$100,000 \times \frac{1}{41} = \frac{100,000}{41}$$

   as mixed numbers (a whole number part and a fractional part).

   d. Use your answers in part (b) to find the decimal representations of the fractional parts of the mixed numbers you found in part (c). Do not use your calculator or do long division; use part (b). Explain your reasoning.

4.3 The Commutative and Associative Properties of Multiplication, Areas of Rectangles, and Volumes of Boxes

Focal Points
Grades 3, 4

In Chapter 3, we saw that the commutative and associative properties of addition are important building blocks for addition, which allow us to calculate sums flexibly. In this section we'll study the commutative and associative properties of multiplication. We'll see why these properties make sense and we'll see how these properties allow us to calculate products flexibly.

There also are geometric ways to think about the commutative and associative properties of multiplication. We can view the commutative property of multiplication in terms of areas of rectangles and we can view the associative property of multiplication in terms of volumes of boxes. In order to examine the commutative and associative properties of multiplication from this geometric perspective, we will first analyze why we can multiply to find the area of a rectangle and to find the volume of a box. Thus, we will identify the origins of the area formula for rectangles and the volume formula for boxes.

Finally, in the Class Activities we will see how the reasoning behind finding areas of rectangles and volumes of boxes applies estimating numbers of things by using multiplication.

The Commutative Property of Multiplication

commutative property of multiplication

The **commutative property of multiplication** says that if A and B are any real numbers, then

$$A \times B = B \times A$$

For example,

$$297 \times 43 = 43 \times 297$$

and

$$1.3 \times 5.7 = 5.7 \times 1.3$$

We assume that this property always holds for *any* pair of numbers, but we can explain why this property makes sense for counting numbers.

Class Activity   *Now Turn to Class Activities Manual*

4F Explaining and Illustrating the Commutative Property of Multiplication, p. 70

If you think of the commutative property of multiplication in terms of groupings, it can seem somewhat mysterious: Why do 3 groups of 5 marbles have the same total number of marbles as 5 groups of 3 marbles? Why do 297 baskets of potatoes with 43 potatoes in each basket contain the same total number of potatoes as 43 baskets of potatoes with 297 potatoes in each basket? Of course, we can calculate 3×5 and 5×3 and see that they are equal, and we can calculate 297×43 and 43×297 and see that they are equal, but by simply calculating, we don't see why the commutative property should hold in other situations as well. A conceptual explanation that will show why this property makes sense will help explain its meaning. We can give a conceptual explanation by working with arrays or with areas of rectangles.

Suppose you have 3 groups with 5 marbles in each group. According to the meaning of multiplication, this is a total of

$$3 \times 5$$

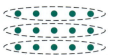
FIGURE 4.10

Grouping marbles
to show that
$3 \times 5 = 5 \times 3$

marbles. You can arrange the 3 groups of marbles so that they form the 3 rows of an array, as pictured on the left in Figure 4.10. But if you now choose the *columns* of the array to be the groups, as on the right in Figure 4.10, then there are 5 groups with 3 marbles in each group. Therefore, according to the meaning of multiplication, there is a total of

$$5 \times 3$$

marbles. But the total number of marbles is the same, either way you count them. Therefore,

$$3 \times 5 = 5 \times 3$$

By imagining pictures like Figure 4.10, it's easy to see why the commutative property of multiplication should hold for counting numbers other than 3 and 5. Even though it would be ridiculously tedious to draw 297 horizontal rows with 43 potatoes in each row, we can imagine such a picture as similar to Figure 4.10. Such a picture, and the argument about changing groups from rows to columns, explains why

$$297 \times 43 = 43 \times 297$$

In these situations, there is nothing special about the numbers 3, 5, 297, and 43. So, if A and B are any counting numbers, there will be a picture similar to Figure 4.10 illustrating that

$$A \times B = B \times A$$

We can also think about the commutative property of multiplication geometrically, in terms of areas of rectangles. In order to do so, let's first examine why we can multiply to find areas of rectangles.

Multiplication, Areas of Rectangles, and the Commutative Property

You probably remember learning the length times width, $L \times W$, formula for areas of rectangles. Why is this formula valid? We can explain this formula by using the meaning of multiplication. But first, what is area?

Area is usually measured in square units. Depending on what we are describing—the page of a book, the floor of a room, a football field, a parcel of land—we usually measure area in square inches, square feet, square yards, square miles, square centimeters, square meters, or square kilometers. A **square inch**, often

square inch

written 1 in², is the area of a square that is 1 inch wide and 1 inch long, as is the square in Figure 4.11. Similarly, one **square foot**, often written 1 ft², is the area of a square that is 1 foot wide and 1 foot long. In general, for any unit of length, one **square unit** is the area of a square that is 1 unit wide and 1 unit long. The **area** of a region, in square units, is the number of 1-unit-by-1-unit squares it takes to cover the region without gaps or overlaps (where squares may be cut apart if necessary).

square foot
square unit

area

Class Activity   *Now Turn to Class Activities Manual*

4G 🏛 Multiplication, Areas of Rectangles, and the Commutative Property, p. 72

FIGURE 4.11

A 1-inch-by-
1-inch square has
area 1 in².

FIGURE 4.12

A-2-inch-by-3-inch
rectangle

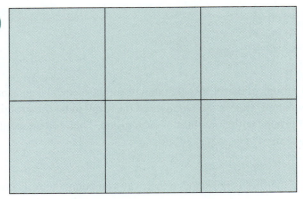

The rectangle in Figure 4.12 is three inches wide and two inches long (or high). Why can its area be calcu-lated by multiplying its length times its width? When we subdivide the rectangle into 1-inch-by-1-inch squares, we have two rows of these squares, and each row has three squares. In other words, the rectangle is made up of 2 groups with 3 squares in each group. Therefore, according to the meaning of multiplication, the rectangle is made up of 2×3, or 6 squares. Because each 1-inch-by-1-inch square has area 1 square inch, the area of the whole rectangle is 2×3 square inches. We can use the same reasoning to determine areas of other rectangles.

In general, if L and W are any whole numbers, then a rectangle that is L units long and W units wide can be subdivided into L rows of 1-unit-by-1-unit squares, with W squares in each row, as indicated in Figure 4.13. In other words, the rectangle consists of L groups, with W squares in each group. Therefore, according to the meaning of multiplication, the rectangle is made of

$$L \times W$$

1-unit-by-1-unit squares. Because each 1-unit-by-1-unit square has area 1 square unit, the entire L-unit-by-W-unit rectangle has area

$$L \times W \text{ square units}$$

This explanation for why the length-times-width formula is valid applies not only when the length and width of the rectangle are whole numbers, but even when the length and width are fractions or decimals. So, a rectangle that is L units by W units has area

$$L \times W \text{ square units}$$

regardless of what kinds of numbers L and W are.

Returning to the commutative property of multiplication, we can see why this property makes sense by using areas of rectangles, as in Class Activity 4G. For example, a rug that is 3 feet by 5 feet can be thought of as being made of 3 rows with 5 squares, each of area 1 square foot, in each row. Therefore, the area of the rug is

$$3 \times 5 \text{ square feet}$$

FIGURE 4.13

A rectangle that is
L units long and
W units wide

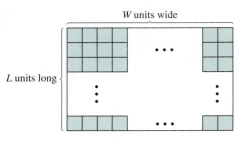

FIGURE 4.14

Using the area of a rectangle to show that $3 \times 5 = 5 \times 3$

On the other hand, the rug can be thought of as made of 5 columns with 3 squares, each of area 1 square foot, in each column, as shown in Figure 4.14. Therefore, the area of the rug is

$$5 \times 3 \text{ square feet}$$

But the area is the same either way you calculate it; therefore,

$$3 \times 5 = 5 \times 3$$

As before, the same line of reasoning will work when other counting numbers replace 3 and 5.

Multiplication and Volumes of Boxes

Not only are *areas* naturally related to multiplication, but *volumes* are as well, as we'll see next.

But first, what is volume? Depending on what you want to measure—a dose of liquid medicine, the size of a compost pile, the volume of coal in a mountain—volume can be measured in cubic inches, cubic feet, cubic yards, cubic miles, cubic centimeters, cubic meters, or cubic kilometers. (Volumes of liquids are also commonly measured in liters, milliliters, gallons, quarts, cups, and fluid ounces.)

One cubic centimeter, often written 1 cm^3, is the volume of a cube that is one centimeter high, one centimeter long, and one centimeter wide. A drawing of such a cube is shown in Figure 4.15, along with a cube of volume 1 cubic inch, 1 inch^3. One cubic yard, often written 1 yd^3, is the volume of a cube that is one yard high, one yard long, and one yard wide. In general, for any unit of length, one **cubic unit**, often written 1 unit^3, is the volume of a cube that is 1 unit high, 1 unit long, and 1 unit wide.

cubic unit

volume

We will be working with volumes of boxes and box shapes. (These shapes are also called rectangular prisms.) The **volume**, in cubic units, of a box or box shape is just the number of 1-unit-by-1-unit-by-1-unit cubes that it would take to fill the box or make the box shape (without gaps or overlaps). You may remember a formula for the volume of a box, but if so, assume for a moment that you don't know this formula. We will see how to derive the formula for the volume of a box from the meaning of multiplication.

Class Activity   *Now Turn to Class Activities Manual*

4H   Ways to Describe the Volume of a Box with Multiplication, p. 72

FIGURE 4.15

Cubes of volume 1 inch^3 and 1 cm^3

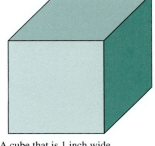

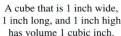

A cube that is 1 inch wide, 1 inch long, and 1 inch high has volume 1 cubic inch.

A cube that is 1 cm wide, 1 cm long, and 1 cm high has volume 1 cubic centimeter.

FIGURE 4.16

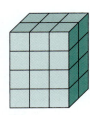

A 4-inch-high,
2-inch-deep,
3-inch-wide box

Suppose you have a box that is 4 inches high, 2 inches long, and 3 inches wide, as pictured in Figure 4.16. What is the volume of this box? If you have a set of building blocks that are all one inch high, one inch long, and one inch wide, then you can use the building blocks to build the box. The number of blocks needed is the volume of the box in cubic inches. We can use multiplication to describe the number of blocks needed by considering the box to be made of 4 horizontal layers, as shown in Figure 4.17. Each layer consists of two rows of three blocks; so, according to the meaning of multiplication, each layer contains 2×3 blocks. There are 4 layers, each containing 2×3 blocks. So, according to the meaning of multiplication, there are

$$4 \times (2 \times 3)$$

blocks making up the box. Therefore, the box has a volume of

$$4 \times (2 \times 3) = 4 \times 6 = 24 \text{ cubic inches}$$

The reasoning used in this example applies generally. Suppose you have a box that is H units high, L units long, and W units wide. If H, L, and W are whole numbers, then as before, you can build such a box out of 1-unit-by-1-unit-by-1-unit cubes. How many cubes does it take? If you consider the box to be made of horizontal layers, then there are H layers. Each layer is made up of L rows of W blocks (or W rows of L blocks) and therefore contains $L \times W$ blocks, according to the meaning of multiplication. There are H layers with $L \times W$ blocks in each layer. Therefore, according to the meaning of multiplication, the box is made out of

$$H \times (L \times W)$$

1-unit-by-1-unit-by-1-unit cubes. Consequently, the box has volume

$$H \times (L \times W) \text{ cubic units}$$

Notice that the order in which the letters H, L, and W appear and the way the parentheses are placed in the expression $H \times (L \times W)$ corresponds to the way the box was divided into groups.

As with areas, it turns out that this height-times-length-times-width formula for the volume of a box remains valid even when the height, length, and width of the box are not whole numbers. The volume of a box that is H units high, L units long, and W units wide is always $H \times (L \times W)$ cubic units.

Next, we examine the use of parentheses in expressions like $H \times (L \times W)$.

FIGURE 4.17

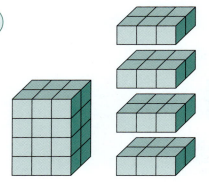

The Associative Property of Multiplication

Recall that the associative property of addition says that, for all real numbers A, B, and C,

$$(A + B) + C = A + (B + C)$$

As we have seen, the associative property of addition gives us flexibility in calculating sums. Likewise, the associative property of multiplication will give us flexibility in calculating products.

associative property of multiplication   The **associative property of multiplication** says that if A, B, and C are any real numbers, then

$$(A \times B) \times C = A \times (B \times C)$$

We assume that this property holds for all real numbers, but as we'll see, we can explain why it makes sense for counting numbers. We will also see how to use the associative property of multiplication.

Class Activity   *Now Turn to Class Activities Manual*

4I   🏺   Explaining the Associative Property, p. 74

Why the Associative Property of Multiplication Makes Sense

Why is the associative property valid? For example, why is

$$(4 \times 2) \times 3 = 4 \times (2 \times 3)?$$

In other words, why is the quantity (4×2) times 3 equal to 4 times the quantity (2×3)? In this example, because specific numbers are involved, we can evaluate the quantities on both sides of the equals sign and see that they really are equal. We get

$$(4 \times 2) \times 3 = 8 \times 3 = 24$$

and

$$4 \times (2 \times 3) = 4 \times 6 = 24$$

Therefore,

$$(4 \times 2) \times 3 = 4 \times (2 \times 3)$$

But when we evaluate the expressions, we show that the associative property holds only in this one special case. Why does it make sense that the associative property of multiplication holds for *all* real numbers?

To explain why the associative property of multiplication makes sense for all counting numbers, we will develop a conceptual explanation for why

$$(4 \times 2) \times 3 = 4 \times (2 \times 3)$$

This explanation will be general in the sense that it will also explain why the equation is true if we were to replace the numbers 4, 2, and 3 with other counting numbers. Our explanation is based on calculating the volume of a box in two different ways.

Previously, we decomposed a 4-inch-high, 2-inch-long, 3-inch-wide box shape into 4 groups of blocks, with 2×3 blocks in each group, as shown on the left of Figure 4.18. When we view it that way, we see that the total number of blocks in the box shape is

$$4 \times (2 \times 3)$$

FIGURE 4.18

Showing that
$4 \times (2 \times 3) =$
$(4 \times 2) \times 3$

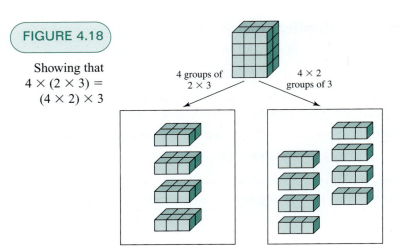

4 groups of
2×3

4×2
groups of 3

On the other hand, the box shape can be decomposed into 4×2 groups, with 3 blocks in each group, as shown on the right of Figure 4.18. According to the meaning of multiplication, there are

$$(4 \times 2) \times 3$$

blocks in the box shape. There are 4×2 groups because the groups of 3 blocks can themselves be arranged into 4 groups of 2. But the total number of blocks in the box shape is the same either way we count them. Therefore,

$$(4 \times 2) \times 3 = 4 \times (2 \times 3)$$

which is what we wanted to establish. Notice that the same argument works when the numbers 4, 2, and 3 are replaced with other counting numbers—only the size of the box will change. This reasoning explains why the associative property of multiplication makes sense for all counting numbers.

Using the Associative and Commutative Properties of Multiplication

How do we use the commutative property of multiplication? The commutative property gives us flexibility by allowing us to calculate in different ways. For example, if there are 15 packages of soap and 2 bars of soap in each package, how many bars of soap are there? One way to determine the total number of bars of soap is to count by twos 15 times.

$$2, 4, 6, 8, 10, 12, 14, 16, 18, 20, 22, 24, 26, 28, 30$$

but a quicker way is to calculate

$$15 + 15 = 30$$

Why can we determine the answer either way? Because of the commutative property of multiplication! The first method determines what 15 twos make, in other words, the first method calculates

$$15 \times 2$$

The second method calculates what two 15s make, in other words, the second method calculates

$$2 \times 15$$

The commutative property of multiplication tells us that we will get the answer either way.

How do we use the associative property of multiplication? One important way is by breaking a number apart and reassociating one of the parts. For example, how can we multiply 7×800 mentally? We can first recall the single-digit multiplication fact $7 \times 8 = 56$ and then multiply that result by 100 to get the answer, 5600. The next equations correspond to this strategy:

$$7 \times 800 = 7 \times (8 \times 100)$$
$$= (7 \times 8) \times 100$$
$$= 56 \times 100 = 5600$$

The equations show that we broke 800 apart into 8×100 and then applied the associative property to reassociate the 8 with the 7 instead of with the 100 at the second equal sign. Several additional examples of this important method of breaking apart and reassociating appear in the next Class Activity. You can also apply this method to some of the problems in this section.

We often apply the associative and commutative properties together. For example, how can we make

$$5 \times 41 \times 2$$

easy to calculate mentally? Since 10 is easy to multiply with, we could multiply the 5 and the 2 first to make 10, then multiply 10 and 41 to make 410. Teachers sometimes like to record this method this way:

$$5 \times 41 \times 2 = ? \qquad 5 \times 41 \times 2 = ? \qquad 5 \times 41 \times 2 = 410$$

Notice that even though we may not see it right away, the teacher's method of recording how to calculate $5 \times 41 \times 2$ actually does involve the commutative and associative properties, as we see from the next equations:

$$5 \times 41 \times 2 = 41 \times 5 \times 2$$
$$= 41 \times 10$$
$$= 410$$

The commutative property was used at the first equal sign and the associative property was used at the second equal sign—to associate the 5 with the 2 instead of with the 41—as well as to simply drop the parentheses altogether.

Class Activity   *Now Turn to Class Activities Manual*

4J   Using the Associative and Commutative Properties of Multiplication, p. 75

4K   Different Ways to Calculate the Total Number of Objects, p. 76

4L   Using Multiplication to Estimate How Many, p. 77

4M   How Many Gumdrops? p. 78

Practice Exercises for Section 4.3

1. State the commutative property of multiplication.

2. Use the meaning of multiplication and a picture to give a conceptual explanation for why $2 \times 4 = 4 \times 2$. Your explanation should be general enough to remain valid if other counting numbers were to replace 2 and 4.

3. Give an example to show how to use the commutative property of multiplication to make a math problem easier to solve mentally.

4. Use the meaning of multiplication to explain why the area of a carpet that is 20 feet long and 12 feet wide is 20×12 square feet.

5. Use the meaning of multiplication to explain why a box that is 3 feet wide, 2 feet long, and 4 feet high has volume $4 \times (2 \times 3)$ cubic feet.

6. One cubic foot of water weighs about 62 pounds. How much will the water weigh in a swimming pool that is 20 feet wide, 30 feet long, and 4 feet deep weigh?

7. State the associative property of multiplication.

8. Give an example of how to use the associative property of multiplication to make a calculation easy to do mentally.

9. Explain how you use the associative property of multiplication when you calculate 7×600 mentally.

10. Use the meaning of multiplication and the idea of decomposing a box shape in two different ways to explain why $4 \times (2 \times 3) = (4 \times 2) \times 3$.

11. Describe how to use the design in Figure 4.19 to explain why

$$5 \times (2 \times 2) = (5 \times 2) \times 2$$

Your explanation should be general, so that it explains why the previous equation is true when you replace the numbers 5, 2, and 2 with other counting numbers, and the design in Figure 4.19 is changed accordingly.

one spiral

FIGURE 4.19   A design of spirals

Answers to Practice Exercises for Section 4.3

1. See text.

2. See Figure 4.20. It shows that the total number of objects in 2 groups with 4 objects in each group (2×4 objects) can also be thought of as made out of 4 groups with 2 objects in each group (4×2 objects). But you have the same number of objects either way you count them; therefore,

$$2 \times 4 = 4 \times 2$$

two groups of 4   four groups of 2

FIGURE 4.20 $2 \times 4 = 4 \times 2$

3. If sponges come 2 to a package and if you have 13 packages of sponges, then the total number of sponges you have is 13×2. To calculate 13×2, you can instead calculate 2×13, according to the commutative property of multiplication. Two groups of 13 is just $13 + 13$, which is easy to calculate mentally as 26.

4. Think of a carpet that is 20 feet long and 12 feet wide as made up of 20 rows of squares with 12 squares in each row, each square being 1 foot wide and 1 foot long. In other words, the carpet consists of 20 groups, with 12 squares in each group. Therefore, according to the meaning of multiplication, the carpet is made of 20×12 squares. The area of a square that is 1 foot wide and 1 foot long is 1 square foot; thus, the area of the carpet is 20×12 square feet.

5. See the explanation in the text. Substitute "feet" where the text says "inches."

6. Because the water in the pool is in the shape of a box that is 4 feet high, 20 feet wide, and 30 feet deep, it can be thought of as

$$4 \times (20 \times 30) = 2400$$

1-foot-by-1-foot-by-1-foot cubes of water. Each of those cubes of water weighs 62 pounds, so the water in the pool will weigh

$$2400 \times 62 \text{ pounds} = 148,800 \text{ pounds}$$

7. See text.

8. See the next practice exercise. See also problems 10, 11, 13, and 14.

9. When we calculate 7×600 mentally, we first calculate 7×6 (a "basic fact") and then we multiply by 100. When we do this, we take the 6 from the 600, which is 6×100, and group the 6 with the 7 instead of with 100. In equation form, we can express this rearrangement as follows:

$$7 \times 600 = 7 \times (6 \times 100)$$
$$= (7 \times 6) \times 100$$
$$= 42 \times 100$$
$$= 4200$$

The associative property of multiplication is used at the second equal sign.

10. See text.

11. See Figure 4.21. On the one hand, we can think of the design as being made up of 5 clusters of spirals, with 2 groups of 2 spirals in each cluster. In this arrangement there are $5 \times (2 \times 2)$ spirals in the design. On the other hand, we can think of the design as made up of 5×2 groups of spirals with 2 spirals in each group. There are 5×2 groups because there are 5 sets of 2 groups. This means there are $(5 \times 2) \times 2$ spirals in the design. There are the same number of spirals, no matter how they are counted. Therefore,

$$5 \times (2 \times 2) = (5 \times 2) \times 2$$

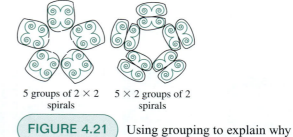

5 groups of 2×2 spirals 5×2 groups of 2 spirals

FIGURE 4.21   Using grouping to explain why $5 \times (2 \times 2) = (5 \times 2) \times 2$

Problems for Section 4.3

1. There are 31 envelopes with 3 stickers in each envelope. Seyong calculates the total number of stickers by counting by threes 31 times. Natasha adds $31 + 31 + 31 = 93$ instead. Discuss the two calculation methods. Are both legitimate? Explain.

2. Here is Amy's explanation for why the commutative property of multiplication is true for counting numbers:

   Whenever I take two counting numbers and multiply them, I always get the same answer as when I multiply them in the reverse order. For example,

   $$6 \times 8 = 48$$
   $$8 \times 6 = 48$$
   $$9 \times 12 = 108$$
   $$12 \times 9 = 108$$
   $$3 \times 15 = 45$$
   $$15 \times 3 = 45$$

   It always works that way; no matter which numbers you multiply, you will get the same answer either way you multiply them.

   Explain why Amy's discussion might not convince a skeptic that the commutative property should always be true for any pair of counting numbers. Then explain why the commutative property of multiplication is valid in another way.

3. Explain why it makes sense to multiply 4×6 to determine the area of a 4-foot-by-6-foot rug in square feet.

4. Use the meaning of multiplication to explain why a box that is 5 inches high, 4 inches wide, and 3 inches long has a volume of

   $$5 \times (4 \times 3) \text{ cubic inches}$$

   Explain the parentheses in the expression $5 \times (4 \times 3)$.

5. 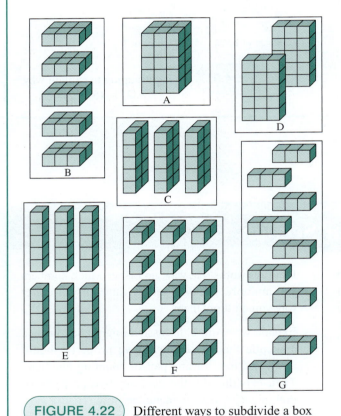 Figure A in Figure 4.22 shows a 5-unit-high, 3-unit-wide, and 2-unit-long box made out of blocks. Figures B through G of Figure 4.22 show different ways of subdividing the box into natural groups of blocks. For each of these ways of subdividing the box, use the meaning of multiplication as we have described it to write an expression for the total number of blocks in the box. Each expression should involve the numbers 5, 3, and 2, the multiplication symbol, and parentheses. Explain why you write your expressions as you do.

FIGURE 4.22   Different ways to subdivide a box

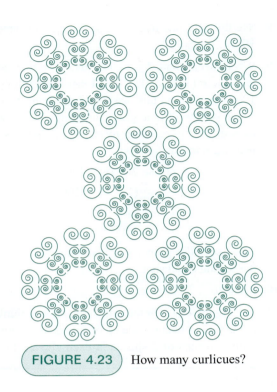

FIGURE 4.23   How many curlicues?

7. Suppose you have 60 pennies arranged into 12 stacks with 5 pennies in each stack. The 12 stacks are arranged into 3 rows with 4 stacks in each row, as shown in Figure 4.24. For each of the following expressions, explain how to see the pennies grouped so that the expression describes the total number of pennies according to that grouping.

Your explanation of the grouping should not involve rearranging the pennies.

a. $3 \times (4 \times 5)$

b. $(3 \times 4) \times 5$

c. $5 \times (3 \times 4)$

d. $4 \times (3 \times 5)$

6. Write three different expressions for the total number of curlicues in Figure 4.23. Each expression should involve only the following: the numbers 2, 3, 5, and 8; the multiplication symbol $\times$ or $\cdot$; and parentheses. For each expression, use the meaning of multiplication as we have described it to explain why your expression represents the total number of curlicues in Figure 4.23.

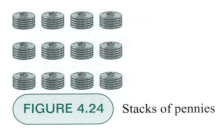

FIGURE 4.24   Stacks of pennies

8. To calculate 3×80 mentally, we can just calculate $3 \times 8 = 24$ and then put a zero on the end to get the answer, 240. Use Figure 4.25 to help you explain why this method of calculation is valid.

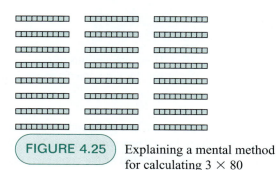

FIGURE 4.25   Explaining a mental method for calculating 3×80

9. Write equations to show how the commutative and associative properties of multiplication are involved when you calculate 40×800 mentally by relying on basic multiplication facts (such as 4×8). Write your equations in the form

$$40 \times 800 = \text{some expression}$$
$$= \vdots$$
$$= \text{some expression}$$

Indicate specifically where the commutative and associative properties of multiplication are used.

10. Explain how to use the associative property of multiplication to make 16×25 easy to calculate mentally. Write equations that show why your method is valid, and show specifically where you have used the associative property of multiplication. Write your equations in the form

$$16 \times 25 = \text{some expression}$$
$$= \vdots$$
$$= \text{some expression}$$

11. Use the associative property of multiplication to make the problem 32×0.25 easy to solve mentally. Write equations to show your use of the associative property of multiplication. Explain how your solution method is related to solving 32×0.25 by thinking in terms of money.

12. Explain how to make the following product easy to calculate mentally (there are five 2s and five 5s in the product):

$$2 \cdot 2 \cdot 2 \cdot 2 \cdot 2 \cdot 5 \cdot 5 \cdot 5 \cdot 5 \cdot 5$$

13. Julia says that it's easy to multiply a number by 4 because you just "double the double." Explain Julia's idea, and explain why it uses the associative property of multiplication.

14. Carmen says that it's easy to multiply even numbers by 5 because you just take half of the number and put a zero on the end. Write equations that incorporate Carmen's method and that demonstrate why her method is valid. Use the case 5×22 for the sake of concreteness. Write your equations in the following form:

$$5 \times 22 = \text{some expression}$$
$$= \text{some expression}$$
$$\vdots$$
$$= 110$$

15. The Browns need new carpet for a room with a rectangular floor that is 35 feet wide and 43 feet long. To save money, they will install the carpet themselves. The carpet comes on a large roll that is 9 feet wide. The carpet store will cut any length of carpet they like, but the Browns must buy the full 9 feet in width.

   a. Draw clear, detailed pictures showing two different ways the Browns could lay their carpet.

   b. For each way of laying the carpet, find how much carpet the Browns will need to buy from the carpet store. Which way is less expensive for the Browns?

16. If a roll of a certain kind of wrapping paper is unrolled, the wrapping paper forms a rectangle that is 3 feet wide and 20 feet long. The wrapping paper is to be covered with an array of ladybugs by repeating the design shown in Figure 4.25. (The arrows and the "3 inches," "2 inches" are not part of the design; they show the dimensions of a portion of the design.) How many ladybugs will be on the wrapping paper? Explain your reasoning.

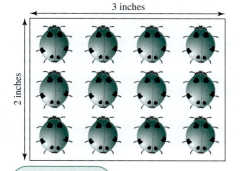

FIGURE 4.25   A ladybug design for wrapping paper

17. A roll of wrapping paper is 30 inches wide. When you unroll the wrapping paper and cut off a portion, you get a rectangular piece of wrapping paper that is 30 inches wide and can have various lengths. The wrapping paper has a design of stars on it. If you were to subdivide the paper into 2-inch-by-2-inch squares, each such square would have 25 stars on it. Use this information about the wrapping paper in parts (a), (b), and (c).

    a. How long a piece of the wrapping paper would you need to get at least 1000 stars? Explain your reasoning.

    b. How long a piece of the wrapping paper would you need to get at least 1,000,000 stars? Explain your reasoning. Realistically, could this length come from a single roll of wrapping paper? Why or why not?

    c. How long a piece of the wrapping paper would you need to get 1,000,000,000 stars? Explain your reasoning. If you had this length of wrapping paper and you wanted to show it to the children in your class, could you roll it out in the school yard? Why or why not?

18. Ms. Dunn's class wants to estimate the number of blades of grass in a lawn that is shaped roughly like a 30-foot-by-40-foot rectangle. The children cut 1-inch-by-1-inch squares out of sheets of paper and place these square holes over patches of grass. Each child cuts the grass from their 1-square-inch patch and counts the number of blades of grass they cut. Inside, the class calculates the average number of blades of grass they cut. This average is 37. Using these data, determine approximately how many blades of grass are in the lawn. Explain your reasoning.

19. Imagine that you are standing on a sandy beach, like the one in Figure 4.26, gazing off into the distance. How many grains of sand might you be looking at? To solve this problem, make reasonable assumptions about how wide the beach is and how far down the length of the beach you can see. Make a reasonable assumption about how many grains of sand are in a very small piece of beach, and explain why your assumption is reasonable. Based on your assumptions, make a calculation that will give you a fairly good estimate of the number of grains of sand that you can see. Explain your reasoning.

FIGURE 4.26 How many grains of sand do you see, looking down a sandy beach?

20. A lot of gumballs are in a glass container. The container is shaped like a box with a square base. When you look down on the top of the container, you see about 50 gumballs at the surface. When you look at one side of the container, you see about 60 gumballs up against the glass. You also notice that there are about 9 gumballs against each vertical edge of the container. Given this information, estimate the total number of gumballs in the container. Explain your reasoning.

21. Estimate how many neatly stacked hundred-dollar bills you could fit in a briefcase that is 20 inches long, 11 inches wide, and 2 inches thick on the inside. Describe your method, and explain why it gives a good estimate.

22. Figure 4.27 shows a grocery store display of cases of soft drinks. The display consists of a large box shape with a "staircase" on top. The display is 7 cases wide and 5 cases deep; it is 4 cases tall in the front and 8 cases tall at the back. How many cases of soft drinks are there in the display? Solve this problem in at least two different ways, and explain your method in each case.

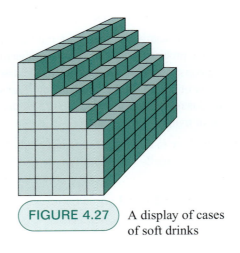

FIGURE 4.27 A display of cases of soft drinks

23. A cube that is 10 inches wide, 10 inches long, and 10 inches high is made out of smaller cubes that are each 1 inch wide, 1 inch long, and 1 inch high. The large cube is then painted on the outside.

    a. How many of the smaller cubes that make up the large cube have paint on them? Explain.

    b. How many of the smaller cubes have paint on exactly two sides? Explain.

    c. How many of the smaller cubes have paint on exactly three sides? Explain.

24. Use the facts that

    1 mile = 1760 yards

    1 yard = 3 feet

    1 foot = 12 inches

    to calculate the number of inches in a mile. Do this in two different ways to illustrate the associative property of multiplication.

25. Investigate the following two questions, and explain your conclusions:

    a. If you make a rectangular garden that is twice as wide and twice as long as a rectangular garden that you already have, how will the area of the larger garden compare with the area of the original garden? (Will the larger garden be twice as big, 3 times as big, 4 times as big, etc.?)

    b. If you make a cardboard box that is twice as wide, twice as tall, and twice as long as a cardboard box that you already have, how will the volume of the larger box compare with the volume of the original box? (Will the larger box be twice as big, 3 times as big, 4 times as big, etc.?)

26. The Better Baking Company is introducing a new line of reduced-fat brownies in addition to its regular brownies. The batter for the reduced-fat brownies contains $\frac{1}{3}$ less fat than the batter for the regular brownies. Both types of brownies will be baked in the same size rectangular pan, which is 24 inches wide and 30 inches long. The bakers cut the regular brownies from this pan by dividing the width into 12 equal segments and by dividing the length into 10 equal segments, so that each regular brownie is 2 inches by 3 inches. Each regular brownie contains 6.3 grams of fat. In addition to using a reduced-fat batter, the Better Baking Company would like to further reduce the amount of fat in their new brownies by making these brownies smaller than the regular ones. You have been contacted to help with this task. Present two different ways to divide the length and width of the pan to produce smaller brownies. In each case, calculate the amount of fat in each brownie and explain the basis for your calculation. The brownies should be of a reasonable size, and there should be no waste left over in the pan after cutting the brownies. The length and width of the brownies do not necessarily have to be whole numbers of inches.

4.4 The Distributive Property

 **Focal Points**
Grades 3, 4

The associative and commutative properties apply in situations when we are adding only (as in Chapter 3) or multiplying only (as in Section 4.3). In this section, we introduce a property that applies to addition and multiplication together—namely, the distributive property of multiplication over addition. The distributive property is the most important and computationally powerful tool in all of arithmetic. It allows for tremendous flexibility in performing mental calculations, and, as we will see in Section 4.6, the distributive property is the foundation of the common longhand multiplication technique.

As with the other properties of arithmetic that we have studied, we will explain why the distributive property makes sense as well as discussing how it is useful in solving problems. Similarly to the commutative property of multiplication, the distributive property can be viewed geometrically in terms of arrays (or areas). As with all the other properties of arithmetic, the distributive property relies on counting a collection of things in two different ways.

Before we discuss the distributive property of multiplication over addition, let's review the conventions for interpreting expressions involving both multiplication and addition.

Expressions Involving Both Multiplication and Addition

Expressions that involve both addition and multiplication must be interpreted suitably, according to the conventions developed by mathematicians. This situation is entirely unlike the situation where only addition, or only multiplication, is involved in an expression. In those situations, parentheses can be dropped safely and adjacent numbers can be combined at will. However, when both multiplication and addition are present in an expression, parentheses cannot generally be dropped without changing the value of the expression. For example,

$$7 + 5 \times 2$$

is *not equal* to

$$(7 + 5) \times 2$$

To properly interpret an expression such as

$$7 + 5 \times 2$$

or

$$5 \times 17 + 9 \times 10^2 - 12 \times 94 + 20 \div 5$$

we need to use the following conventions:

- All powers are calculated first.

- Next, multiplications and divisions are performed from left to right.

- Finally additions and subtractions are performed from left to right.

- Expressions inside parentheses are always evaluated first, using the previous conventions.

Therefore,

$$7 + 5 \times 2 = 7 + 10$$
$$= 17$$

whereas

$$(7 + 5) \times 2 = 12 \times 2$$
$$= 24$$

Similarly,

$$5 \times 17 + 9 \times 10^2 - 12 \times 94 + 20 \div 5$$
$$= 5 \times 17 + 9 \times 100 - 12 \times 94 + 20 \div 5$$
$$= 85 + 900 - 1128 + 4$$
$$= -139$$

Note that the "order of operations" conventions allow us to *interpret* expressions unambiguously, but they do not dictate how to calculate. For example, as we'll see next when we study the distributive property, we could calculate $7 \times (10 + 2)$ by calculating $7 \times 10 = 70$ and adding this result to $7 \times 2 = 14$.

Class Activity   *Now Turn to Class Activities Manual*

4N Order of Operations, p. 79

Why the Distributive Property Makes Sense

distributive
property

The **distributive property** of multiplication over addition says that for all real numbers, A, B, and C,

$$A \times (B + C) = A \times B + A \times C$$

Notice the use of parentheses to group the B and C: The expression

$$A \times (B + C)$$

means A times the *quantity* $B + C$.

As with the other properties of arithmetic that we have studied, we assume that the distributive property holds for all real numbers. However, we can explain why the distributive property makes sense for counting numbers by decomposing arrays of objects or by decomposing rectangles.

Class Activity   *Now Turn to Class Activities Manual*

40   Explaining the Distributive Property, p. 80

Consider an array of dots consisting of 4 horizontal rows, with 7 dots in each row, as shown in Figure 4.28. The shading of the dots shows one way to decompose this array of dots into two smaller arrays. There are two ways of expressing the total number of dots in the array that correspond to this way of decomposing the array. On the one hand, there are 4 rows with 5 + 2 dots in each row, and therefore the total number of dots is

$$4 \times (5 + 2)$$

Notice that parentheses are needed to say 4 times the *quantity* 5 plus 2. On the other hand, the shading decomposes the array of dots into two arrays. The array on the left consists of 4 rows with 5 dots in each row, and thus contains 4×5 dots; the array on the right consists of 4 rows with 2 dots in each row and hence contains 4×2 dots. Therefore, combining these two arrays of dots, we see that there is a total of

$$4 \times 5 + 4 \times 2$$

dots. But the total number of dots is the same, either way you count them. Therefore,

$$4 \times (5 + 2) = 4 \times 5 + 4 \times 2$$

This explains why the distributive property makes sense *in this case*. But notice that the reasoning is general in the sense that if we were to replace the numbers 4, 5, and 2 with other counting numbers, and if we were to correspondingly adjust the array of dots, then our argument would still hold. Therefore, the distributive property makes sense for all counting numbers.

FIGURE 4.28

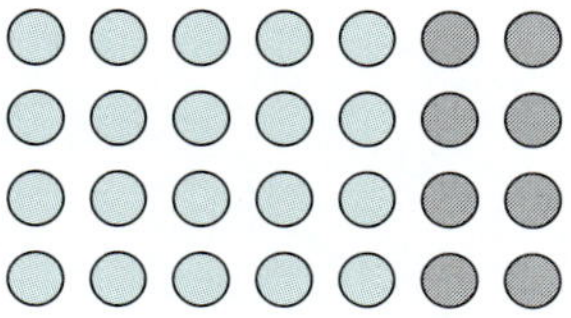

An illustration
of the
distributive
property:
$4 \times (5 + 2) =$
$4 \times 5 + 4 \times 2$

Variations on the Distributive Property

Several useful variations on the distributive property are listed next. We can obtain all these variations from the original distributive property by using the commutative property of multiplication, by using the distributive property repeatedly, or by using the fact that

$$B - C = B + (-C)$$

The Original Distributive Property

$$A \times (B + C) = A \times B + A \times C$$

for all real numbers A, B, and C.

Variation 1

$$(A + B) \times C = A \times C + B \times C$$

for all real numbers A, B, and C.

Variation 2

$$A \times (B + C + D) = A \times B + A \times C + A \times D$$

for all real numbers A, B, C, and D.

Variation 3

$$A \times (B - C) = A \times B - A \times C$$

for all real numbers A, B, and C.

Collectively, we will refer to the original distributive property and all its variations simply as "the distributive property."

Class Activity   *Now Turn to Class Activities Manual*

4P   Using the Distributive Property, p. 81

Using the Distributive Property

Just like the other properties of arithmetic that we have studied, we can often use the distributive property to make mental arithmetic problems easier to solve.

For example, what is an easy way to calculate

$$41 \times 25$$

mentally? We can use the following strategy:

40 times 25 is 1000, plus one more 25 is 1025.

The following sequence of equations corresponds to this strategy and uses the first variation on the distributive property at the second equals sign:

$$41 \times 25 = (40 + 1) \times 25$$
$$= 40 \times 25 + 1 \times 25$$
$$= 1000 + 25$$
$$= 1025$$

What is an easy way to calculate

$$39 \times 25$$

mentally? We can use a strategy similar to the previous one, this time taking away a group of 25 rather than adding it:

40 groups of 25 makes 1000; take away a group of 25, and we are left with 975.

The following sequence of equations corresponds to this strategy and uses the third variation on the distributive property (varied by the first variation) at the second equals sign.

$$\begin{aligned} 39 \times 25 &= (40 - 1) \times 25 \\ &= 40 \times 25 - 1 \times 25 \\ &= 1000 - 25 \\ &= 975 \end{aligned}$$

Class Activity   *Now Turn to Class Activities Manual*

4Q Why Isn't 23 × 23 Equal to 20 × 20 + 3 × 3? p. 82

FOIL and the Extended Distributive Property

If you have studied algebra, then you probably learned the FOIL method for multiplying expressions of the form

$$(A + B) \cdot (C + D)$$

FOIL   FOIL stands for *First, Outer, Inner, Last*, in order to remind you that

$$(A + B) \cdot (C + D) = A \cdot C + A \cdot D + B \cdot C + B \cdot D$$

where $A \cdot C$ is *First*, $A \cdot D$ is *Outer*, $B \cdot C$ is *Inner*, and $B \cdot D$ is *Last*.

Class Activity   *Now Turn to Class Activities Manual*

4R The Distributive Property and FOIL, p. 83

4S Squares and Products Near Squares, p. 84

We can derive FOIL by using the distributive property several times. Hence, FOIL is an extension of the distributive property. To derive FOIL, let's first treat $C + D$ as a single entity; think of $C + D$ as c. Then, by the distributive property,

$$(A + B) \cdot c = A \cdot c + B \cdot c$$

Therefore,

$$(A + B) \cdot \overbrace{(C + D)}^{c} = A \cdot \overbrace{(C + D)}^{c} + B \cdot \overbrace{(C + D)}^{c}$$

Now we can apply the distributive property again, this time to $A \cdot (C + D)$ and to $B \cdot (C + D)$. Stringing all these equations together, we have

$$\begin{aligned} (A + B) \cdot (C + D) &= A \cdot (C + D) + B \cdot (C + D) \\ &= A \cdot C + A \cdot D + B \cdot C + B \cdot D \end{aligned}$$

thereby proving that FOIL is valid.

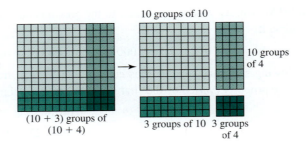

10 groups of 10

10 groups of 4

(10 + 3) groups of (10 + 4)

3 groups of 10   3 groups of 4

FIGURE 4.29   A subdivided rectangle to show that
$(10 + 3) \cdot (10 + 4) = 10 \cdot 10 + 10 \cdot 4 + 3 \cdot 10 + 3 \cdot 4$

We can also see why FOIL makes sense by subdividing arrays or rectangles. Figure 4.29 shows a rectangle made of $10 + 3$ rows with $10 + 4$ small squares in each row. According to the meaning of multiplication, the total number of small squares in the rectangle is

$$(10 + 3) \cdot (10 + 4)$$

As we see by the shading in Figure 4.29, we can subdivide the rectangle into 4 natural parts: 10 groups of 10 small squares, 10 groups of 4 small squares, 3 groups of 10 small squares, and 3 groups of 4 small squares. Therefore, the total number of small squares in the rectangle is

$$10 \cdot 10 + 10 \cdot 4 + 3 \cdot 10 + 3 \cdot 4$$

But the total number of small squares is the same either way you count them. Therefore,

$$(10 + 3) \cdot (10 + 4) = 10 \cdot 10 + 10 \cdot 4 + 3 \cdot 10 + 3 \cdot 4$$

which is the FOIL equation. The same reasoning will apply when the numbers 10, 3, 10, and 4 are replaced with other numbers, and the rectangle is changed accordingly. Thus, we see why FOIL must be valid for all counting numbers.

How do we multiply an expression of the form

$$(A + B) \cdot (C + D + E)?$$

Using our previous reasoning, let's first treat $C + D + E$ as a single entity; think of $C + D + E$ as c. Then, by the distributive property,

$$(A + B) \cdot c = A \cdot c + B \cdot c$$

Therefore,

$$(A + B) \cdot \overbrace{(C + D + E)}^{c} = A \cdot \overbrace{(C + D + E)}^{c} + B \cdot \overbrace{(C + D + E)}^{c}$$

Now we can apply the distributive property again, this time to $A \cdot (C + D + E)$ and to $B \cdot (C + D + E)$. Stringing all these equations together, we have

$$(A + B) \cdot (C + D + E) = A \cdot (C + D + E) + B \cdot (C + D + E)$$
$$= A \cdot C + A \cdot D + A \cdot E + B \cdot C + B \cdot D + B \cdot E$$

Notice that on the right-hand side of the previous equation, each of A and B is multiplied with each of C, D, and E.

Repeating the reasoning of the previous paragraph with other products, we see that FOIL and the previous equation are both special cases of an **extended distributive property**, which tells us that, to multiply a sum by another sum, we can multiply each number in the first sum by each number in the second sum and add all these products.

extended distributive property

Practice Exercises for Section 4.4

1. Does the expression

$$3 \times 4 + 2$$

have a different meaning than the expression

$$3 \times \quad 4 + 2$$

which has a big space between the $\times$ and the 4? Explain.

2. Dana and Sandy are working on

$$8 \times 5 + 20 \div 4$$

Dana says the answer is 15, but Sandy says the answer is 45. Who's right, who's wrong, and why?

3. Mr. Greene has a bag of plastic spiders to give out. After putting 4 spiders in each of 23 bags, he has 8 spiders left. Write an expression using the numbers 4, 23, and 8; the symbols $\times$ (or $\cdot$) and $+$; and parentheses, if needed, for the total number of spiders Mr. Greene had to give out. If you use parentheses, explain why you need them; if you do not use parentheses, explain why you do not need them.

4. State the distributive property.

5. There are 15 goodie bags. Each goodie bag contains 2 pencils, 4 stickers, and 3 wiggly worms. Write two different expressions using the numbers 15, 2, 4, and 3; the symbols $\times$ (or $\cdot$) and $+$; and parentheses, if needed, for the total number of objects in the 15 goodie bags. If you use parentheses

in an expression, explain why you need them; if you do not use parentheses, explain why you do not need them. Use your two expressions to illustrate the distributive property (variation 2).

6. Using a specific example, explain why the distributive property makes sense. Even though you use a specific example, your explanation should be general, in the sense that we should be able to see why it will hold true when other numbers replace your numbers.

7. Give an example of how to use the distributive property to make a calculation easy to carry out mentally. Describe how to carry out the calculation mentally. Write equations showing how the calculation strategy used the distributive property.

8. Compute mentally:

$$97,346 \times 142,349 + 2,654 \times 142,349$$

9. Draw an array that shows why

$$20 \times 19 = 20 \times 20 - 20 \times 1$$

Also, use this equation to help you calculate 20×19 mentally.

10. Draw a subdivided array to show that

$$(10 + 2) \times (10 + 3)$$
$$= 10 \times 10 + 10 \times 3 + 2 \times 10 + 2 \times 3$$

Then write equations that use properties of arithmetic to show why the preceding equation is true.

Answers to Practice Exercises for Section 4.4

1. The expressions

$$3 \times 4 + 2$$

and

$$3 \times \quad 4 + 2$$

have the same meaning. According to the convention on order of operations, multiplication is performed before addition, no matter how big a

space there is following the multiplication symbol. If you want to write 3 times the *quantity* $4 + 2$, then you must use parentheses and write

$$3 \times (4 + 2)$$

2. Sandy is right that the answer is 45. Dana, like many students, just worked from left to right. She did not use the conventions on order of operations: Do multiplication and division first, then addition and

subtraction. According to the conventions on the order of operations,

$$8 \times 5 + 20 \div 4 = 40 + 5$$
$$= 45$$

3. Since there are 23 bags with 4 spiders in each bag, the total number of spiders in bags is 23×4. But there are also 8 more spiders, so the total number of spiders that Mr. Greene has is 8 more than 23×4, which is

$$23 \times 4 + 8$$

We do not need paretheses because, according to the convention on order of operations, we calculate the foregoing expression by multiplying 23 times 4 and then adding 8 to that quantity, which is exactly what we want to express.

4. See text.

5. Since there are 15 goodie bags and each goodie bag has a total of $2 + 4 + 3$ objects in it, one expression for the total number of objects in all the goodie bags is

$$15 \cdot (2 + 4 + 3)$$

We need parentheses in this expression to show that 15 multiplies the entire quantity $2 + 4 + 3$.

Another expression for the total number of objects in the goodie bags is

$$15 \cdot 2 + 15 \cdot 4 + 15 \cdot 3$$

Since there are $15 \cdot 2$ pencils, $15 \cdot 4$ stickers, and $15 \cdot 3$ wiggly worms. Because of the conventions on order of operations, we do not need parentheses here, since we carry out multiplication before addition. Since it's the same total number of objects either way we count them, we have the equation

$$15 \cdot (2 + 4 + 3) = 15 \cdot 2 + 15 \cdot 4 + 15 \cdot 3$$

which illustrates the distributive property (variation 2).

6. See the explanation in the text for why $4 \times (5 + 2) = 4 \times 5 + 4 \times 2$ using the array in Figure 4.28. See also the discussion on why that explanation is a general one. Note in particular that the explanation *does not* rely on evaluating either $4 \times (5 + 2)$ or $4 \times 5 + 4 \times 2$ as equal to the specific number 28. If it did, it would not be a general explanation because we wouldn't be able to tell how the explanation would work if the numbers 4, 5, and 2 in the example were replaced with other counting numbers.

7. We can calculate 7×12 mentally by first calculating $7 \times 10 = 70$, then $7 \times 2 = 14$ and then adding $70 + 14 = 84$. This strategy uses the distributive property, as shown by the next equations:

$$7 \times 12 = 7 \times (10 + 2)$$
$$= 7 \times 10 + 7 \times 2$$
$$= 70 + 14 = 84$$

8. By the distributive property,

$$97{,}346 \times 142{,}349 + 2{,}654 \times 142{,}349$$
$$= (97{,}346 + 2{,}654) \times 142{,}349$$
$$= 100{,}000 \times 142{,}349$$
$$= 14{,}234{,}900{,}000$$

9. Figure 4.30 shows 20 rows of 20 squares. If you take away 20 rows of 1 square—namely, a vertical strip of squares—then you are left with 20 rows of 19 squares, thus illustrating that

$$20 \times 19 = 20 \times 20 - 20 \times 1$$
$$= 400 - 20$$
$$= 380$$

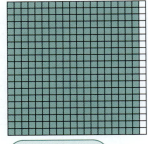

FIGURE 4.30   An array showing that $20 \times 19 = 20 \times 20 - 20 \times 1$

10. The array in Figure 4.31 shows $10 + 2$ groups of $10 + 3$ small squares. According to the meaning of multiplication, there are a total of $(10 + 2) \times (10 + 3)$ small squares in this array. The array is broken into four smaller arrays: 10 groups of 10, 10 groups of 3, 2 groups of 10, and 2 groups of 3. But it's the same number of small squares no matter how you count them; therefore,

$$(10 + 2) \times (10 + 3) = 10 \times 10 + 10 \times 3$$
$$+ 2 \times 10 + 2 \times 3$$

We can also explain this equation by using the distributive property:

$$(10 + 2) \times (10 + 3) = (10 + 2) \times 10$$
$$+ (10 + 2) \times 3$$
$$= 10 \times 10 + 2 \times 10$$
$$+ 10 \times 3 + 2 \times 3$$

The distributive property was used at both equal signs. In fact, it was used twice at the second equals sign.

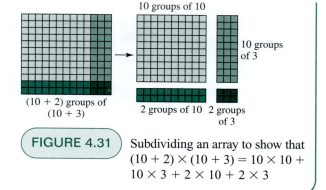

10 groups of 10

10 groups of 3

(10 + 2) groups of (10 + 3)

2 groups of 10   2 groups of 3

FIGURE 4.31 Subdividing an array to show that $(10 + 2) \times (10 + 3) = 10 \times 10 + 10 \times 3 + 2 \times 10 + 2 \times 3$

Problems for Section 4.4

1. Ben and Charles are working on

$$4 + 3 \times 2 \times 10$$

Ben says the answer is 64. Charles says the answer is 140. Who's right, who's wrong, and why?

2. **a.** There are 6 cars traveling together. Each car has 2 people in front and 3 in the back. Write an expression using the numbers 6, 2, and 3; the symbols $\times$ (or $\cdot$) and $+$; and parentheses, if needed, for the total number of people riding in the 6 cars. If you use parentheses, explain why you need them; if you do not use parentheses, explain why you do not need them.

   b. Write a story problem for the expression

$$6 \times 2 + 3$$

3. The children in Mrs. Black's class are arranged as follows: 7 tables have 4 children sitting at each of them; 2 tables have 3 children sitting at each of them. Write an expression using the numbers 7, 4, 2, and 3; the symbols $\times$ (or $\cdot$) and $+$; and parentheses, if needed, for the total number of children in Mrs. Black's class. If you use parentheses, explain why you need them; if you do not use parentheses, explain why you do not need them.

4. Write one story problem for the expression

$$8 \times (4 + 2)$$

and another story problem for the expression

$$8 \times 4 + 2$$

Make clear which is which and why each story problem fits with its expression.

5. There are 6 cars traveling together. Each car has 2 people in front and 3 people in back. Explain how to use this situation to illustrate the distributive property.

6. Draw arrays to help you explain why the equations shown are true, without evaluating any of the products or sums. Explain your answers clearly.

   a. $(5 + 1) \cdot 7 = 5 \cdot 7 + 1 \cdot 7$

   b. $6 \cdot (5 + 2) = 6 \cdot 5 + 6 \cdot 2$

   c. $(10 - 1) \cdot 6 = 10 \cdot 6 - 1 \cdot 6$

7. Explain how to use the distributive property to make 31×25 easy to calculate mentally. Write an equation that corresponds to your strategy. Without drawing all the detail, draw a rough picture of an array that illustrates this calculation strategy.

8. Explain how to calculate 29×20 mentally by using 30×20. Write an equation that uses subtraction and the distributive property and that corresponds to your strategy. Without drawing all the detail, draw a rough picture of an array that illustrates this calculation strategy.

9. Ted thinks that because $10 \times 10 = 100$ and $2 \times 5 = 10$, he should be able to calculate 12×15 by adding $100 + 10$ to get 110. Explain to Ted in two different ways that, even though his method is not correct, his calculations can be part of a correct way to calculate 12×15.

   a. by drawing an array

   b. by writing equations that use the distributive property

10. SunJae is working on the multiplication problem 21 × 34. SunJae says that he can take 1 from the 21 and put it with the 34 to get 20 × 35. SunJae says that this new multiplication problem, 20 × 35, should have the same answer as 21 × 34. Is SunJae right? How might you convince SunJae that his reasoning is or is not correct in a way other than simply showing him the answers to the two multiplication problems?

11. **a.** Use the distributive property several times to show why

    $$(10 + 2) \cdot (10 + 4)$$
    $$= 10 \cdot 10 + 10 \cdot 4 + 2 \cdot 10 + 2 \cdot 4$$

    b. Draw an array to show why

    $$(10 + 2) \cdot (10 + 4)$$
    $$= 10 \cdot 10 + 10 \cdot 4 + 2 \cdot 10 + 2 \cdot 4$$

    c. Relate the steps in your equations in part (a) to your array in part (b).

12. In Section 4.2, we drew pictures of bundled objects to explain why multiplication by 10 shifts the digits in a decimal number. Now use the distributive property, expanded forms, and place value to explain why multiplying a number by 10 moves each digit one place to the left. Use the example 12,345 × 10 to illustrate.

13. **a.** Use an ordinary calculator to calculate 666,666,666 × 999,999,999. Based on the calculator's display, guess how to write the answer in ordinary decimal notation (showing *all* digits).

    b. Now think some more, and determine how to write the product 666,666,666 × 999,999,999 in ordinary decimal notation, showing all its digits, without multiplying longhand or using a calculator or computer.

14. Calculate the product 9,999,999,999 × 9,999,999,999 without using a calculator or computer and without simply multiplying longhand. Give the answer in ordinary decimal notation, showing all its digits. (Both numbers in the product have 10 nines.) Explain your method.

15. Check the following:

    $$11 - 2 = 3 \times 3$$
    $$1111 - 22 = 33 \times 33$$
    $$111111 - 222 = 333 \times 333$$

Continue to find at least three more in the pattern. Does the pattern continue? Now explain why there is such a pattern. Hint: Notice that $1111 - 22 = 11 \times 101 - 11 \times 2$.

16. Determine which of the following two numbers is larger, and explain your reasoning:

    a. $1{,}000{,}000 \times (1 + 2 + 3 + 4 + \cdots + 1{,}000{,}001)$

    b. $1{,}000{,}001 \times (1 + 2 + 3 + 4 + \cdots + 1{,}000{,}000)$

17. The **square** of a number is just the number times itself. For example, the square of 4 is $4 \cdot 4 = 16$.

    a. Find the squares of the numbers 15, 25, 35, 45, ... , 95, 105, 115, 125, ... , 195, 205, and three other whole numbers that end in 5.

    b. Find some patterns in the answers to part (a). Specifically, in all of your answers to part (a), what do you notice about the last two digits, and what do you notice about the number formed by deleting the last two digits?

    c. Use what you discovered in part (b) to predict the squares of 2,005 and 10,005.

    d. Every whole number ending in 5 must be of the form $10A + 5$ for some whole number A. Find the square of $10A + 5$—namely, calculate $(10A + 5) \times (10A + 5)$.

    e. How does your answer to part (d) explain the pattern you found in part (b)?

18. The square of a number is just the number times itself. For example, the square of 4 is 16.

    a. Calculate the squares of 1, 2, ..., 9 and many other whole numbers, including 17, 34, 61, 82, 99, 123, 255, 386, and 728. Record the ones digits in each case. What do you notice? Do any of your squares have a ones digit of 7, for example? Are any other digits missing from the ones digits of squares?

    b. Which digits can never occur as the ones digit of a square of a whole number? Explain why some digits cannot occur as the ones digit of a square of a whole number.

    c. Based on what you've discovered, could the number

    139,787,847,234,329,483

    be the square of a whole number? Why or why not?

4.5 Properties of Arithmetic, Mental Math, and Single-Digit Multiplication Facts

 **Focal Points**
Grades 3, 4

The properties of arithmetic are the building blocks for all of arithmetic. They allow us to break calculations apart into pieces, calculate the pieces, and then put the pieces back together to obtain the full solution. This process of taking apart, analyzing, and putting back together is a powerful technique that is used across all of mathematics: in arithmetic, in geometry, and in algebra—and at all levels. In this section we'll examine how we can combine the properties of arithmetic to take calculations apart. First we'll look at the single-digit multiplication facts and see how students who are learning these facts can build up their knowledge by breaking apart facts they haven't yet mastered into ones they already know. Then we'll examine other mental calculations, including percent calculations. Throughout this section, note how breaking calculations apart is essentially a way of seeing the algebra in the calculations.

The Single-Digit Multiplication Facts

In school, children must develop quick recall of the single-digit multiplication facts from $1 \times 1 =$ to $9 \times 9 = 81$. Fluency with these facts is essential to succeed in mathematics, but the process of becoming fluent is not merely rote memorization.

When children first begin to learn to multiply, they multiply by skip-counting. For example, to multiply 5 by 7, a child will count by 5s seven times:

$$5, 10, 15, 20, 25, 30, 35$$

A visual aid to connect skip-counting with arrays, and thus with multiplication, is shown in Figure 4.32. To skip-count with the aid of an array (which may or may not have the numbers written to the side), children can cover the array with a slip of paper and then uncover more and more of the array as they count.

It may seem that the single-digit multiplication facts should simply be memorized. However, strong programs and accomplished teachers help children learn these multiplication facts by teaching them relationships among the facts. In this way, children learn to organize the facts so that they know them better and learn them more quickly. For example, if a child knows that $7 \times 5 = 35$, and the child knows the commutative property of multiplication, then the child will be able to determine that $5 \times 7 = 35$. The child can then figure out 6×7 by thinking that 6 groups of 7 is one more group of 7 than 5 groups of 7, so 6×7 is 7 more than 35, which is 42.

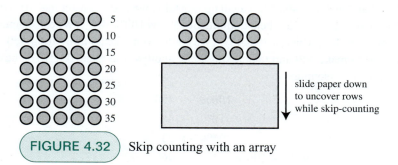

FIGURE 4.32   Skip counting with an array

The next Class Activity will help you think about relationships among the single-digit multiplication facts. It will also help you express those relationships algebraically, with equations, so that you understand how the properties of arithmetic are involved in the relationships. For example, the previous reasoning about deriving 6×7 from 5×7 can be summarized with these equations:

$$6 \times 7 = (5 + 1) \times 7$$
$$= 5 \times 7 + 1 \times 7$$
$$= 35 + 7 = 42$$

Although your students might not write such equations, you may wish to write such equations for them to introduce them to the kind of reasoning and notation they will eventually use in algebra. In algebra, students will multiply polynomials such as

$$(5x^2 + 1) \cdot 7x$$

which they will do with equations that are similar to the ones above, namely:

$$(5x^2 + 1) \cdot 7x = 5x^2 \cdot 7x + 1 \cdot 7x$$
$$= 35x^3 + 7x$$

Class Activity   *Now Turn to Class Activities Manual*

4T 🏛 Using Properties of Arithmetic to Aid the Learning of Basic Multiplication Facts, p. 86

Showing the Algebra in Mental Math by Writing Equations

Sections 4.3 and 4.4 included examples of how to apply properties of arithmetic to make mental calculations easy to do. We used whole numbers in these examples, but the properties of arithmetic apply to all numbers, including percents, for example. In Chapter 2 we reviewed ways of making percent calculations easy to do with the aid of percent tables. Now we'll see how to write equations that fit with the percent table methods, so that we can examine the algebra that is in these methods.

In Chapter 2 we worked with percent calculations by viewing them as problems about equivalent fractions. But we can also think about percent calculations as multiplication problems. Why? Suppose we want to calculate $P\%$ of some quantity Q. Just as "2 of Q" is $2 \cdot Q$ and "3 of Q" is $3 \cdot Q$," so too

$$P\% \text{ of } Q \text{ is } P\% \cdot Q$$

So for example, what is 95% of 460? To calculate mentally, we could reason that 10% of 460 is 46, so 5% of 460 is half of 46, which is 23. Then, thinking of 95% as $100\% - 5\%$, we can calculate 95% of 460 by taking 23 away from 460, which leaves 437. (We could even take 23 away from 460 in two steps by first taking away 20 to make 440 and then taking away another 3 to make 437.) The next percent table summarizes this line of reasoning:

$$100\% \rightarrow 460$$
$$10\% \rightarrow 46$$
$$5\% \rightarrow 23$$
$$95\% = 100\% - 5\% \rightarrow 460 - 23 = 437$$

To show the algebra that is in this mental math strategy, we can write the following equations that fit with the strategy:

$$
\begin{aligned}
95\% \cdot 460 &= (100\% - 5\%) \cdot 460 \\
&= 100\% \cdot 460 - 5\% \cdot 460 \\
&= 460 - \left(\frac{1}{2} \cdot 10\%\right) \cdot 460 \\
&= 460 - \frac{1}{2} \cdot (10\% \cdot 460) \\
&= 460 - \frac{1}{2} \cdot 46 \\
&= 460 - 23 = 437
\end{aligned}
$$

Notice that to formulate these equations, we have to think about the calculation in a big-picture way. These equations show that the distributive property was used at the second equal sign and the associative property of multiplication was used at the fourth equal sign.

One important comment about writing equations that fit with a mental calculation method: We use these equations to show the algebraic structure of the calculations and thus to better understand the nature of the calculations and to go more deeply into their inner workings. In ordinary circumstances, carry out mental calculations in a quick and efficient way—and *mentally*!

Class Activity   *Now Turn to Class Activities Manual*

4U Solving Arithmetic Problems Mentally, p. 88

4V Which Properties of Arithmetic Do These Calculations Use? p. 88

4W Writing Equations That Correspond to a Method of Calculation, p. 90

4X Showing the Algebra in Mental Math, p. 91

Practice Exercises for Section 4.5

1. Write equations that correspond to the following reasoning for determining 5×7:

   I know 2×7 is 14. Then another 2×7 makes 28. And one more 7 makes 35.

   Which properties of arithmetic are used? Explain. Draw an array that corresponds to the reasoning.

2. A child is having difficulty remembering 8×8. Draw two arrays showing how 8×8 is related to other possibly easier facts involving smaller numbers. For each array, write a corresponding equation relating 8×8 to other multiplication facts. Which properties of arithmetic do you use?

3. The string of equations that follows corresponds to a mental method for calculating 45×11. Explain in words why the method of calculation makes sense. Which properties of arithmetic are used, and where are they used?

$$
\begin{aligned}
45 \times 11 &= 45 \times (10 + 1) \\
&= 45 \times 10 + 45 \times 1 \\
&= 450 + 45 \\
&= 495
\end{aligned}
$$

4. Each arithmetic problem in this exercise has a description for solving the problem. In each case, write a string of equations that corresponds to the given description. Identify properties of arithmetic that are used. Write your equations in the following form:

$$
\begin{aligned}
\text{original} &= \text{some expression} \\
&= \vdots \\
&= \text{some expression}
\end{aligned}
$$

a. *Problem:* What is 6×40?

Solution: 6 times 4 is 24; then you multiply by 10 and get 240.

b. *Problem:* What is 110% of 62?

Solution: 100% of 62 is 62 and 10% of 62 is 6.2, so all together it's 68.2.

c. *Problem:* Calculate the 7% tax on a purchase of $25.

Solution: 10% of $25 is $2.50, so 5% is $1.25. One percent of $25 is 25 cents, so 2% is 50 cents. This means 7% is $1.25 plus 50 cents, which is $1.75

d. *Problem:* What is 45% of 300?

Solution: 50% of 300 is 150. Ten percent of 300 is 30, and half of that is 15, so the answer is 135.

e. *Problem:* Find $\frac{3}{4}$ of 72.

Solution: Half of 72 is 36. Half of 36 is 18. Then, to add 36 and 18, I did 40 plus 14, which is 54.

f. *Problem:* Calculate $59 \cdot 70$.

Solution: 6 times 7 is 42, so 60 times 70 is 4200. Therefore, 59 times 70 is 70 less, which is 4130.

5. For each of the problems that follow, use the distributive property to help make the problem easy to solve mentally. In each case, write a string of equations that correspond to your strategies. Write your equations in the following form:

$$\text{original} = \text{some expression}$$
$$= \vdots$$
$$= \text{some expression}$$

a. Calculate 51% of 140.

b. Calculate 95% of 60.

c. Calculate $\frac{5}{8} \times 280$.

d. Calculate 99% of 80.

6. For each of the following arithmetic problems, use properties of arithmetic to make the problem easy to solve mentally. Write a string of equations that corresponds to your method. Say which properties of arithmetic you use and where you use them.

a. 25×84

b. 49×6

c. 486×5

7. Joey, a second-grader, used the following method to mentally calculate the number of minutes in a day: First, Joey calculated 25 times 6 by finding 4 times 25 plus two times 25, which is 150. Then he subtracted 6 from 150 to get 144. Then he multiplied 144 by 10 to get the answer: 1440. Write a sequence of equations that corresponds to Joey's method and that show why his method is legitimate. What properties of arithmetic are involved?

Answers to Practice Exercises for Section 4.5

1. This distributive property is used at the second equals sign in the following equations:

$$5 \times 7 = (2 + 2 + 1) \times 7$$
$$= 2 \times 7 + 2 \times 7 + 1 \times 7$$
$$= 14 + 14 + 7$$
$$= 28 + 7 = 35$$

The following subdivided array corresponds to the arithmetic because it shows 2 rows of seven, another 2 rows of seven, and then a single row of seven, to make a total of 5 rows of seven.

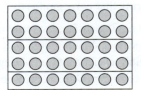

2. See Figure 4.33 for the arrays. The array on the left breaks 8 groups of 8 stars into 2 groups of 4×8. The corresponding equation is

$$8 \times 8 = 2 \times (4 \times 8)$$

This equation uses the associative property of multiplication because the first 8 is broken into 2×4 and this 4 is switched from associating with the 2 to associating with the 8 on the right. We could also write the equation

$$8 \times 8 = 4 \times 8 + 4 \times 8$$

which uses the distributive property instead of the associative property. Notice that if the child knows the fact $4 \times 8 = 32$, then he might be able to add $32 + 32$ quickly to get the correct answer to 8×8.

FIGURE 4.33 Relating 8×8 to other multiplication facts

The array on the right of Figure 4.33 breaks 8 groups of 8 stars into 5 groups of 8 and 3 groups of 8. The corresponding equation is

$$8 \times 8 = 5 \times 8 + 3 \times 8$$

Notice that if the child knows 5×8 and 3×8, then she might be able to add $40 + 24$ quickly to get the correct answer to 8×8.

3. The product 45×11 stands for the total number of objects in 45 groups that have 11 objects in each group. These objects can be broken into 45 groups of 10 and another 45 groups of 1. The number of objects in 45 groups of 10 is 45×10, which is 450, and the number of objects in 45 groups of 1 is 45. In all, that makes $450 + 45$, which is 495 objects. The distributive property is used to say that $45(10 + 1) = 45 \times 10 + 45 \times 1$.

4. **a.** $6 \times 40 = 6 \times (4 \times 10)$

$$= (6 \times 4) \times 10$$
$$= 24 \times 10$$
$$= 240$$

The associative property of multiplication is used at the second equal sign in order to switch the placement of the parentheses and group the 4 with the 6 instead of with the 10.

b. $110\% \times 62 = (100\% + 10\%) \times 62$

$$= 100\% \times 62 + 10\% \times 62$$
$$= 62 + 6.2$$
$$= 68.2$$

The distributive property is used at the second equal sign.

c. $7\% \times 25 = (5\% + 2\%) \times 25$

$$= 5\% \times 25 + 2\% \times 25$$
$$= \frac{1}{2} \times 10\% \times 25 + 2 \times 1\% \times 25$$
$$= \frac{1}{2} \times 2.5 + 2 \times 0.25$$
$$= 1.25 + 0.50$$
$$= 1.75$$

The distributive property is used at the second equal sign. Although parentheses are not shown, the associative property of multiplication is used twice at the fourth equal sign in order to calculate 10% of 25 and 1% of 25 first, and then multiply those by $\frac{1}{2}$ and 2, respectively.

d. $45\% \times 300 = (50\% - 5\%) \times 300$

$$= 50\% \times 300 - 5\% \times 300$$
$$= 150 - \frac{1}{2} \times 10\% \times 300$$
$$= 150 - \frac{1}{2} \times 30$$
$$= 150 - 15$$
$$= 135$$

The distributive property is used at the second equal sign. Although parentheses are not shown, the associative property of multiplication is used at the fourth equal sign to calculate 10% of 300 first, and then find half of that.

e. $\frac{3}{4} \times 72 = \left(\frac{1}{2} + \frac{1}{2} \times \frac{1}{2} \right) \times 72$

$$= \frac{1}{2} \times 72 + \frac{1}{2} \times \frac{1}{2} \times 72$$
$$= 36 + \frac{1}{2} \times 36$$
$$= 36 + 18$$
$$= 36 + (4 + 14)$$
$$= (36 + 4) + 14$$
$$= 40 + 14$$
$$= 54$$

The distributive property is used at the second equal sign. Although parentheses are not shown, the associative property of multiplication is used at the third equal sign to find half of 72 first. Then we take half of that. The associative property of addition was used at the sixth equal sign to change the placement of the parentheses.

f. $59 \cdot 70 = (60 - 1) \cdot 70$

$$= 60 \cdot 70 - 1 \cdot 70$$
$$= (6 \cdot 10) \cdot (7 \cdot 10) - 70$$
$$= (6 \cdot 7) \cdot (10 \cdot 10) - 70$$
$$= 42 \cdot 100 - 70$$
$$= 4200 - 70 = 4130$$

The distributive property is used at the second equal sign. The commutative and associative properties of multiplication are used at the fourth equal sign. Notice that even though the stated solution to the problem did not explicitly mention multiplying 10 by 10 to get 100 and then multiplying 42 by 100, these calculations were used implicitly. So the previous equations actually expand on the stated solution, filling in the implied details.

5. We can decompose 51% as 50% plus 1%. Fifty percent of 140 is half of 140, which is 70. One percent of 140 is 1.4. So 51% of 140 is 70 plus 1.4, which is 71.4. Using equations, we have

a. $51\% \times 140 = (50\% + 1\%) \times 140$
$= 50\% \times 140 + 1\% \times 140$
$= 70 + 1.4 = 71.4$

The distributive property was used at the second equal sign.

b. We can decompose 95% as 100% minus 5%. Now 10% of 60 is 6, so 5% of 60 is half of that, which is 3. Therefore, 95% of 60 is $60 - 3$, which is 57. Using equations, we have

$95\% \times 60 = (100\% - 5\%) \times 60$
$= 100\% \times 60 - 5\% \times 60$
$= 60 - \frac{1}{2} \times 10\% \times 60$
$= 60 - \frac{1}{2} \times 6$
$= 60 - 3 = 57$

The distributive property was used at the second equal sign.

c. Use the fact that $\frac{5}{8} = \frac{4}{8} + \frac{1}{8} = \frac{1}{2} + \frac{1}{8}$.

$\frac{5}{8} \times 280 = \left(\frac{1}{2} + \frac{1}{8}\right) \times 280$
$= \frac{1}{2} \times 280 + \frac{1}{2} \times \frac{1}{2} \times \frac{1}{2} \times 280$
$= 140 + 35 = 175$

The distributive property was used at the second equal sign. The associative property of multiplication could be used at the third equal sign to find $\frac{1}{8}$ of 280 by finding $\frac{1}{2}$ of 280, then $\frac{1}{2}$ of that result, and half of that result.

d. $99\% \times 80 = (100\% - 1\%) \times 80$
$= 100\% \times 80 - 1\% \times 80$
$= 80 - 0.8$
$= 79.2$

The distributive property was used at the second equal sign.

6. a. $25 \times 84 = 25 \times (4 \times 21)$
$= (25 \times 4) \times 21$
$= 100 \times 21$
$= 2100$

The associative property of multiplication was used to rewrite $25 \times (4 \times 21)$ as $(25 \times 4) \times 21$—in other words, to multiply the 4 with the 25 instead of with the 21.

b. $49 \times 6 = (50 - 1) \times 6$
$= 50 \times 6 - 1 \times 6$
$= 300 - 6$
$= 294$

The distributive property was used to rewrite $(50 - 1) \times 6$ as $50 \times 6 - 1 \times 6$.

c. $486 \times 5 = (243 \times 2) \times 5$
$= 243 \times (2 \times 5)$
$= 243 \times 10$
$= 2430$

The associative property of multiplication was used to rewrite $(243 \times 2) \times 5$ as $243 \times (2 \times 5)$.

7. The following sequence of equations shows in detail why Joey's method is valid and how it uses properties of arithmetic:

$24 \times 60 = 24 \times (6 \times 10)$
$= (24 \times 6) \times 10$
$= [(25 - 1) \times 6] \times 10$
$= (25 \times 6 - 1 \times 6) \times 10$
$= [6 \times 25 - 6] \times 10$
$= [(4 + 2) \times 25 - 6] \times 10$
$= [4 \times 25 + 2 \times 25 - 6] \times 10$
$= [100 + 50 - 6] \times 10$
$= (150 - 6) \times 10$
$= 144 \times 10 = 1440$

The associative property of multiplication is used at the second equal sign.

The distributive property is used at the fourth and seventh equal signs.

The commutative property of multiplication is used at the fifth equal sign.

The associative property of addition was essentially used at the ninth equal sign.

Problems for Section 4.5

1. Josh consistently remembers that $7 \times 7 = 49$, but he keeps forgetting 7×8.

   a. Explain to Josh how 7×7 and 7×8 are related. Draw an array to help you show this relationship.

   b. Write an equation relating 7×8 to 7×7. Which property of arithmetic does your equation use? Explain.

2. Demarcus knows his $1\times$, $2\times$, and $3\times$ multiplication tables. He also knows 4×1, 4×2, 4×3, 4×4, and 4×5.

   a. Describe how the three arrays in Figure 4.34 provide Demarcus with three different ways to determine 4×6 from multiplication facts that he already knows. In each case, write an equation that corresponds to the array and that shows how 4×6 is related to other multiplication facts.

FIGURE 4.34 Different ways to think about 4×6

   b. Draw arrays showing two different ways that Demarcus could use the multiplication facts he already knows to determine 4×7. In each case, write an equation that corresponds to the array and that shows how 4×7 is related to other multiplication facts.

   c. Draw arrays showing two different ways that Demarcus could use the multiplication facts he already knows to determine 4×8. In each case, write an equation that corresponds to the array and that shows how 4×8 is related to other multiplication facts.

3. Suppose that a child has learned the following basic multiplication facts:

   • The $\times 1$, $\times 2$, $\times 3$, $\times 4$, and $\times 5$ tables—that is,

$1 \times 1 = 1$	$1 \times 2 = 2$	$1 \times 3 = 3$	$1 \times 4 = 4$	$1 \times 5 = 5$
$2 \times 1 = 2$	$2 \times 2 = 4$	$2 \times 3 = 6$	$2 \times 4 = 8$	$2 \times 5 = 10$
$3 \times 1 = 3$	$3 \times 2 = 6$	$3 \times 3 = 9$	$3 \times 4 = 12$	$3 \times 5 = 15$
$4 \times 1 = 4$	$4 \times 2 = 8$	$4 \times 3 = 12$	$4 \times 4 = 16$	$4 \times 5 = 20$
$5 \times 1 = 5$	$5 \times 2 = 10$	$5 \times 3 = 15$	$5 \times 4 = 20$	$5 \times 5 = 25$
$6 \times 1 = 6$	$6 \times 2 = 12$	$6 \times 3 = 18$	$6 \times 4 = 24$	$6 \times 5 = 30$
$7 \times 1 = 7$	$7 \times 2 = 14$	$7 \times 3 = 21$	$7 \times 4 = 28$	$7 \times 5 = 35$
$8 \times 1 = 8$	$8 \times 2 = 16$	$8 \times 3 = 24$	$8 \times 4 = 32$	$8 \times 5 = 40$
$9 \times 1 = 9$	$9 \times 2 = 18$	$9 \times 3 = 27$	$9 \times 4 = 36$	$9 \times 5 = 45$

   • The squares $1 \times 1 = 1$, $2 \times 2 = 4$, $3 \times 3 = 9, \ldots, 9 \times 9 = 81$.

   For each of the multiplication problems (a) through (c), find at least two different ways that some of the preceding facts, together with properties of arithmetic, could be used to mentally calculate the answer to the problem. Explain your answers, drawing arrays to illustrate how the following problems are related to facts in the given lists:

   a. 6×7 **b.** 7×8 **c.** 6×8

4. For each of the multiplication problems in this exercise, describe a way to make the problem easy to solve mentally. Then write equations that correspond to your method of calculation. Write your equations in the following form:

$$24 \times 25 = \text{some expression}$$
$$= \text{some expression}$$
$$\vdots \quad \vdots$$

   In each case, state which properties of arithmetic you used, and indicate where you used those properties.

   a. 24×25 **b.** 25×48 **c.** 51×6

5. Suppose that the sales tax where you live is 6%. Compare the total amount of sales tax you would pay if you bought a pair of pants and a shirt at the same time, versus if you first bought the pants and then went back to the store and bought the shirt. Write equations to show why the distributive property is relevant to this problem.

6. Clint and Sue went out to dinner and had a nice meal that cost $64.82. With a 7% tax of $4.54, the total came to $69.36. They want to leave a tip of approximately 15% of the cost of the meal (before the tax). Describe a way that Clint and Sue can mentally figure the tip.

7. Your favorite store is having a 10%-off sale, meaning that the store will take 10% off the price of each item you buy. When the clerk rings up your purchases, she takes 10% off the total (before tax), rather than 10% off each item. Will you get the same discount either way? Is there a property of arithmetic related to this? Explain!

8. The exchanges that follow are taken from *Developing Children's Understanding of the Rational Numbers: A New Model and an Experimental Curriculum* by Joan Moss and Robbie Case [53, p. 135]. "*Experimental S1*" and "*Experimental S3*" are two of the fourth-grade students who participated in an experimental curriculum described in the article.

Experimenter: What is 65% of 160?

Experimental S1: Fifty percent (of 160) is 80. I figure 10%, which would be 16. Then I divided by 2, which is 8 (5%) then 16 plus 8 um ... 24. Then I do 80 plus 24, which would be 104.

⋮   ⋮

Experimental S3: Ten percent of 160 is 16; 16 times 6 equals 96. Then I did 5%, and that was 8, so ..., 96 plus 8 equals 104.

For each of the two students' responses, write strings of equations that correspond to the student's method for calculating 65% of 160. State which properties of arithmetic were used and where. (Be specific.)

Write your string of equations in the following form:

$$65\% \times 160 = \text{some expression}$$
$$= \vdots$$
$$= \text{some expression}$$

9. Here is Marco's method for calculating 38×60:

4 times 6 is 24, so 40 times 60 is 2400. Then 2 times 6 is 12, so it's $2400 - 120$, which is 2280.

Write equations that incorporate Marco's method and that also show why his method is valid. Write your equations in the following form:

$$38 \times 60 = \text{some expression}$$
$$= \text{some expression}$$
$$\vdots$$
$$= 2280$$

Which properties of arithmetic did Marco use (knowingly or not), and where?

10. Jenny uses the following method to find 28% of 60,000 mentally:

25% is $\frac{1}{4}$, and $\frac{1}{4}$ of 60 is 15, so 25% of 60,000 is 15,000. One percent of 60,000 is 600, and that times 3 is 1800. So the answer is $15,000 + 1,800$, which is 16,800.

Write a string of equations that calculates 28% of 60,000 and that incorporates Jenny's ideas. Write your equations in the following form:

$$28\% \times 60,000 = \text{some expression}$$
$$= \text{some expression}$$
$$\vdots$$
$$= 16,800$$

11. Use properties of arithmetic to calculate 35% of 440 mentally. Describe your strategy in words, and write a string of equations that corresponds to your strategy. Indicate which properties of arithmetic you used, and where. Be specific. Write your equations in the following form:

$$35\% \times 440 = \text{some expression}$$
$$= \vdots$$
$$= \text{some expression}$$

12. Use the distributive property to make it easy for you to calculate 30% of 240 mentally. Then use the

associative property of multiplication to solve the same problem. In each case, explain your strategy in words, and then write equations that correspond to your strategy. Write your equations in the following form:

$$30\% \times 240 = \text{some expression}$$
$$= \text{some expression}$$
$$\vdots$$

13. Tamar calculated 41×41 as follows:

Four 4s is 16, so four 40s is 160 and forty 40s is 1600. Then forty-one 40s is another 40 added on, which is 1640. So forty-one 41s is 41 more, which is 1681.

   a. Explain briefly why it makes sense for Tamar to solve the problem the way she does. What is the idea behind her strategy?

   b. Write equations that incorporate Tamar's work and that show clearly why Tamar's method calculates the correct answer to 41×41. Which properties of arithmetic did Tamar use (knowingly or not), and where? Be thorough and be specific. Write your equations in the following format:

$$41 \times 41 = \text{some expression}$$
$$= \text{some expression}$$
$$\vdots \quad \vdots$$
$$= 1681$$

14. Here is how Nya solved the problem $\frac{3}{4} \cdot 72$:

Half of 72 is 36. Half of 36 is 18. Then, to add 36 and 18, I did 40 plus 14, which is 54.

Write a string of equations that incorporate Nya's ideas. Which properties of arithmetic did Nya use (knowingly or not), and where? Be thorough and be specific. Write your equations in the following format:

$$\frac{3}{4} \cdot 72 = \text{some expression}$$
$$= \text{some expression}$$
$$\vdots$$
$$= 54$$

15. a. Lindsay calculates two-fifths of 1260 by using the following strategy: First, Lindsay finds half of 1260, which is 630. Then Lindsay subtracts one tenth of 1260, which is 126, from 630 and gives the answer, 504. Discuss the ideas behind

Lindsay's strategy. Then write a string of equations that incorporate Lindsay's strategy and that show why the strategy is valid. What property of arithmetic is involved? Write your equations in the following form:

$$\frac{2}{5} \times 1260 = \text{some expression}$$
$$= \text{some expression}$$
$$\vdots$$
$$= 504$$

   b. Terrell calculates two-fifths of 1260 in the following way: First he multiplies 1260 by 2 to get 2520. Then he multiplies 2520 by 2 to get 5040 and divides this by 10 to get 504. Discuss the idea behind Terrell's strategy. Then write a string of equations that incorporate Terrell's strategy, and that show why the strategy is valid. Write your equations in the form shown previously.

16. Maria is working on the multiplication problem 38×25. Maria says,

4 times 25 is 100, so 40 times 25 is 1000. Now take away 2, so the answer is 998.

Is Maria's method correct or not? If it is correct, write equations that incorporate Maria's work and that show why it's correct. If Maria's reasoning is not correct, work with portions that are right to correct Maria's work, and write a string of equations that corresponds to your corrected method for calculating 38×25. Write your equations in the following form:

$$38 \times 25 = \text{some expression}$$
$$= \text{some expression}$$
$$\vdots$$

17. There is an interesting mental technique for multiplying certain pairs of numbers. The following examples will illustrate how it works:

   • To calculate 32×28, notice that the two factors 28 and 32 are both 2 away—in opposite directions—from 30. To calculate 32×28, do the following:

$$30 \times 30 - 2 \times 2 = 900 - 4$$
$$= 896$$

Notice that you can do this calculation in your head.

- Similarly, to calculate 59×61, notice that both factors are 1 away—in opposite directions—from 60. Then 59×61 is

$$60 \times 60 - 1 \times 1 = 3600 - 1$$
$$= 3599$$

Once again, notice that you can do this mentally.

a. Use the method just shown to calculate 83×77, 195×205, and one other multiplication problem like this that you make up.

b. Now explain why this method works. (*Hint:* A diagram might be helpful. Another approach is to notice that the technique applies to multiplication problems of the form $(A + B) \times (A - B)$.)

18. Try out this next mathematical magic trick. Do the following on a piece of paper:

a. Write the number of days a week you would like to go out (from 1 to 7).

b. Multiply the number by 2.

c. Add 5.

d. Multiply by 50.

e. If you have already had your birthday this year, add 1759 if it is 2009. (Add 1760 if it is 2010, add 1761 if it is 2011, and so on.) If you have not yet had your birthday this year, add 1758 if it is 2009. (Add 1759 if it is 2010, add 1760 if it is 2011, and so on.)

f. Finally, subtract the 4-digit year you were born. You should now have a 3-digit number. If not, try again. If you have a 3-digit number, continue with the following:

The first digit of your answer is your original number (i.e., how many times you want to go out each week). The second two digits are your current age.

Is it magic, or is it math? Explain why the trick works.

4.6 Why the Common Algorithm for Multiplying Whole Numbers Works

Focal Points
Grade 4

In this section, we draw on much of our work in the chapter—multiplication by 10, the commutative and associative properties of multiplication, and the distributive property—to explain why the common algorithm (procedure) for multiplying whole numbers is valid.

The common longhand procedure for multiplying multiple-digit whole numbers is an efficient paper-and-pencil method of calculation. This method is useful because it converts a multiplication problem with numbers that have several digits to many multiplication problems with 1-digit numbers. Therefore, someone who knows the 1-digit multiplication tables (from $1 \times 1 = 1$ to $9 \times 9 = 81$) can multiply any pair of whole numbers by using the longhand multiplication algorithm. But why does this clever method give the correct answer to multiplication problems? What makes it work?

Before we continue, let's first make sense of the questions at the end of the previous paragraph. Notice the distinction between the *meaning* of multiplication and the longhand *procedure* for multiplying. If you have 58 bags of widgets and there are 764 widgets in each bag, then you can ask how many widgets you have in all. There is some specific total number of widgets. But what is this number? According to the meaning of multiplication, the total number of widgets is

$$58 \times 764$$

The longhand multiplication procedure consists of carrying out a number of steps as follows:

Multiply 8×4, write the 2, carry the 3; multiply 8×6, add the carried 3, write the 1, carry the 5, etc.

What is the connection between the meaning of 58×764 and the procedure for calculating 58×764? We need to address the following question:

Why does the common multiplication algorithm for calculating 58×764 give the actual total number of widgets in 58 bags if there are 764 widgets in each bag?

The answer will be explained in this section.

The Partial-Products Multiplication Algorithm

partial-products multiplication algorithm

To understand the common longhand multiplication algorithm, we will work with the **partial-products multiplication algorithm**. The partial-products algorithm is essentially the same as the common multiplication algorithm, but it has the virtue of showing more steps, so it will make analysis easier. A few examples will illustrate how the partial-products algorithm works. In the examples, the arrows and the expressions to the right of the arrows have been added to show how to carry out the algorithm. You do not necessarily have to write these arrows and expressions when you use the algorithm yourself. The common multiplication algorithm, examples of which are shown below, is also known as the **standard multiplication algorithm** and the *standard algorithmic approach* to multiplication.

STANDARD ALGORITHM	PARTIAL-PRODUCTS ALGORITHM
$\overset{4}{3}8$ $\underline{\times\ 6}$ 228	38 $\underline{\times\ 6}$ $48 \quad \leftarrow 6 \times 8$ $\underline{180} \quad \leftarrow 6 \times 30$ $228 \quad \leftarrow \text{add}$

STANDARD ALGORITHM	PARTIAL-PRODUCTS ALGORITHM
$\overset{21}{2}74$ $\underline{\times\ \ 3}$ 822	274 $\underline{\times\ \ 3}$ $12 \quad \leftarrow 3 \times 4$ $210 \quad \leftarrow 3 \times 70$ $\underline{600} \quad \leftarrow 3 \times 200$ $822 \quad \leftarrow \text{add}$

STANDARD ALGORITHM	PARTIAL-PRODUCTS ALGORITHM
$\overset{1}{\underset{}{}}$ $\overset{1}{4}5$ $\underline{\times\ 23}$ 135 $\underline{900}$ 1035	45 $\underline{\times\ 23}$ $15 \quad \leftarrow 3 \times 5$ $120 \quad \leftarrow 3 \times 40$ $100 \quad \leftarrow 20 \times 5$ $\underline{800} \quad \leftarrow 20 \times 40$ $1035 \quad \leftarrow \text{add}$

Relating the Standard and Partial-Products Algorithms

Class Activity   *Now Turn to Class Activities Manual*

4Y The Standard Versus the Partial-Products Multiplication Algorithm, p. 92

Compare the way the standard and partial-products algorithms are used to calculate 58 × 764:

STANDARD ALGORITHM   **PARTIAL-PRODUCTS ALGORITHM**

$$
\begin{array}{r}
^{32}_{53} \\
764 \\
\times\ \ 58 \\
\hline
6{,}112 \quad \leftarrow 32 + 480 + 5{,}600 \\
\\
\\
38{,}200 \quad \leftarrow 200 + 3{,}000 + 35{,}000 \\
\hline
44{,}312
\end{array}
$$

$$
\begin{array}{rl}
764 & \\
\times\quad 58 & \\
\hline
32 & \leftarrow 8 \times 4 \\
480 & \leftarrow 8 \times 60 \\
5{,}600 & \leftarrow 8 \times 700 \\
200 & \leftarrow 50 \times 4 \\
3{,}000 & \leftarrow 50 \times 60 \\
35{,}000 & \leftarrow 50 \times 700 \\
\hline
44{,}312 &
\end{array}
$$

Notice that the standard algorithm condenses the steps in the partial-products algorithm. In this example, three lines of calculations in the partial-products algorithm are condensed to one line in the standard algorithm: The 6,112 produced in the standard algorithm can be obtained by adding the three lines 8 × 4, 8 × 60, and 8 × 700 of the partial-products algorithm; the 38,200 can be obtained by adding 50 × 4, 50 × 60, and 50 × 700. When you multiply and then add carried numbers in the standard algorithm, you are really just combining lines from the partial-products algorithm.

Although the partial-products algorithm is longer than the standard algorithm, it is easier to carry out overall. However, one way that the partial-products algorithm is more difficult is in having to pay attention to place value. In the standard algorithm, you don't have to pay too much attention to place value—as long as you line up the numbers carefully, the algorithm takes care of place value, and you only have to multiply pairs of 1-digit numbers. In the partial-products algorithm, on the other hand, you do have to pay attention to place value. For example, in the last line, you have to be able to calculate 50 × 700, instead of just 5 × 7. Notice in the following equations that relating 50 × 700 to 5 × 7 uses the commutative and associative properties of multiplication:

$$
\begin{aligned}
50 \times 700 &= 5 \times 10 \times 7 \times 100 \\
&= 5 \times 7 \times 10 \times 100 \text{ (by the commutative property of multiplication)} \\
&= 35 \times 1000 \\
&= 35{,}000
\end{aligned}
$$

The associative property of multiplication is used tacitly at the first equal sign in dropping parentheses from (5 × 10) × (7 × 100). The associative property of multiplication is also used tacitly at the third equal sign because the 10 is multiplied by the 100 instead of by the 5 × 7.

Because the partial-products algorithm is an expanded version of the standard multiplication algorithm, we can explain why the standard algorithm correctly calculates the answers to multiplication problems by explaining why the partial-products algorithm correctly calculates the answers to multiplication problems. We can also use the partial-products algorithm to explain why we put zeros in some of the lines of the standard algorithm.

Why We Place Extra Zeros on Some Lines in the Standard Algorithm

We can use the partial-products algorithm to explain why we put an initial zero in the ones column of the second line in the standard multiplication algorithm. (In the previous example, this is the line produced by 5 × 4, 5 × 6, and 5 × 7.) As the partial-products algorithm reveals, that line is actually produced by *multiplying by 50 rather than by 5*. By placing the initial zero in the ones place, all digits are moved one place to the left, which has the same effect as multiplying by 10. The net effect is that this second line is now actually 50 times 764 rather than 5 times 764.

What if the multiplication problem was 258 × 764 instead of 58 × 764? In the standard algorithm, there would be a third line produced by 2 × 4, 2 × 6, and 2 × 7. Before beginning the calculations for this third line, you would put down two zeros. Why two zeros? Because this line should really be produced by multiplying by 200 instead of by 2. By placing zeros in the ones and tens places, all digits are moved two places to the left, which has the same effect as multiplying by 100. The net effect is that the third line is then 200 times 764 instead of 2 times 764.

Why the Algorithms Produce Correct Answers

Class Activity   *Now Turn to Class Activities Manual*

4Z   Why the Multiplication Algorithms Give Correct Answers, Part 1, p. 93

4AA   Why the Multiplication Algorithms Give Correct Answers, Part 2, p. 94

We will now explain why the partial-products algorithm, and, hence, the standard multiplication algorithm, calculates correct answers to multiplication problems. We will use the example 58 × 764 to illustrate. Think of the multiplication problem 58 × 764 as representing the total number of widgets in 58 rows that each contain 764 widgets. The key to explaining why the partial-products algorithm gives the correct answer to 58 × 764 lies in observing that each step in the algorithm counts the number of widgets in a portion of the 58 rows of 764 widgets. The algorithms come from a clever way of subdividing the full collection of widgets into smaller portions whose numbers are easy to calculate as long as you know your 1-digit multiplication tables. At the end of the algorithm, when you add the lines produced by multiplying, you are really just adding the numbers of widgets in each portion to get the total number of widgets.

Figure 4.35 shows 58 rows with 764 widgets in each row, where each in the figure represents a widget. Therefore, according to the meaning of multiplication, Figure 4.35 represents a total of 58 × 764 widgets. Because the numbers 58 and 764 are so large, the diagram doesn't actually show all 58 rows or all 764 widgets in each row.

FIGURE 4.35

Subdividing 58 rows of 764 widgets

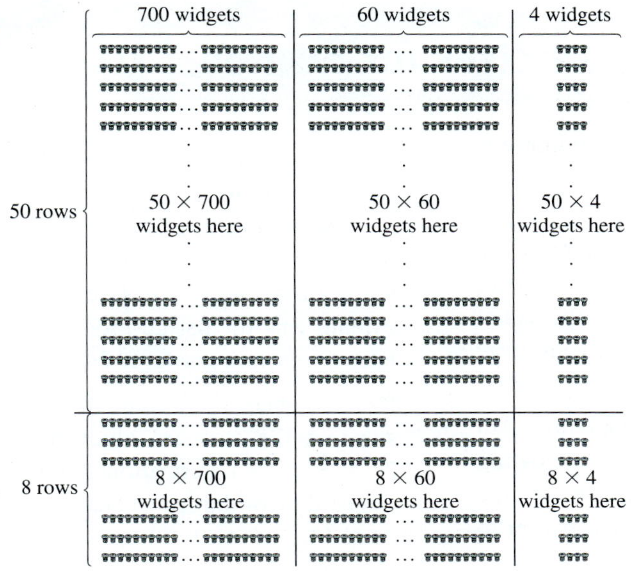

Figure 4.35 illustrates how to subdivide the full collection of widgets so that the number of widgets in each portion corresponds to a line in the partial-products algorithm. For instance, the line 5600 in the partial-products algorithm, which comes from 8×700, is the number of widgets in the piece in the lower left-hand corner, which represents 8 groups of 700 widgets. Similarly, the line 200, which comes from 50×4, is the number of widgets in the piece at the top right-hand corner, which represents 50 groups of 4 widgets. It's the same for all the other lines in the partial-products algorithm—each line in the algorithm corresponds to a portion of the collection of widgets. When you add up the lines in the algorithm, you get the total number of widgets. Therefore, the partial-products algorithm correctly calculates the total number of widgets; in other words, it correctly calculates 58×764. Since the standard algorithm is just a condensation of the partial-products algorithm, it too must correctly calculate answers to multiplication problems.

Another way to explain why the partial-products algorithm, and therefore the standard algorithm, gives the correct answer to 58×764 is to put the numbers 58 and 764 in expanded forms and use properties of arithmetic to calculate 58×764:

$$
\begin{aligned}
58 \times 764 &= (50 + 8) \times (700 + 60 + 4) \\
&= 50 \times (700 + 60 + 4) + 8 \times (700 + 60 + 4) \\
&= 50 \times 700 + 50 \times 60 + 50 \times 4 + 8 \times 700 + 8 \times 60 + 8 \times 4 \\
&= 35{,}000 + 3{,}000 + 200 + 5{,}600 + 480 + 32 \\
&= 44{,}312
\end{aligned}
$$

As the preceding equations show, the six lines produced by the partial-products algorithm are exactly the six products

$$50 \times 700, \quad 50 \times 60, \quad 50 \times 4, \quad 8 \times 700, \quad 8 \times 60, \quad 8 \times 4$$

produced by the use of the distributive property several times, starting with

$$(50 + 8) \times (700 + 60 + 4)$$

Class Activity   *Now Turn to Class Activities Manual*

4BB The Standard Multiplication Algorithm Right Side Up and Upside Down, p. 97

Practice Exercises for Section 4.6

1. Find the last two digits (the ones and hundreds digits) of
$$798{,}312{,}546{,}936 \times 74$$

2. Draw an array on graph paper, and use your array to explain why the partial-products algorithm calculates the correct answer to 24×35. If graph paper is not available, draw a rectangle to represent your array rather than drawing 24 rows with 35 items in each row.

3. Relate the array you drew for Practice Exercise 2 to the steps in the standard algorithm for calculating 24×35. Use this relationship to explain why the standard algorithm calculates the correct answer to 24×35.

4. Solve the multiplication problem 24×35 by writing equations that use expanded forms and the distributive property. Relate your equations to the steps in the partial-products algorithm for calculating 24×35. Use this relationship to explain why the partial-products algorithm calculates the correct answer to 24×35.

5. Relate your equations from Practice Exercise 4 to the steps in the standard algorithm for calculating 24×35. Use this relationship to explain why the standard algorithm calculates the correct answer to 24×35.

6. Cameron wants to calculate 23×23. She says:

$$20 \times 20 = 400 \text{ and } 3 \times 3 = 9,$$

so

$$23 \times 23 = 400 + 9 = 409.$$

Cameron has a good idea, but her reasoning is not completely correct. Compare Cameron's work with the steps in the partial-products algorithm for calculating 23×23. What is missing in Cameron's calculation? Draw an array to show Cameron why she must adjust her calculation.

Answers to Practice Exercises for Section 4.6

1. The last two digits are 64. Notice that you have to find only the last two digits of 36×74 to find this. Why? Think about doing longhand multiplication: Any other contribution will be in the thousands place or higher.

2. Figure 4.36 shows an array consisting of 24 rows of small squares with 35 small squares in each row. According to the meaning of multiplication, there is a total of 24×35 small squares in the array. The shading shows how to subdivide the array into four pieces: one with 20 rows of 30, one with 20 rows of 5, one with 4 rows of 30, and one with 4 rows of 5 small squares. The number of small squares in each of these four pieces is calculated as follows:

$$20 \times 30 = 600$$

$$20 \times 5 = 100$$

$$4 \times 30 = 120$$

$$4 \times 5 = 20$$

But the number of small squares in these four pieces corresponds exactly to the four lines we produce when we use the partial-products algorithm to calculate 24×35:

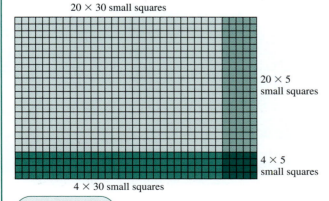

20 × 30 small squares

20 × 5 small squares

4 × 5 small squares

4 × 30 small squares

FIGURE 4.36   An array of 24 rows with 35 small squares in each row subdivided into four parts

$$
\begin{array}{r}
35 \\
\times\ 24 \\
\hline
20 \\
120 \\
100 \\
+\ 600 \\
\hline
840
\end{array}
$$

So the steps in the partial-products algorithm calculate the number of squares in the four parts that make up a full array of 24 rows of 35 squares. When we add the numbers together at the end, we add up the total number of squares in all four parts, and therefore get the total number of small squares in 24 rows of 35 squares, which is 24×35 small squares.

3. The standard algorithm is a condensed version of the partial-products algorithm. The two lines we produce when we multiply 24×35 by using the standard algorithm each come from combining two of the lines in the partial-products algorithm:

$$
\begin{array}{r}
\overset{1}{\underset{2}{}} \\
35 \\
\times\ 24 \\
\hline
140 \\
\\
700 \\
\hline
840
\end{array}
\qquad
\begin{array}{r}
35 \\
\times\ 24 \\
\hline
\left\{ \begin{array}{r} 20 \\ 120 \end{array} \right. \\
\left\{ \begin{array}{r} 100 \\ 600 \end{array} \right. \\
\hline
840
\end{array}
$$

In terms of the array in Figure 4.36, the line 140 in the standard algorithm comes from the bottom 4 rows, which represent 4×35; the line 700 in the standard algorithm comes from the top 20 rows, which represent 20×35. As with the partial-products algorithm, the standard algorithm calculates the total number of small squares in an array of 24 rows with 35 small squares in each row by calculating the number of small squares in parts of the array and adding these totals.

4. $24 \times 35 = (20 + 4) \times (30 + 5)$

$= 20 \times (30 + 5) + 4 \times (30 + 5)$

$= 20 \times 30 + 20 \times 5 + 4 \times 30 + 4 \times 5$

$= 600 + 100 + 120 + 20$

$= 840$

The distributive property was used at the second and third equal signs. The four products, 20×30, 20×5, 4×30, and 4×5, that result from the use of expanded forms and the distributive property to calculate 24×35 are exactly the four products seen in the following calculation using the partial-products algorithm:

$$
\begin{array}{r}
35 \\
\times\ 24 \\
\hline
20 \\
120 \\
100 \\
+\ 600 \\
\hline
840
\end{array}
$$

Therefore, the partial-products algorithm is really just a way of displaying the results produced by using the distributive property when the numbers 24 and 35 are put in expanded form. The partial-products algorithm produces correct calculations because these calculations come from the use of the distributive property.

5. The standard algorithm is a condensed version of the partial-products algorithm. The two lines we produce when we multiply 24×35 by using the standard algorithm each come from combining two of the lines in the partial-products algorithm:

$$
\begin{array}{rr}
\overset{1}{\underset{2}{}} & \\
35 & 35 \\
\times\ 24 & \times\ 24 \\
\hline
140 & \left\{\begin{array}{r}20 \\ 120\end{array}\right. \\
700 & \left\{\begin{array}{r}100 \\ 600\end{array}\right. \\
\hline
840 & 840
\end{array}
$$

Consider the equations we write to calculate 24×35, using expanded forms and the distributive property:

$24 \times 35 = (20 + 4) \times (30 + 5)$

$= 20 \times (30 + 5) + 4 \times (30 + 5)$

$= 20 \times 30 + 20 \times 5 + 4 \times 30 + 4 \times 5$

$= 600 + 100 + 120 + 20$

$= 840$

The two terms in the second line of the equations—namely, $4 \times (30 + 5)$ and $20 \times (30 + 5)$—correspond to the first and second lines, respectively, produced by the standard algorithm—namely, 140 and 700. So the standard algorithm comes from breaking the bottom factor, 24, into expanded form $20 + 4$, and multiplying each of these components by the top factor, 35. The standard algorithm calculates 24×35 by calculating

$$20 \times 35 + 4 \times 35$$

which uses the distributive property.

6. When we calculate 23×23 using the partial-products algorithm, we produce four lines of products:

$$
\begin{array}{r}
23 \\
\times\ 23 \\
\hline
9 \\
60 \\
60 \\
+\ 400 \\
\hline
529
\end{array}
$$

Cameron's calculations produced only two of those lines—the 9 and the 400—so she is missing the two lines of 60. We can see these two missing 60s in Figure 4.37, which shows 23 rows of 23 small squares, subdivided into four portions. The array shows that Cameron has left out 20×3 and 3×20, which correspond to the darkly shaded portions.

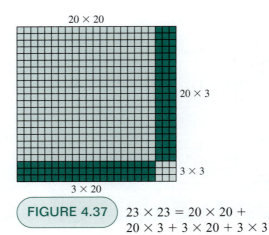

FIGURE 4.37 $23 \times 23 = 20 \times 20 + 20 \times 3 + 3 \times 20 + 3 \times 3$

Problems for Section 4.6

1. Solve the multiplication problem

$$\begin{array}{r} 96 \\ \times\ 8 \\ \hline \end{array}$$

in three different ways: by using the standard algorithm, by using the partial-products algorithm, and by writing the numbers in expanded forms and using properties of arithmetic. For each of the three methods, discuss how the steps in that method are related to the steps in the other methods.

2. Solve the multiplication problem

$$\begin{array}{r} 84 \\ \times\ 76 \\ \hline \end{array}$$

in three different ways: by using the standard algorithm, by using the partial-products algorithm, and by writing the numbers in expanded forms and using properties of arithmetic. Discuss how the steps in the partial-products algorithm are related to the steps in the other methods.

3. Solve the multiplication problem

$$\begin{array}{r} 237 \\ \times\ 43 \\ \hline \end{array}$$

in three different ways: by using the standard algorithm, by using the partial-products algorithm, and by writing the numbers in expanded forms and using properties of arithmetic. Discuss how the steps in the partial-products algorithm are related to the steps in the other methods.

4. When we multiply

$$\begin{array}{r} 37 \\ \times\ 26 \\ \hline \end{array}$$

by using the standard multiplication algorithm, we start the second line by writing a zero:

$$\begin{array}{r} 4 \\ 37 \\ \times\ 26 \\ \hline 222 \\ 0 \end{array}$$

Explain why we place this zero in the second line.

5. **a.** Use the partial-products algorithm to calculate 8×24.

b. Draw an array for 8×24. (You may wish to use graph paper.) Subdivide the array in a natural way so that the parts of the array correspond to the steps in the partial-products algorithm.

c. Solve 8×24 by writing equations that use expanded forms and the distributive property. Relate your equations to the steps in the partial-products algorithm.

6. **a.** Use the partial-products algorithm to calculate 19×28.

b. On graph paper, draw an array for 19×28. If graph paper is not available, draw a rectangle to represent the array rather than drawing 19 rows with 28 items in each row. Subdivide the array in a natural way so that the parts of the array correspond to the steps of the partial-products algorithm.

c. Solve 19×28 by writing equations that use expanded forms and the distributive property. Relate your equations to the steps in the partial-products algorithm.

7. **a.** Use the partial-products and standard algorithms to calculate 27×28.

b. On graph paper, draw an array for 27×28. If graph paper is not available, then draw a rectangle to represent the array rather than drawing 27 rows with 28 items in each row. Subdivide the array in a natural way so that the parts of the array correspond to the steps in the partial-products algorithm.

c. On the array that you drew for part (b), show the parts that correspond to the steps of the standard algorithm.

d. Solve 27×28 by writing equations that use expanded forms and the distributive property. Relate your equations to the steps in the partial-products algorithm.

8. **a.** Draw an array on graph paper, and use your array to explain why the partial-products algorithm calculates the correct answer to 23×27. If graph paper is not available, draw a rectangle to represent the array rather than drawing 23 rows of 27 items.

b. Relate the array you drew for part (a) to the steps in the standard algorithm for calculating 23 × 27. Use this relationship to explain why the standard algorithm calculates the correct answer to 23 × 27.

9. Solve the multiplication problem 23 × 27 by writing equations that use expanded forms and the distributive property. Relate your equations to the steps in the partial-products algorithm for calculating 23 × 27. Use this relationship to explain why the partial-products algorithm calculates the correct answer to 23 × 27.

10. **a.** Use the partial-products and standard algorithms to calculate 43 × 275.

b. Draw a large rectangle to represent an array for 43 × 275. Subdivide the rectangle in a natural way so that the parts of the rectangle correspond to the steps in the partial-products algorithm.

c. On the rectangle that you drew for part (b), show the parts that correspond to the steps in the standard algorithm.

d. Solve 43 × 275 by writing equations that use expanded forms and the distributive property. Relate your equations to the steps in the partial-products algorithm.

11. **a.** Use the standard algorithm to calculate

$$\begin{array}{r} 37 \\ \times\ 24 \\ \hline \end{array}$$

b. On graph paper, draw an array for 24 × 37 (or draw a rectangle to represent such an array). Subdivide the array in a natural way so that the parts of the array correspond to the steps of the standard algorithm in part (a).

c. Write equations that use the distributive property and that correspond to the steps in the standard algorithm in part (a).

d. Use the standard algorithm to calculate

$$\begin{array}{r} 24 \\ \times\ 37 \\ \hline \end{array}$$

e. On graph paper, draw an array for 37 × 24 (or draw a rectangle to represent such an array). Subdivide the array in a natural way so that the parts of the array correspond to the steps of the standard algorithm in part (d). Compare with part (b).

f. Write equations that use the distributive property and that correspond to the steps in the standard algorithm in part (d). Are these the same equations as in part (c)?

g. Other than the distributive property, which you used in parts (c) and (f), which property of arithmetic is relevant to this problem? Explain and discuss.

12. The **lattice method** is a method that some people like to use to multiply. Figure 4.38 shows how to use the lattice method to multiply a 2-digit number by a 2-digit number.

a. Use the lattice method to calculate 38 ×54 and 72 × 83.

b. Use the partial-products algorithm to calculate 38 × 54 and 72 × 83.

c. Explain how the lattice method is related to the partial-products algorithm.

d. Discuss advantages and disadvantages of using the lattice method.

Draw a lattice as shown in the figure. Place the numbers you want to multiply on the top and down the side.

Multiply each digit along the top with each digit along the side. Write the answers in the cells as shown.

Starting at the bottom right, and moving left, then up, add numbers in diagonal strips, carrying as necessary. The answer is 1081.

FIGURE 4.38 Using the lattice method to show that 23 × 47 = 1081

13. The following is a method for multiplying 21 × 23 that relies on repeatedly doubling numbers (by adding them to themselves):

$$\begin{array}{rl}
23 & \\
+\ 23 & \\
\hline
46 & \quad 2 \\
+\ 46 & \\
\hline
92 & \quad 4 \\
+\ 92 & \\
\hline
184 & \quad 8 \\
+\ 184 & \\
\hline
368 & \quad 16
\end{array}$$

$$21 = 16 + 4 + 1$$

$$\begin{array}{r}
368 \\
92 \\
+\ 23 \\
\hline
483
\end{array}$$

The answer, 483, to the multiplication problem 21 × 23, is shown on the bottom of the column on the right.

Notice that, to use this repeated doubling method, you have to be able only to add; you do not need to know how to multiply.

a. Discuss and explain the method of repeated doubling. Address the following questions in your discussion:

i. What is the significance of the column of the numbers 2, 4, 8, 16?

ii. What is the significance of writing 21 = 16 + 4 + 1?

iii. How are the numbers 368, 92, and 23 chosen for the column on the right? Why do we add those numbers?

b. Use the method of repeated doubling to calculate 26 × 35. Show your work.

c. Use the method of repeated doubling to calculate 37 × 51. Show your work.

Chapter Summary and Study Items

Section 4.1 Interpretations of Multiplication

For nonnegative numbers, the product $A \times B$ stands for the total in A groups of B. We can therefore show multiplicative structure in a situation by exhibiting equal groups. We can exhibit equal groups with organized lists, with arrays, and with tree diagrams.

Key skills and understandings:

- Explain why multiplication applies to solve a problem by exhibiting or describing equal groups.

- Write array, ordered pair, and multiplicative comparison story problems for a given multiplication problem (such as 4×6), as well as more ordinary multiplication story problems for which the equal groups are more evident.

Section 4.2 Why Multiplying Numbers by 10 Is Easy in the Decimal System

Because the value of each decimal place is 10 times the value of the place to its right, multiplying a number by 10 moves each digit one place to the left.

Key skills and understandings:

- Describe multiplication by 10 as moving digits one place to the left.

- Use pictures of bundled objects to explain why multiplying a number by 10 moves each digit one place to the left.

- Explain that because $100 = 10 \times 10$, and $1000 = 10 \times 10 \times 10$, multiplying a number by 100 moves each digit 2 places to the left and multiplying a number by 1000 moves each digit 3 places to the left. Explain similar results for other powers of 10.

Section 4.3 The Commutative and Associative Properties of Multiplication, Areas of Rectangles, and Volumes of Boxes

The commutative property of multiplication states that

$$A \times B = B \times A$$

for all real numbers A and B.

The area of a rectangle, in square units, is the number of 1-unit-by-1-unit squares it takes to cover the rectangle without gaps or overlaps. We can multiply $L \times W$ to find the area of an L-by-W rectangle because the rectangle can be subdivided into L rows, each of which contains W 1-unit-by-1-unit squares.

We can explain and illustrate the commutative property of multiplication by subdividing a rectangle into groups of squares in two different ways or by subdividing an array into groups in two different ways. The associative property of multiplications states that

$$(A \times B) \times C = A \times (B \times C)$$

For all real numbers A, B, and C.

The volume of a box, in cubic units, is the number of 1-unit-by-1-unit-by-1-unit cubes it takes to fill or make the box (without gaps or overlaps). We can multiply $H \times (L \times W)$ to find the volume of a box that is H units high, L units long, and W units wide because the box can be subdivided into H layers, each of which contains L rows of W 1-unit-by-1-unit-by-1-unit cubes.

We can explain and illustrate the associative property of multiplication by subdividing a box into groups of cubes in two different ways or by subdividing groups of grouped objects in two different ways.

We use the combination of the commutative and associative properties in many mental calculations, such as in calculating 60×700 by recalling that $6 \times 7 = 42$ and moving 42 three places to the left to make 42,000.

Key skills and understandings:

- Explain why we can multiply to find the area of a rectangle by describing rectangles as subdivided into groups of 1-unit-by-1-unit squares.

- State the commutative property of multiplication, and explain why it makes sense (for counting numbers) by subdividing rectangles or arrays in two different ways.

- Explain why we can multiply to find the volume of a box by describing boxes as subdivided into groups of groups of 1-unit-by-1-unit-by-1-unit cubes.

- State the associative property of multiplication, and explain why it makes sense (for counting numbers) by subdividing boxes two different ways or by subdividing groups of groups of objects in two different ways.

- Give examples of how to use the associative and commutative properties of multiplication in problems and recognize when these properties have been used.

- Apply the methods used in explaining the area formula for rectangles and the volume formula for boxes to estimate numbers of items by multiplying.

Section 4.4 The Distributive Property

The distributive property states that

$$A \times (B + C) = A \times B + A \times C$$

for all real numbers A, B, and C. We can explain why the distributive property is valid by calculating the total number of objects in a subdivided array in two different ways.

We use the distributive property in many calculations. This property is the single most powerful property of arithmetic; in Section 4.6, we see that it is the key idea underlying the multiplication algorithms.

The convention on order of operations governs how we interpret expressions that involve several operations, such as multiplication and addition.

Key skills and understandings:

- State the distributive property, and explain why it makes sense (for counting numbers) by describing the total number of objects in a subdivided array in two different ways. Use simple story situations to explain or illustrate the distributive property.

- Give examples of how to use the distributive property in problems and recognize when the distributive property has been used.

Section 4.5 Properties of Arithmetic, Mental Math, and Single-Digit Multiplication Facts

The commutative and associative properties of addition and multiplication and the distributive property are key in explaining calculations, and they link arithmetic and algebra. Given a description of a mental method of calculation, we can write a sequence of equations that correspond to the method and thus show the "algebra" in the mental math. Many of the basic multiplication facts are related to each other via properties of arithmetic. Students can capitalize on such relationships to aid their learning of the basic facts.

Key skills and understandings:

- Given a mental method of calculation, write a string of equations that correspond to the method and that show which properties of arithmetic were used (knowingly or not).

- Describe how basic multiplication facts are related to other basic facts via properties of arithmetic.

Section 4.6 Why the Common Algorithm for Multiplying Whole Numbers Works

We can explain why the common algorithm (also known as the standard algorithm) for multiplying whole numbers works by considering place value (and multiplication by 10 and 100 and 1000, and so on), and by using the commutative and associative properties of multiplication, and most crucially, the distributive property. The partial-products algorithm is closely related to the standard multiplication algorithm. We can explain the validity of the partial-products algorithm (and therefore also the standard algorithm) by subdividing arrays or by writing numbers in expanded form and using the distributive property.

Key skills and understandings:

- Relate a multiplication problem to an array, and subdivide the array so that the pieces correspond to the lines in the partial-products algorithm. Use the subdivision to explain the validity of the partial-products algorithm. Show the portions of the array that correspond to the lines in the standard algorithm.

- Solve a multiplication problem by writing equations that use expanded forms and the distributive property. Relate the equations to the lines in the partial-products algorithm. Use the relationship to explain why the partial-products algorithm calculates the correct answer to the multiplication problem.

- Understand that multiplication algorithms can be explained in terms of the meaning of multiplication, place value, and properties of arithmetic.

Multiplication of Fractions, Decimals, and Negative Numbers

*I*n *this chapter, we will study multiplication of fractions, decimals, and negative numbers. We will also study the related topics of powers and scientific notation. The procedures for multiplying fractions and decimals are not hard. As a teacher, you should understand not only how to carry out those multiplication procedures, but also why those procedures are the way they are. For example, why do we put the decimal point where we do when we multiply decimals? Why do we multiply fractions by multiplying the numerators and the denominators, even though we don't add fractions by adding the numerators and adding the denominators? We will answer these questions by using the meaning of multiplication, as it applies to fractions and decimals. We will also study fraction and decimal story problems and see why we must pay close attention to the wording of these problems.*

5.1 Multiplying Fractions

F·P **Focal Points**
Grade 6

In this section, we will study the meaning of multiplication of fractions. We will also see why the standard procedure for multiplying fractions gives answers to fraction multiplication problems that agree with what we expect from the meaning of multiplication.

The Meaning of Multiplication for Fractions

At the beginning of Chapter 4, we defined multiplication as follows: If *A* and *B* are nonnegative numbers, then

FIGURE 5.1

There is $3 \times \frac{2}{5}$ of a pie in three groups of $\frac{2}{5}$ of a pie.

represents the total number of objects in A groups if there are B objects in each group. Our shorthand was: $A \times B$ means the total in A groups of B. This definition applies not only to whole numbers, but also to fractions and decimals. As we'll see, though, when fractions are involved, we may prefer to reword slightly for the sake of clarity.

Consider the following examples:

1. $\frac{1}{2} \times 21$ means the total in $\frac{1}{2}$ of a group of 21.

   Or: $\frac{1}{2} \times 21$ means the total number of objects in $\frac{1}{2}$ of a group if there are 21 objects in one whole group.

2. $\frac{1}{2} \times \frac{1}{2}$ means the total in $\frac{1}{2}$ of a group of $\frac{1}{2}$.

   Or: $\frac{1}{2} \times \frac{1}{2}$ means the total number of objects in $\frac{1}{2}$ of a group if there is $\frac{1}{2}$ of an object in one whole group.

3. $3 \times \frac{2}{5}$ means the total in 3 groups of $\frac{2}{5}$. (See Figure 5.1.)

   Or: $3 \times \frac{2}{5}$ means the total number of objects in 3 groups if there is $\frac{2}{5}$ of an object in each group.

What do we mean by the word *object*? In a mathematical context, the word *object* can refer not just to a single thing, but also to a collection of things (as in "the cars in the United States") or to a quantity (as in "one cup of water" or "$20,000"). In the context of fractions, it is also natural to use the term *a whole* instead of the term *object*.

The wording "the total number of objects" must be understood to include the possibility of fractional objects. For example, if the object we're considering is a full jar of peanut butter, then what does the following mean?

The total number of objects in $\frac{1}{2}$ of a group when each group contains $\frac{1}{2}$ of a full jar of peanut butter.

In this case, what we mean by "the total number of objects" is really "the fraction of a full jar of peanut butter" in $\frac{1}{2}$ of $\frac{1}{2}$ of a full jar of peanut butter. This phrasing suggests alternative wording for the meaning of fraction multiplication, which you can use if you find it clearer. The following wording is most appropriate in cases where both factors are proper fractions:

$\frac{A}{B} \times \frac{C}{D}$ means the fraction of an object in $\frac{A}{B}$ of $\frac{C}{D}$ of the object.

For example,

$\frac{1}{2} \times \frac{3}{4}$ means the fraction of a container of yogurt in $\frac{1}{2}$ of $\frac{3}{4}$ of the container of yogurt. (See Figure 5.2.)

The Procedure for Multiplying Fractions

The procedure for multiplying fractions is very easy: Just multiply the numerators, and multiply the denominators. In other words,

$$\frac{A}{B} \cdot \frac{C}{D} = \frac{A \cdot C}{B \cdot D}$$

FIGURE 5.2

$\frac{1}{2}$ of $\frac{3}{4}$ of a container of yogurt

container of yogurt

$\frac{3}{4}$ of the container of yogurt

$\frac{1}{2}$ of the $\frac{3}{4}$ of the container of yogurt

For example, consider the product

$$\frac{2}{3} \cdot \frac{4}{5} = \frac{2 \cdot 4}{3 \cdot 5} = \frac{8}{15}$$

Notice that the procedure can be used even when whole numbers are involved, because a whole number can always be written as a fraction by "putting it over 1." For example,

$$\frac{2}{3} \cdot 17 = \frac{2}{3} \cdot \frac{17}{1} = \frac{2 \cdot 17}{3 \cdot 1} = \frac{34}{3} = 11\frac{1}{3}$$

We can multiply mixed numbers with the fraction multiplication procedure by first converting the mixed numbers to improper fractions. For example,

$$2\frac{3}{4} \cdot 1\frac{2}{3} = \frac{11}{4} \cdot \frac{5}{3} = \frac{55}{12} = 4\frac{7}{12}$$

As with the procedure for multiplying whole numbers, we want to explain why this procedure gives correct answers to fraction multiplication problems. The meaning of fractions and the meaning of multiplication tell us what an expression of the form

$$\frac{A}{B} \cdot \frac{C}{D}$$

means. We must explain why this expression is equal to the fraction

$$\frac{A \cdot C}{B \cdot D}$$

Class Activity   *Now Turn to Class Activities Manual*

5A   The Meaning of Multiplication for Fractions, p. 98

5B Misconceptions with Fraction Multiplication, p. 99

Explaining Why the Procedure for Multiplying Fractions Gives Correct Answers

The procedure for multiplying fractions is easy to carry out. It also seems sensible, because it involves multiplying the numerators and the denominators. However, remember that we don't add fractions by simply adding the numerators and adding the denominators. So, why is the simple procedure of multiplying the numerators and multiplying the denominators valid for fraction multiplication? We will use the meaning of fractions, the meaning of multiplication, and logical reasoning to explain why the simple procedure for multiplying fractions gives correct answers to fraction multiplication problems.

Class Activity   *Now Turn to Class Activities Manual*

5C   Explaining Why the Procedure for Multiplying Fractions Gives Correct Answers, p. 99

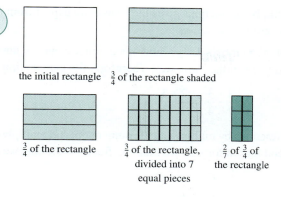

FIGURE 5.3

Finding $\frac{2}{7}$ of $\frac{3}{4}$ of a rectangle

the initial rectangle $\frac{3}{4}$ of the rectangle shaded

$\frac{3}{4}$ of the rectangle $\frac{3}{4}$ of the rectangle, divided into 7 equal pieces $\frac{2}{7}$ of $\frac{3}{4}$ of the rectangle

Consider the multiplication problem

$$\frac{2}{7} \cdot \frac{3}{4}$$

and consider $\frac{2}{7} \cdot \frac{3}{4}$ of a rectangle. By the meaning of multiplication,

$\frac{2}{7} \cdot \frac{3}{4}$ means the fraction of a rectangle in $\frac{2}{7}$ of a group if there is $\frac{3}{4}$ of a rectangle in a whole group.

In other words, $\frac{2}{7} \cdot \frac{3}{4}$ means the fraction of a rectangle in $\frac{2}{7}$ of $\frac{3}{4}$ of the rectangle.

So, starting with a rectangle, first consider $\frac{3}{4}$ of a rectangle, and then consider $\frac{2}{7}$ of the $\frac{3}{4}$ of the rectangle, as shown in Figure 5.3.

Here is the crucial point: We must identify the $\frac{2}{7}$ of the $\frac{3}{4}$ of the rectangle *as a fraction of the original rectangle*. To identify what fraction of the original rectangle is represented by $\frac{2}{7}$ of the $\frac{3}{4}$ of the rectangle, we must put it back inside the original rectangle, as shown in Figure 5.4.

Because the original rectangle was first divided into 4 equal parts, and because each of those parts was then divided into 7 equal parts, the original rectangle has therefore been subdivided into a total of $7 \cdot 4$ small, equal parts. Notice that we multiply $7 \cdot 4$ because there are 7 columns with 4 pieces in each column. Of those $7 \cdot 4$ equal parts, the darkly shaded parts (which represent our $\frac{2}{7}$ of the $\frac{3}{4}$ of the rectangle) form $2 \cdot 3$ parts. Once again, notice that we must multiply $2 \cdot 3$ because there are 2 columns with 3 small pieces in each column that are darkly shaded. So the $\frac{2}{7}$ of the $\frac{3}{4}$ of the rectangle is represented by $2 \cdot 3$ parts out of a total of $7 \cdot 4$ equal parts making up the original rectangle. This means that $\frac{2}{7}$ of the $\frac{3}{4}$ of the rectangle is

$$\frac{2 \cdot 3}{7 \cdot 4}$$

of the rectangle. Therefore,

$$\frac{2}{7} \cdot \frac{3}{4} = \frac{2 \cdot 3}{7 \cdot 4}$$

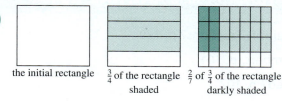

FIGURE 5.4

What fraction is $\frac{2}{7}$ of $\frac{3}{4}$ of a rectangle?

the initial rectangle $\frac{3}{4}$ of the rectangle shaded $\frac{2}{7}$ of $\frac{3}{4}$ of the rectangle darkly shaded

In other words, the procedure for multiplying fractions gives the answer we expect from the meaning of fractions and the meaning of multiplication.

There wasn't anything special about the numbers 2, 3, 7, and 4; any other counting numbers could be substituted, and the argument would work in the same way. Therefore, the explanation given for why

$$\frac{2}{7} \cdot \frac{3}{4} = \frac{2 \cdot 3}{7 \cdot 4}$$

also explains why we multiply other fractions by multiplying the numerators and multiplying the denominators. In other words, we have a good reason for why we multiply fractions the way we do:

$$\frac{A}{B} \cdot \frac{C}{D} = \frac{A \cdot C}{B \cdot D}$$

Notice that, in order to understand fraction multiplication, we had to work with different wholes. In the example, we first took $\frac{3}{4}$ of the original rectangle. Then we took $\frac{2}{7}$ of this $\frac{3}{4}$ portion. We had to treat the $\frac{3}{4}$ of the rectangle as a new whole in its own right in order to take $\frac{2}{7}$ of it. In the end, when we found the portion that was $\frac{2}{7}$ of the $\frac{3}{4}$ of the rectangle, we had to determine what fraction that portion was of the original rectangle. In other words, we had to go back to the original whole again. As always, when we work with fractions, we must pay close attention to the underlying wholes or what the fractions are *of*. When you use pictures to help you solve fraction multiplication problems, you may find it helpful to start by drawing your original reference whole as a reminder.

Why We Need to Know This

As we've seen, the procedure for multiplying fractions is easy, but the explanation for why this procedure gives correct answers to fraction multiplication problems is much harder to grasp. As a teacher, it is especially important for you to have a good conceptual understanding of the mathematics you will teach. This means not just knowing *how*, but also *why* and *when* the various procedures and formulas in mathematics work. Why is this important? Research shows that teachers who have a conceptual understanding of mathematics tend to use a conceptually directed teaching strategy. (See [49, especially pages 38–39].) These teachers teach in a way that encourages understanding. Research also shows that this kind of teaching requires substantial knowledge of subject matter:

> Limited subject matter knowledge restricts a teacher's capacity to promote conceptual learning among students. Even a strong belief of "teaching mathematics for understanding" cannot remedy or supplement a teacher's disadvantage in subject matter knowledge. A few beginning teachers in the procedurally directed group wanted to "teach for understanding." They intended to involve students in the learning process, and to promote conceptual learning that explained the rationale underlying the procedure. However, because of their own deficiency in subject matter knowledge, their conception of teaching could not be realized. [49, page 36]

Class Activity   *Now Turn to Class Activities Manual*

5D When Do We Multiply Fractions? p. 100

5E Multiplying Mixed Numbers, p. 101

5F What Fraction Is Shaded? p. 102

Practice Exercises for Section 5.1

1. What does $\frac{3}{4} \times \frac{1}{2}$ mean? Give an example of a story problem that is solved by calculating $\frac{3}{4} \times \frac{1}{2}$.

2. Use the meaning of fractions and the meaning of multiplication to explain why

$$\frac{2}{7} \cdot \frac{3}{4} = \frac{2 \cdot 3}{7 \cdot 4}$$

3. Anthony is trying to calculate $\frac{5}{7} \cdot 2$. He draws a picture as in Figure 5.5 and concludes from his picture that $\frac{5}{7} \cdot 2 = \frac{10}{14}$, because there are 14 small pieces and 10 of them are shaded. Is Anthony right or not? If not, help Anthony figure out what's wrong with his reasoning.

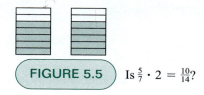

FIGURE 5.5 Is $\frac{5}{7} \cdot 2 = \frac{10}{14}$?

4. Which of the following problems are solved by calculating $\frac{1}{3} \times \frac{2}{3}$, and which are not?

   a. A recipe calls for $\frac{2}{3}$ cup of sugar. You want to make $\frac{1}{3}$ of the recipe. How much sugar should you use?

   b. $\frac{2}{3}$ of the cars at a car dealership have power steering. $\frac{1}{3}$ of the cars at the same car dealership have side-mounted airbags. What fraction of the cars at the car dealership have both power steering and side-mounted airbags?

   c. $\frac{2}{3}$ of the cars at a car dealership have power steering. $\frac{1}{3}$ of those cars that have power steering have side-mounted airbags. What fraction of the cars at the car dealership have both power steering and side-mounted airbags?

   d. Ed put $\frac{2}{3}$ of a bag of candies in a batch of cookies that he made. Ed ate $\frac{1}{3}$ of the batch of cookies. How many candies did Ed eat?

   e. Ed put $\frac{2}{3}$ of a bag of candies in a batch of cookies that he made. Ed ate $\frac{1}{3}$ of the batch of cookies. What fraction of a bag of candies did Ed eat?

5. Maryann has to calculate $\frac{2}{3} \cdot 6\frac{3}{4}$. Rather than convert the $6\frac{3}{4}$ to an improper fraction, she solves the problem this way. She finds $\frac{2}{3}$ of 6, which is 4, and she finds $\frac{2}{3}$ of $\frac{3}{4}$, which is $\frac{1}{2}$. Then Maryann says the answer is $4\frac{1}{2}$. Write equations that correspond to Maryann's work and that explain why her work is correct. Which property of arithmetic is involved?

Answers to Practice Exercises for Section 5.1

1. $\frac{3}{4} \times \frac{1}{2}$ means the fraction of an object in $\frac{3}{4}$ of $\frac{1}{2}$ of the object. A sample story problem: A recipe calls for $\frac{1}{2}$ pound of sea slugs. You decide to make $\frac{3}{4}$ of the recipe. How many pounds of sea slugs will you need?

2. See the text.

3. No, Anthony is not right. Although Anthony has drawn a good representation of the problem, his reasoning is not completely right. 10 pieces are shaded, and these 10 pieces do represent $\frac{5}{7}$ of 2 rectangles. But Anthony must remember that each small piece represents $\frac{1}{7}$ of the original object, which we infer must have been one rectangle. So $\frac{5}{7} \cdot 2$ is the fraction of one rectangle that is shaded. This fraction is $\frac{10}{7}$. When drawing pictures such as Anthony's, it's a good idea to also draw 1 whole object somewhere as a reminder that this is your reference amount.

4. a. Yes, $\frac{1}{3} \cdot \frac{2}{3}$ is the fraction of a cup of sugar you should use, since you will need $\frac{1}{3}$ of $\frac{2}{3}$ cup of sugar.

b. No, from the information given, we don't know if $\frac{1}{3}$ of the cars that have power steering have side-mounted airbags.

c. Yes, $\frac{1}{3}$ of $\frac{2}{3}$ of the cars at the car dealership have both power steering and side-mounted airbags.

d. No, we can't tell how many candies Ed ate, only what fraction of a bag of candies Ed ate.

e. Yes, Ed ate $\frac{1}{3}$ of $\frac{2}{3}$ of a bag of candies.

5.
$$\frac{2}{3} \cdot 6\frac{3}{4} = \frac{2}{3} \cdot \left(6 + \frac{3}{4}\right)$$
$$= \frac{2}{3} \cdot 6 + \frac{2}{3} \cdot \frac{3}{4}$$
$$= 4 + \frac{1}{2}$$
$$= 4\frac{1}{2}$$

The distributive property was used at the second equal sign.

Problems for Section 5.1

1. a. Anita had $\frac{1}{2}$ of a bag of fertilizer left. She used $\frac{3}{4}$ of what was left. What question about Anita's fertilizer can be answered by calculating $\frac{3}{4} \times \frac{1}{2}$?

b. Hermione wants to make 4 batches of a potion. Each batch of potion requires $\frac{2}{3}$ cup of toober pus. What question about the potion will be answered by calculating $4 \times \frac{2}{3}$?

c. There were 6 pieces of pizza left. Tommy ate $\frac{3}{4}$ of them. What question about the pizza can be answered by calculating $\frac{3}{4} \times 6$?

2. a. Discuss the meaning of $\frac{1}{2} \cdot \frac{1}{4}$. Include a simple story problem and a picture in your discussion.

b. Discuss the meaning of $\frac{1}{4} \cdot \frac{1}{2}$. Include a simple story problem and a picture in your discussion.

c. Briefly discuss the difference between parts (a) and (b).

3. Paul used $\frac{3}{4}$ cup of butter in the batch of brownies he made. Paul ate $\frac{1}{6}$ of the batch of brownies. What fraction of a cup of butter did Paul consume when he ate the brownies? Draw pictures to help you solve this problem. Explain in detail how your pictures help you solve the problem.

4. Which of the following are story problems for $\frac{1}{2} \times \frac{1}{3}$, and which are not? Explain your answer in each case.

a. One-third of the children in a class have black hair. One-half of the children in the class have curly hair. How many children in the class have curly black hair?

b. One-third of the children in a class have black hair. One-half of the children who have black hair also have curly hair. How many children in the class have curly black hair?

c. One-third of the children in a class have black hair. One-half of the children who have black hair also have curly hair. What fraction of the children in the class have curly black hair?

d. One-third of the children in a class have black hair. One-half of the children in the class have curly hair. What fraction of the children in the class have curly black hair?

5. Which of the following are story problems for $\frac{3}{4} \times 5$, and which are not? Explain your answer in each case.

a. A cake was cut into pieces of equal size. There are 5 pieces of cake left. John gets $\frac{3}{4}$ of them. What fraction of the cake does John get?

b. A cake was cut into pieces of equal size. There are 5 pieces of cake left. John gets $\frac{3}{4}$ of them. How many pieces of cake does John get?

c. A cake for a class party was cut into pieces of equal size. There are 5 pieces of cake left. Three-quarters of the class still wants cake. What fraction of the cake will be eaten?

d. A cake for a class party was cut into pieces of equal size. There are 5 pieces of cake left. Three-quarters of the class still wants cake. How many pieces does each person get?

6. Consider this story problem about baking brownies:

You are baking brownies for your class. You put white frosting on $\frac{1}{3}$ of the brownies and you put small red hearts on $\frac{1}{4}$ of the brownies. How many brownies have both white frosting and small red hearts on them?

a. Can the brownie problem be solved? If so, solve the problem; if not, explain why not.

b. Is the brownie problem a problem for $\frac{1}{3} \cdot \frac{1}{4}$? If so, explain briefly why it is; if not, modify the problem so that it is a problem for $\frac{1}{3} \cdot \frac{1}{4}$.

c. Is the brownie problem a problem for $\frac{1}{3} + \frac{1}{4}$? If so, explain briefly why it is; if not, modify the problem so that it is a problem for $\frac{1}{3} + \frac{1}{4}$.

7. Explain why it would be easy to interpret the picture in Figure 5.6 incorrectly as showing that $4 \times \frac{3}{5} = \frac{12}{20}$. Explain how to interpret the picture correctly, and explain why your interpretation fits with the meaning of $4 \times \frac{3}{5}$.

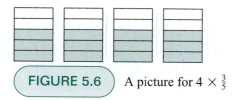

FIGURE 5.6 A picture for $4 \times \frac{3}{5}$

8. Write a story problem for

$$\frac{1}{3} \cdot \frac{1}{4}$$

Use the meaning of fractions, the meaning of multiplication, and pictures to determine the answer to the multiplication problem. Explain your answer.

9. Write a story problem for

$$2 \cdot \frac{3}{5}$$

Use the meaning of fractions, the meaning of multiplication, and pictures to determine the answer to the multiplication problem. Explain your answer.

10. Write a story problem for

$$\frac{2}{3} \cdot 5$$

Use the meaning of fractions, the meaning of multiplication, and pictures to determine the answer to the multiplication problem. Explain your answer.

11. Write a simple story problem for

$$\frac{2}{3} \cdot \frac{4}{5}$$

Use your story problem and pictures to explain why it makes sense that the answer to the fraction multiplication problem is

$$\frac{2 \cdot 4}{3 \cdot 5}$$

In particular, use your pictures to explain why we multiply the numerators and why we multiply the denominators.

12. One serving of Gooey Gushers provides 12% of the daily value of vitamin C. Tim ate $3\frac{1}{2}$ servings of Gooey Gushers.

a. Calculate $3\frac{1}{2} \times 12\%$ without using a calculator. Show your work.

b. What question about the Gooey Gushers can you answer by calculating $3\frac{1}{2} \times 12\%$?

13. a. Write a story problem for $2\frac{1}{2} \times 3\frac{1}{2}$.

b. Use pictures and the meaning of multiplication to solve the problem.

c. Use the distributive property or FOIL to calculate $2\frac{1}{2} \times 3\frac{1}{2}$ by rewriting this product as $(2 + \frac{1}{2}) \times (3 + \frac{1}{2})$.

d. Identify the four terms produced by the distributive property or FOIL [in part (c)] in a picture like the one in part (b).

e. Now write the mixed numbers $2\frac{1}{2}$ and $3\frac{1}{2}$ as improper fractions, and use the standard procedure for multiplying fractions to calculate $2\frac{1}{2} \times 3\frac{1}{2}$. How do you see the product of the numerators in your picture in part (b)? How can you see the product of the denominators in your picture?

14. Manda says that

$$3\frac{2}{3} \times 2\frac{1}{5} = 3 \times 2 + \frac{2}{3} \times \frac{1}{5}$$

Explain why Manda has made a good attempt, but her answer is not correct. Explain how to work with what Manda has already written and modify it to get the correct answer. In other words, don't just start from scratch and show Manda how to do the problem, but rather take what she has already written, use it, and make it mathematically correct. Which property of arithmetic is relevant to correcting Manda's work? Explain.

15. **a.** Write an expression that uses both multiplication and addition (or subtraction) to describe the total fraction of Figure 5.7 that is shaded. (For example, $\frac{5}{7} \cdot \frac{2}{9} + \frac{1}{3}$ is an expression that uses both multiplication and addition). Explain your reasoning. Then determine what fraction of the figure is shaded (in simplest form). You may assume that lengths which appear to be equal really are equal.

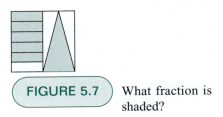

FIGURE 5.7   What fraction is shaded?

b. Draw a figure in which you shade $\frac{1}{4} \cdot \frac{1}{3} + \frac{3}{5} \cdot \frac{1}{3}$ of the figure. Explain your reasoning. Then determine what fraction of the figure is shaded (in simplest form).

16. **a.** Write an expression that uses both multiplication and addition (or subtraction) to describe the total fraction of Figure 5.8 that is shaded. (For example, $\frac{5}{7} \cdot \frac{2}{9} + \frac{1}{3}$ is an expression that uses both multiplication and addition). Explain your reasoning. Then determine what fraction of the figure is shaded (in simplest form). You may assume that lengths which appear to be equal really are equal.

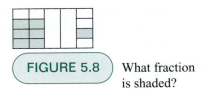

FIGURE 5.8   What fraction is shaded?

b. Draw a figure in which you shade $\frac{5}{7} - \frac{3}{5} \cdot \frac{1}{7}$ of the figure. Explain your reasoning. Then determine what fraction of the figure is shaded (in simplest form).

17. To understand fraction multiplication thoroughly, we must be able to work simultaneously with different wholes. Using the example $\frac{3}{5} \times \frac{3}{4}$, explain why this is so. What are the different wholes that are associated with the fractions in the problem $\frac{3}{5} \times \frac{3}{4}$ (including the answer to the problem)?

18. To understand fraction multiplication thoroughly, we must understand how to divide a whole number of objects into equal parts, such as dividing 8 cookies equally among 3 people. Where do we need to understand how to divide a whole number of objects into equal parts in order to understand $\frac{2}{3} \times \frac{2}{5}$ thoroughly? Be specific.

19. Let's suppose that the liquid in a car's radiator is 75% water and 25% antifreeze. Suppose 30% of the radiator's liquid is drained out and replaced with pure antifreeze. Now what percent of the radiator's liquid is antifreeze? Draw pictures to help you solve this problem. Explain your answer.

20. There are a yellow flask and a red flask. The yellow flask contains more than one cup of yellow paint, and the red flask contains an equal amount of red paint. One cup of the red paint in the red flask is poured into the yellow flask and mixed thoroughly. Then one cup of the paint mixture in the yellow flask is poured into the red flask and mixed thoroughly.

   a. Without doing any calculations, which do you think will be greater, the percentage of yellow paint in the red flask or the percentage of red paint in the yellow flask? Explain your reasoning.

   b. Now use calculations to figure out the answer to the problem in part (a). Work with specific quantities of yellow and red paint. (Remember, both flasks have the same amount at the start, and both contain more than one cup.) Draw pictures to help you calculate the quantities and percentages after the mixing occurs. Do you get a different answer than you got in part (a)? If so, reconcile your answers.

21. Discuss why we must develop an understanding of multiplication that goes beyond seeing it as repeated addition.

22. Suppose you are teaching fraction multiplication. Make the case to your students that multiplication means the same thing whether we are multiplying fractions or whole numbers. Use story problems for 3×4 and $\frac{1}{3} \times \frac{1}{4}$ to illustrate.

5.2 Multiplying Decimals

 **Focal Points**
Grade 6

To multiply (finite) decimals, the standard procedure is first to multiply the numbers without the decimal points, and then to place the decimal point in the answer according to a certain rule. Why is this procedure for multiplying decimals valid? The text, class activities, practice exercises, and problems in this section will help you answer this question in a number of different ways.

The Procedure for Multiplying Decimals

What is the standard procedure for multiplying decimals? We first multiply the numbers without the decimal points. Then we add the number of digits to the right of the decimal points in the numbers we want to multiply, and we put the decimal point in the answer to the product computed without the decimal points that many places from the end. So, to multiply

$$\begin{array}{r} 1.36 \\ \times\ \ 3.7 \\ \hline \end{array}$$

we first multiply as if there were no decimal points:

$$\begin{array}{r} 136 \\ \times\ \ 37 \\ \hline 5032 \end{array}$$

Then we add the number of digits to the right of the decimal points in our two original numbers and put the decimal point that many places from the end in our answer. There are 2 digits behind the decimal point in 1.36 and another 1 digit in 3.7, for a total of $1 + 2 = 3$ digits; so the decimal point goes 3 digits from the end:

$$\begin{array}{r} 1.36 \\ \times\ \ 3.7 \\ \hline 5.032 \end{array}$$

Class Activity   *Now Turn to Class Activities Manual*

5G Multiplying Decimals, p. 102

Using Estimation to Determine Where the Decimal Point Goes

Suppose you have forgotten the rule about where to put the decimal point in the answer to a decimal multiplication problem. One quick way to figure out where to put the decimal point is to think about the sizes of the numbers. For example, 1.36 is between 1 and 2, and 3.7 is between 3 and 4; so 1.36×3.7 must be between $1 \times 3 = 3$ and $2 \times 4 = 8$. So, where will it make sense to put the decimal point in 5032 to get the answer to 1.36×3.7? The numbers 503.2 and 50.32 are far too big. The number 0.5032 is too small. The only number that makes sense is 5.032 because it is between 3 and 8.

Explaining Why the Rule for Placing the Decimal Point is Valid

Class Activity   *Now Turn to Class Activities Manual*

5H Explaining Why We Place the Decimal Point Where We Do When We Multiply Decimals, p. 103

How can we explain why we add the number of places behind the decimal points to determine where to place the decimal point in the answer to a decimal multiplication problem? One way is to compare the decimal multiplication problem with the multiplication problem without the decimal points. For example, how do the multiplication problems

$$0.12 \times 6.24$$

and

$$12 \times 624$$

compare? As indicated in Figure 5.9, to get from 0.12 to 12, we must multiply by ten 2 times; to get from 6.24 to 624, we must multiply by ten 2 times. Therefore, to get from

$$0.12 \times 6.24$$

to

$$12 \times 624$$

we must multiply by ten 2 times and then another 2 times. We know that

$$12 \times 624 = 7488$$

FIGURE 5.9

Comparing
0.12×6.24 and
12×624

$$6.24 \xrightarrow[\times 10 \;\times 10]{\times 10 \;\times 10} 624$$
$$\underline{\times\,0.12} \qquad\qquad \underline{\times\;\;12}$$
$$ \qquad\qquad 7488$$
Therefore,
$$\begin{array}{cc} 6.24 & 624 \\ \underline{\times\,0.12} & \underline{\times\;\;12} \\ 0.7488 & 7488 \end{array}$$
$$\xleftarrow[\div 10 \;\div 10]{\div 10 \;\div 10}$$

Therefore, to get back to the answer to the original problem, 0.12×6.24, we must divide 7488 by ten 2 times and then another 2 times. When we divide by ten 2 times and then another 2 times, we shift the decimal point 2 + 2 places to the left. So we have shown that to solve 0.12×6.24, we must move the decimal point in the answer to 12×624 to the left 2 + 2 places, which is exactly the procedure for decimal multiplication. There wasn't anything special about the numbers we used here—the same line of reasoning applies for any other decimal multiplication problem.

Another way to explain why the standard procedure for multiplying decimals is valid is by writing the decimals as fractions. Then we can use the procedure for multiplying fractions that we have already studied. Every finite decimal can be written as a fraction with a denominator that is a product of 10s. For example,

$$1.25 = \frac{125}{100} = \frac{125}{10 \times 10}$$

$$0.003 = \frac{3}{1000} = \frac{3}{10 \times 10 \times 10}$$

We can use this way of writing the denominators to explain the placement of the decimal point when we multiply decimals.

$$1.25 \times 0.003 = \frac{125}{10 \times 10} \times \frac{3}{10 \times 10 \times 10}$$

$$= \frac{125 \times 3}{(10 \times 10) \times (10 \times 10 \times 10)}$$

These equations tell us that to calculate the answer to 1.25×0.003, we should calculate the answer to 125×3 and divide by ten 2 times and then another 3 times. When we divide by ten 2 times and then another 3 times, we move the decimal point 2 places and then 3 places to the left. In other words, to calculate the answer to 1.25×0.003, we should calculate the answer to 125×3 and then move the decimal point a total of $2 + 3 = 5$ places to the left, which is exactly the standard decimal multiplication procedure.

Class Activity   *Now Turn to Class Activities Manual*

51 Decimal Multiplication and Areas of Rectangles, p. 104

Practice Exercises for Section 5.2

1. The product $1.35 \times 7.2 = 9.72$, but shouldn't the answer have 3 digits to the right of its decimal point? Why doesn't it?

2. Write a story problem for 8.3×2.15.

3. Suppose you multiply a decimal that has 5 digits to the right of its decimal point by a decimal that has 2 digits to the right of its decimal point. Explain why you put the decimal point $5 + 2$ places from the end of the product without the decimal points.

4. Use the meaning of multiplication to explain why we can multiply to find the area of a 2.4-unit-by-1.6-unit rectangle. Show how to subdivide the rectangle and recombine pieces to determine the area of the rectangle.

Answers to Practice Exercises for Section 5.2

1. The rule says to find the product without the decimal points (i.e., 135×72) and then move the decimal point in that answer $2 + 1 = 3$ places to the left. But

$$135 \times 72 = 9720$$

ends in a zero. That's why that third digit doesn't appear in the answer.

2. You bought 8.3 gallons of gas. Gas cost $2.15 per gallon. How much did you pay?

3. Let's work with a particular example to illustrate, say, 1.23456×1.23. To get from 1.23456 to 123456, we must multiply by ten 5 times. To get from 1.23 to 123, we must multiply by ten 2 times. So, to get from

$$1.23456 \times 1.23$$

to

$$123456 \times 123$$

we must multiply by ten 5 times and then another 2 times. Therefore, to get from the answer to

$$123456 \times 123$$

back to the answer to

$$1.23456 \times 1.23$$

we must divide by ten 5 times and then another 2 times.

When we divide the answer to

$$123456 \times 123$$

by ten 5 times and then another 2 times, we move the decimal point to the left $5 + 2$ places. The explanation works the same way for any other decimals with 5 digits and 2 digits to the right of their decimal points.

4. Starting with a 1-unit-by-1-unit square, Figure 5.10 shows 1.6 of the square, then 2 groups of 1.6 of the square, and finally 2.4 groups of the 1.6 of the square. So, according to the meaning of multiplication, the 2.4-unit-by-1.6-unit rectangle in Figure 5.10 consists of 2.4×1.6 squares and therefore has area 2.4×1.6 square units. In the 2.4 groups of the 1.6 of the square, there are 2 whole squares, a bunch of strips, each of which is $\frac{1}{10}$ of a square, and a bunch of small squares, each of which is $\frac{1}{100}$ of a square.

If we take the large rectangle apart, we can recombine it as shown in Figure 5.11. We see that the rectangle is made of 3 full 1-unit-by-1-unit squares, $\frac{8}{10}$ of a square, and $\frac{4}{100}$ of a square. Therefore, the total area of the 2.4-unit-by-1.6-unit rectangle is 3.84 square units.

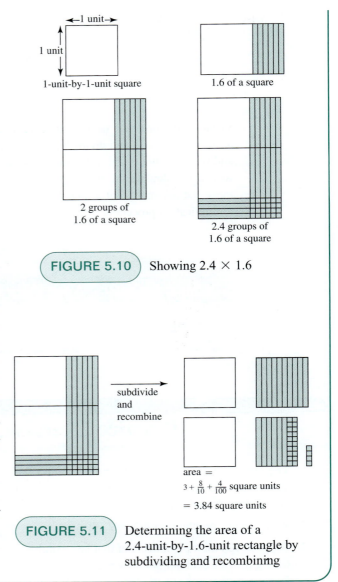

FIGURE 5.10   Showing 2.4×1.6

FIGURE 5.11   Determining the area of a 2.4-unit-by-1.6-unit rectangle by subdividing and recombining

Problems for Section 5.2

1. Write and solve your own story problem for 14.3×1.39.

2. Write a story problem for 1.3×2.79. Solve your problem without using a calculator.

3. Leah is working on the multiplication problem 2.43×0.148. Ignoring the decimal points, Leah multiplies 243×148 and gets the answer 35964. But Leah can't remember the rule about where to put the decimal point in this answer to get the correct answer to 2.43×0.148. Explain how Leah can use reasoning about the sizes of the numbers to determine where to put the decimal point.

4. Ron used a calculator to determine that

$$0.35 \times 2.4 = 0.84$$

Ron wants to know why the rule about adding up the number of places to the right of the decimal point doesn't work in this case. Why aren't there $2 + 1 = 3$ digits to the right of the decimal point in the answer? Is Ron correct that the rule about adding the number of places to the right of the decimal points is not correct in this case? Explain.

5. When we multiply 0.48×3.9, we first multiply as if the decimal points were not there:

$$
\begin{array}{r}
3.9 \\
\times\ .48 \\
\hline
312 \\
1560 \\
\hline
1872
\end{array}
$$

Then we place a decimal point in 1872 to get the correct answer. Explain clearly why it makes sense to put the decimal point where we do.

6.  Suppose you multiply a decimal that has 2 digits to the right of its decimal point by a decimal that has 3 digits to the right of its decimal point. Explain why you put the decimal point $2 + 3$ places from the end of the product calculated without the decimal points.

7. Suppose you multiply a decimal that has 2 digits to the right of its decimal point by a decimal that has 4 digits to the right of its decimal point. Explain why you put the decimal point $2 + 4$ places from the end of the product calculated without the decimal points.

8. Suppose you multiply a decimal that has M digits to the right of its decimal point by a decimal that has N digits to the right of its decimal point. Explain why you put the decimal point $M + N$ places from the end of the product calculated without the decimal points.

9. Explain how to write 1.89 and 3.57 as improper fractions whose denominators are products of 10s. Then use fraction multiplication to explain where to place the decimal point in the solution to 1.89×3.57. Show how to use the products of 10s in the denominators to explain why we add the number of digits behind the decimal points in 1.89 and 3.57.

10. *A shampoo problem:* A bottle contained 25.4 fluid ounces of shampoo. Katie used 0.25 of the bottle. How much shampoo is left?

    a. Is the shampoo problem a story problem for 0.25×25.4, is it a story problem for $25.4 - 0.25$, or is it not a story problem for either of these? Explain.

    b. Write a new shampoo story problem for 0.25×25.4 and write a new shampoo story problem for $25.4 - 0.25$. Make clear which is which.

11. Show how to subdivide the rectangle in Figure 5.12 and recombine the pieces to determine the area of the rectangle. Calculate 1.4×2.8, and verify that it produces the correct area.

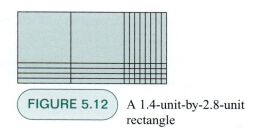

FIGURE 5.12 A 1.4-unit-by-2.8-unit rectangle

12. Show how to subdivide a 1.7-unit-by-3.1-unit rectangle and recombine the pieces to determine the area of the rectangle. Calculate 1.7×3.1, and verify that it produces the correct area.

5.3 Multiplying Negative Numbers

F·P Focal Points
Grade 7

So far, we have examined multiplication only for numbers that are not negative. What is the meaning of multiplication for negative numbers? In this section, we will examine several different ways to understand multiplication of negative numbers.

Class Activity   *Now Turn to Class Activities Manual*

5J Patterns with Multiplication and Negative Numbers, p. 106

5K Explaining Multiplication with Negative Numbers (and 0), p. 106

What should 3×-2 mean? We know that 3×2 stands for the total number of objects in 3 groups if there are 2 objects in each group. If we try to use the same interpretation for 3×-2, then we have the following:

3×-2 is the total number of objects in 3 groups if there are -2 objects in each group.

What does this mean? One way to interpret this sensibly is to think of "-2 objects in each group" as meaning "each group *owes* 2 objects." For example, if Lakeisha, Mary, and Jayna each *owe* \$2.00, then we can think of each girl as *having* -2 dollars, and all together, the 3 girls have 3×-2 dollars. Notice that with this interpretation, it makes sense to say

$$3 \times -2 = -6$$

because all together, the 3 girls collectively owe 6 dollars, as represented by the -6. In general, if A and B are positive numbers, then we can define $A \times -B$ to be $-(A \times B)$; that is,

$$A \times -B = -(A \times B) \tag{5.1}$$

This definition is consistent with the interpretation of negative numbers as amounts owed. Equation 5.1 gives us the familiar rule:

A positive times a negative is negative.

What about a negative number times a positive number, such as

$$-2 \times 3?$$

We could perhaps interpret this product as the total number of objects in 2 owed groups with 3 objects in each group. Similarly, we could interpret a negative number times a negative number, such as

$$-4 \times -5$$

as the total number of objects in 4 owed groups, with each group owing 5 objects. But these interpretations seem difficult to grasp. A better way to interpret these multiplication problems is as the change in the amount of money we will have if we give away checks or bills. (See Class Activity 5L.)

There is another way we can understand how to multiply negative numbers, which draws upon the properties of arithmetic. We have seen why the distributive property and the commutative and associative properties of addition and multiplication make sense for counting numbers. These properties of arithmetic are fundamental and describe the algebraic structure of numbers and operations. We have seen that these properties of arithmetic underlie mental calculations and standard calculation algorithms. Because the properties of arithmetic are so important and fundamental, we should expect them to hold for any number system that is an extension of the counting numbers. In particular, the properties of arithmetic should continue to hold for all the integers, positive and negative.

Let us now assume that the properties of arithmetic we have studied hold not only for positive numbers, but for negative numbers as well. We will see that this assumption determines how we multiply negative numbers.

How do we multiply a negative number with a positive number, such as

$$-2 \times 3?$$

According to the commutative property of multiplication,

$$-2 \times 3 = 3 \times -2,$$

which must be equal to -6, according to Equation 5.1 and the previous discussion about how to interpret 3×-2. In general, if A and B are any positive numbers, then because we are assuming that the commutative property of multiplication holds, we must define $-A \times B$ to be $-(A \times B)$. That is,

$$-A \times B = -(A \times B) \tag{5.2}$$

This equation gives us the following familiar rule:

> A negative times a positive is negative.

Finally, how do we multiply a negative number with a negative number, as in

$$-4 \times -5?$$

We are assuming that the distributive property holds for both positive and negative numbers. Therefore,

$$\begin{aligned}
(-4 \times -5) + (4 \times -5) &= (-4 + 4) \times -5 \\
&= 0 \times -5 \\
&= 0
\end{aligned}$$

The preceding equations tell us that -4×-5 and 4×-5 add to zero. Since $4 \times -5 = -20$ (by Equation 5.1), -4×-5 and -20 add to zero. But 20 is the only number which, when added to -20, yields zero. Therefore, -4×-5 must equal 20. In general, if A and B are any positive numbers, then if we assume that the distributive property holds, we must define $-A \times -B$ to be $A \times B$; that is,

$$-A \times -B = A \times B \tag{5.3}$$

This equation gives us the following familiar rule:

> A negative times a negative is positive.

Class Activity *Now Turn to Class Activities Manual*

5L Using Checks and Bills to Interpret Multiplication with Negative Numbers, p. 107

5M Does Multiplication Always Make Larger? p. 108

Practice Exercises for Section 5.3

1. Explain why it makes sense that $3 \times -2 = -6$ by interpreting negative numbers as amounts owed.

2. Given that $3 \times -2 = -6$, use a property of arithmetic to explain why it makes sense that $-2 \times 3 = 6$.

3. Given that $4 \times -5 = -20$, use a property of arithmetic to explain why it makes sense that $-4 \times -5 = 20$.

Answers to Practice Exercises for Section 5.3

1., 2., 3. See text. Pages 208, 209.

Problems for Section 5.3

1. a. Explain why it makes sense that $5 \times -2 = -10$ by interpreting negative numbers as amounts owed.

   b. Given that $5 \times -2 = -10$, use a property of arithmetic to explain why it makes sense that $-2 \times 5 = -10$.

   c. Given that $5 \times -2 = -10$, use a property of arithmetic to explain why it makes sense that $-5 \times -2 = 10$.

2. For which numbers, N, is $N \times 2$ greater than 2? In other words, for which numbers, N, is

$$N \times 2 > 2?$$

Investigate this question as follows:

   a. Without using a calculator, multiply 2 by the given numbers. Show your work. In each case, determine whether the resulting product is greater than 2 or not.

$$2.85, \quad 1.25, \quad 0.95, \quad 0.03 \quad -0.7 \quad -3.5$$

$$3\frac{1}{2}, \quad 1\frac{1}{8}, \quad \frac{7}{8}, \quad \frac{2}{5}, \quad -\frac{2}{3}, \quad -2\frac{3}{4}$$

   b. Based on your answers in part (a) and on the meaning and rules of multiplication, describe the collection of all numbers, N, for which $N \times 2$ is greater than 2.

3. For which numbers, N, is $N \times -2$ greater than -2? In other words, for which numbers, N, is

$$N \times -2 > -2?$$

Investigate this question as follows:

   a. Without using a calculator, multiply -2 by the given numbers. Show your work. In each case, determine whether the resulting product is greater than -2 or not.

$$3.15, \quad 1.01, \quad 0.85, \quad 0.002, \quad -0.3, \quad -4.2$$

$$4\frac{3}{5}, \quad 1\frac{2}{3}, \quad \frac{9}{10}, \quad \frac{3}{5}, \quad -\frac{7}{8}, \quad -1\frac{1}{8}$$

   b. Based on your answers in part (a) and on the meaning and rules of multiplication, describe the collection of all numbers, N, for which $N \times -2$ is greater than -2.

5.4 Powers and Scientific Notation

F·P Focal Points
Grade 8

Many scientific applications require the use of very large or very small numbers: Distances between stars are huge, whereas molecular distances are tiny. These kinds of numbers can be cumbersome to write in ordinary decimal notation. Therefore, a special notation called scientific notation is often used to write such numbers. When a number is in scientific notation, we can see its order of magnitude—how big or small it is—at a glance.

Use a calculator to multiply

$$123{,}456{,}789 \times 987{,}654{,}321$$

How is the answer displayed? The answer is probably displayed in one of the following forms:

$$1.2193263\ 17$$

or

$$1.2193263\ E\ 17$$

Both of these displays represent

$$1.2193263 \times 10^{17}$$

which is in scientific notation. Scientific notation involves multiplying by powers of 10, such as 10^{17}. And next we discuss powers.

Powers

A convenient notation for writing an expression such as

$$10 \times 10 \times 10 \times 10 \times 10 \times 10$$

which is six tens multiplied together, is

$$10^6$$

So,

$$1,000,000 = 10 \times 10 \times 10 \times 10 \times 10 \times 10 = 10^6$$

The expression 10^6 is read "ten to the sixth power" or just "ten to the sixth," and we can refer to an expression like 10^6 as a **power of ten**. The number 6 in 10^6 is called the **exponent** of 10^6.

Power of ten

exponent

More generally, if A is any real number and B is any counting number, then A^B stands for B As multiplied together:

$$A^B = \underbrace{A \times A \times \cdots \times A}_{B \text{ times}}$$

For example,

$$2^5 = 2 \times 2 \times 2 \times 2 \times 2 = 32$$

As with powers of 10, we read A^B as "A to the Bth power" or "A to the B," and B is called the exponent of A^B.

Table 5.1 shows powers of ten from 10^1 to 10^{12}. Notice the correlation between the number of zeros and the exponent on the 10 that is visible in Table 5.1. For example, 10,000 is written as a 1 followed by 4 zeros; it also can be written as 10^4, where the exponent on the 10 is 4. Similarly, 1,000,000 is written as a 1 followed by 6 zeros and it can also be written as 10^6.

What about decimal places to the right of the decimal point? As before, there is special notation to write one tenth, one hundredth, one thousandth, and so on, as powers of 10. This time, the powers are *negative*, as in the following:

$$\frac{1}{10} = \frac{1}{10^1} = 10^{-1}$$
$$\frac{1}{100} = \frac{1}{10^2} = 10^{-2}$$
$$\frac{1}{1000} = \frac{1}{10^3} = 10^{-3}$$
$$\frac{1}{10,000} = \frac{1}{10^4} = 10^{-4}$$
$$\vdots \qquad \vdots \qquad \vdots$$

TABLE 5.1 Powers of ten

Ten Hundred	$10 = 10^1$ $100 = 10 \times 10 = 10^2$
Thousand Ten thousand Hundred thousand	$1000 = 10 \times 10 \times 10 = 10^3$ $10{,}000 = 10 \times 10 \times 10 \times 10 = 10^4$ $100{,}000 = 10 \times 10 \times 10 \times 10 \times 10 = 10^5$
Million Ten million Hundred million	$1{,}000{,}000 = 10 \times 10 \times 10 \times 10 \times 10 \times 10 = 10^6$ $10{,}000{,}000 = 10^7$ $100{,}000{,}000 = 10^8$
Billion Ten billion Hundred billion	$1{,}000{,}000{,}000 = 10^9$ $10{,}000{,}000{,}000 = 10^{10}$ $100{,}000{,}000{,}000 = 10^{11}$
Trillion	$1{,}000{,}000{,}000{,}000 = 10^{12}$

Since $1/10 = 0.1$ and $1/100 = 0.01$ and $1/1000 = 0.001$, and so on, the negative exponents fit the following pattern of positive exponents:

$$
\begin{array}{rcl}
10{,}000 & = & 10^4 \\
1{,}000 & = & 10^3 \\
100 & = & 10^2 \\
10 & = & 10^1 \\
1 & = & 10^0 \\
0.1 & = & 10^{-1} \\
0.01 & = & 10^{-2} \\
0.001 & = & 10^{-3} \\
0.0001 & = & 10^{-4}
\end{array}
$$

Notice that it makes sense to *define* 10^0 to be the number one—that clearly fits with the pattern of moving the decimal point one place to the left (moving down the left-hand column of numbers) and lowering the exponent on the 10 by one (moving down the column on the right). There is another good reason to define 10^0 to be 1, which you will see at the end of the next activity.

Class Activity   *Now Turn to Class Activities Manual*

5N Multiplying Powers of 10, p. 109

Scientific Notation

scientific notation A number is in **scientific notation** if it is written as a decimal that has exactly one nonzero digit to the left of the decimal point, multiplied by a power of ten. So a number is in scientific notation if it is written in the form

$$\#.\#\#\#\#\# \times 10^{\#}$$

where the # to the left of the decimal point is not zero and where any number of digits can be displayed to the right of the decimal point. Thus, neither

$$121.93263 \times 10^{15}$$

nor

$$12.193263 \times 10^{16}$$

is in scientific notation, but

$$1.2193263 \times 10^{17}$$

is in scientific notation.

Other than 0, every real number can be expressed in scientific notation. To write a number in scientific notation, think about how multiplication by powers of 10 works. For example, how do we express

$$847{,}930{,}000$$

in scientific notation? We need to find an exponent that will make the following equation true:

$$847{,}930{,}000 = 8.4793 \times 10^{?}$$

The decimal point in 8.4793 must be moved 8 places to the right to get 847,930,000; therefore, we should multiply 8.4793 by 10^8. In other words,

$$847{,}930{,}000 = 8.4793 \times 10^{8}$$

How do we write

$$0.0000345$$

in scientific notation? We need to find an exponent that will make the following equation true:

$$0.0000345 = 3.45 \times 10^{?}$$

The decimal point in 3.45 must be moved 5 places to the left to get 0.0000345; therefore, we must multiply 3.45 by

$$0.00001 = \frac{1}{10^5} = 10^{-5}$$

in order to get 0.0000345. So

$$0.0000345 = 3.45 \times 10^{-5}$$

Here are a few more examples:

Ordinary decimal notation	Scientific notation
12	1.2×10^{1}
123	1.23×10^{2}
1234	1.234×10^{3}
12345	1.2345×10^{4}
1234.5	1.2345×10^{3}
123.45	1.2345×10^{2}
12.345	1.2345×10^{1}
1.2345	1.2345 or 1.2345×10^{0}
.12345	1.2345×10^{-1}
.012345	1.2345×10^{-2}
.0012345	1.2345×10^{-3}

Although scientific notation is mainly used in scientific settings, it is common to use a variation of scientific notation when discussing large numbers in more common situations, such as when budgets or populations are concerned. For example, an amount of money may be described as $3.5 billion, which means

$$\$3.5 \times \text{(one billion)}$$

or

$$\$3.5 \times 1{,}000{,}000{,}000$$

Two other ways to express $ 3.5 billion are

$$\$3.5 \times 10^9$$

and

$$\$3{,}500{,}000{,}000$$

The form

$$\$3.5 \text{ billion}$$

is probably more quickly and easily grasped by most people than any of the other forms.

By the way, what is a big number—or a small number? Children sometimes ask questions such as the following:

Is 100 a big number? What about 1000, is that a big number?

The answer is, "It depends." We wouldn't think of 1000 grains of sand as a lot of sand, but we might consider 10 pages of homework to be a lot.

Speaking of big numbers, there is one special number name that many children find amusing and fascinating: a googol. A **googol** is the name for the number whose decimal representation is a 1 followed by one hundred zeros:

googol

$$\underbrace{10000000000 \ldots 0000000000}_{100 \text{ zeros}}$$

A googol is so large that, according to current theories in physics, it is even larger than the number of atoms in the universe. Another very large number that has a special name is a googolplex. **A googolplex** is the name for the number whose decimal representation is a 1 followed by a googol zeros:

googolplex

$$\underbrace{10000000000 \ldots 0000000000}_{a \text{ googol zeros}}$$

It is very difficult to comprehend such a number.

Class Activity   *Now Turn to Class Activities Manual*

5O Scientific Notation versus Ordinary Decimal Notation, p. 110

5P How Many Digits Are in a Product of Counting Numbers? p. 111

5Q Explaining the Pattern in the Number of Digits in Products, p. 112

Practice Exercises for Section 5.4

1. Write the following numbers as powers of 10:

   a. 10,000,000,000,000

   b. 0.1

   c. 0.000001

   d. 1

   e. 10

   f. the number whose decimal representation is a 1 followed by 200 zeros

   g. a googol

   h. a googolplex

2. We call 10^6 a million, we call 10^9 a billion, and we call 10^{12} a trillion. What are 10^{13} and 10^{14} called? If we call 10^{15} a thousand trillion and 10^{18} a million trillion, then what are 10^{19}, 10^{20}, and 10^{21} called?

3. Explain why $10^A \times 10^B = 10^{A+B}$ is always true whenever A and B are counting numbers.

4. Write the following numbers in scientific notation:

   a. 153,293,043,922

   b. 0.00000321

   c. $(2.398 \times 10^{15}) \times (3.52 \times 10^9)$

   d. $(5.9 \times 10^{15}) \times (8.3 \times 10^9)$

5. Write 1.2 trillion in ordinary decimal notation and in scientific notation.

6. The astronomical unit (AU) is used to measure distances. One AU is the average distance from the earth to the sun, which is 92,955,630 miles. We are 2 billion AU from the center of the Milky Way galaxy (our galaxy). How many miles are we from the center of the Milky Way galaxy? Give your answer in scientific notation. Explain how the meaning of multiplication applies to this problem.

7. Sam uses a calculator to multiply 666,666 × 7,777,777. The calculator's answer is displayed as follows:

   $$5.1851795 \text{ E } 12$$

   So Sam writes

   $$666{,}666 \times 7{,}777{,}777 = 5{,}185{,}179{,}500{,}000$$

   Is Sam's answer correct or not? If not, why not?

Answers to Practice Exercises for Section 5.4

1. a. $10{,}000{,}000{,}000{,}000 = 10^{13}$

   b. $0.1 = 10^{-1}$

   c. $0.000001 = 10^{-6}$

   d. $1 = 10^0$

   e. $10 = 10^1$

   f. The number whose decimal representation is a 1 followed by 200 zeros can be written 10^{200}.

   g. A googol can be written 10^{100}.

   h. A googlpex can be written 10^{googol} or $10^{(10^{100})}$.

2. Since 10^{12} is a trillion, 10^{13} is ten trillion and 10^{14} is a hundred trillion. Since 10^{18} is a million trillion, 10^{19} is ten-million trillion, 10^{20} is a hundred-million trillion, and 10^{21} is a billion trillion.

3. The expression $10^A \times 10^B$ stands for A tens multiplied by B tens. If we multiply A tens with B tens,

   then all together, we multiply $A + B$ tens. Therefore,

   $$10^A \times 10^B = 10^{(A+B)}$$

   We can also explain this relationship with the following equations:

   $$10^A \times 10^B$$
   $$= \underbrace{10 \times 10 \times \cdots \times 10}_{A \text{ times}} \times \underbrace{10 \times 10 \times \cdots \times 10}_{B \text{ times}}$$
   $$= \underbrace{10 \times 10 \times \cdots \times 10 \times 10 \times 10 \times \cdots \times 10}_{A + B \text{ times}}$$

4. a. $1.53293043922 \times 10^{11}$

   b. 3.21×10^{-6}

   c. 8.44096×10^{24}

   d. 4.897×10^{25}

5. Ordinary decimal notation: 1,200,000,000,000. Scientific notation: 1.2×10^{12}.

6. Since we are 2 billion AU from the center of the galaxy and each AU is 92,955,630 miles, the number of miles from the earth to the center of the galaxy is the total number of objects in 2 billion groups (each AU is a group) when there are 92,955,630 objects in each group (each object is 1 mile). Therefore, according to the meaning of multiplication, we are

$$2 \text{ billion} \times 92{,}955{,}630 = 2 \times 10^{9} \times 92{,}955{,}630$$
$$= 185{,}911{,}260 \times 10^{9}$$
$$= 1.86 \times 10^{8} \times 10^{9}$$
$$= 1.86 \times 10^{17}$$

miles from the center of the galaxy.

7. No, Sam's answer is not correct. When the calculator displays the answer to $666{,}666 \times 7{,}777{,}777$ as

$$5.1851795 \text{ E } 12$$

this stands for

$$5.1851795 \times 10^{12}$$

However, the calculator is forced to round its answer because it can display only so many digits on its screen. Therefore, although it is true that

$$5.1851795 \times 10^{12} = 5{,}185{,}179{,}500{,}000$$

this is not the exact answer to $666{,}666 \times 7{,}777{,}777$. Instead, it is the answer to $666{,}666 \times 7{,}777{,}777$ rounded to the nearest hundred-thousand.

Problems for Section 5.4

1. Write the following numbers as powers of 10:

   a. 0.000001

   b. 10,000,000

   c. the number whose decimal representation is a 1 followed by 50 zeros

   d. the number whose decimal representation is a decimal point followed by 10 zeros, followed by a 1

   e. the number whose decimal representation is a decimal point followed by 50 zeros, followed by a 1

2. a. The winnings of a lottery were $250 million. Write 250 million in ordinary decimal notation and in scientific notation.

   b. A company's revenues are $15 billion. Write 15 billion in ordinary decimal notation and in scientific notation.

3. Write the following numbers in scientific notation:

   a. 201,348,761,098

   b. 0.000000078

   c. $(2.4 \times 10^{12}) \times (8.6 \times 10^{11})$

   d. $(6.1 \times 10^{13}) \times (9.2 \times 10^{8})$

4. A calculator might display the answer to

$$555{,}555 \times 6{,}666{,}666$$

as

$$3.7037 \text{ E } 12$$

   a. What does the calculator's display mean?

   b. What information about the solution to $555{,}555 \times 6{,}666{,}666$ can you obtain from the calculator's display? Can you determine the exact answer to $555{,}555 \times 6{,}666{,}666$ in ordinary decimal notation? Why or why not? Can you determine how many digits are in the ordinary decimal representation of the product $555{,}555 \times 6{,}666{,}666$? Explain.

5. Tanya says that the ones digit of 2^{59} is a 5 because her calculator's display for 2^{59} reads

$$5.76460752303 \text{ E } 17$$

and there is a 5 in the ones place. Is Tanya right? Why or why not?

6. Is 2×10^{7} equal to 2^{7}? Is 1×10^{9} equal to 1^{9}? If not, explain the distinctions between the expressions.

7. Suppose you have a calculator that displays no more than 12 digits. Describe how to use the distributive property and the calculator to multiply

$$8 \times 123{,}456{,}123{,}456$$

showing the answer in ordinary decimal notation. Do not just calculate the product with the longhand multiplication algorithm. Make efficient use of the calculator.

8. Let's say that you want to write the answer to 179,234,652 × 437,481,694 as a whole number in ordinary decimal notation, showing all its digits. Find a way to use a calculator that displays no more than 12 digits to help you do this in an efficient way. Do not just multiply longhand. Explain your technique, and explain why it works.

9. Suppose you multiply a 6-digit number by an 8-digit number. How many digits will the product have? The following problem will help you answer this question:

a. When you write a 6-digit number and an 8-digit number in scientific notation, what will the exponents on the 10s be? Explain.

b. Write

$$(1.3 \times 10^5) \times (2.5 \times 10^7)$$

and

$$(7.9 \times 10^5) \times (8.3 \times 10^7)$$

in scientific notation.

c. Suppose you write a 6-digit number and an 8-digit number in scientific notation. If you multiply these numbers and write the product in scientific notation, what will the exponent on the 10 be when the product is expressed in scientific notation? Explain why your answer is in an "either…or…" form. (*Hint:* See part (b).)

d. Use your answer to part (c) to answer the following question: When you multiply a 6-digit number by an 8-digit number, how many digits will the product have?

10. Light travels at a speed of about 300,000 kilometers per second.

a. How far does light travel in one day? Give your answer both in scientific notation and in ordinary decimal notation. Explain your work.

b. How far does light travel in one year? Give your answer both in scientific notation and in ordinary decimal notation, rounded to the nearest hundred-billion kilometers. Explain your work. The distance that light travels in one year is called a *light year*.

11. According to scientific theories, the solar system formed between 5 and 6 billion years ago. Light travels 186,282 miles per second. How far has the light from the forming solar system traveled in 5.5 billion years? Give your answer in scientific notation.

12. Suppose that a laboratory has one gram of a radioactive substance that has a half-life of 100 years. "A half-life of 100 years" means that, no matter what amount of the radioactive substance one starts with, after 100 years, only half of it will be left. So, after the first 100 years, only half a gram would be left, and after another 100 years, only one quarter of a gram would be left.

a. How many hundreds of years will it take until there is less than one hundred millionth of a gram left of the laboratory's radioactive substance? (Give a whole number of hundreds of years.)

b. How many hundreds of years will it take until there is less than one billionth of a gram left of your radioactive substance? (Give a whole number of hundreds of years.)

Chapter Summary and Study Items

Section 5.1 Multiplying Fractions

The meaning of $A \times B$ as the total in A groups of B applies not only to whole numbers, but also to fractions and decimals. To multiply fractions, we multiply the numerators and denominators. We can explain why this procedure makes sense by working with fractions of fractions.

Key skills and understandings:

- Write story problems for a given fraction multiplication problem, and solve the problems by using simple reasoning (such as with the aid of a picture).

- Use simple story problems and pictures to explain why the procedure for multiplying fractions makes sense.

- Given a story problem, determine whether it can be solved by fraction multiplication.

Section 5.2 Multiplying Decimals

To multiply decimals, the standard procedure is to multiply the numbers without the decimal points present. Then we add the number of digits to the right of the decimal points in the two original numbers and put the decimal point in the answer that many places from the end. We can explain why we place the decimal point where we do in the answer to a decimal multiplication problem by recognizing that we have multiplied and then divided by powers of 10 in order to shift decimal points suitably. We can also explain decimal multiplication in terms of fraction multiplication and in terms of areas of rectangles.

Key skills and understandings:

- Explain why we put the decimal point where we do when we multiply decimals.

- Use estimation to determine where to place the decimal point in a decimal multiplication problem.

- Describe decimal multiplication in terms of area.

- Write story problems for a given decimal multiplication problem.

Section 5.3 Multiplying Negative Numbers

We can explain why a positive number times a negative number is negative by considering negative numbers as representing owed amounts. By looking at patterns, we can see why our rules for multiplying with negative numbers are reasonable. A more thorough way to explain the way we multiply with negative numbers is to use the properties of arithmetic.

Key skills and understandings:

- Explain why a positive number times a negative number should be negative by considering negative numbers as owed amounts.

- Use properties of arithmetic to explain the rules for multiplying with negative numbers.

Section 5.4 Powers and Scientific Notation

Scientific notation is a way to write numbers by using powers of 10. Very large numbers and very small numbers are usually put in scientific notation in order to work with them and grasp them more easily.

Key skills and understandings:

- Put numbers that are in ordinary decimal notation in scientific notation, and put numbers that are in scientific notation in ordinary decimal notation.

- Use scientific notation in calculations.

Division

*I*n *this chapter, we will study division of different kinds of numbers: whole numbers, fractions, decimals, and integers. We can use division to solve a wide variety of problems, such as determining how many pencils each child will get if 100 pencils are divided equally among 22 children, determining how many servings of rice we can make from 5 cups of rice if one serving is $\frac{2}{3}$ cup of rice, or determining which paint mixture will be more yellow: one made by mixing 4 parts yellow paint with 25 parts white paint or one made by mixing 3 parts yellow paint with 16 parts white paint. On the surface, these are very different kinds of problems, but all can be solved with division.*

As with multiplication, we will study the meaning of division, and we will analyze why the various division procedures are valid based on the meaning of division. We will also study the link between division and fractions.

6.1 Interpretations of Division

 **Focal Points**
Grades 3, 4

What does division mean? Think about a simple story problem you might give your students to help them understand what $15 \div 3$ means. Does an example like giving out 15 cookies to 3 children come to mind? There is another quite different but equally valid way to think about the meaning of division. In this section, we will discuss these two interpretations of division. Just as every subtraction problem can be rewritten as an addition problem, every division problem can be rewritten as a multiplication problem. The two interpretations of division arise from this connection between division and multiplication.

Have you ever wondered why we are not allowed to divide by zero? Is there a reason behind this "law"? In this section we'll see that the link between division and multiplication gives us a mathematical reason for why we can't divide by zero.

FIGURE 6.1

(a) How many groups?   (b) How many in each group?

Two
interpretations
of 8 ÷ 2

Division Notation and Terminology

Division is denoted in three standard ways:

$$A \div B, \quad A/B, \quad \text{and} \quad B\overline{)A}$$

quotient

dividend

divisor

All three are read "*A* divided by *B*." In a division problem $A \div B$, the result is called the **quotient**, the number *A* is called the **dividend**, and the number *B* is called the **divisor**.

What Division Means

Class Activity   *Now Turn to Class Activities Manual*

6A What Does Division Mean? p. 114

A simple example will show that there are two distinct ways to interpret the meaning of division. How would you draw a simple picture to show your students what 8 ÷ 2 means? Your drawing might look like either part (a) or part (b) of Figure 6.1.

In (a) of Figure 6.1, 8 ÷ 2 is represented by 8 objects divided into *groups of* 2. With this interpretation we're thinking of 8 ÷ 2 = ? as asking how many groups there will be if 8 objects are divided into groups with 2 objects in each group. Whereas in (b) of Figure 6.1, 8 ÷ 2 is represented by 8 objects divided into 2 *groups*. With the second interpretation, we're thinking of 8 ÷ 2 = ? as asking how many objects will be in each group if 8 objects are divided equally among 2 groups.

The "How Many Groups?" Interpretation   If *A* and *B* are nonnegative numbers, and *B* is not 0, then, according to the "how many groups?" interpretation of division,

(Exact) division

$A \div B$

$A \div B$ means the number of groups that are formed when *A* objects are divided into groups with *B* objects in each group.

With this "how many groups?" interpretation of division, the problem

$$A \div B = ?$$

is equivalent to the problem

$$? \times B = A$$

In other words, with this interpretation, $A \div B$ means "the number of *B*s that are in *A*." This interpretation of division is sometimes called the measurement model of division or the subtractive model of division.

With the "how many groups?" interpretation of division, 18 ÷ 6 = 3, because if you have 18 objects and you divide these into groups of 6, then there will be 3 groups, as shown in Figure 6.2.

FIGURE 6.2

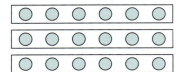

Using the "how many groups?" approach to show 18 ÷ 6 = 3

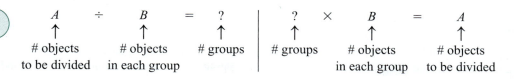

FIGURE 6.3

"How many
groups?"
division

Figure 6.3 shows that in "how many groups?" division problems, the divisor is the number of objects in each group and the quotient is the number of groups. The dividend is the number of objects to be divided (which will also be the case in our second interpretation of division).

The "How Many in Each Group?" Interpretation If A and B are nonnegative numbers and B is not 0, then, according to the "how many in each group?" interpretation of division,

(Exact)
division $A \div B$

$A \div B$ means the number of objects that are in each group when A objects are divided equally among B groups.

With this interpretation of division, the problem

$$A \div B = ?$$

is equivalent to the problem

$$B \times ? = A$$

This interpretation of division is sometimes called the partitive model of division or the sharing model of division.

With this interpretation of division we conclude that $18 \div 6 = 3$ because if you divide 18 objects equally among 6 groups, then there will be 3 objects in each group, as seen in Figure 6.4.

Figure 6.5 shows that in "how many in each group?" division problems, the divisor is the number of groups and the quotient is the number of objects in each group. The dividend is the number of objects to be divided, as it was with our other interpretation of division.

Distinguishing "How Many Groups?" from "How Many in Each Group?" Problems If you can't determine what kind of division problem a problem is, it may help to reformulate the problem in terms of multiplication. For example, consider this problem:

> 7 cups of snodgrass filled 3 identical containers full. How many cups of snodgrass are needed to fill one container full?

We can reformulate this problem with the equation

$$3 \quad \times \quad (\text{cups of snodgrass in one container}) \quad = \quad 7 \text{ cups}$$

groups   amount in each group   total amount

which we can solve by dividing $7 \div 3$. Since there are 3 groups and the "cups of snodgrass in one container" is the amount in each group, the problem is a "how many in each group?" problem.

Here is another example:

> Beezlebugs cost $3 per pound. You have $14 to spend on beezlebugs. How many pounds of beezlebugs can you buy?

FIGURE 6.4

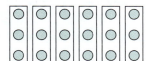

Using the "how
many in each
group?"
approach to show
$18 \div 6 = 3$

FIGURE 6.5

"How many in each group? division

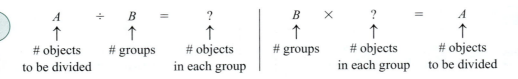

The wording "$3 per pound" means that each pound is worth $3, so we can think of each pound as a "group of $3". We can, therefore, reformulate the beezlebug problem with the equation

which we can solve by dividing $14 \div 3$. Since the "# pounds beezlebugs" stands for the number of groups and since there are $3 in each group, the problem is a "how many groups?" problem.

Even though you probably won't teach your students to distinguish "how many groups?" story problems from "how many in each group?" story problems, it's important for *you* to distinguish them so that you can give your students a mix of both types.

Class Activity   *Now Turn to Class Activities Manual*

6B   Division Story Problems, p. 115

Relating the "How Many Groups?" and "How Many in Each Group?" Interpretations   Why do we get the same answer to division problems, regardless of which interpretation of division we use? By viewing division in terms of multiplication, we can explain why a division problem will have the same answer, regardless of whether the "how many groups?" or the "how many in each group?" interpretation of division is used. The key lies in the commutative property of multiplication. According to the "how many groups?" interpretation, solving $A \div B = ?$ is equivalent to solving $? \times B = A$. But according to the commutative property of multiplication, $? \times B = B \times ?$. Therefore, solving $? \times B = A$ is equivalent to solving $B \times ? = A$, which is equivalent to solving $A \div B = ?$ with the "how many in each group?" interpretation of division. Thus, we will always get the same answer to a division problem, regardless of which way we interpret division.

Array, Area, and Other Division Story Problems

In Chapter 4 we studied array, ordered pair, and multiplicative comparison story problems, and we saw why areas of rectangles are calculated by multiplying. In each of these cases, it is possible to describe the situation in terms of equal groups and, therefore, in terms of multiplication. By viewing division problems as multiplication problems that have a known product, one known factor, and one unknown factor, we can create division problems that concern arrays, areas, multiplicative comparisons, and ordered pairs.

For example, consider these problems (drawings to accompany the first three problems are shown in Figure 6.6):

1. There are 54 cans arranged in an array that has 6 rows. How many cans are in each row?

2. A rectangular room has area 54 square meters. One side of the room is 6 meters long. How long is an adjacent side?

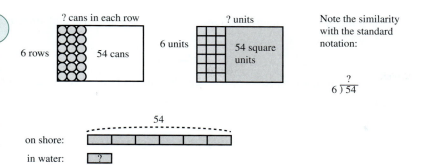

FIGURE 6.6

Pictorial representations of array, area, and multiplicative comparison division problems

3. There were 54 penguins on shore, and that was 6 times as many penguins as were in the water. How many penguins were in the water?

4. In a game, when you roll a 6-sided number cube and then pick a card, there are 54 different possible outcomes. How many different cards are there that you can pick?

Each of these problems can be viewed as a problem for

$$6 \times ? = 54 \quad \text{or} \quad ? \times 6 = 54$$

Therefore, each of these problems can be solved by calculating $54 \div 6 = 9$.

Pictorial representations for array and area division problems are similar to the division notation of the form $6\overline{)54}$, as indicated in Figure 6.6.

Why We Can't Divide by Zero

When we defined division, $A \div B$, we said that the divisor, B, should not be 0. Why not? The next class activity will help you explain this in several ways. One way is to translate a division problem with divisor 0 into an equivalent multiplication problem.

Class Activity *Now Turn to Class Activities Manual*

6C Why Can't We Divide by Zero? p. 116

Division with Negative Numbers

So far we have defined division for nonnegative numbers. How can we make sense of division problems like

$$-18 \div 3 = ?$$

or

$$18 \div -3 = ?$$

We could perhaps talk about dividing $18 that is owed among 3 people. Another way is to rewrite a division problem with negative numbers as a multiplication problem. For example, the problem

$$-18 \div 3 = ?$$

is equivalent to

$$? \times 3 = -18$$

(if we use the "how many groups?" interpretation). Since $-6 \times 3 = -18$, and since no other number times 3 is -18, it follows that

$$-18 \div 3 = -6$$

By considering how multiplication works when negative numbers are involved, we can see that the following rules apply to division problems involving negative numbers:

$$\text{negative} \div \text{positive} = \text{negative}$$
$$\text{positive} \div \text{negative} = \text{negative}$$
$$\text{negative} \div \text{negative} = \text{positive}$$

Practice Exercises for Section 6.1

1. For each of the following story problems, write the corresponding numerical division problem and decide which interpretation of division is involved (the "how many groups?" or the "how many in each group?").

   a. There are 5235 tennis balls that are to be put into packages of 3. How many packages of balls can be made?

   b. If 50 fluid ounces of mouthwash costs $5, then what is the price per fluid ounce of this mouthwash?

   c. If 50 fluid ounces of mouthwash costs $5, then how much of this mouthwash is worth $1?

   d. If you have a full 50-fluid-ounce bottle of mouthwash and you use 2 fluid ounces per day, then how many days will this mouthwash last?

   e. 1 yard is 3 feet. How long is an 84-foot-long stretch of sidewalk in yards?

   f. 1 gallon is 16 cups. How many gallons are 64 cups of lemonade?

   g. If you drive 315 miles at a constant speed and it takes you 5 hours, then how fast did you go?

2. Write two story problems for $300 \div 12$, one for each of the two interpretations of division.

3. a. Write a multiplicative comparison story problem for $18 \div 3$.

   b. Write an area story problem for $18 \div 3$.

4. Josh says that $0 \div 5$ doesn't make sense because if you have nothing to divide among 5 groups, then you won't be able to put anything in the 5 groups. Which interpretation of division is Josh using? What does Josh's reasoning actually say about $0 \div 5$?

5. Roland says,

   I have 23 pencils to give out to my students, but I have no students. How many pencils should each student get? There's no possible answer because I can't give out 23 pencils to my students if I have no students.

   Write a division problem that corresponds to what Roland said. Which interpretation of division does this use? What does Roland's reasoning say about the answer to this division problem?

6. Katie says,

   I have no candies to give out and I'm going to give each of my friends 0 candies. How many friends do I have? I could have 10 friends, or 50 friends, or 0 friends—there's no way to tell.

   Write a division problem that corresponds to what Katie said. Which interpretation of division does this use? What does Katie's reasoning say about the answer to this division problem?

7. Use the "how many groups?" interpretation of division to explain why $1 \div 0$ is not defined.

8. Explain why $0 \div 0$ is not defined (or indeterminate) by rewriting $0 \div 0 = ?$ as a multiplication problem.

Answers to Practice Exercises for Section 6.1

1. a. $5235 \div 3$. This uses the "how many groups?" interpretation. 1745 packages can be made.

b. $5 \div 50$. This uses the "how many in each group?" interpretation. Each ounce of mouthwash represents a group. We want to divide \$5 equally among these groups. Each fluid ounce costs \$0.10.

c. $50 \div 5$. This uses the "how many in each group?" interpretation. Each dollar represents a group. We want to divide 50 ounces among the 5 groups. \$1 buys 10 fluid ounces.

d. $50 \div 2$. This uses the "how many groups?" interpretation. Each 2 fluid ounces is a group. We want to know how many of these groups are in 50 fluid ounces. The mouthwash will last for 25 days.

e. $84 \div 3$. This uses the "how many groups?" interpretation. Each 3 feet is a group (a yard). We want to know how many of these groups are in 84 feet. There are 26 yards in 84 feet.

f. $64 \div 16$. This uses the "how many groups?" interpretation. Each 16 cups is a group (a gallon). We want to know how many of these groups are in 64 cups. The answer is 4 gallons.

g. $315 \div 5 = 63$ miles per hour. This uses the "how many in each group?" interpretation. Divide the 315 miles equally among the 5 hours. Each hour represents a group. In each hour, you drove $315 \div 5$ miles. This means your speed was 63 miles per hour.

2. A good example for the "how many groups?" interpretation is "how many feet are in 300 inches?" Because each foot is 12 inches, this problem can be interpreted as "how many 12s are in 300?" An example for the "how many in each group?" interpretation is "300 snozzcumbers will be divided equally among 12 hungry boys. How many snozzcumbers does each hungry boy get?"

3. a. Kaya saved \$18 and that is 3 times as much as her little sister, Ana, saved. How much did Ana save?

b. A rectangular poster is to have an area of 18 square feet. If the poster will be 3 feet wide, how long should it be?

4. Josh is using the "how many in each group?" interpretation of division. Josh's statement can be reinterpreted as this: If you have 0 objects and you divide them equally among 5 groups, then there will be 0 objects in each group. Therefore, $5 \times 0 = 0$ and so $0 \div 5 = 0$.

5. The division problem that corresponds to what Roland said is this: $23 \div 0 = ?$. Written in terms of multiplication it is $0 \times ? = 23$. Roland is using the "how many in each group?" interpretation of division. Each group is represented by a student. Roland wants to divide 23 objects equally among 0 groups. But as Roland says, this is impossible to do. Therefore, $23 \div 0$ is undefined.

6. The division problem that corresponds to what Katie said is this: $0 \div 0 = ?$. Written in terms of multiplication it is $? \times 0 = 0$. Katie is using the "how many groups?" interpretation of division. Each group is represented by a friend. There are 0 objects to be distributed equally, with 0 objects in each group. But from this information, there is no way to determine how many groups there are. There could be *any* number of groups—50, or 100, or 1000. So the reason that $0 \div 0$ is undefined (or indeterminate) is there isn't *one unique answer*.

7. With the "how many groups?" interpretation, $1 \div 0$ means the number of groups there are when 1 object is divided into groups with 0 objects in each group. But if you put 0 objects in each group, then there's no way to distribute the 1 object—it can never be distributed among the groups, no matter how many groups there are. In other words, $? \times 0 = 1$ cannot be solved. Therefore, $1 \div 0$ is not defined.

8. Any division problem $A \div B = ?$ can be rewritten in terms of multiplication, namely, as either $? \times B = A$ or as $B \times ? = A$. Therefore, $0 \div 0 = ?$ means the same as $? \times 0 = 0$ or $0 \times ? = 0$. But *any number* times 0 is zero, so the ? can stand for any number. Since there isn't one unique answer to $0 \div 0 = ?$, we say that $0 \div 0$ is undefined (or indeterminate).

Problems for Section 6.1

1.  For each of the following story problems, write the corresponding numerical division problem, state which interpretation of division is involved (the "how many groups?" or the "how many in each group?") and solve the problem:

   a. If 252 rolls are to be put in packages of 12, then how many packages of rolls can be made?

   b. If you have 506 stickers to give out equally to 23 children, then how many stickers will each child get?

   c. Given that 1 gallon is 8 pints, how many gallons of water is 48 pints of water?

   d. If your car used 12 gallons of gasoline to drive 360 miles, then how many miles per gallon did your car get?

   e. If you drove 177 miles at a constant speed and if it took you 3 hours, then how fast were you going?

   f. Given that 1 foot is 12 inches, how many feet long is an 84-inch-long board? in feet?

2. Write two story problems for $63 \div 7$, one for each of the two interpretations of division. Solve each problem.

3. a. Write an array problem for $21 \div 3$. Draw a simple picture that a child could use to determine the answer to the problem.

   b. Write an area problem for $21 \div 3$. Draw a simple picture that a child could use to determine the answer to the problem.

4. a. Write a multiplicative comparison problem for $21 \div 3$.

   b. Draw a strip diagram to accompany your problem in part (a).

c. Reformulate your problem in part (a), but this time use fraction language. (If you used fraction language in part (a), then reformulate your problem without using fraction language.)

d. Discuss why a child might mistakenly attempt to solve your problem in part (a)—or its reformulation in part (c)—by multiplying 21 by 3 instead of by dividing 21 by 3. Explain how drawing a strip diagram might help the child determine how to solve the problem.

5. a. Is $0 \div 5$ defined or not? Write a story problem for $0 \div 5$, and use your story problem to discuss whether or not $0 \div 5$ is defined.

   b. Is $5 \div 0$ defined or not? Write a story problem for $5 \div 0$, and use your story problem to discuss whether or not $5 \div 0$ is defined.

6. a. Is $0 \div 3$ defined or not? Explain your reasoning.

   b. Is $3 \div 0$ defined or not? Explain your reasoning.

7. Write and solve one story problem for $35 \div 70$ and another for $70 \div 35$. Say which is which.

8. a. Use the meaning of powers of ten to show how to write the following expressions as a single power of ten (i.e., in the form $10^{\text{something}}$):

   i. $10^5 \div 10^2$ ii. $10^6 \div 10^4$ iii. $10^7 \div 10^6$

   b. In each of (i), (ii), and (iii), compare the exponents involved. In each case, what is the relationship among the exponents?

   c. Explain why it is always true that $10^A \div 10^B = 10^{A-B}$ when A and B are counting numbers and A is greater than B.

6.2 Division and Fractions and Division with Remainder

F·P Focal Points
Grade 5

When we *multiply* two whole numbers, the product is always a whole number. So it's interesting that when we *divide* two whole numbers, the quotient might not be a whole number. Furthermore, when the quotient is not a whole number, we have different options for how to express the answer to the division problem. In this section, we'll study the different kinds of answers we can give to whole number division problems. We'll see that if the division problem arose from a story problem, we should take the context into account when deciding which answer to the numerical division problem to use. We'll also see how the different answers to a division problem are related. Especially important in this relationship is the link between division and fractions, which we examine first.

Fractions and Division

What is the relationship between fractions and division? Have you noticed that the same notation is often used for division as well as for fractions? Instead of writing the "divided by" symbol, $\div$, we sometimes write $2 \div 5$ as 2/5. But the expression 2/5 also stands for the fraction $\frac{2}{5}$, so the notation we use equates fractions with division. However, according to our definitions of fractions and of division, the expressions

$$\frac{2}{5} \quad \text{and} \quad 2 \div 5$$

have the following *different* meanings:

- $\frac{2}{5}$ of a pie is the amount of pie formed by 2 pieces when the pie is divided into 5 equal pieces.
- $2 \div 5$ is the amount of pie each person will receive if 2 (identical) pies are divided equally among 5 people (using the "how many in each group?" interpretation).

Notice the difference: $\frac{2}{5}$ refers to *2 pieces of pie*, whereas $2 \div 5$ refers to dividing *2 pies*. But is it the same amount of pie either way? Before you read on, please do the next Class Activity.

Class Activity *Now Turn to Class Activities Manual*

6D 🏺 Relating Fractions and Division, p. 117

Let's return to the case of dividing 2 pies equally among 5 people. To do so, you can divide each pie into 5 equal pieces and give each person one piece from each pie, as shown in Figure 6.7. One person's share of pie consists of 2 pieces of pie, and each of those pieces is $\frac{1}{5}$ of a pie. (That is, each piece comes from a pie that has been divided into 5 equal pieces.) So each of the 5 people sharing the 2 pies gets $\frac{2}{5}$ of a pie. Thus, one person's share of pie can be described in two ways: as $2 \div 5$ of a pie and as $\frac{2}{5}$ of a pie. Therefore, both ways of describing a person's share of pie must be equal, and we have

$$2 \div 5 = \frac{2}{5}$$

The preceding discussion applies equally well when other whole numbers replace 2 and 5 (except that 5 should not be replaced with 0). So, in general, if A and B are whole numbers and B is not 0, then $A \div B$ really is equal to $\frac{A}{B}$.

FIGURE 6.7

Explaining why $2 \div 5 = \frac{2}{5}$ by dividing 2 pies equally among 5 people and determining that one person's share is $\frac{2}{5}$ of a pie.

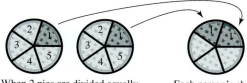

When 2 pies are divided equally among 5 people, each person's share is $2 \div 5$ of a pie. → Each person's share is $\frac{2}{5}$ of a pie.

Fractions with Negative Numerators or Denominators How can we make sense of fractions such as $\frac{-13}{5}$, $\frac{13}{-5}$, or $\frac{-13}{-5}$? Earlier, we explained why, when A and B are counting numbers,

$$\frac{A}{B} = A \div B$$

We can use this relationship between fractions and division to *define* fractions that have negative numerators, negative denominators, or both. Therefore, if A and B are whole numbers and B is not zero, then

$$\frac{-A}{B} = (-A) \div B = -(A \div B) = -\frac{A}{B}$$

$$\frac{A}{-B} = A \div (-B) = -(A \div B) = -\frac{A}{B}$$

$$\frac{-A}{-B} = (-A) \div (-B) = A \div B = \frac{A}{B}$$

So,

$$\frac{-13}{5} = -\frac{13}{5}$$

$$\frac{13}{-5} = -\frac{13}{5}$$

$$\frac{-13}{-5} = \frac{13}{5}$$

Exact Division versus Division with Remainder

Consider the division problem $23 \div 4$. Depending on the context, any one of the following three answers could be most appropriate:

$$23 \div 4 = 5.75$$

$$23 \div 4 = 5\frac{3}{4}$$

$$23 \div 4 = 5, \text{ remainder } 3$$

The first two answers fit with division as we have been interpreting it so far—exact division. For example, to divide \$23 equally among 4 people, each person should get $23 \div 4 = \$5.75$. Or, if you have 23 cups of flour, and a batch of cookies requires 4 cups of flour, then you can make $23 \div 4 = 5\frac{3}{4}$ batches of cookies (assuming that it is possible to make $\frac{3}{4}$ of a batch of cookies). But if you have 23 pencils to divide equally among 4 children, then it doesn't make sense to give each child 5.75 or $5\frac{3}{4}$ pencils. Instead, it is best to give each child 5 pencils and keep the remaining 3 pencils in reserve. How does this answer, 5, remainder 3, fit with division as we have defined it?

In order to get the third answer, 5, remainder 3, we need to interpret division in a slightly different manner than we have so far. For each of the two main interpretations of division, there is an alternative formulation, which allows for a remainder. In these alternative formulations, we seek a quotient that is a *whole number*.

If A and B are whole numbers, and B is not zero, then

division with remainder, how many in each group?

$A \div B$ is the largest whole number of objects that are in each group when A objects are divided equally among B groups. The **remainder** is the number of objects left over (i.e., that can't be placed in a group). This is the "**how many in each group?**" interpretation, with remainder.

division with remainder, how many groups?

$A \div B$ is the largest whole number of groups that can be made when A objects are divided into groups with B objects in each group. The **remainder** is the number of objects left over (i.e., that can't be placed in a group). This is the "**how many groups?**" interpretation, with remainder.

The traditional notation for division with remainder is to write "equations" of the form

$$23 \div 4 = 5, \text{remainder } 3 \quad \text{or} \quad 23 \div 4 = 5, \text{R } 3$$

However, there are problems with this notation, especially with using the equal sign in this way. Before you read on, please do the next Class Activity.

Class Activity *Now Turn to Class Activities Manual*

6E Division with Remainder Notation, p. 117

This traditional notation, $23 \div 4 = 5$, R 3, should be interpreted as

$$23 = 5 \times 4 + 3 \quad \text{or} \quad 23 = 4 \times 5 + 3$$

How is division with remainder connected with exact division? We'll study this question next.

Connecting Whole-Number-with-Remainder Answers to Mixed Number Answers How are the whole-number-with-remainder answer and the mixed number answer to a whole number division problem related? Given the division problem $29 \div 6$, the whole-number-with remainder answer is 4 remainder 5, and its (exact) mixed number answer is $4\frac{5}{6}$. The remainder is 5, the divisor is 6, and the fractional part of the mixed number answer, $4\frac{5}{6}$ is $\frac{5}{6}$, which is in the form

$$\frac{\text{remainder}}{\text{divisor}}$$

This will be the case in general. We can see why by supposing we have 29 pizzas to be divided equally among 6 classrooms. How many pizzas does each classroom get? Since $29 \div 6 = 4$, remainder 5, each classroom will get 4 pizzas, and there will be 5 pizzas left over. Instead of just giving away the remaining 5 pizzas, we might want to divide these 5 pizzas into 6 equal parts. When 5 pizzas are divided equally among 6 classrooms, how much pizza does each classroom get? Each classroom gets

$$5 \div 6 = \frac{5}{6}$$

of a pizza, according to the work we did previously, in which we equated fractions and division, and as we also see in Figure 6.8.

So, instead of giving each class 4 pizzas and having 5 pizzas left over, we can give each class $4\frac{5}{6}$ pizzas. In other words,

$$29 \div 6 = 4, \text{remainder } 5$$

is equivalent to

$$29 \div 6 = 4\frac{5}{6}$$

FIGURE 6.8

Dividing
5 pizzas among
6 classrooms

 Each class gets $\frac{5}{6}$ of a pizza.

Notice that in the mixed number version, the remainder 5 becomes the numerator of the fractional part of the answer, whereas the denominator is the divisor, 6.

The same reasoning works with other numbers. If A and B are whole numbers, then

$$A \div B = Q, \text{ remainder } R$$

is equivalent to

$$A \div B = Q\frac{R}{B}$$

In the mixed number answer to $A \div B$, the fractional part of the answer is the remainder R over the divisor B.

In the next Class Activity, you will interpret whole-number-with-remainder and mixed number answers to division problems in the context of story problems. You'll see that you should pick the form of the answer to a numerical division problem suitably in order to best answer a related story problem. Even then, the answer to the story problem is not necessarily identical to the answer to the numerical division problem.

Class Activity   *Now Turn to Class Activities Manual*

6F What to Do with the Remainder? p. 118

Using Division to Convert Improper Fractions to Mixed Numbers In Chapter 3, we discussed how to turn mixed numbers into improper fractions by viewing a mixed number as the sum of its whole number part and its fractional part. Now we will go the other way around and turn improper fractions into mixed numbers. We do so by viewing an improper fraction in terms of division.

For example, to convert the improper fraction $\frac{27}{4}$ to a mixed number, write $\frac{27}{4}$ in terms of division, and calculate the mixed number answer to the corresponding division problem:

$$\frac{27}{4} = 27 \div 4$$

$$27 \div 4 = 6 \text{ remainder } 3$$

Thus,

$$\frac{27}{4} = 27 \div 4 = 6\frac{3}{4}$$

Another way to explain why the preceding procedure for turning $\frac{27}{4}$ into a mixed number makes sense is to notice that $\frac{27}{4}$ stands for 27 copies of parts of an object that has been divided into 4 equal parts. Each group of 4 parts makes a whole. When we divide 27 by 4, we find that there are 6 groups of 4 and 3 left over. In terms of the fraction $\frac{27}{4}$, the 6 groups of 4 parts make 6 wholes. The 3 remaining parts make $\frac{3}{4}$ of a whole because each of the parts is one fourth.

The same reasoning works with other numbers. If A and B are whole numbers, and if

$$A \div B = Q, \text{ remainder } R$$

then

$$\frac{A}{B} = Q\frac{R}{B}$$

When we write a fraction as a mixed number, the fractional part of the mixed number is the remainder over the divisor.

Practice Exercises for Section 6.2

1. Use the meaning of fractions and the meaning of division, from the "how many in each group?" viewpoint, to explain why $3 \div 10 = \frac{3}{10}$. Your explanation should be general, in the sense that you could see why $3 \div 10 = \frac{3}{10}$ would still be true if other numbers were to replace 3 and 10.

2. Describe how the whole number with remainder and mixed number answers to $14 \div 3$ are related. Use a simple story problem to help you explain this relationship.

3. For each of the following story problems, write the corresponding numerical division problem. Also, interpret the meaning of the whole-number-with-remainder answer and the meaning of the mixed number answer to the division problem in terms of the story problem and its answer.

   a. You have 27 pints of soup and a bunch of containers that hold 4 pints. How many containers will you need to hold all the soup?

   b. How long will it take you to drive 105 miles at a steady speed of 45 miles per hour?

   c. You drove 135 miles at a steady speed and it took you 2 hours. How fast were you going?

4. If March 5th is a Wednesday, then why do you know right away that March 12th, 19th, 26th, and April 2nd are Wednesdays? Explain this, using mathematics and knowledge of our calendar system.

5. Suppose that today is Friday. What day of the week will it be 36 days from today? What day of the week will it be 52 days from today? What about 74 days from today? Explain how division with a remainder is relevant to these questions.

Answers to Practice Exercises for Section 6.2

1. If there are 3 (identical) pies to be divided equally among 10 people, then according to the meaning of division (with the "how many in each group?" interpretation) each person will get $3 \div 10$ of a pie. To divide the pies, you can divide each pie into 10 pieces and give each person 1 piece from each of the 3 pies. One person's share is shown shaded in Figure 6.9. Each person then gets 3 pieces, where each piece is $\frac{1}{10}$ of a pie. That is, each piece is 1 part when the pie is divided into 10 equal parts. Therefore, each person gets $\frac{3}{10}$ of a pie, according to the meaning of fractions.

FIGURE 6.9 The shaded portion is one person's share when 3 pies are divided equally among 10 people.

Because each person's share can be described both as $3 \div 10$ of a pie and as $\frac{3}{10}$ of a pie, it follows that $\frac{3}{10} = 3 \div 10$.

2. The whole-number-with-remainder answer to $14 \div 3$ is 4, remainder 2. The mixed number answer is $4\frac{2}{3}$. In general, to get the fractional part of the mixed number answer, you make a fraction whose numerator is the remainder and whose denominator is the divisor. To see why this makes sense, consider an example. If you have 14 cookies to be divided equally among 3 people, then you could give each person 4 cookies and have 2 cookies left over, or you could take those 2 remaining cookies and divide them equally among the 3 people. You could do this by dividing each cookie into 3 equal parts and giving each person 2 parts (as you could show in a picture like Figure 6.8). In this way, each person would get $4\frac{2}{3}$ cookies.

3. a. 27 ÷ 4. The whole-number-with-remainder answer to 27 ÷ 4 is 6, remainder 3. In terms of the story problem, this means that you can fill 6 containers completely full and will then have 3 pints remaining. So to answer the story problem, you will need another container, for a total of 7 containers, to hold all the soup. The mixed number answer to 27 ÷ 4 is $6\frac{3}{4}$. In terms of the story problem, this means that you can fill 6 containers completely full and fill the 7th container $\frac{3}{4}$ full. Again, the answer to the story problem is therefore 7.

b. 105 ÷ 45. The whole-number-with-remainder answer to 105 ÷ 45 is 2, remainder 15. In terms of the story problem, this means that after 2 full hours of driving, you will still have another 15 miles to go. Notice that this answer doesn't tell us exactly how long it will take to drive the full distance, only that it will take between 2 and 3 hours. The mixed number answer to 105 ÷ 45 is $2\frac{15}{45} = 2\frac{1}{3}$. In terms of the story problem, this means it will take $2\frac{1}{3}$ hours to drive the 105 miles. Since $\frac{1}{3}$ of an hour is 20 minutes, it will take 2 hours, 20 minutes.

c. 135 ÷ 2. The whole-number-with-remainder answer to 135 ÷ 2 is 67, remainder 1. In terms of the story problem, this means you went 67 whole miles in each hour, and another part of a mile in each hour. The whole-number-with-remainder answer to the numerical division problem is really not appropriate or illuminating for solving the story problem. The mixed number answer to 135 ÷ 2 is $67\frac{1}{2}$. In terms of the story problem, this means that you drove $67\frac{1}{2}$ miles each hour, in other words, that you were driving at a speed of $67\frac{1}{2}$ miles per hour.

4. Every seven days after Wednesday is another Wednesday. March 12th, 19th, 26th, and April 2nd are 7, 14, 21, and 28 days after March 5th, and 7, 14, 21, and 28 are multiples of 7, so these days are all Wednesdays, too.

5. Every seven days after a Friday is another Friday. So 35 days after today (assumed to be a Friday) is another Friday. Therefore, 36 days from today is a Saturday, because it is 1 day after a Friday. Notice that we really only needed to find the remainder of 36 when divided by 7 in order to determine the answer. 52 ÷ 7 = 7, remainder 3, so 52 days from today will be 3 days after a Friday (because the 7 groups of 7 get us to another Friday). Three days after a Friday is a Monday. 74 ÷ 7 = 10, remainder 4, so 74 days from today will be 4 days after a Friday, which is a Tuesday.

Problems for Section 6.2

1. Use the meaning of fractions and the meaning of division from the "how many in each group?" viewpoint, to explain in your own words why $3 \div 7 = \frac{3}{7}$. Your explanation should be general, in the sense that you could see why $3 \div 7 = \frac{3}{7}$ would still be true if other numbers were to replace 3 and 7.

2. Describe how to get the mixed-number answer to 23 ÷ 5 from the whole-number-with-remainder answer. Explain why your method makes sense by interpreting it in terms of a simple story problem.

3. a. Write a simple "how many in each group? story problem for 17 ÷ 5 for which the answer "3, remainder 2" is appropriate. Explain what the answer "3, remainder 2" means in the context of the story problem.

b. Write a simple "how many groups?" story problem for 17 ÷ 5 for which the answer "3, remainder 2" is appropriate. Explain what the answer "3, remainder 2" means in the context of the story problem.

c. Write a simple "how many in each group?" story problem for 17 ÷ 5 for which the answer $3\frac{2}{5}$ is appropriate. Explain what the answer $3\frac{2}{5}$ means in the context of the story problem.

d. Write a simple "how many groups?" story problem for $17 \div 5$ for which the answer $3\frac{2}{5}$ is appropriate. Explain what the answer $3\frac{2}{5}$ means in the context of the story problem.

4. [icon] Write and solve four different story problems for $21 \div 4$.

a. In the first story problem, the answer should be best expressed as 5, remainder 1. Explain why this is the best answer. Interpret the meaning of "5, remainder 1" in the context of the story problem.

b. In the second story problem, the answer should be best expressed as $5\frac{1}{4}$. Explain why this is the best answer. What does the $\frac{1}{4}$ mean in the context of the story problem?

c. In the third story problem, the answer should be best expressed as 5.25. Explain why this is the best answer.

d. The answer to the fourth story problem should be 6 (even though $21 \div 4 \neq 6$). Explain why this is the best answer.

5. For each of the problems that follow, write the corresponding numerical division problem and solve the problem. Determine the best form (or forms) of the answer: a mixed number, a decimal, a whole number with remainder, or a whole number that is not equal to the solution of the numerical division problem. Briefly explain your answers.

a. For purposes of maintenance, a 58-mile stretch of road will be divided into 3 equal parts. How many miles of road are in each part?

b. For purposes of maintenance, a 58-mile stretch of road will be divided into sections of 15 miles. Each section of 15 miles will be the responsibility of a particular road crew. How many road crews are needed?

c. You have 75 pencils to give out to a class of 23 children who insist on fair distribution. How many pencils will each child get?

d. You have 7 packs of chips to share equally among 4 hungry people. How many packs of chips will each person get?

6. Explain how to solve the next problems with division. Which interpretation of division do you use?

a. What day of the week will it be 50 days from today?

b. What day of the week will it be 60 days from today?

c. What day of the week will it be 91 days from today?

d. What day of the week will it be 365 days from today?

7. [icon] In your own words, describe a procedure for turning an improper fraction, such as $\frac{19}{4}$, into a mixed number, and explain why this procedure makes sense.

8. Halloween (October 31) of 2008 was on a Friday, which was great for kids.

a. How can you use division to determine what day of the week Halloween was on in 2009? (The year 2009 was not a leap year, so Halloween of 2009 was 365 days from Halloween of 2009.)

b. After 2009, when are the next two times that Halloween falls on either a Friday or a Saturday? Again, use mathematics to determine this. Explain your reasoning. (The years 2012, 2016, 2020, etc. are leap years, so they have 366 days instead of 365.)

9. Must there be at least one Friday the 13th in every year? Use division to answer this question. (You may answer only for years that aren't leap years.) To get started on solving this problem, answer the following: If January 13th falls on a Monday, then what day of the week will February 13th, March 13th, and so forth, fall on? Use division with remainder to answer these questions. Now consider what will happen if January 13th falls on a Tuesday, a Wednesday, and so on.

10. Presidents' Day is the third Monday in February. In 2010, Presidents' Day was on February 15. What is the date of Presidents' Day in 2011? Use mathematics to solve this problem without looking at a calendar. Explain your reasoning clearly.

11. I'm thinking of a number. When you divide it by 2, it has remainder 1; when you divide it by 3, it has remainder 1; when you divide by 4, 5, or 6, it always has remainder 1. The number I am thinking of could just be the number 1 (because $1 \div 2 = 0$, remainder 1; $1 \div 3 = 0$, remainder 1; $1 \div 4 = 0$, remainder 1, etc.). Find at least one other such number that is greater than 1.

12. I'm thinking of a number. When you divide it by 12, it has remainder 2; and when you divide it by 16, it also has remainder 2. The number I am thinking of could be the number 2 because $2 \div 12 = 0$, remainder 2 and $2 \div 16 = 0$, remainder 2. Find at least three other such numbers that are greater than 2. How are these numbers related?

13. Three robbers have just acquired a large pile of gold coins. They go to bed, leaving their faithful servant to guard it. In the middle of the night, the first robber gets up, gives 2 gold coins from the pile to the servant as hush money, divides the remaining pile of gold evenly into 3 parts, takes 1 part, forms the remaining 2 parts back into a single pile and goes back to bed. A little later, the second robber gets up, gives 2 gold coins from the remaining pile to the servant as hush money, divides the remaining pile evenly into 3 parts, takes 1 part, forms the remaining 2 parts back into a single pile and goes back to bed. A little later, the third robber gets up and does the very same thing. In the morning, when they count up the gold coins, there are 100 of them left. How many were in the pile originally? Explain your answer.

14. A year that is not a leap year has 365 days. (Leap years have 366 days and generally occur every four years.) There are 7 days in a week and 52 whole weeks in a year. How many whole weeks are there in 3 years? How many whole weeks are there in 7 years? Is the number of whole weeks in 3 years three times the number of whole weeks in 1 year? Is the number of whole weeks in 7 years seven times the number of whole weeks in 1 year? Explain the discrepancy!

15. The text explained why $A \div B = \frac{A}{B}$ with the "how many in each group?" view of division. In this problem, you will explain why $A \div B = \frac{A}{B}$ with the "how many groups?" view of division. Write a "how many groups?" story problem for $3 \div 5$, and explain why the solution is $\frac{3}{5}$. Will this work generally, if different counting numbers replace 3 and 5? Explain.

16. The text presented one way to explain why $A \div B = \frac{A}{B}$. This problem will help you explain in a different way why we can express fractions in terms of division.

  a. Use the meaning of multiplication, the meaning of fractions, and the meaning of division to explain why

  $$\frac{1}{10} \cdot 3 = 3 \div 10$$

  b. Use the equation in part (a), the commutative property of multiplication, and the meaning of fractions to explain why

  $$3 \div 10 = \frac{3}{10}$$

6.3 Why the Common Long Division Algorithm Works

Focal Points
Grade 5

In Chapter 4, we studied the connection between the meaning of multiplication and the standard longhand multiplication procedure. We explained why the multiplication procedure gives answers to multiplication problems that agree with the meaning of multiplication. In this section, we will do the same for division. Think about what a remarkable human achievement the discovery of the long division procedure is: using it we can solve any whole number division problem. Even so, students sometimes wonder why we should study the division algorithm when we can simply use a calculator to calculate answers to division problems. As we'll see in this section, to understand why the division algorithm works, students must think about the meaning of division and they must reason about place value and properties of arithmetic. When the focus is on reasoning and sense-making, the study of the common algorithms of arithmetic can provide students with the opportunity to deepen their understanding of some of the most fundamental ideas in mathematics.

Class Activity   *Now Turn to Class Activities Manual*

6G Can We Use Properties of Arithmetic to Divide?, p. 119

Solving Division Problems Without the Common Longhand Procedure

To analyze why the common (also called "standard") long division procedure works, we will first study how to solve division problems without the procedure and without a calculator.

Class Activity   *Now Turn to Class Activities Manual*

6H Dividing Without Using a Calculator or Long Division, p. 120

How can we divide without using a longhand division procedure? In the following example, Vanessa is a student in a combined third-and fourth-grade class ([19, p. 69]):

Problem 1: Jesse has 24 shirts. If he puts 8 of them in each drawer, how many drawers does he use?

Vanessa wrote, "$24 - 8 = 16$, $16 - 8 = 8$, $8 - 8 = 0$," and then circled "3" for the answer.

Problem 2: If Jeremy needs to buy 36 cans of seltzer water for his family and they come in packs of 6, how many packs should he buy?

This time Vanessa added, "$6 + 6 = 12$, $12 + 12 = 24$, $24 + 6 = 30$, $30 + 6 = 36$,…"

In the first problem, Vanessa starts with 24 and repeatedly *subtracts* 8s until she reaches 0. In the second problem, Vanessa repeatedly *adds* 6s until she reaches 36. (The 12s are 2 sixes, and the 24 is 4 sixes because it came from 2 sixes plus 2 sixes.) As we'll see later in this section, Vanessa's approach in the first problem is used in the standard division procedure. Vanessa's second approach is equally valid.

Consider another example, in which there are remainders. Suppose that we want to put 110 candies into packages of 8 candies each. How many packages can we make, and how many candies will be left over? To solve this problem, we must calculate $110 \div 8$, which we can do in the following way:

10 packages will use up $10 \times 8 = 80$ candies. Then there will be $110 - 80 = 30$ candies left. Three packages will use $3 \times 8 = 24$ candies. Then there will be only $30 - 24 = 6$ candies left. So we can make $10 + 3 = 13$ packages, and 6 candies will be left over.

The equations to express this reasoning are

$$110 - 10 \times 8 = 30$$
$$30 - 3 \times 8 = 6$$

These equations can be condensed to the single equation

$$110 - 10 \times 8 - 3 \times 8 = 6 \tag{6.1}$$

According to the distributive property, we can rewrite this equation as

$$110 - 13 \times 8 = 6 \tag{6.2}$$

Since 6 is less than 8,

$$110 \div 8 = 13, \text{ remainder } 6$$

To solve this problem we used the idea of repeatedly *subtracting* multiples of 8 from 110 in order to calculate 110 ÷ 8. Another approach is to repeatedly *add* multiples of 8 until we get as close to 110 as possible without going over. The "repeated subtraction" reasoning is only slightly different from the "repeated addition" reasoning given in the following:

> Ten packages use 10 × 8 = 80 candies. Three packages use 3 × 8 = 24 candies. So far, we have used 80 + 24 = 104 candies in 10 + 3 = 13 packages. Six more candies make 110, so we can make 13 packages of candies with 6 candies left over.

As before, we can write the following single equation to express this reasoning; it involves addition rather than subtraction:

$$10 \times 8 + 3 \times 8 + 6 = 110 \tag{6.3}$$

According to the distributive property, we can rewrite this equation as

$$13 \times 8 + 6 = 110 \tag{6.4}$$

This equation shows that 110 ÷ 8 = 13, remainder 6.

As these examples show, to divide, we can repeatedly subtract multiples of the divisor or we can repeatedly add multiples of the divisor.

The Scaffold Method of Long Division

As we just saw, we can calculate 110 ÷ 8 by repeatedly subtracting multiples of 8. Repeated subtraction is the basis of the common long division procedure. In order to understand the common long division procedure, we will work with a modification of it called the **scaffold method**. The scaffold method is less efficient than the common long division procedure, but it allows greater flexibility in calculating and it is a stepping-stone to the common algorithm.

Table 6.1 shows how to calculate 4581 ÷ 7 by using the common longhand procedure and by using the scaffold method. Both methods arrive at the conclusion that 4581 divided by 7 is 654 with remainder 3.

To use the scaffold method to calculate 4581 ÷ 7 and to understand why this method works, let's use the "how many groups" interpretation of division and think of 4581 ÷ 7 as the largest whole number of sevens in 4581, or in other words, the largest whole number that we can multiply 7 by without going

TABLE 6.1 Common long division versus the scaffold method of long division

Common Method	Scaffold Method
	4 ← How many 7s are in 31?
	50 ← How many tens of 7s are in 381?
654	600 ← How many hundreds of 7s are in 4581?
7)4581	7)4581
−42	−4200 ← 600 sevens
38	381 ← what is left over after subtracting 600 sevens
−35	−350 ← 50 sevens
31	31 ← what is left over after subtracting another 50 sevens
−28	−28 ← 4 sevens
3	3 ← what is left over after subtracting another 4 sevens

over 4581. With that in mind, examine the scaffold method shown in Table 6.1 (on the right). To carry out the scaffold method, we begin by asking how many hundreds of sevens are in 4581. We can start with hundreds because 1000 sevens is 7000, which is already greater than 4581. There are 600 sevens in 4581 because $600 \times 7 = 4200$, but 700 sevens would be 4900, which is greater than 4581. We subtract the 600 sevens, namely 4200, from 4581, leaving 381. Now we ask how many tens of sevens are in 381. There are 50 sevens in 381 because $50 \times 7 = 350$, but $60 \times 7 = 420$ is greater than 381. We subtract the 50 sevens—namely, 350—from 381, leaving 31. Finally, we ask how many sevens are in 31. There are 4, leaving 3 as a remainder. To determine the answer to $4581 \div 7$ we add the numbers at the top of the scaffold:

$$600 + 50 + 4 = 654$$

Therefore,

$$4581 \div 7 = 654, \text{ remainder } 3$$

Another way to write the scaffold method is to write the numbers 600, 50, 4 to the *side* of the scaffold instead of at the top, as shown in Table 6.2. However, some teachers have found this notation to be confusing for some immigrant students because it conflicts with similar looking notation used in some countries in which the *divisor* is written to the right.

TABLE 6.2
Another way of writing the scaffold method

$$
\begin{array}{r|l}
7\overline{)4581} & \\
-4200 & 600 \\
\hline
381 & \\
-350 & 50 \\
\hline
31 & \\
-28 & 4 \\
\hline
3 &
\end{array}
$$

In carrying out the scaffold method, we started with 4581, subtracted 600 sevens, subtracted another 50 sevens, subtracted another 4 sevens, and in the end, 3 were left over. In other words,

$$4581 - 600 \times 7 - 50 \times 7 - 4 \times 7 = 3 \tag{6.5}$$

Notice that all together, starting with 4581, we subtracted a total of 654 sevens and were left with 3. We come to the same conclusion when we first rewrite Equation (6.5) as

$$4581 - (600 \times 7 + 50 \times 7 + 4 \times 7) = 3$$

and then apply the distributive property to get

$$4581 - (600 + 50 + 4) \times 7 = 3$$

or

$$4581 - 654 \times 7 = 3 \tag{6.6}$$

Equation (6.6) tells us that when we take 654 sevens away from 4581, we are left with 3. Therefore, we can conclude that 654 is the largest whole number of 7s in 4581, and therefore $4581 \div 7 = 654$, remainder 3.

In general, why does the scaffold method of division give correct answers to division problems, based on the meaning of division? When you solve a division problem $A \div B$ by using the scaffold method, you start with the number A, and you repeatedly subtract multiples of B (numbers times B) until a number remains that is less than B. Since you subtracted as many Bs as possible, when you add the total number of Bs that were subtracted, that is the largest whole number of Bs that are in A. What's left over is the remainder. Therefore, the answer provided by the scaffold method to calculate $A \div B$ gives the answer that we expect for $A \div B$ based on the meaning of division.

Class Activity   *Now Turn to Class Activities Manual*

6I 🏺 Why the Scaffold Method of Long Division Works, p. 122

We used the "how many groups?" interpretation of division to make sense of the scaffold method, but can the "how many in each group" interpretation also be used? Table 6.3 shows that the answer is yes.

TABLE 6.3 Using the scaffold method of long division with the "how many in each group?" viewpoint to divide 4581 objects equally among 7 groups

```
        4    ←    From 31, how many individuals can we put in each group?
       50    ←    From 381, how many tens can we put in each group?
      600    ←    From 4581, how many hundreds can we put in each group?
  7)4581
  − 4200    ←    Put 600 in each group.
      381    ←    left over after putting 600 in each group
  − 350     ←    Put another 50 in each group.
       31    ←    left over after putting 50 in each group
  − 28      ←    Put another 4 in each group.
        3    ←    left over after putting 4 in each group
```

Class Activity   *Now Turn to Class Activities Manual*

6J Using the Scaffold Method Flexibly, p. 123

Using the Scaffold Method Flexibly Some teachers like to introduce long division with the scaffold method, using it as a stepping-stone to the standard long division algorithm. Why? The scaffold method is flexible. It allows students at different levels of computational fluency to carry out long division successfully and in a way that makes sense to them. As Table 6.4 indicates, students can learn to become more efficient until they are doing the standard version of the scaffold method (the version shown in Table 6.1), which parallels the standard division algorithm.

Comparing the Scaffold Method with Common Long Division Looking back at Table 6.1, we see that the scaffold and common long division methods are very similar and accomplish the same thing. The scaffold method works with entire numbers (the entire dividend), rather than just portions of numbers (e.g., 381 rather than 38). When you use the scaffold method, you have to keep track of place value. For example, in the first step, you ask, "How many *hundreds* of sevens are in 4581?" whereas, for the common method, you ask only, "How many sevens are in 45?" The common method is really just an abbreviated version of the scaffold method. Therefore, if we can use the meaning of division to explain why the scaffold method gives correct answers to division problems, then we will also know why the common longhand method gives correct answers to division problems.

TABLE 6.4 Using the scaffold method of long division flexibly

Hundreds and tens are easy to work with, but this is inefficient:	Fewer steps make this more efficient; 5s and 2s are easy to work with:	Efficient use of the scaffold method parallels the standard algorithm:
$8\overline{)5072}$	4	4
-800 100	10	30
4272	20	600
-800 100	100	$8\overline{)5072}$
3472	500	-4800
-800 100	$8\overline{)5072}$	272
2672	-4000	-240
-800 100	1072	32
1872	-800	-32
-800 100	272	0
1072	-160	
-800 100	112	
272	-80	
-80 10	32	
192	-32	
-80 10	0	
112		
-80 10		
32		
-32 4		
0		

Why the Standard Long Division Algorithm Works

What is the logic behind the standard longhand division algorithm? To understand the logic of this algorithm we'll need to focus on several elements: what division means; how to break numbers apart by place value; how to work with a number part by part, thereby applying properties of arithmetic (even if only implicitly); and how the values of adjacent places are related.

The standard long division algorithm is especially nicely interpreted from the "how many in each group?" viewpoint. To do so, consider the number we want to divide (the dividend) as a number of objects bundled into ones, tens, hundreds, thousands, and so on. Consider again the case of dividing 4581 objects equally among 7 groups, this time thinking of the objects as bundled into 4 thousands, 5 hundreds, 8 tens, and 1 individual object. At the first step of long division we ask how many hundreds of objects we can put in each group. We can put 6 hundreds in each of the 7 groups, using 42 hundreds, as we see in the first step of the standard long division in Table 6.5. Then $45 - 42 = 3$ hundreds remain. We must unbundle these 3 hundreds in order to subdivide them. Unbundled, the 3 hundreds become 30 tens. We can combine these 30 tens with the 8 tens we have in the 4581 objects. So, we now have

TABLE 6.5 Using the standard method of long division with the "how many in each group?" viewpoint to divide 4581 objects equally among 7 groups

```
      654
  7)4581
   -42      ← Each group gets 6 hundreds.
    38      ← 3 hundreds and 8 tens = 38 tens remain.
   -35      ← Each group gets another 5 tens.
    31      ← 3 tens and 1 one = 31 ones remain.
   -28      ← Each group gets another 4 ones.
     3      ← 3 ones remain.
```

38 tens to distribute among the 7 groups. This process of unbundling and combining corresponds to "bringing down" the 8 next to the 3 from 45 − 42 in the long division process. From the 38 tens we now have, we can give each of the 7 groups 5 tens, leaving 38 − 35 = 3 tens. We must unbundle these 3 tens in order to subdivide them. Unbundled, the 3 tens become 30 ones. We can combine these 30 ones with the 1 one we have in the 4581 objects. So we now have 31 ones to distribute among the 7 groups. Again, the process of unbundling and combining corresponds to "bringing down" the 1 next to the 3 from 38 − 35 in the long division process. From the 31 ones, each group gets another 4 individual objects, leaving 31 − 28 = 3 objects remaining. All together, each of the 7 groups got 654 objects, and 3 objects were left over, as before.

In the next Class Activities, you will see some pictorial methods that students can use to reason about and make sense of the standard long division algorithm. These methods can be placed side by side with the usual way we write the standard division algorithm to help students understand the standard notation. From a mathematical perspective, the steps in an algorithm are what constitute the algorithm. So the pictorial methods in the Class Activities actually *are* the standard algorithm—they are just another way of recording the algorithm on paper.

Class Activity   *Now Turn to Class Activities Manual*

6K Interpreting the Standard Division Algorithm as Dividing Bundled Toothpicks, p. 124

6L Interpreting the Standard Division Algorithm in Terms of Area, p. 126

6M Student Errors in Using the Division Algorithm, p. 127

Calculating Decimal Answers to Whole Number Division Problems

If a whole number division problem doesn't have a whole number quotient, then we have several options for describing the answer to the division problem: We can describe it as a whole number quotient with a remainder; we can describe it as a mixed number (or fraction); or we can describe it as a decimal. We discussed whole-number-with-remainder answers and mixed number (and fraction) answers to division problems in Section 6.2. To get the decimal answer to a whole number division problem, all we need to do is extend the standard division algorithm into smaller decimal places.

For example, suppose there is $243 that is to be divided equally among 7 people. In this case, neither a whole-number-with-remainder answer nor a mixed number answer to 243 ÷ 7 is practical. Instead,

TABLE 6.6 Calculating the decimal number answer to 243 ÷ 7 by viewing the division algorithm as dividing $243 equally among 7 people in stages that fit with the decimal system: doling out tens, then ones, then dimes, then pennies

```
      34.71
 7)243.000
  −21          ← Each person gets 3 tens.
   33          ← 3 tens and 3 ones = 33 ones remain.
  −28          ← Each person gets 4 ones.
    50         ← 5 ones = 50 dimes (tenths) remain.
   −49         ← Each person gets 7 dimes (tenths).
    10         ← 1 dime (tenth) = 10 pennies (hundredths) remain.
    −7         ← Each person gets 1 penny (hundredth).
     3         ← 3 pennies (hundredths) remain.
```

a decimal answer is most appropriate. Table 6.6 shows how to use the standard division algorithm to calculate this decimal answer to the hundredths place and how to interpret each step in the process in terms of doling out money in stages. The stages fit with the decimal system: First dole out tens, then ones, then dimes (tenths of a dollar), then pennies (hundredths of a dollar).

From the calculations of Table 6.6, we conclude that when we divide $243 equally among 7 people, each person gets $34.71. But there will still be 3 cents left over, which cannot be divided further because we do not have denominations smaller than one-hundredth of a dollar. In an abstract setting, or in a context where thousandths, ten-thousandths, and so on make sense, we could continue the long division process, dividing the remaining 3 hundredths to as many decimal places as we needed. We would find that

$$243 \div 7 = 34.7142\ldots$$

Class Activity   *Now Turn to Class Activities Manual*

6N   Interpreting the Calculation of Decimal Answers to Whole Number Division Problems in Terms of Money, p. 127

6O Errors in Decimal Answers to Division Problems, p. 127

Using Division to Calculate Decimal Representations of Fractions

We've seen how to use the long division algorithm to calculate the decimal number answer to a whole number division problem. We can use this procedure to express a fraction whose numerator and denominator are whole numbers as a decimal by equating the fraction with division:

$$\frac{A}{B} = A \div B$$

For example, how can we express the fraction $\frac{5}{16}$ as a decimal? First note that

$$\frac{5}{16} = 5 \div 16$$

Then use long division to calculate $5 \div 16$ as a decimal:

$$
\begin{array}{r}
0.3125 \\
16\overline{)5.0000} \\
-48 \\
\hline
20 \\
-16 \\
\hline
40 \\
-32 \\
\hline
80 \\
-80 \\
\hline
0
\end{array}
$$

Therefore,

$$\frac{5}{16} = 0.3125$$

Similarly, we can calculate

$$\frac{1}{12} = 1 \div 12 = 0.083333333\ldots,$$

$$\frac{2}{7} = 2 \div 7 = 0.285714285714\ldots$$

The decimal representations of these last two fractions have infinitely many digits to the right of the decimal point, unlike the decimal representation of $\frac{5}{16} = 0.3125$, which has only 4 digits to the right of its decimal point.

By equating fractions with division, and by using long division, we can express any fraction whose numerator and denominator are whole numbers as a decimal.

Class Activity   *Now Turn to Class Activities Manual*

6P Using Division to Calculate Decimal Representations of Fractions, p. 128

Using Subdivided Squares to Determine Decimal Representations of Fractions

In simple cases we can use diagrams to determine the decimal representations of fractions. These diagrams can help us better understand the relationship between decimals and fractions.

Figure 6.10 shows that

$$\frac{1}{4} = 0.25$$

in the following way: Consider a large square (made up of 100 small squares) as representing 1. To show $\frac{1}{4}$ of the large square, we must divide the large square into 4 equal parts. One of those 4 equal parts represents $\frac{1}{4}$ of the large square. Therefore, we can represent $\frac{1}{4}$ of the large square by the shaded portion of Figure 6.10, which consists of 2 vertical strips of 10 small squares, and 5 more small squares. Because each vertical strip of 10 small squares represents $\frac{1}{10}$ of the large square and each small square represents $\frac{1}{100}$ of the large square, Figure 6.10 shows that

$$\frac{1}{4} = 2 \cdot \frac{1}{10} + 5 \cdot \frac{1}{100} = 0.25$$

FIGURE 6.10

The shaded area is $\frac{1}{4}$ of the large square. Therefore, $\frac{1}{4} = 2 \cdot \frac{1}{10} + 5 \cdot \frac{1}{100} = 0.25$

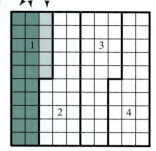

Division with Multi-Digit Divisors

When we use a division algorithm with a divisor that has two or more digits, we must estimate quotients during the process. Students should be aware that even if they round numbers correctly and make sensible estimates, they may still need to revise their estimates during the division process. The next class activity examines rounding and division.

Class Activity *Now Turn to Class Activities Manual*

6Q Rounding to Estimate Solutions to Division Problems, p. 128

Practice Exercises for Section 6.3

1. Use the scaffold method to calculate $31\overline{)73125}$. Interpret the steps in the scaffold in terms of the following story problem:

   There are 73,125 beads that will be put into bags with 31 beads in each bag. How many bags of beads can be made, and how many beads will be left over?

2. Here is one way to calculate $239 \div 9$:

   Ten nines is 90. Another 10 nines makes 180. Five more nines makes 225. One more nine makes 234. Five ones more makes 239. All together that's 26 nines, with 5 left over. So the answer is 26, remainder 5.

   Write a single equation, in the form of Equation (6.1) or Equation (6.3), that incorporates this reasoning. Use your equation and the distributive property to write another equation, like Equation (6.2) or Equation (6.4), that shows the answer to $239 \div 9$.

3. Here is one way to calculate $2687 \div 4$:

   Five hundred fours is 2000. That leaves 687. One hundred fours is 400. That leaves 287. Another 50 fours is 200. Now 87 are left. Another 20 fours is 80. Now there are 7 left, and we can get one more 4 out of that with 3 left. All together there were $500 + 100 + 50 + 20 + 1 = 671$ fours in 2687 with 3 left over, so $2687 \div 4 = 671$, remainder 3.

   a. Write a scaffold that corresponds to the reasoning in the problem.

   b. Calculate $2687 \div 4$, using a scaffold with fewer steps than your scaffold for part (a).

   c. Even though your scaffold in part (a) uses more steps than necessary, is it still based on sound reasoning? Explain.

4. Calculate $2950 \div 13$ without using a calculator or a long division procedure.

5. Calculate $1000 \div 27$ without using a calculator or a long division procedure.

6. Here is one way to calculate $320 \div 17$:

> 20 seventeens is 340. So 19 seventeens is 17 less, which is 323. Therefore, 18 seventeens must be 306. So the answer is 18, remainder 14.

Write a scaffold that corresponds to this reasoning.

7. Show how to use the standard long division algorithm to calculate $4581 \div 7$. Interpret each step in the algorithm in terms of dividing 4581 toothpicks equally among 7 groups, where the toothpicks are bundled according to the decimal system. In particular, be sure to explain how to interpret the "bringing down" steps in terms of the bundles.

8. Show how to use the standard long division algorithm to calculate the decimal answer to $20 \div 11$ to the hundredths place. Interpret each step in terms of dividing \$20 equally among 11 people.

9. Show how to use the standard long division algorithm to write the fraction $\frac{5}{8}$ as a decimal.

10. Use the large square in Figure 6.11 (which is subdivided into 100 small squares) to help you explain why the decimal representation of $\frac{1}{8}$ is 0.125.

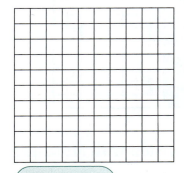

FIGURE 6.11   A square subdivided into 100 smaller squares

11. Marina says that since $300 \div 50 = 6$, that means you can put 300 marbles into 6 groups with 50 marbles in each group. Marina thinks that knowing $300 \div 50 = 6$ should help her calculate $300 \div 55$. At first she tries $300 \div 50 + 300 \div 5$, but that answer seems too big. Is there some other way to use the fact that $300 \div 50 = 6$ in order to calculate $300 \div 55$?

Answers to Practice Exercises for Section 6.3

1.
$$
\begin{array}{r}
8 \\
50 \\
300 \\
2000 \\
\hline
31\overline{)73125} \\
-62000 \\
\hline
11125 \\
-9300 \\
\hline
1825 \\
-1550 \\
\hline
275 \\
-248 \\
\hline
27
\end{array}
$$
so $73125 \div 31 = 2358$, remainder 27.

If we make 2000 bags of beads, we will use $2000 \times 31 = 62{,}000$ beads, which will leave 11,125 beads remaining from the original 73,125. If we make another 300 bags of beads, we will use $300 \times 31 = 9300$ beads, which leaves 1825 beads remaining. If we make another 50 bags of beads, we will use 1550 beads, leaving 275 beads. Finally, if we make another 8 bags of beads, we will use 248 beads, leaving only 27 beads, which is not enough for another bag. All together, we made $2000 + 300 + 50 + 8 = 2358$ bags of beads, and 27 beads are left over.

2. $10 \times 9 + 10 \times 9 + 5 \times 9 + 1 \times 9 + 5 = 239$

So, by the distributive property,

$$(10 + 10 + 5 + 1) \times 9 + 5 = 239,$$

Therefore,

$$26 \times 9 + 5 = 239,$$

which means that

$$239 \div 9 = 26,$$

remainder 5.

3. a.

```
        1
       20
       50
      100
      500
   4)2687    So 2687 ÷ 4 = 671, remainder 3.
    −2000
      687
     −400
      287
     −200
       87
      −80
        7
       −4
        3
```

b.

```
        1
       70
      600
   4)2687    So 2687 ÷ 4 = 671, remainder 3.
    −2400
      287
     −280
        7
       −4
        3
```

c. Even though the scaffold in part (a) uses more steps than necessary, it is still based on sound reasoning. Instead of subtracting the full 600 fours from 2687, the scaffold subtracted the same number of fours in two steps instead of one: first subtracting 500 fours and then subtracting another 100 fours. If we think in terms of putting 2687 cookies into packages of 4, we are first making 500 packages and then making another 100 packages instead of making 600 packages straight away. Similarly, instead of subtracting the full 70 fours from 287, the scaffold subtracted 50 fours and then another 20 fours. All together, we are still finding the same total number of fours in 2687, just in a slightly less efficient way than in the scaffold in part (b).

4. There are many ways you could do this. Here is one way: 100 thirteens is 1300, so 200 thirteens is 2600. Taking 2600 away from 2950 leaves 350. Another 20 thirteens is 260, leaving 90. Five thirteens make 65. Now we have 25 left. We can only get one more

13, and 12 will be left. All together we have $200 + 20 + 5 + 1 = 226$ thirteens and 12 are left, so $2950 ÷ 13 = 226$, remainder 12.

5. There are many ways you could do this. Here is one way: The product $10 × 27 = 270$ and $20 × 27 = 540$, so $30 × 27 = 540 + 270 = 810$. Five 27s must be half of 270, which is 135. Therefore,

$$35 × 27 = (30 + 5) × 27$$
$$= 30 × 27 + 5 × 27$$
$$= 810 + 135$$
$$= 945.$$

Two 27s are 54, and adding that to 945 makes 999. Therefore, $37 × 27 = 999$, so $37 × 27 + 1 = 1000$, and this means that $1000 ÷ 27 = 37$, remainder 1.

6.

```
       −1
       −1
       20
   17)320
    −340
     −20
   −(−17)
       −3
   −(−17)
       14
```

7. See text.

8. Using the standard algorithm, we find that $20 ÷ 11 = 1.81 … :$

```
       1.81
   11)20.00
     −11      ← Each person gets $1.
       90     ← $9 = 90 dimes remain.
      −88     ← Each person gets 8 dimes.
       20     ← 2 dimes = 20 pennies remain.
      −11     ← Each person gets 1 penny.
        9     ← 9 pennies remain.
```

If we think in terms of dividing $20 equally among 11 people, then at the first step we ask how many ones we should give each person. Each of the 11 people will get $1, leaving $20 − $11 = $9 remaining. If we trade each dollar for 10 dimes, we will have 90 dimes. We can now give each of the 11 people 8 dimes, using a total of 88 dimes, leaving 2 dimes left. If we trade each dime for 10 pennies, we will have 20 pennies. Each person gets 1 penny, leaving $20 − 11 = 9$ pennies. Therefore, each person gets $1.81, and $20 \div 11 = 1.81\ldots$. The decimal representation of $20 \div 11$ continues to have digits in the thousandths place, the ten-thousandths place, and so on, but we can't use our money interpretation for these places because we don't have denominations smaller than 1 penny.

9. The following long division calculation shows that $\frac{5}{8} = 0.625$:

$$
\begin{array}{r}
0.625 \\
8\overline{)5.000} \\
-48 \\
\hline
20 \\
-16 \\
\hline
40 \\
-40 \\
\hline
0
\end{array}
$$

10. By thinking of the large square as representing 1, each small square then represents $\frac{1}{100}$. To divide the large square into 8 equal pieces, first make 8 strips of 10 small squares, as indicated in Figure 6.12. Each of the 8 equal pieces into which we want to divide the large square gets one of these strips. Then there are still 20 small squares left to be divided into 8 equal pieces. Make 8 groups of 2 small squares from these 20 small squares. Each of the 8 equal pieces gets 2 of these small squares. That leaves 4 small squares left to be divided into 8 equal pieces. Dividing each small square in half makes 8 half-squares. All together, we obtain $\frac{1}{8}$ of the big square by collecting these various parts: one strip of 10 small squares, 2 small squares, and one half of a

small square, as shown in Figure 6.12. Notice that since 1 small square is $\frac{1}{100}$ of the large square, $\frac{1}{2}$ of a small square is $\frac{1}{200} = \frac{5}{1000}$ of the large square. Therefore,

$$
\frac{1}{8} = 1 \cdot \frac{1}{10} + 2 \cdot \frac{1}{100} + 5 \cdot \frac{1}{1000}
$$
$$
= 0.125
$$

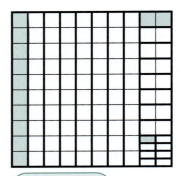

FIGURE 6.12   Showing $\frac{1}{8} = 0.125$

11. Marina is right that $300 \div 50 + 300 \div 5$ won't work—it divides the 300 marbles into groups *twice*, so it gives too many groups. Also, if you get 6 groups when you put 50 marbles in each group, then you must get fewer groups when you put more marbles in each group. If you think about removing one of those 6 groups of marbles, then you could take the 50 marbles in that group and from them, put 5 marbles in each of the remaining 5 groups. Then you'd have 5 groups of 55 marbles, and you'd have 25 marbles left over. Therefore, $300 \div 55 = 5$, remainder 25.

Problems for Section 6.3

1. a. Calculate $4215 \div 6$ and $62{,}635 \div 32$ in two ways: with the common division method and with the scaffold method.

   b. Compare the common division method and the scaffold method. How are the methods alike? How are the methods different? What are advantages and disadvantages of each method?

2. a. Use the scaffold method to calculate $793 \div 4$.

   b. Interpret the steps in your scaffold in terms of the following story problem: If you have 793 cookies and you want to put them in packages of 4, how many packages will there be, and how many cookies will be left over?

c. Write a single equation, like Equation 6.5, that incorporates the steps of your scaffold. Use your equation and the distributive property to write another equation, like Equation (6.6), that shows the answer to 793 ÷ 4.

Relate this last equation to portions in the scaffold method.

3. 🏺 **a.** Use the standard long division algorithm to calculate 1875 ÷ 8.

b. Interpret each step in your calculation in part (a) in terms of the following problem: You have 1875 toothpicks bundled into 1 thousand, 8 hundreds, 7 tens, and 5 individual toothpicks. If you divide these toothpicks equally among 8 groups, how many toothpicks will each group get?

4. 🏺 **a.** Use the standard long division algorithm to determine the decimal number answer to 2893 ÷ 6 to the hundredths place.

b. Interpret each step in your long division calculation in part (a) in terms of dividing $2893 equally among 6 people.

5. 🏺 Tamarin calculates 834 ÷ 25 in the following way:

I know that four 25s make 100, so I counted 4 for each of the 8 hundreds. This gives me 32. Then there is one more 25 in 34, but there will be 9 left. So the answer is 33, remainder 9.

a. Explain Tamarin's method in detail, and explain why her method is legitimate. (Do not just state that she gets the correct answer; explain why her method gives the correct answer.) Include equations as part of your explanation.

b. Use Tamarin's method to calculate 781 ÷ 25.

6. Felicia is working on the following problem: There are 730 balls to be put into packages of 3. How many packages can be made, and how many balls will be left over? Here are Felicia's ideas:

One hundred packages of balls will use 300 balls. After another 100 packages, we will have used up 600 balls. Another 30 packages will use another 90 balls, for a total of 690 balls used. Ten more packages brings us to 720 balls used. Three more packages will bring us to 729 balls used. Then there is 1 ball left over. All together we could make 100 + 100 + 30 + 10 + 3 = 243 packages of balls with 1 ball left over.

Write equations like Equations (6.3) and (6.4) that correspond to Felicia's work. Explain how your equations show that 730 ÷ 3 = 243, remainder 1.

7. Rodrigo calculates 650 ÷ 15 in the following way:

$$
\begin{array}{ll}
150 & \leftarrow 10 \\
+\,150 & \leftarrow 10 \\
\hline
300 & \leftarrow 20
\end{array}
\qquad
\begin{array}{ll}
300 & \leftarrow 20 \\
+\,300 & \leftarrow 20 \\
\hline
600 & \leftarrow 40
\end{array}
$$

$$
\begin{array}{ll}
600 & \leftarrow \ 40 \\
+\,30 & \leftarrow \ \ 2 \\
\hline
630 & \leftarrow 42 \\
+\,15 & \\
\hline
645 & \leftarrow 43 \\
+\,5 & \leftarrow \text{left} \\
\hline
650 &
\end{array}
\qquad 43\ R\ 5
$$

a. Explain why Rodrigo's method makes sense. It may help you to work with a story problem for 650 ÷ 15.

b. Write equations that correspond to Rodrigo's work and that demonstrate that 650 ÷ 15 = 43, remainder 5.

8. 🏺 Meili calculates 1200 ÷ 45 in the following way:

$$
\begin{array}{l}
\begin{array}{r}
45 \\
\times\,10 \\
\hline
450
\end{array}
\quad
\begin{array}{ll}
450 & \leftarrow 10 \\
+\,450 & \leftarrow 10 \\
\hline
900 & \\
+\ \ 90 & \leftarrow 2 \\
\hline
990 & \\
+\ \ 90 & \leftarrow 2 \\
\hline
1080 & \\
+\ \ 90 & \leftarrow 2 \\
\hline
1170 & \\
+\ \ 30 & \leftarrow \text{left} \\
\hline
1200 &
\end{array}
\end{array}
$$

10 + 10 + 2 + 2 + 2 = 26

26 R 30

a. Explain why Meili's strategy makes sense. It may help you to work with a story problem for 1200 ÷ 45.

b. Write equations that correspond to Meili's work and that demonstrate that 1200 ÷ 45 = 26, remainder 30.

9. Describe how the whole-number-with-remainder and the decimal answer to 23 ÷ 6 are related and explain why this relationship holds.

10. Show how to use long division to determine the decimal representation of $\frac{1}{37}$.

11. Describe how to use either dimes and pennies or a subdivided square to determine the tenths and hundredths places in the decimal representation of $\frac{1}{11}$. Explain your reasoning.

12. Describe how to use either dimes and pennies or a subdivided square to determine the tenths and hundredths places in the decimal representation of $\frac{1}{9}$. Explain your reasoning.

13. **a.** Use the standard long division algorithm to determine the decimal representation of $\frac{1}{9}$ to the ten-thousandths place.

    b. Interpret the steps to the hundredths place in part (a) in terms of dividing $1 equally among 9 people.

14. Describe how to use either dimes and pennies or a subdivided square to determine the tenths and hundredths places in the decimal representation of $\frac{1}{7}$. Explain your reasoning.

15. **a.** Use long division to determine the decimal representation of $\frac{1}{7}$ to 7 decimal places.

    b. Interpret the steps to the hundredths place in part (a) in terms of dividing $1 equally among 7 people.

16. Jessica calculates that $7 \div 3 = 2$, remainder 1. When Jessica is asked to write her answer as a decimal, she simply puts the remainder 1 behind the decimal point:

$$7 \div 3 = 2.1$$

Is Jessica correct or not? If not, explain why not, and explain to Jessica in a concrete way why the correct answer makes sense.

17. Use some or all of the multiplication facts

$$2 \times 35 = 70$$
$$10 \times 35 = 350$$
$$20 \times 35 = 700$$

repeatedly to calculate $2368 \div 35$ without using a calculator. Explain your method.

18. Use the two multiplication facts, $30 \times 12 = 360$ and $12 \times 12 = 144$, to calculate $500 \div 12$ without the use of a calculator, common long division, or the scaffold method of division. Use both multiplication facts, and explain your method. Give your answer as a whole number with a remainder. Explain your method.

19. Calculate $623 \div 8$ without using a calculator or any longhand method of division. Show your work. Then briefly describe your reasoning.

20. Calculate $2000 \div 75$ without using a calculator or any longhand method of division. Show your work. Then briefly describe your reasoning.

21. **a.** Write and solve a simple story problem for $1200 \div 30$.

    b. Use the situation of your story problem in part (a) to help you solve $1200 \div 31$ without a calculator or long division by modifying your solution to $1200 \div 30$.

    c. Use the situation of your story problem in part (a) to help you solve $1200 \div 29$ without a calculator or long division by modifying your solution to $1200 \div 30$.

22. **a.** Write and solve a simple story problem for $630 \div 30$.

    b. Use the situation of your story problem in part (a) to help you solve $630 \div 31$ without a calculator or long division by modifying your solution to $630 \div 30$.

    c. Use the situation of your story problem in part (a) to help you solve $630 \div 29$ without a calculator or long division by modifying your solution to $630 \div 30$.

23. **a.** Suppose you want to estimate

$$459 \div 38$$

by rounding 38 up to 40. Both

$$440 \div 40$$

and

$$480 \div 40$$

are easy to calculate mentally. Use reasoning about division to determine which division problem, $440 \div 40$ or $480 \div 40$, should give you a better estimate to $459 \div 38$. Then check your answer by solving the division problems.

    b. Suppose you want to estimate

$$459 \div 42$$

by rounding 42 down to 40. As before, both $440 \div 40$ and $480 \div 40$ are easy to calculate mentally. Use reasoning about division to determine which division problem, 440 ÷ 40 or 480 ÷ 40, should give you a better estimate to 459 ÷ 42. Then check your answer by solving the division problems.

c. Suppose you want to estimate 632 ÷ 58. What is a good way to round the numbers 632 and 58 so that you get an easy division problem which will give a good estimate to 632 ÷ 58? Explain, drawing on what you learned from parts (a) and (b).

d. Suppose you want to estimate 632 ÷ 62. What is a good way to round the numbers 632 and 62 so that you get an easy division problem which will give a good estimate to 632 ÷ 62? Explain, drawing on what you learned from parts (a) and (b).

24. Bob wants to estimate 1893 ÷ 275. He decides to round 1893 up to 2000. Bob says that since he rounded 1893 up, he'll get a better estimate if he rounds 275 in the opposite direction, namely, down to 250 rather than up to 300. Therefore, Bob says that 1893 ÷ 275 is closer to 8 (which is 2000 ÷ 250) than to 7, because 2000 ÷ 300 is a little less than 7. This problem will help you investigate whether or not Bob's reasoning is correct.

a. Which of 2000 ÷ 250 and 2000 ÷ 300 gives a better estimate to 1893 ÷ 275? Can Bob's reasoning (just described) be correct?

b. Use the meaning of division (either of the two main interpretations) to explain the following: When estimating the answer to a division problem $A \div B$, if you round A up, you will generally get a better estimate if you also round B up rather than down. Draw diagrams to aid your explanation.

c. Now consider the division problem 1978 ÷ 205. Suppose you round 1978 up to 2000. In this case, will you get a better estimate to 1978 ÷ 205 if you round 205 up to 250 (and compute 2000 ÷ 250) or if you round 205 down to 200 (and compute 2000 ÷ 200)? Reconcile this with your findings in part (b).

25. Suppose a student uses the scaffold method as shown in Table 6.7.

a. Write another scaffold for 743,425 ÷ 365 that uses fewer steps.

b. Even though the student's scaffold for 743,425 ÷ 365 uses more steps than your scaffold in part (a), is it still based on sound reasoning? Explain.

TABLE 6.7 A student's scaffold for 743,425 ÷ 365

```
              1
              5
             10
             20
           1000
           1000
      _____
  365)743425        so, 743425 ÷ 365 = 1000 + 1000 + 20 + 10 + 5 + 1
    −365000                          = 2036 remainder 285
      _____
     378425
    −365000
      _____
      13425
     −7300
      _____
       6125
      −3650
      _____
       2475
      −1825
      _____
        650
       −365
      _____
        285
```

26. A student calculates $6998 \div 7$ as follows, and concludes that $6998 \div 7 = 1000 - 1$, remainder 5, which is 999, remainder 5:

$$
\begin{array}{r}
-1 \\
1000 \\
7\overline{)6998} \\
-7000 \\
\hline
-2 \\
-(-7) \\
\hline
5
\end{array}
$$

Even though it is not conventional to use negative numbers in a scaffold, explain why the student's method corresponds to legitimate reasoning. Write a simple story problem for $6998 \div 7$, and use your story problem to discuss the reasoning that corresponds to the scaffold.

27. When you divide whole numbers using an ordinary calculator, the answer is displayed as a decimal. But what if you want the answer as a whole number with a remainder? Here's a method to determine the remainder with an ordinary calculator, illustrated with the example of $236 \div 7$.

Use the calculator to divide the whole numbers:

$$236 \div 7 = 33.71428\ldots$$

Subtract the whole number part of the answer:

$$33.71428\ldots - 33 = 0.71428\ldots$$

Multiply the resulting decimal by the divisor (which was 7):

$$0.71428\ldots \times 7 = 4.99999$$

Round the resulting number to the nearest whole number. This is your remainder. In this example the remainder is 5. So $236 \div 7 = 33$, remainder 5.

a. Using a calculator, solve at least two more division problems with whole numbers, and use the preceding method to find the answer as a whole number with a remainder. Check that your answers are correct.

b. Now solve the same division problems you did in part (a), except this time give your answers as mixed numbers instead of as decimals.

c. Use the mixed number version to help you explain why the calculator method for determining the remainder works.

6.4 Fraction Division from the "How Many Groups?" Perspective

Focal Points
Grade 6

Have you ever wondered why we "invert and multiply" to divide fractions? In this section, we will see why this standard procedure for dividing fractions gives answers to fraction division problems that agree with what we expect from the meaning of division. We will also study story problems that can be solved with fraction division.

We'll start by recalling the "invert and multiply" or "multiply by the reciprocal" procedure for fraction division. Then we'll examine what fraction division means from the "how many groups?" perspective. Finally, we'll consider several related ways why the fraction division procedure is valid.

You might wonder why we need several different explanations for why a mathematical fact is true. One reason is that teachers should know multiple explanations in order to reach more students effectively. But explanations are also important in mathematics not only because they prove that mathematical facts really are true, but because they show how ideas are connected and related to each other. In mathematics, the paths through which ideas are linked and interconnected are interesting in their own right.

The "Invert and Multiply" or "Multiply by the Reciprocal" Procedure

The procedure for dividing fractions is straightforward. To divide fractions, such as

$$\frac{3}{4} \div \frac{2}{3} \quad \text{and} \quad 6 \div \frac{2}{5}$$

we can use the familiar "invert and multiply" method in which we invert the divisor and multiply by it, as in

$$\frac{3}{4} \div \frac{2}{3} = \frac{3}{4} \cdot \frac{3}{2} = \frac{3 \cdot 3}{4 \cdot 2} = \frac{9}{8} = 1\frac{1}{8}$$

and

$$6 \div \frac{2}{5} = \frac{6}{1} \div \frac{2}{5} = \frac{6}{1} \cdot \frac{5}{2} = \frac{6 \cdot 5}{1 \cdot 2} = \frac{30}{2} = 15$$

reciprocal   Another way to describe this "invert and multiply" method for dividing fractions is in terms of the reciprocal of the divisor. The **reciprocal** of a fraction $\frac{C}{D}$ is the fraction $\frac{D}{C}$. In order to divide fractions, we multiply by the reciprocal of the divisor. So, in general,

$$\frac{A}{B} \div \frac{C}{D} = \frac{A}{B} \cdot \frac{D}{C} = \frac{A \cdot D}{B \cdot C}$$

As an interesting special case, notice that we can apply the fraction division procedure to whole number division because every whole number is equal to a fraction (e.g., $2 = \frac{2}{1}$ and $3 = \frac{3}{1}$). Therefore,

$$2 \div 3 = \frac{2}{1} \div \frac{3}{1} = \frac{2}{1} \cdot \frac{1}{3} = \frac{2 \cdot 1}{1 \cdot 3} = \frac{2}{3}$$

Notice that this result, $2 \div 3 = \frac{2}{3}$, agrees with our finding earlier in this chapter on the link between division and fractions, namely, that $A \div B = \frac{A}{B}$.

The "How Many Groups?" Interpretation

Class Activity   *Now Turn to Class Activities Manual*

6R   "How Many Groups?" Fraction Division Problems, p. 129

With the "how many groups?" interpretation of division, $8 \div 2$ means the number of groups we can make when we divide 8 objects into groups with 2 objects in each group. In other words, $8 \div 2$ tells us how many groups of 2 we can make from 8.

Similarly, with the "how many groups?" interpretation of division,

$$\frac{8}{3} \div \frac{2}{3}$$

tells us how many groups of $\frac{2}{3}$ we can make from $\frac{8}{3}$. For example, suppose you are making popcorn balls and each popcorn ball requires $\frac{2}{3}$ cup of popcorn. If you have $2\frac{1}{3} = \frac{8}{3}$ cup of popcorn, then how many popcorn balls can you make? In this case you want to divide $\frac{8}{3}$ cup of popcorn into groups (balls) so that there is $\frac{2}{3}$ cup of popcorn in each group, as indicated in Figure 6.13. According to the "how many groups?" interpretation of division, you can make

$$\frac{8}{3} \div \frac{2}{3}$$

popcorn balls.

FIGURE 6.13

A "how many groups?" story problem for $\frac{8}{3} \div \frac{2}{3}$

1 cup popcorn

$\frac{2}{3}$ cup popcorn makes one popcorn ball

How many popcorn balls can we make from $\frac{8}{3}$ cups popcorn?

$\frac{8}{3} \div \frac{2}{3}$ popcorn balls

Another way to analyze "how many groups?" fraction division is to reformulate fraction division problems as equivalent multiplication problems. From the "how many groups?" viewpoint,

$$8 \div 2 = ? \text{ corresponds to } ? \times 2 = 8$$

since we are asking how many groups of 2 are in 8. Therefore, similarly,

$$\frac{8}{3} \div \frac{2}{3} = ? \text{ corresponds to } ? \times \frac{2}{3} = \frac{8}{3}$$

and we are asking how many $\frac{2}{3}$s are in $\frac{8}{3}$.

Class Activity   *Now Turn to Class Activities Manual*

6S Dividing Fractions by Dividing the Numerators and Dividing the Denominators, p. 131

Using the "How Many Groups?" Interpretation to Explain Why "Invert and Multiply" Is Valid

There are many ways to explain why the "invert and multiply" procedure for dividing fractions is valid. One way, which we study next, involves reasoning about "how many groups?" fraction division problems. A side benefit of this way of reasoning is that it actually develops *another* valid procedure for dividing fractions. As we'll see, this other method consists of first giving the two fractions a common denominator and then dividing the numerators.

Consider the division problem

$$\frac{2}{3} \div \frac{1}{2}$$

The following is a story problem for this division problem:

How many $\frac{1}{2}$ cups of water are in $\frac{2}{3}$ cup of water?

Or, said another way,

How many times will we need to fill a $\frac{1}{2}$ cup measuring cup with water and pour it into a container that holds $\frac{2}{3}$ cup of water in order to fill the container?

From the diagram in Figure 6.14 we can say right away that the answer to this problem is "one and a little more" because one-half cup clearly fits in two-thirds of a cup, but then a little more is still needed to fill the two-thirds of a cup. But what is this "little more"? Remember the original question: How many $\frac{1}{2}$ cups of water are in $\frac{2}{3}$ cup of water? The answer should be of the form "so and so many $\frac{1}{2}$ cups of water." This means that we need to express this "little more" as *a fraction of $\frac{1}{2}$ cup of water*. How can we do that? By subdividing both the $\frac{1}{2}$ and the $\frac{2}{3}$ into common parts—namely, by using common denominators.

FIGURE 6.14

$\frac{2}{3} \div \frac{1}{2} = ?$ How many $\frac{1}{2}$ cups of water are in $\frac{2}{3}$ cup?

When we give $\frac{1}{2}$ and $\frac{2}{3}$ the common denominator of 6, then, as on the right of Figure 6.14, and as in Figure 6.15, the $\frac{1}{2}$ cup of water is made out of 3 parts (3 sixths cup of water), and the $\frac{2}{3}$ cup of water is made out of 4 parts (4 sixths cup of water), so the "little more" we were discussing in the previous paragraph is just one of those parts. Since $\frac{1}{2}$ cup is 3 parts, and the "little more" is 1 part, the "little more" is $\frac{1}{3}$ of the $\frac{1}{2}$ cup of water. This explains why $\frac{2}{3} \div \frac{1}{2} = 1\frac{1}{3}$: There's an entire $\frac{1}{2}$ cup plus another $\frac{1}{3}$ of the $\frac{1}{2}$ cup in $\frac{2}{3}$ cup of water.

To summarize, we are considering the fraction division problem $\frac{2}{3} \div \frac{1}{2}$ in terms of the story problem "how many $\frac{1}{2}$ cups of water are in $\frac{2}{3}$ cup of water?" If we give $\frac{1}{2}$ and $\frac{2}{3}$ the common denominator of 6, then we can rephrase the problem as "how many $\frac{3}{6}$ cup are in $\frac{4}{6}$ cup?" But in terms of Figures 6.14 and 6.15, this is equivalent to the problem "how many 3s are in 4?" which is the problem $4 \div 3 = ?$, whose answer is $\frac{4}{3} = 1\frac{1}{3}$.

The approach that we just developed actually provides us with another general procedure for dividing fractions! The reasoning we used was general and can be applied in exactly the same way to other fractions. Notice the process for dividing fractions that we developed: Starting with a fraction division problem, we first gave the fractions a common denominator; then we just divided the numerators of these new fractions, disregarding the common denominators. The same line of reasoning will work for any fraction division problem,

$$\frac{A}{B} \div \frac{C}{D}$$

Thinking logically, as before, and interpreting $\frac{A}{B} \div \frac{C}{D}$ as "how many $\frac{C}{D}$ cups of water are in $\frac{A}{B}$ cups of water?", we can conclude that

$$\frac{A}{B} \div \frac{C}{D} = \frac{A \cdot D}{B \cdot D} \div \frac{B \cdot C}{B \cdot D} = (A \cdot D) \div (B \cdot C) = \frac{A \cdot D}{B \cdot C}$$

FIGURE 6.15

$\frac{2}{3} \div \frac{1}{2} = ?$
How many $\frac{1}{2}$ are in $\frac{2}{3}$?

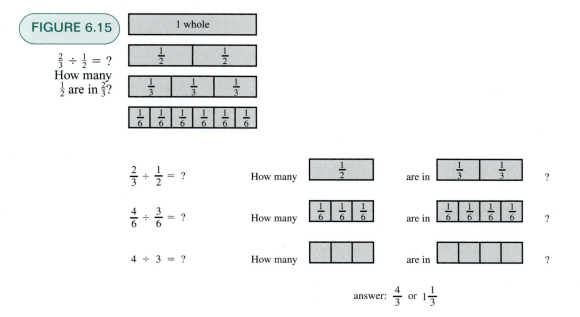

answer: $\frac{4}{3}$ or $1\frac{1}{3}$

The final expression, $\frac{A \cdot D}{B \cdot C}$, is the answer provided by the "invert and multiply" procedure for dividing fractions. Therefore, we know that the "invert and multiply" procedure gives answers to division problems that agree with what we expect from the meaning of division.

Viewing Division Problems as Unknown Factor Multiplication Problems to Explain Why "Multiply by the Reciprocal" Is Valid

Another way to explain why the usual fraction division procedure is valid is to use the important link between division and multiplication. Recall that every division problem is equivalent to a multiplication problem with an unknown factor (actually two multiplication problems). Thus,

$$S \div T = ?$$

is equivalent to

$$? \div T = S$$

(or $T \cdot ? = S$). So, for example,

$$1 \div \frac{C}{D} = ? \quad \text{is equivalent to} \quad ? \cdot \frac{C}{D} = 1$$

What can we put in place of the ? to make either of these equations true? Notice that the reciprocal of $\frac{C}{D}$, namely $\frac{D}{C}$, will make the second equation true:

$$\frac{D}{C} \cdot \frac{C}{D} = \frac{D \cdot C}{C \cdot D} = 1 \tag{6.7}$$

Equation (6.7) shows that when you multiply a fraction by its reciprocal, you get 1. Another way to say this is that the reciprocal of a fraction is the **multiplicative inverse** of the fraction.

In general, it is the fact that the reciprocal of a fraction is its multiplicative inverse that makes the "multiply by the reciprocal" procedure for fraction division work. Let's see why that is so. The equation

$$\frac{A}{B} \div \frac{C}{D} = ? \quad \text{is equivalent to} \quad ? \cdot \frac{C}{D} = \frac{A}{B}$$

What can we put in place of the ? to make either of these equations true? If we put

$$\frac{A}{B} \cdot \frac{D}{C}$$

in for the ? in the second equation, the $\frac{D}{C}$ and $\frac{C}{D}$ will multiply together to make 1, thus leaving us with $\frac{A}{B}$, which makes the equation true:

$$\left(\frac{A}{B} \cdot \frac{D}{C} \right) \cdot \frac{C}{D} = \frac{A}{B} \cdot \left(\frac{D}{C} \cdot \frac{C}{D} \right) = \frac{A}{B}$$

Therefore,

$$\frac{A}{B} \cdot \frac{D}{C}$$

really is the solution to the division problem

$$\frac{A}{B} \div \frac{C}{D} = ?$$

and we have developed another line of reasoning for why the "multiplying by the reciprocal" procedure for fraction division is valid.

Practice Exercises for Section 6.4

1. Write a "how many groups?" story problem for $1 \div \frac{5}{7}$. Use the story problem and a simple picture to help you solve the problem.

2. Annie wants to solve the division problem $\frac{3}{4} \div \frac{1}{2}$ by using the following story problem:

   I need $\frac{1}{2}$ cup of chocolate chips to make a batch of cookies. How many batches of cookies can I make with $\frac{3}{4}$ of a cup of chocolate chips?

   $\frac{1}{2}$ cup makes one batch.   } $\frac{1}{4}$ cup is left.

   FIGURE 6.16   How many batches of cookies can be made with $\frac{3}{4}$ cup of chocolate chips if 1 batch requires $\frac{1}{2}$ cup of chocolate chips?

   Annie draws a diagram like the one in Figure 6.16. Explain why it would be easy for Annie to misinterpret her diagram as showing that $\frac{3}{4} \div \frac{1}{2} = 1\frac{1}{4}$. How should Annie interpret her diagram so as to conclude that $\frac{3}{4} \div \frac{1}{2} = 1\frac{1}{2}$?

3. Write a simple "how many groups?" story problem for $\frac{5}{2} \div \frac{2}{3}$ and use the story problem to help you explain why you can solve the division problem by first giving the fractions a common denominator and then dividing the numerators.

4. Use the fact that we can rewrite the division problem $\frac{7}{11} \div \frac{3}{5} = ?$ as a multiplication problem with an unknown factor to explain why the division problem can be solved by multiplying $\frac{7}{11}$ by the reciprocal of $\frac{3}{5}$.

Answers to Practice Exercises for Section 6.4

1. A simple "how many groups?" story problem for $1 \div \frac{5}{7}$ is "how many $\frac{5}{7}$ cup of water are in 1 cup of water?" Figure 6.17 shows 1 cup of water and shows $\frac{5}{7}$ cup of water shaded. The shaded portion is divided into 5 equal parts, and the full cup is 7 of those parts. Thus, the full cup is $\frac{7}{5}$ of the shaded part, and there are $\frac{7}{5}$ of $\frac{5}{7}$ of a cup of water in 1 cup of water, so $1 \div \frac{5}{7} = \frac{7}{5}$.

1 cup $\frac{5}{7}$ of a cup   Each piece is $\frac{1}{5}$ of the shaded portion.

FIGURE 6.17   Showing why $1 \div \frac{5}{7} = \frac{7}{5}$ by considering how many $\frac{5}{7}$ cup of water are in 1 cup of water.

2. Annie's diagram shows that she can make 1 full batch of cookies from her $\frac{3}{4}$ cup of chocolate chips and that $\frac{1}{4}$ cup of chocolate chips will be left over.

Because $\frac{1}{4}$ cup of chocolate chips is left over, it would be easy for Annie to misinterpret her picture as showing $\frac{3}{4} \div \frac{1}{2} = 1\frac{1}{4}$. The answer to the problem is supposed to be the number of *batches* Annie can make. In terms of batches, the remaining $\frac{1}{4}$ cup of chocolate chips makes $\frac{1}{2}$ of a batch of cookies. We can see this because 2 quarter-cup sections make a full batch, so each quarter-cup section makes $\frac{1}{2}$ of a batch of cookies. Thus, by interpreting the remaining $\frac{1}{4}$ cup of chocolate chips in terms of batches, we see that Annie can make $1\frac{1}{2}$ batches of chocolate chips, thereby showing that $\frac{3}{4} \div \frac{1}{2} = 1\frac{1}{2}$, not $1\frac{1}{4}$.

3. Story problem: One serving of juice is $\frac{2}{3}$ of a cup. If you drink $2\frac{1}{2} = \frac{5}{2}$ cups of juice, how many servings will that be? To solve the problem, we need to find how many $\frac{2}{3}$ are in $\frac{5}{2}$, which, by giving the fractions common denominators, is the same as finding how many $\frac{4}{6}$ are in $\frac{15}{6}$. In other words, to solve the problem we need to find how many groups of 4 sixths are in 15 sixths. But to find how

many groups of 4 sixths are in 15 sixths we just need to find how many 4s are in 15, which is $15 \div 4 = 3\frac{3}{4}$. So to solve the division problem, we first rewrote it in terms of common denominators:

$$\frac{5}{2} \div \frac{2}{3} = \frac{15}{6} \div \frac{4}{6}$$

and then we divided the numerators:

$$15 \div 4 = 3\frac{3}{4}$$

4. Work with the unknown factor multiplication problem

$$? \times \frac{3}{5} = \frac{7}{11}$$

See the text for one explanation. Another explanation is indicated by Class Activity 6S. This explanation works by changing $\frac{7}{11}$ into an equivalent fraction so that the multiplication problem will "work." If we change $\frac{7}{11}$ to the equivalent fraction

$$\frac{7 \times 3 \times 5}{11 \times 3 \times 5}$$

then to solve

$$? \times \frac{3}{5} = \frac{7 \times 3 \times 5}{11 \times 3 \times 5}$$

the ? should be

$$\frac{7 \times 5}{11 \times 3}$$

which is equal to

$$\frac{7}{11} \times \frac{5}{3}$$

which is $\frac{7}{11}$ times the reciprocal of $\frac{3}{5}$.

Problems for Section 6.4

1. A bread problem: If one loaf of bread requires $1\frac{1}{4}$ cups of flour, then how many loaves of bread can you make with 10 cups of flour? (Assume that you have enough of all other ingredients on hand.)

   a. Solve the bread problem by drawing a diagram. Explain your reasoning.

   b. Write a division problem that corresponds to the bread problem. Solve the division problem by "inverting and multiplying." Verify that your solution agrees with your solution in part (a).

2. A measuring problem: You are making a recipe that calls for $\frac{2}{3}$ cup of water, but you can't find your $\frac{1}{3}$ cup measure. You can, however, find your $\frac{1}{4}$ cup measure. How many times should you fill your $\frac{1}{4}$ cup measure to measure $\frac{2}{3}$ cup of water?

   a. Solve the measuring problem by drawing a diagram. Explain your reasoning.

   b. Write a division problem that corresponds to the measuring problem. Solve the division problem

by "inverting and multiplying." Verify that your solution agrees with your solution in part (a)

3. Write a "how many groups?" story problem for $4 \div \frac{2}{3}$, and solve your problem in a simple and concrete way without using the "invert and multiply" procedure. Explain your reasoning. Verify that your solution agrees with the solution you obtain by using the "invert and multiply" procedure.

4. Write a "how many groups?" story problem for $5\frac{1}{4} \div 1\frac{3}{4}$, and solve your problem in a simple and concrete way without using the "invert and multiply" procedure. Explain your reasoning. Verify that your solution agrees with the solution you obtain by using the "invert and multiply" procedure.

5. Jose and Mark are making cookies for a bake sale. Their recipe calls for $2\frac{1}{4}$ cups of flour for each batch. They have 5 cups of flour. Jose and Mark realize that they can make two batches of cookies and that there will be some flour left. Since the

recipe doesn't call for eggs, and since they have plenty of the other ingredients on hand, they decide they can make a fraction of a batch in addition to the two whole batches. But Jose and Mark have a difference of opinion. Jose says that

$$5 \div 2\frac{1}{4} = 2\frac{2}{9}$$

so he says that they can make $2\frac{2}{9}$ batches of cookies. Mark says that two batches of cookies will use up $4\frac{1}{2}$ cups of flour, leaving $\frac{1}{2}$ left, so they should be able to make $2\frac{1}{2}$ batches. Mark draws the picture in Figure 6.18 to explain his thinking to Jose.

Discuss the boys' mathematics: What's right, what's not right, and why? If anything is incorrect, how could you modify it to make it correct?

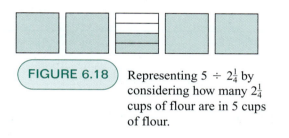

FIGURE 6.18 Representing $5 \div 2\frac{1}{4}$ by considering how many $2\frac{1}{4}$ cups of flour are in 5 cups of flour.

6. Marvin has 11 yards of cloth to make costumes for a play. Each costume requires $1\frac{1}{2}$ yards of cloth.

   a. Solve the following two problems:

   i. How many costumes can Marvin make, and how much cloth will be left over?

   ii. What is $11 \div 1\frac{1}{2}$?

   b. Compare and contrast your answers in part (a).

7. A laundry problem: You need $\frac{3}{4}$ cup of laundry detergent to wash one full load of laundry. How many loads of laundry can you wash with 5 cups of laundry detergent? (Assume that you can wash fractional loads of laundry.)

   a. Solve the laundry problem by drawing a simple picture. Explain your reasoning.

   b. Write a division problem that corresponds to the laundry problem. Solve the division problem by "inverting and multiplying." Verify that your solution agrees with your solution in part (a).

8. Write a "how many groups?" story problem for $2 \div \frac{3}{4}$, and solve your problem in a simple and concrete way without using the "invert and multiply" procedure. Explain your reasoning. Verify that your solution agrees with the solution you obtain by using the "invert and multiply" procedure.

9. Write a "how many groups?" story problem for $\frac{1}{3} \div \frac{1}{4}$, and solve your problem in a simple and concrete way without using the "invert and multiply" procedure. Explain your reasoning. Verify that your solution agrees with the solution you obtain by using the "invert and multiply" procedure.

10. Write a "how many groups?" story problem for $\frac{1}{2} \div \frac{2}{3}$, and solve your problem in a simple and concrete way without using the "invert and multiply" procedure. Explain your reasoning. Verify that your solution agrees with the solution you obtain by using the "invert and multiply" procedure.

11. Write a simple "how many groups?" story problem for $\frac{1}{3} \div \frac{1}{2}$. Use the story problem together with a simple picture to help you explain in your own words why you can solve the division problem by first giving the fractions a common denominator and then dividing the numerators.

12. **a.** By rewriting the division problem

   $$\frac{18}{33} \div \frac{9}{11} = ?$$

   as a multiplication problem with an unknown factor, explain why it is valid to divide the fractions as follows:

   $$\frac{18}{33} \div \frac{9}{11} = \frac{18 \div 9}{33 \div 11} = \frac{2}{3}$$

   b. Give an example of a numerical fraction division problem that can be made easy to solve by using the method of dividing the numerators and dividing the denominators that is demonstrated in part (a).

13. Use the fact that we can rewrite the division problem $\frac{7}{10} \div \frac{2}{3} = ?$ as a multiplication problem with an unknown factor to explain in your own words why the division problem can be solved by multiplying $\frac{7}{10}$ by the reciprocal of $\frac{2}{3}$.

14. Fraction division story problems involve the simultaneous use of different wholes. Solve the paint problem that follows in a simple and concrete way without using the "invert and multiply" procedure. Describe how you must work simultaneously with different wholes in solving the problem.

A paint problem: You need $\frac{3}{4}$ of a bottle of paint to paint a poster board. You have $3\frac{1}{2}$ bottles of paint. How many poster boards can you paint?

15. An article by Dina Tirosh, [86], discusses some common errors in division. The following problems are based on some of the findings of this article:

a. Tyrone says that $\frac{1}{2} \div 5$ doesn't make sense because 5 is bigger than $\frac{1}{2}$ and you can't divide a smaller number by a bigger number. Give Tyrone an example of a sensible story problem for $\frac{1}{2} \div 5$. Solve your problem, and explain your solution.

b. Kim says that $4 \div \frac{1}{3}$ can't be equal to 12 because when you divide, the answer should be smaller. Kim thinks the answer should be $\frac{1}{12}$ because that is less than 4. Give Kim an example of a story problem for $4 \div \frac{1}{3}$, and explain why it makes sense that the answer really is 12, not $\frac{1}{12}$.

16. Give an example of either a hands-on activity or a story problem for elementary school children that is related to a fraction division problem (even if the children wouldn't think of the activity or problem as fraction division). Write the fraction division problem that is related to your activity or story problem. Describe how the children could solve the problem by using logical thinking aided by physical actions or by drawing pictures.

6.5 Fraction Division from the "How Many in One Group?" Perspective

Focal Points
Grade 6

In this section we'll continue to study fraction division, this time from the "how many in each (or one) group?" viewpoint. Often "how many in one group?" fraction division story problems don't seem like division problems at all, but rather like proportion problems—which in fact they are. We'll see that the "multiply by the reciprocal" procedure for fraction division is especially nicely explained from the "how many in one group?" perspective. Finally, we'll consider a variety of story problems and we'll pay special attention to distinguishing fraction division story problems from fraction multiplication story problems. In particular, we'll distinguish dividing *in* half from dividing *by* one half.

The "How Many in One Group?" Interpretation

Class Activity *Now Turn to Class Activities Manual*

6T "How Many in One Group?" Fraction Division Problems, p. 132

With the "how many in each group?" interpretation of division, $12 \div 3$ means the number of objects in each group when we distribute 12 objects equally among 3 groups. In other words, $12 \div 3$ is the number of objects in one group if we use 12 objects to evenly fill 3 groups. When we work with fractions, it often helps to think of "how many in each group?" division story problems as asking "how many are in *one whole group*?", and it helps to think of *filling* groups or part of a group. So in the context of fractions, we will usually refer to the "how many in each group?" interpretation as "how many in *one* group?"

With the "how many in one group?" interpretation of division,

$$\frac{3}{4} \div \frac{1}{2}$$

is the number of objects in one group when we distribute $\frac{3}{4}$ of an object equally among $\frac{1}{2}$ of a group. A clearer way to say this is "$\frac{3}{4} \div \frac{1}{2}$ is the number of objects (or fraction of an object) in one whole group when $\frac{3}{4}$ of an object fills $\frac{1}{2}$ of a group." For example, suppose you pour $\frac{3}{4}$ pint of blueberries into a container and this fills $\frac{1}{2}$ of the container. How many pints of blueberries will it take to fill the whole container? In this case, $\frac{3}{4}$ pint of blueberries fills (i.e., is distributed equally among) $\frac{1}{2}$ of a group (a container). So, according to the "how many in one group?" interpretation of division, the number of pints of blueberries in one whole group (one full container) is

$$\frac{3}{4} \div \frac{1}{2}$$

One way to better understand fraction division story problems is to think about replacing the fractions in the problem with whole numbers. For example, if you have 6 pints of blueberries and they fill 2 containers, then how many pints of blueberries are in each container? We solve this problem by dividing $6 \div 2$, according to the "how many in each group?" interpretation. If we replace the 6 pints with $\frac{3}{4}$ pint and the 2 containers with $\frac{1}{2}$ container, we solve the problem in the same way as before: $6 \div 2$ now becomes $\frac{3}{4} \div \frac{1}{2}$ as indicated in Figure 6.19.

Another way to analyze "how many in one group?" fraction division is to reformulate fraction division problems as equivalent multiplication problems. From the "how many in one group?" viewpoint,

$$6 \div 2 = ? \text{ corresponds to } 2 \times ? = 6$$

since we are asking 2 of how many make 6. (2 of how many pints make 6 pints?) Therefore, similarly,

$$\frac{3}{4} \div \frac{1}{2} = ? \text{ corresponds to } \frac{1}{2} \times ? = \frac{3}{4}$$

and we are asking $\frac{1}{2}$ of how much makes $\frac{3}{4}$. In the case of the blueberries where $\frac{1}{2}$ of a container is filled and this amount is $\frac{3}{4}$ of a pint of blueberries, $\frac{1}{2}$ of the number of pints in a full container is $\frac{3}{4}$ pint. In other words,

$$\frac{1}{2} \times \text{number of pints in full container} = \frac{3}{4}$$

Therefore,

$$\text{number of pints in full container} = \frac{3}{4} \div \frac{1}{2}$$

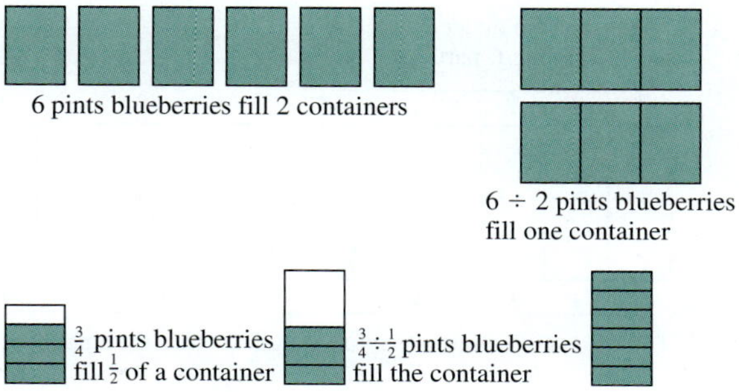

FIGURE 6.19

"How many in one group?" story problems for $6 \div 2$ and $\frac{3}{4} \div \frac{1}{2}$

6 pints blueberries fill 2 containers

$6 \div 2$ pints blueberries fill one container

$\frac{3}{4}$ pints blueberries fill $\frac{1}{2}$ of a container

$\frac{3}{4} \div \frac{1}{2}$ pints blueberries fill the container

Class Activity   *Now Turn to Class Activities Manual*

6U Using Double Number Lines to Solve "How Many in One Group?" Division Problems, p. 134

Using the "How Many in One Group?" Interpretation to Explain Why "Invert and Multiply" Is Valid

In Section 6.4, we saw how to use the "how many groups?" interpretation of division to explain why the "invert and multiply" procedure for fraction division is valid. We can also use the "how many in one group?" interpretation for the same purpose. This interpretation has the advantage of showing us directly why we can multiply by the reciprocal of the divisor in order to divide fractions.

Consider the following "how many in one group?" story problem for $\frac{1}{2} \div \frac{3}{5}$:

> You used $\frac{1}{2}$ can of paint to paint $\frac{3}{5}$ of a wall. How many cans of paint will it take to paint the whole wall?

This is a "how many in one group?" problem because we can think of the paint as "filling" $\frac{3}{5}$ of the wall. Or we could realize that if it takes 6 cans of paint for 3 walls (of the same size), then one wall takes $6 \div 3$ cans of paint, and this uses the "how many in one group?" view of division. So the same must be true when we replace 6 with $\frac{1}{2}$ and 3 with $\frac{3}{5}$. We can also see that this is a division problem by writing the corresponding equation:

$$\frac{3}{5} \cdot (\text{amount to paint the whole wall}) = \frac{1}{2}$$

Therefore,

$$\text{amount to paint the whole wall} = \frac{1}{2} \div \frac{3}{5}$$

We will now see why it makes sense to solve this division problem by multiplying $\frac{1}{2}$ by the reciprocal of $\frac{3}{5}$, namely, by $\frac{5}{3}$. Let's focus on the wall to be painted, as shown in Figure 6.20. Think of dividing the wall into 5 equal sections, 3 of which you painted with the $\frac{1}{2}$ can of paint. If you used $\frac{1}{2}$ can of paint to paint 3 sections, then each of the 3 sections required $\frac{1}{2} \div 3$ or $\frac{1}{2} \times \frac{1}{3}$ cans of paint. To determine how much paint you will need for the whole wall, multiply the amount you need for one section by 5. So you can

$$\frac{1}{2} \div \frac{3}{5} = ?$$

$\frac{1}{2}$ can paint covers $\frac{3}{5}$ of a wall.

? cans paint cover 1 whole wall.

$$\frac{A}{B} \div \frac{C}{D} = ?$$

$\frac{A}{B}$ cans paint cover $\frac{C}{D}$ of a wall.

? cans paint cover 1 whole wall.

FIGURE 6.20

The amount of paint needed for the whole wall is $\frac{5}{3}$ of the $\frac{1}{2}$ can used to cover $\frac{3}{5}$ of the wall.

The $\frac{1}{2}$ can of paint is divided equally among 3 parts.

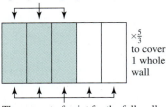

$\times \frac{5}{3}$ to cover 1 whole wall

The amount of paint for the full wall is 5 times the amount in one part.

The $\frac{A}{B}$ cans are divided equally among C parts.

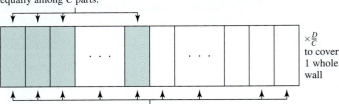

$\times \frac{D}{C}$ to cover 1 whole wall

The amount of paint for the full wall is D times the amount in one part.

TABLE 6.8
To solve $\frac{1}{2} \div \frac{3}{5}$ or $\frac{A}{B} \div \frac{C}{D}$, determine how much paint to use for a whole wall if $\frac{1}{2}$ can of paint covers $\frac{3}{5}$ of the wall or if $\frac{A}{B}$ of a can of paint covers $\frac{C}{D}$ of the wall.

$\frac{1}{2} \div \frac{3}{5} = ?$				$\frac{A}{B} \div \frac{C}{D} = ?$			
Use	$\frac{1}{2}$ can paint	for	$\frac{3}{5}$ of the wall.	Use	$\frac{A}{B}$ can paint	for	$\frac{C}{D}$ of the wall.
	$\downarrow \div 3$ or $\times \frac{1}{3}$		$\downarrow \div 3$ or $\times \frac{1}{3}$		$\downarrow \div C$ or $\times \frac{1}{C}$		$\downarrow \div C$ or $\times \frac{1}{C}$
Use	$\frac{1}{6}$ can paint	for	$\frac{1}{5}$ of the wall.	Use	$\frac{A}{B} \times \frac{1}{C}$ can paint	for	$\frac{1}{D}$ of the wall.
	$\downarrow \times 5$		$\downarrow \times 5$		$\downarrow \times D$		$\downarrow \times D$
Use	$\frac{5}{6}$ can paint	for	1 whole wall.	Use	$\frac{A}{B} \times \frac{D}{C}$ can paint	for	1 whole wall.
in one step:				**in one step:**			
Use	$\frac{1}{2}$ can paint	for	$\frac{3}{5}$ of the wall.	Use	$\frac{A}{B}$ can paint	for	$\frac{C}{D}$ of the wall.
	$\downarrow \times \frac{5}{3}$		$\downarrow \times \frac{5}{3}$		$\downarrow \times \frac{C}{D}$		$\downarrow \times \frac{D}{C}$
Use	$\frac{5}{6}$ can paint	for	1 whole wall.	Use	$\frac{A}{B} \times \frac{D}{C}$ can paint	for	1 whole wall.
	so $\frac{1}{2} \div \frac{3}{5} = \frac{1}{2} \times \frac{5}{3}$				so $\frac{A}{B} \div \frac{C}{D} = \frac{A}{B} \times \frac{D}{C}$		

determine the amount of paint you need for the whole wall by multiplying the $\frac{1}{2}$ can of paint by $\frac{1}{3}$ and then multiplying that result by 5, as summarized in Table 6.8. But to multiply a number by $\frac{1}{3}$ and then multiply it by 5 is the same as multiplying the number by $\frac{5}{3}$. Therefore, we can determine the number of cans of paint you need for the whole wall by multiplying $\frac{1}{2}$ by $\frac{5}{3}$:

$$\frac{1}{2} \cdot \frac{5}{3} = \frac{5}{6}$$

Another way to summarize this reasoning is to observe that the whole wall is $\frac{5}{3}$ of the part painted by $\frac{1}{2}$ can of paint, so the whole will take $\frac{5}{3}$ as much paint, which is $\frac{1}{2} \times \frac{5}{3}$ cans of paint.

This is exactly the "invert and multiply" procedure for dividing $\frac{1}{2} \div \frac{3}{5}$. It shows that you will need $\frac{5}{6}$ can of paint for the whole wall.

The preceding argument works when other fractions replace $\frac{1}{2}$ and $\frac{3}{5}$, as summarized on the right in Figure 6.20 and Table 6.8, thereby explaining why

$$\frac{A}{B} \div \frac{C}{D} = \frac{A}{B} \cdot \frac{D}{C}$$

In other words, to divide fractions, multiply the dividend by the reciprocal of the divisor.

Attending Closely to the Wording of Story Problems: Dividing *by* $\frac{1}{2}$ versus Dividing *in* $\frac{1}{2}$

In mathematics, language is used much more precisely and carefully than in everyday conversation. This is one source of difficulty in learning mathematics. For example, consider the following two phrases:

dividing *by* $\frac{1}{2}$

dividing *in* $\frac{1}{2}$

You may feel that these two phrases mean the same thing; however, mathematically, they do not. To divide a number—say, 5—by $\frac{1}{2}$ means to calculate $5 \div \frac{1}{2}$. Remember that we read $A \div B$ as A divided by B.

We would divide 5 by $\frac{1}{2}$ if we wanted to know how many half cups of flour are in 5 cups of flour, for example. (Notice that there are 10 half cups of flour in 5 cups of flour, not $2\frac{1}{2}$.)

On the other hand, to divide a number *in* half means to find half of that number. So to divide 5 in half means to find $\frac{1}{2}$ of 5. One half of 5 means $\frac{1}{2} \times 5$. So dividing in $\frac{1}{2}$ is the same as dividing by 2.

When you examine the story problems in the next Class Activity, pay close attention to the wording in order to decide if they are division problems or not. When it comes to story problems, there is simply no substitute for careful reading and close attention to meaning!

Class Activity   *Now Turn to Class Activities Manual*

6V Are These Division Problems? p. 135

Practice Exercises for Section 6.5

1. Write a "how many in one group?" story problem for $2 \div \frac{3}{4}$. Use the situation of the story problem to help you explain why you can solve the division problem by multiplying 2 by the reciprocal of $\frac{3}{4}$.

2. Which of the following are solved by the division problem $\frac{3}{4} \div \frac{1}{2}$? For those that are, which interpretation of division is used? For those which are not, determine how to solve the problem, if it can be solved.

  a. $\frac{3}{4}$ of a bag of jelly worms make $\frac{1}{2}$ cup. How many cups of jelly worms are in one bag?

  b. $\frac{3}{4}$ of a bag of jelly worms make $\frac{1}{2}$ cup. How many bags of jelly worms does it take to make one cup?

  c. You have $\frac{3}{4}$ of a bag of jelly worms and a recipe that calls for $\frac{1}{2}$ cup of jelly worms. How many batches of your recipe can you make?

  d. You have $\frac{3}{4}$ cup of jelly worms and a recipe that calls for $\frac{1}{2}$ cup of jelly worms. How many batches of your recipe can you make?

  e. If $\frac{3}{4}$ pound of candy costs $\frac{1}{2}$ of a dollar, then how many pounds of candy should you be able to buy for 1 dollar?

  f. If you have $\frac{3}{4}$ pound of candy and you divide the candy in $\frac{1}{2}$, then how much candy will you have in each portion?

  g. If $\frac{1}{2}$ pound of candy costs \$1, then how many dollars should you expect to pay for $\frac{3}{4}$ pound of candy?

3. Frank, John, and David earned \$14 together. They want to divide it equally, except that David should only get a half share, since he did half as much work as either Frank or John did (and Frank and John worked equal amounts). Write a division problem to find out how much Frank should get. Which interpretation of division does this story problem use?

4. Bill leaves a tip of \$4.50 for a meal. If the tip is 15% of the cost of the meal, then how much did the meal cost? Write a division problem to solve this. Which interpretation of division does this story problem use?

5. Compare the arithmetic needed to solve the following problems:

  a. What fraction of a $\frac{1}{3}$ cup measure is filled when we pour in $\frac{1}{4}$ cup of water?

  b. What is one quarter of $\frac{1}{3}$ cup?

  c. How much more is $\frac{1}{3}$ cup than $\frac{1}{4}$ cup?

  d. If $\frac{1}{4}$ cup of water fills $\frac{1}{3}$ of a plastic container, then how much water will the full container hold?

6. It takes an assembly line $\frac{1}{2}$ hour to produce enough boxes of widgets to fill a truck $\frac{2}{3}$ full. At that rate, how long does it take the assembly line to produce enough boxes of widgets to fill the truck completely full?

Explain how to use the double number line in Figure 6.21 to help you solve this problem.

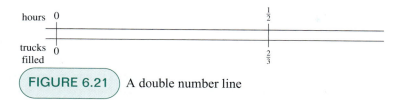

FIGURE 6.21 A double number line

Answers to Practice Exercises for Section 6.5

1. Here is a "how many in one group?" story problem for $2 \div \frac{3}{4}$. If 2 tons of dirt fills a truck $\frac{3}{4}$ full, then how many tons of dirt will be needed to fill the truck completely full?

We can see that this is a "how many in one group?" type of problem because the 2 tons of dirt fills $\frac{3}{4}$ of a group (the truck) and we want to know the amount of dirt in 1 whole group. Or we could realize that if 6 tons of dirt fill 3 trucks, then $6 \div 3$ tons of dirt fill one truck, and this uses the "how many in one group?" view of division. So the same must be true when we replace 6 with 2 and 3 with $\frac{3}{4}$.

Figure 6.22 shows a truck bed divided into 4 equal parts with 3 of those parts filled with dirt. Since the 3 parts are filled with 2 tons of dirt, each of the 3 parts must contain $\frac{1}{3}$ of the 2 tons of dirt. To fill the truck completely, 4 parts, each containing $\frac{1}{3}$ of the 2 tons of dirt are needed. "4 parts, each containing $\frac{1}{3}$" means $\frac{4}{3}$, so to fill the truck completely, $\frac{4}{3}$ of the 2 tons of dirt are needed. Therefore, the truck takes $2 \times \frac{4}{3}$ tons of dirt to fill it. So $2 \div \frac{3}{4} = 2 \times \frac{4}{3}$.

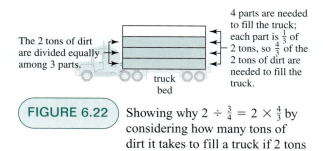

The 2 tons of dirt are divided equally among 3 parts.

4 parts are needed to fill the truck; each part is $\frac{1}{3}$ of 2 tons, so $\frac{4}{3}$ of the 2 tons of dirt are needed to fill the truck.

truck bed

FIGURE 6.22 Showing why $2 \div \frac{3}{4} = 2 \times \frac{4}{3}$ by considering how many tons of dirt it takes to fill a truck if 2 tons fills it $\frac{3}{4}$ full

2. a. This problem can be rephrased as "if $\frac{1}{2}$ cup of jelly worms fill $\frac{3}{4}$ of a bag, then how many cups fill a whole bag?" Therefore, this is a "how many in one group?" division problem illustrating $\frac{1}{2} \div \frac{3}{4}$, not $\frac{3}{4} \div \frac{1}{2}$. Since $\frac{1}{2} \div \frac{3}{4} = \frac{1}{2} \cdot \frac{4}{3} = \frac{2}{3}$, there are $\frac{2}{3}$ cup of jelly worms in a whole bag.

b. This problem is solved by $\frac{3}{4} \div \frac{1}{2}$, according to the "how many in one group?" interpretation. A group is a cup, and each object is a bag of jelly worms.

c. This problem can't be solved because you don't know how many cups of jelly worms are in $\frac{3}{4}$ of a bag.

d. This problem is solved by $\frac{3}{4} \div \frac{1}{2}$, according to the "how many groups?" interpretation. Each group consists of $\frac{1}{2}$ cup of jelly worms.

e. This problem is solved by $\frac{3}{4} \div \frac{1}{2}$, according to the "how many in one group?" interpretation. This is because you can think of the problem as saying that $\frac{3}{4}$ pound of candy fills $\frac{1}{2}$ of a group and you want to know how many pounds fills 1 whole group.

f. This problem is solved by $\frac{3}{4} \times \frac{1}{2}$, not $\frac{3}{4} \div \frac{1}{2}$. It is dividing *in* half, not dividing *by* half.

g. This problem is solved by $\frac{3}{4} \div \frac{1}{2}$, according to the "how many groups?" interpretation because you want to know how many $\frac{1}{2}$ pounds are in $\frac{3}{4}$ pound. Each group consists of $\frac{1}{2}$ pound of candy.

3. If we consider Frank and John as each representing one group, and David as representing half of a group, then the $14 should be distributed equally among $2\frac{1}{2}$ groups. Therefore, this is a "how many in one group" division problem. Each group should get

$$14 \div 2\frac{1}{2} = 14 \div \frac{5}{2} = 14 \cdot \frac{2}{5} = \frac{28}{5}$$

$$= 5\frac{3}{5} = 5\frac{6}{10} = 5.60$$

dollars. Therefore, Frank and John should each get $5.60, and David should get half of that, which is $2.80.

4. According to the "how many in one group?" interpretation, the problem is solved by $4.50 ÷ 0.15 because $4.50 fills 0.15 of a group and we want to know how much is in 1 whole group. So the meal cost

$$\$4.50 \div 0.15 = \$4.50 \div \frac{15}{100} = \$4.50 \cdot \frac{100}{15}$$

$$= \frac{\$450}{15} = \$30$$

5. Each problem, except for the first and last, requires different arithmetic to solve it.

a. This is asking, "$\frac{1}{4}$ equals what times $\frac{1}{3}$?" We solve this by calculating $\frac{1}{4} \div \frac{1}{3}$, which is $\frac{3}{4}$. We can also think of this as a division problem with the "how many groups?" interpretation because we want to know how many $\frac{1}{3}$ cup are in $\frac{1}{4}$ cup. According to the meaning of division, this is $\frac{1}{4} \div \frac{1}{3}$.

b. This is asking, "What is $\frac{1}{4}$ of $\frac{1}{3}$?" We solve this by calculating $\frac{1}{4} \times \frac{1}{3} = \frac{1}{12}$.

c. This is asking, "What is $\frac{1}{3} - \frac{1}{4}$?" The answer is $\frac{1}{12}$, which happens to be the same answer as in part (b), but the arithmetic to solve it is different.

d. Since $\frac{1}{4}$ cup of water fills $\frac{1}{3}$ of a plastic container, the full container will hold 3 times as much water, or $3 \times \frac{1}{4} = \frac{3}{4}$ of a cup. We can also think of this as a division problem with the "how many in one group?" interpretation. $\frac{1}{4}$ cup of water is put into $\frac{1}{3}$ of a group. We want to know how much is in one group. According to the meaning of division, it's $\frac{1}{4} \div \frac{1}{3}$, which again is equal to $\frac{3}{4}$.

6. To determine how long it will take to produce enough widgets to fill the truck completely full we must find the location on the top number line that is above 1, as indicated on the first double number line in Figure 6.23. If we give the bottom number line tick marks at every third, then $\frac{2}{3}$ is the second tick mark and $1 = \frac{3}{3}$ is the third tick mark. The number above $\frac{1}{3}$ must be half of $\frac{1}{2}$, which is $\frac{1}{4}$, as shown on the second double number line in Figure 6.23. So on the top number line, the tick marks are spaced $\frac{1}{4}$ apart. The number above 1 is $\frac{3}{4}$, so it will take $\frac{3}{4}$ of an hour to produce enough widgets to fill the truck completely full.

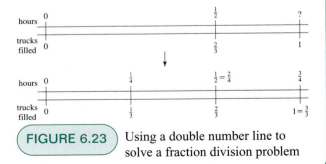

FIGURE 6.23   Using a double number line to solve a fraction division problem

Problems for Section 6.5

1. Write a "how many in one group?" story problem for $4 \div \frac{1}{3}$, and use your story problem to explain why it makes sense to solve $4 \div \frac{1}{3}$ by "inverting and multiplying"—in other words, by multiplying 4 by $\frac{3}{1}$.

2. Write a "how many in one group?" story problem for $4 \div \frac{2}{3}$, and use your story problem to explain why it makes sense to solve $4 \div \frac{2}{3}$ by "inverting and multiplying"—in other words, by multiplying 4 by $\frac{3}{2}$.

3. Write a "how many in one group?" story problem for $9 \div \frac{3}{4}$, and use your story problem to explain why it makes sense to solve $9 \div \frac{3}{4}$ by "inverting and multiplying"—in other words, by multiplying 9 by $\frac{4}{3}$.

4. Write a "how many in one group?" story problem for $\frac{1}{2} \div \frac{3}{4}$, and use your story problem to explain why it makes sense to solve $\frac{1}{2} \div \frac{3}{4}$ by "inverting and multiplying"—in other words, by multiplying $\frac{1}{2}$ by $\frac{4}{3}$.

5. Write a "how many in one group?" story problem for $1 \div 2\frac{1}{2}$, and use your story problem to explain why it makes sense to solve $1 \div 2\frac{1}{2}$ by "inverting and multiplying."

6. If $1\frac{1}{2}$ cups of a cereal weigh 2 pounds, how much does 1 cup of the cereal weigh? Solve this problem with the aid of a double number line, explaining your reasoning.

7. It took a mule $\frac{2}{3}$ of an hour to go $\frac{4}{5}$ of a mile. At that rate, how long would it take the mule to go 1 mile? How far would the mule go in an hour? Solve these problems with the aid of double number lines, explaining your reasoning.

8. It took $1\frac{1}{3}$ cans of paint to paint $\frac{2}{5}$ of a room. At that rate, how many cans of paint will it take to paint the whole room? What fraction of the room can you paint with 1 can of paint? Solve these problems with the aid of a double number line, explaining your reasoning.

9. Write a story problem for $\frac{3}{4} \times \frac{1}{2}$ and another story problem for $\frac{3}{4} \div \frac{1}{2}$. (Make clear which is which.) In each case, use elementary reasoning about the story situation to solve your problem. Explain your reasoning.

10. Sam picked $\frac{1}{2}$ gallon of blueberries. Sam poured the blueberries into one of his plastic containers and noticed that the berries filled the container $\frac{2}{3}$ full. Solve the following problems in any way you like without using a calculator, and explain your reasoning in detail:

    a. How many of Sam's containers will 1 gallon of blueberries fill? (Assume that Sam has a number of containers of the same size.)

    b. How many gallons of blueberries does it take to fill Sam's container completely full?

11. A road crew is building a road. So far, $\frac{2}{3}$ of the road has been completed and this portion of the road is $\frac{3}{4}$ of a mile long. Solve the following problems in any way you like without using a calculator, and explain your reasoning in detail:

    a. How long will the road be when it is completed?

    b. When the road is 1 mile long, what fraction of the road will be completed?

12. Will has mowed $\frac{2}{3}$ of his lawn, and so far it's taken him 45 minutes. For each of the following problems, solve the problem in two ways: (1) by using elementary reasoning about the story situation and (2) by interpreting the problem as a division problem (say whether it is a "how many groups?" or a "how many in one group?" type of problem) and by solving the division problem using standard paper-and-pencil methods. Do not use a calculator. Verify that you get the same answer both ways.

    a. How long will it take Will to mow the entire lawn (all together)?

    b. What fraction of the lawn can Will mow in an hour?

13. Grandma's favorite muffin recipe uses $1\frac{3}{4}$ cups of flour for one batch of 12 muffins. For each of the problems (a) through (c), solve the problem in two ways: (1) by using elementary reasoning about the story situation and (2) by interpreting the problem as a division problem (say whether it is a "how many groups?" or a "how many in one group?" type of problem) and by solving the division problem using standard paper-and-pencil methods. Do not use a calculator. Verify that you get the same answer both ways.

    a. How many cups of flour are in 1 muffin?

    b. How many muffins does 1 cup of flour make?

    c. If you have 3 cups of flour, then how many batches of muffins can you make? (Assume that you can make fractional batches of muffins and that you have enough of all the ingredients.)

14. Tran writes the following story problem for $1\frac{3}{4} \div \frac{1}{2}$:

    There are $1\frac{3}{4}$ pizzas left. Tran eats half of the pizza that is left. How many pieces of pizza did Tran eat?

Tran shows the picture in Figure 6.24, saying that it shows that

$$1\frac{3}{4} \div \frac{1}{2} = 3\frac{1}{2}$$

a. Even though Tran gets the correct answer to the division problem, explain why his story problem is not a correct story problem for $1\frac{3}{4} \div \frac{1}{2}$.

b. Write a correct story problem for $1\frac{3}{4} \div \frac{1}{2}$ that is about $1\frac{3}{4}$ pizzas.

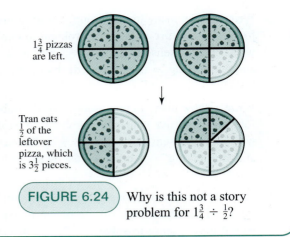

$1\frac{3}{4}$ pizzas are left.

Tran eats $\frac{1}{2}$ of the leftover pizza, which is $3\frac{1}{2}$ pieces.

FIGURE 6.24   Why is this not a story problem for $1\frac{3}{4} \div \frac{1}{2}$?

6.6 Dividing Decimals

 **Focal Points**
Grade 6

Why do we divide decimals the way we do? When we divide finite decimals, such as

$$2.35\overline{)3.714}$$

the standard procedure is to move the decimal point in both the divisor (2.35) and the dividend (3.714) the same number of places to the right until the divisor becomes a whole number:

$$235\overline{)371.4}$$

Then a decimal point is placed above the decimal point in the new dividend (371.4):

$$235\overline{)371.4}$$

From then on, division proceeds just as if we were doing the whole number division problem $3714 \div 235$, except that a decimal point is left in place for the answer. Why does this procedure of shifting the decimal points make sense? In this section, we will explain in several different ways why we can shift the decimal points as we do. Our technique for shifting decimal points in division problems also allows us to calculate efficiently with large numbers, such as those in the millions, billions, and trillions. But first we consider the two interpretations of division for decimals.

Class Activity   *Now Turn to Class Activities Manual*

6W   Quick Tricks for Some Decimal Division Problems, p. 136

6X   Decimal Division, p. 136

The Two Interpretations of Division for Decimals

The two interpretations that we have used for whole numbers and for fractions also apply to decimals.

The "How Many Groups?" Interpretation With the "how many groups?" interpretation of division, $35 \div 7$ means the number of groups we can make when we divide 35 objects into groups with 7 objects in each group. For example, if meat costs \$7 per pound and you have \$35, then how many pounds of meat can you buy?

Similarly, with the "how many groups?" interpretation of division,

$$14.5 \div 2.45$$

means the number of groups we can make when we divide 14.5 objects into groups with 2.45 objects in each group. For example, suppose gas costs \$2.45 per gallon and you have \$14.50 to spend on gas. How many gallons of gas can you buy? We want to know how many groups of \$2.45 are in \$14.50, so this is a "how many groups?" problem for $14.5 \div 2.45$.

The "How Many in One (Each) Group?" Interpretation With the "how many in one (or each) group?" interpretation of division, $35 \div 7$ means the number of objects in one group if 35 objects are divided equally among 7 groups. For example, if 7 identical action figures cost \$35, then how much did one (or each) action figure cost?

Similarly, with the "how many in one group?" interpretation of division,

$$9.48 \div 5.3$$

means the number of objects in one group if 9.48 objects are divided equally among 5.3 groups. You can also think of the 9.48 objects as "filling" 5.3 groups. For example, if you bought 5.3 pounds of peaches for \$9.48, then how much did 1 pound of peaches cost? The \$9.48 is distributed equally among or "fills" 5.3 groups, and we want to know how many dollars are in one group. Or, if you bought 5 pounds of peaches for \$10, then 1 pound of peaches would cost $\$10 \div 5$, and this is a "how many in one group?" division problem. So the same must be true when we replace 10 with 9.48 and 5 with 5.3. Therefore, this is a "how many in one group?" problem for $9.48 \div 5.3$.

Next we turn to different methods for explaining why we move the decimal point as we do when we divide decimals.

Explaining the Shifting of Decimal Points by Multiplying and Dividing by the Same Power of 10

One way to explain why it is valid to shift the decimal points in a division problem is as follows: When we shift the decimal points in both the divisor and the dividend the same number of places, we have multiplied and divided by the same power of 10, thereby replacing the original problem with an equivalent problem that can be solved by previous methods. Recall that multiplying a decimal number by a power of 10 shifts the digits in the number and therefore can be thought of as shifting the decimal point. Consider the following example:

$$2.35\overline{)3.714}$$

Because

$$2.35 \times 100 = 235$$
$$3.714 \times 100 = 371.4$$

it follows that

$$
\begin{array}{cc}
\times 100 & \div 100 \\
\downarrow & \downarrow
\end{array}
$$
$$3.714 \div 2.35 = (3.714 \times 100) \div (2.35 \times 100)$$
$$= 371.4 \div 235$$

Since we multiplied and divided by the same number—namely, 100—the original problem $371.4 \div 235$ is equivalent to the new problem $371.4 \div 235$. In other words, the two problems have the identical answer. We can give the same explanation by writing the division problems in fraction form:

$$\frac{3.714}{2.35} = \frac{3.714}{2.35} \times \frac{100}{100}$$

$$= \frac{3.714 \times 100}{2.35 \times 100}$$

$$= \frac{371.4}{235}$$

So the shifting of decimal points is really just a way to replace the problem

$$3.714 \div 2.35$$

with the equivalent problem

$$371.4 \div 235$$

Explaining the Shifting of Decimal Points by Changing from Dollars to Cents

If the decimals in a division problem have no digits in places lower than the hundredths place, then we can use money to interpret the division problem. For example, we can interpret

$$1.27\overline{)4.5}$$

as follows:

    $4.5 \div 1.27$ is the number of pounds of plums we can buy for $4.50 if plums cost $1.27 per pound.

(Note that we are using the "how many groups?" interpretation of division because we want to know how many groups of $1.27 are in $4.50.) If we think in terms of cents rather than dollars, then an equivalent way to rephrase the preceding statement is as follows:

    $4.5 \div 1.27$ is the number of pounds of plums we can buy for 450 cents if plums cost 127 cents per pound.

But this last statement also describes the division problem

$$127\overline{)450}$$

Therefore, the problems $4.5 \div 1.27$ and $450 \div 127$ are equivalent. In other words, they must have the same answer. We can buy the same number of pounds of plums whether we are thinking in terms of dollars or in terms of cents. So, when we shift the decimal points of 1.27 and 4.5 two places to the right in the division problem $1.27\overline{)4.5}$ to obtain the problem $127\overline{)450}$, we have simply replaced the original problem with a problem that has the same answer and can be solved with the techniques we learned for whole number division.

Explaining the Shifting of Decimal Points by Changing the Unit

A more general way to explain why we can shift the decimal points in a decimal division problem is to think about representing the division problem with bundled objects or with pictures representing bundled objects. For example, consider the division problem

$$0.29\overline{)1.7}$$

We can interpret this division problem as asking, "How many 0.29 are in 1.7?" from a "how many groups?" viewpoint. The diagram in Figure 6.25 gives us a way to picture this division problem by interpreting the

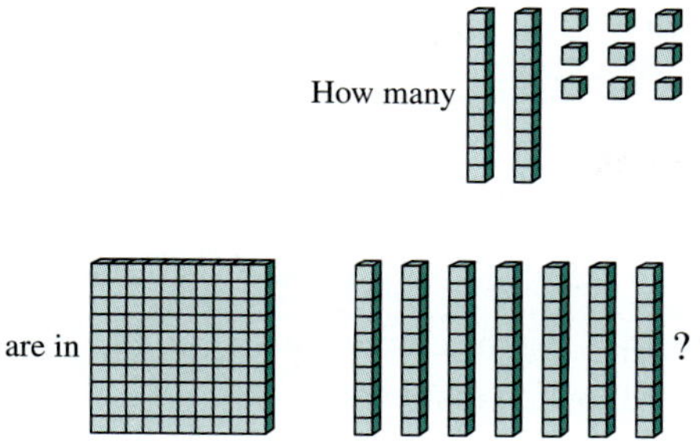

FIGURE 6.25

Representing
$1.7 \div 0.29$

How many

are in

?

smallest square as representing $\frac{1}{100}$. But if we interpret the smallest square as representing 1 instead of $\frac{1}{100}$, then Figure 6.25 asks "how many 29s are in 170?" which is the division problem

$$29\overline{)170}$$

The answer must be the same, no matter which way we interpret the diagram; therefore, the division problems $1.7 \div 0.29$ and $170 \div 29$ are equivalent problems. Notice that from the diagram we can estimate that the answer is between 5 and 6 because 5 copies of the squares on the top will be fewer squares than on the bottom, but 6 copies of the squares on the top will be more squares than those on the bottom.

By allowing the smallest square in Figure 6.25 to represent different values, we see that this diagram can represent infinitely many different division problems, as indicated in Table 6.9. Therefore, all of these division problems must have the same answer.

Deciding Where to Put the Decimal Point by Estimating the Size of the Answer

As a way of checking your work, or as a backup if you forget what to do about the decimal points when dividing decimals, you can often estimate to determine where to put the decimal point in the answer to a decimal division problem.

TABLE 6.9
Different ways
to interpret
Figure 6.25

If the Small Square Represents	Then Figure 7.27 Represents
$\vdots$	$\vdots$
$\dfrac{1}{10,000}$	$0.017 \div 0.0029$
$\dfrac{1}{1000}$	$0.17 \div 0.029$
$\dfrac{1}{100}$	$1.7 \div 0.29$
$\dfrac{1}{10}$	$17 \div 2.9$
1	$170 \div 29$
10	$1700 \div 290$
100	$17,000 \div 2900$
$\vdots$	$\vdots$
1 million	(170 million) $\div$ (29 million)
$\vdots$	$\vdots$

For example, suppose that you want to calculate

$$0.7\overline{)2.53}$$

but instead you calculate

$$\begin{array}{r} 36.14 \\ 7\overline{)253} \end{array}$$

How should you move the decimal point in 36.14 to get the correct answer to 2.53 ÷ 0.7? By shifting the decimal point in 36.14, you can get any of the following numbers:

$$\ldots 0.03614,\ 0.3614,\ 3.614,\ 36.14,\ 361.4,\ 3614,\ \ldots$$

Which number in this list is a reasonable answer to 2.53 ÷ 0.7? You can reason that 2.53 is close to 3 and 0.7 is close to 1, so the answer to 2.53 ÷ 0.7 should be close to 3 ÷ 1 = 3. Therefore, the correct answer should be 3.614.

Dividing Numbers in the Millions, Billions, and Trillions

Earlier, we explained why we can shift the decimal point in both the divisor and the dividend the same number of places to obtain an equivalent division problem. To divide decimals, we shift the decimal points to the right, thereby turning the decimal division problem into a whole number division problem. But we can also shift the decimal points to the left. By shifting decimal points to the left, we can replace a division problem involving large numbers with an equivalent division problem that involves smaller numbers. For example, to calculate

$$(170 \text{ million}) \div (29 \text{ million}) = 170,000,000 \div 29,000,000$$

we can shift the decimal points (which are not shown, but are to the right of the final 0s in each number) 6 places to the left. We thereby obtain the equivalent problem

$$170 \div 29$$

which doesn't involve so many zeros. In this way, we can make division problems involving numbers in the millions, billions, and trillions easier to solve.

What do we do if a problem mixes trillions, billions, and millions? For example, how much money will each person get if \$1.3 billion is divided equally among 8 million people? If we express 1.3 billion in terms of millions, then both numbers will be in terms of millions. Since 1 billion is 1000 million,

$$1.3 \text{ billion} = 1.3 \times 1000 \text{ million} = 1300 \text{ million}$$

Therefore,

$$(1.3 \text{ billion}) \div (8 \text{ million}) = (1300 \text{ million}) \div (8 \text{ million}) = 1300 \div 8$$

Since 1300 ÷ 8 = 162.5, each person will get \$162.50.

Practice Exercises for Section 6.6

1. Theresa needs to cut a piece of wood 0.33 inches thick, or just a little less thick. Theresa's ruler shows subdivisions of $\frac{1}{32}$ of an inch. How many $\frac{1}{32}$ of an inch thick should Theresa cut her piece of wood? What type of division problem is this?

2. Describe a quick way to mentally calculate 0.11 ÷ 0.125 by thinking in terms of fractions.

3. Write a "how many groups?" story problem for 0.35 ÷ 1.45.

4. Write a "how many in one group?" story problem for 0.35 ÷ 1.45.

5. Without calculating the answers, give two different explanations for why the two division problems

$$0.65\overline{)4.3} \quad \text{and} \quad 65\overline{)430}$$

must have the same answer.

6. Show how to calculate 2.3 ÷ 0.008 without a calculator.

7. Explain how to calculate

$$(\text{4 billion}) \div (\text{2 million})$$

mentally.

8. Use the idea of multiplying and dividing by the same number to explain why the problems

$$170{,}000{,}000 \div 29{,}000{,}000$$

and

$$170 \div 29$$

are equivalent.

9. Use Figure 6.25 to explain why the problems

$$170{,}000{,}000 \div 29{,}000{,}000$$

and

$$170 \div 29$$

are equivalent.

Answers to Practice Exercises for Section 6.6

1. $0.33 \div \frac{1}{32} = 10.56$, so Theresa should cut her piece of wood $\frac{10}{32}$ of an inch thick. This uses the "how many groups?" interpretation of division, exact division (although in the end we round down because Theresa needs to cut her piece of wood a little less thick than the actual answer). Each $\frac{1}{32}$ of an inch represents one group. We want to know how many of these groups of $\frac{1}{32}$ of an inch are in 0.33 inches.

2. Observe that $0.125 = \frac{1}{8}$. To divide by a fraction, we can "invert and multiply." So, to divide by $\frac{1}{8}$, we multiply by 8. Therefore, $0.11 \div 0.125 = 0.11 \div \frac{1}{8} = 0.11 \times 8 = 0.88$.

3. "If apples cost \$1.45 per pound, then how many pounds of apples can you buy for \$0.35?" is a "how many groups?" story problem for $0.35 \div 1.45$, because you want to know how many groups of \$1.45 are in \$0.35.

4. "If you paid \$0.35 for 1.45 liters of water, then how much does 1 liter of water cost?" is a "how many in one group?" story problem for $0.35 \div 1.45$, because \$0.35 is distributed equally among 1.45 liters and you want to know how many dollars are "in" 1 liter. Or you could notice that if you paid \$6 for 2 liters of water, then 1 liter would cost $\$6 \div 2$, and this uses the "how many in each group?" view of division. So the same must be true when we replace \$6 with \$0.35 and 2 liters with 1.45 liters.

5. Explanation 1: When we shift the decimal points in the numbers 0.65 and 4.3 two places to the right, we have really just multiplied and divided the problem $4.3 \div 0.65$ by 100, thereby arriving at an equivalent division problem. In equations,

$$
\begin{array}{cc}
\times 100 & \div 100 \\
\downarrow & \downarrow
\end{array}
$$
$$4.3 \div 0.65 = (4.3 \times 100) \div (0.65 \times 100)$$
$$= 430 \div 65$$

or

$$
4.3 \div 0.65 = \frac{4.3}{0.65} = \frac{4.3}{0.65} \times \frac{100}{100}
$$
$$
= \frac{4.3 \times 100}{0.65 \times 100}
$$
$$
= \frac{430}{65} = 430 \div 65
$$

Explanation 2: Thinking in terms of money, we can interpret $4.3 \div 0.65 = ?$ as asking, "How many groups of \$0.65 are in \$4.30?" If we phrase the question in terms of cents instead of dollars, it becomes "How many groups of 65 cents are in 430 cents?", which we can solve by calculating $430 \div 65$. The answer must be the same either way we ask the question. Therefore, $4.3 \div 0.65$ and $430 \div 65$ are equivalent problems.

We can give a third explanation by using a diagram like Figure 6.25, in which we ask how many groups of 6 ten-sticks and 5 small squares are in 4 hundred-squares and 3 ten-sticks. By interpreting the small square as $\frac{1}{100}$, we find that the diagram represents $4.3 \div 0.65$. By interpreting the small square as 1, we see that the diagram represents $430 \div 65$. The answer must be the same either way we interpret it, so once again, we conclude that $4.3 \div 0.65$ and $430 \div 65$ are equivalent problems.

6. To calculate $2.3 \div 0.008$, we can move both decimal points 3 places to the right and calculate $2300 \div 8$ instead:

$$
\begin{array}{r}
287.5 \\
8\overline{)2300.} \\
-16 \\
\hline
70 \\
-64 \\
\hline
60 \\
-56. \\
\hline
40 \\
-40 \\
\hline
0
\end{array}
$$

7. Since 1 billion is 1000 million, 4 billion is 4000 million. So,

$$(4 \text{ billion}) \div (2 \text{ million}) =$$

$$(4000 \text{ million}) \div (2 \text{ million}) = 4000 \div 2 = 2000$$

Mentally, we just think that 4000 million divided by 2 million is the same as 4000 divided by 2.

8. If we divide the numbers 170 million and 29 million in the division problem $170{,}000{,}000 \div 29{,}000{,}000$ by 1 million, then we will have divided and multiplied by 1 million to obtain the new problem $170 \div 29$. Since we divided and multiplied by the same number, the new problem and the old problem are equivalent. In other words, they have the same answer. In equation form,

$$(170 \text{ million}) \div (29 \text{ million})$$
$$\times 1 \text{ million} \quad \div 1 \text{ million}$$
$$\downarrow \qquad\qquad \downarrow$$
$$= (170 \times 1 \text{ million}) \div (29 \times 1 \text{ million})$$
$$= 170 \div 29 = 5.86$$

In fraction form,

$$(170 \text{ million}) \div (29 \text{ million}) = \frac{170 \times 1 \text{ million}}{29 \times 1 \text{ million}}$$

$$= \frac{170}{29} \times \frac{1 \text{ million}}{1 \text{ million}} = \frac{170}{29}$$

$$= 170 \div 29$$

9. As shown in Table 6.9, if we let the small square in Figure 6.25 represent 1 million, then Figure 6.25 represents

$$170{,}000{,}000 \div 29{,}000{,}000$$

But if we let the small square represent 1, then Figure 6.25 represents

$$170 \div 29$$

We must get the same answer either way we interpret the diagram; therefore, the two division problems $170{,}000{,}000 \div 29{,}000{,}000$ and $170 \div 29$ are equivalent.

Problems for Section 6.6

1. Sue needs to cut a piece of wood 0.4 of an inch thick, or just a little less thick. Sue's ruler shows sixteenths of an inch. How many sixteenths of an inch thick should Sue cut her piece of wood? What type of division problem is this?

2. a. Write a "how many groups?" story problem for $5.6 \div 1.83$.

    b. Write a "how many in one group?" story problem for $5.6 \div 1.83$.

    c. Write a story problem (any type) for $0.75 \div 2.4$.

3.   **a.** Calculate $28.3 \div 0.07$ to the hundredths place without a calculator. Show your work.

    b. Describe the standard procedure for determining where to put the decimal point in the answer to $28.3 \div 0.07$.

    c. Explain in two different ways why the placement of the decimal point that you described in part (b) is valid.

4. a. Calculate 16.8 ÷ 0.35 to the hundredths place without a calculator. Show your work.

b. Describe the standard procedure for determining where to put the decimal point in the answer to 16.8 ÷ 0.35.

c. Explain in two different ways why the placement of the decimal point that you described in part (b) is valid.

5. Ramin must calculate 8.42 ÷ 3.6 longhand, but he can't remember what to do about decimal points. Instead, Ramin solves the division problem 842 ÷ 36 longhand and gets the answer 23.38. Ramin knows that he must shift the decimal point in 23.38 somehow to get the correct answer to 8.42 ÷ 3.6. Explain how Ramin could reason about the sizes of the numbers to determine where to put the decimal point.

6. a. Draw a diagram like Figure 6.25, and use your diagram to help you explain why the division problems 2.15 ÷ 0.36 and 215 ÷ 36 are equivalent.

b. Use a diagram to help you explain why the division problems $\frac{7}{8} ÷ \frac{3}{8}$ and 7 ÷ 3 are equivalent.

c. Discuss how parts (a) and (b) are related.

7. A federal debt problem: If the federal debt is $6.8 billion, and if this debt were divided equally among 290 million people, then how much would each person owe? Describe how to estimate mentally the answer to the federal debt problem, and explain briefly why your strategy makes sense.

8. A tax cut problem: If a 1.3 trillion dollar tax cut were divided equally among 290 million people over a 10-year period, then how much would each person get each year? Assume that you have only a very simple calculator that cannot use scientific notation and that displays at most 8 digits. Describe how to use such a calculator to solve the tax cut problem, and explain why your solution method is valid.

9. Light travels at about 300,000 kilometers per second. If Pluto is 6 billion kilometers away from us, then how long does light from Pluto take to reach us? Explain why you can calculate the way you do.

10. A new star was discovered. The star is about 75 trillion kilometers away from us. Light travels at about 300,000 kilometers per second. How long does light from the new star take to reach us? Explain why you can calculate the way you do.

11. Light travels at a speed of about 300,000 kilometers per second. The distance that light travels in one year is called a *light year*. The star Alpha Centauri is 4.34 light years from earth. How many years would it take a rocket traveling at 6000 kilometers per hour to reach Alpha Centauri? Solve this problem, and explain how you use the meanings of multiplication and division in solving it.

12. Susan has a 5-pound bag of flour and an old recipe of her grandmother's calling for 1 kilogram of flour. She reads on the bag of flour that it weighs 2.26 kilograms. She also reads on the bag of flour that 1 serving of flour is about $\frac{1}{4}$ cup and that there are about 78 servings in the bag of flour.

a. Based on the information given, how many cups of flour should Susan use in her grandmother's recipe? Solve this problem, and explain how you use the meanings of multiplication and division in solving it.

b. How can Susan measure this amount of flour as precisely as possible if she has the following measuring containers available: 1 cup, $\frac{1}{2}$ cup, $\frac{1}{3}$ cup, $\frac{1}{4}$ cup measures, 1 tablespoon? Remember that 1 cup = 16 tablespoons.

13. In ordinary language, the term *divide* means "partition and make smaller," as in "Divide and conquer."

a. In mathematics, does dividing always make smaller? In other words, if you start with a number N and divide it by another number M, is the resulting quotient $N ÷ M$ necessarily less than N? Explain briefly.

b. For which positive numbers, M, is 10 ÷ M less than 10? Determine all such positive numbers M. Use the meaning of division (either interpretation) to explain why your answer is correct.

14. When Mary converted a recipe from metric measurements to U.S. customary measurements, she discovered that she needed 8.63 cups of flour. Mary has a 1-cup measure, a $\frac{1}{2}$-cup measure, a $\frac{1}{4}$-cup measure, and a measuring tablespoon, which is $\frac{1}{16}$ cup. How should Mary use her measuring implements to measure the 8.63 cups of flour as accurately and efficiently as possible? Explain your reasoning.

15. Suppose you need to know how many thirty-secondths ($\frac{1}{32}$) of an inch 0.685 inch is (rounded to the nearest thirty-secondth of an inch). Explain why the following method is a legitimate way to solve this problem:

- Calculate $0.685 \times 32 = 21.92$, and round the result to the nearest whole number, namely, 22.

- Use your result from the previous step, 22, to form the fraction $\frac{22}{32}$. Then 0.685 inch is $\frac{22}{32}$ inch rounded to the nearest thirty-secondth of an inch.

Don't just verify that this method gives the right answer; explain why it works.

16. Sarah is building a carefully crafted cabinet and calculates that she must cut a certain piece of wood 33.33 inches long. Sarah has a standard tape measure that shows subdivisions of one-sixteenth ($\frac{1}{16}$) of an inch. How should Sarah measure 33.33 inches with her tape measure, using the closest sixteenth of an inch? (How many whole inches and how many sixteenths of an inch?) Explain your reasoning.

Chapter Summary and Study Items

Section 6.1 Interpretations

There are two interpretations of division: the "how many groups?" interpretation and the "how many in each group?" interpretation. With the "how many groups?" interpretation of division, $A \div B$ means the number of Bs that are in A—that is, the number of groups when A objects are divided into groups with B objects in each group. With the "how many in each group?" interpretation of division, $A \div B$ means the number of objects in each group when A objects are divided equally among B groups. Every division problem can be reformulated as a multiplication problem. With the "how many groups?" interpretation,

$$A \div B = ? \text{ is equivalent to } ? \times B = A$$

With the "how many in each group?" interpretation,

$$A \div B = ? \text{ is equivalent to } B \times ? = A$$

Key skills and understandings:

- Write and recognize whole number division story problems for both interpretations of division.

- Explain why we can't divide by 0, but why we *can* divide 0 by a nonzero number.

- Use division to solve problems.

Section 6.2 Division and Fractions and Division with Remainder

Division and fractions are related by $A \div B = \frac{A}{B}$. Given a whole number division problem, the exact quotient might not be a whole number. In this case, the quotient can be expressed as a mixed number or fraction or as a decimal, or it can be expressed as a whole number with a remainder.

Key skills and understandings:

- Explain why it makes sense that the result of dividing $A \div B$ is $\frac{A}{B}$.

- Write and recognize whole number division story problems that are best answered exactly, with either a decimal or mixed number, or answered with a whole number and a remainder.

- In story problems, interpret quotients and remainders appropriately, and recognize the distinction between solving a numerical division problem that is related to a story problem and solving the story problem.

- Describe how the whole-number-with-remainder answer to a division problem is related to the mixed number answer, and explain why this relationship holds.

Section 6.3 Why the Common Long Division Algorithm Works

We can solve whole number division problems in a primitive way by repeatedly subtracting the divisor or repeatedly adding the divisor. The scaffold method is a variation on the standard division algorithm that works with entire numbers instead of just portions of them. The scaffold method allows for flexibility in calculation. We can explain why the scaffold method is valid by using the distributive property to "collect up" multiples of the divisor that were subtracted during the division process. We can interpret each step in the standard division algorithm by viewing numbers as representing objects bundled according to place value. During common long division, we repeatedly unbundle amounts and combine them with the amount in the next-lower place.

Key skills and understandings:

- Use the scaffold method of division, and interpret the process from either the "how many groups?" or the "how many in each group?" point of view.

- Use the standard long division algorithm, and interpret the process from the "how many in each group?" point of view. Explain the "bringing down" steps in terms of unbundling the remaining amount and combining it with the amount in the next-lower place.

- Understand and use nonstandard methods of division.

- Use the standard division algorithm to write fractions as decimals and to give decimal answers to division problems.

Section 6.4 Fraction Division from the "How Many Groups?" Perspective

We can interpret fraction division from the "how many groups?" (as well as from the "how many in one group?") viewpoint. To divide fractions, we simply "invert and multiply"; in other words, we multiply by the reciprocal. We can explain why this procedure for fraction division works by writing a fraction division problem as an equivalent multiplication problem.

Key skills and understandings:

- Write and recognize fraction division story problems for the "how many groups?" interpretation of division.

- Solve fraction division story problems with the aid of pictures and tables as well as numerically. Know how to interpret pictures appropriately.

- Explain why the "invert and multiply" procedure is a valid method of dividing fractions.

Section 6.5 Fraction Division from the "How Many in One Group?" Perspective

We can interpret fraction division from the "how many in one group?" (as well as the "how many groups?") viewpoint. The "invert and multiply" or "multiply by the reciprocal" procedure is especially nicely explained from this perspective.

Key skills and understandings:

- Write and recognize fraction division story problems for the "how many in one group?" as well as from the "how many groups?" interpretations.

- Solve division problems with the aid of pictures, tables, and double number lines as well as numerically. Know how to interpret pictures appropriately.

- Use the "how many in one group?" interpretation of division to explain why the "invert and multiply" procedure for dividing fractions is valid.

Section 6.6 Dividing Decimals

We can interpret decimal division from the "how many groups?" as well as from the "how many in one group?" viewpoint. To divide decimals, we move the decimal points in both the divisor and dividend the same number of places so that the divisor becomes a whole number; then we proceed as with whole number division. There are several ways to explain this process of shifting the decimal points in decimal division.

Key skills and understandings:

- Write decimal division story problems for both interpretations of division.

- Explain in several different ways why we move the decimals points the way we do when we divide decimals, and calculate decimal divisions (without a calculator).

- Use estimation to determine the location of the decimal point in a decimal division problem.

- Explain why many division problems are equivalent, such as $6000 \div 2000$, $600 \div 200$, $60 \div 20$, $6 \div 2$, $0.6 \div 0.2$, and so on.

Combining Multiplication and Division: Proportional Reasoning

Many situations in daily life—as well as in science, medicine, and business—require the use of ratios, rates, and proportions. For example, in cooking; if we increase or decrease a recipe, we usually keep the ratios of the various ingredients the same. Any time we use percents, we are using a ratio. When pharmacists mix drugs, they pay careful attention to the ratios of the ingredients. When a car company lists the miles per gallon that a car gets, it is using a rate, when a store lists the price per pound of some food, it is using a rate. When you increase or decrease the quantities in a recipe, you use a proportion. The situations to which ratios, rates, and proportions apply are virtually limitless.

When it comes to solving problems that involve ratios, you are probably familiar and comfortable with a certain routine: Set up a proportion, cross-multiply, and solve the resulting equation—you know the drill! In this chapter we will delve more deeply into ratios, rates, and proportions to develop a more comprehensive sense of their meaning. We'll analyze when ratios apply and when they don't apply. Perhaps most importantly, we'll see that reasoning about ratios, rates, and proportions is really just reasoning about multiplication and division in combination. So to understand ratios, rates, and proportions, we must understand what multiplication and division mean and what kinds of problems these operations solve. At the same time, our work with ratios, rates, and proportions can be a way to strengthen and deepen our understanding of multiplication and division.

7.1 The Meanings of Ratio, Rate, and Proportion

Focal Points
Grades 6, 7

What are ratios, rates, and proportions? In this section, we'll define what these terms mean. In the same way that there are two ways to interpret the meaning of division ("how many groups?" and "how many in each group?"), there are two related ways to interpret the meaning of ratio. Sometimes different amounts of quantities are mixed in the same ratio. This type of situation gives rise to the idea of equivalent ratios and the notion of proportion.

Ratios and Rates

Ratios arise naturally when we make mixtures of substances. Before you read on, please do the next Class Activity.

> **Class Activity**   *Now Turn to Class Activities Manual*
>
> **7A** Comparing Mixtures, p. 138

The Class Activity asked you to compare different mixtures of red punch and lemon-lime soda. When we mix red punch and lemon-lime soda, we get a new drink, but the specific flavor of this new drink depends on the ratio of red punch to lemon-lime soda. Similarly, when we mix yellow paint and blue paint we get green paint, but the specific shade of green paint that we get depends on the ratio of yellow paint to blue paint. But what exactly is a ratio?

ratio ("how many groups?" definition) A **ratio** describes a specific type of relationship between two quantities. If the ratio of flour to milk in a muffin recipe is 7 to 2, then it means that for every 7 cups of flour we use, we must use 2 cups of milk, as shown in Figure 7.1. Equivalently, for every 2 cups of milk we use, we must use 7 cups of flour. In general, to say that two quantities are in a *ratio* of *A* to *B* means that for every *A* units of the first quantity there are *B* units of the second quantity. Equivalently, for every *B* units of the second quantity there are *A* units of the first quantity. With this perspective, we are thinking of quantities in a ratio of *A* to *B* as made of a number of groups or batches, where each group or batch consists of *A* units of the first quantity and *B* units of the second quantity.

There is another slightly different way to think about ratios. Consider again the muffin recipe in which the ratio of flour to milk is 7 to 2. We interpreted this ratio as meaning that for every 7 cups of flour we use, we must use 2 cups of milk. Here is another way to interpret that ratio. If we divide the flour into 7 parts of equal volume, we must use 2 parts of that same volume of milk, as shown in Figure 7.2. For

FIGURE 7.1

The ratio 7 to 2 viewed as groups of 7 and groups of 2

7 cups of flour,   2 cups of milk

14 cups of flour,   4 cups of milk

21 cups of flour,   6 cups of milk

28 cups of flour,   8 cups of milk

FIGURE 7.2

The ratio 7 to 2 viewed as 7 groups and 2 groups

7 groups and 2 groups

example, if we use 7 pints of flour, we should use 2 pints of milk; if we use 7 pails of flour, we should use 2 pails of milk. No matter what the size of the container we use, if we use 7 containers worth of flour, we should use 2 containers worth of milk. In general, with this point of view, to say that two quantities are in a **ratio** of *A* to *B* means that if we divide the first quantity into *A* equal parts, then the second quantity consists of *B* parts of the same size. Notice that in the case of the muffin recipe, "parts of the same size" meant that the parts were of equal *volume*. In other contexts, "parts of the same size" could mean that the parts are of equal weight or equal length, for example. With this perspective, we are thinking of quantities in a ratio of *A* to *B* as made of *A* parts of the first quantity and *B* parts of the second quantity.

ratio ("how many in each ratio group?" definition)

The two different points of view about ratios correspond to the two different points of view about division. The first view of ratio corresponds to the "how many groups?" view of division, because we view the quantities as consisting of multiple groups of the same size (such as groups of 7 cups of flour and groups of 2 cups of milk). The second view of ratio corresponds to the "how many in each group?" view of division, because we put the quantities into a fixed number of groups of the same size (such as 7 groups of flour and 2 groups of milk, all groups being of the same size).

rate

A **rate** is a ratio between two quantities that are measured in different units. For example, speed, which is the ratio of distance traveled to elapsed time, is a rate because distance is measured in miles, kilometers, or other similar units, whereas time is measured in hours, minutes, or seconds. If you are traveling at a speed of 60 mph, that means that you are traveling 60 miles for every hour that you travel. When we work with rates, we use the first point of view about ratios. **Ratio notation** can take several forms:

ratio notation

- We can use words, as in the ratio of flour to milk is 7 to 2.
- We can use a colon, as in the ratio of flour to milk is 7:2.
- We can use a fraction, as in the ratio of flour to milk is 7/2 (or $\frac{7}{2}$).

Later in this chapter, we will see why it makes sense to write ratios as fractions.

Equivalent Ratios, Proportions, and Ratio Tables

Suppose that to make 1 batch of muffins we need 7 cups of flour and 2 cups of milk. If we make several batches of muffins, the ratio of total flour used to total milk used is still 7 to 2 (because for every 7 cups of flour we use 2 cups of milk). On the other hand, the actual amounts of flour and milk that are needed describe a different ratio, such as 14 to 4 (if we make 2 batches). Thus, the ratios 7 to 2 and 14 to 4 are **equivalent**; in other words, they are the same ratio. The statement that two ratios are equivalent is a **proportion**. So the equation 7:2 = 14:4 is an example of a proportion.

equivalent ratios
proportion

An easy way to begin to work with ratios and proportions is to use a ratio table. A ratio table is just a table that lists equivalent ratios. If we make several batches of muffins or half a batch of muffins, we will need the amounts of flour and milk shown in Table 7.1. How do we determine the entries in this table?

TABLE 7.1 A ratio table for batches of muffins

# Batches	1	2	3	4	5	$\frac{1}{2}$	N
# Cups flour	7	14	21	28	35	$3\frac{1}{2}$	$N \cdot 7$
# Cups milk	2	4	6	8	10	1	$N \cdot 2$

TABLE 7.2 Finding entries in a ratio table by repeatedly adding

# Batches	1		2		3		4		5
# Cups flour	7	$\xrightarrow{+7}$	14	$\xrightarrow{+7}$	21	$\xrightarrow{+7}$	28	$\xrightarrow{+7}$	35
# Cups milk	2	$\xrightarrow{+2}$	4	$\xrightarrow{+2}$	6	$\xrightarrow{+2}$	8	$\xrightarrow{+2}$	10

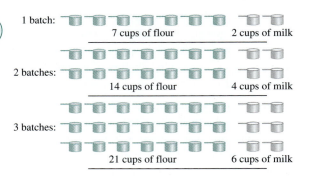

FIGURE 7.3

Different combinations of flour and milk that are in the ratio 7 to 2

1 batch: 7 cups of flour   2 cups of milk

2 batches: 14 cups of flour   4 cups of milk

3 batches: 21 cups of flour   6 cups of milk

At first, we might start filling in the table by simply adding 7 cups of flour and 2 cups of milk every time we make an additional batch, as indicated in Table 7.2 and Figure 7.3.

But we can use simple reasoning about multiplication to determine the entries in Table 7.1. For example, to make 4 batches of muffins, we will need 4 times as much flour and 4 times as much milk as in 1 batch. To make $\frac{1}{2}$ of a batch of muffins, we will need $\frac{1}{2}$ as much flour and $\frac{1}{2}$ as much milk, so we'll need $\frac{1}{2} \cdot 7 = 3\frac{1}{2}$ cups of flour and $\frac{1}{2} \cdot 2 = 1$ cup of milk. In general, to make N batches of muffins, we will need N times as much flour and milk as in 1 batch, so we will need $N \cdot 7$ cups of flour and $N \cdot 2$ cups of milk. This reasoning is summarized in Table 7.3.

In N batches of muffins the ratio of flour to milk is still 7 to 2 because for every 7 cups of flour, we use 2 cups of milk, no matter what N is (as long as it's positive). But on the other hand, if we make N batches of muffins, we will use $N \cdot 7$ cups of flour and $N \cdot 2$ cups of milk, so we can also describe the ratio of flour to milk as $N \cdot 7$ to $N \cdot 2$. Thus, the two ratios are equivalent, so that we have the following proportion:

$$7:2 = N \cdot 7:N \cdot 2$$

In general, the preceding reasoning tells us this:

The ratio A to B (where A and B are positive) is equivalent to the ratio $N \cdot A$ to $N \cdot B$, so that

$$A:B = N \cdot A:N \cdot B$$

is a proportion (if N is positive).

Notice that we made essentially this same statement about equivalent fractions in Chapter 2.

TABLE 7.3 Finding entries in a ratio table by multiplying

# Batches	1		4	1		$\frac{1}{2}$	1		N
# Cups flour	7	$\xrightarrow{\times 4}$	28	7	$\xrightarrow{\times \frac{1}{2}}$	$3\frac{1}{2}$	7	$\xrightarrow{\times N}$	$N \cdot 7$
# Cups milk	2	$\xrightarrow{\times 4}$	8	2	$\xrightarrow{\times \frac{1}{2}}$	1	2	$\xrightarrow{\times N}$	$N \cdot 2$

FIGURE 7.4

Orange paint created by mixing yellow paint and red paint in a ratio of 3 to 1

TABLE 7.4 Relating amounts of yellow and red paint when the ratio of yellow paint to red paint is 3 to 1

# Containers red	1	N	1	$\frac{1}{3} \cdot M$
	$\downarrow \times 3$	$\downarrow \times 3$	$\uparrow \times \frac{1}{3}$	$\uparrow \times \frac{1}{3}$
# Containers yellow	3	$3 \cdot N$	3	M

The previous paragraph tells us that we can create an equivalent ratio by multiplying the entries in a ratio by the same positive number N, which does not need to be a whole number—it can be a fraction or a decimal.

There is another useful way to create equivalent ratios. Let's consider a shade of orange paint created by mixing red paint with yellow paint in a ratio of 1 to 3, as shown in Figure 7.4. If we use 1 container of red paint, then we must use 3 containers of yellow paint to make this shade of orange (where all containers are the same size). This means that however much red paint we use, we must use 3 times as much yellow paint, and however much yellow paint we use, we must use $\frac{1}{3}$ as much red paint to create equivalent ratios of red to yellow paint. This reasoning is summarized in Table 7.4.

Now if we want to make the orange paint by using $2\frac{1}{2}$ gallons of red paint, then we should use

$$3 \cdot 2\frac{1}{2} = 7\frac{1}{2}$$

gallons of yellow paint, because we need 3 times as much yellow paint as red paint. If we want to make the orange paint by using 2 gallons of yellow paint, then we should use

$$\frac{1}{3} \cdot 2 = \frac{2}{3}$$

gallons of red paint, because we need $\frac{1}{3}$ as much red paint as yellow paint. This reasoning is summarized in Table 7.5.

Notice that we could use the reasoning we used for Table 7.3 in this case as well, as indicated in Table 7.6. In the next section of this chapter, we will apply the reasoning about ratios and multiplication that we have been studying to solve proportions in several different ways.

Class Activity   *Now Turn to Class Activities Manual*

7B Using Strip Diagrams, Ratio Tables, and Double Number Lines to Describe Equivalent Ratios, p. 139

7C Using Ratio Tables to Compare Two Ratios, p. 140

TABLE 7.5 Finding amounts of yellow or red paint when the ratio of yellow paint to red paint is 3 to 1

# Gallons red	1	$2\frac{1}{2}$		1	$\frac{2}{3}$
	$\downarrow \times 3$	$\downarrow \times 3$		$\uparrow \times \frac{1}{3}$	$\uparrow \times \frac{1}{3}$
# Gallons yellow	3	$7\frac{1}{2}$		3	2

TABLE 7.6 Finding amounts of yellow or red paint when the ratio of yellow paint to red paint is 3 to 1

# Gallons red	1	$\xrightarrow{\times 2\frac{1}{2}}$	$2\frac{1}{2}$	1	$\xrightarrow{\times \frac{2}{3}}$	$\frac{2}{3}$
# Gallons yellow	3	$\xrightarrow{\times 2\frac{1}{2}}$	$7\frac{1}{2}$	3	$\xrightarrow{\times \frac{2}{3}}$	2

Practice Exercises for Section 7.1

1. Describe in two different ways what it means to say that liquid soap and rose water are mixed in a ratio of 5 to 2.

2. Lavender oil and bergamot can be mixed in a ratio of 2 to 3. Give three different examples of quantities of lavender oil and bergamot that are mixed in the same ratio. Explain why your examples really are in the same ratio.

3. Give an example of how to use a double number line to show equivalent rates.

4. Which of the following two mixtures will be more lemony?

   • 2 tablespoons of lemon juice mixed in 3 cups of water

   • 5 tablespoons of lemon juice mixed in 7 cups of water

   Use ratio tables to solve this problem. Explain your reasoning.

Answers to Practice Exercises for Section 7.1

1. One way: The mixture is made of 5 parts liquid soap and 2 parts rose water, where all the parts are the same size, but can be any size. Visually, we can show this description with a strip diagram, as in Figure 7.5. We can think of each part in the strip diagram as "filled with" any (common) amount such as 5 drops, or $\frac{1}{2}$ cup, or 3 gallons, or any other amount.

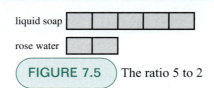

liquid soap

rose water

FIGURE 7.5 The ratio 5 to 2

Another way: For every 5 teaspoons (say) of liquid soap that are in the mixture, there are 2 teaspoons of rose water in the mixture. (A ratio table nicely displays this way of thinking about the ratio.) With this point of view, we think of the mixture as made of a certain number of batches, where 1 batch is made of 5 teaspoons liquid soap and 2 teaspoons rose water.

2. The strip diagram in Figure 7.6 can help picture this ratio. If each part is filled with 5 drops, then there will be 10 drops lavender oil and 15 drops bergamot. If each part is filled with $\frac{1}{2}$ cup, then there will be 1 cup lavender oil and $1\frac{1}{2}$ cups bergamot. If each part is filled with $\frac{1}{3}$ of a cup, then there will be $\frac{2}{3}$ of a cup of lavender oil and 1 cup bergamot. All of these pairs are in the same ratio because all apply to the same strip diagram that defines the ratio of 2 to 3. In each example there are 2 parts lavender oil and 3 parts bergamot, where all parts in that example are the same size.

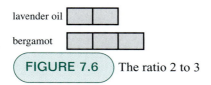

FIGURE 7.6 The ratio 2 to 3

3. Figure 7.7 shows a double number line with equivalent rates of price per volume (of juice, for example).

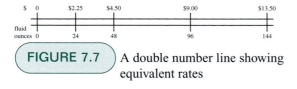

FIGURE 7.7 A double number line showing equivalent rates

4.

Mixture 1								
tbs Lemon juice	2	4	6	8	10	12	14	16	18
Cups water	3	6	9	12	15	18	21	24	27

Mixture 2								
tbs Lemon juice	5	10	15	20	25	30	35	40	45
Cups water	7	14	21	28	35	42	49	56	63

The mixtures within a ratio table all taste the same because they are just multiple batches of the first mixture. So we can compare any pair in the mixture 1 table with any pair in the mixture 2 table to compare the original two mixtures. In the first ratio table, 10 tablespoons of lemon juice requires 15 cups of water, whereas in the second mixture, the same number of tablespoons of lemon juice requires only 14 cups of water. Therefore, the second mixture is more lemony because it has less water for the same amount of lemon than the first mixture does. You can also compare the two mixtures when they have the same amount of water.

Problems for Section 7.1

1. Draw a strip diagram and use it to explain in your own words what it means to say that phosphate and nitrogen are mixed in a ratio of 2 to 7 in a fertilizer.

2. Use a ratio table to explain in your own words what it means to say that phosphate and nitrogen are mixed in a ratio of 2 to 7 in a fertilizer.

3. John is driving at a constant speed of 10 miles every 12 minutes. Draw a double number line and use it to show at least three other equivalent rates.

Include at least one example that involves one or more decimals, fractions, or mixed numbers. Explain briefly why the rates are equivalent.

4. Use ratio tables to compare the two paint mixtures in Class Activity 7A, part 2, in *two ways*. Explain your reasoning clearly in your own words.

5. Without using a calculator, explain how you can use reasoning and mental arithmetic to determine which of the following two laundry detergents is a

better buy (i.e., has a lower rate of dollars per load washed):

- a box of laundry detergent that washes 80 loads and costs $12.75

- a box of laundry detergent that washes 36 loads and costs $6.75

6. You can make grape juice by mixing 1 can of frozen grape juice concentrate with 3 cans of water. Use simple reasoning with multiplication and division to find at least four other ways of mixing grape juice concentrate with water so that the result will be in the same ratio and therefore taste the same. At least two of your ways should involve numbers that are not whole numbers.

7. You can make a pink paint by mixing $2\frac{1}{2}$ cups white paint with $1\frac{3}{4}$ cups red paint. Use simple reasoning with multiplication and divison to find at least four other ways of mixing white paint and red paint in the same ratio to make the same shade of pink paint.

8. Allie, Benton, and Cathy are planning to mix red and yellow paint. They are considering which of the two following paint mixtures will make a more yellow paint:

- a mixture of 3 parts red to 5 parts yellow

- a mixture of 4 parts red to 6 parts yellow

Allie says that both paints will look the same because to make the second mixture you just add 1 part of each color to the first mixture. Because you add the same amount of each color, the second mixture should look the same as the first mixture. Benton says that the second mixture should be more yellow than the first because it uses more yellow than the first mixture. Cathy says that both paints should look the same because each uses 2 parts more yellow than red.

a. Discuss the students' ideas. Is their reasoning valid or not?

b. Which paint will be more yellow, and why? Solve this problem in two different ways, explaining in detail why you can solve the problem the way you do.

7.2 Solving Proportion Problems by Reasoning with Multiplication and Division

F·P Focal Points
Grade 6

When you think of solving proportions, you may think of the method in which you set two fractions equal to each other, cross-multiply these fractions, and then solve the resulting equation. However, we can also solve proportions by using multiplication, division, and simple logical reasoning. When students solve proportion problems this way, they have an opportunity to think more deeply about the operations of multiplication and division and when to apply these operations. Perhaps this is why the mathematics textbooks used in Singapore at 6th grade (see [78], volume 6A) and the textbooks used in Japan at 6th grade (see [36], volume 6A) do not teach the "cross-multiplication" method; instead they have children solve proportion problems by reasoning about multiplication and division.

In this section we will study three different methods for reasoning about multiplication and division to solve ratio and rate problems. The first method involves finding the amount in one part first, then using that value to find the unknown amount. The second method involves comparing like quantities, and the third method involves comparing unlike quantities. We'll also see that sometimes the comparisons we would like to make are more easily accomplished by breaking them into several

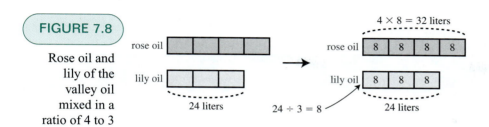

FIGURE 7.8

Rose oil and lily of the valley oil mixed in a ratio of 4 to 3

steps. Especially useful is the method of "going through 1," in which we first find a unit rate to solve the problem.

The Method of Finding the Amount in One Part

The first method is relevant when we think of the quantities in a ratio as consisting of certain numbers of parts. It's especially useful to apply when we can use a strip diagram to represent the ratio. To apply this way of reasoning, we find the value of one part and then use that value to find the unknown amount. Let's see how it works in an example.

Problem: At a perfume factory, the workers want to mix rose oil and lily of the valley oil in a ratio of 4 to 3. How much rose oil will the workers need to mix with 24 liters of lily of the valley oil?

If we think of the perfume mixture as consisting of 4 parts rose oil and 3 parts lily of the valley oil, as pictured in the strip diagram in Figure 7.8, then the 3 parts of lily of the valley oil must consist of 24 liters. Since all parts are the same size, each part must contain $24 \div 3 = 8$ liters. The rose oil consists of 4 parts, so the workers should use $4 \times 8 = 32$ liters of rose oil.

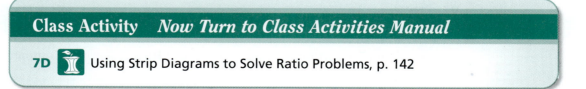

Class Activity *Now Turn to Class Activities Manual*

7D   Using Strip Diagrams to Solve Ratio Problems, p. 142

The Methods of Comparing Like Quantities and Comparing Unlike Quantities

Let's think about the methods of comparing like quantities and comparing unlike quantities in the context of a problem.

> *Problem:* Mr. Arias used 4 gallons of gas to drive 100 miles. Assuming that Mr. Arias continues to get the same gas mileage (in other words, assuming that the rate of miles driven to gas used continues to remain the same), how far will Mr. Arias be able to drive with 20 gallons of gas?

This problem involves two amounts of gas and it involves two distances. Each distance is associated with a certain amount of gas that is needed to drive that distance. In the method of *comparing like quantities,* we'll compare gas to gas and distance to distance. Think of these comparisons as "external" comparisons because we're comparing across different driving scenarios. On the other hand, in the method of *comparing unlike quantities,* we'll compare gas to distance within one driving scenario and apply that comparison to another driving scenario. Think of these comparisons as "internal" comparisons because we're comparing quantities that are linked to each other within one scenario.

TABLE 7.7 Solving a proportion by using reasoning about multiplication and division and comparing like quantities

	Problem		1st Step		2nd Step	
Miles	100	?	100	?	$100 \xrightarrow{\times 5} 500$	
Gallons	4	20	$4 \xrightarrow{\times 5} 20$		$4 \xrightarrow{\times 5} 20$	
	$\dfrac{100}{4} = \dfrac{?}{20}$		$\dfrac{100}{4} \underset{\times 5}{=} \dfrac{?}{20}$		$\dfrac{100}{4} \overset{\times 5}{\underset{\times 5}{=}} \dfrac{500}{20}$	

Comparing Like Quantities Since 20 gallons of gas is 5 times as much as 4 gallons of gas, Mr. Arias will be able to drive 5 times as far, so he will be able to drive $5 \times 100 = 500$ miles. This line of reasoning is summarized in Table 7.7 (in both tabular and fraction form) and shown pictorially with the double number line in Figure 7.9.

Comparing Unlike Quantities If we think of dividing the 100 miles equally among the 4 gallons of gas, we see that each gallon of gas will take Mr. Arias 25 miles. So the number of miles that Mr. Arias can drive is always 25 times the number of gallons of gas he has. Since Mr. Arias has 20 gallons of gas, he can drive $25 \times 20 = 500$ miles. This line of reasoning is summarized in two different ways in Table 7.8 (in both tabular and fraction forms).

Class Activity   *Now Turn to Class Activities Manual*

7E Solving Proportions by Reasoning about How Quantities Compare, p. 143

"Going Through 1" to Make Numerical Calculations Easier

So far in this section, the numbers we've used have been easy to work with. But the three methods we've studied for reasoning about multiplication and division to solve ratio and rate problems can be used with any numbers—even in "messy" cases. When more difficult comparisons are involved, it helps to break the calculations into steps by "going through 1."

FIGURE 7.9

Solving a proportion by using reasoning about multiplication and division on a double number line and comparing like quantities

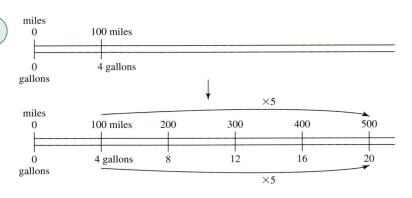

TABLE 7.8 Solving a proportion by using reasoning about multiplication and division and comparing unlike quantities

	Problem		1st Step		2nd Step	
Miles	100	?	100	?	100	500
			↑×25		↑×25	↑×25
Gallons	4	20	4	20	4	20
	$\frac{100}{4} = \frac{?}{20}$		×25 $\left(\frac{100}{4} = \frac{?}{20}\right.$		×25 $\left(\frac{100}{4} = \frac{500}{20}\right)$×25	

Class Activity   *Now Turn to Class Activities Manual*

7F "Going Through 1", p. 144

Let's consider a problem to see how to "go through 1."

> *Problem:* If we want to use 4 cups of flour in a muffin recipe in which the ratio of flour to milk is 7 to 2, how much milk should we use?

The ratio of flour to milk is 7 to 2, so we know 7 cups of flour require 2 cups of milk. We want to use 4 cups of flour, but it seems hard to get from 7 cups to 4 cups. Instead, let's go through 1 cup first, so we first ask how much milk we would need for 1 cup of flour. Once we find out, we'll multiply that amount by 4 to learn how much milk we'll need for 4 cups of flour. If 7 cups of flour takes 2 cups of milk, then we can think of the 2 cups of milk as divided equally among the 7 cups of flour. Therefore, each cup of flour takes

$$2 \div 7 = \frac{2}{7}$$

of a cup of milk. Now if we want to use 4 cups of flour, then we need to use 4 groups of $\frac{2}{7}$ cup of milk. In other words, we need to use

$$4 \cdot \frac{2}{7} = \frac{8}{7} = 1\frac{1}{7}$$

cups of milk. Table 7.9 summarizes this reasoning.

TABLE 7.9 Solving a proportion by using reasoning about multiplication and division and "going through 1"

	Problem		Solution				
Cups flour	7	4	7	$\xrightarrow{\div 7}$	1	$\xrightarrow{\times 4}$	4
Cups milk	2	?	2	$\xrightarrow{\div 7}$	$\frac{2}{7}$	$\xrightarrow{\times 4}$	$1\frac{1}{7}$

TABLE 7.10 Solving a proportion by using reasoning about multiplication and division

	Easier Problem		Solution		
Cups flour	7	21	7	$\xrightarrow{\times 3}$	21
Cups milk	2	?	2	$\xrightarrow{\times 3}$	6
	Our Problem		Solution		
Cups flour	7	4	7	$\xrightarrow{\times \frac{4}{7}}$	4
Cups milk	2	?	2	$\xrightarrow{\times \frac{4}{7}}$	$1\frac{1}{7}$

Notice that in solving this muffin recipe problem by "going through 1" we actually solved a "how many in one group?" division problem first, and then we multiplied. In general, we can view solving proportions as combining division and multiplication.

Instead of going through 1 cup of flour, we actually could have gone straight from 7 cups to 4 cups, as indicated in Table 7.10. In the next section we'll see that when we "go through 1" we are actually calculating a unit rate, and that unit rates connect ratios with fractions.

Class Activity   *Now Turn to Class Activities Manual*

7G More Ratio Problem Solving, p. 145

Practice Exercises for Section 7.2

1. A soda mixture can be made by mixing cola and lemon-lime soda in a ratio of 4 to 3. Explain how to solve each of the following problems about the soda mixture by using logical reasoning about multiplication and division in two ways—with the aid of a ratio table and with the aid of a strip diagram:

   a. How much lemon-lime soda should you use to make the soda mixture with 48 cups of cola? How much soda mixture will this make?

   b. How much cola and how much lemon-lime soda should you use to make 105 cups of the soda mixture?

2. Traveling at a constant speed, a race car goes 15 miles every 6 minutes. Use simple, logical reasoning supported by a double number line to help you determine the answers to the next questions.

   a. How far does the race car go in 15 minutes?

   b. How long does it take the race car to go 20 miles?

3. The ratio of Quint's CDs to Chris's CDs was 7 to 3. After Quint gave 6 CDs to Chris, they had an equal number of CDs. How many CDs did Chris have at first? Explain how to solve this problem with the aid of a strip diagram.

4. Cali mixed $3\frac{1}{2}$ cups of red paint with $4\frac{1}{2}$ cups of yellow paint to make an orange paint. How many cups of red paint and how many cups of yellow paint will Cali need to make 12 cups of the same shade of orange paint? Solve this problem by using only simple reasoning about multiplication or division (or both). Explain your reasoning.

5. To make a punch you mixed $\frac{1}{4}$ cup grape juice concentrate with $1\frac{1}{2}$ cups bubbly water. If you want to make the same punch with 2 cups of bubbly water, then how many cups of grape juice concentrate

should you use? Solve this problem by using only simple reasoning about multiplication or division (or both). Explain your reasoning.

6. In order to reconstitute a medicine properly, a pharmacist must mix 10 milliliters (mL) of a liquid medicine for every 12 mL of water. If one dose must contain $2\frac{1}{2}$ mL of medicine, then how many milliliters of medicine/water mixture provides one dose of the medicine? Solve this problem by using only simple reasoning about multiplication or division (or both). Explain your reasoning.

Answers to Practice Exercises for Section 7.2

1. **a.** The ratio of cola to lemon-lime soda is 4 to 3, so for every 4 cups of cola, you must use 3 cups of lemon-lime soda. To make the mixture with 48 cups cola, you'll need 12 groups of 4 cups of cola, since $48 \div 4 = 12$. So you'll also need 12 groups of 3 cups of lemon-lime soda, which is $12 \times 3 = 36$ cups of lemon-lime soda. All together, this will make $12 \times (4 + 3) = 84$ cups of soda mixture. Table 7.11a summarizes this reasoning.

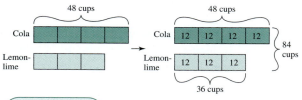

FIGURE 7.11   Solving a proportion by using reasoning about multiplication and division and a strip diagram

TABLE 7.11a Solving a proportion by using reasoning about multiplication and division and a ratio table

Cups cola	4	$\xrightarrow{\times 12}$ 48	4	$\xrightarrow{\times 12}$ 48
Cups lemon-lime	3	?	3	$\xrightarrow{\times 12}$ 36
Total cups	7	?	7	$\xrightarrow{\times 12}$ 84

Another way to think about the problem is to view the soda mixture as made of 4 parts cola and 3 parts lemon-lime soda, as shown in the strip diagrams in Figure 7.11. Since you want the 4 parts of cola to be 48 cups, each part must be $48 \div 4 = 12$ cups. Therefore, the 3 parts lemon-lime soda are $3 \times 12 = 36$ cups. The full mixture consists of 7 parts, which is therefore $7 \times 12 = 84$ cups.

b. Once again, for every 4 cups of cola, you must use 3 cups of lemon-lime soda. Combined, 4 cups of cola and 3 cups of lemon-lime soda make 7 cups of soda mixture. To make 105 cups of soda mixture, you must use 15 groups of 7 cups of mixture since $105 \div 7 = 15$. Therefore, you must also use 15 groups of 4 cups of cola, or $15 \times 4 = 60$ cups of cola, and 15 groups of 3 cups of lemon-lime soda, or $15 \times 3 = 45$ cups of lemon-lime soda. Table 7.11b summarizes this reasoning.

TABLE 7.11b Solving a proportion by using reasoning about multiplication and division and a ratio table

Cups cola	4	?	4	$\xrightarrow{\times 15}$ 60
Cups lemon-lime	3	?	3	$\xrightarrow{\times 15}$ 45
Total cups	7	$\xrightarrow{\times 15}$ 105	7	$\xrightarrow{\times 15}$ 105

Another way to think about the problem is to view the soda mixture as made of a total of 7 parts, of which 4 parts are cola and 3 parts are lemon-lime soda, as indicated in the strip diagrams in Figure 7.12. Since you want the 7 parts of soda mixture to be 105 cups, each part must be 105 ÷ 7 = 15 cups. Therefore, the 4 parts of cola are 4 × 15 = 60 cups, and the 3 parts of lemon-lime soda are 3 × 15 = 45 cups.

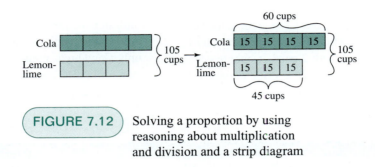

FIGURE 7.12   Solving a proportion by using reasoning about multiplication and division and a strip diagram

2. **a.** Since the race car goes 15 miles every 6 minutes, it goes 30 miles in 12 minutes and 45 miles in 18 minutes, as shown in the double number line in Figure 7.13. Fifteen minutes is halfway between 12 and 18 minutes, so the race car must go halfway between 30 and 45 miles in 15 minutes. Since 45 − 30 = 15 and 15 ÷ 2 = 7.5, the race car goes 30 + 7.5 = 37.5 miles in 15 minutes.

b. Since the race car goes 15 miles every 6 minutes, it goes 30 miles in 12 minutes, and 45 miles in 18 minutes, as shown in the double number line in Figure 7.14. Twenty miles is $\frac{1}{3}$ of the way between 15 miles and 30 miles. So it must take the race car the time that is $\frac{1}{3}$ of the way between 6 and 12 minutes to go 20 miles. Since 12 − 6 = 6 and 6 ÷ 3 = 2, one-third of the way between 6 and 12 minutes is 6 + 2 = 8 minutes.

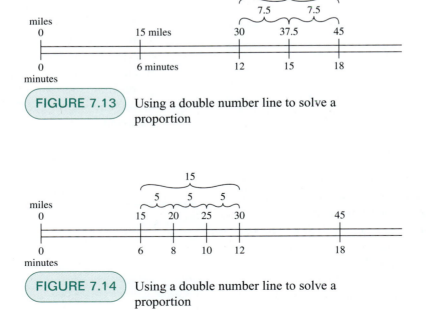

FIGURE 7.13   Using a double number line to solve a proportion

FIGURE 7.14   Using a double number line to solve a proportion

3. Since the ratio of Quint's CDs to Chris's CDs is 7 to 3, Quint's CDs can be divided into 7 parts and Chris's CDs can be divided into 3 parts, where all parts are the same size, as shown in the strip diagram in Figure 7.15. Since the boys have the same number of CDs in the end, and since Quint has 4 more parts than Chris at first, two of Quint's parts must go to Chris. Those two parts are 6 CDs, so each part consists of 3 CDs. Therefore, Quint had $7 \times 3 = 21$ CDs to start with, and Chris had $3 \times 3 = 9$ CDs to start with.

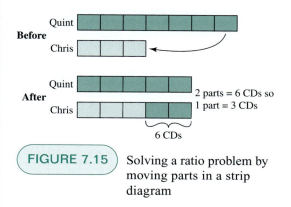

FIGURE 7.15 Solving a ratio problem by moving parts in a strip diagram

4. The $3\frac{1}{2}$ cups of red paint and $4\frac{1}{2}$ cups of yellow paint combine to make $3\frac{1}{2} + 4\frac{1}{2} = 8$ cups orange paint. Since 12 cups is $1\frac{1}{2}$ times as much as 8 cups, Cali will need $1\frac{1}{2}$ times as much red paint and yellow paint. Therefore, Cali will need

$$1\frac{1}{2} \cdot 3\frac{1}{2} = \frac{3}{2} \cdot \frac{7}{2} = \frac{21}{4} = 5\frac{1}{4}$$

cups of red paint and

$$1\frac{1}{2} \cdot 4\frac{1}{2} = \frac{3}{2} \cdot \frac{9}{2} = \frac{27}{4} = 6\frac{3}{4}$$

cups of yellow paint. This reasoning is summarized in Table 7.12.

TABLE 7.12 Solving a proportion by using reasoning about multiplication and division and a ratio table

Cups red	$3\frac{1}{2}$	?	$3\frac{1}{2} \xrightarrow{\times 1\frac{1}{2}} 5\frac{1}{4}$
Cups yellow	$4\frac{1}{2}$	?	$4\frac{1}{2} \xrightarrow{\times 1\frac{1}{2}} 6\frac{3}{4}$
Total cups orange	$8 \xrightarrow{\times 1\frac{1}{2}} 12$		$8 \xrightarrow{\times 1\frac{1}{2}} 12$

5. You can reason that $1\frac{1}{2}$ cups is 6 times as much as $\frac{1}{4}$ cup, so you use 6 times as much bubbly water as grape juice concentrate. To find out how much concentrate to use for 2 cups bubbly water ask, "Six times what is equal to 2 cups?" The answer is $2 \div 6 = \frac{2}{6} = \frac{1}{3}$ cup, so you should use $\frac{1}{3}$ of a cup of grape juice concentrate for 2 cups bubbly water. This reasoning is summarized on the left in Table 7.13.

Another way to reason is to "go through 1" cup of bubbly water by first going through a whole number of cups of bubbly water. If you use twice as much bubbly water, which is 3 cups, you should also use twice as much concentrate, which is $\frac{1}{2}$ of a cup. So, if you wanted to use 1 cup of bubbly water, which is $\frac{1}{3}$ as much as 3 cups, then you should use $\frac{1}{3}$ of $\frac{1}{2}$ cup of grape juice concentrate, namely $\frac{1}{6}$ of a cup. Then to use 2 cups of bubbly water, which is twice as much as 1 cup, you should use twice as much as $\frac{1}{6}$ of a cup of grape juice concentrate, namely, $\frac{2}{6} = \frac{1}{3}$ cups. This reasoning is summarized on the right in Table 7.13.

TABLE 7.13 Using reasoning about multiplication and division to solve a proportion

Comparing Unlike Quantities				Comparing Like Quantities and "Going Through 1"			
Cups bubbly water $\quad 1\frac{1}{2} \quad 2 \quad 1\frac{1}{2} \quad 2$				Cups bubbly water $\quad 1\frac{1}{2} \xrightarrow{\times 2} 3 \xrightarrow{\div 3} 1 \xrightarrow{\times 2} 2$			
$\uparrow \times 6 \qquad \uparrow \times 6 \; \uparrow \times 6$							
Cups concentrate $\quad \frac{1}{4} \quad ? \quad \frac{1}{4} \quad \frac{1}{3}$				Cups concentrate $\quad \frac{1}{4} \xrightarrow{\times 2} \frac{1}{2} \xrightarrow{\div 3} \frac{1}{6} \xrightarrow{\times 2} \frac{1}{3}$			

6. *Comparing Like Quantities:* Since $2\frac{1}{2}$ mL is $\frac{1}{4}$ of 10 mL, the pharmacist will also need $\frac{1}{4}$ as much water, which is $\frac{1}{4} \cdot 12 = 3$ mL of water. The medicine and water combined makes $2\frac{1}{2} + 3 = 5\frac{1}{2}$ mL of medicine/water mixture. This reasoning is summarized on the left in Table 7.14.

Comparing Like Quantities by "Going through 1": To make the mixture with 1 mL of medicine, which is $\frac{1}{10}$ as much as 10 mL of medicine, the pharmacist will also need $\frac{1}{10}$ as much water, namely, $\frac{1}{10} \cdot 12 = \frac{6}{5} = 1\frac{1}{5}$ mL of water, thus making $1 + 1\frac{1}{5} = 2\frac{1}{5}$ mL of medicine/water mixture. Then to make the mixture with $2\frac{1}{2}$ mL of medicine, which is $2\frac{1}{2}$ times as much as 1 mL, the pharmacist must use $2\frac{1}{2}$ times as much water, namely, $2\frac{1}{2} \cdot 1\frac{1}{5} = \frac{5}{2} \cdot \frac{6}{5} = 3$ mL of water, thus making $5\frac{1}{2}$ mL of medicine/water mixture. This reasoning is summarized below in Table 7.14.

TABLE 7.14 Solving a proportion by using reasoning about multiplication and division and a ratio table

Comparing Like Quantities				Comparing Like Quantities by "Going Through 1" First				
mL Medicine	10	$\xrightarrow{\div 4}$	$2\frac{1}{2}$	mL Medicine	10	$\xrightarrow{\div 10}$ 1	$\xrightarrow{\times 2\frac{1}{2}}$	$2\frac{1}{2}$
mL Water	12	$\xrightarrow{\div 4}$	3	mL Water	12	$\xrightarrow{\div 10}$ $1\frac{1}{5}$	$\xrightarrow{\times 2\frac{1}{2}}$	3
mL Total	22	$\xrightarrow{\div 4}$	$5\frac{1}{2}$	mL Total	22	$\xrightarrow{\div 10}$ $2\frac{1}{5}$	$\xrightarrow{\times 2\frac{1}{2}}$	$5\frac{1}{2}$

Problems for Section 7.2

1. Walking at a constant speed, a person walks $\frac{3}{4}$ of a mile every 12 minutes. Use simple, logical reasoning supported by a double number line to help you determine the answers to the next questions.

   a. How far does the person walk in 30 minutes?

   b. How long does it take the person to walk $2\frac{1}{2}$ miles?

2. In a terrarium, the ratio of grasshoppers to crickets is $6:5$. There are 48 grasshoppers. How many crickets are there? Explain how to solve this problem in two ways: with the aid of a strip diagram and with the aid of a ratio table. In each case, be sure to discuss the reasoning involved in the method; in other words, explain why the method makes sense.

3. At a zoo, the ratio of king penguins to emperor penguins is $2:3$. In all, there are 45 king and emperor penguins combined. How many emperor penguins are at the zoo? Explain how to solve this problem in two ways: with the aid of a strip diagram and with the aid of a ratio table. In each case, be sure to discuss the reasoning involved in the method; in other words, explain why the method makes sense.

4. On a farm, the ratio of grey goats to white goats is $2:5$. There are 100 grey goats. How many white goats are there? Explain how to solve this problem in two ways: with the aid of a strip diagram and with the aid of a ratio table. In each case, be sure to discuss the reasoning involved in the method; in other words, explain why the method makes sense.

5. ⟨icon⟩ An orange-lemon juice mixture can be made by mixing orange juice and lemonade in a ratio of 5 to 2. Explain how to solve each of the following problems about the juice mixture by using logical reasoning about multiplication and division in two ways—with the aid of a ratio table and with the aid of a strip diagram:

   a. How much orange juice and how much lemonade should you use to make the juice mixture with 75 cups of orange juice? How much juice mixture will this make?

   b. How much orange juice and how much lemonade should you use to make 140 cups of the juice mixture?

   c. How much orange juice and how much lemonade should you use to make the juice mixture with 15 cups of lemonade? How much juice mixture will this make?

   d. How much orange juice and how much lemonade should you use to make 10 cups of the juice mixture?

6. (This problem refers to part 5 (a).) Create two new problems for your students by changing the ratio 5 to 2 and the number, 75, of cups of juice mixture. One problem should be about the same level of difficulty as part 5 (a), and the other problem should be harder. Say which problem is which and why. Explain how to solve each problem in two different ways.

7. Brad made some punch by mixing $\frac{1}{2}$ of a cup of grape juice with $\frac{1}{4}$ cup of sparkling water. Brad really likes his punch, so he decides to make a larger batch of it by using the same ratio.

   a. Brad wants to make 6 cups of his punch. How much grape juice and how much sparkling water should Brad use? Explain how to use reasoning about quantities to solve this problem in two different ways.

   b. Now Brad wants to make 4 cups of his punch. How much grape juice and how much sparkling water should Brad use? Explain how to use reasoning about quantities to solve this problem in two different ways.

8. To make grape juice by using frozen juice concentrate, you must mix the frozen juice concentrate with water in a ratio of 1 to 3. How much frozen juice concentrate and how much water should you use to make $1\frac{1}{2}$ cups of grape juice? Solve this problem in two different ways, explaining your reasoning in each case.

9. You can make a soap bubble mixture by combining 2 tablespoons water with 1 tablespoon liquid dishwashing soap and 4 drops of corn syrup. Using the same ratios, how much liquid dishwashing soap and how many drops of corn syrup should you use if you want to make a soap bubble mixture with 5 tablespoons of water? Solve this problem using only simple reasoning about multiplication or division (or both). Explain your reasoning.

10. You can make concrete by mixing 1 part cement with 2 parts pea gravel and 3 parts sand. Using the same ratios, how much cement and how much pea gravel should you use if you want to make concrete with 8 cubic feet of sand? Solve this problem using only simple reasoning about multiplication or division (or both). Explain your reasoning.

11. a. John was paid $250 for $3\frac{3}{4}$ hours of work. At that rate, how much should John make for $2\frac{1}{2}$ hours of work? Explain how to solve this problem by reasoning about quantities.

    b. John was paid $250 for $3\frac{3}{4}$ hours of work. At that rate, how long should John be willing to work for $100? Explain how to solve this problem by reasoning about quantities.

12. The ratio of Frank's marbles to Huang's marbles is 3 to 2. After Frank gives $\frac{1}{2}$ of his marbles to another friend, Frank has 30 fewer marbles than Huang. How many marbles does Huang have?

    a. Explain how to solve the problem with the aid of a strip diagram.

    b. Create an easier problem for your students by changing the ratio, 3 to 2, to a different ratio and by changing the number of marbles, 30, to a different number of marbles. Make sure the problem has a sensible answer. Explain how to solve the problem.

    c. Create a problem of about the same level of difficulty as the original problem by changing the ratio, 3 to 2, to a different ratio and by changing the number of marbles, 30, to a different number of marbles. Make sure the problem has a sensible answer. Explain how to solve the problem.

13. Asia and Taryn each had the same amount of money. After Asia spent $14 and Taryn spent $22, the ratio of Asia's money to Taryn's money was 4 to 3. How much money did each girl have at first?

   a. Explain how to solve the problem with the aid of a strip diagram.

   b. Create a harder problem for your students by changing the ratio, 4 to 3, to a different ratio and by changing the dollar amounts, $14 and $22, to different dollar amounts. Explain how to solve the problem.

   c. Create a problem of about the same level of difficulty as the original problem by changing the ratio, 4 to 3, to a different ratio and by changing the dollar amounts, $14 and $22, to different dollar amounts. Explain how to solve the problem.

14. An aquarium contained an equal number of horseshoe crabs and sea stars. After 15 horseshoe crabs were removed and 27 sea stars were removed, the ratio of horseshoe crabs to sea stars was 5 : 3. How many horseshoe crabs and sea stars were there at first? Solve this problem and explain your reasoning.

15. Marge made light blue paint by mixing $2\frac{1}{2}$ cups of blue paint with $1\frac{3}{4}$ cups of white paint. Homer poured another cup of white paint into Marge's paint mixture. How many cups of blue paint should Marge add to bring the paint back to its original shade of light blue (mixed in the same ratio as before)? Use the most elementary reasoning you can to solve this problem. Explain your reasoning.

16. A batch of lotion was made at a factory by mixing 1.3 liters of ingredient A with 2.7 liters of ingredient B in a mixing vat. By accident, a worker added an extra 0.5 liters of ingredient A to the mixing vat. How many liters of ingredient B should the worker add to the mixing vat so that the ingredients will be in the original ratio? Use the most elementary reasoning you can to solve this problem. Explain your reasoning.

17. If a $\frac{3}{4}$ cup serving of snack food gives you 60% of your daily value of calcium, then what percent of your daily value of calcium is in $\frac{1}{2}$ cup of the snack food? Solve the problem with the aid of a picture or a table. Explain your reasoning.

18. If $6000 is 75% of a company's budget for a project, then what percent of the budget is $10,000?

Solve the problem with the aid of a picture or a table. Explain your reasoning.

19. Amy mixed 2 tablespoons of chocolate syrup in $\frac{3}{4}$ cup of milk to make chocolate milk. To make chocolate milk that is mixed in the same ratio and therefore tastes the same as Amy's, how much chocolate syrup will you need for 1 gallon of milk? Express your answer by using appropriate units. Explain your reasoning. Recall that 1 gallon = 4 quarts, 1 quart = 2 pints, 1 pint = 2 cups, 1 cup = 8 fluid ounces, 1 fluid ounce = 2 tablespoons.

20. A 5-gallon bucket filled with water is being pulled from the ground up to a height of 20 feet at a rate of 2 feet every 15 seconds. The bucket has a hole in it, so that water leaks out of the bucket at a rate of 1 quart ($\frac{1}{4}$ gallon) every 3 minutes. How much of the water will be left in the bucket by the time the bucket gets to the top? Solve this problem by using logical thinking and by using the most elementary reasoning you can. Explain your reasoning clearly.

21. Use the meanings of multiplication and division to solve the following problems:

   a. Suppose you drive 4500 miles every half year in your car. At the end of $3\frac{3}{4}$ years, how many miles will you have driven?

   b. Mo used 128 ounces of liquid laundry detergent in $6\frac{1}{2}$ weeks. If Mo continues to use laundry detergent at this rate, how much will he use in a year?

   c. Suppose you have a 32-ounce bottle of weed killer concentrate. The directions say to mix $2\frac{1}{2}$ ounces of weed killer concentrate with enough water to make a gallon. How many gallons of weed killer will you be able to make from this bottle?

22. Buttercup the gerbil drank $\frac{2}{3}$ of a bottle of water in $1\frac{1}{2}$ days. Assuming Buttercup continues to drink water at the same rate, how many bottles of water will Buttercup drink in 5 days? Use multiplication and division to solve this problem. Explain in detail why you can use multiplication when you do and why you can use division when you do.

23. If you used $2\frac{1}{2}$ truck loads of mulch for a garden that covers $\frac{3}{4}$ acre, then how many truck loads of mulch should you order for a garden that covers $3\frac{1}{2}$ acres? (Assume that you will spread the mulch

at the same rate as before.) Use multiplication and division to solve this problem. Explain in detail why you can use multiplication when you do and why you can use division when you do.

24. If $2\frac{1}{2}$ pints of jelly filled $3\frac{1}{2}$ jars, then how many jars will you need for 12 pints of jelly? Will the last jar of jelly be completely full? If not, how full will it be? (Assume that all jars are the same size.) Use multiplication and division to solve this problem.

Explain in detail why you can use multiplication when you do and why you can use division when you do.

25. A standard bathtub is approximately $4\frac{1}{2}$ ft long, 2 ft wide, and 1 ft deep. If water comes out of a faucet at the rate of $2\frac{1}{2}$ gallons each minute, how long will it take to fill the bathtub $\frac{3}{4}$ full? Use the fact that 1 gallon of water occupies 0.134 cubic feet.

7.3 Connecting Ratios and Fractions

 **Focal Points**
Grade 7

Ratios are often written as fractions, but we defined ratios and fractions in different ways. So why is it legitimate to equate ratios and fractions as is commonly done? And why is the common method of solving proportions (i.e., by setting two fractions equal to each other, then cross-multiplying and solving the resulting equation) a legitimate way to solve proportions? We will answer these questions in this section. The key connection between ratios and fractions is the notion of a unit rate, and we equate unit rates when we solve a proportion by setting two fractions equal to each other.

Unit Rates Connect Ratios and Fractions

Let's now see how every ratio of quantities is naturally associated with a fraction via a unit rate. Consider a bread recipe in which the ratio of flour to water is 14 to 5, so that if we use 14 cups of flour, we will need 5 cups of water. If we think of dividing those 14 cups of flour equally among the 5 cups of water, then, according to the meaning of division (the "how many in each group?" view), each cup of water goes with

$$14 \div 5 = \frac{14}{5}$$

cups of flour. So in this recipe, there are $\frac{14}{5}$ cups of flour *per cup of water*. In other words, there are $\frac{14}{5}$ cups of flour *for 1 cup of water*. In this way, the ratio 14 to 5 of flour to water is naturally associated with the fraction $\frac{14}{5}$, which tells us the number of cups of flour per cup of water, or the *unit rate* of cups of flour per cup of water.

unit rate   In general, given a ratio A to B relating two (nonzero) quantities, the fraction $\frac{A}{B}$ is the **unit rate** that tells us how many units there are of the first quantity for 1 unit of the second quantity. (Similarly, the fraction $\frac{B}{A}$ is the unit rate that tells us how many units there are of the second quantity for 1 unit of the first quantity.)

So when we write a fraction to stand for a ratio, we should really interpret that fraction as a unit rate. For example, if blue and yellow paint are mixed in a ratio of

2 to 3

and if the fraction

$$\frac{2}{3}$$

is written to stand for that ratio, then the fraction should really be interpreted as the ratio

$$\frac{2}{3} \text{ to } 1$$

which is equivalent to the original 2 to 3 ratio.

Class Activity   *Now Turn to Class Activities Manual*

7H Connecting Ratios, Fractions, and Division with Unit Rates, p. 146

The Logic Behind Solving Proportions by Cross-Multiplying Fractions

You are probably familiar with the technique of solving proportions by cross-multiplying. Why is this a valid technique for solving proportions? We will examine this now. Before you read on, please do the next Class Activity.

Class Activity   *Now Turn to Class Activities Manual*

7I Solving Proportions by Cross-Multiplying Fractions, p. 147

Why is the method of solving proportions by setting two fractions equal to each other and cross-multiplying valid? Consider a light-blue paint mixture made with $\frac{1}{4}$ cup blue paint and 4 cups white paint. How much blue paint will we need if we want to use 6 cups white paint and if we are using the same ratio of blue paint to white paint to make the same shade of light-blue paint? A common method for solving such a problem is to set up the following proportion in fraction form:

$$\frac{\frac{1}{4}}{4} = \frac{A}{6}$$

Here, A represents the as-yet-unknown amount of blue paint we will need for 6 cups of white paint. We then cross-multiply to get

$$6 \cdot \frac{1}{4} = 4 \cdot A$$

Therefore,

$$A = \left(6 \cdot \frac{1}{4} \right) \div 4 = 1\frac{1}{2} \div 4 = \frac{3}{8}$$

so that we must use $\frac{3}{8}$ cups of blue paint for 6 cups of white paint.

Let's analyze the preceding steps. First, when we set the two fractions

$$\frac{\frac{1}{4}}{4}$$

and

$$\frac{A}{6}$$

equal to each other, why can we do that, and what does it mean? If we think of the fractions as representing division—that is,

$$\frac{1}{4} \div 4$$

and

$$A \div 6$$

then each of these expressions stands for the number of cups of blue paint per cup of white paint. In other words, each fraction stands for the same unit rate. We want to use the same amount of blue paint per cup of white paint either way; therefore, the two fractions should be equal to each other, or

$$\frac{\frac{1}{4}}{4} = \frac{A}{6} \tag{7.1}$$

Next, why do we cross-multiply? We can cross-muliply because two fractions are equal exactly when their "cross-multiples" are equal. Recall that the method of cross-multiplying is really just a shortcut for giving fractions a common denominator. If we give the fractions in Equation 7.1 the common denominator $6 \cdot 4$, which is the product of the two denominators, then we can replace Equation 7.1 with the proportion

$$\frac{\frac{1}{4} \cdot 6}{4 \cdot 6} = \frac{A \cdot 4}{6 \cdot 4} \tag{7.2}$$

In terms of the paint mixture, both sides of this proportion now refer to 24 cups of paint, instead of 4 cups and 6 cups of paint, as in Equation 7.1. But two fractions that have the same denominator are equal exactly when their numerators are equal. Since the denominators of the fractions in Equation 7.2 are equal, since $4 \cdot 6 = 6 \cdot 4$, the proportion will be solved exactly when the numerators are equal—namely, when

$$\frac{1}{4} \cdot 6 = A \cdot 4 \tag{7.3}$$

Therefore, we can solve the proportion in Equation 7.1 by solving Equation 7.3, which was obtained by cross-multiplying.

Practice Exercises for Section 7.3

1. Jose mixed 3 cups of blue paint with 4 cups of red paint to make a purple paint. For each of the following fractions and division problems, interpret the fraction or the division problem in terms of Jose's paint mixture, and explain why your interpretation makes sense:

   $\frac{3}{7}$ or $3 \div 7$; $\frac{4}{7}$ or $4 \div 7$; $\frac{3}{4}$ or $3 \div 4$;

   $\frac{4}{3}$ or $4 \div 3$; $\frac{7}{3}$ or $7 \div 3$; $\frac{7}{4}$ or $7 \div 4$

2. Which of the following two mixtures will be more salty?

   • 3 tablespoons of salt mixed in 4 cups of water

   • 4 tablespoons of salt mixed in 5 cups of water

   Solve this problem in two different ways: by comparing fractions and by using a ratio table. Explain your reasoning in each case.

Answers to Practice Exercises for Section 7.3

1. Out of the total 7 cups of purple paint, 3 cups are blue, so $\frac{3}{7}$ of the paint mixture is blue. Thinking in terms of division, if we imagine the blue paint divided equally among the 7 cups of purple paint, then there is $3 \div 7 = \frac{3}{7}$ cups of blue paint in each cup of purple paint. Similarly, 4 out of 7 cups of the purple paint are red, so $\frac{4}{7}$ of the paint is red. Imagining the red paint divided equally among the 7 cups of purple paint, there are $4 \div 7 = \frac{4}{7}$ cups of red paint in each cup of purple paint.

Since there are 3 cups of blue paint and 4 cups of red paint in the mixture, if we think of dividing the blue paint equally among the 4 cups of red paint, then there are $3 \div 4 = \frac{3}{4}$ cups of blue paint for each cup of red paint. By the same logic, there are $4 \div 3 = \frac{4}{3} = 1\frac{1}{3}$ cups of red paint for each cup of blue paint.

If we think of dividing the 7 cups of purple paint equally among the 3 cups of blue paint, then there are $7 \div 3 = \frac{7}{3} = 2\frac{1}{3}$ cups of purple paint for each cup of blue paint. By the same logic, there are

$7 \div 4 = \frac{7}{4} = 1\frac{3}{4}$ cups of purple paint for each cup of red paint.

2. With fractions: If we think of the 3 tablespoons of salt in the first mixture as being divided equally among the 4 cups of water, then each cup of water contains $3 \div 4 = \frac{3}{4}$ tablespoons of salt. Similarly, each cup of water in the second mixture contains $4 \div 5 = \frac{4}{5}$ tablespoons of salt. Since $\frac{4}{5} = 0.8$ and $\frac{3}{4} = 0.75$, and since $0.8 > 0.75$, the second mixture contains more salt per cup of water. Thus, it is more salty.

With ratio table (table not shown): If we make 5 batches of the first mixture and 4 batches of the second mixture, then both will contain 20 cups of water. The first mixture will contain $5 \times 3 = 15$ tablespoons of salt, and the second mixture will contain $4 \times 4 = 16$ tablespoons of salt. Since both mixtures contain the same amount of water, but the second mixture contains 1 more tablespoon of salt than the first, the second mixture must be more salty.

Problems for Section 7.3

1. Pat mixed 2 cups of blue paint with 5 cups of yellow paint to make a green paint. For each of the following fractions and division problems, interpret the fraction or the division problem in terms of Pat's paint mixture, and explain why your interpretation makes sense. Use the definition of fraction from Chapter 2; do not simply refer to the fractions as ratios.

$\frac{2}{5}$ or $2 \div 5$; $\frac{5}{2}$ or $5 \div 2$; $\frac{2}{7}$ or $2 \div 7$;

$\frac{7}{2}$ or $7 \div 2$; $\frac{5}{7}$ or $5 \div 7$; $\frac{7}{5}$ or $7 \div 5$

2. a. Which of the following two mixtures will have a stronger lime flavor?

   • 2 parts lime juice concentrate mixed in 5 parts water

   • 4 parts lime juice concentrate mixed in 7 parts water

Solve this problem in two different ways: with ratio tables and by comparing fractions. Explain your reasoning in each case. In particular, be sure to explain what the factions mean in the context of the mixtures.

   b. A student might say that the second mixture has a stronger flavor than the first mixture because the numbers for the second mixture are greater (in other words, $4 > 2$ and $7 > 5$). Even if the conclusion is correct, is the student's reasoning valid? Explain why or why not.

3. a. Snail A moved 6 feet in 7 hours. Snail B moved 7 feet in 8 hours. Both snails moved at constant speeds. Which snail went faster? Solve this problem in two different ways, explaining in detail why you can solve the problem the way you do. In particular, if you use fractions in your

explanation, be sure to explain how the fractions are relevant.

b. A student might say that snail B moved faster than snail A because the numbers for snail B are greater (in other words, $7 > 6$ and $8 > 7$). Even if the conclusion is correct, is the student's reasoning valid? Explain why or why not.

4.  A dough recipe calls for 3 cups of flour and $1\frac{1}{4}$ cups of water. You want to use the same ratio of flour to water to make a dough with 10 cups of flour. How much water should you use?

a. Solve this problem by setting up a proportion in which you set two fractions equal to each other.

b. Interpret the two fractions that you set equal to each other in part (a) in terms of the recipe. Explain why it makes sense to set these two fractions equal to each other.

c. Why does it make sense to cross-multiply the two fractions in part (a)? What is the logic behind the procedure of cross-multiplying?

d. Now solve the problem of how much water to use for 10 cups of flour in a different way, by using the most elementary reasoning you can. Explain your reasoning clearly.

5. A recipe that serves 6 people calls for $1\frac{1}{2}$ cups of rice. How much rice will you need to serve 8 people (assuming that the ratio of people to cups of rice stays the same)?

a. Solve this problem by setting up a proportion in which you set two fractions equal to each other.

b. Interpret the two fractions that you set equal to each other in part (a) in terms of the recipe. Explain why it makes sense to set these two fractions equal to each other.

c. Why does it make sense to cross-multiply the two fractions in part (a)? What is the logic behind the procedure of cross-multiplying?

d. Now solve the same problem in a different way by using logical thinking and by using the most elementary reasoning you can. Explain your reasoning clearly.

7.4 When You Can Use a Proportion and When You Cannot

F·P **Focal Points**
Grade 7

Some problems may seem as if they should be solved with a proportion, but in fact they cannot be solved that way. In this brief section, we'll look at a few examples of such problems and note the distinction between problems that can and can't be solved with a proportion.

How can we tell if a proportion is appropriate to use? A proportion is a statement that two ratios are equal (equivalent) to each other. In terms of ratio tables, this means that the two pairs of quantities could both appear in the same ratio table. But think about how ratio tables work: If we double one quantity, we also have to double the other quantity. If we multiply one quantity by 3, then we also have to multiply the other quantity by 3. If we divide one quantity by 2, then we also have to divide the other quantity by 2, and so on. In some cases, we might work with quantities that are related, and we might make a table to show the relationship. However, not every table is a *ratio table* that shows a set of equivalent ratios.

Class Activity *Now Turn to Class Activities Manual*

7J Can You Always Use a Proportion?, p. 148

7K Who Says You Can't Do Rocket Science?, p. 150

Practice Exercises for Section 7.4

1. Suppose that a logging crew can cut down 5 acres of trees every 2 days. Assume that the crew works at a steady rate. Solve problems (a) and (b) using logical thinking and using the most elementary reasoning you can. Explain your reasoning clearly.

   a. How many days will it take the crew to cut 8 acres of trees? Give your answer as a mixed number.

   b. Now suppose there are *three* logging crews that all work at the same rate as the original one. How long will it take these 3 crews to cut down 10 acres of trees?

2. If 3 people take 2 days to paint 5 fences, how long will it take 2 people to paint 1 fence? (Assume that the fences are all the same size and the painters are all equally good workers who work at a steady rate.) Can this problem be solved by setting up the following proportion to find how long it will take 2 people to paint 5 fences?

$$\frac{3 \text{ people}}{2 \text{ days}} = \frac{2 \text{ people}}{x \text{ days}}$$

Solve the problem by thinking logically about the situation. Explain your reasoning clearly.

Answers to Practice Exercises for Section 7.4

1. a. Because the crew cuts 5 acres every 2 days, it will cut half as much in 1 day, namely, $2\frac{1}{2}$ acres. To determine how many days it will take to cut 8 acres, we must determine how many groups of $2\frac{1}{2}$ are in 8, which is $8 \div 2\frac{1}{2} = 3\frac{1}{5}$ days.

   b. In part (a) we saw that 1 crew cuts $2\frac{1}{2}$ acres per day. Therefore, 3 crews will cut 3 times as much per day, which is $3 \cdot 2\frac{1}{2} = 7\frac{1}{2}$ acres per day. To determine how many days it will take to cut 10 acres, we must figure out how many groups of $7\frac{1}{2}$ are in 10. This is solved by the division problem $10 \div 7\frac{1}{2}$. Because $10 \div 7\frac{1}{2} = 10 \div \frac{15}{2} = \frac{10}{1} \cdot \frac{2}{15} = 1\frac{1}{3}$, we conclude that it will take the 3 crews $1\frac{1}{3}$ days to cut 10 acres.

2. The proportion

$$\frac{3 \text{ people}}{2 \text{ days}} = \frac{2 \text{ people}}{x \text{ days}}$$

is not valid for this situation, because when more people are painting, it will take less time to paint a fence. It is not the case that for each group of 3 people there are, it will take 2 days to paint a fence. Therefore, the relationship between the number of people and the number of days it takes to paint a fence is not a ratio.

Thinking logically, if 3 people take 2 days to paint 5 fences, then those 3 people will take $2 \div 5 = \frac{2}{5}$ of a day to paint just 1 fence (dividing the 2 days equally among the 5 fences). If just 1 person were painting, it would take 3 times as long to paint the fence, namely, $3 \cdot \frac{2}{5} = \frac{6}{5}$ of a day (which is $1\frac{1}{5}$ days). With 2 people painting, it will take half as much time to paint the fence, namely, $\frac{3}{5}$ of a day.

Problems for Section 7.4

1. Suppose that you have two square garden plots: One is 10 feet by 10 feet and the other is 15 feet by 15 feet. You want to cover both gardens with a 1-inch layer of mulch. If the 10-by-10 garden took $3\frac{1}{2}$ bags of mulch, could you calculate how many bags of mulch you'd need for the 15-by-15 garden

by setting up the following proportion

$$\frac{3\frac{1}{2}}{10} = \frac{x}{15}$$

Explain clearly why or why not. If the answer is no, is there another proportion that you could set up? It may help you to draw pictures of the gardens.

2. If you can rent 5 DVDs for 5 nights for $5, then at that rate, how much should you expect to pay to rent 1 DVD for 1 night? Solve this problem by using logical thinking and the most elementary reasoning you can. Explain your reasoning clearly.

3. If a crew of 3 people takes $2\frac{1}{2}$ hours to clean a house, then how long should a crew of 2 take to clean the same house? Assume that all people in the cleaning crew work at the same steady rate. Solve this problem by using logical thinking and the most elementary reasoning you can. Explain your reasoning clearly.

4.  If 6 men take 3 days to dig 8 ditches, then how long would it take 4 men to dig 10 ditches? Assume that all the ditches are the same size and take equally long to dig, and that all the men work at the same steady rate. Solve this problem by using logical thinking and the most elementary reasoning you can. Explain your reasoning clearly.

5. A candy factory has a large vat into which workers pour chocolate and cream. Each ingredient flows into the vat from its own special hose, and each ingredient comes out of its hose at a constant rate. Workers at the factory know that it takes 20 minutes to fill the vat with chocolate from the chocolate hose, and it takes 15 minutes to fill the vat with cream from the cream hose. If workers pour both chocolate and cream into the vat at the same time (each coming full tilt out of its own hose), how long will it take to fill the vat? Before you find an exact answer to this problem, find an approximate answer, or find a range, such as "between ... and ... minutes." Explain your reasoning.

6. Jay and Mark run a lawn-mowing service. Mark's mower is twice as big as Jay's, so whenever they both mow, Mark mows twice as much as Jay in a given time period. When Jay and Mark are working together, it takes them 4 hours to cut the lawn of an estate. How long would it take Mark to mow the lawn by himself? How long would it take Jay to mow the lawn by himself? Explain your answers.

7. If liquid pouring at a steady rate from hose A takes 15 minutes to fill a vat, and liquid pouring at a steady rate from hose B takes 10 minutes to fill the same vat, then how long will it take for liquid pouring from both hose A and hose B to fill the vat? Solve this problem by using logical thinking and the most elementary reasoning you can. Explain your reasoning clearly.

8. Suppose that there are 400 pounds of freshly picked tomatoes and that 99% of their weight is water. After one day, the same tomatoes only weigh 200 pounds due to evaporation of water. (The tomatoes consist of water and solids. Only the water evaporates; the solids remain.)

   a. How many pounds of solids are present in the tomatoes? (Notice that this is the same when they are freshly picked as after one day.)

   b. Therefore, when the tomatoes weigh 200 pounds, what percent of the tomatoes is water?

   c. Is it valid to use the following proportion to solve for the percent of water, x, in the tomatoes when they weigh 200 pounds?

   $$\frac{0.99}{400} = \frac{x}{200}$$

   If not, why not?

7.5 Percent Revisited: Percent Increase and Decrease

F·P **Focal Points**
Grade 7

When a quantity increases or decreases, determining the amount of change is a simple matter of subtraction. However, in many situations, the actual value of the increase or decrease is less informative than the *percent* that this increase or decrease represents. For example, suppose that this year, there are 50 more children at Barrow Elementary School than there were last year. If Barrow Elementary had only 100 children last year, then 50 additional children is a very large increase. On the other hand, if Barrow Elementary had 500 children last year, an increase of 50 children is less significant. In this section, we will study increases and decreases in quantities as *percents* rather than as fixed values.

Class Activity   *Now Turn to Class Activities Manual*

7L How Should We Describe the Change?, p. 150

---

percent increase

If the value of a quantity goes up, then the increase in the quantity, figured as a percent of the original, is the **percent increase** of the quantity. If a piece of furniture costs $279 last week, and this week the same piece of furniture costs $319, then the price went up by $40. The percent increase of the cost of the furniture is the percent represented by this $40 increase over the original cost, $279. What percent of $279 is $40? Let $P\%$ be this percentage. Then

$$\frac{P}{100} = \frac{40}{279}$$

So, $P\%$ is about 14%; therefore, the price of the furniture increased by about 14%.

percent decrease

When a quantity decreases in value, there is the notion of a **percent decrease**. It is the decrease, figured as a percent of the original. If Lower Heeltoe got 215 inches of rainfall in 2009 and only 193 inches of rainfall in 2010, then Lower Heeltoe got 22 inches less rain in 2010 than in 2009. What percent of 215 inches is 22 inches? Let $P\%$ be this percentage. Then

$$\frac{P}{100} = \frac{22}{215}$$

So, $P\%$ is about 10%; therefore, the amount of rainfall in Lower Heeltoe decreased by about 10% from 2009 to 2010.

Two Methods for Calculating Percent Increase or Decrease

We can choose from two methods to find a percent increase or decrease when we know the original and final amounts.

Class Activity   *Now Turn to Class Activities Manual*

7M   Calculating Percent Increase and Decrease, p. 151

The First Method The first method for calculating a percent increase or percent decrease was the method we used in the previous examples. This method relies directly on the definitions of percent increase and percent decrease. In general, the method works as follows: Suppose a quantity changes from an amount *A*—the reference amount—to an amount *B*:

$$A \qquad \rightarrow \qquad B$$
$$\text{reference amount} \qquad \text{changed amount}$$

To calculate the percent increase or decrease in the quantity, we use the following procedure:

1. Calculate the *change*, *C*, in the quantity, namely, either $B - A$ or $A - B$, whichever is positive (or 0).

2. Calculate the percent that the change *C* is of the reference amount *A* (using any method from Section 2.6). The result is the percent increase or decrease of the quantity from *A* to *B*.

For example, if D.J.'s weight dropped from 250 pounds to 225 pounds, by what percent did D.J.'s weight decrease? D.J.'s weight decreased by 25 pounds (250 – 225), so we must find what percent 25 is of 250. Twenty-five is 10% of 250, so D.J.'s weight decreased by 10%.

On the other hand, if Terrence's weight went up from 225 pounds to 250 pounds, by what percent did Terrence's weight increase? Terrence's weight increased by 25 pounds (250 – 225), so we must find what percent 25 is of 225. Solving

$$\frac{25}{225} = \frac{P}{100}$$

We find that Terrence's weight increased by about 11%.

The Second Method The second method for calculating percent increase or decrease is usually more efficient than the first method. Suppose that a quantity increases from an amount *A* to a larger amount *B*. If we calculate *B* as a percent of *A*, we will find that it is more than 100%. This makes sense because *B* is more than *A*, so *B* must represent more than 100% of *A*. *The amount by which B, calculated as a percent of A, exceeds 100% is the percent increase from A to B.* On the left of Figure 7.16, we see that *B* is 135% of *A*. Therefore, the increase from *A* to *B* is 35%.

To summarize, the following steps describe how to calculate the percent increase from *A* to a larger amount *B*:

1. Calculate *B* as a percent of *A*.

2. Subtract 100%. The result is the percent increase of the quantity from *A* to *B*.

For example, suppose that the population of a city increases from 35,000 people to 44,000 people. What is the percent increase in the population? First, let's calculate what percent 44,000 is of 35,000. Solving

$$\frac{44,000}{35,000} = \frac{P}{100}$$

We find that 44,000 is about 126% of 35,000. Subtracting 100%, we determine that the population increased by about 26%.

FIGURE 7.16

Calculating percent increase and decrease

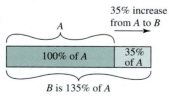

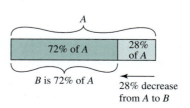

Similarly, if a quantity decreases from an amount A to a smaller amount B, then if we calculate B as a percent of A, we will find that it is less than 100%. This makes sense because B is less than A, so B must represent less than 100% of A. *The amount that B, calculated as a percent of A, is under 100% is the percent decrease from A to B.* On the right of Figure 7.16, we see that B is 72% of A. Therefore, the decrease from A to B is 28% (namely, 100% − 72%).

To summarize, the following steps describe how to calculate the percent decrease from A to a smaller amount B:

1. Calculate B as a percent of A.

2. Subtract this percent from 100%. The result is the percent decrease of the quantity from A to B.

For example, suppose that the price of a computer dropped from $899 to $825. What is the percent decrease in the price of the computer? First, let's calculate what percent 825 is of 899. Solving

$$\frac{825}{899} = \frac{P}{100}$$

We find that 825 is about 92% of 899. Since 100% − 92% = 8%, the price of the computer decreased by about 8%.

Two Methods for Calculating Amounts When the Percent Increase or Decrease Is Given

There are also two methods for calculating a new amount when we know the original amount and the percent increase or decrease. The second method also allows us to calculate the original amount if we know the final amount and the percent increase or decrease.

Class Activity   *Now Turn to Class Activities Manual*

7N Calculating Amounts from a Percent Increase or Decrease, p. 152

The First Method When an amount and its percent increase are given, we can calculate the new amount by calculating the increase and adding this increase to the original amount. Similarly, when an amount and its percent decrease are given, we can calculate the new amount by calculating the decrease and subtracting this decrease from the original amount.

If a table costs $187 and if the price of the table goes up by 6%, then what will the new price of the table be? We find that 6% of $187 is

$$0.06 \times \$187 = \$11$$

Thus, the new price of the table will be

$$\$187 + \$11 = \$198$$

On the other hand, if a table costs $187 and the price of the table goes down by 6%, then the new price of the table will be

$$\$187 - \$11 = \$176$$

The Second Method The second method for calculating amounts when percent increases or decreases are given uses the same idea as the second method for calculating percent increases and decreases. Unlike the first method, we can use this method to calculate the *initial amount* when we know the percent increase or decrease and the *final amount*.

FIGURE 7.17

Calculating amounts from the percent increase or percent decrease

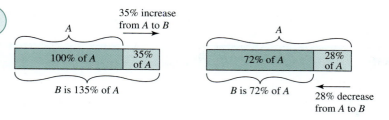

On the left of Figure 7.17, note that if an amount A increases by 35% to an amount B, then

$$B \text{ is } 135\% \text{ of } A$$

In general, if an amount A increases by $P\%$ to an amount B, then

$$B \text{ is } (100 + P)\% \text{ of } A$$

So, if a table costs \$187 and if the price of the table goes up by 6%, then the new price of the table will be $(100 + 6)\% = 106\%$ of \$187, which is

$$1.06 \times \$187 = \$198$$

The new price of the table after the 6% increase is \$198.

Notice that we can also solve the following type of problem, where the percent increase and the final amount are known, and the initial amount is to be calculated: If the price of gas went up 5% and is now \$2.15 per gallon, then how much did gas cost before this increase? If A was the initial price per gallon of gas before the increase, then 105% of A is \$2.15. Therefore,

$$1.05 \times A = \$2.15$$

So,

$$A = \frac{\$2.15}{1.05} = \$2.05$$

Hence, gas cost \$2.05 per gallon before the 5% increase. Table 7.15 shows how to use a percent table to carry out this calculation.

The situation is similar when the percent decrease is given. On the right of Figure 7.17, we see that if an amount A decreases by 28% to an amount B, then

$$B \text{ is } 72\% \text{ of } A$$

because $100\% - 28\% = 72\%$. In general, if an amount A decreases by $P\%$ to an amount B, then

$$B \text{ is } (100 - P)\% \text{ of } A$$

TABLE 7.15 Calculating the initial price of gas before a 5% increase by calculating 1% of the initial price and then 100% of the initial price

			After a 5% increase, the price of gas is \$2.15.
105%	→	\$2.15	Therefore, 105% of the initial price is \$2.15.
1%	→	$\$2.15 \div 105 = \$0.0205$	So, 1% of the initial price is \$0.0205.
100%	→	$100 \times \$0.0205 = \$2.05$	Thus, 100% of the initial price of gas is \$2.05.

TABLE 7.16 Calculating the initial price of gas before a 5% decrease by calculating 1% of the initial price and then 100% of the initial price

		After a 5% decrease, the price of gas is $2.15.
95% →	$2.15	Therefore, 95% of the initial price is $2.15.
1% →	$2.15 ÷ 95 = $0.0226	So 1% of the initial price is $0.0226.
100% →	100 × $0.0226 = $2.26	Thus, 100% of the initial price of gas is $2.26.

So, if a table costs $187 and if the price of the table goes down by 6%, then the new price of the table will be $(100 - 6)\% = 94\%$ of $187, which is

$$0.94 \times \$187 = \$176$$

Therefore, after the 6% decrease in price, the new price of the table is $176.

Once again, we can also solve the following type of problem, where the percent decrease and the final amount are known, and the initial amount is to be calculated: If the price of gas went down 5% and is now $2.15 per gallon, then how much did gas cost before this decrease? If A was the initial price of gas before the decrease, then 95% of A is $2.15. Therefore,

$$0.95 \times A = \$2.15$$

So,

$$A = \frac{\$2.15}{0.95} = \$2.26$$

Then, gas cost $2.26 per gallon before the 5% decrease. Table 7.16 shows how to use a percent table to carry out this calculation.

The Importance of the Reference Amount

When you find a percent increase or decrease, make sure you calculate the increase or decrease as a percent *of the reference amount*, namely, the amount for which you want to know the percent increase or decrease. The reference amount is the whole or 100% for the problem. Let's say the cost of a Dozey-Chair increases from $200 to $300. Then the new price is 50% more than the old price, but the old price is only 33% less than the new price. In both percent calculations, the change is $100. The different percentages come about because of the different reference amounts or wholes that are used. In the first case, the reference amount is $200, and $100 is 50% of $200. In the second case, the reference amount is $300, and $100 is only 33% of $300.

Class Activity   *Now Turn to Class Activities Manual*

7O Percent *of* versus Percent Increase or Decrease, p. 154

7P Percent Problem Solving, p. 155

7Q Percent Change and the Commutative Property of Multiplication, p. 156

Practice Exercises for Section 7.5

1. Last year, Ken had 2.5 tons of sand in his sand pile. This year, Ken has 3.5 tons of sand in his sand pile. By what percent did Ken's sand pile increase from last year to this year? First solve the problem by drawing a picture. Explain how your picture helps you solve the problem. Then solve the problem numerically.

2. Last year's profits were $16 million, but this year's profits are only $6 million. By what percent did profits decrease from last year to this year? First solve the problem by drawing a picture. Explain how your picture helps you solve the problem. Then solve the problem numerically.

3. If sales taxes are 6%, then how much should you charge for an item so that the total cost, including tax, is $35?

4. A pair of shoes has just been reduced from $75.95 to $30.38. Fill in the blanks:

   a. The new price is _____% less than the old price.

   b. The new price is _____% of the old price.

   c. The old price is _____% higher than the new price.

   d. The old price is _____% of the new price.

5. John bought a piece of land adjoining the land he owns. Now John has 25% more land than he did originally. John plans to give 20% of his new, larger amount of land to his daughter. Once John does this, how much land will John have in comparison with the amount he had originally? Draw a picture to help you solve this problem. Then solve the problem numerically, assuming, for example, that John starts with 100 acres of land.

6. The population of a certain city increased by 2% from 2006 to 2007 and then decreased by 2% from 2007 to 2008. By what percent did the population of the city change from 2006 to 2008? Did the population increase, decrease, or stay the same? Make a guess first, then calculate the answer carefully.

7. There are two boxes of chocolate. The chocolate in the second box weighs 15% more than the chocolate in the first box. There are 3 more ounces of chocolate in the second box than in the first. How much does the chocolate in the two boxes together weigh? Explain your reasoning.

8. There are two vats of grape juice. After 20% of the juice in the first vat was poured into the second vat, the first vat had 2 times as much juice as the second vat. By what percent did the amount of juice in the second vat increase when the juice was poured into it? Explain your reasoning.

Answers to Practice Exercises for Section 7.5

1. Using a picture, we see that each strip in Figure 7.18 represents 20% of last year's sand pile. Since two additional strips have been added since last year, that is a 40% increase.

   To calculate the percent increase numerically, let's first calculate 3.5 as a percent of 2.5.

   $$\frac{3.5}{2.5} = 1.4 = 140\%$$

   Subtracting 100%, we see that the percent increase is 40%.

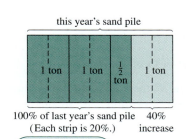

this year's sand pile

| 1 ton | 1 ton | $\frac{1}{2}$ ton | 1 ton |

100% of last year's sand pile   40%
(Each strip is 20%.)   increase

FIGURE 7.18 Last year's and this year's sand pile

2. Using a picture, we see that each strip in Figure 7.19 represents $2 million. As the picture shows, the decrease in profits is $\frac{1}{8}$ more than 50%. Since $\frac{1}{8} = 12.5\%$, the profits decreased by 50% + 12.5%, namely, by 62.5%.

last year's profits of $16 million

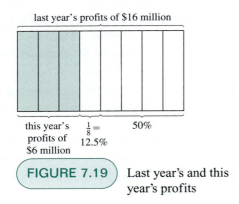

this year's profits of $6 million $\frac{1}{8} = 12.5\%$ 50%

FIGURE 7.19   Last year's and this year's profits

To calculate the percent decrease numerically, we should determine what percent the decrease, 10, is of 16.

$$\frac{10}{16} = 0.625 = \frac{62.5}{100} = 62.5\%$$

So, the profits went down by 62.5%.

Another way to calculate the percent decrease is to calculate what percent 6 is of 16 first, and then subtract this from 100%.

$$\frac{6}{16} = 0.375 = \frac{37.5}{100} = 37.5\%$$

Therefore, the percent decrease is 100% − 37.5%, which is 62.5%.

3. The sales tax increases the amount that the customer pays by 6%. Therefore, if P represents the price of the item, 106% of P must equal $35. Thus,

$$1.06 \times P = \$35$$

So,

$$P = \$35 \div 1.06 = \$33.02$$

If the price of the item is $33.02, then with a 6% sales tax, the total cost to the customer is $35.

Table 7.17 shows how to solve the problem with a percent table.

TABLE 7.17 A percent table for calculating the amount which becomes $35 when 6% is added

106%	→	$35
1%	→	$35 ÷ 106 = $0.3302
100%	→	$0.3302 × 100 = $33.02

4. **a.** The new price is <u>60%</u> less than the old price.

   b. The new price is <u>40%</u> of the old price.

   c. The old price is <u>150%</u> higher than the new price.

   d. The old price is <u>250%</u> of the new price.

5. See Figure 7.20. If John starts with 100 acres of land and gets 25% more, then he will have 125 acres. We find that 20% of 125 acres is 25 acres, so if John gives 20% of his new amount of land away, he will have 125 − 25 = 100 acres of land, which is the amount he started with.

1. John's original plot of land.

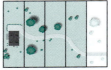

2. Now John has 25% more land.

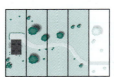

3. 20% of John's new, larger amount of land is represented by the unshaded part.

4. When John gives 20% of his larger plot of land away, he's left with the amount he started with.

FIGURE 7.20   Solving the land problem with a picture

6. Although you might have guessed that the population stayed the same, it actually decreased by 0.04%. Try this on a city of 100,000, for example. You can tell that the population will have to decrease because when the population goes back down by 2%, this 2% is *of a larger number* than the original population.

7. Since 15% of the chocolate in the first box is 3 ounces, the percent table in Table 7.18 shows that the chocolate in the first box weighs 20 ounces. Therefore, the chocolate in the second box weighs 20 + 3 = 23 ounces and the chocolate in the two boxes together weighs 20 + 23 = 43 ounces.

TABLE 7.18 A percent table for the chocolate in box 1

15%	→	3 ounces
5%	→	1 ounce
100%	→	20 ounces

8. See Figure 7.21. The contents of the first vat are represented by a long strip. Since 20% of the juice is poured out, and since 20% = $\frac{1}{5}$, the strip is broken into 5 equal parts. One of those parts will be poured into the second vat, leaving 4 parts remaining. Since the 4 remaining parts must be 2 times the contents of the second vat after pouring, the second vat must contain 2 parts of juice after pouring. One of those parts was poured in. Therefore, the other part was there originally. Since an equal amount of juice was poured in as was there originally, the amount of juice in the second vat increased by 100%.

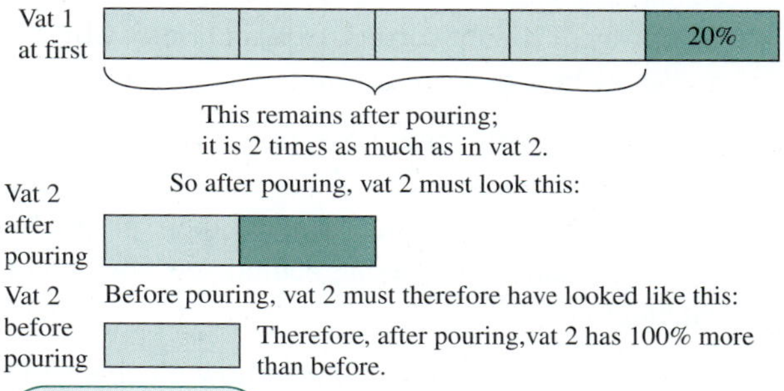

FIGURE 7.21 Percent change in juice in vats

Problems for Section 7.5

1. Last year, the population of South Skratchankle was 48,000. This year, the population is 60,000. By what percent did the population increase? First solve the problem by drawing a picture. Explain how your picture helps you solve the problem. Then solve the problem numerically.

2. Last year's sales were $7.5 million. This year's sales are only $6 million. By what percent did sales decrease from last year to this year? First solve the problem by drawing a picture. Explain how your picture helps you solve the problem. Then solve the problem numerically.

3. Jayna and Lisa are comparing the prices of two boxes of cereal of equal weight. Brand A costs $3.29 per box, and brand B costs $2.87 per box. Jayna calculates that

$$\frac{\$3.29}{\$2.87} = 1.15$$

Lisa calculates that

$$\frac{\$2.87}{\$3.29} = 0.87$$

a. Use Jayna's calculation to make *two* correct statements comparing the prices of brand A and brand B with percentages. Explain your reasoning.

b. Use Lisa's calculation to make *two* correct statements comparing the prices of brand A and brand B with percentages. Explain your reasoning.

4. Are the two problems that follow solved in the same way? Are the answers the same? Solve both and compare your solutions. Explain your reasoning.

a. A television that originally cost $500 is marked down by 25%. What is its new price?

b. Last week a store raised the price of a television by 25%. The new price is $500. What was the old price?

5. The price of play equipment for the school has just been reduced by 25%. The new, reduced price of the play equipment is $400. Bob says he can find the original price (before the reduction) in the following way:

> First I noticed that 25% is $\frac{1}{4}$. Then I found $\frac{1}{4}$ of $400, which is $100. Next I added $400 and $100, so the original price was $500.

Is Bob's method correct or not? If it's correct, say so and also explain how to solve the problem in another way. If it's not correct, explain briefly why not and show how to solve the problem correctly.

6. How much should Swanko Jewelers charge now for a necklace if they want the necklace to cost $79.95 when they reduce their prices by 60%? Explain the reasoning behind your method of calculation.

7. If sales taxes are 7%, then how much should you charge for an item if you want the total cost, including tax, to be $15? Explain the reasoning behind your method of calculation.

8. Connie and Benton bought identical plane tickets, but Benton spent more than Connie (and Connie did not get her ticket for free).

   a. If Connie spent 25% less than Benton, then did Benton spend 25% more than Connie? If not, then what percent more than Connie did Benton spend? Explain.

   b. If Benton spent 25% more than Connie, then did Connie spend 25% less than Benton? If not, then what percent less than Benton did Connie spend? Explain.

9. Of the five statements that follow, which have the same meaning? In other words, which of these statements could be used interchangeably (in a news report, for example)? Explain your answers.

   a. The price increased by 53%.

   b. The price increased by 153%.

   c. The new price is 153% of the old price.

   d. The new price is higher than the old price by 53%.

   e. The old price was 53% less than the new price.

10. Write a paragraph explaining the difference between a 150% increase in an amount and 150% of an amount. Give examples to illustrate.

11. The SuperDiscount store is planning a "35%-off sale" in two weeks. This week, a pair of pants costs $59.95.

   a. Suppose SuperDiscount raises the price of the pants by 35% this week, and then two weeks from now, lowers the price by 35%. How much will the pants cost two weeks from now? Explain your method of calculation. Explain why it makes sense that the pants won't return to their original price of $59.95 two weeks from now. (A picture may help you explain.)

   b. By what percent does SuperDiscount need to raise the price of the pants this week, so that two weeks from now, when it lowers the price by 35%, the pants will return to the original price of $59.95? Explain your method of calculation.

12. Every week, DollarDeals lowers the price of items it has in stock by 10%. Suppose that the price of an item has been lowered twice, each time by 10% of that week's price. Explain why it makes sense that the total discount on the item is *not* 20%, even though the price has been lowered twice by 10% each time. A picture may help you explain. What percent is the total discount?

13. One box of cereal contains 12% more cereal than another box. The larger box contains 3 more ounces of cereal than the smaller box. How much does the cereal in each box weigh? Explain your reasoning.

14. One box of cereal contains 25% more cereal than another box. Together, the cereal in both boxes weighs 54 ounces. How much does the cereal in each box weigh? Explain your reasoning.

15. In a box of chocolate candies, 40% of the candies are dark chocolate; the rest are milk chocolate. There are 6 more milk chocolate candies than dark chocolate candies. In all, how many chocolate candies are in the box? Explain your reasoning.

16. In a box of chocolate candies, 30% of the candies are dark chocolate; the rest are milk chocolate. What percent more milk chocolate candies are in the box than dark chocolate candies? Explain why we can't solve this problem by subtracting 70% − 30%. Then solve the problem correctly, explaining your reasoning.

17. There are two elementary schools in a county. After 10% of the children at the first school were moved to the second school, both schools had the same number of children. By what percent did the number of children at the second school increase when the children from the first school were added? Explain your reasoning.

18. There are two middle schools in a county. The first middle school had 20% more children than the second middle school. Then 10% of the children at the second middle school left the county. What percent more children are now at the first middle school than at the second one? Explain your reasoning.

19. One school had 10% more children than another school. After 18 children moved from one school to the other, both schools had the same number of children. How many children are in the two schools together? Explain your reasoning.

20. Sue and Tonya started the same job at the same time and earned identical salaries. After one year, Sue got a 5% raise and Tonya got a 6% raise. The following year, the situation was reversed: Sue got a 6% raise and Tonya got a 5% raise. After the first year, Tonya's salary was higher than Sue's, of course, but whose salary was higher after both raises? Solve this problem by using the fact that if someone's salary goes up by 5%, then that person's new salary is 1.05 times the old salary (similarly for a 6% raise, of course). Using this method for calculating the women's salaries, explain how the commutative property of multiplication is relevant to comparing the salaries after both raises.

21. Suppose that the sales tax is 7%, and suppose that some towels are on sale at a 20% discount. When you buy the towels, you pay 7% tax on the discounted price. What if you were to pay 7% tax on the full price, but you got a 20% discount on the price *including* the tax? Would you pay more, less, or the same amount? Explain how the commutative property of multiplication is relevant to this question.

22. A dress is marked down 25%, and then it is marked down 20% from the discounted price.

    a. By what percent is the dress marked down after both discounts?

    b. Does the dress cost the same, less, or more than if the dress were marked down 45% from the start? Explain how you can determine the answer to this question without doing any calculating.

    c. If the dress were marked down 20% first and then 25%, would you get a different answer to part (a)? Explain how the commutative property of multiplication is relevant to this question.

23. Frank's Jewelers runs the following advertisement: "Come to our 40%-off sale on Saturday. We're not like the competition, who raise prices by 30% and then have a 70%-off sale!"

    a. If two items start off with the same price, which gives you the lower price in the end: taking off 40% or raising the price by 30% and then taking off 70% (of the raised price)?

    b. Consider the same problem more generally, with other numbers. For example, if you raise prices by 20% and then take off 50% (of the raised price), how does this compare with taking 30% off of the original price? If you raise a price by 30% and then lower the raised price by 30%, how does that compare with the original price? Try at least two other pairs of percentages by which to raise and then lower a price. Describe what you observe.

    Predict what happens in general: If you raise a price by A% and then take B% off of the raised price, does that have the same result as if you had lowered the original price by $(B - A)$%? If not, which produces the lower final price?

    c. Use the distributive property or FOIL to explain the pattern you discovered in part (b). Remember that to raise a price by 15%, for example, you multiply the price by $1 + 0.15$, whereas to lower a price by 15%, you multiply the price by $1 - 0.15$.

24. According to the 2000 Census, from 1990 to 2000 the population of Clarke County, Georgia, increased by 15.86% and the population of adjacent Oconee County increased by 48.85%.

    a. Can we calculate the percent increase in the total population of the two-county Clarke/Oconee area from 1990 to 2000 by adding 15.86% and 48.85%? Why or why not?

    b. Use the census data in the following table to calculate the percent increase in the total population of the two-county Clarke/Oconee area from 1990 to 2000:

County	1990 Population	2000 Population
Clarke	87,594	101,489
Oconee	17,618	26,225

25. The following information about two different snack foods is taken from their packages:

Snack	Serving Size	Calories in One Serving	Total Fat in One Serving
Small crackers	28 g	140 calories	6 g
Chocolate hearts	40 g	220 calories	13 g

Suppose we want to compare the amount of fat in these foods. One way to compare the fat is to compare the amount of fat in one serving of each food; another way is to compare the amount of fat in some fixed number of calories of the foods, such as 1 calorie or 100 calories; yet another way is to compare the amount of fat in a fixed number of grams of the foods, such as 1 gram or 100 grams. Solve the following problems by comparing the small crackers and chocolate hearts in different ways:

a. Explain how to interpret the information in the table so that the chocolate hearts have 117% more fat than the small crackers.

b. Explain how to interpret the information in the table so that the chocolate hearts have 52% more fat than the small crackers.

c. Explain how to interpret the information in the table so that the chocolate hearts have 38% more fat than the small crackers.

26. In 2000, Washington County had a total population—urban and rural populations combined—of 200,000. From 2000 to 2010, the rural population of Washington County went up by 4%, and the urban population of Washington County went up by 8%.

a. Based on the preceding information, make a reasonable guess for the percent increase of the total population of Washington County from 2000 to 2010. Based on your guess, what do you expect the total population of Washington County to have been in 2010?

b. Make up three very different examples for the rural and urban populations of Washington County in 2000 (i.e., pick pairs of numbers that add to 200,000). For each example, calculate the total population in 2010, and calculate the percent increase in the total population of Washington County from 2000 to 2010. Compare these answers with your answers in part (a).

c. If you had only the data given at the beginning of the problem (the 4% and 8% increases and the total population of 200,000 in 2000), would you be able to say exactly what the total population of Washington County was in 2010? Could you give a range for the total population of Washington County in 2010? In other words, could you say that the total population must have been between certain numbers in 2010? If so, what is this range of numbers?

27. Suppose you owe $1000 on your credit card and that at the end of each month an additional 1.4708% of the amount you owe is added on to the amount you owe. (This is what would happen if your credit card charged an annual percentage rate of 17.6496%.) Let's assume that you do not pay off any of this debt or the interest that is added to it. Let's also assume that you don't add on any more debt (other than the interest that you are charged).

a. How much will you owe after 6 months? after one year? after two years?

b. Use a calculator to calculate $(1.014708)^{48}$, and notice that this number is just a little larger than 2. Based on this calculation, what will happen to your debt every 4 years? Explain.

c. Suppose you didn't pay off your debt for 40 years. Using part (b), determine approximately how much you owe.

Chapter Summary and Study Items

Section 7.1 The Meanings of Ratio, Rate, and Proportion

A ratio or a rate describes a specific type of relationship between quantities. There are two definitions of ratio, each corresponding to one of the definitions of division. Equivalent ratios can be shown in strip diagrams, in ratio tables, and on double number lines. A proportion is the statement that two ratios are equal.

Key skills and understandings:

- Using a strip diagram as a support, explain what it means for two quantities to be in a certain ratio.
- Using a ratio table as a support, explain what it means for two quantities to be in a certain ratio.
- Given a ratio, find equivalent ratios.
- Use ratio tables to compare mixtures.

Section 7.2 Solving Proportion Problems by Reasoning with Multiplication and Division

Problems about ratios and rates can be solved in several different ways by using multiplication and division, while reasoning about how quantities compare.

Key skills and understandings:

- Solve ratio problems by using multiplication and division in the process of reasoning about how quantities compare. Use strip diagrams, ratio tables, or double number lines as supports.

Section 7.3 Connecting Ratios and Fractions

Ratios are connected to fractions (and division) by unit rates. In the standard method of solving proportions, we first set two fractions equal to each other; we can do so because we are equating the unit rates.

Key skills and understandings:

- Understand that if two quantities are in a ratio of A to B, then $\frac{A}{B}$ is a unit rate; namely, it is the amount of the first quantity per unit amount of the second quantity.
- Identify unit rates and explain what they mean.
- Explain why this method of solving proportions is valid: setting two fractions equal to each other, then cross-multiplying, then solving the resulting equation.

Section 7.4 When You Can Use a Proportion and When You Cannot

Some problems that involve related quantities can't be solved with a proportion.

Key skill and understandings:

- Recognize problems that should not be solved with a proportion.

Section 7.5 Percent Revisited: Percent Increase and Decrease

If the value of a quantity goes up, then to find the increase in the quantity, we subtract. But to find the *percent* increase, we must figure the increase as a percent of the original. If the value of a quantity goes down, then to find the decrease in the quantity, we subtract. But to find the *percent* decrease, we must figure the decrease as a percent of the original. We can also calculate a percent increase by calculating what percent the increased amount is of the original amount and subtracting 100%. Similarly, we can calculate a percent decrease by calculating what percent the decreased amount is of the original amount and subtracting this from 100%. We can also calculate quantities from a given percent increase or decrease.

Key skills and understandings:

- Calculate percent increase and percent decrease in several different ways, and explain why the calculation methods make sense.
- Calculate quantities from a given percent increase or decrease, and explain why the calculation method makes sense.
- Distinguish between percent increase/decrease and percent *of*.
- Solve problems involving percent increase or decrease.

Number Theory

In this chapter, we will study some of the theory of numbers—including factors, multiples, prime numbers, and even and odd numbers—and some quick tests for determining whether a number is divisible by another number. Most elementary school curricula do not emphasize these topics; however, most children learn about these topics in the upper elementary grades. Even young children can, and often do, learn about even and odd numbers.

8.1 Factors and Multiples

In this section, we study situations where counting numbers decompose into products of counting numbers.

Definitions of Factors and Multiples

divisible
divides
If A and B are counting numbers and if $A \div B$ is a counting number with no remainder, then we say that A is **divisible** by B or **evenly divisible** by B, or that B **divides** A. For example, 12 is divisible by 4, and 5 divides 30.

factor
If A and B are counting numbers, then we say that B is a **factor** or a **divisor** of A if there is a counting number C such that

$$A = B \times C$$

So 7 is a factor of 21 because

$$21 = 7 \times 3$$

The numbers 3, 1, and 21 are also factors of 21, and these are the only other factors of 21. Notice that the concepts of divisibility and of factor are almost identical: A counting number B is a factor of a counting number A exactly when A is divisible by B.

multiple
If A and B are counting numbers, then we say that A is a **multiple** of B if there is a counting number C such that

$$A = C \times B$$

So 44 is a multiple of 11 because

$$44 = 4 \times 11$$

Similarly, the numbers 55 and 66 are also multiples of 11. The list of multiples of 11 is infinitely long:

$$11, 22, 33, 44, 55, 66, \ldots$$

To find all the multiples of a counting number we simply multiply the number by 1, by 2, by 3, by 4, by 5, and so on. We can never explicitly list all the multiples of a number because the list is infinitely long.

We commonly use the concepts of factors and multiples in the context of the counting numbers. However, in some cases, it is useful to include 0 as well. In these cases, it is acceptable to describe 0 as divisible by every counting number, or as a multiple of every whole number.

Notice that the concepts of factors and multiples are closely linked: A counting number A is a multiple of a counting number B exactly when B is a factor of A. For this reason it's easy to get the concepts of factors and multiples confused. Remember that the factors of a number are the numbers you get from writing the original number as a product (i.e., from "breaking down" the number by dividing it). On the other hand, the multiples of a number are the numbers you get by multiplying the number by counting numbers (i.e., by "building up" the number by multiplying it).

To summarize, if A, B, and C are counting numbers, and if

$$A = B \times C$$

then

- A is a multiple of B
- A is a multiple of C
- B is a factor of A
- C is a factor of A

Class Activity   *Now Turn to Class Activities Manual*

8A Factors, Multiples, and Rectangles, p. 157

8B Problems about Factors and Multiples, p. 158

Finding Factors

Class Activity   *Now Turn to Class Activities Manual*

8C Finding All Factors, p. 159

One way to find all the factors of a counting number is to divide the number by all the counting numbers smaller than it, to see which ones divide the number evenly, without a remainder. To make the work efficient, keep track of the quotients, because the whole number quotients will also be factors. For example, to find all the factors of 40, divide 40 by 1, by 2, by 3, by 4, by 5, and so on, recording those numbers that divide 40 and recording the corresponding quotients:

$$1, 40 \quad \text{(because } 1 \times 40 = 40)$$
$$2, 20 \quad \text{(because } 2 \times 20 = 40)$$
$$4, 10 \quad \text{(because } 4 \times 10 = 40)$$
$$5, 8 \quad \text{(because } 5 \times 8 = 40)$$

After dividing 40 by 1 through 8 and verifying that 1, 2, 4, 5, and 8 are the only factors of 40 up to 8, you don't have to check if 9, 10, 11, 12, and so on divide 40. Why not? Any counting number larger than 8 that divides 40 would have a quotient less than 5 and would already have to appear as one of the factors listed previously. So we can conclude that the full list of factors of 40 is

$$1, 2, 4, 5, 8, 10, 20, 40$$

Notice that every counting number except 1 must have at least two distinct factors, namely, 1 and itself.

to factor

So far we've used the word *factor* as a noun. However, the word *factor* can also be used as a verb. If A is a counting number, then **to factor** A means to write A as a product of two or more counting numbers, each of which is less than A. So we can factor 18 as 9 times 2:

$$18 = 9 \times 2$$

But notice that we can factor 18 even further, because we can factor 9 as 3 times 3, so

$$18 = 3 \times 3 \times 2$$

Class Activity   *Now Turn to Class Activities Manual*

8D Do Factors Always Come in Pairs? p. 159

Practice Exercises for Section 8.1

1. Describe an organized, efficient method to find all the factors of a counting number, and explain why this method finds all the factors. Illustrate with example 72.

2. What are the multiples of the number 72? Explain.

3. Write a story problem such that solving the problem will require finding all the factors of 54. Solve the problem.

4. Write a story problem such that solving the problem will require finding many multiples of 4. Solve the problem.

Answers to Practice Exercises for Section 8.1

1. The factors of the number 72 are the counting numbers that divide 72. To find these numbers, we can simply divide 72 by smaller counting numbers in order, checking to see which ones divide 72. So we can divide 72 by 1, by 2, by 3, by 4, by 5, and so on. To work efficiently, we can record the factors we

find and the corresponding quotients as follows:

1, 72	(because $1 \times 72 = 72$)
2, 36	(because $2 \times 36 = 72$)
3, 24	(because $3 \times 24 = 72$)
4, 18	(because $4 \times 18 = 72$)
6, 12	(because $6 \times 12 = 72$)
8, 9	(because $8 \times 9 = 72$)

Once we have checked the counting numbers up to 9 to see if they divide 72, we won't have to continue dividing 72 by counting numbers greater than 9. This is because any counting number greater than 9 that divides 72 must have a quotient that is less than 9 and therefore must already occur in the column on the right in the previous list. Hence, we can conclude that the list of factors of 72 is

$$1, 2, 3, 4, 6, 8, 9, 12, 18, 24, 36, 72$$

2. The multiples of 72 are the numbers that can be expressed as 72 times another counting number. They are

$$72, 144, 216, 288, \ldots$$

because these numbers are

$$72 \times 1, \quad 72 \times 2, \quad 72 \times 3, \quad 72 \times 4, \ldots$$

3. There are 54 members of a band. What are the ways of arranging the band members into equal rows? The ways of factoring 54 into a product of two counting numbers are

$$1 \times 54, \quad 2 \times 27, \quad 3 \times 18, \quad 6 \times 9$$

and the reverse factorizations,

$$54 \times 1, \quad 27 \times 2, \quad 18 \times 3, \quad 9 \times 6$$

So the band can be arranged in 1 row of 54 people, 2 rows of 27 people, 3 rows of 18 people, 6 rows of 9 people, 9 rows of 6 people, 18 rows of 3 people, 27 rows of 2 people, or 54 rows of 1 person.

4. Tawanda has painted many dry noodles and will put strings through them to make necklaces that are completely filled with noodles, as in Figure 8.1. Each noodle is 4 centimeters long. What are the lengths of the necklaces Tawanda can make?

Whenever Tawanda makes a necklace from these noodles, the length of the necklace will be the number of noodles she used times 4 centimeters. Therefore, this problem is almost the same as the problem of finding the multiples of 4. In Tawanda's case, some of the multiples of 4 would make necklaces that are too short to be practical, and some multiples of 4 would not work because Tawanda doesn't have enough noodles. The multiples of 4 are

$$4, 8, 12, 16, 20, 24, \ldots$$

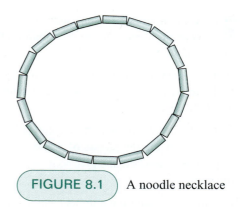

FIGURE 8.1 A noodle necklace

Problems for Section 8.1

1. Johnny says that 3 is a multiple of 6 because you can arrange 3 cookies into 6 groups by putting $\frac{1}{2}$ of a cookie in each group. Discuss Johnny's idea in detail: In what way does he have the right idea about what the term *multiple* means, and in what way does he need to modify his idea?

2. Manuela is looking for all the factors of 90. So far, Manuela has divided 90 by all the counting numbers from 1 to 10, listing those numbers that divide 90

and listing the corresponding quotients. Here is Manuela's work so far:

$$1, 90 \qquad 1 \times 90 = 90$$
$$2, 45 \qquad 2 \times 45 = 90$$
$$3, 30 \qquad 3 \times 30 = 90$$
$$5, 18 \qquad 5 \times 18 = 90$$
$$6, 15 \qquad 6 \times 15 = 90$$
$$9, 10 \qquad 9 \times 10 = 90$$
$$10, 9 \qquad 10 \times 9 = 90$$

Should Manuela keep checking numbers to see if any numbers larger than 10 divide 90, or can Manuela stop dividing at this point? If so, why? What are all the factors of 90?

3. Show how to find all the factors of the following numbers in an efficient manner. Explain why you can stop checking for factors when you do.

   a. 63

   b. 75

   c. 126

4. 🏺 a. Write a story problem such that solving your problem will require finding all the factors of 48. Solve your problem.

   b. Write a story problem such that solving your problem will require finding several multiples of 15. Solve your problem.

5. a. If A and B are counting numbers and B is a factor of A, how are the factors of A and B related? Explain your answer, and give some examples to illustrate.

   b. If A and B are counting numbers and A is a multiple of B, how are the multiples of A and B related? Explain your answer, and give some examples to illustrate.

6. 🏺 a. Write a problem about a realistic situation that involves the concept of factors. Solving your problem should involve finding all the factors of a number. Solve your problem.

   b. Write a problem about a realistic situation that involves the concept of multiples. Solving your problem should involve finding multiples of a number. Solve your problem.

7. Solve the next two problems, and determine whether the answers are different or not. Explain why or why not.

   a. Josh has 1159 bottle caps in his collection. In how many different ways can Josh arrange his bottle-cap collection into groups so that the same number of bottle caps are in each group and so that there are no bottle caps left over (i.e., not in a group)?

   b. How many different rectangles can be made whose side lengths, in centimeters, are counting numbers and whose area is 1159 square centimeters?

8. If A, B, and C are counting numbers and both A and B are multiples of C, what can you say about $A + B$? Explain why your answer is always true, and give some examples to illustrate. Which property of arithmetic is relevant to this problem?

9. 🔄 At school there is a long line of closed lockers, numbered 1 to 1000 in order. Outside, 1000 students are lined up, waiting to come in and open or close locker doors. The first student in line opens every locker door. The second student in line closes the doors of lockers that are multiples of 2. The third student in line changes the doors of lockers that are multiples of 3. (The student opens the doors that are closed and closes the doors that are open.) The fourth student in line changes the doors of lockers that are multiples of 4. (The student opens the doors that are closed and closes the doors that are open.) Students keep coming in and changing the locker doors, continuing in the pattern that the Nth student changes the lockers that are multiples of N. Which lockers are open after all 1000 students have gone through changing locker doors? Explain your answer.

8.2 Greatest Common Factor and Least Common Multiple

What do gears, different types of cicadas, and the spirograph drawing toy all have in common? All have aspects that are related to greatest common factors and least common multiples! In Section 8.1, we discussed factors and multiples of individual numbers. If we have two or more counting numbers in mind, we can consider the factors that the two numbers have in common and the multiples that the two numbers have in common. By considering common multiples and common factors of two or more counting numbers, we arrive at the concepts of *greatest common factor* and *least common multiple*, which we will discuss in this section. Greatest common factors and least common multiples can be useful when working with fractions.

> ## Class Activity   *Now Turn to Class Activities Manual*
>
> **8E**   Finding Commonality, p. 160

Definitions of GCF and LCM

greatest common factor GCF

If you have two or more counting numbers, then the **greatest common factor**, abbreviated **GCF**, or **greatest common divisor**, abbreviated **GCD**, of these numbers is the greatest counting number that is a factor of all the given counting numbers. For example, what is the GCF of 12 and 18? The factors of 12 are

$$1, 2, 3, 4, 6, 12$$

and the factors of 18 are

$$1, 2, 3, 6, 9, 18$$

Therefore, the common factors of 12 and 18 are the numbers that are common to the two lists, namely,

$$1, 2, 3, 6$$

The greatest of these numbers is 6; therefore, the greatest common factor of 12 and 18 is 6.

least common multiple LCM

Similarly, if you have two or more counting numbers, then the **least common multiple**, abbreviated **LCM**, of these numbers is the least counting number that is a multiple of all the given numbers. For example, what is the LCM of 6 and 8? The multiples of 6 are

$$6, 12, 18, 24, 30, 36, 42, 48, 54, 60, 66, 72, 78, \ldots$$

and the multiples of 8 are

$$8, 16, 24, 32, 40, 48, 56, 64, 72, 80, \ldots$$

Therefore, the common multiples of 6 and 8 are the numbers that are common to the two lists, namely,

$$24, 48, 72, \ldots$$

The least of these numbers is 24; therefore, the least common multiple of 6 and 8 is 24.

Although there are other methods for calculating GCFs and LCMs, we can always use the definition of these concepts to calculate GCFs and LCMs.

The Slide Method for Finding GCFs and LCMs

> ## Class Activity   *Now Turn to Class Activities Manual*
>
> **8F** The Slide Method, p. 162

To determine the GCF of counting numbers, we can always use the definition of GCF: We find all the factors of the numbers, determine which factors the numbers have in common, and select the greatest of those common factors. Similarly, we can use the definition of LCM to determine the LCM of counting numbers. Another quicker way to determine GCFs and LCMs is sometimes called the "slide method."

TABLE 8.1 Using the slide method to calculate the GCF and LCM of 8800 and 10,000

Step 1:	10	8800, 10,000
		880, 1000

Step 2:	10	8800, 10,000
	10	880, 1000
		88, 100

Step 3:	10	8800, 10,000
	10	880, 1000
	2	88, 100
		44, 50

Step 4:	10	8800, 10,000
	10	880, 1000
	2	88, 100
	2	44, 50
		22, 25

Conclusion:

GCF $= 10 \times 10 \times 2 \times 2 = 400$

LCM $= 10 \times 10 \times 2 \times 2 \times 22 \times 25 = 220,000$

Table 8.1 shows the steps for one way to use the slide method to determine the GCF and LCM of 8800 and 10,000. Table 8.2 shows the final result of a more efficient way to use the slide method for determining the GCF and LCM of 8800 and 10,000.

Observe that to use the slide method, you repeatedly find common factors. You write these common factors on the left, and you write the quotients that result from dividing by the common factors on the right. The "slide" stops when the resulting quotients on the right no longer have any common factor except 1. The GCF is then the product of the factors down the left-hand side of the slide, and the LCM is the product of the factors down the left-hand side of the slide *and* the numbers in the last row of the slide.

Notice that we can use slides flexibly. For example, in Table 8.1 we found the common factor 10 of 8800 and 10,000, and at the next step we found another common factor of 10. But in Table 8.2 we found the common factor 100 in one step, rather than taking two steps ($10 \times 10 = 100$).

Why does the slide method give us the GCF? Essentially, it's because we repeatedly take out common factors until there are no more common factors left (except 1). When we multiply all the common factors we collected, that product should form the greatest common factor. By listing successive quotients on the right-hand side of the slide, we ensure that we don't repeat a common factor that has already been accounted for.

Why does the slide method give us the LCM? This is harder to explain, and we will see only roughly why it works. Notice that the product of the numbers on the left in a slide and one of the numbers in the bottom row is equal to the number above it at the top. For example, referring to Table 8.2, we see that

TABLE 8.2 Using the slide method to calculate the GCF and LCM of 8800 and 10,000 in fewer steps

100	8800, 10,000
4	88, 100
	22, 25

GCF $= 100 \times 4 = 400$

LCM $= 100 \times 4 \times 22 \times 25 = 220,000$

$100 \times 4 \times 22 = 8800$ and $100 \times 4 \times 25 = 10,000$. So the product of the numbers on the left and the numbers in the bottom row of a slide must be a multiple of the initial numbers. Because we don't repeat the factors that the two initial numbers have in common (namely, the factors on the left), we get the smallest possible multiple of the initial numbers.

Class Activity   *Now Turn to Class Activities Manual*

8G Problems Involving Greatest Common Factors and Least Common Multiples, p. 162

8H Spirograph Flower Designs, p. 164

8I Relationships Between the GCF and the LCM and Explaining the Flower Designs, p. 165

Using GCFs and LCMs with Fractions

Class Activity   *Now Turn to Class Activities Manual*

8J Using GCFs and LCMs with Fractions, p. 166

We often use GCFs and LCMs when working with fractions.

To put a fraction in simplest form, we can divide the numerator and denominator of the fraction by the GCF of the numerator and the denominator. For example, to put

$$\frac{24}{36}$$

in simplest form, we can first determine that the greatest common factor of 24 and 36 is 12. Since $24 = 2 \cdot 12$ and $36 = 3 \cdot 12$, we have

$$\frac{24}{36} = \frac{2 \cdot 12}{3 \cdot 12} = \frac{2}{3}$$

So $\frac{2}{3}$ is the simplest form of $\frac{24}{36}$.

It is not necessary to determine the greatest common factor of the numerator and denominator of a fraction in order to put the fraction in simplest form. Instead, we can keep simplifying the fraction until it can no longer be simplified. For example, we could use the following steps to simplify $\frac{24}{36}$:

$$\frac{24}{36} = \frac{12 \cdot 2}{18 \cdot 2} = \frac{12}{18} = \frac{6 \cdot 2}{9 \cdot 2} = \frac{6}{9} = \frac{2 \cdot 3}{3 \cdot 3} = \frac{2}{3}$$

We can use the LCM of the denominators of two fractions to add the fractions. In order to add fractions, we must first find a common denominator. Although any common denominator will do, we may wish to work with the least common denominator. Because common denominators of two fractions must be multiples of both denominators, the least common denominator is the least common multiple of the two denominators. For example, to add

$$\frac{1}{6} + \frac{3}{8}$$

we need a common denominator that is a multiple of 6 and of 8. The least common multiple of 6 and 8 is 24. Using this least common denominator, we calculate

$$\frac{1}{6} + \frac{3}{8} = \frac{1 \cdot 4}{6 \cdot 4} + \frac{3 \cdot 3}{8 \cdot 3} = \frac{4}{24} + \frac{9}{24} = \frac{13}{24}$$

To add $\frac{1}{6} + \frac{3}{8}$ we could also have used the common denominator $6 \cdot 8 = 48$. In this case,

$$\frac{1}{6} + \frac{3}{8} = \frac{1 \cdot 8}{6 \cdot 8} + \frac{3 \cdot 6}{8 \cdot 6} = \frac{8}{48} + \frac{18}{48} = \frac{26}{48}$$

Because we did not use the least common denominator, the resulting sum is not in simplest form. We need the following additional step to put the answer in simplest form:

$$\frac{26}{48} = \frac{13 \cdot 2}{24 \cdot 2} = \frac{13}{24}$$

When we use a common denominator that is not the least common multiple of the denominators, the resulting sum will not be in simplest form. However, even if we use the least common denominator when adding two fractions, the resulting fraction may not be in simplest form. For example,

$$\frac{1}{2} + \frac{1}{6} = \frac{3}{6} + \frac{1}{6} = \frac{4}{6}$$

but $\frac{4}{6}$ is not in simplest form even though we used the least common denominator when adding.

Practice Exercises for Section 8.2

1. Use the definition of LCM to calculate the least common multiple of 9 and 12.

2. Use the definition of GCF to calculate the greatest common factor of 36 and 63.

3. Use the definition of GCF to calculate the greatest common factor of 16 and 27.

4. Write a story problem such that solving the problem requires finding the GCF of 100 and 75. Solve the problem.

5. Write a story problem such that solving the problem requires finding the LCM of 12 and 5. Solve the problem.

6. Show how to use the slide method to find the GCF and LCM of 224 and 392.

7. Show how to use the slide method to find the GCF and LCM of 12,375 and 16,875.

Answers to Practice Exercises for Section 8.2

1. The multiples of 9 are

    9, 18, 27, 36, 45, 54, 63, 72, 81, ...

   and the multiples of 12 are

    12, 24, 36, 48, 60, 72, 84, ...

   Therefore, the common multiples of 9 and 12 are

    36, 72, ...

   The least of these is 36, which is thus the least common multiple of 9 and 12.

2. The factors of 36 are

    1, 2, 3, 4, 6, 9, 12, 18, 36

   and the factors of 63 are

    1, 3, 7, 9, 21, 63

Therefore, the common factors of 36 and 63 are

$$1, 3, 9$$

The greatest of these numbers is 9; hence, the greatest common factor of 36 and 63 is 9.

3. The factors of 16 are

$$1, 2, 4, 8, 16$$

and the factors of 27 are

$$1, 3, 9, 27$$

The only common factor of 16 and 27 is 1, so this must be the greatest common factor of 16 and 27.

4. If you have 100 pencils and 75 small notebooks, what is the largest group of children that you can give all the pencils and all the notebooks to so that each child gets the same number of pencils and each child gets the same number of notebooks, and so that no pencils or notebooks are left over?

If each child gets the same number of pencils and no pencils are left over, then the number of children you can give the pencils to must be a factor of 100. Similarly, the number of children that you can give the notebooks to must be a factor of 75. Since you are looking for the largest number of children to give the pencils and notebooks out to, this will be the greatest common factor of 100 and 75. The factors of 100 are

$$1, 2, 4, 5, 10, 20, 25, 50, 100$$

The factors of 75 are

$$1, 3, 5, 15, 25, 75$$

The common factors of 100 and 75 are

$$1, 5, 25$$

Therefore, the GCF of 100 and 75 is 25, so the largest group of children you can give the pencils and notebooks out to is 25.

5. If pencils come in packages of 12 and small notebooks come in packages of 5, what is the smallest number of pencils and notebooks you can buy so that you can match each pencil with a notebook and so that no pencils or notebooks will be left over? Explain why you can solve the problem the way you do.

The total number of pencils you buy will be a multiple of 12, and the total number of notebooks you buy will be a multiple of 5. You want both of these multiples to be equal and to be as small as possible. Therefore, the number of pencils and notebooks you will buy will be the least common multiple of 12 and 5. The multiples of 12 are

$$12, 24, 36, 48, 60, 72, 84, 96, 108, \ldots$$

The multiples of 5 are

$$5, 10, 15, 20, 25, 30, 35, 40, 45, 50, 55, 60, 65, \ldots$$

The only common multiples that we see so far in these lists is 60, so this is the LCM of 12 and 5. Therefore, you should buy 5 packs of pencils and 12 packs of notebooks for a total of 60 pencils and 60 notebooks.

6.

2	224,	392
2	112,	196
2	56,	98
7	28,	49
	4,	7

GCF $= 2 \times 2 \times 2 \times 7 = 56$
LCM $= 2 \times 2 \times 2 \times 7 \times 4 \times 7 = 1568$

7.

5	12,375,	16,875
5	2475,	3375
5	495,	675
3	99,	135
3	33,	45
	11,	15

GCF $= 5 \times 5 \times 5 \times 3 \times 3 = 1125$
LCM $= 5 \times 5 \times 5 \times 3 \times 3 \times 11 \times 15 = 185,625$

Problems for Section 8.2

1. Why do we not talk about a *greatest* common multiple and a *least* common factor?

2. Show how to use the definition of GCF to determine the greatest common factor of 27 and 36.

3. Show how to use the definition of GCF to determine the greatest common factor of 30 and 77.

4. Show how to use the definition of LCM to determine the least common multiple of 44 and 55.

5. Show how to use the definition of LCM to determine the least common multiple of 7 and 8.

6. Show how to use the slide method to determine the GCF and LCM of 2880 and 2400.

7. Show how to use the slide method to determine the GCF and LCM of 144 and 2240.

8. Show how to use the slide method to determine the GCF and LCM of 360 and 1344.

9. Show all the details in the following calculations:

a. Put $\frac{90}{126}$ in simplest form by first determining the greatest common factor of 90 and 126. Show the details of determining this greatest common factor.

b. Put $\frac{90}{126}$ in simplest form without first determining the greatest common factor of 90 and 126.

10. Show all the details in the following calculations:

a. Add $\frac{5}{12} + \frac{1}{8}$ by using the least common denominator. Show the details of determining the least common denominator. Give the answer in simplest form.

b. Add $\frac{5}{12} + \frac{1}{8}$ by using the common denominator $12 \cdot 8$. Give the answer in simplest form.

11. Write a story problem that requires calculating the least common multiple of 45 and 40 to solve. Solve your problem, explaining how the least common multiple is involved.

12. Write a story problem that requires calculating the greatest common factor of 45 and 40 to solve. Solve your problem, explaining how the greatest common factor is involved.

13. Suppose you are teaching children about least common multiples and you use only the following examples to illustrate the concept:

$$4 \quad \text{and} \quad 7$$
$$3 \quad \text{and} \quad 10$$
$$5 \quad \text{and} \quad 12$$
$$8 \quad \text{and} \quad 9$$

What misconception about the LCM might the children develop? Explain why. What other kinds of examples should the children see? Describe some examples that you think are good, and explain why you think they are good.

14. Kwan and Clevere are playing drums together, making a steady beat. Kwan beats his drum hard

on beats that are multiples of 8. Clevere beats his drum hard on beats that are multiples of 12. Find the first 4 beats on which both Kwan and Clevere will beat their drum hard. Use mathematical terms to describe the first beat, and all the beats, on which both Kwan and Clevere beat their drum hard. Explain.

15. At the zoo, the birds must be fed 12 cups of sunflower seeds and 20 cups of millet seed every day. The zoo keepers would like to find a container for scooping both the sunflower seeds and the millet seeds with. The container should allow the zoo keepers to scoop out the proper amount of sunflower seeds and millet seeds by filling the container completely a certain number of times. What is the largest possible container the zoo keepers could use? Use mathematical terms to describe the size of this container. Explain.

16. There are periodical cicadas with 13-year life cycles and other periodical cicadas with 17-year life cycles. The ones with 13-year life cycles become active adults every 13 years, and the ones with 17-year life cycles become active adults every 17 years. Suppose that one year, both the 13-year cicadas and the 17-year cicadas are active adults simultaneously. Find the next four times that both kinds of cicadas will be active adults simultaneously. Use mathematical terms to describe the next time, and all the times, when both kinds of cicadas will be active adults simultaneously. Explain.

17. In a clothing factory, a worker can sew 18 Garment A seams in a minute and 30 Garment B seams per minute. If the factory manager wants to complete equal numbers of Garments A and B every minute, how many workers should she hire for each type of garment? Give three different possibilities, and find the smallest number of workers the manager could hire. Explain your answers.

18. Keiko has a rectangular piece of fabric that is 48 inches wide and 72 inches long. Keiko wants to cut her fabric into identical square pieces, leaving no fabric remaining. She wants the side lengths of the squares to be whole numbers of inches.

a. Draw rough sketches indicating three different ways that Keiko could cut her fabric into squares.

b. Keiko decides that she wants her squares to be as large as possible. How big should Keiko make her squares? Explain your reasoning.

Draw a sketch showing how Keiko should cut the squares from her fabric.

19. A large gear will be used to turn a smaller gear. The large gear will make 300 revolutions per minute. The smaller gear must make 1536 revolutions per minute. How many teeth could each gear have? Give three different possibilities, and find the smallest

number of teeth each gear could have. Explain your reasoning.

20. A large gear is used to turn a smaller gear. The large gear has 60 teeth and makes 12 revolutions per minute; the small gear has 24 teeth. How fast does the small gear turn (i.e., how many revolutions per minute does it make)? Explain your reasoning.

8.3 Prime Numbers

In this section, we will discuss the prime numbers, which can be considered the building blocks of the counting numbers. Interestingly, although prime numbers might seem to be only of theoretical interest, they actually have important practical applications to encryption. So, for example, when you use a secure Web site on the Internet, prime numbers are involved.

prime numbers

primes

The **prime numbers**, or simply, the **primes**, are the counting numbers other than 1 that are divisible only by 1 and themselves. In other words, the prime numbers are the counting numbers that cannot be factored in a nontrivial way. For example, 2, 3, 5, and 7 are prime numbers, but 6 is not because $6 = 2 \times 3$.

The prime numbers are considered to be the *building blocks of the counting numbers*. Why? Because it turns out that every counting number greater than or equal to 2 is either a prime number or can be factored as a product of prime numbers. For example,

$$145 = 29 \times 5$$
$$2009 = 41 \times 7 \times 7$$
$$264,264 = 13 \times 11 \times 11 \times 7 \times 3 \times 2 \times 2 \times 2$$

In this way, all counting numbers from 2 onward are "built" from prime numbers. Furthermore, except for rearranging the prime factors, there is only one way to factor a counting number that is not prime into a product of prime numbers. In other words, it is not possible for a product of prime numbers to be equal to a product of different prime numbers. For example, it couldn't happen that $5 \times 7 \times 23 \times 29$ is equal to a product of some different prime numbers, say, involving 11 or 13 or any primes other than 5, 7, 23, and 29. The fact that every counting number greater than 1 can be factored as a product of prime numbers in a unique way as just described is not at all obvious (and beyond the scope of this book to explain).

Fundamental Theorem of Arithmetic

It is a theorem called the **Fundamental Theorem of Arithmetic**.

By the way, students sometimes wonder why 1 is not included as a prime number. We don't include 1 as a prime number because if we did, we would have to restate the Fundamental Theorem of Arithmetic in a complicated way.

The prime numbers are to the counting numbers as atoms are to matter—basic and fundamental. And just as many physicists study atoms, so too, many mathematicians study the prime numbers. How do we find prime numbers, and how can we tell if a number is prime? We will consider these questions next.

The Sieve of Eratosthenes

Sieve of Eratosthenes

How can we find prime numbers? Eratosthenes (275–195 B.C.) of ancient Greece, discovered a method for finding and listing prime numbers, called the **Sieve of Eratosthenes**. To use the Sieve of Eratosthenes, list all the whole numbers from 2 up to wherever you want to stop looking for prime

TABLE 8.3 Using the Sieve of Eratosthenes to find prime numbers

	2	3	4̶	5	6̶	7	8̶	9̶	1̶0̶
11	1̶2̶	13	1̶4̶	1̶5̶	1̶6̶	17	1̶8̶	19	2̶0̶
2̶1̶	2̶2̶	23	2̶4̶	2̶5̶	2̶6̶	2̶7̶	2̶8̶	29	3̶0̶

numbers. The list in Table 8.3 goes from 2 to 30, so we will look for all the primes up to 30. Next, carry out the following process of circling and crossing out numbers:

1. On the list, circle the number 2, and then cross out every *other* number after 2. (So cross out 4, 6, 8, etc., the multiples of 2.)

2. Then circle the next number that has not been crossed out, 3, and cross out every third number after 3—even if it has already been crossed out. (So cross out 6, 9, 12, etc., the multiples of 3. Notice that you can do this "mechanically," by repeatedly counting, "1, 2, 3," and crossing out on "3.")

3. Circle the next number that has not been crossed out, 5, and cross out every fifth number after 5—even if it has already been crossed out. (So cross out 10, 15, 20, etc., the multiples of 5. Again, notice that you can do this "mechanically," by repeatedly counting, "1, 2, 3, 4, 5," and crossing out on "5.")

4. Continue in this way, going back to the beginning of the list, circling the next number N that has not been crossed out and then crossing out every Nth number after it until every number in the list has either been circled or crossed out.

The circled numbers in the list are the prime numbers.

Class Activity   *Now Turn to Class Activities Manual*

8K   The Sieve of Eratosthenes, p. 167

Why are the circled numbers produced by the Sieve of Eratosthenes the prime numbers in the list? If a number is circled, then we did not cross it out, using one of the previously circled numbers. The numbers we cross out are exactly the multiples of a circled number (beyond that circled number). For example, when we circle 3 and cross out every third number after 3, we cross out the multiples of 3 beyond 3, namely, 6, 9, 12, 15, and so on. Therefore, the circled numbers are the numbers that are not multiples of any smaller number (other than 1). Hence, a circled number is divisible only by 1 and itself and so is a prime number.

The Trial Division Method for Determining Whether a Number Is Prime

How can you tell whether a number, such as 239, is a prime number? You could use the Sieve of Eratosthenes to find all prime numbers up to 239, but that would be slow. A faster way is to divide your number by 2. If it's not divisible by 2, divide it by 3; if it's not divisible by 3, divide it by 5; if it's not divisible by 5, divide it by 7, and so on, dividing by consecutive prime numbers. If you ever find a prime number that divides your number, then your number is not prime; otherwise, it is prime.

trial division   This method for determining whether a number is prime is called the method of **trial division**.

Class Activity   *Now Turn to Class Activities Manual*

8L   The Trial Division Method for Determining Whether a Number Is Prime, p. 168

Why is trial division a valid way to determine whether a number is prime? Why do we have to check only *prime* numbers to see if they divide the number in question? Why don't we have to check other numbers, such as 4, 6, 8, and 9, which are not prime numbers, to see if they divide the given number? If a number is divisible by 4, for example, then it must also be divisible by 2; if a number is divisible by 6, then it must also be divisible by 2 and by 3. In general, if a number is divisible by some counting number other than 1 and itself, then, by the Fundamental Theorem of Arithmetic, that divisor is either a prime number or can be factored into a product of prime numbers, and each of those prime factors must also divide the number in question. So, to determine whether a number is prime, we have to find out only whether any prime numbers divide it.

When we use trial division to determine whether a number is prime, how do we know when to stop dividing by primes? Consider the example of 283. To see if 283 is prime or not, divide it by 2, by 3, by 5, by 7, and so on, dividing by consecutive prime numbers.

We find that none of these prime numbers divide 283, but how do we know when we can stop dividing? Let's look at the quotients that result when we divide 283 by prime numbers in order:

$$283 \div 2 = 141.5$$
$$283 \div 3 = 94.33\ldots$$
$$283 \div 5 = 56.6$$
$$283 \div 7 = 40.43\ldots$$
$$283 \div 11 = 25.73\ldots$$
$$283 \div 13 = 21.77\ldots$$
$$283 \div 17 = 16.65\ldots$$

Notice that as we divide by larger prime numbers, the quotients get smaller. Up to 13, the quotient is larger than the divisor, but when we divide 283 by 17, the quotient, 16.6, is smaller than the divisor. Now imagine that we were to continue dividing 283 by larger prime numbers: by 19, by 23, by 29, and so on. The quotients would get smaller and would be less than 17. So if a prime number larger than 17 were to divide 283, the corresponding quotient would be a whole number less than 17 and thus would have a prime number divisor less than 17. But we already checked all the prime numbers up to 17 and found that none of them divide 283. Therefore, we can already tell that 283 is prime.

In general, to determine whether a number is prime by the trial division method, divide the number by consecutive prime numbers starting at 2. Suppose that none of the prime numbers divide your number. If you reach a point when the quotient becomes smaller than the prime number by which you are dividing, then the previous reasoning tells us your number must be prime.

Of course, it may happen that the number you are checking is not prime. For example, consider 899. When you divide 899 by 2, by 3, by 5, by 7, and so on, dividing by consecutive prime numbers, you will find that none of these primes divides 899 until you get to 29. Then you will discover that $899 = 29 \times 31$, so 899 is not prime.

Factoring Into Products of Prime Numbers

The prime numbers are the building blocks of the counting numbers because every counting number greater than 1 that is not already a prime number can be factored into a product of prime numbers. How do we write numbers as products of prime numbers?

factor tree   A convenient way to factor a counting number into a product of prime numbers is to create a factor tree. A **factor tree** is a diagram like the one in Figure 8.2, which shows how to factor a number, how to factor the factors, how to factor the factors of factors, and so on, until prime numbers are reached.

Starting with a given counting number, such as 600, you can create a factor tree as follows: Factor 600 in any way you can think of—for example,

$$600 = 6 \times 100$$

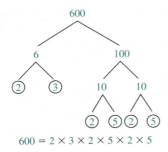

FIGURE 8.2

A factor tree
for 600

$600 = 2 \times 3 \times 2 \times 5 \times 2 \times 5$

Record this factorization below the number 600, as shown in Figure 8.2. Then proceed by factoring each of 6 and 100 separately, and recording these factorizations below each number. Continue until all your factors are prime numbers. Then the numbers at the bottom of the branches of your factor tree are the prime factors of your original number. From Figure 8.2, we determine that

$$600 = 2 \times 3 \times 2 \times 5 \times 2 \times 5$$

You might want to rearrange your factorization by placing identical primes next to each other:

$$600 = 2 \times 2 \times 2 \times 3 \times 5 \times 5$$

Using the notation of exponents, you can also write

$$600 = 2^3 \times 3 \times 5^2$$

Even if you thought of a different way to factor the number initially or along the way, you will still wind up with the same prime factors in the end. This is what the Fundamental Theorem of Arithmetic tells us.

What if we don't immediately see a way to factor a number? For example, how can we factor 26,741? In this case, we try to divide it by consecutive prime numbers until we find a prime number that divides it evenly. We find that 26,741 is not divisible by 2, by 3, by 5, or by 7, but is divisible by 11:

$$26{,}741 = 11 \times 2431$$

Now we need to factor 2431. It is not divisible by 2, by 3, by 5, or by 7. (If it were, 26,741 would also be divisible by one of these.) But we still must find out whether 2431 is divisible by 11, and sure enough, it is:

$$2431 = 11 \times 221$$

Now we determine whether 221 is divisible by 11. It isn't. So we see if 221 is divisible by the next prime, 13. It is:

$$221 = 13 \times 17$$

Collecting the prime factors we have found, we conclude that

$$26{,}741 = 11 \times 11 \times 13 \times 17$$

The factor tree in Figure 8.3 records how we factored 26,741.

Class Activity   *Now Turn to Class Activities Manual*

8M Factoring Into Products of Primes, p. 168

How Many Prime Numbers Are There?

Mathematicians through the ages have been fascinated by prime numbers. There are many interesting facts and questions about them. For instance, does the list of prime numbers go on forever—does it ever come to a stop? Ponder this for a moment before you read on.

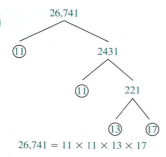

FIGURE 8.3

A factor tree
for 26,741

$26{,}741 = 11 \times 11 \times 13 \times 17$

Do you have a definite opinion about this or are you unsure? If you do have a definite opinion, can you give convincing evidence for it? This is not so easy.

It turns out that the list of prime numbers does go on forever. How do we know this? It certainly isn't possible to demonstrate this by listing all the prime numbers. It is not at all obvious why there are infinitely many prime numbers, but there is a beautiful line of reasoning that proves this.

In about 300 B.C., Euclid, another mathematician who lived in ancient Greece, wrote the most famous mathematics book of all time, *The Elements*. In it appears an argument that proves there are infinitely many prime numbers [25, Proposition 20 of book IX]. The following is a variation of that argument, presented in more modern language:

Suppose that

$$P_1, P_2, P_3, \ldots, P_n$$

is a list of *n* prime numbers. Consider the number

$$P_1 \cdot P_2 \cdot P_3 \cdot \ldots \cdot P_n + 1$$

which is just the product of the list of prime numbers with the number 1 added on. Then this new number is not divisible by any of the primes $P_1, P_2, P_3, \ldots, P_n$ on our list, because each of these leaves a remainder of 1 when our new number is divided by any of these numbers. Therefore, either $P_1 \cdot P_2 \cdot P_3 \cdot \ldots \cdot P_n + 1$ is a prime number that is not on our list of primes or, when we factor our number $P_1 \cdot P_2 \cdot P_3 \cdot \ldots \cdot P_n + 1$ into a product of prime numbers, each of these prime factors is a new prime number that is not on our list. Hence, no matter how many prime numbers we started with, we can always find new prime numbers, and it follows that there must be infinitely many prime numbers.

Practice Exercises for Section 8.3

1. What is a prime number?

2. What is the Sieve of Eratosthenes, and why does it work?

3. Explain how to use trial division to determine whether 283 is prime. Why do you have to divide only by prime numbers? When can you stop dividing and why?

4. For each of the listed numbers, determine whether it is prime. If it is not prime, factor the number into a product of prime numbers.

   a. 1081

   b. 1087

   c. 269,059

   d. 2081

   e. 1147

5. Given that $550 = 2 \cdot 5 \cdot 5 \cdot 11$, find all the factors of 550, and explain your reasoning.

Answers to Practice Exercises for Section 8.3

1. See text.

2. See text.

3. See text.

4. **a.** $1081 = 23 \times 47$.

   b. 1087 is prime.

   c. $269{,}059 = 7 \times 7 \times 17 \times 17 \times 19$.

   d. 2081 is prime.

   e. $1147 = 31 \times 37$.

5. 1 and 550 are factors of 550. Given any other factor of 550, when this factor is factored, it must be a product of some of the primes in the list 2, 5, 5, 11. (If not, you would get a different way to factor 550 into a product of prime numbers, which would contradict the Fundamental Theorem of Arithmetic.) Therefore, the factors of 550 are

$$1, 550, 2, 5, 11, 2 \cdot 5 = 10, 2 \cdot 11 = 22,$$
$$5 \cdot 5 = 25, 5 \cdot 11 = 55, 2 \cdot 5 \cdot 5 = 50,$$
$$2 \cdot 5 \cdot 11 = 110, 5 \cdot 5 \cdot 11 = 275$$

Problems for Section 8.3

1. For which counting numbers, N, greater than 1, is there only one rectangle whose side lengths, in inches, are counting numbers and whose area, in square inches, is N? Explain.

2. ⚗ Use trial division to determine whether 251 is prime. In your own words, explain why you have to divide only by prime numbers and why you can stop dividing when you do.

3. For each of the numbers in (a) through (d), determine whether it is a prime number. If it is not a prime number, factor the number into a product of prime numbers.

   a. 8303

   b. 3719

   c. 3721

   d. 80,000

4. Given that $792 = 2^3 \cdot 3^2 \cdot 11$, find all the factors of 792, and explain your reasoning.

5. **a.** Given that $1440 = 2^5 \cdot 3^2 \cdot 5$ and $4536 = 2^3 \cdot 3^4 \cdot 7$, find the GCF and LCM of 1440 and 4536, and write the GCF and LCM as products of prime numbers. Explain your reasoning.

   b. Given that $374{,}375{,}925 = 3^4 \cdot 5^2 \cdot 7^5 \cdot 11$ and $93{,}767{,}625 = 3^7 \cdot 5^3 \cdot 7^3$, find the GCF and LCM of 374,375,925 and 93,767,625, and write the GCF and LCM as products of prime numbers. Explain your reasoning.

   c. Describe in general how the prime factorizations of the GCF and LCM of two counting numbers are related to the prime factorizations of the two numbers, and explain why this is so.

6. Without calculating the number $19 \times 23 + 1$, explain why this number is not divisible by 19 or by 23.

7. Following Euclid's proof that there are infinitely many primes, and starting the list with 5 and 7, you form the number $5 \times 7 + 1 = 36$, and you get the new prime numbers 2 and 3 because $36 = 2 \times 2 \times 3 \times 3$. Which new prime numbers would you get if your starting list were the following?

   a. 2, 3, 5

   b. 2, 3, 5, 7

   c. 3, 5

8.4 Even and Odd

The study of even and odd numbers is fertile ground for investigating and exploring math. This is an area of math where it's not too hard to find interesting questions to ask about what is true, to investigate these questions, and to explain why the answers are right. Even young children can do this. In this section, in addition to asking and investigating questions about even and odd numbers, we will examine the familiar method for determining whether a number is even or odd.

How would you answer the following questions?

- What does it *mean* for a counting number to be even?

- How can you *tell* if a counting number is even?

Did you give different answers? Then why do these two different characterizations describe *the same numbers*?

even The meaning of *even* is this: A counting number is **even** if it is divisible by 2—in other words, if there is no remainder when you divide the number by 2. There are various equivalent ways to say that a counting number is even:

- A counting number is even if it is divisible by 2—in other words, if there is no remainder when you divide the number by 2.

- A counting number is even if it can be factored into 2 times another counting number (or another counting number times 2).

- A counting number is even if you can divide that number of things into 2 equal groups with none left over.

- A counting number is even if you can divide that number of things into groups of 2 with none left over. (See Figure 8.4.)

odd If a counting number is not even, then we call it **odd**. Each of the ways previously described of saying what it means for a counting number to be even can be modified to say what it means for a counting number to be odd:

- A counting number is odd if it is not divisible by 2—in other words, if there is a remainder when you divide the number by 2.

- A counting number is odd if it cannot be factored into 2 times another counting number (or another counting number times 2).

- A counting number is odd if there is one thing left over when you divide that number of things into 2 equal groups.

- A counting number is odd if there is one thing left over when you divide that number of things into groups of 2. (See Figure 8.4.)

If you have a particular number in mind, such as 237,921, how do you *tell* if the number is even or odd? Chances are, you don't actually divide 237,921 by 2 to see if there is a remainder or not. Instead, you look at the ones digit. In general, if a counting number has a 0, 2, 4, 6, or 8 in the ones place, then it is even; if it has a 1, 3, 5, 7, or 9 in the ones place, then it is odd. So 237,921 is odd because there is a 1 in the ones place. But *why* is this a valid way to determine that there will be a remainder when you divide 237,921 by 2?

Class Activity *Now Turn to Class Activities Manual*

8N   Why Can We Check the Ones Digit to Determine Whether a Number is Even or Odd? p. 169

FIGURE 8.4

Even and odd

even ‖ ‖ ‖ ‖ ‖ ‖ ‖ ‖

odd ‖ ‖ ‖ ‖ ‖ ‖ ‖ ‖ ‖ |

To explain why we can determine whether a counting number is even or odd by looking at the ones place, think about representing counting numbers with bundled toothpicks. For example, we represent 351 with 3 bundles of 100, 5 bundles of 10, and 1 individual toothpick; we represent 4736 with 4 bundles of 1000, 7 bundles of 100, 3 bundles of 10, and 6 individual toothpicks. A counting number is even if that number of toothpicks can be divided into groups of 2 with none left over; the number is odd if there is a toothpick left over. So consider dividing bundled toothpicks that represent a counting number into groups of 2. To divide these toothpicks into groups of 2, we could divide each of the bundles of 10, of 100, of 1000, and so on, into groups of 2. Each bundle of 10 toothpicks can be divided into 5 groups of 2. Each bundle of 100 toothpicks can be divided into 50 groups of 2. Each bundle of 1000 toothpicks can be divided into 500 groups of 2, and so on, for all the higher places. For each place from the 10s place on up, each bundle of toothpicks can always be divided into groups of 2 with none left over. Therefore, the full number of toothpicks can be divided into groups of 2 with none left over exactly when the number of toothpicks that are in the ones place can be divided into groups of 2 with none left over. This is why we have to check only the ones place of a counting number to see if the number is even or odd.

Class Activity   *Now Turn to Class Activities Manual*

8O   Questions about Even and Odd Numbers, p. 170

8P   Extending the Definitions of Even and Odd, p. 171

Practice Exercises for Section 8.4

1. Explain why it is valid to determine whether a counting number is divisible by 2 by seeing if its ones digit is 0, 2, 4, 6, or 8.

2. If you add an odd number and an even number, what kind of number do you get? Explain why your answer is always correct.

3. If you multiply an even number and an even number, what kind of number do you get? Explain why your answer is always correct.

Answers to Practice Exercises for Section 8.4

1. See text.

2. If you add an odd number to an even number, the result is always an odd number. One way to explain why is to think about putting together an odd num-

ber of blocks and an even number of blocks, as in Figure 8.5. The odd number of blocks can be divided into groups of 2 with 1 block left over. The even number of blocks can be divided into groups of

2 with none left over. When the two block collections are joined together, there will be a bunch of groups of 2, and the 1 block that was left over from the odd number of blocks will still be left over. Therefore, the sum of an odd number and an even number is odd.

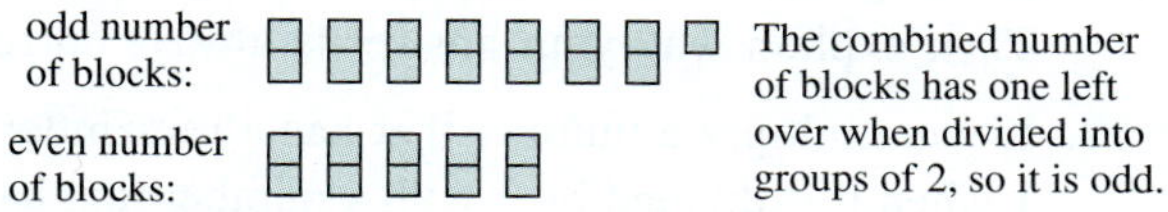

odd number of blocks:

even number of blocks:

The combined number of blocks has one left over when divided into groups of 2, so it is odd.

FIGURE 8.5   Adding an odd number to an even number

3. If you multiply an even number and an even number, the result is always an even number. One way to explain why is to think about creating an even number of groups of blocks, where each group of blocks contains the same even number of blocks, as in Figure 8.6. The total number of blocks is the product

of the number of groups with the number of blocks in each group. Because each group of blocks can be divided into groups of 2 with none left over, the whole collection of blocks can be divided into groups of 2 with none left over. Therefore, an even number times an even number is always an even number. (Notice that this argument actually shows that any counting number times an even number is even.)

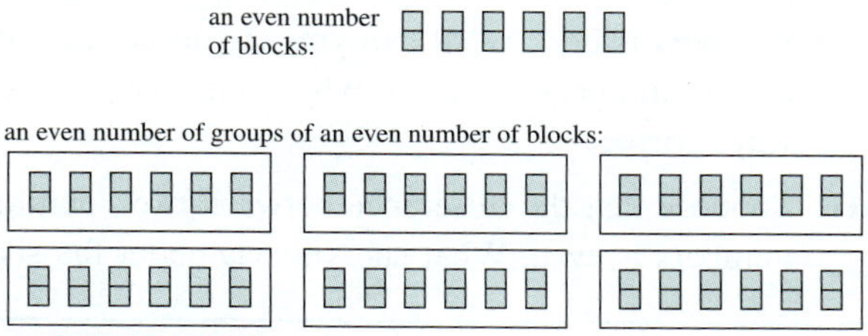

an even number of blocks:

an even number of groups of an even number of blocks:

FIGURE 8.6   Multiplying an even number with an even number

Problems for Section 8.4

1. Describe a way that the children in Mrs. Verner's kindergarten class can tell if there are an even number or an odd number of children present in the class without counting.

2. Explain why an odd counting number can always be written in the form $2N + 1$ for some whole number N.

3. For each of the two designs in Figure 8.7, explain how you can tell whether the number of dots in the design is even or odd without determining the number of dots in the design.

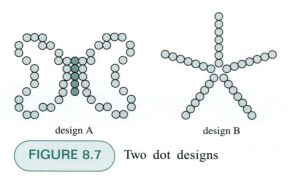

design A   design B

FIGURE 8.7   Two dot designs

4. a. Without determining the number of dots in the following design (Figure 8.8), determine whether this number is even or odd. Explain.

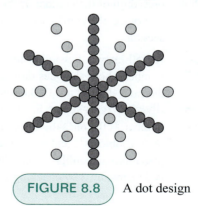

FIGURE 8.8   A dot design

   b. Give some examples of things where you can tell right away that their number is even or odd without actually counting them.

5. If you add an even number and an even number, what kind of number do you get? Explain why your answer is always correct.

6. If you multiply an odd number and an odd number, what kind of number do you get? Explain why your answer is always correct.

7. If you multiply an even number by 3 and then add 1, what kind of number do you get? Explain why your answer is always correct.

8. If you multiply an odd number by 3 and then add 1, what kind of number do you get? Explain why your answer is always correct.

9. Suppose that the difference between two counting numbers is odd. What can you say about the sum of the numbers? Explain why your answer is always correct.

10. Suppose that the difference between two counting numbers is even. What can you say about the sum of the numbers? Explain why your answer is always correct.

11. If you add a number that has a remainder of 1 when it is divided by 3 to a number that has a remainder of 2 when it is divided by 3, then what is the remainder of the sum when you divide it by 3? Investigate this question by working examples. Then explain why your answer is always correct.

12. If you multiply a number that has a remainder of 1 when it is divided by 3 with a number that has a remainder of 2 when it is divided by 3, then what is the remainder of the product when you divide it by 3? Investigate this question by working examples. Then explain why your answer is always correct.

8.5 Divisibility Tests

How can you tell if one counting number is divisible by another counting number? In the previous section, we saw a way to explain why we can look at the ones digit to tell whether a counting number is even or odd. Are there easy ways to tell if a counting number is divisible by other numbers, such as 3, 4, or 5? Yes, there are, and we will study some of them in this section.

divisibility test

A method for *checking* or *testing* to see if a counting number is divisible by another counting number, without actually carrying out the division, is called a **divisibility test**.

divisibility test for 2

The **divisibility test for 2** is just the familiar test to see if a counting number is even or odd: Check the ones digit of the number. If the ones digit is 0, 2, 4, 6, or 8, then the number is divisible by 2 (it is even); otherwise, the number is not divisible by 2 (it is odd). This divisibility test was discussed in the previous section.

divisibility test for 10

A well-known divisibility test is the **divisibility test for 10**. The test is as follows: Given a counting number, check the ones digit of the number. If the ones digit is 0, then the number is divisible by 10; if the ones digit is not 0, then the number is not divisible by 10. Another familiar divisibility test is the

divisibility test for 5

divisibility test for 5. The test is as follows: Given a counting number, check the ones digit of the number. If the ones digit is 0 or 5, then the original number is divisible by 5; otherwise, it is not divisible by 5. Problems 2 and 3 at the end of this section focus on why these divisibility tests are valid. The explanations are similar to the explanation developed in the previous section for why we can see if a number is divisible by 2 by checking its ones digit.

> **Class Activity**   *Now Turn to Class Activities Manual*
>
> **8Q** The Divisibility Test for 3, p. 171

divisibility test for 3

The **divisibility test for 3** is as follows: Given a counting number, add the digits of the number. If the sum of the digits is divisible by 3, then the original number is also divisible by 3; otherwise, the original number is not divisible by 3. This divisibility test makes it easy to determine whether a large whole number is divisible by 3. For example, is 127,358 divisible by 3? To answer this, all we have to do is add all the digits of the number:

$$1 + 2 + 7 + 3 + 5 + 8 = 26$$

Because 26 is not divisible by 3, the original number 127,358 is also not divisible by 3. Similarly, is 111,111 divisible by 3? Because the sum of the digits is 6,

$$1 + 1 + 1 + 1 + 1 + 1 = 6$$

and because 6 is divisible by 3, the original number 111,111 must also be divisible by 3.

Why does the divisibility test for 3 work? In other words, why is it a valid way to determine whether a counting number is divisible by 3? As with the divisibility test for 2 (the test for even or odd), let's consider counting numbers as represented by bundled toothpicks. A counting number is divisible by 3 exactly when that number of toothpicks can be divided into groups of 3 with none left over. So consider dividing bundled toothpicks that represent a counting number into groups of 3. To divide these toothpicks into groups of 3, we could start by dividing each of the bundles of 10, of 100, of 1000, and so on into groups of 3. Each bundle of 10 can be divided into 3 groups of 3 with 1 left over. Each bundle of 100 can be divided into 33 groups of 3 with 1 left over. Each bundle of 1000 can be divided into 333 groups of 3 with 1 left over. For each place from the 10s place on up, each bundle of toothpicks can always be divided into groups of 3 with 1 left over. Now think of collecting all those leftover toothpicks from each bundle and joining them with the individual toothpicks. How many toothpicks will that be? If the original number was 248, then there will be 2 leftover toothpicks from the 2 bundles of 100, as indicated in Figure 8.9. There will be 4 leftover toothpicks from the 4 bundles of 10. (Even though we could make another group of 3 from these, let's not, because we want to explain why we can add the digits of the number to determine if the number is divisible by 3. So we want to work with the digit 4, not 1.) If we combine the leftover toothpicks with the 8 individual toothpicks, then in all, that is

$$2 + 4 + 8$$

toothpicks. The case is similar for any other counting number: If we combine the individual toothpicks with all the leftover toothpicks that result from dividing each bundle of 10, 100, 1000, and so on into groups of 3, then the number of toothpicks we will have is given by the sum of the digits of the number. All the other toothpicks have already been put into groups of 3, so if we can put these leftover toothpicks into groups of 3, then the original number is divisible by 3. If we can't put these leftover toothpicks into groups of 3 with none left, then the original number is not divisible by 3. In other words, we can determine whether a number is divisible by 3 by adding its digits and seeing if that number is divisible by 3.

FIGURE 8.9

Explaining the divisibility test for 3 by dividing individual bundles into groups of 3 and collecting the remainders

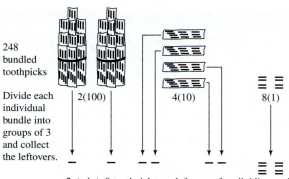

248 bundled toothpicks

Divide each individual bundle into groups of 3 and collect the leftovers.

2(100)   4(10)   8(1)

2 + 4 + 8 toothpicks are left over after dividing each individual bundle into groups of 3.

We can express the previous explanation for why the divisibility test for 3 is valid in a more algebraic way. Let's restrict our explanation to 4-digit counting numbers. Such a number is of the form $ABCD$. Now consider the following equations:

$$ABCD = A \cdot 1000 + B \cdot 100 + C \cdot 10 + D$$
$$= (A \cdot 999 + B \cdot 99 + C \cdot 9) + (A + B + C + D)$$
$$= (A \cdot 333 + B \cdot 33 + C \cdot 3) \cdot 3 + (A + B + C + D)$$

The first equation expresses $ABCD$ in its expanded form. After the second = sign, we have separated the leftover toothpicks that result from dividing the bundles of 1000, 100, and 10 into groups of 3, and we have joined these leftovers with the individual toothpicks to make $A + B + C + D$ toothpicks. The expression after the third = sign shows that the toothpicks other than these remaining $A + B + C + D$ can be divided evenly into groups of 3—namely, into $A \cdot 333 + B \cdot 33 + C \cdot 3$ groups of 3. Therefore, the original $ABCD$ toothpicks can be divided into groups of 3 with none left over exactly when the remaining $A + B + C + D$ toothpicks can be divided into groups of 3 with none left over.

divisibility test for 9 The **divisibility test for 9** is similar to the divisibility test for 3: Given a counting number, add the digits of the number. If the sum of the digits is divisible by 9, then the original number is also divisible by 9; otherwise, the original number is not divisible by 9. To explain why this test is a valid way to see if a number is divisible by 9, use the same argument as for the divisibility test for 3, suitably modified for 9 instead of 3. You are asked to explain this divisibility test in Problem 9.

divisibility test for 4 The **divisibility test for 4** is as follows: Given a whole number, check the number formed by its last two digits. For example, given 123,456,789, check the number 89. If the number formed by the last two digits is divisible by 4, then the original number is divisible by 4; otherwise, it is not. So, because 89 is not divisible by 4, according to the divisibility test for 4, the number 123,456,789 is also not divisible by 4. You are asked to explain this divisibility test in Practice Exercise 2.

Practice Exercises for Section 8.5

1. Use the divisibility test for 3 to determine which of the following numbers are divisible by 3:

   a. 125,389,211,464

   b. 111,111,111,111,111

   c. 123,123,123,123,123

   d. 101,101,101,101,101

2. Explain why the divisibility test for 4 described in the text is a valid way to determine if a counting number is divisible by 4.

3. In your own words, explain why the divisibility test for 3 is a valid way to determine if a counting number is divisible by 3.

Answers to Practice Exercises for Section 8.5

1. a. The sum of the digits of 125,389,211,464 is $1 + 2 + 5 + 3 + 8 + 9 + 2 + 1 + 1 + 4 + 6 + 4 = 46$, which is not divisible by 3. Therefore, 125,389,211,464 is not divisible by 3.

   b. The sum of the digits of 111,111,111,111,111 is 5×3, which is divisible by 3. Therefore, 111,111,111,111,111 is divisible by 3.

   c. The sum of the digits of 123,123,123,123,123 is 5×6, which is divisible by 3. Therefore, 123,123,123,123,123 is divisible by 3.

   d. The sum of the digits of 101,101,101,101,101 is 5×2, which is not divisible by 3. Therefore, 101,101,101,101,101 is not divisible by 3.

2. A counting number is divisible by 4 exactly when that number of toothpicks can be divided into groups of 4 with none left over. Consider representing a counting number physically with bundled toothpicks, and consider dividing the bundles of toothpicks, from the hundreds place on up, into groups of 4. Each bundle of 100 toothpicks can be divided into 25 groups of 4. Each bundle of 1000 toothpicks can be divided into 250 groups of 4. Each bundle of 10,000 toothpicks can be divided into 2500 groups of 4, and so on, for all the higher places. For each place from the 100s place on up, each bundle of toothpicks can always be divided evenly into groups of 4 with none left over. Therefore, the full number of toothpicks can be divided into groups of 4 with none left over exactly when the number of toothpicks that are together in the tens and ones places can be divided into groups of 4 with none leftover.

3. See text.

Problems for Section 8.5

1. Use the divisibility test for 3 to determine whether the following numbers are divisible by 3:

   a. 7,591,348

   b. 777,777,777

   c. 157,157,157

   d. 241,241,241,241

2. According to the divisibility test for 10, to determine whether a counting number is divisible by 10, you have to check only its ones digit. If the ones digit is 0, then the number is divisible by 10. Otherwise, it is not. Give a clear and complete explanation for why this divisibility test is a valid way to determine whether a number is divisible by 10, using your own words.

3. According to the divisibility test for 5, to determine whether a counting number is divisible by 5, you have to check only its ones digit. If the ones digit is 0 or 5, then the number is divisible by 5. Otherwise, it is not. Give a clear and complete explanation for why this divisibility test is a valid way to determine whether a number is divisible by 5, using your own words.

4. Beth knows the divisibility test for 3. Beth says that she can tell just by looking, and without doing any calculations at all, that the number

   999,888,777,666,555, 444,333, 222,111

   is divisible by 3. How can Beth do that? Explain why it's not just a lucky guess.

5. What are all the different ways to choose the ones digit, A, in the number $572435A$, so that the number will be divisible by 3? Explain your reasoning.

6. Sam used his calculator to calculate

   $$123,123,123,123,123 \div 3$$

   Sam's calculator displayed the answer as

   4.1041041041E13

   Sam says that because the calculator's answer is not a whole number, the number 123,123,123,123 is not evenly divisible by 3. Is Sam right? Why or why not? How do you reconcile this with Sam's calculator's display? Discuss.

7. Explain how to modify the divisibility test for 3 so that you can determine the remainder of a counting number when it is divided by 3 without dividing the number by 3. Explain why your method for determining the remainder when a number is divided by 3 is valid. Illustrate your method by determining the remainder of 8,127,534 when it is divided by 3 without actually dividing.

8. For each of the numbers in (a) through (d), verify that the divisibility test for 9 accurately predicts which numbers are divisible by 9. That is, for each number, determine whether it is divisible by 9 by using long division or a calculator. Then see if the sum of the digits of the number is divisible by 9. The two conclusions should agree.

   Example: 52,371 is divisible by 9 because 52,371 ÷ 9 is the whole number 5819, with no remainder. The sum of the digits of 52,371, namely, 5 + 2 + 3 + 7 + 1 = 18, is also divisible by 9.

   a. 1,827,364,554,637,281

   b. 2,578,109

   c. 777,777

   d. 777,777,777

9. **a.** Give a clear and complete explanation for why the divisibility test for 9 is a valid way to determine whether a three-digit counting number *ABC* is divisible by 9, using your own words.

    b. Relate your explanation in part (a) to the following equations:

$$ABC = A \cdot 100 + B \cdot 10 + C$$
$$= (A \cdot 99 + B \cdot 9) + (A + B + C)$$
$$= (A \cdot 11 + B \cdot 1) \cdot 9 + (A + B + C)$$

10. **a.** What are all the different ways to choose the ones digit, *A*, in 271854*A* so that the number will be divisible by 9? Explain your reasoning.

    b. What are all the different ways to choose the tens digit, *A*, and the ones digit, *B*, in the number 631872*AB* so that the number will be divisible by 9? Explain your reasoning.

11. **a.** Find a divisibility test for 25; in other words, find a way to determine if a counting number is divisible by 25 without actually dividing the number by 25.

    b. Explain why your divisibility test for 25 is a valid way to determine whether a counting number is divisible by 25.

12. **a.** Find a divisibility test for 8. In other words, find a way to determine if a counting number is divisible by 8 without actually dividing the number by 8.

    b. Explain why your divisibility test for 8 is a valid way to determine whether a counting number is divisible by 8.

13. **a.** Is it true that a whole number is divisible by 6 exactly when the sum of its digits is divisible by 6? Investigate this by considering a number of examples. State your conclusion.

    b. How could you determine whether the number

      111,222,333,444,555,666,777,888,999,000

is divisible by 6 without using a calculator or doing long division? Explain! (*Hint:* $6 = 2 \times 3$.)

14. Investigate the questions in the following parts (a) and (b) by considering a number of examples.

    a. If a whole number is divisible both by 6 and by 2, is it necessarily divisible by 12?

    b. If a whole number is divisible both by 3 and by 4, is it necessarily divisible by 12?

    c. Based on your examples, what do you think the answers to the questions in (a) and (b) should be?

    d. Based on your answer to part (c), determine whether

        3,321,297,402,348,516

is divisible by 12 without using a calculator or doing long division. Explain your reasoning.

15. **a.** If you add 2 consecutive counting numbers (such as 47, 48), will the resulting sum always, sometimes, or never be divisible by 2? Explain why your answer is true.

    b. If you add 3 consecutive counting numbers (such as 47, 48, 49), will the resulting sum always, sometimes, or never be divisible by 3? Explain why your answer is true.

    c. If you add 4 consecutive counting numbers (such as 47, 48, 49, 50), will the resulting sum always, sometimes, or never be divisible by 4? Explain why your answer is true.

    d. If you add 5 consecutive counting numbers (such as 47, 48, 49, 50, 51), will the resulting sum always, sometimes, or never be divisible by 5? Explain why your answer is true.

    e. If you add *N* consecutive counting numbers, will the resulting sum always, sometimes, or never be divisible by *N*? Explain why your answer is true.

8.6   Rational and Irrational Numbers

How are fractions and decimals related? We saw in Section 6.3 that every fraction whose numerator and denominator are whole numbers can be written as a decimal by dividing the denominator into the numerator. What about the other way around? If we have a decimal, is it the decimal representation of a fraction? In this section, we will analyze the surprising fact that not every decimal comes from a fraction. In fact, the decimals that come from fractions are of two special types: repeating or terminating. We will then see that every decimal that either repeats or terminates can be written as a fraction, and we will see how to write such decimals as fractions. In particular, we will note that the number 0.999999 . . . , where the 9s repeat forever, is equal to 1. Finally, in this section, we will consider why numbers such as $\sqrt{2}$ and $\sqrt{3}$ cannot be written as a fraction, and we'll look at a surprising consequence of this fact reflected in the way pattern tiles—which even the very youngest children play with—behave.

Before we continue, let's review the standard terminology that mathematicians use when discussing fractions versus decimals.

rational numbers The **rational numbers** are those numbers that can be expressed as a fraction or the negative of a fraction. Note that we can write the fraction $\frac{1}{3}$ as the decimal 0.33333 . . . ; these are two different ways to write the same number, and this number is rational no matter which way we write it.

real numbers The **real numbers** are those numbers that can be expressed as a decimal number or the negative of a decimal number. Included among the real numbers are those decimals that have infinitely many nonzero entries to the right of the decimal point.

irrational A real number is called **irrational** if it is not rational, in other words, if it cannot be written as a fraction or the negative of a fraction.

Every fraction can be expressed as a decimal, so the following question is perfectly natural: Are there any irrational numbers or are all real numbers rational? As we will see in this section, there are indeed irrational numbers. The discovery of irrational numbers was shocking and disconcerting to people in ancient times. We will also see that there is an exact characterization of the decimals that represent rational numbers.

Class Activity   *Now Turn to Class Activities Manual*

8R   Decimal Representations of Fractions, p. 173

Decimal Representations of Fractions: Terminating or Repeating

Recall from Chapter 6 that we can use longhand division to write a fraction whose numerator and denominator are whole numbers as a decimal. When we write a fraction as a decimal, the decimal representation is one of two types. For example, consider the decimal representations of the fractions in Table 8.4.

terminating The fractions on the right in Table 8.4 have decimal representations that are **terminating** because these decimals only have finitely many nonzero digits. The fractions on the left in Table 8.4 have decimal

repeating representations that are **repeating** because it turns out that each of these has a single digit or a fixed string of digits that repeats forever. Notice that the repeating portion of the decimal need not begin right after the decimal point, as in the decimal representations of $\frac{1}{12}$ and $\frac{1}{888}$.

To indicate that a digit or a string of digits repeats forever, we can write a bar above the repeating portion. For example,

$$0.\overline{3} \qquad \text{means } 0.333333\ldots, \text{ where the 3s repeat forever}$$
$$0.00\overline{126} \qquad \text{means } 0.00126126126\ldots, \text{ where the 126s repeat forever}$$

We will now use the long division process to explain why the decimal representation of every fraction whose numerator and denominator are whole numbers must either terminate or repeat. Consider a proper fraction $\frac{A}{B}$, where A and B are whole numbers and A is less than B. When we carry out the long division process to calculate the decimal representation of $\frac{A}{B}$, we will get various remainders along the way. For example, when we divide $1 \div 7$, we get the remainders

$$1, 3, 2, 6, 4, 5, 1, 3, \ldots$$

TABLE 8.4 Decimal representations of fractions are of two types: repeating or terminating

$\frac{5}{9} = 0.555555555555555\ldots$	$\frac{1}{4} = 0.25$
$\frac{1}{7} = 0.142857142857142\ldots$	$\frac{5}{8} = 0.625$
$\frac{41}{333} = 0.123123123123123\ldots$	$\frac{1}{80} = 0.0125$
$\frac{1}{12} = 0.083333333333333\ldots$	$\frac{41}{100} = 0.41$
$\frac{1}{888} = 0.001126126126126\ldots$	$\frac{12,345}{100,000} = 0.12345$

and when we divide $1 \div 8$, we get the remainders

$$1, 2, 4, 0$$

as shown in Table 8.5. There are three key points to observe about the longhand division process for writing the fraction as a decimal:

1. If we get a remainder of 0, then the decimal representation terminates at that point (i.e., all subsequent digits in the decimal representation of the fraction are 0).

2. If we get a remainder that we got before, the decimal representation will repeat from there on.

3. We can get only remainders that are less than the denominator of the fraction. For example, in finding the decimal representation of $\frac{1}{7}$ we could get only the remainders 0, 1, 2, 3, 4, 5, or 6. (In fact, in that case we get all those remainders except for 0.)

TABLE 8.5 Using long division, we see that decimal representations of fractions either repeat or terminate

repeats		terminates	
↓		↓	
0.1428571		0.125	
7)1.0000000	remainder 1	8)1.000	remainder 1
−7		−8	
30	remainder 3	20	remainder 2
−28		−16	
20	remainder 2	40	remainder 4
−14		−40	
60	remander 6	0	remander 0
−56			terminates
40	remainder 4		
−35			
50	remainder 5		
−49			
10	remainder 1		
−7	repeats		
3	remainder 3		

Now imagine carrying out the longhand division process to find the decimal representation of a proper fraction $\frac{A}{B}$. Suppose that we have found $B - 1$ digits to the right of the decimal point. If any of the remainders were 0, then the decimal representation terminates. Now suppose that none of the remainders we found were 0. Each of the $B - 1$ digits we found in the decimal representation was obtained from a remainder by putting a 0 behind the remainder and dividing B into the resulting number. Because of item 3, and because we are assuming that we didn't get 0 as a remainder, the remainders must be among the numbers

$$1, 2, \ldots, B - 1$$

When we calculate the Bth digit to the right of the decimal point, we use the Bth remainder, which either is 0 or must also be among the numbers

$$1, 2, \ldots, B - 1$$

If the Bth remainder is 0, then the decimal representation terminates at that point. But if the Bth remainder is not 0, then since there are only $B - 1$ distinct nonzero remainders that we can possibly get, there must be 2 equal remainders among the first B remainders. When we get a remainder that has appeared before, the decimal representation repeats. So, in this case, the decimal representation of our fraction is repeating. Therefore, for any proper fraction, $\frac{A}{B}$, where A and B are whole numbers, the decimal representation of $\frac{A}{B}$ either terminates or repeats after at most $B - 1$ places to the right of the decimal point.

We can use the conclusion we just derived to show that there are decimal numbers that are not the decimal representation of any fraction whose numerator and denominator are whole numbers. For example, consider the decimal number

$$0.28228222822228222228\ldots$$

where the pattern of putting more and more 2s in between 8s continues forever. Notice that even though there is a pattern to this decimal, it does not have a *fixed string* of digits that repeats forever. Therefore, this decimal number is neither terminating nor repeating, so it cannot be the decimal representation of a fraction whose numerator and denominator are whole numbers. Thus, the set of real numbers, which consists of all decimal numbers, is strictly larger than the set of rational numbers, which consists of all fractions whose numerator and denominator are whole numbers and the negatives of these fractions.

Using Subdivided Squares to See Repeating Decimal Representations of Fractions

Students sometimes work with large squares that have been subdivided into 100 smaller squares (or even into 1000 tiny rectangles) in order to make sense of decimals. If the large square in Figure 8.10 represents 1, then each of the small squares represents one hundredth, or 0.01, and a strip of 10 small squares represents one tenth, or 0.1. To use a subdivided square to represent the fraction $\frac{1}{3}$ as a decimal (to the hundredths place), we first need to divide the square into 3 equal parts. One of those parts will represent $\frac{1}{3}$, so we need to determine how many small squares it takes to make that part. Figure 8.10 shows how to subdivide the square into 3 equal parts. One part consists of all the pieces labeled 1, a second part consists of all the pieces labeled 2, and a third part consists of all the pieces labeled 3.

FIGURE 8.10

The shaded area is $\frac{1}{3}$ of the large square. Therefore, $\frac{1}{3} = 3 \cdot \frac{1}{10} + 3 \cdot \frac{1}{100} + \ldots = 0.33\ldots$

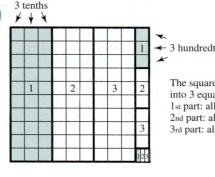

3 tenths

1 ← 3 hundredths

The square is divided into 3 equal parts:
1st part: all pieces labeled "1"
2nd part: all pieces labeled "2"
3rd part: all pieces labeled "3"

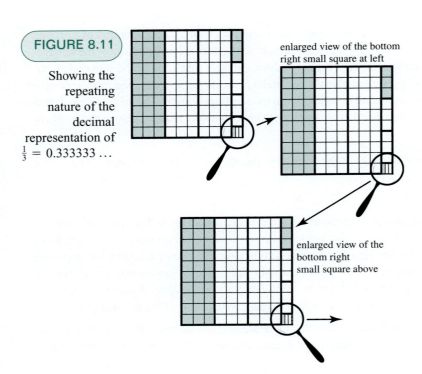

FIGURE 8.11

Showing the repeating nature of the decimal representation of $\frac{1}{3} = 0.333333\ldots$

enlarged view of the bottom right small square at left

enlarged view of the bottom right small square above

Interestingly, and perhaps surprisingly, we can use the subdivided square to *see* the repeating nature of the decimal representation of $\frac{1}{3}$, as indicated in Figure 8.11. Although it is not an important goal for students to be able to use a subdivided square to see why a decimal representation is repeating, some students may find this visual aspect of repeating decimal representations intriguing and appealing.

Writing Repeating and Terminating Decimals as Fractions

As we have seen, if we start with a fraction whose numerator and denominator are whole numbers, its decimal representation is necessarily either terminating or repeating. So if a decimal is neither terminating nor repeating, then it is definitely not the decimal representation of a fraction whose numerator and denominator are whole numbers. But if we have a decimal that is either terminating or repeating, is it necessarily the decimal representation of a fraction whose numerator and denominator are whole numbers? The answer is yes, as we will see.

Class Activity   *Now Turn to Class Activities Manual*

8S Writing Terminating and Repeating Decimals as Fractions, p. 176

Why can every terminating or repeating decimal be written as a fraction whose numerator and denominator are whole numbers? First, it is easy to write a terminating (i.e., finite) decimal as a fraction by using a denominator that is an appropriate power of 10. For example,

$$0.12345 = \frac{12{,}345}{100{,}000} \qquad 1.2345 = \frac{12345}{10{,}000}$$

$$0.2004 = \frac{2004}{10{,}000} \qquad 12.8 = \frac{128}{10}$$

Think of these fractions in terms of division; that is,

$$\frac{2004}{10{,}000} = 2004 \div 10{,}000$$

TABLE 8.6 Facts we can use to write repeating decimals as fractions

$$\frac{1}{9} = 0.\overline{1} = 0.111111\ldots \qquad \frac{1}{9999} = 0.\overline{0001} = 0.00010001\ldots$$

$$\frac{1}{99} = 0.\overline{01} = 0.010101\ldots \qquad \frac{1}{99{,}999} = 0.\overline{00001} = 0.0000100001\ldots$$

$$\frac{1}{999} = 0.\overline{001} = 0.001001\ldots \qquad \vdots$$

From that point of view, we are simply choosing a denominator so that when we divide, the decimal point will go in the correct place.

Now let's see how to write a repeating decimal as a fraction whose numerator and denominator are whole numbers. By using long division we can determine the useful facts in Table 8.6. We can use these facts to write repeating decimals as fractions as shown in Table 8.7. If we have a repeating decimal whose repeating pattern doesn't start right after the decimal point, then we first shift the decimal point by dividing by a suitable power of 10. We then add on the nonrepeating part, as demonstrated in Table 8.8.

Another Method for Writing Repeating Decimals as Fractions

Another method for writing a repeating decimal as a fraction involves shifting the decimal point by multiplying by a suitable power of 10 and then subtracting so that the resulting quantity is a whole number. For example, let N stand for the repeating decimal $0.123123123\ldots = 0.\overline{123}$, so that

$$N = 0.\overline{123} = 0.123123123\ldots$$

Then

$$1000N = 123.123123123\ldots$$

We can subtract N from $1000N$ in two different ways:

$$
\begin{array}{rl}
1000N & 123.123123123\ldots \\
-N & -0.123123123\ldots \\
\hline
999N & 123
\end{array}
$$

TABLE 8.7 Writing repeating decimals as fractions, using facts from Table 8.6

$$0.\overline{12} = 12 \times 0.010101\ldots = 12 \times \frac{1}{99} = \frac{12}{99}$$

$$0.\overline{123} = 123 \times 0.001001001\ldots = 123 \times \frac{1}{999} = \frac{123}{999}$$

$$0.\overline{1234} = 1234 \times 0.000100010001\ldots = 1234 \times \frac{1}{9999} = \frac{1234}{9999}$$

TABLE 8.8 Writing repeating decimals whose repeating part doesn't start right after the decimal point as fractions

$$0.00\overline{12} = \frac{1}{1000} \times 0.121212\ldots = \frac{1}{1000} \times \frac{12}{99} = \frac{12}{99{,}000}$$

$$0.987\overline{12} = 0.987 + 0.00\overline{12} = \frac{987}{1000} + \frac{12}{99{,}000} = \frac{97{,}725}{99{,}000}$$

Therefore,

$$999N = 123$$

and

$$N = \frac{123}{999}$$

Thus,

$$0.\overline{123} = \frac{123}{999}$$

Notice that by multiplying N by 1000 we were able to arrange for the decimal portions to cancel when we subtracted. If we had multiplied by 10 or 100, this canceling would not have happened.

If the repeating part of the decimal does not start right after the decimal point, then we can modify the previous method by shifting with additional powers of 10. For example, let

$$N = 0.98\overline{123} = 0.98123123123\ldots$$

Then

$$100,000N = 98,123.123123123\ldots$$

and

$$100N = 98.123123123\ldots$$

Now subtract $100N$ from $100.000N$ in the following two different ways:

$$
\begin{array}{ll}
100,000N & 98,123.123123123\ldots \\
\underline{-100N} & \underline{-98.123123123\ldots} \\
99,900N & 98,025
\end{array}
$$

Therefore,

$$99,900N = 98,025$$

and

$$N = \frac{98,025}{99,900}$$

Thus,

$$0.98\overline{123} = \frac{98,025}{99,900}$$

Notice that by multiplying by 100,000 and by 100, we arranged for the decimal portions to cancel when we subtracted.

Class Activity   *Now Turn to Class Activities Manual*

8T What is 0.9999...? p. 177

The Surprising Fact That 0.99999... = 1

We can apply either of the two methods for writing a repeating decimal as a fraction to the repeating decimal

$$0.\overline{9} = 0.99999999\ldots$$

When we do this, we reach the following surprising conclusion:

$$0.\overline{9} = 9 \times 0.111111111\ldots = 9 \times \frac{1}{9} = \frac{9}{9} = 1$$

Similarly, let

$$N = 0.\overline{9} = 0.999999\ldots$$

Then

$$10N = 9.999999\ldots$$

Now subtract N from $10N$ in the following two ways:

$$
\begin{array}{rr}
10N & 9.999999\ldots \\
-N & -0.999999\ldots \\
\hline
9N & 9
\end{array}
$$

Therefore,

$$9N = 9$$

and

$$N = 1$$

Thus,

$$0.\overline{9} = 1$$

Although it may seem like 0.999999... should be just a little bit less than 1, it is in fact *equal to* 1.

The Irrationality of the Square Root of 2 and Other Square Roots

square root   Given a positive number N, the **square root** of N, denoted $\sqrt{N}$, is the positive number S such that

$$S^2 = N$$

For example, $\sqrt{2}$ is the positive number such that

$$(\sqrt{2})^2 = 2$$

Square roots of numbers provide many examples of irrational numbers—that is, numbers that cannot be written as fractions whose numerator and denominator are whole numbers. For example, we will see why $\sqrt{3}$ is irrational. In Class Activity 8U you will show that $\sqrt{2}$ is irrational. Note, however, that square roots of some numbers are rational. For example, $\sqrt{4}$ is rational because

$$\sqrt{4} = 2 = \frac{2}{1}$$

so $\sqrt{4}$ can be written as a fraction whose numerator and denominator are whole numbers.

Class Activity   *Now Turn to Class Activities Manual*

8U The Square Root of 2, p. 179

If you use a calculator to find $\sqrt{3}$, your calculator's display might read

$$1.73205080757$$

Of course, a calculator can display only a certain number of digits, so the calculator is not telling you that $\sqrt{3}$ is exactly equal to 1.73205080757. Rather, it is telling you the first few digits in the decimal representation of $\sqrt{3}$ (although the last digit may be rounded). When we look at the calculator's display, it appears that the decimal representation of $\sqrt{3}$ does not repeat or terminate, but we can *never tell for sure* by looking at a finite portion of a decimal representation whether or not the decimal repeats or terminates. The decimal might not start to repeat until after 100 places, or the decimal might appear to repeat at first, but in fact not repeat. So to determine whether a number such as $\sqrt{3}$ is rational or irrational, we must do something other than look at some digits in the decimal representation.

We will now show that $\sqrt{3}$ is irrational by supposing that we could write $\sqrt{3}$ as a fraction whose numerator and denominator are counting numbers and by deducing that this could not be the case. So suppose that

$$\sqrt{3} = \frac{A}{B}$$

where A and B are counting numbers. Then

$$3 = \left(\frac{A}{B}\right)^2$$

so

$$3 = \frac{A^2}{B^2}$$

By multiplying both sides of this equation by B^2, we have

$$3 \cdot B^2 = A^2$$

Now imagine factoring A and B into products of prime numbers. Then A^2 has an even number of prime factors because it has twice as many prime factors as A does. For example, if A were 42, then

$$A = 2 \cdot 3 \cdot 7$$

and

$$A^2 = 2 \cdot 3 \cdot 7 \cdot 2 \cdot 3 \cdot 7$$

which has 6 prime factors, twice as many as A. Similarly, $3 \cdot B^2$ has an odd number of prime factors because it has one more than twice as many prime factors as B. For example, if B were 35, then

$$B = 5 \cdot 7$$

and

$$3 \cdot B^2 = 3 \cdot 5 \cdot 7 \cdot 5 \cdot 7$$

which has 5 prime factors, one more than twice as many as A.

But if $3 \cdot B^2$ has an odd number of prime factors and A^2 has an even number of prime factors, then it cannot be the case that

$$3 \cdot B^2 = A^2$$

because of the uniqueness of factorization into products of prime numbers (the Fundamental Theorem of Arithmetic).

Therefore, it cannot be the case that

$$\sqrt{3} = \frac{A}{B}$$

where A and B are counting numbers. Thus, $\sqrt{3}$ is irrational.

As abstract an idea as the irrationality of $\sqrt{3}$ may seem to be, it is actually reflected in pattern tiles, which even children in PreK play with! You'll see how in the next Class Activity.

Class Activity   *Now Turn to Class Activities Manual*

8V Pattern Tiles and the Irrationality of the Square Root of 3, p. 180

⚡ Proof by Contradiction

The method of proof that we just used to show that $\sqrt{3}$ is irrational is called **proof by contradiction**. Proof by contradiction works as follows: Assume that the statement you want to prove is *false*. Argue logically until you arrive at a contradiction. Therefore, conclude that your assumption (that the statement you want to prove is false) cannot be true. So the statement you want to prove must in fact be true.

Here is how we used proof by contradiction to prove that $\sqrt{3}$ is irrational. We first assumed that $\sqrt{3}$ is *not* irrational; in other words, we assumed that we could write $\sqrt{3}$ as a fraction, $\frac{A}{B}$, where A and B are whole numbers. We reasoned logically and showed that in that case,

$$3 \cdot B^2 = A^2$$

would be true. We also used logical reasoning to show that the prime factorization of $3 \cdot B^2$ would have an odd number of factors and the prime factorization of A^2 would have an even number of factors. But this contradicts the equation $3 \cdot B^2 = A^2$. Therefore, it must be false that $\sqrt{3}$ is *not* irrational, and so $\sqrt{3}$ must be irrational.

Proof by contradiction is an interesting method of proof because it relies on believing that if we can prove that the statement "Statement S is false" is itself false, then statement S must be true. In the past, this method of proof was disputed by some mathematicians.

Practice Exercises for Section 8.6

1. In your own words, explain why the decimal representation of a fraction must either terminate or repeat.

2. Write the following as fractions whose numerator and denominator are whole numbers. Explain your reasoning.

   a. $0.00\overline{577}$

   b. $1.24\overline{3}$

   c. $13.2\overline{83}$

3. What is the 103rd digit to the right of the decimal point in the decimal representation of $\frac{13}{101}$? Explain your answer.

4. What is another way to write $3.45\overline{9}$ as a decimal? Explain your answer.

5. We have seen that $0.\overline{9} = 1$. Does this mean that there are other decimal representations for the numbers $0.\overline{8}$, $0.\overline{7}$, and $0.\overline{6}$?

6. Find at least two fractions whose numerator and denominator are whole numbers and whose decimal representation begins 0.349205986.

7. Are the following numbers rational or irrational?

   a. 0.2522522252252252225252252225... where the pattern of a 2 followed by a 5, two 2s followed

   by a 5, and three 2s followed by a 5 continues to repeat.

   b. 0.2522522522225222252222225... where the pattern of placing more and more 2s between 5s continues.

8. Suppose you have two decimals and each one is either repeating or terminating. Is it possible that the product of these two decimals could be neither repeating nor terminating?

9. Use your calculator to find the decimal representation of $\frac{1}{29}$. When you look at the calculator's display, does the number look as if its decimal representation either repeats or terminates? Does the number in fact have a repeating or terminating decimal representation?

10. a. Find the square root of 5 on a calculator. Make a guess: Does it look like it is rational, or does it look like it is irrational? Why?

    b. Can you tell for sure whether or not $\sqrt{5}$ is rational just by looking at your calculator's display? Explain your answer.

11. Is $\sqrt{1.96}$ rational or irrational?

12. Is the square root of $1.\overline{7}$ rational or irrational?

13. In your own words, explain why $\sqrt{3}$ is irrational.

Answers to Practice Exercise for Section 8.6

1. See text.

2. **a.** Since

$$\frac{1}{999} = 0.\overline{001}$$

it follows that

$$0.\overline{577} = \frac{577}{999}$$

Dividing by 100, we have

$$0.00\overline{577} = \frac{577}{99,900}$$

b. Since

$$0.\overline{3} = \frac{3}{9} = \frac{1}{3}$$

dividing by 100, we have

$$0.00\overline{3} = \frac{1}{300}$$

Since

$$1.24 = \frac{124}{100}$$

it follows that

$$1.24\overline{3} = 1.24 + 0.00\overline{3} = \frac{124}{100} + \frac{1}{300} = \frac{373}{300}$$

c. Since

$$0.\overline{83} = \frac{83}{99}$$

dividing by 10, we have

$$0.0\overline{83} = \frac{83}{990}$$

Since

$$13.2 = \frac{132}{10}$$

it follows that

$$13.2\overline{83} = 13.2 + 0.0\overline{83} = \frac{132}{10} + \frac{83}{990} = \frac{13,151}{990}$$

3. Since

$$\frac{13}{101} = 0.\overline{1287} = 0.128712871287\ldots$$

has a string of 4 digits that repeats, every 4th digit is a 7. So the 4th, the 8th, the 12th, … digit is a 7. Since 100 is divisible by 4, the 100th digit is a 7. The 101st digit is then a 1, the 102nd is a 2, and the 103rd digit is an 8.

4. We know that $0.\overline{9} = 1$. Dividing by 100, we see that $0.00\overline{9} = 0.01$. Therefore,

$$3.45\overline{9} = 3.45 + 0.00\overline{9}$$
$$= 3.45 + 0.01$$
$$= 3.46$$

5. No, it turns out that there is no other way to write decimal representations for these numbers. We can, however, write these numbers as the fractions $\frac{8}{9}, \frac{7}{9}$, and $\frac{6}{9} = \frac{2}{3}$.

6. There are many examples. The fraction

$$\frac{349,205,986}{1,000,000,000}$$

is one example. So is

$$\frac{349,205,986}{999,999,999}$$

(This one is repeating.) So is

$$\frac{3,492,059,861,544}{10,000,000,000,000}$$

(This one is terminating.)

7. **a.** $0.252252225252252225252252225\ldots = 0.\overline{252252225}$

is a repeating decimal, so it is a rational number.

b. $0.2522522252222522222252222225\ldots$

is irrational because it is neither repeating nor terminating.

8. No. To see why not, notice that your two decimals can both be written as fractions whose numerator and denominator are whole numbers because they are repeating or terminating. When you multiply fractions whose numerator and denominator are whole numbers, the result is again a fraction whose numerator and denominator are whole numbers. Therefore, the product of the two decimals must have either a repeating or a terminating decimal representation.

9. On a calculator, the decimal representation looks like it is neither terminating nor repeating. But this

is just because a calculator can display only so many digits. In fact, since $\frac{1}{29}$ is a fraction whose numerator and denominator are whole numbers, its decimal representation must be either terminating or repeating. (In fact, it is repeating.)

10. a. The calculator display of the decimal representation of the square root of 5 does not appear to be either terminating or repeating, so it looks like the square root of 5 is irrational. But this is not a proof.

b. No, you definitely can't tell for sure whether the square root of 5 is irrational just from looking at the calculator's display. Conceivably, its decimal representation might end after 200 digits. Or it

might start to repeat after 63 digits. You can't tell for sure what's going to happen just by looking at the calculator. (It turns out that the square root of 5 is irrational. The proof is the same as for the square root of 3 and the square root of 2.)

11. $\sqrt{1.96} = \sqrt{\dfrac{196}{100}} = \dfrac{14}{10}$

So the square root of 1.96 is rational.

12. $1.\overline{7} = 1 + \dfrac{7}{9} = \dfrac{16}{9}$

So the square root of $1.\overline{7}$ is $\frac{4}{3}$, which is rational.

13. See text.

Problems for Section 8.6

1. a. Use long division to determine the decimal representation of $\frac{1}{37}$. Determine whether the decimal representation repeats or terminates; if it repeats, describe the string of repeating digits.

b. Use long division to determine the decimal representation of $\frac{1}{74}$. Determine whether the decimal representation repeats or terminates; if it repeats, describe the string of repeating digits.

2. a. Use long division to determine the decimal representation of $\frac{1}{101}$. Determine whether the decimal representation repeats or terminates; if it repeats, describe the string of repeating digits.

b. Use long division to determine the decimal representation of $\frac{1}{505}$. Determine whether the decimal representation repeats or terminates; if it repeats, describe the string of repeating digits.

3. Write the following decimal numbers as fractions whose numerator and denominator are whole numbers. Explain your reasoning (you need not write the fractions in simplest form):

a. $0.\overline{56}$, $0.000\overline{56}$, $1.111\overline{56}$

b. $0.9\overline{87}$, $0.009\overline{87}$, $0.409\overline{87}$

c. $1.234\overline{567}$

4. What is the 100th digit to the right of the decimal point in the decimal representation of $\frac{2}{37}$? Explain your reasoning.

5. What is another way to write 1.824 as a decimal? Explain your reasoning.

6. Without actually determining the decimal representation of $\frac{1}{47}$, explain in your own words why its decimal representation must either terminate or repeat after at most 46 decimal places to the right of the decimal point.

7. Give an example of an irrational number, and explain in detail why your number is irrational.

8. In your own words, prove that the square root of 5 is irrational, using the same ideas as we did to prove that the square root of 3 is irrational.

9. Show how to find the exact decimal representation of

$0.\overline{027} \times 0.\overline{18}$
$$= 0.027027027\ldots \times 0.18181818\ldots$$

without using a calculator. Is the product a repeating decimal? If so, what are the repeating digits? *Hint:* Write the numbers in a different form.

10. Carl's calculator displays only 10 digits. Carl punches some numbers and some operations on his calculator and shows you his calculator's display: 0.034482758. Carl has to figure out whether the number whose first 9 digits behind the decimal point are displayed on his calculator is rational or irrational.

a. If you had to make a guess, what would you guess about Carl's number: Do you think it is rational or irrational? Why?

b. Find a fraction of whole numbers whose first nine digits behind the decimal point in its decimal representation agree with Carl's calculator's display. (You do not have to put the fraction in simplest form.)

c. What if Carl's number was actually

0.0344837580034483758003448375800034483 758…,

where the pattern of putting in more and more zeros before a block of 34483758 continues forever? In that case, is Carl's number rational or irrational? Why?

d. Without any knowledge of how Carl got his number to display on his calculator, is it possible to say for sure whether or not it is rational? Explain.

11. Fran has a calculator that shows at most 10 digits. Fran punches some numbers and operations on her calculator and shows you the display: 0.232323232.

a. If you had to make a guess, do you think Fran's number is rational or irrational? Why?

b. Find two different rational numbers whose decimal representation begins 0.232323232.

c. Find an irrational number whose decimal representation begins 0.232323232.

12. Tyrone used a calculator to solve a problem. The calculator gave Tyrone the answer 0.217391304. Without any further information, is it possible to tell if the answer to Tyrone's problem can be written as a fraction whose numerator and denominator are whole numbers? Explain.

13. It is a fact that

$$513{,}239 \times 194{,}841 = 99{,}999{,}999{,}999.$$

Use the preceding multiplication fact to find the exact decimal representations of

$$\frac{1}{513{,}239}$$

and

$$\frac{1}{194{,}841}$$

(Your answer should not just show the first bunch of digits, as on a calculator, but should make clear what *all* the digits in the decimal representation are.) If the decimal representation is repeating, describe the string of digits that repeat. Explain your reasoning. *Hint:* Think about rewriting the fractions, using the denominator 99,999,999,999.

14. Suppose you have a fraction of the form $\frac{1}{A}$, where A is a counting number. Also, suppose that the decimal representation of this fraction is repeating, that the string of repeating digits starts right after the decimal point, and that this string consists of 5 digits.

a. Explain why $\frac{1}{A}$ is equal to a fraction with denominator 99,999.

b. Factor 99,999 as a product of prime numbers.

c. Use parts (a) and (b) to help you find all the prime numbers A such that $\frac{1}{A}$ has a repeating decimal representation with a 5-digit string of repeating digits.

15. Suppose you have a counting number N that divides evenly into 9,999,999 (in other words 9,999,999 ÷ N is a counting number), but N is not 1, 3, or 9. What can you say about the decimal representation of $\frac{1}{N}$? Explain your reasoning.

16. a. For each of the fractions in the first column of Table 8.9, factor the denominator of the fraction as a product of prime numbers.

TABLE 8.9 Decimal representations of various fractions

$\frac{1}{12} = 0.0833333333333333$	$\frac{1}{4} = 0.25$
$\frac{1}{11} = 0.0909090909090909$	$\frac{2}{5} = 0.4$
$\frac{113}{33} = 3.424242424242424$	$\frac{37}{8} = 4.625$
$\frac{491}{550} = 0.8927272727272727$	$\frac{17}{50} = 0.34$
$\frac{14}{37} = 0.3783783783783783$	$\frac{1}{125} = 0.008$
$\frac{35}{101} = 0.3465346534653465$	$\frac{9}{20} = 0.45$
$\frac{1}{41} = 0.0243902439024390$	$\frac{19}{32} = 0.59375$
$\frac{5}{7} = 0.7142857142857142$	$\frac{1}{3200} = 0.0003125$
$\frac{1}{14} = 0.0714285714285714$	$\frac{1}{64{,}000} = 0.000015625$
$\frac{1}{21} = 0.0476190476190476$	$\frac{1}{625} = 0.0016$

b. For each of the fractions in the second column of Table 8.9, factor the denominator of the fraction as a product of prime numbers.

c. Contrast the prime factors that you found in part (a) with the prime factors that you found in part (b). Contrast the decimal representations of the fractions in the first column of Table 8.9 with the decimal representations of the fractions in the second column of Table 8.9. Based on your findings, develop a conjecture (an educated guess) about decimal representations of fractions.

17. **a.** Use a calculator to calculate the decimal representation of

$$\frac{1}{5^{100}}$$

b. Based on your calculator's display, is it possible to tell whether or not $\frac{1}{5^{100}}$ has a repeating decimal representation or a terminating decimal representation? Why or why not?

c. Find a fraction that is equal to $\frac{1}{5^{100}}$ and that has a denominator that is a power of 10.

d. Based on part (c), determine whether $\frac{1}{5^{100}}$ has a repeating or a terminating decimal representation. Explain your reasoning.

18. **a.** Use a calculator to calculate the decimal representation of

$$\frac{1}{7^{100}}$$

b. Based on your calculator's display, is it possible to tell whether or not $\frac{1}{7^{100}}$ has a repeating decimal representation or a terminating decimal representation? Why or why not?

c. Can there be a fraction of whole numbers that is equal to $\frac{1}{7^{100}}$ and that has a denominator that is a power of 10? Why or why not?

d. Based on part (c), determine whether $\frac{1}{7^{100}}$ has a repeating or a terminating decimal representation. Explain your reasoning.

19. **a.** Suppose that a fraction $\frac{A}{B}$ where A and B are counting numbers has a terminating decimal representation and that $\frac{A}{B}$ is in simplest form. Explain why B must divide a power of 10.

b. Suppose that a fraction $\frac{A}{B}$ where A and B are counting numbers has a terminating decimal

representation and that $\frac{A}{B}$ is in simplest form. Show that when B is factored into a product of prime numbers, the only prime numbers that can appear are 2 and 5.

c. Given a counting number B, suppose that when B is factored into a product of prime numbers, the only prime numbers that appear are 2, or 5, or both 2 and 5. Let A be another counting number. Explain why $\frac{A}{B}$ must have a terminating decimal representation.

d. Does

$$\frac{1}{2^{100}}$$

have a repeating or a terminating decimal representation? Explain your answer.

e. Does

$$\frac{1}{3^{100}}$$

have a repeating or a terminating decimal representation? Explain your answer.

20. **a.** Find the decimal representations of the fractions that follow. In each case, compare the number of digits in the repeating string and the *denominator minus one*. (Compare the number of digits in the repeating string of $\frac{1}{7}$ and the number 6, compare the number of digits in the repeating string of $\frac{1}{11}$ and the number 10, etc.) Find a general relationship (other than the fact that the number of digits in the repeating string is less than the denominator minus one). Here are the fractions:

$$\frac{1}{7}, \quad \frac{1}{11}, \quad \frac{1}{13}, \quad \frac{1}{37}, \quad \frac{1}{41}, \quad \frac{1}{73},$$
$$\frac{1}{101}, \quad \frac{1}{137}, \quad \frac{1}{239}, \quad \frac{1}{271}$$

b. The denominators of the fractions in part (a) are prime numbers. Now look at lots of examples of other fractions of the form $\frac{1}{A}$, where A is a whole number that is not a prime number, but where the decimal representation of $\frac{1}{A}$ is repeating. In each case, compare the number of digits in the repeating string and the denominator minus one, and see if the relationship you found in part (a) still holds all the time.

8.7 Looking Back at the Number Systems

In the previous section we saw that the rational numbers (fractions and their negatives) and the real numbers (decimals and their negatives) are related but not the same: Every rational number is a real number, but there are real numbers that are not rational numbers. So the system of real numbers is larger than the system of rational numbers. In this section we take a final look at the number systems—at how they fit together and at the differences among them. We can think of the number systems as fitting together in a hierarchy, with more "advanced" number systems arising from more "basic" number systems. From this perspective, the more advanced systems allow to solve more problems.

First, what is a number system? A number system is a collection of numbers that fit together in a natural way. The most important number systems are the counting numbers, the whole numbers, the integers, the rational numbers, and the real numbers. (You may also have studied the complex numbers, which we do not discuss in this book.)

The most basic system of numbers is the counting numbers (also called the natural numbers),

$$1, 2, 3, 4, 5, \ldots$$

But the counting numbers alone don't allow us to describe situations where there are none of something. The whole numbers,

$$0, 1, 2, 3, 4, 5, \ldots$$

is a number system that solves the problem of not being able to describe "none" by including the number 0.

Given any two whole numbers, when we add them, we still get a whole number. In other words, the whole numbers are "closed under addition." But what about when we subtract two whole numbers? If we subtract 5 from 8, we get the whole number 3. But we can't subtract 8 from 5 within the system of whole numbers. In other words,

$$5 - 8 = ?$$

cannot be solved in the system of whole numbers. So in order to carry out all possible subtractions with whole numbers, we need a larger number system to work in. The integers,

$$\ldots -4, -3, -2, -1, 0, 1, 2, 3, 4, \ldots$$

is a number system that contains the whole numbers and that solves the problem of not always being able to subtract within the whole numbers. Given any two integers, when we add them or subtract them, we still get an integer. In other words, the integers are "closed under addition and subtraction."

Given any two integers, when we multiply them, we still get an integer. But what about when we divide two integers? If we divide 6 by 3, we get the integer 2. But we can't divide 3 by 6 within the integers. In other words,

$$3 \div 6 = ?$$

cannot be solved in the integers. So in order to carry out all possible divisions with integers, we need a larger number system to work in. The rational numbers, which consist of all the fractions and their negatives, is a number system that contains the integers and that solves the problem of not always being able to divide within the integers. (Why do the rational numbers contain the integers? Because every integer is a rational number: given an integer, N or $-N$, we can write it as $\frac{N}{1}$ or $-\frac{N}{1}$.) Given any two rational numbers, when we add, subtract, multiply, or divide them, we still get a rational number. In other words, the rational numbers are "closed under addition, subtraction, multiplication, and division."

There are some problems, however, that involve only rational numbers but can't be solved within the rational numbers. For example: Given a positive rational number, A, find the side lengths of a square that has area A. If A is 9, we can solve this problem within the rational numbers because a square of side length 3 units has area 9 square units. But if A is 2, for example, then we can't solve this problem within the rational numbers. As we saw in the previous section, $\sqrt{2}$ is irrational, so there is no rational number x such that

$$x^2 = 2$$

and therefore there is no square that has rational side lengths and area 2. However, $\sqrt{2}$ can be expressed as a decimal. The real numbers, which consist of all the decimals and their negatives, is a number system that contains the rational numbers. As with the rational numbers, the real numbers are "closed under addition, subtraction, multiplication, and division."

In summary, the relationships among the number systems are as follows:

- Every counting number is a whole number.

- Every whole number is an integer.

- Every integer is a rational number because if A is an integer, then you can also write it as $\frac{4}{1}$, which shows that it is a rational number.

- Every rational number is a real number because if you have a fraction $\frac{4}{B}$, you can divide B into A to get a decimal representation of the number.

But none of these kinds of numbers are the same:

- Some real numbers are not rational numbers. For example, $\sqrt{2}$ is a real number that is not a rational number.

- Some rational numbers are not integers. For example, $\frac{1}{2}$ is a rational number, but it is not an integer.

- Some integers are not whole numbers. For example, -1 is an integer, but it is not a whole number.

- 0 is a whole number that is not a counting number.

Figure 8.12 shows the relationships between the various sets of numbers. All the numbers that we are working with—whole numbers, integers, rational numbers (fractions), and real numbers (decimal numbers)—are real numbers. So, the real numbers unify all the kinds of numbers we have discussed.

FIGURE 8.12

Relationships among the number systems

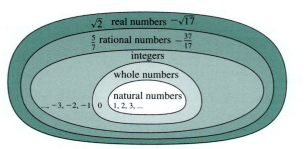

Problems for Section 8.7

1. a. Give your own example of a math problem that is formulated with whole numbers but does not have a whole-number solution. Explain briefly why your problem fulfills the request.

b. Give your own example of a math problem that is formulated with integers but does not have an integer solution. Explain briefly why your problem fulfills the request.

c. Give an example of a math problem that is formulated with rational numbers but does not have a rational solution. Explain briefly why your problem fulfills the request.

2. In your own words, discuss how the whole numbers, the integers, the rational numbers, and the real numbers are related.

Chapter Summary and Study Items

Section 8.1 Factors and Multiples

If A, B, and C are counting numbers and $A = B \times C$, then B and C are factors of A and A is a multiple of B and of C. Also, A is divisible by B and by C, and B and C divide A.

Key skills and understandings:

- State the meaning of the terms *factor* and *multiple*.

- Given a counting number, find all its factors in an efficient way and explain why the method finds all the factors; list several multiples of the number.

- Write and solve story problems that can be solved by finding all the factors of a number. Write and solve story problems that can be solved by finding several multiples of a number.

Section 8.2 Greatest Common Factor and Least Common Multiple

The GCF of two (or more) counting numbers is the greatest counting number that is a common factor of the numbers. The LCM of two (or more) counting numbers is the least counting number that is a common multiple of the numbers. In addition to using the definitions to find the GCF and LCM of counting numbers, we can also use the slide method in which we repeatedly divide by common factors. GCFs and LCMs are often used in fraction arithmetic, although their use is not necessary.

Key skills and understandings:

- State the meaning of GCF and LCM.

- Use the definitions to determine GCFs and LCMs. Use the slide method to determine GCFs and LCMs.

- Write and solve story problems that can be solved by finding a GCF. Write and solve story problems that can be solved by finding a LCM.

- Apply GCFs and LCMs to fraction arithmetic, but understand that the arithmetic can also be done without explicitly using GCFs and LCMs.

Section 8.3 Prime Numbers

The prime numbers are the counting numbers other than 1 that are divisible only by 1 and themselves. We can use the Sieve of Eratosthenes to find all the prime numbers up to a given number. To determine if a given counting number is prime, we can use the method of trial division, in which we check to see if consecutive prime numbers divide the given number. The Fundamental Theorem of Arithmetic states that every counting number other than 1 either is a prime number or can be factored in a unique way into a product of prime numbers. Factor trees can help find the prime factorization of a counting number. There are infinitely many prime numbers.

Key skills and understandings:

- State the meaning of prime number.

- Use the Sieve of Eratosthenes, and explain why it produces a list of prime numbers.

- Use trial division to determine if a given counting number is prime. Explain why you have to divide only by prime numbers when using trial division. Explain why you can stop dividing when you do.

- Given a counting number, determine whether it is prime, or if not, factor it into a product of prime numbers.

Section 8.4 Even and Odd

A counting number is even if it is divisible by 2, or in other words, if there is no remainder when the number is divided by 2. A counting number is odd if it is not divisible by 2, or in other words, if there is a remainder when it is divided by 2. To check if a counting number is even or odd, we have to look only at the ones digit of the number. We can explain why this way of checking if a number is even or odd is valid by considering place value and by recognizing that tens, hundreds, thousands, and so on can all be divided into groups of 2 with none left over.

Key skills and understandings:

- State the meaning of even and odd.

- Explain why it is valid to check if a number is even or odd by examining the ones place of the number.

- Solve problems about even and odd numbers.

Section 8.5 Divisibility Tests

A divisibility test is a quick way to check if a counting number is divisible by a given number without actually dividing. In addition to the divisibility test for 2, there are commonly used divisibility tests for 3, 4, 5, 9, and 10 (plus other divisibility tests). We can explain the divisibility tests by considering place value and by considering what happens when tens, hundreds, thousands, and so on are divided by the number in question.

Key skills and understandings:

- Describe how to use the divisibility tests for 2, 3, 4, 5, 9, and 10, use the divisibility tests, and explain why these tests work.

Section 8.6 Rational and Irrational Numbers

If $\frac{A}{B}$ is a fraction where A and B are counting numbers, then the decimal representation of $\frac{A}{B}$ either terminates or repeats. Conversely, every terminating or repeating decimal can be written as a fraction $\frac{A}{B}$ where A and B are integers. Therefore, the rational numbers are exactly those numbers whose decimal representation either terminates or repeats. Surprisingly, the rational number $0.\overline{9}$ is equal to 1. The square root of 2 and the square root of 3 (along with many other square roots) are irrational numbers.

Key skills and understandings:

- Understand that the rational numbers (which were defined as fractions whose numerator and denominator are integers) are the same as those numbers that have a terminating or repeating decimal representation.

- Use long division to explain why fractions of counting numbers have decimal representations that either terminate or repeat.

- Given a terminating or repeating decimal, write it as a fraction.

- Explain why $0.\overline{9} = 1$.

- Prove that various square roots such as $\sqrt{2}$ and $\sqrt{3}$ are irrational.

Section 8.7 Looking Back at the Number Systems

The basic number systems are the counting numbers, the whole numbers, the integers, the rational numbers, and the real numbers. Each one of these number systems is contained inside the next one, and the next number system can be thought of as arising from the need to solve problems.

Key skills and understandings:

- Know what the counting numbers (natural numbers), the whole numbers, the integers, the rational numbers, and the real numbers are and how these number systems are related.

- Give examples of problems that can be formulated within a number system but require a larger number system to solve.

Algebra

Algebra is the study of the structure of arithmetic and of ways to use the structure of arithmetic to describe situations and to solve problems. Using algebra, we can study growing or repeating patterns and we can solve many problems that can be formulated with an equation. Although children usually do not study algebra formally until the end of middle school or in high school, the foundations for learning algebra must be laid in elementary school. We began to study the structure of arithmetic in Chapters 3 and 4 when we learned about the commutative, associative, and distributive properties and their importance to standard computational procedures and to mental calculation methods. In this chapter, we continue our study of algebra by studying patterns, functions, and some elementary ways to solve equations. Algebra is a deep subject that continues far beyond the content of this book. Our focus here is on aspects of algebra that will be most useful to elementary school teaching.

The National Council of Teachers of Mathematics has made recommendations concerning the learning of algebra (see [53]): Instructional programs should enable all students from prekindergarten through grade 12 to

- *understand patterns, relations, and functions;*
- *represent and analyze mathematical situations and structures using algebraic symbols;*
- *use mathematical models to represent and understand quantitative relationships;*
- *analyze change in various contexts.*

For the NCTM's specific algebra recommendations for grades PreK–2, grades 3–5, grades 6–8, and grades 9–12, go to www.pearsonhighered.com/beckmann.

9.1 Mathematical Expressions and Formulas

 **Focal Points**
Grade 6

Essential skills in algebra include creating and evaluating mathematical expressions, applying formulas, and manipulating and solving equations. In this section and the next, we define some of the basic terms used in algebra and we study some of the ways these fundamental concepts are applied.

Variables, Expressions, and Formulas

variable A **variable** is a letter or other symbol that stands for any number within a specified (or understood) set of numbers. For example, if we say "a circle of radius r has area πr^2," then r is a variable that stands for any positive real number. In this context, it wouldn't make sense for r to be negative, so it's simply understood that r is restricted to positive numbers.

mathematical expression
expression A **mathematical expression**, or simply, **expression**, is a meaningful string of numbers, operations (such as $+$, $-$, $\times$, $\div$, or raising to a power, but not $=$), and possibly variables. For example,

$$3 + 5 \times 2 - 4^3$$

and

$$2x + 3y + 7$$

are expressions. A convention in writing expressions with variables is to drop multiplication symbols between numbers and variables or variables and variables. For example,

$$2x \text{ means } 2 \cdot x$$

and

$$3xy \text{ means } 3 \cdot x \cdot y$$

Also, we usually write numbers before variables when numbers and variables are multiplied in an expression. So instead of writing $x \cdot 2$ or $x2$, we prefer to write $2x$. Since multiplication is commutative, we can always write expressions so that the numbers precede the variables.

Even very young children in elementary school learn about simple mathematical expressions, such as

$$2 + 3$$
$$3 \times 5$$
$$35 \div 7$$

Older elementary school children learn to work with more complex expressions, such as

$$5 \cdot 12 + 7$$
$$\frac{3 \cdot 10}{5 \cdot 9}$$

formula A **formula** is an expression involving variables. A commonly used expression that involves variables is usually called a formula rather than an expression. For example, we talk about the formula for the area of a rectangle of length L and width W,

$$L \times W$$

or the formula for the circumference of a circle of radius r,

$$2\pi r$$

To apply algebra in solving story problems, we must be able to write expressions or formulas for given situations. For example, suppose that Corinna has two collections of postcards on her wall. One collection is arranged in 5 rows with 6 postcards in each row. The other collection is arranged in 10 rows with 2 postcards in each row. Then the following expression stands for the total number of postcards that Corinna has on her wall:

$$5 \cdot 6 + 10 \cdot 2$$

The expression shows the $5 \cdot 6$ postcards from the 5 groups of 6 added to the $10 \cdot 2$ from the 10 groups of 2 postcards. Notice that we do not need to put parentheses around $5 \cdot 6$ and $10 \cdot 2$ because of the conventions of order of operations. (See Section 4.4.)

Evaluating Expressions and Formulas

evaluate an expression   When we have an expression that does not involve variables, we usually want to evaluate it. To **evaluate an expression** that does not involve variables means to determine the number to which the expression is equal. We usually summarize the result in an equation. For example, to evaluate the expression

$$5 \cdot 6 + 10 \cdot 2$$

for the total number of Corinna's postcards, we carry out the indicated operations, following the conventional order of operations. (See Section 4.4). We find that

$$5 \cdot 6 + 10 \cdot 2 = 50$$

Sometimes, we may prefer to string together several equations and evaluate the expression in several steps:

$$5 \cdot 6 + 10 \cdot 2 = 30 + 20$$
$$= 50$$

evaluate a formula   To **evaluate a formula** or an expression that involves variables means to replace the variables with numbers and to determine the number to which the resulting expression is equal. For example, to evaluate

$$3x + 4y$$

at $x = 6$ and $y = 7$, we replace x with 6 and y with 7. We find that

$$3 \cdot 6 + 4 \cdot 7 = 18 + 28$$
$$= 46$$

Note that when a variable occurs several times in an expression, we must replace *all* occurrences of the variable with the same number when evaluating the expression. For example, to evaluate

$$3x^2 + 5x + 1$$

at $x = 2$, we replace each occurrence of x with 2 to obtain

$$3 \cdot 2^2 + 5 \cdot 2 + 1 = 3 \cdot 4 + 10 + 1$$
$$= 12 + 10 + 1$$
$$= 23$$

When we evaluate an expression that has two or more different variables, the values we use for the different variables can be different or they can be the same. Previously we evaluated $3x + 4y$ at $x = 6$ and $y = 7$, but we can also evaluate this same expression at $x = 6$ and $y = 6$:

$$3 \cdot 6 + 4 \cdot 6 = 18 + 24$$
$$= 42$$

Even though the variables x and y are different, we can still use the same numbers for them when we evaluate the expression.

Class Activity   *Now Turn to Class Activities Manual*

9A   Writing Expressions and a Formula for a Flower Pattern, p. 185

9B   Expressions for Dot and Star Designs, p. 187

9C   Expressions with Fractions, p. 188

Evaluating Expressions with Fractions Efficiently

To be able to use algebra effectively, students must become proficient at evaluating expressions. To evaluate expressions correctly and efficiently is just a matter of using the correct order of operations, using the properties of arithmetic, and working with fractions and mixed numbers according to the rules we have studied in previous chapters.

Class Activity   *Now Turn to Class Activities Manual*

9D Evaluating Expressions with Fractions Efficiently and Correctly, p. 188

When an expression involves fractions, we can sometimes find ways to make evaluating the expression easy. For example, to evaluate

$$\frac{8}{57} \cdot \frac{57}{61}$$

we could write

$$\frac{8}{57} \cdot \frac{57}{61} = \frac{8 \cdot 57}{57 \cdot 61} = \frac{456}{3477}$$

We would then put $\frac{456}{3477}$ in its simplest form, $\frac{8}{61}$. However, a more efficient way to evaluate $\frac{8}{57} \cdot \frac{57}{61}$ is to "cancel" the 57s, namely, to write

$$\frac{8}{57} \cdot \frac{57}{61} = \frac{8}{\cancel{57}} \cdot \frac{\cancel{57}}{61} = \frac{8}{61}$$

Why is this canceling valid? Canceling is just shorthand for several steps, such as the following, that use fraction multiplication, the commutative property of multiplication, and equivalent fractions:

$$\frac{8}{57} \cdot \frac{57}{61} = \frac{8 \cdot 57}{57 \cdot 61} = \frac{8 \cdot 57}{61 \cdot 57} = \frac{8}{61} \cdot \frac{57}{57} = \frac{8}{61}$$

Rather than write all these steps, we simply cancel the 57s. In effect, the 57s are just "pulled out" to make $\frac{57}{57}$, which is 1.

How can we make

$$\frac{21}{50} \cdot \frac{1}{35}$$

easy to evaluate? In this case, a factor 7 in the 35 and a factor 7 in the 21 cancel:

$$\frac{\overset{3}{\cancel{21}}}{50} \cdot \frac{1}{\underset{5}{\cancel{35}}} = \frac{3 \cdot 1}{50 \cdot 5} = \frac{3}{250}$$

Again, we can write equations to explain why this canceling is valid:

$$\frac{21}{50} \cdot \frac{1}{35} = \frac{21 \cdot 1}{50 \cdot 35} = \frac{3 \cdot 7}{50 \cdot 5 \cdot 7} = \frac{3}{50 \cdot 5} \cdot \frac{7}{7} = \frac{3}{250}$$

In effect, the 7s are just "pulled out" to make $\frac{7}{7}$, which is 1.

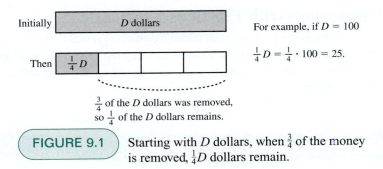

For example, if $D = 100$

$$\frac{1}{4}D = \frac{1}{4} \cdot 100 = 25.$$

$\frac{3}{4}$ of the D dollars was removed,
so $\frac{1}{4}$ of the D dollars remains.

FIGURE 9.1 Starting with D dollars, when $\frac{3}{4}$ of the money is removed, $\frac{1}{4}D$ dollars remain.

Formulating Expressions That Arise in Story Problems

Class Activity   *Now Turn to Class Activities Manual*

9E Expressions for Story Problems, p. 190

Expressions often arise from the situation in a story problem. For example, suppose there are D dollars in a bank account initially. Then $\frac{3}{4}$ of the money in the account is removed and another \$150 is added. At that point, how much money is in the account? This situation gives rise to this expression:

$$\frac{1}{4}D + 150$$

Why? When $\frac{3}{4}$ of the money in the account is removed, $\frac{1}{4}$ of the money remains, so $\frac{1}{4}D$ dollars remain as shown in Figure 9.1.

When \$150 is added, there are then

$$\frac{1}{4}D + 150$$

dollars in the account.

In problem situations, we usually find an expression in order to formulate an equation that can then be solved to solve the problem. We discuss formulating and solving equations in the next three sections.

Practice Exercises for Section 9.1

1. Write an expression that uses addition and multiplication to describe the total number of dots in Figure 9.2.

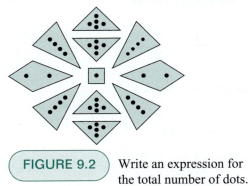

FIGURE 9.2 Write an expression for the total number of dots.

2. Write two expressions for the area of the floor plan in Figure 9.3. (Assume all the angles are right angles.) Explain. Then explain in two different ways why your expressions are equal.

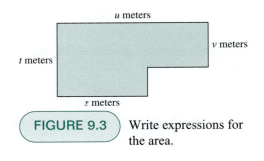

FIGURE 9.3 Write expressions for the area.

3. Write an expression using multiplication and addition (or subtraction) for the fraction of the area of the rectangle in Figure 9.4 that is shaded. Explain briefly. You may assume that all parts that appear to be the same size really are the same size.

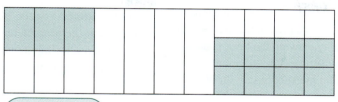

FIGURE 9.4   What fraction is shaded?

4. Show how to make the expression
$$\frac{37}{80} \cdot \frac{40}{37} + \frac{17}{30} \cdot \frac{15}{34}$$
easy to evaluate without a calculator.

5. Write equations that use rules about operating with fractions, as well as properties of arithmetic, to show why it is valid to cancel to evaluate the expression
$$\frac{35 + 21}{63}$$
as follows:
$$\frac{\overset{5}{\cancel{35}} + \overset{3}{\cancel{21}}}{\underset{9}{\cancel{63}}} = \frac{5 + 3}{9} = \frac{8}{9}$$

6. Explain why it is *not valid* to evaluate
$$\frac{35 \cdot 21}{63}$$
by canceling as follows:
warning, incorrect: $\dfrac{\overset{5}{\cancel{35}} \cdot \overset{3}{\cancel{21}}}{\underset{9}{\cancel{63}}} = \dfrac{5 \cdot 3}{9} = \dfrac{15}{9} = 1\dfrac{2}{3}$

7. Goo-Young has lollipops to distribute among goodie bags. When Goo-Young tries to put L lollipops in each of 6 goodie bags, she is 1 lollipop short. Write a formula for the total number of lollipops Goo-Young has in terms of L. Explain briefly.

8. Initially, there were L liters of gasoline in a can.

   a. If $\frac{1}{3}$ of the gasoline in the can is poured out and then another $\frac{1}{2}$ liter of gasoline is poured into the can, then how much gasoline is in the can? Write a formula in terms of L. Evaluate your formula when $L = 3$ and when $L = 3\frac{1}{2}$.

   b. If $\frac{1}{2}$ liter of gasoline is poured into the can and then $\frac{1}{3}$ of the gasoline in the can is poured out, how much gasoline is in the can? Write a formula in terms of L. Is your formula the same as in part (a)? Evaluate your formula when $L = 3$ and when $L = 3\frac{1}{2}$.

9. There were P pounds of apples for sale in the produce area of a store. After $\frac{2}{5}$ of the apples were sold, a clerk brought out another 30 pounds of apples. Then $\frac{1}{3}$ of all the apples in the produce area were sold. Write a formula in terms of P for the number of pounds of apples that are in the produce area now.

Answers to Practice Exercises for Section 9.1

1. There are 4 triangles (arranged vertically in the center) that each contain 5 dots. These 4 triangles contribute 4 · 5 dots. There are 4 triangles (the elongated ones at the corners) that each contain 4 dots. These 4 triangles contribute 4 · 4 dots. There are 2 quadrilaterals (at the left and right) that each contain 2 dots. These quadrilaterals contribute 2 · 2 dots. The square in the center contributes 1 more dot. All together, the number of dots in the figure is
$$4 \cdot 5 + 4 \cdot 4 + 2 \cdot 2 + 1$$

2. In order to find expressions for the area of the floor plan, we should first determine the unspecified side lengths. Since all the angles are right angles, the length of the long vertical side of the floor plan must be equal to the sum of the lengths of the two shorter vertical sides. Similarly, the length of the long horizontal side must be equal to the sum of the lengths of the two shorter horizontal sides. Therefore, the length of the unspecified vertical side is $t - v$ meters

and the length of the unspecified horizontal side is $u - s$ meters.

Figure 9.5 shows two ways to subdivide the floor plan into two rectangles. We can use these subdivisions to find the area of the floor plan. (We could also find an expression for the area by viewing the floor plan as a large t-by-u rectangle with a $t - v$-by-$u - s$ rectangular piece missing.) Using the subdivision on the left, the area is

$$ts + v(u - s)$$

square meters. Using the subdivision on the right, we find that the area is

$$(t - v)s + vu$$

square meters. Since both of these expressions represent the area of the floor plan, they must be equal. Another way to see why the expressions are equal is to apply properties of arithmatic.

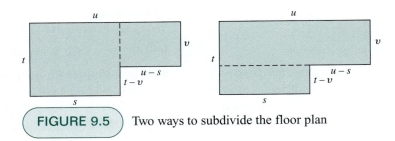

FIGURE 9.5   Two ways to subdivide the floor plan

3. The fraction of the rectangle that is shaded is

$$\frac{1}{2} \cdot \frac{3}{11} + \frac{2}{3} \cdot \frac{4}{11}$$

if we view the rectangle as consisting of 11 identical vertical strips, then the shaded portion on the left is $\frac{1}{2}$ of 3 strips, so it is $\frac{1}{2}$ of $\frac{3}{11}$ of the rectangle and therefore

$$\frac{1}{2} \cdot \frac{3}{11}$$

of the rectangle. The shaded portion on the right is $\frac{2}{3}$ of 4 strips, so it is $\frac{2}{3}$ of $\frac{4}{11}$ of the rectangle and therefore

$$\frac{2}{3} \cdot \frac{4}{11}$$

of the rectangle. All together, the fraction of the area of the rectangle that is shaded is $\frac{1}{2} \cdot \frac{3}{11} + \frac{2}{3} \cdot \frac{4}{11}$.

4. The 37s cancel each other. The 40 in the numerator cancels with 40 in the denominator, leaving 2 in the denominator. The 17 in the numerator cancels with a 17 in the denominator, leaving 2 in the denominator. The 15 in the numerator cancels with a 15 in the denominator, leaving 2 in the denominator. We can thus write

$$\frac{\overset{1}{\cancel{37}}}{\underset{2}{\cancel{80}}} \cdot \frac{\overset{1}{\cancel{40}}}{\underset{1}{\cancel{37}}} + \frac{\overset{1}{\cancel{17}}}{\underset{2}{\cancel{30}}} \cdot \frac{\overset{1}{\cancel{15}}}{\underset{2}{\cancel{34}}} = \frac{1}{2} + \frac{1}{4} = \frac{3}{4}$$

5. We can use the distributive property to "pull out" a 7 in the numerator. We can also "pull out" a 7 in the

denominator to make a $\frac{7}{7}$, which is 1. All together, we have

$$\frac{35 + 21}{63} = \frac{(5 + 3) \cdot 7}{9 \cdot 7} = \frac{5 + 3}{9} \cdot \frac{7}{7} = \frac{8}{9} \cdot 1 = \frac{8}{9}$$

6. In the expression

$$\frac{35 \cdot 21}{63} = \frac{(5 \cdot 7) \cdot (3 \cdot 7)}{9 \cdot 7}$$

the 7 in the denominator can cancel only one of the 7s in the numerator, not both of them. So it is correct to write

$$\frac{35 \cdot 21}{63} = \frac{35 \cdot (3 \cdot 7)}{9 \cdot 7} = \frac{35 \cdot 3}{9} \cdot \frac{7}{7} = \frac{35}{3} = 11\frac{2}{3}$$

7. If Goo-Young were to put L lollipops in each bag, then she'd have $6L$ lollipops in all. But since she is 1 lollipop short, she only has

$$6L - 1$$

lollipops in all. We could also write the formula

$$5L + (L - 1)$$

for the number of lollipops. This formula is valid because there are L lollipops in each of 5 bags, but the 6th bag contains $L - 1$ lollipops.

8. **a.** When $\frac{1}{3}$ of the gasoline in the can is poured out, $\frac{2}{3}$ of the gasoline in the can remains. Therefore, after pouring out the gasoline, there are $\frac{2}{3}L$ liters

remaining in the can. When another $\frac{1}{2}$ liter is poured in, there are

$$\frac{2}{3}L + \frac{1}{2}$$

liters of gasoline in the can. When $L = 3$, the number of liters of gasoline remaining in the can is

$$\frac{2}{3} \cdot 3 + \frac{1}{2} = 2\frac{1}{2}$$

When $L = 3\frac{1}{2}$, the number of liters of gasoline remaining in the can is

$$\frac{2}{3} \cdot 3\frac{1}{2} + \frac{1}{2} = \frac{2}{3} \cdot \frac{7}{2} + \frac{1}{2} = \frac{17}{6} = 2\frac{5}{6}$$

b. When $\frac{1}{2}$ liter of gasoline is poured into the can, there are $L + \frac{1}{2}$ liters in the can. When $\frac{1}{3}$ of this amount is poured out, $\frac{2}{3}$ of the amount remains in the can. Therefore, after pouring out $\frac{1}{3}$ of the gasoline in the can, there are

$$\frac{2}{3}\left(L + \frac{1}{2}\right)$$

liters left in the can. We can use the distributive property to rewrite this formula for the number of liters of gasoline in the can as

$$\frac{2}{3}L + \frac{2}{3} \cdot \frac{1}{2} = \frac{2}{3}L + \frac{1}{3}$$

Notice that this formula, $\frac{2}{3}L + \frac{1}{3}$, is different from the formula $\frac{2}{3}L + \frac{1}{2}$ in part (a). When $L = 3$, the number of liters of gasoline remaining in the can is

$$\frac{2}{3} \cdot 3 + \frac{1}{3} = 2\frac{1}{3}$$

When $L = 3\frac{1}{2}$, the number of liters of gasoline remaining in the can is

$$\frac{2}{3} \cdot 3\frac{1}{2} + \frac{1}{3} = \frac{2}{3} \cdot \frac{7}{2} + \frac{1}{3} = \frac{7}{3} + \frac{1}{3} = 2\frac{2}{3}$$

9. When $\frac{2}{5}$ of the P pounds of apples were sold, $\frac{3}{5}$ of the apples remain. So the number of pounds of apples remaining at this point is $\frac{3}{5}P$. When 30 more pounds are brought in, there are $\frac{3}{5}P + 30$ pounds of apples. When $\frac{1}{3}$ of these apples are sold, $\frac{2}{3}$ of the apples remain. Notice that this is $\frac{2}{3}$ of the $\frac{3}{5}P + 30$ pounds of apples. Therefore, there are now

$$\frac{2}{3} \cdot \left(\frac{3}{5}P + 30\right)$$

pounds of apples set out. Using the distributive property, we can also write this formula as

$$\frac{2}{5}P + 20$$

Problems for Section 9.1

1. Write an expression that uses addition and multiplication to describe the total number of stars in Figure 9.6.

FIGURE 9.6 Write an expression for this picture.

2. Write at least three different expressions for the total number of small squares that Design 1 in Figure 9.7 is made of. Each expression should fit with the way the squares are arranged in the design and should

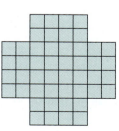
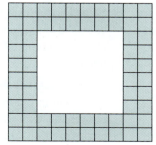

Design 1   Design 2

FIGURE 9.7 Write expressions for the total number of small squares.

involve multiplication and either addition or subtraction. In each case, explain briefly why the expression describes the total number of small squares in the design (without counting the squares or evaluating the expression).

3. Write at least three different expressions for the total number of small squares that Design 2 in Figure 9.7 is made of. Each expression should fit with the way the squares are arranged in the design and should involve multiplication and either addition or subtraction. In each case, explain briefly why the expression describes the total number of small squares in the design (without counting the squares or evaluating the expression).

4. a. Draw a design made of small circles (or another shape) so that the total number of circles in the design is

$$3 \cdot 2 + 3 \cdot 5$$

and so that you can tell this expression stands for the correct number of circles in the design without counting the circles or evaluating the expression. Explain briefly.

  b. Write another expression for the total number of circles in your design in part (a). Explain briefly. (Your expression should not consist of a single number.) Discuss connections to material you have studied previously (such as in Chapter 4).

5. a. Draw a design made of small circles (or another shape) so that the total number of circles in the design is

$$7 \cdot 8 - 2 \cdot 3$$

and so that you can tell this expression stands for the correct number of circles in the design without counting the circles or evaluating the expression. Explain briefly.

  b. Write another expression for the total number of circles in your design in part (a). Explain briefly. (Your expression should not consist of a single number.)

6. a. Write at least two expressions for the area of the floor plan in Figure 9.8. Explain briefly how to obtain the expressions. (Assume that all the angles are right angles.)

  b. Write two expressions for the perimeter of the floor plan in Figure 9.8. Explain briefly how to obtain the expressions.

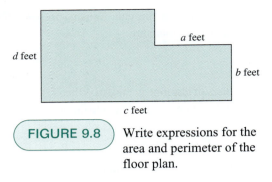

FIGURE 9.8   Write expressions for the area and perimeter of the floor plan.

  c. Explain in two different ways why your expressions in part (a) are equal. One way should use properties of arithmetic, and the other should use the fact that the area is the same no matter how you calculate it.

7. a. Write at least two expressions for the area of the floor plan in Figure 9.9. Explain briefly how to obtain the expressions. (Assume that all the angles are right angles.)

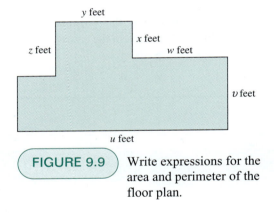

FIGURE 9.9   Write expressions for the area and perimeter of the floor plan.

  b. Write two expressions for the perimeter of the floor plan in Figure 9.9. Explain briefly how to obtain the expressions.

  c. Explain in two different ways why your expressions in part (a) are equal. One way should use properties of arithmetic, and the other should use the fact that the area is the same no matter how you calculate it.

8. Write at least two expressions for the total number of 1-cm-by-1-cm-by-1-cm cubes it would take to build a 6-cm-tall prism (tower) over Base 1 in Figure 9.10. Each expression should use both multiplication and addition (or subtraction). Explain briefly how to obtain the expressions.

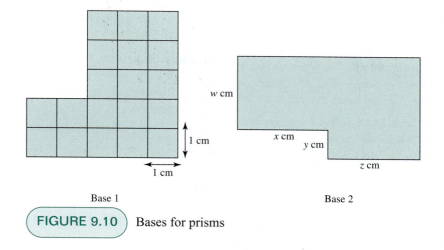

Base 1   Base 2

FIGURE 9.10   Bases for prisms

9. **a.** Write at least two expressions for the total number of 1-cm-by-1-cm-by-1-cm cubes it would
take to build a v-cm-tall prism (tower) over Base 2
in Figure 9.10. (Assume that all the angles are
right angles.) Each expression should be in
terms of v, w, x, y, and z. Explain briefly how to
obtain the expressions.

b. Explain in two different ways why your expressions in part (a) are equal. One way should use
properties of arithmetic, and the other should
use the fact that the number of cubes is the same
no matter how you calculate it.

10. Write an expression using multiplication and addition (or subtraction) for the fraction of the area of
Rectangle 1 in Figure 9.11 that is shaded. Explain
briefly. You may assume that all parts that appear to
be the same size really are the same size.

11. Write an expression using multiplication and addition (or subtraction) for the fraction of the area
of Rectangle 2 in Figure 9.11 that is shaded. Explain briefly. You may assume that all parts that
appear to be the same size really are the same
size.

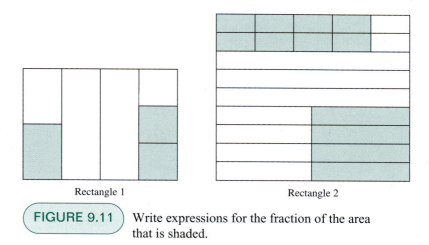

Rectangle 1   Rectangle 2

FIGURE 9.11   Write expressions for the fraction of the area
that is shaded.

12. Shade

$$\frac{2}{3} \cdot \frac{2}{5} + \frac{1}{2} \cdot \frac{1}{5}$$

of a rectangle in such a way that you can tell the
correct amount is shaded without evaluating the
expression. Explain briefly.

13. Shade

$$\frac{3}{4} - \frac{1}{3} \cdot \frac{1}{2}$$

of a rectangle in such a way that you can tell the
correct amount is shaded without evaluating the
expression. Explain briefly.

14. Tyrone has a marble collection that he wants to divide among some bags. When Tyrone puts M marbles into each of 8 bags, he has 3 marbles left over. Write a formula for the total number of marbles Tyrone has in terms of M. Evaluate your formula when $M = 3$ and when $M = 4$.

15. Kenny has stickers to distribute among some goodie bags. When Kenny tries to put 5 stickers into each of G goodie bags, he is 2 stickers short of filling the last bag. Write a formula for the total number of stickers Kenny has in terms of G. Explain.

16. Ted is saving for a toy. Ted has no money now, but he figures that if he saves $\$D$ per week for the next 4 weeks he will have $\$1$ more than he needs for the toy. Write a formula for the cost of the toy Ted wants to buy in terms of D. Explain.

17. Show how to make the expression

$$\frac{21}{13} \cdot \frac{13}{56} + \frac{17}{36} \cdot \frac{18}{34}$$

easy to evaluate without using a calculator. Explain briefly.

18. Show how to make the expression

$$\left(3 - \frac{1}{11}\right) \div \left(\frac{1}{5} + \frac{1}{11}\right)$$

easy to evaluate without using a calculator. Explain briefly.

19. Write equations that use rules about operating with fractions and possibly also properties of arithmetic to show why it is valid to cancel to evaluate the expression

$$\frac{22 \cdot 3}{77}$$

as follows:

$$\frac{\overset{2}{\cancel{22}} \cdot 3}{\underset{7}{\cancel{77}}} = \frac{2 \cdot 3}{7} = \frac{6}{7}$$

20. Explain why it is *not valid* to evaluate

$$\frac{22 + 3}{77}$$

using the canceling shown in the following step:

warning, incorrect: $\dfrac{\overset{2}{\cancel{22}} + 3}{\underset{7}{\cancel{77}}} = \dfrac{2 + 3}{7} = \dfrac{5}{7}$

Show how to evaluate $\frac{22 + 3}{77}$ correctly. (Is canceling possible or not?)

21. Write equations that use rules about operating with fractions as well as properties of arithmetic to show why it is valid to cancel to evaluate the expression

$$\frac{15}{25 + 10}$$

as follows:

$$\frac{\overset{3}{\cancel{15}}}{\underset{5}{\cancel{25}} + \underset{2}{\cancel{10}}} = \frac{3}{5 + 2} = \frac{3}{7}$$

22. If you pay $\$7$ for a night light that costs $\$0.50$ per year in electricity to operate, then how much will you have spent on the night light if you operate it continuously for Y years? Write a formula in terms of Y. Evaluate your formula for the cost of the night light when $Y = 4$ and when $Y = 5\frac{1}{2}$. Explain your work.

23. There were Q quarts of liquid in a container. First, $\frac{3}{4}$ of the liquid in the container was removed. Then another $\frac{1}{2}$ quart was poured into the container. Write an expression in terms of Q for the number of quarts of liquid in the container at the end. Explain your work.

24. Martin had M dollars initially. First, Martin spent $\frac{1}{3}$ of his money. Then Martin spent $\frac{3}{4}$ of what was left. Finally, Martin spent another $\$20$. Write an expression in terms of M for the final amount of money that Martin has. Explain.

25. At a store, the price of an item is $\$P$. Consider three different scenarios:

First scenario: Starting at the price $\$P$, the price of the item was raised by $A\%$. Then the new price was raised by $B\%$.

Second scenario: Starting at the price $\$P$, the price of the item was raised by $B\%$. Then the new price was raised by $A\%$.

Third scenario: Starting at the price $\$P$, the price of the item was raised by $(A + B)\%$.

For each scenario, write an expression for the final price of the item. Are any of the final prices the same? Explain.

9.2 Equations

 **Focal Points**
Grades 6, 7, 8

Equations are a major idea in mathematics. They are used throughout mathematics—in all topics and at every level. Different types of equations arise in different contexts, and we use these equations in different ways.

equation   An **equation** is a statement that an expression or number is equal to another expression or number. For example, the following statements are equations:

$$4x = 8$$
$$3x + 5 = 2x + 6$$
$$y = 4x + 1$$
$$3 + 2 = 5$$
$$3 + 2 = 4 + 1$$

Unlike expressions, equations involve an equals (=) sign, which always means "equals" or "is equal to."

Equations That Are Used in Calculations

Most commonly, we use equations in the process of calculating the answer to an arithmetic problem—in other words, in evaluating an expression. For example, to evaluate the expression $2 + 3 \cdot 4$ we might write

$$2 + 3 \cdot 4 = 2 + 12 = 14$$

in order to conclude that the expression $2 + 3 \cdot 4$ is equal to the number 14. Or we might concatenate several equations to show a method of calculation, as in

$$99 + 17 = 99 + (1 + 16) \tag{9.1}$$
$$= (99 + 1) + 16 \tag{9.2}$$
$$= 100 + 16 = 116 \tag{9.3}$$

to conclude that $99 + 17 = 116$. Of course, there are also simple equations that record the result of a calculation directly, such as

$$3 + 2 = 5$$

It is very important, however, that children do not develop a view of the equals sign as a "calculate the answer" symbol. Students who view the equals sign this way will not be able to make sense of solving algebraic equations later on. Instead, students must understand that the equals sign tells us that the left and right side of the equation stand for the same number.

Equations That Are True for All Values of the Variables

We also use equations to describe general relationships, such as the properties of arithmetic. For example, the commutative property of multiplication states that for all real numbers A and B,

$$A \cdot B = B \cdot A$$

Similarly, the commutative property of addition, the associative properties of addition and multiplication, and the distributive property, which states that

$$A \cdot (B + C) = A \cdot B + A \cdot C$$

for all real numbers A, B, and C, are examples of equations that are true for all values of the variables involved.

Equations that are true for all values of the variables are useful because they tell us that we can replace an expression with another equivalent expression. In Equation 9.1, we replaced

$$99 + (1 + 16)$$

with

$$(99 + 1) + 16$$

by applying the associative property of addition. We can apply the distributive property to replace an expression of the form

$$A \cdot B + A \cdot C$$

where A, B, and C are either numbers or other expressions in their own right, with the expression

$$A \cdot (B + C)$$

In this way, we can pick whichever expression is most useful for our purposes. For example, if we want to show that

$$2 \cdot 8134277 + 2 \cdot 906123$$

is even, we can apply the distributive property to rewrite this expression as

$$2 \cdot (8134277 + 906123)$$

This new expression is in the form "2 times some number," so by definition, it represents an even number.

Think of equations that are always true as giving us two different ways to describe an expression, and therefore giving us flexibility in calculating answers and solving problems. The next Class Activities will help you develop some interesting and useful equations that are true for all counting numbers.

Class Activity   *Now Turn to Class Activities Manual*

9F How Many High-Fives?, p. 191

9G Sums of Counting Numbers, p. 192

9H Sums of Odd Numbers, p. 193

9I Equations Arising from Rectangular Designs, p. 194

Equations That Describe How Quantities Are Related

Another use of equations is to describe a relationship that holds between specific quantities. For example, if C is the number of cats in a room and P is the number of cat paws in the room, then the equation

$$P = 4C$$

states that the number of cat paws in the room is 4 times the number of cats in the room.

Several precautions are important in formulating equations that describe relationships among quantities. One is making sure that the equation is formulated correctly. For example, students might be tempted to translate

there are 4 times as many paws as cats

as

warning, incorrect: $4P = C$

by translating the sentence word by word into an equation without attending carefully to its meaning. Figure 9.12 shows why the correct equation is $P = 4C$.

FIGURE 9.12

There are 4 times as many cat paws as cats, so $P = 4C$.

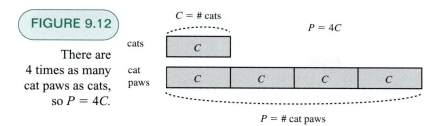

The error of writing an incorrect equation may also arise because students aren't thinking of the letters P and C as standing for *numbers*. Teachers and students should take care to define the variables they use and to make clear that these variables stand for numbers. For example, in a problem about how much lemon juice to use in a recipe, students might write

warning, incorrect: L = lemon juice

using the letter L as an abbreviation for lemon juice, instead of as representing a *number* of units of lemon juice, as in

L = # of cups of lemon juice

Class Activity   *Now Turn to Class Activities Manual*

9J Equations about Related Quantities, p. 196

Equations That Must Be Solved to Solve a Problem

A common way of using equations in algebra is for solving story problems, such as this:

Carlos had some blocks. After he gave 5 blocks to Tanya, he gave half of his remaining blocks to Sam. At that point, Carlos had 18 blocks left. How many blocks did Carlos have at first?

If x stands for the number of blocks that Carlos had at first, then we can formulate the equation

$$\frac{1}{2}(x - 5) = 18$$

To solve the story problem, we'll want to solve the equation; in other words, we'll want to find out what number for x makes the equation true.

In the next section we'll discuss techniques for solving equations. Then in Section 9.4, we'll discuss how to formulate and solve story problems with strip diagrams and how to connect the strip diagrams to standard algebraic methods.

Class Activity   *Now Turn to Class Activities Manual*

9K Equations for Story Problems, p. 197

Practice Exercises for Section 9.2

1. Write two different expressions for the total number of small squares in Figure 9.13. Each expression should use either multiplication or addition, or both. Write an equation by setting these two expressions equal to each other.

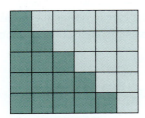

FIGURE 9.13   Write an equation for this picture.

2. If there are 25 people at a party and everyone "clinks" glasses with everyone else, how many "clinks" will there be? Describe *two different ways* to solve this problem, and explain why these methods are valid.

3. Calculate $1 + 2 + 3 + 4 + \cdots + 500$ by finding and evaluating another expression that equals this sum. Explain your method.

4. **a.** Using the four square patterns in Figure 9.14, find a way to express the following sums in terms of multiplication and subtraction:

$$2 + 4$$
$$2 + 4 + 6$$
$$2 + 4 + 6 + 8$$
$$2 + 4 + 6 + 8 + 10$$

Write the associated equations.

   b. Based on your results in part (a), predict the sum

$$2 + 4 + 6 + 8 + \cdots + 98$$

   c. Find an equation whose left-hand side is

$$2 + 4 + 6 + 8 + \cdots + 2N$$

Explain why your equation is true for all counting numbers N.

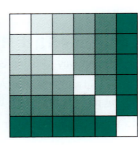

FIGURE 9.14   Sums of even numbers

5. Draw, label, and shade a rectangle so that it gives rise to the equation

$$(x + 3) \cdot (y + 4) = xy + 4x + 3y + 12$$

Explain briefly.

6. To make orange paint, you need to mix yellow and red paint together. For a certain shade of orange paint, you need to use $2\frac{1}{2}$ times as much yellow paint as red paint. Write an equation that corresponds to this situation, taking care to define your variables clearly and correctly.

7. Let P be the initial population of a town. After 10% of the town leaves, another 1400 people move in to the town. The new population is 5000. Write an equation, involving P, that corresponds to this situation.

Answers to Practice Exercises for Section 9.2

1. On the one hand, since Figure 9.13 consists of 5 rows of small squares with 6 squares in each row, the total number of small squares in the figure is $5 \cdot 6$. On the other hand, the darkly shaded portion of the figure consists of

$$1 + 2 + 3 + 4 + 5$$

small squares, which we see by adding the number of squares in the 1st, 2nd, 3rd, 4th, and 5th rows. The lightly shaded portion of the figure is a rotated copy of the darkly shaded portion. Therefore, the unshaded portion has the same number of small squares, so the total number of small squares in the figure is

$$2 \cdot (1 + 2 + 3 + 4 + 5)$$

Thus, the figure gives us the equation

$$2 \cdot (1 + 2 + 3 + 4 + 5) = 5 \cdot 6$$

2. One way to solve this problem is to add

$$24 + 23 + 22 + \cdots + 4 + 3 + 2 + 1 = 300$$

to find the total number of "clinks." To see why, imagine the 25 people lined up in a row. The 25th person in the row could go down the row and clink glasses with all other 24 people (and then stand in the corner). The 24th person in the row could go down the row and clink glasses with the other 23 people left in the row (and then stand in the corner also). The 23rd person could clink with the remaining 22, and so on, until the 2nd person clinks with the first. Counting up all the clinks, you get the previous sum $24 + 23 + \cdots + 2 + 1$.

Another way to solve this problem is to calculate

$$(25 \times 24) \div 2 = 300$$

This method is valid because each of the 25 people clinks with 24 other people. This makes for 25×24 clinks—except that each clink has been counted *twice* this way. Why? Because when Anna clinks with Beatrice, it's the same as Beatrice clinking with Anna, but the 25×24 figure counts those as 2 separate clinks. So, to get the correct count, divide 25×24 by 2.

3. If you think about a large 500-unit-by-501-unit rectangle that is subdivided it into small squares and is shaded in a staircase pattern (as in Figure 9.13), then the number of darkly shaded squares can be expressed in two ways. On the one hand, adding up the darkly shaded squares row by row, there are

$$1 + 2 + 3 + \cdots + 500$$

darkly shaded, small squares. On the other hand, the darkly shaded squares take up half the rectangle, and there are $500 \cdot 501$ small squares in the rectangle, so there are

$$\frac{1}{2}(500 \cdot 501) = 125{,}250$$

darkly shaded, small squares. It's the same number either way you count, so the sum of the first 500 counting numbers must also be 125,250.

4. a. There are $2 + 4$ small, shaded squares in the first square pattern. The shaded squares fill up all but 3 of the $3 \cdot 3$ squares in the pattern. Therefore,

$$2 + 4 = 3^2 - 3$$

Similarly, there are $2 + 4 + 6$ small, shaded squares in the second square pattern. The shaded squares fill up all but 4 of the $4 \cdot 4$ squares in the pattern. Therefore,

$$2 + 4 + 6 = 4^2 - 4$$

Using the same reasoning for the 3rd and 4th square patterns, we see that

$$2 + 4 + 6 + 8 = 5^2 - 5$$
$$2 + 4 + 6 + 8 + 10 = 6^2 - 6$$

b. Assuming that the pattern we found in part (a) continues, it should be the case that

$$2 + 4 + 6 + 8 + \cdots + 98 = 50^2 - 50 = 2450$$

Why $50^2 - 50$? We can find the expressions on the right-hand side of the equations in part (a) by taking half of the last term in the series (on the left-hand side of the equation) and adding 1. Half of 98 is 49; adding 1, we get 50.

c. If we imagine a square made of $N + 1$ rows with $N + 1$ small squares in each row, and if we imagine shading the small squares as in Figure 9.14, then counting the squares of each color, we will find

$$2 + 4 + 6 + \cdots + 2N$$

small, shaded squares. These small, shaded squares fill up all but $N + 1$ squares of the $N + 1$

by $N + 1$ square pattern. Therefore, there are

$$(N + 1)^2 - (N + 1)$$

small, shaded squares. Since it's the same number of small, shaded squares either way you count them,

$$2 + 4 + 6 + \cdots + 2N = (N + 1)^2 - (N + 1)$$

Notice that since

$$(N + 1)^2 - (N + 1) = N^2 + 2N + 1 - N - 1$$
$$= N^2 + N$$

we can also say that

$$2 + 4 + 6 + \cdots + 2N = N^2 + N$$

5. See Figure 9.15. Overall, the rectangle has dimensions $x + 3$ by $y + 4$ and therefore has area $(x + 3) \cdot (y + 4)$ square units. The 4 portions making up the rectangle have area xy (top left), $4x$ (top right), $3y$ (bottom left), and 12 (bottom right) square units. Therefore, all together, the area of the rectangle is $xy + 4x + 3y + 12$ square units. Since it's the same area either way we calculate it, we have illustrated the equation

$$(x + 3) \cdot (y + 4) = xy + 4x + 3y + 12$$

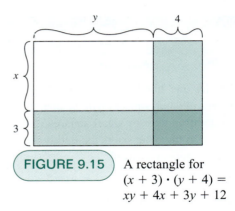

FIGURE 9.15 A rectangle for $(x + 3) \cdot (y + 4) = xy + 4x + 3y + 12$

6. If you let Y be the number of liters of yellow paint that you will use and you let R be the number of liters of red paint you will use, then an equation relating the two is

$$Y = 2.5R$$

7. After 10% of the town leaves, 90% of the town remains. So, after the 10% leave, there are $0.9P$ people in the town. When another 1400 people move in, the population rises to $0.9P + 1400$. We are given that this new population is 5000. Therefore,

$$0.9P + 1400 = 5000$$

Problems for Section 9.2

1. Write two different expressions for the total number of small squares in design (a) of Figure 9.16. Each expression should use either multiplication or addition, or both. Write an equation by setting these two expressions equal to each other. Explain briefly.

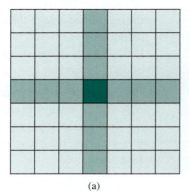

(a)

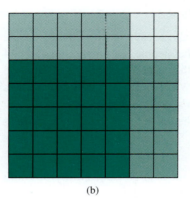

(b)

FIGURE 9.16 Write equations for these square patterns.

2. Write two different expressions for the total number of small squares in design (b) of Figure 9.16. Each expression should use either multiplication or addition, or both. Write an equation by setting these two expressions equal to each other. Explain briefly.

3. By determining the total number of small squares in design (a) in Figure 9.17 in two different ways, find an equation that the design gives rise to. (One side of the equation should correspond to the shading, the other side should correspond to the overall rectangular shape.) Explain briefly.

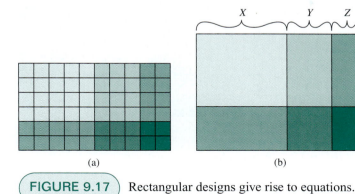

(a)   (b)

FIGURE 9.17 Rectangular designs give rise to equations.

4. a. Find an equation that design (b) in Figure 9.17 gives rise to. One side of the equation should be

$$(X + Y + Z) \cdot (S + T)$$

Explain briefly how the design gives rise to your equation.

b. Explain how to obtain your equation in part (a) from properties of arithmetic.

5. Draw, label, and shade a rectangle so that it gives rise to the equation

$$(A + B + C) \cdot (X + Y + Z) = AX + AY + AZ \\ + BX + BY + BZ \\ + CX + CY + CZ$$

Explain briefly.

6. Draw, label, and shade a square so that it gives rise to the equation

$$(A + B)^2 = A^2 + 2AB + B^2$$

Explain briefly.

7. a. Write two different expressions, each using only the numbers 9, 5, and 3, for the unknown length (?) on the left of Figure 9.18. One of your expressions should involve parentheses. Write an equation by setting these two expressions equal to each other. Explain briefly.

b. Write two different expressions, for the unknown length (?) on the right of Figure 9.18. One of your expressions should involve parentheses. Write an equation by setting these two expressions equal to each other. Explain briefly.

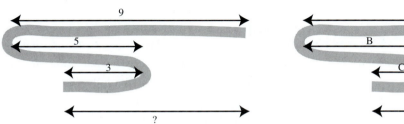
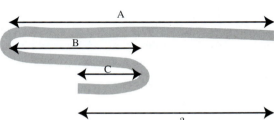

FIGURE 9.18 Write an equation arising from the unknown length.

8. a. For each of the 4 rectangular patterns in Figure 9.19, find two different ways to express the total number of small squares that the pattern is made of. In each case, write an associated equation.

b. Based on your work in part (a), write an equation whose left-hand side is the sum

$$1 + 5 + 9 + 13 + \cdots + 93 + 97$$

Use the right-hand side of the equation to evaluate this sum. Explain your reasoning.

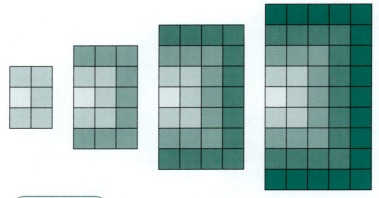

FIGURE 9.19 Patterns for some equations

9. An auditorium has 50 rows of seats. The first row has 20 seats, the second row has 21 seats, the third row has 22 seats, and so on, each row having one more seat than the previous row. How many seats are there all together?

   a. Write two different expressions for the total number of seats in the auditorium. One expression should use addition; the other should involve multiplication. Write an equation that uses these two expressions.

   b. Solve the auditorium problem, explaining your reasoning.

10. Max was asked to write an equation corresponding to the following situation:

    There are 2 times as many pencils in Antrice's pencil box as in Ben's pencil box.

    Max responded thus:

    A = Antrice B = Ben

    $2A = B$

    Discuss Max's work, describing and correcting any errors he has made. Explain why your corrections are correct.

11. A company sells $3\frac{1}{4}$ times as many tissues as napkins. Write an equation that corresponds to this situation. Explain why your equation is correct. Be sure to define your variables with care.

12. Write a story problem for the equation

    $$6x = y$$

    Explain briefly why your story problem fits with this equation.

13. **a.** Try this mathematical magic trick several times.

    • Write down a secret number, which can be any whole number.

    • Multiply by 5.

    • Add 2.

    • Multiply by 4.

    • Add 2.

    • Multiply by 5.

    • Subtract 50.

    • Drop the two zeros at the end.

    Do you recognize the result?

   b. Let S stand for a secret number that you begin with in part (a). Write an expression in terms of S that shows the result of applying the steps in part (a). Apply properties of arithmetic to this expression to explain why the end result of the steps in part (a) is always predictable.

14. Bonnie had B stickers. First, Bonnie gave away $\frac{1}{6}$ of her stickers. Then Bonnie got 12 more stickers, and she has 47 stickers. Write an equation, involving B, that corresponds to this situation. Explain your work.

15. There are F cups of flour in a bag. First, $\frac{3}{8}$ of the flour in the bag was used. Then $\frac{1}{5}$ of the remaining flour was used. At that point, there were 10 cups of flour in the bag. Write an equation, involving F, that corresponds to this situation. Explain your work.

16. Write a story problem for the equation

$$(x + 15) - \frac{1}{3}(x + 15) = 50$$

Explain briefly why your story problem fits with this equation.

17. a. Write a story problem for the equation

$$\frac{4}{5}x - 25 = 35$$

Explain briefly why your story problem fits with this equation.

b. Write a story problem for the equation

$$\frac{4}{5}(x - 25) = 35$$

Explain briefly why your story problem fits with this equation.

9.3 Solving Equations

 **Focal Points**
Grades 6, 7

A major part of algebra is solving equations. In this section, we focus on the fundamental ideas we need to solve equations. We will apply these ideas to solve some of the simplest kinds of equations.

Solutions of Equations

We must first ask what it means to solve an equation and what the solutions of an equation are. Some equations are true; for example, the equations

$$3 + 2 = 5$$
$$6 \times 3 = 3 \times 6$$
$$10 = 2 \times 5$$

are true. Some equations are false; for example, the equations

$$1 = 0$$
$$3 + 2 = 6$$

are false. Most equations with variables are true for some values of the variable and false for other values of the variable. For example, the equation

$$3 + x = 5$$

is true when $x = 2$, but is false for all other values of x.

solve an equation
solutions

To **solve an equation** involving variables means to determine those values for the variables which make the equation true. The values for the variables which make the equation true are called the **solutions** to the equation. To solve the equation

$$3x = 12$$

means to find all those values for x for which $3x$ is equal to 12. Only $3 \cdot 4$ is equal to 12; 3 times any number other than 4 is not equal to 12. So, there is only one solution to $3x = 12$, namely, $x = 4$.

Even young children in elementary school learn to solve simple equations. For example, first- or second-graders could be asked to fill in the box to make the following equation true:

$$5 + \square = 7$$

One source of difficulty in solving equations is understanding that the equals sign does not mean "calculate the answer." For example, when children are asked to fill in the box to make the equation

$$5 + 3 = \square + 2$$

true, many will fill in the number 8 because $5 + 3 = 8$. In order to understand how to solve equations, we must understand that we want to make the expressions to the left and right of the equal sign equal to

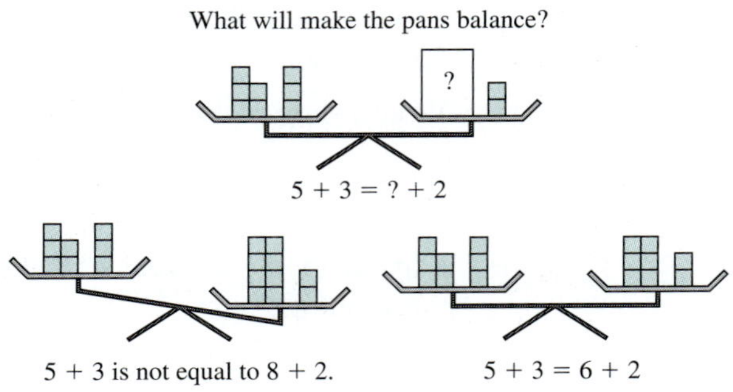

FIGURE 9.20

Viewing the
equation
$5 + 3 = ? + 2$
in terms of a
pan balance

What will make the pans balance?

$5 + 3 = ? + 2$

$5 + 3$ is not equal to $8 + 2$.

$5 + 3 = 6 + 2$

each other. One helpful piece of imagery for understanding equations is a pan balance, as shown in Figure 9.20. If we view the equation $5 + 3 = \square + 2$ in terms of a pan balance, each side of the equation corresponds to a side of the pan balance. To solve the equation means to make the pans balance rather than tilt to one side or the other.

Solving Equations Using Number Sense

Although there are general techniques for solving equations, sometimes we can solve an equation just by thinking about what the equation means and using "number sense." For example, to solve

$$2x = 6$$

we can think: "2 times what number is 6?" The solution is 3, which we could also find by dividing:

$$x = 6 \div 2 = 3$$

To solve

$$x + 3 = 7$$

we can think: "what number plus 3 is 7?" The solution is 4, which we could also find by subtracting:

$$x = 7 - 3 = 4$$

The next Class Activity asks you to solve equations by using your understanding of numbers and operations—number sense.

Class Activity *Now Turn to Class Activities Manual*

9L Solving Equations Using Number Sense, p. 198

Solving Equations Using Algebra

How do we solve equations? We just saw that in some cases we can use number sense to determine the solutions to equations.

But how do we solve an equation like

$$5x + 2 = 3x + 8?$$

The general strategy for solving equations is to change the equation into a new equation that has the same solutions and is easy to solve. We can change an equation into a new equation that has the same solutions by doing any of the following:

1. Use properties of arithmetic or valid ways of operating with numbers (including fractions) to change an expression on either side of the equals sign into an equal expression. For example, we can change the equation

$$3x + 4x + 2 = 1$$

to the equation

$$7x + 2 = 1$$

because, according to the distributive property,

$$3x + 4x = (3 + 4)x = 7x$$

2. Add the same quantity to both sides of the equation or subtract the same quantity from both sides of the equation. For example, we can change the equation

$$6x - 2 = 3$$

to the equation

$$6x - 2 + 2 = 3 + 2$$

which becomes

$$6x = 5$$

by using item (1).

3. Multiply both sides of the equation by the same nonzero number, or divide both sides of the equation by the same nonzero number. For example, we can change the equation

$$5x = 3$$

to the equation

$$x = \frac{3}{5}$$

by dividing both sides of the equation by 5. Remember that $5x$ stands for $5 \cdot x$, so when we divide this by 5, we get x.

Why do items (1), (2), and (3) change an equation into a new equation that has the same solutions as the original one? When we use item (1), we don't really change the equation at all; we just write the expression on one side of the equals sign in a different way. Therefore, when we use item (1), the new equation must have the same solutions as the original equation.

To see why item (2) should produce a new equation that has the same solutions as the original one, think in terms of a pan balance. If the two pans are balanced and you add the same amount to both sides or take the same amount away from both sides, then the pans will remain balanced. The same is true with equations.

To see why item (3) should produce a new equation that has the same solutions as the original one, think again in terms of a pan balance. If the two pans are balanced and you put twice as much or three times as much, and so on, on each pan, then the pans will remain balanced. Similarly, if the two pans are balanced and you take $\frac{1}{2}$ or $\frac{1}{3}$, and so on, of each pan, then the pans will remain balanced.

Thus, the strategy for solving an equation such as

$$5x + 2 = 3x + 8$$

is to use items (1), (2), and (3) to change the equation into an equation of the form

$$x = a$$

where a is a number. This equation has the same solutions as the original one. And we can see that this equation has solution a. We can find the equation $x = a$ by getting all the terms with x on one side of the equation and getting all the numbers on the other side. Figure 9.21 shows how to solve $5x + 2 = 3x + 8$ this way. Each step is also illustrated in terms of a pan balance. Each small square on the pan balance represents 1. Each larger square represents an x. To solve the equation means to determine the value of x that will make the equation true. In terms of the pan balance, we want to determine how many small squares should fill a box with an x to make the pans balance. We see from Figure 9.21 that the solution is $x = 3$. In other words, 3 small squares should fill a square with an x.

FIGURE 9.21

Solving
$5x + 2 = 3x + 8$
with equations
and with a pan
balance

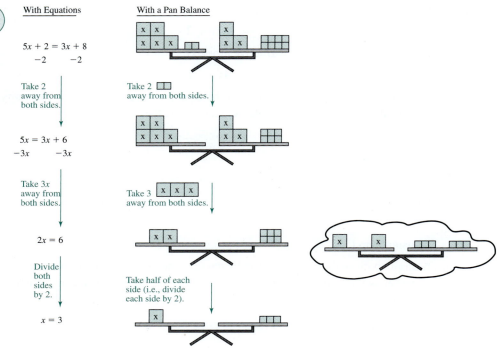

With Equations

$5x + 2 = 3x + 8$
$-2 \qquad -2$

Take 2
away from
both sides.

$5x = 3x + 6$
$-3x \qquad -3x$

Take $3x$
away from
both sides.

$2x = 6$

Divide
both
sides
by 2.

$x = 3$

With a Pan Balance

Take 2 ▢▢
away from both sides.

Take 3 x x x
away from both sides.

Take half of each
side (i.e., divide
each side by 2).

Class Activity   *Now Turn to Class Activities Manual*

9M Solving Equations Algebraically and with a Pan Balance, p. 199

One final comment about the pan balance view of equations: This view of equations provides an excellent concrete way to solve simple equations and to understand why we solve equations the way we do. However, many algebra story problems, especially those involving fractions or percentages, are not readily formulated in terms of a pan balance. Furthermore, equations involving negative numbers do not fit well with pan balances. So pan balances are best used as a starting point for solving equations.

Practice Exercises for Section 9.3

1. Use number sense (your understanding of numbers and operations) to solve

$$\frac{18}{7} = \frac{18 \cdot 2y}{12 \cdot 7}$$

Explain your reasoning. Do not use standard algebraic techniques for solving the equation.

2. Solve $5x + 2 = 2x + 17$ in two ways: with equations and with pictures of a pan balance. Relate the two methods.

Answers to Practice Exercises for Section 9.3

1. The solution is $y = 6$. Since

$$\frac{18 \cdot 2y}{12 \cdot 7} = \frac{18 \cdot 2y}{7 \cdot 12} = \frac{18}{7} \cdot \frac{2y}{12}$$

we must find a y which makes

$$\frac{2y}{12} = 1$$

Which is the case when $y = 6$.

2. See Figure 9.22. On the pan balance, each small square represents 1 and each larger square, filled with an x, represents the unknown amount x. At the end, we find that x must stand for 5 small squares in order to make the pans balance. Therefore, the solution of the equation is 5.

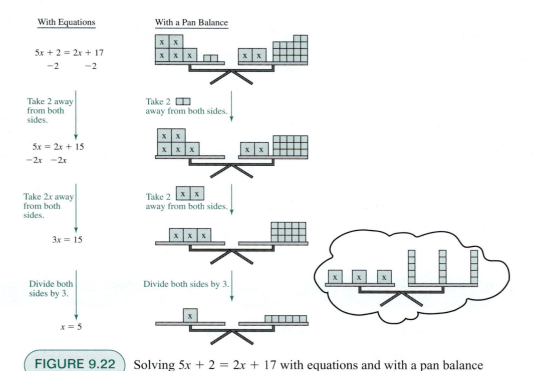

With Equations

$$5x + 2 = 2x + 17$$
$$-2 -2$$

Take 2 away from both sides.

$$5x = 2x + 15$$
$$-2x -2x$$

Take $2x$ away from both sides.

$$3x = 15$$

Divide both sides by 3.

$$x = 5$$

With a Pan Balance

Take 2 away from both sides.

Take 2 away from both sides.

Divide both sides by 3.

FIGURE 9.22 Solving $5x + 2 = 2x + 17$ with equations and with a pan balance

Problems for Section 9.3

1. Use number sense to solve

$$487 + 176 = x + 490$$

Explain your reasoning. Do not use standard algebraic techniques for solving the equation.

2. Use number sense to solve
$$333 \cdot 213 = 111A$$

Explain your reasoning. Do not use standard algebraic techniques for solving the equation.

3. Use number sense to solve

$$C + 34 \cdot 8 = 28 \cdot 34$$

Explain your reasoning. Do not use standard algebraic techniques for solving the equation.

4. Use number sense to solve

$$87 \cdot 14 = y + 13 \cdot 86$$

Explain your reasoning. Do not use standard algebraic techniques for solving the equation.

5. Write an equation that can be solved by using number sense. Explain how to solve your equation by number sense.

6. Solve

$$4x + 5 = x + 8$$

in two ways: with equations and with pictures of a pan balance. Relate the two methods.

7. Solve

$$3x + 2 = x + 8$$

in two ways: with equations and with pictures of a pan balance. Relate the two methods.

9.4 Solving Algebra Story Problems with Strip Diagrams and with Algebra

F·P Focal Points
Grades 6, 7

In this section, we will see how drawing strip diagrams can make some algebra story problems easy to solve without using variables. But we will also note that there is a close connection between the strip-diagram approach and the standard algebraic approach to formulating and solving story problems. The strip-diagram-drawing method of solving problems is shown in textbooks used in grades 3 through 6 in Singapore. Of the 38 nations studied in the Third International Mathematics and Science Study, children in Singapore scored highest in math. (See [57].)

Before reading on, do the next Class Activity.

Class Activity   *Now Turn to Class Activities Manual*

9N How Many Pencils Were There?, p. 199

If you did the Class Activity, you might have solved the pencil problem by drawing and analyzing a picture. Or you might have solved the problem by setting up and solving an equation. If you did both, perhaps you were even able to relate the two approaches. Now let's study a way to use simple pictures—strip diagrams—to solve some algebra and other story problems. In most cases, a strip diagram can even help formulate a story problem algebraically, with an equation. And sometimes the steps we use in solving a problem with a picture correspond to the steps we use when we solve the problem algebraically by manipulating equations.

Class Activity   *Now Turn to Class Activities Manual*

9O Solving Story Problems with Strip Diagrams and with Equations, p. 200

FIGURE 9.23

How much were each of the 4 equal payments?

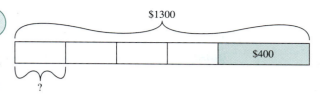

model diagrams
bar diagrams

In the elementary school mathematics textbooks used in Singapore (see [78] and [79]), narrow strips, such as the ones shown in Figure 9.23 and in the previous Class Activity, are often used to represent quantities. These strip diagrams, also known as **model diagrams** and **bar diagrams**, are an excellent aid for reasoning about the quantities in a problem.

The next three problems demonstrate how to analyze a strip diagram to solve a problem and how to relate the strip diagram to equations.

Problem 1: A Payment Plan Problem

The Brown family is buying a TV that costs $1300. After the Browns make 4 equal payments, they still owe $400 on the TV. How much was each payment?

Figure 9.23 shows a strip diagram for this problem. The total strip represents the price of the TV. The 4 small strips represent the 4 equal payments, and the darkly shaded strip represents the $400 left to pay. From the strip diagram we can see that $1300 − $400 = $900 is the amount to be split among the 4 equal payments. So each payment was $900 ÷ 4 = $225.

We can also formulate this problem algebraically, with an equation. Let x be the amount of each payment. Then the 4 payments are $4x$. Since the TV costs $1300, after making the 4 payments, the Browns still owe

$$1300 - 4x$$

dollars. Since they still owe $400, we have the equation

$$1300 - 4x = 400$$

If we add $4x$ to both sides, we obtain the equation

$$1300 = 400 + 4x$$

which fits nicely with the strip diagram in Figure 9.23, where one small white strip represents x. If we subtract 400 from both sides of the equation, we obtain the equation

$$1300 - 400 = 4x \quad \text{after simplifying: } 900 = 4x$$

Dividing both sides by 4, we have

$$\frac{900}{4} = x \quad \text{after simplifying: } 225 = x$$

So $x = 225$ is the solution, and each payment was $225. Notice that the calculations we performed when solving the problem algebraically are the same as those we performed when using the strip diagram: In both cases we subtracted 400 from 1300 to obtain 900, and then we divided 900 by 4 to obtain 225.

Problem 2: Tea in Two Urns

70 cups of tea are poured into two urns. The first urn contains 6 more cups of tea than the second urn. How many cups of tea are in the second urn?

Figure 9.24 shows a strip diagram for this problem. Each strip represents the tea in an urn. From the diagram we see that if we subtract 6 cups from 70 cups, then the remaining cups of tea—namely, 64 cups—must be divided equally in two. So the second urn contains 64 ÷ 2 = 32 cups of tea, and the first urn contains 32 + 6 = 38 cups.

FIGURE 9.24

How much tea
is in each
urn?

We can also formulate this problem algebraically, with an equation. Let x be the number of cups of tea in the second urn. So x corresponds to the strip at the bottom of Figure 9.24. Then there are $x + 6$ cups of tea in the first urn, which corresponds to the long strip at the top of Figure 9.24. All together, there are $(x + 6) + x = 2x + 6$ cups of tea in the urn. Since there are 70 cups of tea in all, we have the equation

$$2x + 6 = 70$$

Notice that we can also see this equation directly from Figure 9.24, since there are two long strips that each represent x cups and a smaller strip representing 6 cups. All together, these make 70 cups. Returning to the equation $2x + 6 = 70$, if we subtract 6 from both sides, we obtain the equation

$$2x = 64$$

Dividing both sides by 2, we have

$$x = 32$$

So there are 32 cups of tea in the second urn. Notice that the calculations we performed when solving the problem algebraically are the same as those we performed when using the strip diagram: In both cases we subtracted 6 from 70 to obtain 64, and then we divided 64 by 2 to obtain 32.

Class Activity   *Now Turn to Class Activities Manual*

9P Modifying Problems, p. 202

9Q Solving Story Problems, p. 204

Practice Exercises for Section 9.4

Some of these Practice Exercises were inspired by problems in the 4th-, 5th-, and 6th-grade mathematics texts used in Singapore (see [78] and [79], volumes 4A–6B).

1. Dante has twice as many jelly beans as Carlos. Natalia has 4 more jelly beans than Carlos and Dante together. All together, Dante, Carlos, and Natalia have 58 jelly beans. How many jelly beans does Carlos have? Solve this problem in two ways: with the aid of a diagram and with equations. Explain both solution methods, and discuss how they are related.

2. A school has two hip-hop dance teams, the Jelani team and the Zuri team. There are $1\frac{1}{2}$ times as many students on the Jelani team as on the Zuri team. All together, there are 65 students on the two teams combined. How many students are on the Jelani team, and how many are on the Zuri team? Solve this problem in two ways: with the aid of a diagram and with equations. Explain both solution methods, and discuss how they are related.

3. At first, only $\frac{1}{3}$ of the students who were going on the band trip were on the bus. After another 21 students got on the bus, $\frac{4}{5}$ of the students who were going on the band trip were on the bus. How many students were going on the band trip? Solve this problem in two ways: with the aid of a diagram and with equations. Explain both solution methods, and discuss how they are related.

Answers to Practice Exercises for Section 9.4

1. We see from Figure 9.25 that the total number of jelly beans consists of 6 equal groups of jelly beans and 4 more jelly beans. If we take the 4 away from 58, we have 54 jelly beans that must be distributed equally among the 6 groups. Therefore, each group has $54 \div 6 = 9$ jelly beans. Since Carlos has one group of jelly beans, he has 9 jelly beans. (Dante has $2 \cdot 9 = 18$ jelly beans, and Natalia has $9 + 18 + 4 = 31$ jelly beans.)

To solve the problem with equations, let x be the number of jelly beans that Carlos has. Since Dante has twice as many jelly beans as Carlos, Dante has $2x$ jelly beans. Since Natalia has 4 more than Carlos and Dante combined, Natalia has $x + 2x + 4$ jelly beans, which we can also write as $3x + 4$ jelly beans. We can also see these relationships in Figure 9.25. Carlos's strip represents x jelly beans. Since Dante has 2 such strips, he has $2x$ jelly beans. Natalia has 3 such strips and 4 more jelly beans, so she has $3x + 4$ jelly beans.

The total number of jelly beans is Carlos's, Dante's, and Natalia's combined, which is

$$x + 2x + 3x + 4$$

By combining the x's, we can rewrite this expression as

$$6x + 4$$

We are given that the total number of jelly beans is 58, so we have the equation

$$6x + 4 = 58$$

We can also deduce this equation from Figure 9.25 because there are 6 strips that each represent x jelly beans and another small strip representing 4 more jelly beans; the combined amount must be 58, so $6x + 4 = 58$.

To solve $6x + 4 = 58$, first subtract 4 from both sides to obtain

$$6x = 58 - 4 \quad \text{after simplifying: } 6x = 54$$

Then divide both sides by 6 to obtain

$$x = \frac{54}{6} \quad \text{after simplifying: } x = 9$$

Therefore, Carlos has 9 jelly beans. Notice that the arithmetic we performed to solve the equation $6x + 4 = 58$ is the same as the arithmetic we performed to solve the problem with the diagram: We first subtracted 4 from 58, and then we divided the resulting amount, 54, by 6.

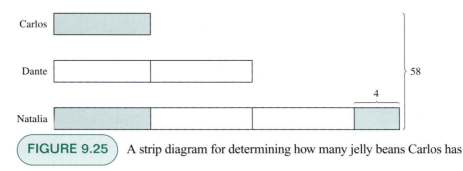

FIGURE 9.25 A strip diagram for determining how many jelly beans Carlos has

2. Figure 9.26 shows that there are $1\frac{1}{2}$ times as many students on the Jelani team as on the Zuri team and that there are 65 students in all. Notice that the total number of students is represented by 5 half-strips, each of

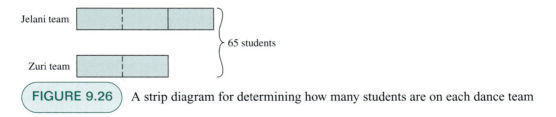

FIGURE 9.26 A strip diagram for determining how many students are on each dance team

which is the same size. The 65 students must be divided equally among these 5 portions, so each portion contains $65 \div 5 = 13$ students. The Jelani team consists of 3 of those half portions, so there are $3 \times 13 = 39$ students on the Jelani team. Since the Zuri team consist of 2 half portions, there are $2 \times 13 = 26$ students on the Zuri team.

To solve the problem with equations, let Z stand for the number of students on the Zuri team. Then the number of students on the Jelani team is $1\frac{1}{2} \cdot Z$, or $\frac{3}{2}Z$. Therefore, the two teams combined have $Z + \frac{3}{2}Z$ students. Since this number of students is 65, we obtain the equation

$$Z + \frac{3}{2}Z = 65$$

Simplifying, we have

$$\left(1 + \frac{3}{2}\right)Z = 65$$

after further simplifying: $\frac{5}{2}Z = 65$

Notice that we could also formulate this last equation from Figure 9.26, as the total number of students is represented by the 5 half-segments. Therefore, the total number of students is $\frac{5}{2}$ of the number of students on the Zuri team, so $\frac{5}{2}Z = 65$. By multiplying both sides of the equation $\frac{5}{2}Z = 65$ by $\frac{2}{5}$ (or by dividing both sides by $\frac{5}{2}$), we obtain

$$Z = 65 \cdot \frac{2}{5} = \overset{13}{\cancel{65}} \cdot \frac{2}{\cancel{5}} = 13 \cdot 2 = 26$$

So there are 26 students on the Zuri team. Therefore, there must be

$$1\frac{1}{2} \cdot 26 = \frac{3}{2} \cdot \overset{13}{\cancel{26}} = 3 \cdot 13 = 39$$

students on the Jelani team.

With both solution methods, we divided the total number of students, 65, by 5 and multiplied

the result, 13, by 2 to obtain the number of students on the Zuri team. We also multiplied the 13 by 3 to obtain the number of students on the Jelani team.

3. The full strip in Figure 9.27 represents all the students going on the band trip. The dark portion represents the students who were on the bus initially, and the lighter shaded portion represents the additional 21 students who got on the bus. This lighter shaded portion must be the difference between $\frac{4}{5}$ and $\frac{1}{3}$ of the students going on the trip. Since the fractions $\frac{4}{5}$ and $\frac{1}{3}$ have 15 as a common denominator, it makes sense to break the total number of students into 15 equal parts. Because

$$\frac{4}{5} - \frac{1}{3} = \frac{12}{15} - \frac{5}{15} = \frac{7}{15}$$

the 21 additional students take up 7 parts. So each part consists of $21 \div 7 = 3$ students. Since there are 15 parts making the whole group of students going on the band trip, there are $15 \times 3 = 45$ students going on the band trip.

To solve the problem with equations, let x be the number of students going on the band trip. In terms of the diagram in Figure 9.27, x stands for the whole long strip. Then $\frac{1}{3}x$ is the number of students on the bus at first. After 21 students get on the bus, there are

$$\frac{1}{3}x + 21$$

students on the bus. But this number of students is also equal to $\frac{4}{5}x$ because now $\frac{4}{5}$ of the students going on the trip are on the bus. Therefore, we have the equation

$$\frac{1}{3}x + 21 = \frac{4}{5}x$$

Notice that we can also see this equation from the relationship between the shaded portions in Figure 9.27.

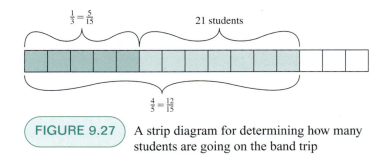

$$\frac{1}{3} = \frac{5}{15} \qquad\qquad \text{21 students}$$

$$\frac{4}{5} = \frac{12}{15}$$

FIGURE 9.27 A strip diagram for determining how many students are going on the band trip

Subtracting $\frac{1}{3}x$ from both sides of the previous equation, we have

$$21 = \frac{4}{5}x - \frac{1}{3}x$$

So

$$21 = \left(\frac{4}{5} - \frac{1}{3}\right)x \quad \text{after simplifying: } 21 = \frac{7}{15}x$$

After multiplying both sides by $\frac{15}{7}$ (or dividing both sides by $\frac{7}{15}$), we have

$$\frac{15}{7} \cdot \overset{3}{\cancel{21}} = \frac{\cancel{15}}{7} \cdot \frac{7}{\cancel{15}}x \quad \text{after simplifying: } 45 = x$$

Therefore, $x = 45$, and there were 45 students going on the band trip.

We perform the same arithmetic by using both solution methods. With each method, we calculated $\frac{4}{5} - \frac{1}{3}$. When we solved the problem by using a diagram, we divided $21 \div 7$ to find the number of students in each $\frac{1}{15}$ part of the total students. Then we multiplied that result, 3, by the total number of parts making up all the students—namely, 15. Similarly, using the algebraic solution method, we had to calculate

$$\frac{15}{7} \cdot 21$$

When we used canceling, we also divided 21 by 7 and multiplied the result, 3 by 15.

Problems for Section 9.4

Some of these problems were inspired by problems in the 4th-, 5th-, and 6th-grade mathematics texts used in Singapore (see [78] and [79], volumes 4A–6B).

1. A candy bowl contained 723 candies. Some of the candies were red, and the rest were green. There were twice as many green candies as red candies. How many red candies were in the candy bowl? Solve this problem in two ways: with the aid of a diagram and with equations. Explain both solution methods, and discuss how they are related.

2. Sarah spends $\frac{2}{5}$ of her monthly income on rent. Sarah's rent is $750. What is Sarah's monthly income? Solve this problem in two ways: with the aid of a diagram and with equations. Explain both solution methods, and discuss how they are related.

3. After Joey spent $\frac{3}{8}$ of his money on an electronic game player, he had $360 left. How much money did Joey have before he bought the electronic game player? How much did the electronic game player cost? Solve this problem in two ways: with the aid of a diagram and with equations. Explain both solution methods, and discuss how they are related.

4. Katie made some chocolate truffles. Katie gave $\frac{1}{4}$ of her truffles to Leanne. Then Katie gave $\frac{1}{2}$ of the remaining truffles to Jeff. At that point, Katie had 18 truffles left. How many chocolate truffles did Katie have at first? Solve this

problem in two ways: with the aid of a diagram and with equations. Explain both solution methods, and discuss how they are related.

5. a. Suppose you want to modify Problem 4 about Katie's chocolate truffles by changing the number 18 to a different number. Which numbers could you select to replace the 18 in the problem and still have a sensible problem (without changing anything else in the problem)? Explain.

   b. Change the fraction $\frac{1}{2}$ in Problem 4 about Katie's chocolate truffles so that the problem will become easier to solve. You may change the number 18 as well. Show how to solve the new problem, and explain why it is easier to solve than the original problem.

6. Write a problem like Problem 4 in which a fraction of a collection of objects are removed and then a fraction of the remaining objects are removed and a certain number of objects remain. Solve your problem in two ways: with the aid of a diagram and with equations. Explain both solution methods, and discuss how they are related.

7. By washing cars, Beatrice and Hannah raised $2000 total. If Beatrice raised $600 more than Hannah, how much money did Hannah raise? Solve this problem in two ways: with the aid of a diagram and with equations. Explain both solution methods, and discuss how they are related.

8. There are 1700 bottles of water arranged into 4 groups. The second group has 20 more bottles than the first group. The third group has twice as many bottles as the second group. The fourth group has 10 more bottles than the third group. How many bottles are in each group? Solve this problem in two ways: with the aid of a diagram and with equations. Explain both solution methods, and discuss how they are related.

9. Write a problem like Problem 8 in which a certain number of objects are divided into a number of unequal groups. Solve your problem in two ways: with the aid of a diagram and with equations. Explain both solution methods, and discuss how they are related.

10. One day a coffee shop sold 360 cups of coffee, some regular coffee and the rest decaffeinated. The shop sold 3 times as many cups of regular coffee as cups of decaffeinated coffee. The shop made 100 cups of regular coffee with foamed milk. How many cups of regular coffee did the shop make without foamed milk? Solve this problem in two ways: with the aid of a diagram and with equations. Explain both solution methods, and discuss how they are related.

11. A 1-pound-5-ounce piece of cheese is divided into 2 pieces. The larger piece weighs twice as much as the smaller piece. How much does each piece weigh? Solve this problem in two ways: with the aid of a diagram and with equations. Explain both solution methods, and discuss how they are related.

12. A hot dog stand sold $\frac{3}{8}$ of its hot dogs before noon and $\frac{1}{2}$ of its hot dogs after noon. The stand sold 280 hot dogs in all. How many hot dogs were left? How many hot dogs were there at first? Solve this problem in two ways: with the aid of a diagram and with equations. Explain both solution methods, and discuss how they are related.

13. Write a problem like Problem 12 in which the combined fractional amounts of a quantity are equal to a given quantity and the fractions in the problem have different denominators. Solve your problem in two ways: with the aid of a diagram and with equations. Explain both solution methods, and discuss how they are related.

14. Julie bought donuts for a party—$\frac{1}{6}$ of the donuts were jelly donuts, $\frac{1}{3}$ of the donuts were cinnamon donuts, and $\frac{5}{6}$ of the remainder were glazed donuts. Julie bought 20 glazed donuts. How many donuts did Julie buy in all? Solve this problem in two ways: with the aid of a diagram and with equations. Explain both solution methods, and discuss how they are related.

15. Shauntay caught twice as many fireflies as Robert. Jessica caught 10 more fireflies than Robert. All together, Shauntay, Robert, and Jessica caught 150 fireflies. How many fireflies did Jessica catch? Solve this problem in two ways: with the aid of a diagram and with equations. Explain both solution methods, and discuss how they are related.

16. At a frog exhibit, $\frac{3}{5}$ of the frogs are bullfrogs, $\frac{2}{3}$ of the remainder are tree frogs, and the rest are river frogs. There are 36 bullfrogs in the exhibit. How many river frogs are there? Solve this problem in two ways: with the aid of a diagram and with equations. Explain both solution methods, and discuss how they are related.

17. (See Problem 14 in Section 9.2.) Bonnie had some stickers. First, Bonnie gave away $\frac{1}{6}$ of her stickers. When Bonnie gets 12 more stickers, she has 47 stickers. Determine how many stickers Bonnie had at the start. Explain your solution.

18. (See Problem 15 in Section 9.2.) There was some flour in a bag. First, $\frac{3}{8}$ of the flour in the bag was used. Then $\frac{1}{5}$ of the remaining flour was used. At that point, there were 10 cups of flour in the bag. Determine how much flour was in the bag at the start. Explain your solution.

19. Not every algebra story problem is naturally modeled with either a pan balance or with a strip diagram. Solve the following problem in any way that makes sense, and explain your solution:

Kenny has stickers to distribute among some goodie bags. When Kenny tries to put 5 stickers in each goodie bag, he is 2 stickers short. But when Kenny puts 4 stickers in each bag, he has 3 stickers left over. How many goodie bags does Kenny have? How many stickers does Kenny have?

20. Not every algebra story problem is naturally modeled with either a pan balance or with a strip diagram. Solve the following problem in any way that makes sense, and explain your solution:

Frannie has no money now, but she plans to save $D every week so that she can buy a toy she would like to have. Frannie figures that if she saves her money for 3 weeks, she will have $1 less than she needs to buy the toy, but if she saves her money for 4 weeks, she will have $3 more than she needs to buy the toy. How much is Frannie saving every week, and how much does the toy Frannie plans to buy cost?

21. In a group of ladybugs there were 10% more male ladybugs than female ladybugs. There were 8 more male ladybugs than female ladybugs. How many ladybugs were there in all? Solve this problem in two ways: with the aid of a diagram and with equations. Explain both solution methods, and discuss how they are related.

22. A farmer planted 65% of his farmland with onions, 20% of his land with cotton, and the rest with soybeans. The soybeans are on 90 acres of land. How many acres of farmland does the farmer have? Solve this problem in two different ways, one of which involves solving an equation. Explain both solution methods.

23. Some exercise equipment is on sale for 20% off. The sale price is $340. What was the price of the exercise equipment before the discount? Solve this problem in two different ways, one of which involves solving an equation. Explain both solution methods.

24. Clevere makes 30% more money than Shane. If Clevere makes $8450 per month, how much does Shane make per month? Solve this problem in two different ways, one of which involves solving an equation. Explain both solution methods.

25. Two stores had a combined total of 210 teddy bears. After the first store sold $\frac{1}{2}$ of its teddy bears and the second store sold $\frac{1}{3}$ of its teddy bears, each store had the same number of teddy bears left. How many teddy bears were sold from the two stores combined? Solve this problem in two different ways, one of which involves solving an equation. Explain both solution methods.

26. If $\frac{2}{3}$ of a cup of juice gives you 120% of your daily value of vitamin C, what percent of your daily value of vitamin C will you get in $\frac{1}{2}$ cup of the juice? Solve this problem in two different ways, one of which involves solving an equation. Explain both solution methods.

27. Erika is weaving a wool blanket. So far, she has used 10 balls of wool and has completed $\frac{5}{8}$ of the blanket. How many more balls of wool will Erika need to finish the blanket? Solve this problem in two different ways, one of which involves solving an equation. Explain both solution methods.

28. Jane had a bottle filled with juice. At first, Jane drank $\frac{1}{5}$ of the juice in the bottle. After 1 hour, Jane drank $\frac{1}{4}$ of the juice remaining in the bottle. After another 2 hours, Jane drank $\frac{1}{3}$ of the remaining juice in the bottle. At that point, Jane checked how much juice was left in the bottle: There was $\frac{2}{3}$ of a cup left. No other juice was added to or removed from the bottle. How much juice was in the bottle originally? Solve this problem, and explain your solution.

29. A flock of geese on a pond was being observed continuously. At 1:00 P.M., $\frac{1}{5}$ of the geese flew away. At 2:00 P.M., $\frac{1}{8}$ of the geese that remained flew away. At 3:00 P.M., 3 times as many geese as had flown away at 1:00 P.M. flew away, leaving 28 geese on the pond. At no other time did any geese arrive or fly away. How many geese were in the original flock? Solve this problem, and explain your solution.

9.5 Sequences

Most people, including young children, enjoy discovering and experimenting with growing and repeating patterns. But patterns aren't just fun. By thinking about how to describe them generally, we begin to develop algebraic reasoning. In this section, we will study growing and repeating patterns from a mathematical point of view by studying sequences.

sequence   A **sequence** is a list of items occurring in a specified order. Sequences of physical objects or figures can give rise to sequences of numbers, which we can study mathematically.

The following example shows a simple way that sequences can arise. Austin is making "trains" out of snap cubes (snap-together cubes). Each train has an engine, which is made of 4 white snap cubes, and one or more train cars. Each train car is made of 3 yellow and 2 red snap cubes. Austin makes a sequence of trains: The first train consists of an engine and 1 train car; the second train consists of an engine and 2 train cars; the third train consists of an engine and 3 train cars; and so on, as indicated in Figure 9.28.

FIGURE 9.28

A sequence of snap cube "trains"

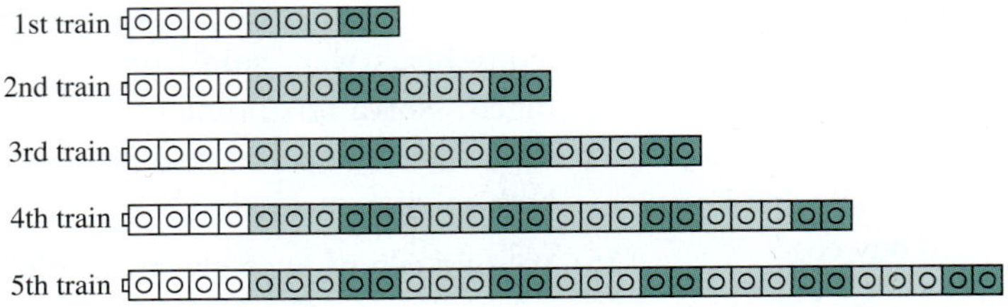

1st train
2nd train
3rd train
4th train
5th train

We see that Austin's sequence of trains gives rise to several sequences of numbers by considering the red snap cubes, the yellow snap cubes, and the total number of snap cubes in the trains. The numbers of red snaps cubes in the trains are

$$2, \quad 4, \quad 6, \quad 8, \quad 10, \ldots$$

the numbers of yellow snap cubes in the trains are

$$3, \quad 6, \quad 9, \quad 12, \quad 15, \ldots$$

and the total numbers of snap cubes in the trains are

$$9, \quad 14, \quad 19, \quad 24, \quad 29 \ldots$$

Of course, Austin will eventually run out of snap cubes and will be able to make only so many trains. But if we imagine an infinite supply of snap cubes, we can imagine each of these sequences going on forever.

Arithmetic Sequences

arithmetic sequence

The sequences of the numbers of red and yellow snap cubes in Austin's trains and the total number of snap cubes in Austin's trains are examples of arithmetic sequences. We create an **arithmetic sequence** by starting with any number as the first entry. Each subsequent entry is obtained from the previous entry by adding or subtracting the same fixed number, as shown in Figure 9.29.

Class Activity   *Now Turn to Class Activities Manual*

9R   Arithmetic Sequences of Numbers Corresponding to Sequences of Figures, p. 204

9S How Are Formulas for Arithmetic Sequences Related to the Way Sequences Start and Grow?, p. 207

9T   Explaining Formulas for Arithmetic Sequences, p. 208

Every arithmetic sequence is described by a simple formula. This formula is related to the sequence in a special way. Let's consider the example of the arithmetic sequence

$$9, \quad 14, \quad 19, \quad 24, \quad 29 \ldots$$

FIGURE 9.29

Arithmetic sequences are produced by repeatedly adding or subtracting the same amount.

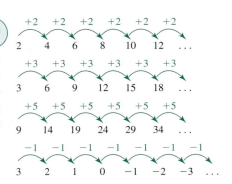

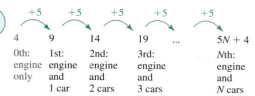

FIGURE 9.30

We obtain the formula $5N + 4$ by starting with the 0th entry and repeatedly adding 5.

of the total number of snap cubes in Austin's trains. This sequence starts with 9 and increases by 5. That is, each entry is 5 more than the previous entry. Imagine this sequence going on forever, so that there is a 100th entry, a 1000th entry, and so on; for each counting number, N, there is an Nth entry in the sequence. Is there a formula in terms of N for this Nth entry? Yes, there is, namely,

$$5N + 4$$

Although you might be able to guess this formula, there is a systematic way to find it, to explain why it makes sense, and to relate it to the growth of the sequence.

To explain why the formula $5N + 4$ for the Nth entry in the foregoing sequence makes sense, think about how to obtain each entry. Notice that the first entry, 9, was obtained by adding one 5 to 4, because a single train car made of 5 snap cubes was added to an engine made of 4 snap cubes. The second entry, 14, was obtained by adding two 5s to 4 because 2 train cars were added to an engine. The third entry, 19, was obtained by adding three 5s to 4. In general, the Nth entry is obtained by adding N 5s to 4. Therefore, the Nth entry is $5N + 4$. So, it's like starting with the 0th entry (just the engine, no train cars), which is 4, and adding N 5s to get the Nth entry, as indicated in Figure 9.30.

The reasoning to find the formula for the Nth entry of any other arithmetic sequence is similar. Consider a "0th entry" that you obtain by subtracting the increase amount from the initial entry. Then the Nth entry will be obtained by adding the increase amount N times to the 0th entry. Therefore, the Nth entry is

$$(\text{increase amount}) \cdot N + (\text{0th entry})$$

Geometric Sequences

We just saw that arithmetic sequences are obtained by repeatedly adding (or subtracting) the same amount. Geometric sequences are similar, but are obtained by repeatedly *multiplying* (or dividing) by the same amount. We create a **geometric sequence** by starting with any number as the first entry. Each subsequent entry is obtained from the previous entry by multiplying (or dividing) by the same fixed number, as shown in Figure 9.31.

geometric sequence

Class Activity   *Now Turn to Class Activities Manual*

9U Geometric Sequences, p. 210

FIGURE 9.31

Geometric sequences are produced by repeatedly multiplying or dividing by the same amount.

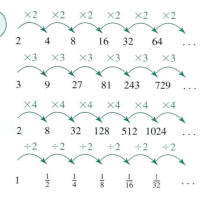

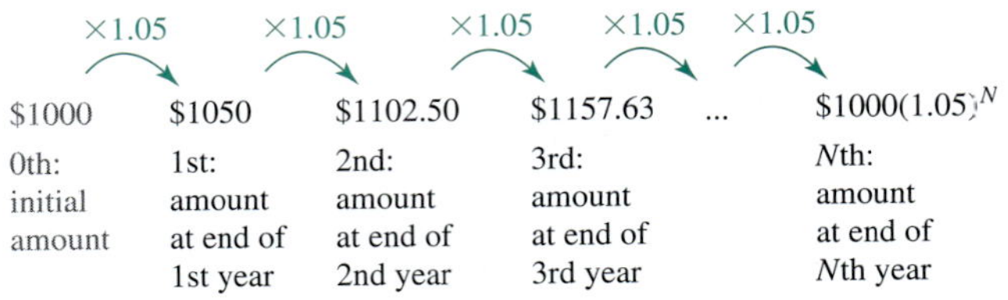

FIGURE 9.32

$\times 1.05$ $\times 1.05$ $\times 1.05$ $\times 1.05$ $\times 1.05$

$\$1000$	$\$1050$	$\$1102.50$	$\$1157.63$	...	$\$1000(1.05)^N$
0th:	1st:	2nd:	3rd:		Nth:
initial	amount	amount	amount		amount
amount	at end of	at end of	at end of		at end of
	1st year	2nd year	3rd year		Nth year

The geometric sequence of the amounts of money in a bank account after N years when the initial amount is $1000 and the interest rate is 5%.

Geometric sequences have important practical applications. For example, suppose that $1000 is deposited in a bank account that pays 5% interest annually. If no money other than the interest is added or removed, then at the end of each year, the amount of money in the account will be 5% more than the amount at the end of the previous year. Therefore, the amount each year is obtained from the amount the previous year by multiplying by 1.05, as shown in Figure 9.32. (The 1.05 is 1 +.05, which is 100% and an additional 5%; see Section 7.5.) So the amounts of money in the account at the end of each year form a geometric sequence.

As with arithmetic sequences, geometric sequences can also be described by formulas. Consider the sequence

$$1050, 1102.50, 1157.63, 1215.51 \ldots$$

which is the dollar amount in the bank account described earlier after 1 year, 2 years, 3 years, and so on. What is a formula for the number of dollars in the account after N years? To find this formula, think about how to obtain each entry in the sequence by starting with the initial $1000 in the account. After 1 year, there is

$$1.05 \times 1000$$

dollars in the account. After another year, this amount is multiplied by 1.05, so after 2 years there is

$$1.05 \times 1.05 \times 1000 = (1.05)^2 \cdot 1000$$

dollars in the account. After another year, this amount is multiplied by 1.05 again, so after 3 years there is

$$1.05 \times 1.05 \times 1.05 \times 1000 = (1.05)^3 \cdot 1000$$

dollars in the account. Continuing in this way, after N years there is

$$\underbrace{1.05 \times 1.05 \times \cdots \times 1.05}_{N \text{ times}} \times 1000 = (1.05)^N \cdot 1000$$

dollars in the account.

The reasoning to find the Nth entry of any other geometric sequence is similar. Let's call the factor by which each entry in a geometric sequence is multiplied in order to obtain the next entry the "ratio" of the geometric sequence. Consider a "0th entry" that you obtain from the first entry by dividing by the ratio of the geometric sequence. Then the Nth entry will be obtained by multiplying the 0th entry by the ratio N times. Therefore, the Nth entry is

$$\underbrace{(\text{ratio}) \cdot (\text{ratio}) \cdot \cdots \cdot (\text{ratio})}_{N \text{ times}} \cdot (\text{0th entry})$$

or

$$(\text{ratio})^N \cdot (\text{0th entry})$$

Other Sequences

> **Class Activity** *Now Turn to Class Activities Manual*
>
> **9V** Repeating Patterns, p. 211

FIGURE 9.33

A sequence
of shapes

Some of the sequences that young children encounter first in math generally consist of shapes or colors that are repeated in a pattern, such as the sequence of shapes in Figure 9.33. With such a sequence, we can ask which shape will be in the 100th entry, for example. Assume that the sequence continues with repetitions of a square followed by a circle, followed by a triangle. Then, as there are 3 shapes in the pattern, we can answer this question by dividing 100 by 3. Since

$$100 \div 3 = 33, \text{ remainder } 1$$

there are 33 repetitions of a square followed by a circle followed by a triangle in the entries 1 through $33 \times 3 = 99$, as indicated in Figure 9.34. The 100th entry is then a square.

Some numerical sequences are neither arithmetic nor geometric. For example, the sequence

$$1, 4, 9, 16, 25, 36, \dots$$

is neither arithmetic nor geometric. For each of the six entries shown, the Nth entry is N^2.

Fibonacci sequence   One fascinating sequence is the **Fibonacci sequence,** which is the sequence

$$1, 1, 2, 3, 5, 8, 13, 21, 34, \dots$$

Each entry in the sequence is obtained by adding the previous two entries. The Fibonacci sequence has surprising connections to nature and art. For example, the number of petals on many flowers is a Fibonacci number (i.e., an entry in the Fibonacci sequence). If you look carefully at pine cones and pineapples, you will see "swirls" going in two directions. The number of swirls in the two directions are usually consecutive Fibonacci numbers such as 5 and 8. Rectangles whose width and length are in the ratio of consecutive Fibonacci numbers, such as 3 to 5 or 5 to 8, are pleasing to the eye and often used in art and architecture. To find links to Web sites about the Fibonacci sequence, go to *www.pearsonhighered.com/beckmann.*

Class Activity   *Now Turn to Class Activities Manual*

9W Comparing and Contrasting Sequences, p. 213

9X The Fibonacci Sequence in Nature and Art, p. 214

Rules for Sequences

Class Activity   *Now Turn to Class Activities Manual*

9Y What's the Next Entry? p. 215

FIGURE 9.34

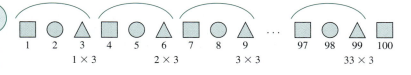

Sequence
repeating in a
pattern of 3

A common mathematics problem for children in elementary school is to determine the next several entries in a sequence when the first few entries are given. For example, what are the next three entries in the sequence

$$3, 5, 7, \ldots?$$

Be aware that if a rule or formula for the sequence has not been given, and if the sequence has not been specified as of a certain type (such as arithmetic, geometric, or repeating), there are always several different ways to continue the sequence. For example, we could continue the previous sequence like this:

$$3, 5, 7, 9, 11, 13 \ldots$$

viewing the sequence as an arithmetic sequence that increases by 2. Or we could continue the sequence like this:

$$3, 5, 7, 3, 5, 7, \ldots$$

and we could say that a rule for the sequence is to repeat the numbers 3, 5, 7 in order. Yet another possibility is to continue the sequence like this:

$$3, 5, 7, 11, 13, 17, \ldots$$

and we could say that the sequence consists of the odd prime numbers in order. Technically, the next three entries could be *any* three numbers, although the most interesting sequences are ones that have definite rules and patterns.

Practice Exercises for Section 9.5

1. Figure 9.35 shows a sequence of figures made of small squares. Assume that the sequence continues by adding 2 darkly shaded squares to the top of a figure in order to get the next figure in the sequence.

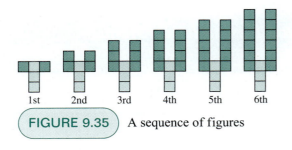

1st   2nd   3rd   4th   5th   6th

FIGURE 9.35   A sequence of figures

a. Write the first 6 entries in the sequence whose entries are the number of small squares making up the figures.

b. Find a formula for the number of small squares making up the Nth figure in the sequence. Explain why your formula makes sense by relating it to the structure of the figures.

c. Will there be a figure in the sequence that is made of 100 small squares? If yes, which one? If no, why not? Determine the answer to these questions in two ways: with algebra and in a way that a student in elementary school might be able to.

d. Will there be a figure in the sequence that is made of 199 small squares? If yes, which one; if no, why not? Determine the answer to these questions in two ways: with algebra and in a way that a student in elementary school might be able to.

2. Consider the arithmetic sequence whose first few entries are

$$1, 4, 7, 10, 13, 16, \ldots$$

Find a formula for the Nth entry in this sequence, and explain in detail why your formula is valid.

3. Draw the first 5 figures in a sequence of figures whose Nth entry is made of $3N + 2$ small circles. Describe how subsequent figures in your sequence would be formed.

4. Consider the geometric sequence whose first few entries are

$$6, 12, 24, 48, 96, 192, \ldots$$

Find a formula for the Nth entry in this sequence, and explain in detail why your formula is valid.

5. Describe four different rules for determining the next entries in the following sequence. Give the next three entries in the sequence for each rule.

$$3, 7, 15, \ldots$$

Answers to Practice Exercises for Section 9.5

1. a. The number of squares in the figures are

$$5, 7, 9, 11, 13, 15, \ldots$$

b. The Nth pattern in the sequence is made of $2N + 3$ small squares. We can explain why this formula is valid by relating it to the structure of the figures. Each figure is made of 3 small, lightly shaded squares and 2 "prongs" of darkly shaded squares. In the Nth figure, each prong is made of N squares. So in all, there are $2N + 3$ small squares in the Nth figure.

c. A student in elementary school might realize that the number of squares making up the figures is always odd. Since 100 is even, it could not be the number of small squares that a figure in the sequence is made of. Or the student might try to find how many squares are in each prong if there were a figure made of 100 squares. There would have to be $100 - 3 = 97$ squares in the two prongs, so each prong would have to have $97 \div 2$ squares. Since 97 is not evenly divisible by 2, there can't be such a figure.

To answer the questions with algebra, notice that we want to know if there is a counting number, N, such that

$$2N + 3 = 100$$

Subtracting 3 from both sides, we have the new equation

$$2N = 97$$

which has the same solution as the original equation. Dividing both sides by 2, we have

$$N = \frac{97}{2}$$

But $\frac{97}{2}$ is not a counting number, so there is no such figure.

d. A student in elementary school might try to find the number of squares in each prong. There must be $199 - 3 = 196$ squares in the two prongs combined. Therefore, each prong must have $196 \div 2 = 98$ squares. So the 98th figure in the sequence will be made of 199 small squares.

To answer the questions with algebra, notice that we want to solve the equation

$$2N + 3 = 199$$

Subtracting 3 from both sides, we have the new equation

$$2N = 196$$

which has the same solutions as the original equation. Dividing both sides by 2, we have

$$N = 98$$

So the 98th figure is made of 199 small squares.

2. The entries in the sequence increase by 3 each time. If we imagine a 0th entry preceding the first entry, this 0th entry would have to be $1 - 3 = -2$. Starting at the 0th entry, the 1st entry is obtained by adding one 3, the second entry is obtained by adding two 3s, the third entry is obtained by adding three 3s, and so on, as indicated in Figure 9.36. The Nth entry is obtained by adding N 3s to -2, so the Nth entry is

$$3N + (-2)$$

which we can also write as

$$3N - 2$$

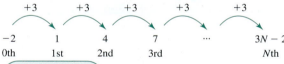

| +3 | +3 | +3 | +3 | +3 |

| -2 | 1 | 4 | 7 | $\cdots$ | $3N - 2$ |
| 0th | 1st | 2nd | 3rd | | Nth |

FIGURE 9.36 We obtain the formula $3N - 2$ by starting with the 0th entry and repeatedly adding 3.

3. See Figure 9.37. Of course, many other sequences of figures are also possible. Each figure is formed by adding one circle to each of the three "prongs" of the previous figure.

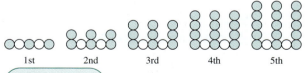

1st   2nd   3rd   4th   5th

FIGURE 9.37 A sequence of figures whose Nth entry is made of $3N + 2$ small circles.

4. We multiply each entry in the sequence by 2 to obtain the next entry. If we imagine a 0th entry preceding the 1st entry, this 0th entry would have to be 6 divided by 2, namely, 3. Starting at the 0th entry, the

1st entry is obtained by multiplying by 2 once, the second entry is obtained by multiplying by 2 twice, the third entry is obtained by multiplying by 2 three times, and so on, as indicated in Figure 9.38. The Nth entry is obtained by multiplying N 2s with 3, so the Nth entry is

$$3 \cdot 2^N$$

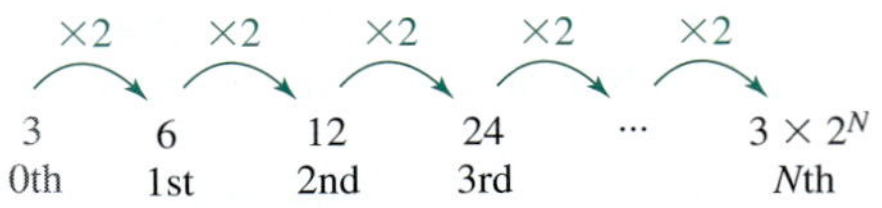

$$
\begin{array}{cccccc}
\times 2 & \times 2 & \times 2 & \times 2 & \times 2 & \\
3 & 6 & 12 & 24 & \cdots & 3 \times 2^N \\
\text{0th} & \text{1st} & \text{2nd} & \text{3rd} & & N\text{th}
\end{array}
$$

FIGURE 9.38   We obtain the formula $3 \cdot 2^N$ by starting with the 0th entry and repeatedly multiplying by 2.

5. One rule is simply to continue to repeat the sequence 3, 7, 15 in order. In this case, the first 6 entries in the sequence are

$$3, 7, 15, 3, 7, 15, \ldots$$

Another rule is that each entry is obtained from the previous entry by doubling and adding 1. In this case, the first 6 entries are

$$3, 7, 15, 31, 63, 127, \ldots$$

A third rule is that the next entries are obtained by adding 4, then adding 8, then adding 16, then adding 32, and so on, so that each time, twice as much is added as the time before. In this case, the first 6 entries are

$$3, 7, 15, 31, 63, 127, \ldots$$

which are the same entries as in the previous rule. A fourth rule is that the next entries are obtained by adding 4, then adding 8, then adding 12, then adding 16, and so on, so that each time, 4 more are added than the time before. In this case, the first 6 entries are

$$3, 7, 15, 27, 43, 63, \ldots$$

Problems for Section 9.5

1. Figure 9.39 shows a sequence of figures made of small circles. Assume that the sequence continues by the addition of one circle to each of the 5 "arms" of a figure in order to get the next figure in the sequence.

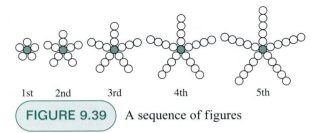

1st   2nd   3rd   4th   5th

FIGURE 9.39   A sequence of figures

a. Find a formula for the number of circles making up the Nth figure in the sequence. Explain why your formula makes sense by relating it to the structure of the figures.

b. Will there be a figure in the sequence that is made of 100 circles? If yes, which one? If no, why not? Determine the answer to these questions in two ways: with algebra and in a way that a student in elementary school might be able to.

c. Will there be a figure in the sequence that is made of 206 circles? If yes, which one? If no, why not? Determine the answer to these questions in two ways: with algebra and in a way that a student in elementary school might be able to.

2. Consider the arithmetic sequence whose first few entries are

$$6, 11, 16, 21, 26, 31, \ldots$$

a. Determine the 100th entry in the sequence, and explain why your answer is correct.

b. Find a formula for the Nth entry in this sequence, and explain in detail why your formula is valid.

c. Is 1000 an entry in the sequence? If yes, which entry? If no, why not? Determine the answer to these questions in two ways: with algebra and in a way that a student in elementary school might be able to.

d. Is 201 an entry in the sequence? If yes, which entry? If no, why not? Determine the answer to these questions in two ways: with algebra and in a way that a student in elementary school might be able to.

3. Consider the arithmetic sequence whose first few entries are

$$3, 7, 11, 15, 19, 23, \ldots$$

a. Determine the 50th entry in the sequence, and explain why your answer is correct.

b. Find a formula for the Nth entry in this sequence, and explain in detail why your formula is valid.

c. Is 207 an entry in the sequence? If yes, which entry? If no, why not? Determine the answer to these questions in two ways: with algebra and in a way that a student in elementary school might be able to.

d. Is 100 an entry in the sequence? If yes, which entry? If no, why not? Determine the answer to these questions in two ways: with algebra and in a way that a student in elementary school might be able to.

4. a. Draw the first 5 figures in a sequence of figures whose Nth entry is made of $4N + 1$ small circles. Describe how subsequent figures in your sequence would be formed.

b. Is there a figure in your sequence in part (a) that is made of 889 small circles? If yes, which one? If no, why not? Determine the answer to these questions in two ways: with algebra and in a way that a student in elementary school might be able to.

c. Is there a figure in your sequence in part (a) that is made of 150 small circles? If yes, which one? If no, why not? Determine the answer to these questions in two ways: with algebra and in a way that a student in elementary school might be able to.

5. a. Draw the first 5 figures in a sequence of figures whose Nth entry is made of $6N + 3$ small circles. Describe how subsequent figures in your sequence would be formed.

b. Is there a figure in your sequence in part (a) that is made of 102 small circles? If yes, which one? If no, why not? Determine the answer to these

questions in two ways: with algebra and in a way that a student in elementary school might be able to.

c. Is there a figure in your sequence in part (a) that is made of 333 small circles? If yes, which one? If no, why not? Determine the answer to these questions in two ways: with algebra and in a way that a student in elementary school might be able to.

6. Consider an arithmetic sequence whose third entry is 10 and whose fifth entry is 16. Use the most elementary reasoning you can to find the first and second entries of the sequence. Explain your reasoning.

7. Consider an arithmetic sequence whose fourth entry is 2 and whose seventh entry is 4. Use the most elementary reasoning you can to find the first and second entries of the sequence. Explain your reasoning.

8. Consider an arithmetic sequence whose fifth entry is 1 and whose tenth entry is 7. Use the most elementary reasoning you can to find the first and second entries of the sequence. Explain your reasoning.

9. The Widget Company sells boxes of widgets by mail order. The company charges a fixed amount for shipping, no matter how many boxes of widgets are ordered. All boxes of widgets are the same size and cost the same amount. The company is out of state, so there are no taxes charged. You find out that the total cost (including shipping) for 6 boxes of widgets is $27, and the total cost (including shipping) for 10 boxes of widgets is $41. Find the cost of shipping and the price of one box of widgets. Also, find a formula for the total cost (including shipping) of N boxes of widgets. Explain your reasoning.

10. Consider the geometric sequence whose first few entries are

$$2, 10, 50, 250, 1250, 6250, \ldots$$

Find a formula for the Nth entry in this sequence, and explain in detail why your formula is valid.

11. Consider the geometric sequence whose first few entries are

$$\frac{1}{2}, \frac{1}{4}, \frac{1}{8}, \frac{1}{16}, \frac{1}{32}, \frac{1}{64}, \ldots$$

Find a formula for the *N*th entry in this sequence, and explain in detail why your formula is valid.

12. Consider the geometric sequence whose first few entries are

$$\frac{1}{6}, \frac{1}{12}, \frac{1}{24}, \frac{1}{48}, \frac{1}{96}, \frac{1}{192}, \dots$$

Find a formula for the *N*th entry in this sequence, and explain in detail why your formula is valid.

13. Consider the geometric sequence whose first few entries are

$$\frac{2}{3}, \frac{2}{9}, \frac{2}{27}, \frac{2}{81}, \frac{2}{243}, \frac{2}{729}, \dots$$

Find a formula for the *N*th entry in this sequence, and explain in detail why your formula is valid.

14. Parts (a) and (b) of this problem are similar to some of the problems in *Navigating Through Algebra in Prekindergarten–Grade 2* by the National Council of Teachers of Mathematics [61]. Assume that the pattern of a square followed by 2 circles and a triangle continues to repeat in the sequence of shapes in Figure 9.40 and that the numbers below the shapes indicate the position of each shape in the sequence.

FIGURE 9.40   A repeating pattern of shapes

a. What shape will be above the number 150? Explain how you can tell.

b. How many circles will there be above the numbers 1 through 160? Explain your answer.

c. Consider the numbers that are below the squares. What kind of sequence do these numbers form: an arithmetic sequence, a geometric sequence, or neither? Why? If it is either an arithmetic sequence or a geometric sequence, give a formula for the *N*th entry.

d. Consider the numbers that are below the triangles. What kind of sequence do these numbers form: an arithmetic sequence, a geometric sequence, or neither? Why? If it is either an arithmetic sequence or a geometric sequence, give a formula for the *N*th entry.

e. Consider the numbers that are below the circles. What kind of sequence do these numbers form: an arithmetic sequence, a geometric sequence, or neither? Why? If it is either an arithmetic sequence or a geometric sequence, give a formula for the *N*th entry.

15. Parts (a) and (b) for this problem are similar to some of the problems in *Navigating Through Algebra in Prekindergarten–Grade 2* by the National Council of Teachers of Mathematics [61]. Assume that the pattern of a circle followed by 3 squares and a triangle continues to repeat in the sequence of shapes in Figure 9.41 and that the numbers below the shapes indicate the position of each shape in the sequence.

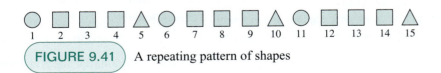

FIGURE 9.41   A repeating pattern of shapes

a. What shape will be above the number 999? Explain how you can tell.

b. How many squares will there be above the numbers 1 through 200? Explain your answer.

c. Consider the numbers that are below the triangles. What kind of sequence do these numbers form: an arithmetic sequence, a geometric sequence, or neither? Why? If it is either an arithmetic sequence or a geometric sequence, give a formula for the *N*th entry.

d. Consider the numbers that are below the circles. What kind of sequence do these numbers form: an arithmetic sequence, a geometric sequence, or neither? Why? If it is either an arithmetic sequence or a geometric sequence, give a formula for the Nth entry.

e. Consider the numbers that are below the squares. What kind of sequence do these numbers form: an arithmetic sequence, a geometric sequence, or neither? Why? If it is either an arithmetic sequence or a geometric sequence, give a formula for the Nth entry.

16. Assume that the repeating pattern of 4 squares followed by a circle and a triangle shown in Figure 9.42 continues to the right. Discuss whether the following reasoning is valid:

Since there are 8 squares above the numbers 1–10, there will be 24 times as many squares above the numbers 1–240. So there will be $24 \times 8 = 192$ squares above 1–240.

If the reasoning is not valid, explain why not and find a different way to determine the number of squares above the numbers 1–240.

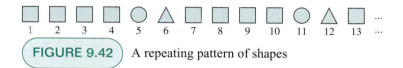

$$\boxed{\text{FIGURE 9.42}} \quad \text{A repeating pattern of shapes}$$

17. What day of the week will it be 1000 days from today? Use math to solve this problem. Explain your answer.

18. Assume that the first day of school is a Wednesday and that school runs Monday through Friday every week with no days off. What day of the week will the 100th school day be? Use math to solve this problem. Explain your answer.

19. Describe four different rules for determining the next entries in the sequence that follows. Give the next 3 entries in the sequence for each rule.

$$5, 7, 11, \ldots$$

20. Describe four different rules for determining the next entries in the sequence that follows. Give the next 3 entries in the sequence for each rule.

$$1, 2, 4, \ldots$$

21. Describe four different rules for determining the next entries in the sequence that follows. Give the next 3 entries in the sequence for each rule.

$$1, 4, 9, \ldots$$

22. Suppose you owe $500 on a credit card that charges you 1.6% interest on the amount you owe at the end of every month. Assume that you do not pay off any of your debt and that you do not add any more debt other than the interest you are charged.

a. Explain why you will owe

$$(1.016)^N \cdot 500$$

dollars at the end of N months.

b. Use a calculator to determine how much you will owe after 2 years.

23. Suppose you put $1000 in an account whose value will increase by 6% every year. Assume that you do not take any money out of this account and that you do not put any money into this account other than the interest it earns. Find a formula for the amount of money that will be in the account after N years. Explain in detail why your formula is valid.

24. Suppose a scientist puts a colony of bacteria weighing 1 gram in a large container. Assume that these bacteria reproduce in such a way that their number doubles every 20 minutes.

a. How much will the colony of bacteria weigh after 1 hour? after 8 hours? after 1 day? after 1 week? In each case, explain your reasoning.

b. Is it plausible that bacteria in a container could double every 20 minutes for more than a few hours? Why or why not?

25. a. Use a calculator to compute 2^{161}. Are you able to tell what the ones digit of this number is from the calculator's display? Why or why not?

b. Determine the ones digits of each of the numbers $2^1, 2^2, 2^3, 2^4, 2^5, \ldots, 2^{17}$. Describe the pattern in the ones digits of these numbers.

c. Using the pattern that you discovered in part (b), predict the ones digit of 2^{161}. Explain your reasoning clearly.

9.6 Series

series

Have you ever wondered how banks calculate loan payments? Loan payments are calculated by the use of formulas for mathematical series. A **series** is a sum of the numbers in a sequence.

Just as there are arithmetic and geometric sequences, there are arithmetic and geometric series. There are also series that are neither arithmetic nor geometric.

Arithmetic Series

arithmetic series

An **arithmetic series** is a sum of consecutive entries in an arithmetic sequence. For example, if we add the first 100 entries in the arithmetic sequence

$$1, 2, 3, 4, 5, 6, \ldots$$

we have the arithmetic series

$$1 + 2 + 3 + 4 + 5 + \cdots + 97 + 98 + 99 + 100$$

If you solved the practice exercises in Section 9.2, you may recall finding a quick way to calculate such a sum.

Geometric Series

Geometric series are of great practical importance, as they are used in calculating loan payments, for example. Although we will not be determining loan payments, we will discuss formulas for adding geometric series, which are the basis for various financial formulas and are widely used in mathematics.

geometric series

A **geometric series** is a sum of some or all of the consecutive entries in a geometric sequence. For example,

$$1 + 2 + 4 + 8 + 16 + 32 + 64$$

and

$$\frac{1}{10} + \frac{1}{100} + \frac{1}{1000} + \frac{1}{10,000} + \cdots$$

are geometric series. Notice that the second series is an infinite sum. It may seem surprising that we can consider an infinite sum of numbers, but observe that

$$\frac{1}{10} + \frac{1}{100} + \frac{1}{1000} + \frac{1}{10,000} + \cdots$$

is really just another way to write the decimal

$$0.1111111111 \ldots = 0.\overline{1}$$

Class Activity   *Now Turn to Class Activities Manual*

9Z Sums of Powers of Two, p. 216

9AA An Infinite Geometric Series, p. 218

Consider a finite geometric series such as

$$1 + 3 + 9 + 27 + 81 + 243 + 729 + 2187$$

Although we could add the numbers one by one to determine the sum, there is a quicker way: Let S stand for the sum of the series, so that

$$S = 1 + 3 + 9 + 27 + 81 + 243 + 729 + 2187$$

If we multiply the series by 3, many of the terms will be the same as in the original series (because each term in the series is 3 times the previous term):

$$3S = 3 \cdot (1 + 3 + 9 + 27 + 81 + 243 + 729 + 2187)$$

$$= 3 + 9 + 27 + 81 + 243 + 729 + 2187 + 6561$$

We get the last equation by the distributive property. If we subtract S from $3S$, we have the following two different ways we can write the result, as an expression in terms of S and as a series:

in terms of S: as a series:

$$
\begin{array}{r}
3S \\
- S \\
\hline
2S
\end{array}
\qquad
\begin{array}{r}
3 + 9 + 27 + 81 + 243 + 729 + 2187 + 6561 \\
- 1 - 3 - 9 - 27 - 81 - 243 - 729 - 2187 \\
\hline
- 1 \hspace{4.5cm} + 6561
\end{array}
$$

So

$$2S = 6560$$

and

$$S = 6560 \div 2 = 3280$$

So the sum

$$1 + 3 + 9 + 27 + 81 + 243 + 729 + 2187 = 3280$$

Notice how we got this answer: We took the next term that would go in the series—namely, $3 \times 2187 = 6561$—we subtracted the initial term, which is 1, and we divided the result by $3 - 1 = 2$. The reasoning that we used previously will show that for any finite geometric series,

$$\text{sum of finite geometric series} = \frac{\text{next term} - \text{first term}}{\text{ratio} - 1}$$

where *ratio* stands for the number that each term is multiplied by to get the next term in the series.

We can use the same method to determine sums of infinite geometric series. For example, consider the infinite series

$$\frac{1}{10} + \frac{1}{100} + \frac{1}{1000} + \frac{1}{10,000} + \cdots$$

Let S stand for its sum. The ratio of this series is $\frac{1}{10}$ (i.e., we multiply a term in the series by $\frac{1}{10}$ to get the next term in the series). So if we multiply S by $\frac{1}{10}$, many of the terms in S and $\frac{1}{10}S$ will match:

$$\frac{1}{10}S = \frac{1}{10}\left(\frac{1}{10} + \frac{1}{100} + \frac{1}{1000} + \cdots\right)$$

$$= \frac{1}{100} + \frac{1}{1000} + \frac{1}{10,000} + \cdots$$

assuming that there is an "infinite distributive property." When we subtract $\frac{1}{10}S$ from S, many of the terms will cancel. If we subtract $\frac{1}{10}S$ from S, we can write the result in the following two ways, in terms of S and in terms of the series:

in terms of S: as a series:

$$
\begin{array}{r}
S \\
-\frac{1}{10}S \\
\hline
\frac{9}{10}S
\end{array}
\qquad
\begin{array}{r}
\frac{1}{10} + \frac{1}{100} + \frac{1}{1000} + \frac{1}{10,000} + \cdots \\
- \frac{1}{100} - \frac{1}{1000} - \frac{1}{10,000} + \cdots \\
\hline
\frac{1}{10}
\end{array}
$$

Therefore,

$$\frac{9}{10}S = \frac{1}{10}$$

So

$$S = \frac{1}{10} \div \frac{9}{10} = \frac{1}{9}$$

and

$$\frac{1}{10} + \frac{1}{100} + \frac{1}{1000} + \frac{1}{10,000} + \ldots = \frac{1}{9}$$

Notice how we got this answer: We took the first term in the series and divided it by 1 minus the ratio. The reasoning that we used previously will show that for any geometric series,

$$\text{sum of infinite geometric series} = \frac{\text{first term}}{1 - \text{ratio}}$$

Notice also that the infinite sum $\frac{1}{10} + \frac{1}{100} + \frac{1}{1000} + \ldots$ is just another way to write

$$0.11111111111 \ldots = 0.\overline{1}$$

We showed previously that

$$0.\overline{1} = \frac{1}{9}$$

The reasoning we just used is similar to the reasoning used in Section 8.6 to show how to write repeating decimals as fractions.

Class Activity *Now Turn to Class Activities Manual*

9BB Making Payments Into an Account, p. 219

Practice Exercises for Section 9.6

1. Suppose that at the beginning of every month, you make a payment of $100 into an account that earns 0.5% interest per month. (That is, the value of the account at the end of the month is 0.5% higher than it was at the beginning of the month.)

   a. Find a series that describes the amount of money in the account at the beginning of the Nth month, right after the $100 payment for that month has been added.

   b. Use the techniques described in this section to find a formula for the total amount of money in the account at the beginning of the Nth month, right after the Nth payment into the account.

   c. Determine how much will be in the account at the beginning of the 24th month, right after the 24th payment into the account.

Answers to Practice Exercises for Section 9.6

1. a. At the beginning of the first month there is $100. At the end of the first month there are therefore

$$(1.005)100$$

dollars in the account, because the amount is increased by 0.5%. (See Section 7.5.) At the beginning of the second month, another $100 is added to this amount, so then there are

$$100 + (1.005)\,100$$

dollars in the account. At the end of the second month there is 1.005 times as much as this in the account; therefore, there are

$$(1.005) \cdot (100 + (1.005)100) = (1.005)100 + (1.005)^2 100$$

dollars in the account, by the distributive property. At the beginning of the third month another \$100 is added to this amount, so then there are

$$100 + (1.005)100 + (1.005)^2 100$$

dollars in the account. Continuing in this way, we see that at the beginning of the Nth month, after the Nth payment is added, there are

$$100 + (1.005)100 + (1.005)^2 100 + (1.005)^3 100 + \cdots + (1.005)^{(N-1)} 100$$

dollars in the account.

b. Let S be the sum of the geometric series in part (a). So

$$S = 100 + (1.005)100 + (1.005)^2 100 + (1.005)^3\, 100 + \cdots + (1.005)^{(N-1)} 100$$

Then, by the distributive property,

$$(1.005)S = (1.005)100 + (1.005)^2 100 + (1.005)^3 100 + \cdots + (1.005)^N 100$$

When we subtract $(1.005)\, S - S$, we can write the answer in two ways: in terms of S and as a series. In terms of S,

$$(1.005)S - S = (0.005)S$$

In terms of a series, most of the terms in $(1.005)S - S$ cancel:

$$
\begin{array}{l}
(1.005)100 + (1.005)^2 100 + \cdots + (1.005)^{(N-1)} 100 + (1.005)^N 100 \\
-100 - (1.005)100 - (1.005)^2 100 - \cdots - (1.005)^{(N-1)} 100 \\
\hline
-100 \hspace{6cm} + (1.005)^N 100
\end{array}
$$

Therefore,

$$(0.005)S = (1.005)^N 100 - 100$$

and it follows that

$$S = \frac{(1.005)^N\, 100 - 100}{0.005}$$

c. According to the formula in part (b), at the beginning of 24 months, there will be

$$\frac{(1.005)^{24} 100 - 100}{0.005} = \$2543.20$$

in the account.

Problems for Section 9.6

1. Use the method described in this section to determine the sum of the geometric series

$$1 + 4 + 16 + 64 + 256 + 1024 + 4096$$

without adding the terms. Explain your work.

2. Use the method described in this section to determine the sum of the geometric series

$$1 + 4 + 4^2 + 4^3 + \cdots + 4^N$$

Explain your work.

3. Use the method described in this section to determine the sum of the geometric series

$$1 + 5 + 25 + 125 + 625 + 3125 + 15{,}625$$

without adding the terms. Explain your work.

4. Use the method described in this section to determine the sum of the geometric series

$$1 + 5 + 5^2 + 5^3 + \cdots + 5^N$$

Explain your work.

5. Use the method described in this section to determine the sum of the infinite geometric series

$$1 + \frac{1}{3} + \left(\frac{1}{3}\right)^2 + \left(\frac{1}{3}\right)^3 + \left(\frac{1}{3}\right)^4 + \cdots$$

Explain your work.

6. Use the method described in this section to determine the sum of the infinite geometric series

$$1 + \frac{1}{4} + \left(\frac{1}{4}\right)^2 + \left(\frac{1}{4}\right)^3 + \left(\frac{1}{4}\right)^4 + \cdots$$

Explain your work.

7. **a.** Use the method described in this section to determine the sum of the infinite geometric series

$$7\left(\frac{1}{10}\right) + 7\left(\frac{1}{10}\right)^2 + 7\left(\frac{1}{10}\right)^3 + 7\left(\frac{1}{10}\right)^4 + \cdots$$

Explain your work.

b. Rephrase your result in part (a) as a result about a repeating decimal.

8. **a.** Use the method described in this section to determine the sum of the infinite geometric series

$$12\left(\frac{1}{100}\right) + 12\left(\frac{1}{100}\right)^2 + 12\left(\frac{1}{100}\right)^3 + 12\left(\frac{1}{100}\right)^4 + \cdots$$

Explain your work.

b. Rephrase your result in part (a) as a result about a repeating decimal.

9. Suppose that at the beginning of every month, you make a payment of $100 into an account that earns 0.75% interest per month. (That is, the value of the account at the end of the month is 0.75% higher than it was at the beginning of the month.)

a. Find a series that describes the amount of money in the account at the beginning of the Nth month, right after the $100 payment for that month has been added.

b. Use the techniques described in this section to find a formula for the amount of money in the account at the beginning of the Nth month, right after the Nth payment into the account.

c. Determine how much will be in the account at the beginning of the 36th month, right after the 36th payment into the account.

9.7 Functions

In this section, we extend our study of sequences in Section 9.5 to the study of mathematical functions. We will define the concept of function and consider the four main ways that functions can be represented: with words, with formulas, with tables, and with graphs. Functions are one of the cornerstones of mathematics and science. By studying and using functions, mathematicians and scientists have been able to describe and predict a variety of phenomena, such as motions of planets, the distance a rocket will travel, or the dose of medicine that a patient needs. The study of functions goes far beyond what we discuss in this section. But even at the most elementary level, functions provide a way to organize, represent, and study information.

function   A **function** is a rule that assigns one output to each allowable input. Most commonly, these "inputs" and "outputs" are numbers. The set of allowable inputs is called the **domain** of the function. The set of allowable outputs is called the **range** of the function.

Texts for children often portray functions as machines, as in Figure 9.43. With this imagery, it's easy to imagine putting a number into a machine; the machine somehow transforms the input number into

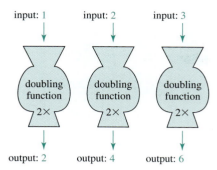

FIGURE 9.43

Viewing functions as machines

FIGURE 9.44

A sequence of shapes

a new output number. For example, consider the *doubling function*: to each input number, the assigned output is twice the input. To the input 1, the doubling function assigns output 2; to the input 2, the doubling function assigns the output 4; to the input 3, the doubling function assigns the output 6; and so on.

We can consider every sequence a function by assigning to each counting number, N, the output that is the Nth entry in the sequence. For example, the sequence

$$10, 12, 14, 16, 18, \ldots$$

corresponds to the function that takes the input 1 to the output 10, the input 2 to the output 12, the input 3 to the output 14, and so on. The sequence in Figure 9.44 corresponds to the function that takes the input 1 to a square, the input 2 to a circle, the input 3 to a triangle, the input 4 to a square, and so on.

It can be cumbersome to describe functions in words by repeatedly referring to inputs and outputs. Therefore, there are several standard ways of displaying functions: with tables, with graphs, and with formulas. These methods can help make functions comprehensible. Now let's look at these ways of representing functions.

Representing Functions with Tables

A table provides an easily understood, organized way to display a function. To display a function in a table, make two columns: the column on the left consisting of various input values, the column on the right consisting of the corresponding output values. Some people like to call such a table a t-chart because you can make such a table by first drawing a large T.

For example, consider the function that assigns to each one of five girls, her height. This function is fully displayed in the table on the left in Table 9.1.

Of course, we can also consider the height function that assigns to *each person in the world,* his or her height. In this case, it would be neither informative nor realistically possible to make a table for this function that would show all inputs and their outputs.

The table on the right in Table 9.1 represents the doubling function. This table does not show all possible inputs and outputs and therefore does not give complete information about the function. But an incomplete table can still be useful for summarizing a function.

TABLE 9.1 Tables for two functions: A height function and a doubling function

Height Function		Doubling Function	
Input (Girl)	Output (Height)	Input	Output
Kaitlyn	53 inches	1	2
Lameisha	56 inches	2	4
Sarah	49 inches	3	6
Manuela	50 inches	4	8
Kelli	54 inches	5	10
		$\vdots$	$\vdots$

In the case of the doubling function, we can make different choices for the inputs that we want to allow. We can decide to allow only counting numbers as inputs for the doubling function. In this case, the doubling function corresponds exactly to the sequence

$$2, 4, 6, 8, 10, \ldots$$

because we assign to the Nth place in the sequence the number $2N$.

If we decide that we want to allow all real numbers as inputs for the doubling function, then the doubling function encompasses more than just the sequence

$$2, 4, 6, 8, 10, \ldots$$

For example, the doubling function assigns to the input 3.25, the output 6.5, but 6.5 is not in the sequence

$$2, 4, 6, 8, 10, \ldots$$

Tables are easy to read and to understand, but they rarely display complete information about a function. Another good way to display a function is to use a graph.

Representing Functions with Graphs

The graph of a function displays the function in a picture. By graphing a function we can often display more inputs and outputs than we could display in a table, and we can often discern trends and patterns in the function that are less obvious in a table. Like tables, however, graphs usually do not display complete information about a function. Graphs of functions are displayed in coordinate planes.

coordinate
plane

axes

origin

Coordinate Planes A **coordinate plane** is a plane, together with two perpendicular number lines in the plane that meet at the location of 0 on each number line. The two number lines are called the **axes** of the coordinate plane (singular. axis). Traditionally, one number line is displayed horizontally and the other is displayed vertically. The horizontal axis is often called the *x-axis*, and the vertixal axis is often called the *y-axis*. The point where the *x*- and *y*-axes meet is called the **origin.**

The main feature of a coordinate plane is that *the location of every point in the plane can be specified by referring to the two axes.* This works in the following way: A pair of numbers, such as (4.5, 3), corresponds to the point in the plane that is located where a vertical line through 4.5 on the horizontal axis meets a horizontal line through 3 on the vertical axis, as shown in Figure 9.45. This point is designated (4.5, 3); or we say that 4.5 and 3 are the **coordinates** of the point. Specifically, 4.5 is the first coordinate, or *x*-coordinate, of the point and 3 is the second coordinate, or *y*-coordinate, of the point.

coordinates

You can use your fingers to locate the point (4.5, 3) by putting your right index finger on 4.5 on the horizontal axis and your left index finger on 3 on the vertical axis, and sliding the right index finger

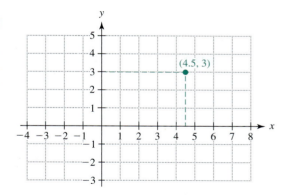

FIGURE 9.45

Locating (4.5, 3) in a coordinate plane

vertically upward and the left index finger horizontally to the right until your two fingers meet. Some people like to use two sheets of paper to locate points in a coordinate plane by holding the edge of one sheet parallel to the *y*-axis and the edge of the other sheet parallel to the *x*-axis.

Figure 9.46 shows the coordinates of several different points in a coordinate plane. Notice that the coordinates of *a point can include negative numbers.*

graph of a function   **The Graph of a Function**   A function whose inputs and outputs are *numbers* has a graph. The **graph of a function** consists of all those points in a coordinate plane whose second coordinate is the output of the first coordinate. For example, the graph on the left of Figure 9.47 shows the graph of the doubling function when only the counting numbers are allowed to be inputs. In this case, the graph of the doubling function consists of the points

$$(1, 2), (2, 4), (3, 6), (4, 8), (5, 10), \ldots$$

For each point on the graph, the second coordinate is twice the first coordinate.

The graph on the right of Figure 9.47 shows the graph of the doubling function when we allow all real numbers as inputs. In this case, the graph consists of all points of the form

$$(N, 2N)$$

where N can be any real number. For example, in this case

$$(3.25, 6.5)$$

is on the graph of the doubling function, and so is

$$(-1.3, -2.6)$$

The graph on the left in Figure 9.47 consists of isolated points, but the graph on the right in Figure 9.47 consists of so many points that the points fill in a solid line.

Drawing the Graph of a Function   To draw the graph of a function, it often helps to display the function in a table first. For example, suppose that a seed is planted and the height of the plant growing from the seed is recorded every day after planting. This situation gives rise to a *height* function, such as the

FIGURE 9.46

Points in a coordinate plane

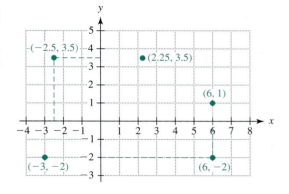

FIGURE 9.47

Graph of the
doubling function
with different
allowable inputs
(domains)

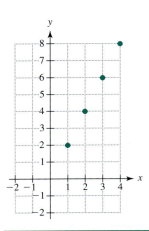

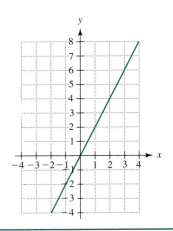

Graph of the doubling function with only counting numbers as inputs

Graph of the doubling function when all real numbers are allowed as inputs

hypothetical one shown in the table in Figure 9.48. Each input and corresponding output produces a point in a coordinate plane. The collection of all such points in a coordinate plane forms the graph of the function.

The points obtained from the table for the function can then be plotted in a coordinate plane, as on the right in Figure 9.48. Notice that in this graph, the points obtained from the table have been connected by a curve. Why does it make sense to connect the points? In between the times that the plant's height was measured, the plant continued to grow. For example, between days 7 and 8, the plant grew from 3 cm to 5 cm. So, $7\frac{1}{2}$ days after planting, the plant was probably about 4 cm tall. Therefore, the point (7.5, 4), or a point that is very close to this point, is also on the graph of the plant height function. For every time between 7 days after planting and 8 days after planting, the plant had some height between 3 cm and 5 cm, and this yields points on the graph that lie between the points (7, 3) and (8, 5) and that connect these two points with a curve.

Note the following two features of the graph in Figure 9.48 that help make the graph easy to see and to interpret:

- The scales on the two axes are different: The distances between numbers on the horizontal axis are smaller than the distances between numbers on the vertical axis. This is acceptable

FIGURE 9.48

A table for a plant
height function
yields points that
can be plotted in a
coordinate plane
to make a graph of
the function.

Plant Height Function		yields the point
Input # of Days Since Planting	Output Plant Height (in cm)	
1	0	(1, 0)
2	0	(2, 0)
3	0	(3, 0)
4	0	(4, 0)
5	0.5	(5, 0.5)
6	1.5	(6, 1.5)
7	3	(7, 3)
8	5	(8, 5)
9	5.5	(9, 5.5)
10	6	(10, 6)
11	6	(11, 6)
12	6	(12, 6)

Graph of the plant height function

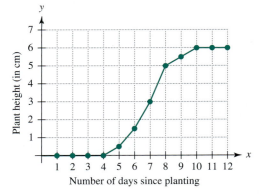

because the horizontal and vertical axes represent different kinds of quantities: The horizontal axis represents time and the vertical axis represents height. When graphing a function, choose scales that make the function easy to see and interpret.

- The axes are labeled. The horizontal axis is labeled "Number of days since planting," and the vertical axis is labeled "Plant height (in cm)." Labeling the axes helps a reader understand and interpret a graph when the function arises from a real or realistic situation.

Interpreting Graphs of Functions Once you learn how to interpret graphs of functions, you can quickly and easily see qualitative aspects of functions. Look at the graph in Figure 9.48. "Read" it from left to right. Moving from left to right along the graph represents time passing: How far a point on the graph is to the right corresponds to how many days have passed since the seed was planted. The height of a point on the graph above the horizontal axis represents the height of the plant at that time (i.e., at the time represented by the horizontal location of the point). Thus, following the graph from left to right, we see that the plant doesn't sprout until after 4 days. Between the 6th and 8th days, the plant grows more quickly than at other times. We see this because the graph rises steeply during those times, meaning that over a short period of time, a lot of growth has occurred. Between days 8 and 10, the plant grows less rapidly; the height doesn't increase as much over those 2 days as it did over the previous 2 days. After 10 days, the plant has stopped growing—its height no longer increases.

Class Activity   *Now Turn to Class Activities Manual*

9CC   Interpreting Graphs of Functions, p. 220

9DD   Are These Graphs Correct? p. 223

Representing Functions with Formulas

Some functions can be described by formulas; others cannot. To give a formula for a function means to give a formula for the output of the function in terms of the input of the function. For example, consider the doubling function. The output is always 2 times the input. Therefore, if the input is x, the output is $2x$, and so the doubling function is described by the formula

$$2x$$

If we give the doubling function the name "D," then a standard way to describe the formula for the doubling function is to write

$$D(x) = 2x$$

This is shorthand for saying that when x is put into the function D, the output is $2x$.

Not all functions have formulas. For example, consider the plant height function we just studied. There is no formula to describe the height of the plant in terms of the number of days since the seed was planted.

If you have a formula for a function, then you can make a table and a graph for the function by using the formula. For example, consider the function, f, which is given by the formula

$$f(x) = x^2 - x - 2$$

To make a table for the function, remember what it means to say that $f(x) = x^2 - x - 2$. It means that when x is put into the function, the output is $x^2 - x - 2$. So, if 1 is put into the function, the output is

$$1^2 - 1 - 2 = 1 - 1 - 2 = -2$$

If 2 is put into the function, the output is

$$2^2 - 2 - 2 = 4 - 2 - 2 = 0$$

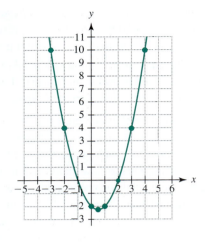

FIGURE 9.49

A table and graph
for the function
$f(x) = x^2 - x - 2$

f	
Input	**Output**
−3	10
−2	4
−1	0
0	−2
0.5	−2.25
1	−2
2	0
3	4
4	10

If 3 is put into the function, the output is

$$3^2 - 3 - 2 = 9 - 3 - 2 = 4$$

and so on. Therefore, the function f has the table given on the left in Figure 9.49.

We can use the table in Figure 9.49 to graph the function f, as shown on the right in Figure 9.49. Assuming that all real numbers are allowed as inputs for f, the graph of f will be a curve that connects the points from the table. If you were to plot many more points, you would see that this curve is smooth, as shown in Figure 9.49.

In more advanced mathematics, a central topic of study is the relationship between the nature of a formula for a function and the nature of the graph of the function.

Practice Exercises for Section 9.7

1. Plot the following points in a coordinate plane:

   $(4, -2.5)$, $(-4, 2.5)$, $(3, 2.75)$, $(-3, -2.75)$

2. For each of the following descriptions, draw the graph of an associated pollen count function for which the input is time elapsed since the beginning of the week and the output is the pollen count at that time. In each case, explain why you draw your graph in the shape that you do.

   a. At the beginning of the week, the pollen count rose sharply. Later in the week, the pollen count continued to rise, but more slowly.

   b. The pollen count fell steadily during the week.

   c. At the beginning of the week, the pollen count fell slowly. Later in the week, the pollen count fell more rapidly.

3. A company that sells juice is interested in a profit function whose inputs are the possible prices that a bottle of juice could sell for and whose outputs are the profits the company will make when juice is sold at that price. The graph of this profit function is shown in Figure 9.50.

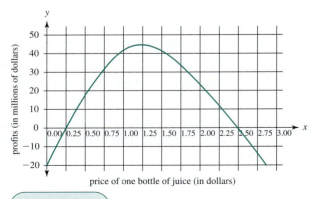

FIGURE 9.50   A profit function for juice sales

a. Based on the graph in Figure 9.50, if juice is sold for $2.00 per bottle, approximately what will the company's profit be?

b. Based on the graph in Figure 9.50, which price for one bottle of juice will result in maximal profit?

c. Where does the graph cross the x-axis, and what is the significance of those points?

d. What is the significance of the points on the graph that lie above the x-axis? below the x-axis?

4. Geneva went for a bike ride. She started off riding at a constant speed on level ground. After a while she went down a hill, going faster and faster. But as soon as she got to the bottom of the hill, there was another hill for Geneva to ride up. Geneva had built up speed coming down the other hill, so she started up the hill going fast. But soon Geneva was going more and more slowly. When she got to the top of the hill, Geneva stopped for a long rest.

Describe three different functions that fit with this story about Geneva's bike ride. In each case, sketch a graph that could be the graph of the function. Indicate how these graphs fit with the story.

5. Fill in the blanks so that the next points lie on the graph of the function f defined by $f(x) = -2x + 1$. Explain.

$$(3, \underline{}), (\underline{}, -13), (a, \underline{}), (\underline{}, b)$$

Answers to Practice Exercises for Section 9.7

1. See Figure 9.51.

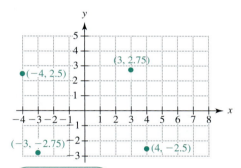

FIGURE 9.51   Points in a coordinate plane

2. See Figure 9.52. Moving from left to right along the graph corresponds to time passing during the week.

a. Graph *a* goes up steeply at first (reading it from left to right) because for each of the first few days of the week, the pollen count is significantly higher than it was the day before. Thus, the graph slopes steeply upward at first. But later in the week, toward the right of the graph, the graph must go up less steeply because for each day later in the week, the pollen count is only a little higher than it was the day before.

b. Graph *b* slopes downward (reading from left to right) and shows that for each day that goes by, the pollen count drops by the same amount.

c. Graph *c* goes down slowly at first (reading from left to right) because for each of the first few days of the week, the pollen count is only a little

FIGURE 9.52

Pollen counts

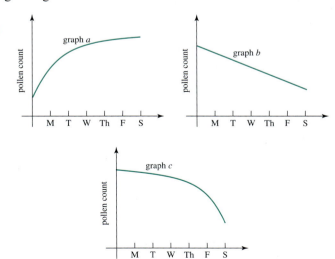

less than it was the day before. Thus, the graph slopes gently downward at first. But later in the week, toward the right of the graph, the graph must go down more steeply because for each day later in the week, the pollen count is significantly lower than it was the day before. Thus, toward the right, the graph slopes steeply downward.

3. **a.** If juice sold for $2.00 per bottle, the company's profit would be about $24 million. We can see this from the graph because when the input (*x*-coordinate) is 2.00, the output (*y*-coordinate) appears to be about 24, since the point (2.00, 24) seems to be on (or nearly on) the graph, as shown in Figure 9.53.

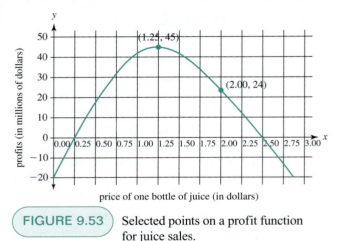

FIGURE 9.53   Selected points on a profit function for juice sales.

b. The maximal profit will occur when the graph reaches its highest point, which appears to be approximately at the point (1.25, 45), as we see in Figure 9.53. So when a bottle of juice sells for $1.25, the company will make a maximal profit of $45 million.

c. The graph crosses the *x*-axis at the points (0.25, 0) and (2.50, 0). At these points the profit is 0. So if juice sells for $0.25 or for $2.50 per bottle, the company breaks even, but does not make a profit.

d. The points on the graph that lie above the *x*-axis correspond to prices for a bottle of juice for which the company makes a profit. This is because at these points, the profits (i.e., outputs) are greater than 0. If the company sells a bottle

of juice for anywhere between $0.25 and $2.50, the company will make a profit.

The points on the graph that lie below the *x*-axis correspond to prices for a bottle of juice for which the company loses money. This is because at these points, the profits (i.e., outputs) are less than 0. For example, if the company sets the price of a bottle of juice either less than $0.25 or greater than $2.50, then the company will lose money.

4. One function that corresponds to Geneva's bike ride is a distance function. The input for the distance function is the time that has elapsed since Geneva started her bike ride. The output is the total distance Geneva has traveled at that time. A possible graph of this function is at the top of Figure 9.54.

A second function that corresponds to Geneva's bike ride is a speed function. The input for the speed function is the time that has elapsed since Geneva started her bike ride. The output is the speed at which Geneva is riding at that time. A possible graph of this function is on the bottom left of Figure 9.54.

A third function that corresponds to Geneva's bike ride is a height function. The input for the height function is the time that has elapsed since Geneva started her bike ride. The output is the height that Geneva is above sea level at that time. A possible graph of this function is on the bottom right of Figure 9.54.

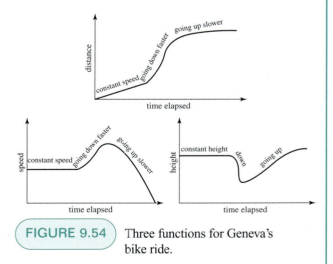

FIGURE 9.54   Three functions for Geneva's bike ride.

5. When the input is 3, the output of the function *f* is $f(3) = -2 \cdot 3 + 1 = -6 + 1 = -5$. So $(3, -5)$ is on the graph of the function.

When the output is -13, the input of the function f is a number x such that $f(x) = -13$. So

$$-2x + 1 = -13$$

Solving for x, we find that $x = 7$. So $(7, -13)$ is on the graph of the function. To check, $f(7) = -2 \cdot 7 + 1 = -14 + 1 = -13$.

When the input is a, the output of the function f is $f(a) = -2a + 1$. So $(a, -2a + 1)$ is on the graph of the function.

When the output is b, the input of the function f is a number x such that $f(x) = b$. So

$$-2x + 1 = b$$

We can solve this equation for x:

$$-2x + 1 = b \quad \text{subtract 1 from both sides}$$
$$-2x = b - 1 \quad \text{divide both sides by } -2$$
$$x = \frac{b - 1}{-2} = \frac{1 - b}{2}$$

So $(\frac{1-b}{2}, b)$ is on the graph of the function. To check,

$$f\left(\frac{1 - b}{2}\right) = -2\left(\frac{1 - b}{2}\right) + 1$$
$$= -(1 - b) + 1$$
$$= -1 + b + 1 = b$$

Problems for Section 9.7

1. Consider the *doubling plus one* function: The output is two times the input, plus one. For example, if the input is 3, the output is 7. Determine which of the following points are on the graph of the doubling plus one function:

   $(2, 5), \quad (2, 8), \quad (-1, -1), \quad (3, 4),$
   $(-2, -3), \quad (1.5, 4), \quad (0.5, 1.5)$

   Explain your reasoning.

2. Consider the *squaring* function: The output is the input times itself. For example, if the input is 3, the output is 9. Determine which of the following points are on the graph of the squaring function:

   $(6, 36), \quad (4, 8), \quad (-2, 4), \quad (-3, -6),$
   $(5, 25), \quad (6, 12), \quad (-6, 36)$

   Explain your reasoning.

3. Consider the function f defined by $f(x) = x^2 - 2x$.

   a. Make a table in which you show the outputs for the following inputs:

      $$-2, -1, 0, 1, 2, 3, 4$$

   b. Plot the points that are associated with the inputs and outputs you found in part (a).

   c. Based on the points you found in part (b), draw the graph of the function $f(x) = x^2 - 2x$ as best you can.

4. For each of the following descriptions, draw a possible graph of an associated temperature function, for which the input is time elapsed since the beginning of the day, and the output is the temperature at that time. In each case, explain why you draw your graph in the shape that you do.

   a. At the beginning of the day, the temperature dropped sharply. Later on, the temperature continued to drop, but more gradually.

   b. The temperature rose steadily at the beginning of the day, remained stable in the middle of the day, and fell steadily later in the day.

   c. The temperature rose quickly at the beginning of the day until it reached its peak. Then the temperature fell gradually throughout the rest of the day.

5. Sally invests $5000 in stock. This situation gives rise to a stock value function for which the input is the time elapsed since investing and the output is the value of Sally's stock. The graph of this function is shown in Figure 9.55.

   a. Compare the rate of growth of the value of Sally's stock in the first 10 months versus in the next 20 months. During which time period does the value of the stock grow more rapidly? Explain how the graph shows you this.

b. What happens to Sally's stock between 30 months and 40 months? Explain your answer.

c. What happens to Sally's stock between 45 months and 60 months? Explain your answer.

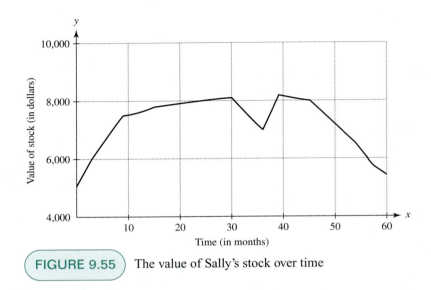

FIGURE 9.55 The value of Sally's stock over time

---

6. Consider the following scenario: Over a period of 5 years, the level of water in a lake drops slowly at first, then much more rapidly. During the next 5 years, the water level in the lake remains stable. Over the next 3 years, the water level rises slowly, and after another 2 years of more rapid increase in water level, the lake is back to its original level from 15 years before.

Describe a function that this scenario gives rise to. Sketch a graph that could be the graph of this function. Explain how features of your graph correspond to events described in the previous scenario.

7. A patient receives a drug. This situation gives rise to a drug function for which the input is the time elapsed since the drug was administered and the output is the amount of the drug in the patient's blood (in milligrams per milliliter of blood). The graph of this drug function is shown in Figure 9.56.

a. Give the approximate coordinates of the point *A*, and discuss the significance of this point in terms of the scenario about the patient and the drug.

b. Give the approximate coordinates of the point *B*, and discuss the significance of this point in terms of the scenario about the patient and the drug.

c. Overall, discuss what the graph tells you about the patient and the drug.

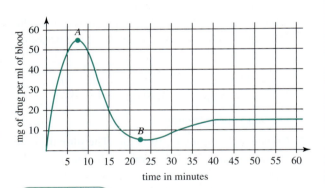

FIGURE 9.56 The graph of a drug function

8. An object is dropped from the top of a building. At first the object falls slowly, but as it continues to fall, it falls faster and faster. The speed at which the object falls increases at a steady rate.

One function that arises from this scenario is the height function, whose input is the time elapsed since the object was dropped and whose output is the height of the object above the ground at that time. Another function that arises from this scenario is the speed function, whose input is the time elapsed since the object was dropped and whose output is the speed at which the object is falling at that time. Identify graphs of these two functions from among the graphs shown in Figure 9.57 and explain your choices.

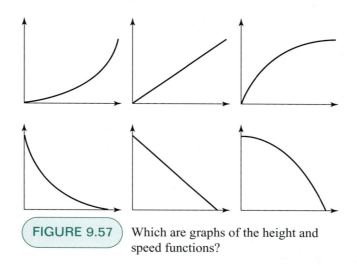

FIGURE 9.57   Which are graphs of the height and speed functions?

---

9.  A skydiver jumps out of a plane. At first, the skydiver falls faster and faster, but then the sky-diver reaches a point where he falls with a constant speed. After falling at a constant speed for a while, the skydiver's parachute opens and the skydiver falls at a slower constant speed until he lands safely on the ground.

   Describe two different functions that fit with this story about a skydiver. In each case, sketch a graph that could be the graph of the function. Indicate how these graphs fit with the story.

10. Water is being poured at a steady rate into a round vase whose vertical cross-section is shown in Figure 9.58. One function that arises from this situation is the volume function, whose input is the time elapsed since water began pouring into the vase and whose output is the volume of water in the vase at that time. Another function that arises from this situation is the height function, whose input is the time elapsed since water began pouring

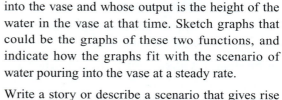

FIGURE 9.58   The cross-section of a vase

into the vase and whose output is the height of the water in the vase at that time. Sketch graphs that could be the graphs of these two functions, and indicate how the graphs fit with the scenario of water pouring into the vase at a steady rate.

11. Write a story or describe a scenario that gives rise to a function. Describe the function (taking care to describe the inputs and outputs clearly), sketch a possible graph of the function, and indicate how your graph fits with your story or scenario.

12. Draw *two different* graphs of two different functions that have the following three properties:
    - When the input is 1, the output is 3.
    - When the input is 2, the output is 6.
    - When the input is 3, the output is 9.

13. Find *two different* formulas for two different functions that have the following two properties:
    - When the input is 0, the output is 1.
    - When the input is 1, the output is 2.

14. During a 6-hour rainstorm, rain fell at a rate of 1 inch every hour for the first 3 hours. During the next hour, $\frac{1}{2}$ inch of rain fell. During the last 2 hours, rain fell at a rate of $\frac{1}{4}$ inch every hour.

    This scenario gives rise to a rainfall function in which the input is the time elapsed since the rainstorm began and the output is the amount of rain that fell during the storm up to that time.

a. Make a table for the rainfall function.

b. Draw a graph of the rainfall function.

c. Is there a single formula for the rainfall function?

15. In the country of Taxo, each household pays taxes on its annual household income as determined by the following:

• A household with income between $0 and $20,000 pays no tax.

• For each $1 of income beyond $20,000, and up to $50,000, the household must pay $0.20 in tax.

• For each $1 of income beyond $50,000, and up to $100,000, the household must pay $0.40 in tax.

• For each $1 of income beyond $100,000 the household must pay $0.60 in tax.

This situation gives rise to a tax function in which the input is annual household income and the output is the taxes paid.

a. Make a table for the tax function.

b. Draw a graph of the tax function.

c. Is there a single formula for the tax function?

9.8 Linear Functions

 Focal Points
Grade 8

This section, about linear functions, will allow us to tie together two topics we have studied: arithmetic sequences and ratios.

linear function A **linear function** is a function whose graph lies on a line in a coordinate plane. There are several other ways to characterize linear functions; first we will see how linear functions are related to proportions and to arithmetic sequences.

Class Activity   *Now Turn to Class Activities Manual*

9EE A Function Arising from Proportions, p. 224

9FF Arithmetic Sequences as Functions, p. 225

Proportional Relationships Are Linear Functions

Suppose two quantities are in a ratio of 3 to 5. For example, suppose that to make juice, for every 3 parts of juice concentrate, we must mix in 5 parts water to make (8 parts) juice. A juice function arises naturally from this situation. This function's input is the number of cups of juice concentrate we could use; its output is the number of cups of water we should mix with that amount of concentrate. Figure 9.59 shows a table and graph for this juice function.

Notice that the table in Figure 9.59 is just a ratio table, of the type we studied in Chapter 7. Notice also that the graph is a line which goes through the origin (the point (0, 0)). Notice also how the table and the graph reflect the ratio 3 to 5. Whenever the input increases by 3, the output increases by 5. Starting at a point on the graph, if we move to the right 3 units, we must move up 5 units to stay on the graph. Also, the points on the graph give rise to all ratios that are equivalent to the ratio 3 to 5.

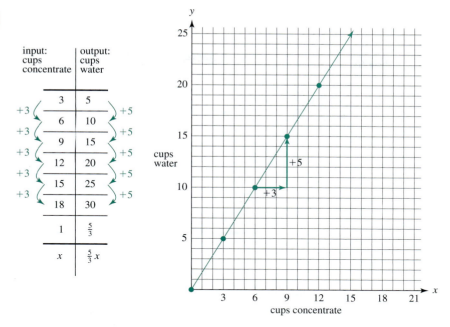

A function arising from the ratio 3 to 5 of juice concentrate to water

What about a formula for the juice function? Since for every 3 cups concentrate we must use 5 cups water, for 1 cup concentrate we should use $\frac{1}{3}$ as much water—namely, $\frac{5}{3}$ cups water. So for x cups concentrate, we must use x times as much water, namely, $\frac{5}{3} \times$ cups water, which gives us a formula for the juice function.

What we have seen in the juice example holds generally. In general, given a pair of quantities that are in a fixed ratio (such as 3 to 5), there is an associated function that arises from all equivalent ratios; the graph of this function is a line through the origin. So proportional relationships provide many examples of linear functions.

Arithmetic Sequences Are Linear Functions

Arithmetic sequences also provide examples of linear functions. For example, suppose that the Tropical Orchid Company sells orchids by mail order. The company charges $7.00 for shipping plus $3.00 shipping for each orchid ordered. This situation gives rise to an arithmetic sequence in which the first entry is the cost of shipping 1 orchid, the second entry is the cost of shipping 2 orchids, the third entry is the cost of shipping 3 orchids, and so on, as shown in Figure 9.60. Each entry in the sequence is $3 more than the previous entry because each additional orchid costs an extra $3 in shipping. Based on our work on arithmetic sequences, the Nth entry is

$$3N + 7$$

dollars. In other words, the cost for shipping N orchids is $3N + 7$ dollars.

Now let's view the orchid shipping sequence as a function whose input is the number of orchids to be shipped and whose output is the cost of shipping that many orchids. We can make a table for this

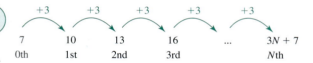

An arithmetic sequence of shipping costs of orchids if there is a $7 fee plus $3 for each orchid

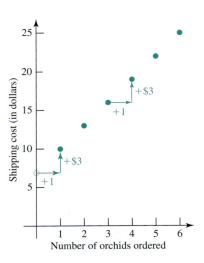

FIGURE 9.61

A table and a graph of the orchid shipping sequence

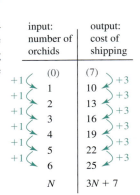

input: number of orchids	output: cost of shipping
(0)	(7)
1	10
2	13
3	16
4	19
5	22
6	25
N	3N + 7

function, and we can graph this function, as in Figure 9.61. Notice that the arithmetic nature of the orchid shipping sequence is reflected in its table in the following way: Whenever the input goes up by 1, the output goes up by a fixed amount, namely, 3. In other words, the ratio of the change in input to the corresponding change in output is always the same. Notice also that the graph of the orchid shipping sequence appears to lie on a straight line.

In fact, the following properties are true for all arithmetic sequences:

1. When the sequence is viewed as a function, the ratio of the change in input to the corresponding change in output is always the same.

2. There are numbers m and b so that the Nth entry in the sequence is $mN + b$.

3. The graph of the sequence (viewed as a function) lies on a line.

Viewed as functions, arithmetic sequences are linear functions whose input set (domain) is restricted to counting numbers rather than allowing all real numbers as inputs. We will now see that all linear functions have properties (1) and (2)—once property (2) is suitably rephrased.

Properties of Linear Functions

Consider a linear function. By definition, its graph is a line. We can use this line to form many different right triangles that have a horizontal and a vertical side, as shown in Figure 9.62. Because the horizontal lines are all parallel, the angles they form with the graph of the function are all the same. Since they are right triangles, corresponding angles of these triangles must be equal; so, as we'll see in Chapter 14, all such triangles are similar. Therefore, the ratios of the lengths of the horizontal sides to the lengths of the vertical sides are equal for all these triangles. The lengths of horizontal sides of these triangles represent "changes in inputs" because the horizontal axis represents inputs; the lengths of vertical sides of these triangles represent the corresponding "changes in outputs" because the vertical axis represents outputs. Therefore, for a linear function, the ratio of the change in input to the corresponding change in output is always the same.

We have seen that for linear functions, the ratio of the change in input to the change in output is always the same. We will now use this fact to deduce that every linear function has a formula of a certain type. Let b be the location on the y-axis where the graph of the function crosses the y-axis. Suppose that when the input is increased by 1, the output increases by m. Because the ratio of the change in input to the corresponding change in output is always the same, if the input increases by x, the output increases by mx. Therefore, when the input is x, the output is

$$mx + b$$

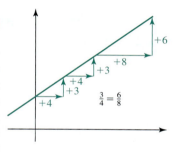

FIGURE 9.62

For linear functions, the ratio of the change in input to the corresponding change in output is always the same due to similar triangles.

as indicated in Figure 9.63. So every linear function has a formula of the form

$$f(x) = mx + b$$

slope   for some numbers m and b. The number m is called the **slope** of the line; the number b is called the
y-intercept   **y-intercept** of the line.

To summarize, for any linear function, the following properties are true:

1.   The ratio of the change in input to the corresponding change in output is always the same.

2.   There are numbers m and b so that the linear function is described by the formula $f(x) = mx + b$.

3.   The number b in part (2) is the output for the input 0. The number m in part (2) is the increase in output when the input increases by 1.

Notice that part (3) tells us that the ratio in part 1 is 1 to m.

Class Activity   *Now Turn to Class Activities Manual*

9GG Analyzing the Way Functions Change, p. 227

9HH Story Problems for Linear Functions, p. 228

9II   Deriving the Formula for Temperature in Degrees Fahrenheit in Terms of Degrees Celsius, p. 229

FIGURE 9.63

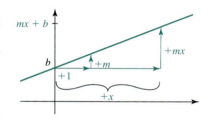

For a linear function, if the output increases by m when the input increases by 1, then the output increases by mx when the input increases by x

Practice Exercises for Section 9.8

1. A nightlight costs $7.00 to buy and uses $0.75 of electricity per year to operate. What function does this situation give rise to? Make a table for this function, sketch a graph of this function, and find a formula for this function.

2. A juice mixture can be made by mixing juice and lemon-lime soda in a ratio of 3 to 2.

   a. Describe a function that is associated with this situation. Write a formula for the function and explain why the function has this formula. Sketch the graph of the function.

   b. Discuss how the formula for the function in part (a), the graph of the function, and the 3-to-2 ratio (or a different, related ratio) are related to each other.

3. Consider this arithmetic sequence:

   $$2, 5, 8, 11, \ldots$$

   a. Describe a function that is associated with this arithmetic sequence. Write a formula for this function and explain why the function has this formula. Sketch the graph of this function.

   b. Discuss how the components of the formula in part (a) are related to the arithmetic sequence and to the graph in part (a).

4. For each of the tables that follow, determine whether it could be the table for a linear function. If so, find a formula for the linear function. If not, explain why not.

f	
Input	Output
0	2
1	7
2	12
3	17
4	22

g	
Input	Output
2	5
4	10
6	15
8	20
10	25

h	
Input	Output
1	2
2	4
3	8
4	16
5	32

5. Assuming that the following tables are for linear functions, fill in the blank outputs and find formulas for these functions:

Input	Output
0	
1	
2	
3	4
4	
5	8
6	

Input	Output
0	
1	
2	3
3	
4	
5	
6	6

Input	Output
−3	7
−2	
−1	
0	
1	−1
2	
3	

Answers to Practice Exercises for Section 9.8

1. The situation gives rise to a nightlight function for which the input is the number of years since purchase and the output is the cost of operating the nightlight for that amount of time (where this cost includes the cost of buying the nightlight). The following is a table for the nightlight function:

Nightlight Function	
Input Years Since Purchase	output Total Cost of Operating
0	$7.00
1	$7.75
2	$8.50
3	$9.25
4	$10.00
5	$10.75

Because the nightlight costs $0.75 for each year of operation, it costs $0.75x$ dollars to operate for x years. Adding this amount to the purchase price of $7.00, we have it costs a total of

$$7 + 0.75x$$

dollars to operate the nightlight for x years. A sketch of the graph of the nightlight function is in Figure 9.64.

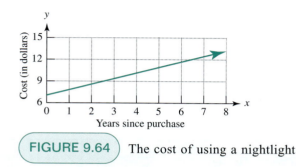

FIGURE 9.64 The cost of using a nightlight

2. **a.** One function is the function whose inputs are amounts of lemon-lime soda and whose outputs are the corresponding amount of juice that will be needed to mix with that amount of lemon-lime soda. If we call this function f, then it has the formula given by

$$f(x) = \frac{3}{2}x$$

This formula is valid because for every 2 cups of lemon-lime soda, we need 3 cups of juice. So for 1 cup of soda we need half as much juice, namely, $\frac{3}{2}$ cups of juice. So for x cups of soda, we need x times as much juice, namely, $\frac{3}{2}x$ cups of juice, which gives the formula for the function. The graph of this function (not shown) is a line through the origin with slope $\frac{3}{2}$. (Another function has the same inputs but the outputs are the total amount of juice mixture that is made. Two more functions have amounts of juice as inputs and either the corresponding amount of lemon-lime soda as outputs or the total amount of juice mix created as outputs.)

b. The 3-to-2 ratio shows up in the graph this way: Every time we go over 2 units to the right, we go up 3 units. In the formula for the function, the 3-to-2 ratio shows up as the unit rate, $\frac{3}{2}$ of cups of juice per cup of lemon-lime soda. This unit rate is also the slope of the graph: Whenever we go to the right 1 unit on the graph, we go up $\frac{3}{2}$ units.

3. **a.** The natural function associated with this sequence is the function whose inputs are the counting numbers and whose outputs are the entries of the sequence in those positions. In other words, if we call this function S, then $S(N)$ is the Nth entry of the sequence. To find a formula for the Nth entry, use the reasoning of Section 9.5. Since the sequence starts at 2 and increases by 3, the 0th entry of the sequence would be $2 - 3 = -1$. The Nth entry is "N jumps of 3 from -1," which is $-1 + 3N$. Therefore, the function S is given by the formula $S(N) = -1 + 3N$. The graph of this function (not shown) consists of isolated points on a line.

b. The 0th entry of the sequence is the y-intercept of the graph of the function (i.e., the y-value where the graph hits the *y*-axis). The amount by which the sequence increases, 3, is slope of the graph, namely, whenever we go to the right 1 unit, we go up 3 units.

4. The first table is a table of the linear function with formula

$$f(x) = 5x + 2$$

This formula fits with the table because whenever the input increases by 1, the output increases by 5. For a linear function with these data, when the input increases by *x*, the output must increase by 5*x*. When the input is 0, the output is 2. So when the input is *x*, the output must be 5*x* more than 2, which is $5x + 2$.

The second table is a table of the linear function

$$g(x) = \frac{5}{2}x$$

This formula fits with the table because whenever the input increases by 2, the output increases by 5. For a linear function with these data, the ratio of increase in input to increase in output is 2 to 5. This is the same as the ratio 1 to $\frac{5}{2}$. So, when the input increases by 1, the output increases by $\frac{5}{2}$. Therefore, if the input increases by *x*, the output must increase by $\frac{5}{2}x$. When the input is 0, the output is 0, so when the input is *x*, the output is $\frac{5}{2}x$ more than 0, which is $\frac{5}{2}x$.

The third table is not a table for a linear function because when the input increases by 1, the output increases by different amounts. For example, in going from the input 1 to the input 2, the output increases by 2 from 2 to 4. But in going from the input 2 to the input 3, the output increases by 4, from 4 to 8.

5. Formulas for each function are shown at the bottom of each table.

Input	Output
0	−2
1	0
2	2
3	4
4	6
5	8
6	10
x	2*x* − 2

Input	Output
0	1.5
1	2.25
2	3
3	3.75
4	4.5
5	5.25
6	6
x	0.75*x* + 1.5

Input	Output
−3	7
−2	5
−1	3
0	1
1	−1
2	−3
3	−5
x	−2*x* + 1

Problems for Section 9.8

1. A long-distance phone company charges $3 for a phone call plus $0.75 for each minute of the call. What function does this situation give rise to? Make a table for this function, sketch a graph of this function, and find a formula for this function.

2. For each of the tables that follow, determine whether it could be the table for a linear function. If so, find a formula for the linear function. If not, explain why not.

f	
Input	Output
0	3
1	6
2	9
3	12
4	15

g	
Input	Output
0	3
2	6
4	9
6	12
8	15

h	
Input	Output
0	1
1	2
2	5
3	10
4	17

3. For each of the tables that follow, determine whether it could be the table for a linear function. If so, find a formula for the linear function. If not, explain why not.

f	
Input	Output
0	3
1	7
2	11
3	15
4	19

g	
Input	Output
2	4
4	7
6	10
8	13
10	16

h	
Input	Output
0	−1
1	0
2	3
3	8
4	15

4. Assuming that the following tables are for linear functions, fill in the blank outputs and find formulas for these functions:

Input	Output
0	
1	
2	7
3	
4	10
5	
6	

Input	Output
0	
1	−2
2	
3	
4	
5	
6	2

Input	Output
−3	13
−2	
−1	
0	
1	
2	−2
3	

5. Consider a function that has the following properties:

- When the input is 0, the output is 2.
- Whenever the input increases by 1, the output increases by 4.

   a. Make a table and draw the graph of a function that has the given properties.

   b. Explain how the properties in the two bulleted items are reflected in the graph of the function.

   c. Find a formula for a function that has the given properties.

6. Consider a function that has the following properties:

- When the input is 2, the output is 4.

- Whenever the input increases by 3, the output increases by 5.

a. Make a table and draw the graph of a function that has the given properties.

b. Explain how the properties in the two bulleted items are reflected in the graph of the function.

c. Find a formula for a function that has the given properties.

7. Starting from milepost 100, John drove at a steady 60 miles per hour down the straight road. Consider the distance function whose input is the time elapsed since John was at milepost 100, and whose output is the distance that John is from milepost 100 at that time.

a. Make a table and draw a graph for the distance function.

b. Find a formula for the distance function, and explain why your formula is valid.

8. Scientists determined that for each inch of rain that falls in a certain area, 50,000 gallons of water will flow into the area's storm sewer system.

a. Describe a function that arises from this situation.

b. Make a table and draw the graph of your function.

c. Find a formula for your function, and explain why your formula is valid.

9. For each dollar spent at Athens stores, Athens receives $0.01 in sales tax.

a. Describe a function that arises from this situation.

b. Make a table and draw the graph of your function.

c. Find a formula for your function, and explain why your formula is valid.

10. At one stage in a factory's candy production, cocoa is mixed with sugar. For every 20 pounds of cocoa, 13 pounds of sugar are needed.

a. Describe a function that arises from this situation.

b. Make a table and draw the graph of your function.

c. Find a formula for your function, and explain why your formula is valid.

11. Sherri's monthly bill for phone service is $20. If Sherri calls her sister in France, then she pays an additional $0.15 per minute of phone call.

a. Describe a function that arises from this situation.

b. Make a table and draw a graph of your function in part (a).

c. Find a formula for your function in part (a), and explain why your formula is valid.

12. A school bought 100 rolls of wrapping paper for a total of $150. As a fund-raiser, the school will sell the rolls of wrapping paper for $6 each. This situation gives rise to a profit function whose input is the number of rolls sold and whose output is the total profit the school makes when that many rolls are sold.

a. Make a table and draw a graph for the profit function.

b. Find a formula for the profit function, and explain why your formula is valid.

c. Where does the graph of the profit function cross the x-axis? What is the significance of this point in terms of the school's fund-raiser?

13. For the spring fling, a school bought 200 slices of pizza for a total of $100. At the spring fling, pizza will be sold for $2.50 per slice.

a. Describe a function that arises from this spring fling scenario.

b. Make a table and draw a graph of your function in part (a).

c. Find a formula for your function in part (a), and explain why your formula is valid.

d. Discuss useful information concerning the spring fling that your graph shows.

14. The Gizmo store sells gizmos. You pay $2 for the first gizmo and $0.75 for each additional gizmo. This situation gives rise to a gizmo function whose inputs are the numbers of gizmos you could buy and whose outputs are how much you would pay for that many gizmos.

a. Make a table and draw a graph of the gizmo function.

b. Find a formula for the gizmo function, and explain why your formula is valid.

15. Write a story problem which gives rise to a function, f, that has the formula $f(x) = 0.35x + 2.50$. Explain why your function has that formula.

16. Write a story problem that gives rise to a function, g, having the formula $g(x) = \frac{2}{3}x$.

Chapter Summary and Study Items

Section 9.1 Mathematical Expressions and Formulas

An expression is a meaningful string of numbers, operations, and possibly variables. Expressions arise in a variety of scenarios and provide a way to formulate a scenario mathematically.

Key skills and understandings:

- Formulate expressions arising from scenarios.
- Evaluate expressions and formulas efficiently. Use cancellation correctly when evaluating expressions involving fractions.

Section 9.2 Equations

An equation is a statement that an expression or number is equal to another expression or number. Equations involve an equals ($=$) sign. Different kinds of equations arise in different contexts. Some equations show the result of a calculation. Some equations are true for all values of the variables. Some equations describe how quantities are related. And some equations must be solved to solve a problem.

Key skills and understandings:

- Formulate equations arising from scenarios.

Section 9.3 Solving Equations

To solve an equation means to determine values for the variables that make the equation true. Sometimes we can solve an equation just by applying our sense of numbers and operations. However, a general strategy for solving equations is to change the equation into a new equation that has the same solution and is easy to solve.

Key skills and understandings:

- Solve appropriate equations by using number sense.
- Solve equations algebraically, and describe how the method can be viewed in terms of a pan balance if appropriate

Section 9.4 Solving Algebra Story Problems with Strip Diagrams and with Algebra

Many algebra story problems can be solved with the aid of strip diagrams. Often, when we use strip diagrams to solve a story problem, our method parallels the algebraic method of setting up and solving an equation with variables.

Key skills and understandings:

- Set up and solve story problems with the aid of strip diagrams as well as algebraically, and relate the two.

Section 9.5 Sequences

A sequence is a list of items (usually numbers or shapes) occurring in a specified order. Arithmetic sequences are produced by repeatedly adding (or subtracting) the same fixed number. Geometric sequences are produced by repeatedly multiplying (or dividing) by the same fixed number. Arithmetic sequences and geometric sequences both have formulas that describe their Nth entry. Repeating sequences are frequently studied in elementary school.

Key skills and understandings:

- Given an arithmetic sequence, find a formula for the Nth entry, and explain why this formula is valid in terms of the way the sequence grows.

- Given a geometric sequence, find a formula for the *N*th entry, and explain why this formula is valid in terms of the way the sequence grows.
- Solve problems about growing or repeating sequences of shapes.

Section 9.6 Series

A series is a sum of the numbers in a sequence. An arithmetic (respectively, geometric) series is a sum of the numbers in an arithmetic (respectively, geometric) sequence.

Key skills and understandings:

- Find formulas for sums of finite or infinite geometric series, and explain why the formulas are valid.

Section 9.7 Functions

A function is a rule that assigns one output to each allowable input. We can describe or represent functions with words, with tables (which usually don't convey complete information), with graphs (which usually don't convey complete information), and with formulas (although not every function has a formula).

Key skills and understandings:

- Given a description of a function in words, draw a graph that could be the graph of the function. Explain how the graph fits with the description of the function.
- Given the graph of a function, describe aspects or features of the function. Explain how the graph informs you about those aspects or features.

Section 9.8 Linear Functions

A linear function is a function whose graph lies on a line in a coordinate plane. Proportional relationships and arithmetic sequences both give rise to linear functions. For any linear function, the ratio of change in input to change in corresponding output is always the same. There are numbers m and b such that the linear function is described by the formula $f(x) = mx + b$; the number b is the output for the input 0 and the number m is the increase in output when the input increases by 1.

Key skills and understandings:

- Given two quantities that vary proportionally, describe the associated function, make a table for the function, verify that the graph of this function lies on a line, and find a formula for the function. Discuss relationships among the given ratio, the table and graph for the function, and the formula for the function.
- Given an arithmetic sequence, describe the associated function, make a table for the function, verify that the graph of this function lies on a line, find a formula for the function, and explain why the formula is valid. Discuss relationships among the arithmetic sequence, the table and graph for the function, and the formula for the function.
- Given information about a function, determine if the function is linear or not.
- Given a scenario that gives rise to a linear function, describe the function in words, make a table and graph for the function, find a formula for the function, and explain why the formula is valid.

Geometry

Geometry is the study of space and shapes in space. The word **geometry** comes from the Greek and means measurement of the earth (geo = earth, metry = measurement). Mathematicians of ancient Greece developed fundamental concepts of geometry to answer basic questions about the earth and its relationship to the sun, the moon, and the planets: How big is the earth? How far away is the moon? How far away is the sun? The geometers of ancient Greece had to be able to visualize the earth and the heavenly bodies in space. They had to extract the relevant relationships from their mental pictures, and they had to analyze these pictures mathematically. The methods the ancient Greeks developed are still useful today for solving modern problems in construction, road building, medicine, and other disciplines.

We begin this chapter by visualizing three-dimensional situations and making two-dimensional drawings to help analyze phenomena. We then study two-dimensional shapes. We will focus on analyzing features and properties of shapes, on constructing shapes, and on relating shapes.

Traditionally, very little geometry has been taught in elementary school in the United States. Therefore, you may feel that you won't need to know much geometry in order to teach mathematics in elementary school. However, the recommendations of the National Council of Teachers of Mathematics include a strong component in geometry (see [53]): Instructional programs should enable all students from prekindergarten through grade 12 to

- analyze characteristics and properties of two- and three-dimensional geometric shapes and develop mathematical arguments about geometric relationships;

- specify locations and describe spatial relationships, using coordinate geometry and other representational systems;

- apply transformations and use symmetry to analyze mathematical situations;

- use visualization, spatial reasoning, and geometric modeling to solve problems.

For the NCTM's specific geometry recommendations for grades PreK–2, grades 3–5, grades 6–8, and grades 9–12, go to www.pearsonhighered.com/beckmann.

10.1 Visualization

One important reason to study geometry is that it promotes the ability to visualize and mentally manipulate objects in space. This is a necessary skill for many professions. For example, a surgeon or dentist must be able to visualize the steps in and outcomes of an operation, a carpenter must be able to see different designs in his or her mind's eye, an architect must be able to visualize many possibilities for a building that satisfies certain design criteria, and a clothes designer must be able to visualize how pieces of fabric will fit together to make a garment.

According to the report "What Work Requires of Schools" of the U.S. Department of Labor [76], being able to "see things in the mind's eye" is a foundational skill for solid job performance. You will want to foster this skill in your students, so you must first foster it in yourself.

Another important skill for understanding and analyzing phenomena in the world around us is to make and use two-dimensional drawings of three-dimensional situations. The most helpful drawings are ones that are simple, yet capture just enough useful information to allow us to analyze the situation. In order to use a drawing effectively, we must be able to go back and forth between the real three-dimensional situation and the simplified two-dimensional drawing. To go back and forth that way requires visualization.

This brief section consists of a collection of Class Activities and problems that can help you begin to improve your visualization skills. For some people, visualization comes naturally; for others, it is more of a struggle. No matter where you find yourself in this spectrum, you can improve your visualization skills by working at it.

The next Class Activity asks you to visualize lines in a plane and planes in space.

Class Activity   *Now Turn to Class Activities Manual*

10A Visualizing Lines and Planes, p. 231

The terms *point*, *line*, and *plane* are usually considered primitive, undefined terms. Even so, we can describe how to visualize points, lines, and planes (see Figure 10.1):

- To visualize a **point**, think of a tiny dot, such as the period at the end of a sentence. A point is an idealized version of a dot, having no size or shape.

- To visualize a **line**, think of an infinitely long, stretched string that has no beginning or end. A line is an idealized version of such a string, having no thickness.

- To visualize a **plane**, think of an infinite flat piece of paper that has no beginning or end. A plane is an idealized version of such a piece of paper, having no thickness.

FIGURE 10.1

Points, lines, planes. Arrow heads indicate that the line or ray extends indefinitely in that direction.

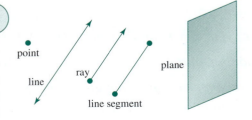

line segment Related to lines are line segments and rays, as pictured in Figure 10.1. A **line segment** is the part of a line lying between two points on a line. These two points are called the **endpoints** of the line segment. Think of a line segment as having both a beginning and an end, even though both points are called **ray** endpoints. A **ray** is the part of a line lying on one side of a point on the line. Think of a ray as having a beginning, but no end.

The next two Class Activities ask you to visualize the earth, sun, and moon in space and to use and make simple drawings to help you analyze phenomena that you experience every day.

Class Activity *Now Turn to Class Activities Manual*

10B The Rotation of the Earth and Time Zones, p. 232

10C Explaining the Phases of the Moon, p. 234

Practice Exercises for Section 10.1

1. Figure 10.2 shows a diagram of the earth as seen from outer space looking down on the North Pole, which is marked N. How does the sun appear to people at locations A, B, C, and D? Use your answers, together with the fact that the sun rises in the east and sets in the west, to explain what time of day it is at A, B, C, and D.

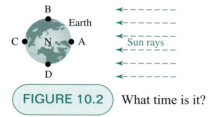

FIGURE 10.2   What time is it?

2. Imagine floating in outer space above the North Pole. Looking down on the earth, which way is the earth rotating, clockwise or counterclockwise? Use your results in Practice Exercise 4 to explain your answer.

   If you were floating above the South Pole instead of the North Pole, would the rotation look different?

3. Figure 10.3 shows four different positions of the earth and moon as seen from outer space, floating above the North Pole (not to scale). The moon rotates around the earth in the direction indicated. In which of diagrams A, B, C, and D is the moon waxing (getting bigger); in which is it waning (getting smaller)? Explain your answers.

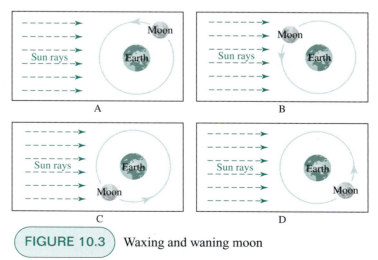

FIGURE 10.3   Waxing and waning moon

Answers to Practice Exercises for Section 10.1

1. Picture yourself at point A in Figure 10.2. At this location, the sun's rays are directly overhead, which means it must be about noon. Location C is not receiving any sunlight and is directly opposite the location marked noon, so it must be about midnight at point C. Now picture yourself at points B and D. At both locations, the sun is on the horizon because the sun's rays just graze the surface of the earth there. Therefore, it must be sunrise or sunset at B and D. How can you tell which is which? Picture yourself standing at point D and facing north. The sun is on your right, which is to the east. Since the sun rises in the east, it is sunrise at point D. Now picture yourself standing at point B and facing north. Now the sun is on your left, which is to the west. Since the sun sets in the west, it is sunset at point B.

2. According to the results of the previous practice exercise, noon, midnight, sunrise, and sunset are at the locations indicated in Figure 10.4. Because the day progresses from midnight to sunrise to noon to sunset, the earth must be rotating counterclock-

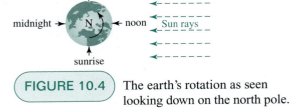

FIGURE 10.4   The earth's rotation as seen looking down on the north pole.

wise when viewed from above the North Pole. However, when viewed from above the South Pole, the earth rotates in the *clockwise* direction. You can see why the sense of rotation is reversed by doing the following: Hold a small ball between your thumb and index finger. Let your index finger represent the North Pole and let your thumb represent the South Pole. Now rotate the ball counterclockwise when viewed looking down on your index finger (North Pole). Keep rotating the ball in the same way, but now look down on your thumb (South Pole). Notice that the same rotation now appears clockwise.

3. See Figure 10.5. The shaded portions in the moon show what part of the moon is visible from earth. The portion of the moon that is visible from earth is the portion that is facing earth *and also* illuminated by the sun. The dashed lines on the moon help you see the part of the moon that is illuminated by the sun and the part of the moon that is facing earth. In diagram A, the moon is just past being full. (The moon is full when it is opposite the sun, with the earth in between.) As the moon moves from the configuration in A to the configuration shown in B and beyond, less of the moon becomes visible. Therefore, the moon is waning in A and B. After the configuration shown in B, the moon becomes new (when it is between the sun and earth) and then begins to become visible again, as in C. In D, even more of the moon is visible than in C. So the moon is waxing in C and D.

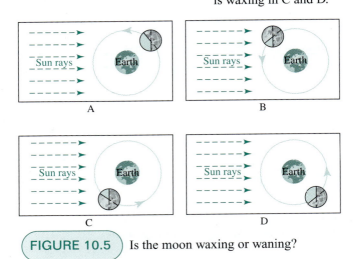

FIGURE 10.5   Is the moon waxing or waning?

Problems for Section 10.1

1. The photo of the earth in Figure 10.6 was taken by NASA. Identify some of the landforms shown. Use the landforms and the position of the shadow to explain how you can determine whether it is sunrise or sunset at the edge of the shadow.

FIGURE 10.6   A photograph taken by NASA

2. ▓ Figure 10.7 shows one possible arrangement of the earth, moon, and sun as seen from outer space (not to scale). In this arrangement, what does the moon look like to people on earth who can see it? Is the moon waxing or waning? Explain your answers.

FIGURE 10.7   In what phase is the moon?

3. Figure 10.8 shows one possible arrangement of the earth, moon, and sun as seen from outer space (not to scale). In this arrangement, what does the moon look like to people on earth who can see it? Is the moon waxing or waning? Explain your answers.

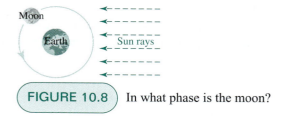

FIGURE 10.8   In what phase is the moon?

4. ▓ Figure 10.9 shows two possible arrangements of the earth, moon, and sun as seen from outer space (not to scale). In these arrangements, what does the moon look like to people on the earth who can see it? What is the relevance of the sentence, "Sun rays are not in the plane of the page; they come from slightly above (or below) this plane," which appears at the top of Figure 10.9? Explain your answers.

Sun rays are not in the plane of the page; they come from slightly above (or below) this plane.

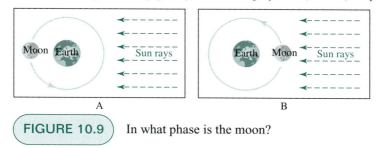

FIGURE 10.9   In what phase is the moon?

5. Figure 10.10 shows one possible arrangement of the earth, moon, and sun as seen from outer space, shown to scale. In this arrangement, what does the moon look like to people on the earth who can see it? Assuming that the moon orbits counterclockwise around the earth in the plane of the page, is the moon waxing or waning? Explain your answers.

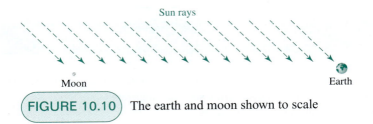

Sun rays

Moon   Earth

FIGURE 10.10 The earth and moon shown to scale

6. Figure 10.11 shows one possible configuration of the earth, moon, and sun as seen from outer space, shown to scale. In this configuration, what does the moon look like to people on the earth who can see it? Assuming that the moon orbits counterclockwise around the earth in the plane of the page, is the moon waxing or waning? Explain your answers.

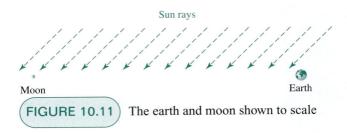

Sun rays

Moon   Earth

FIGURE 10.11 The earth and moon shown to scale

7. **a.** Figure 10.12 shows the earth as seen from outer space, looking down on the North Pole (labeled N). What does the sun look like to a person standing at point P? Therefore, what time of day is it at point P? Use the fact that the sun rises in the east and sets in the west in your answer.

b. If a person at point P in Figure 10.12 can see the moon, and if the moon is neither new nor full, then is the moon waxing or is it waning? Explain how you can tell from the diagram.

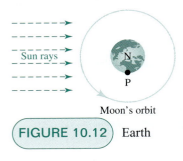

Sun rays

Moon's orbit

FIGURE 10.12 Earth

8. **a.** Figure 10.13 shows the earth as seen from outer space, looking down on the North Pole (labeled N). What does the sun look like to a person standing at point P? Therefore, what time of day is it at point P? Use the fact that the sun rises in the east and sets in the west in your answer.

b. If a person at point P in Figure. 10.13 can see the moon, and if the moon is neither new nor full, then is the moon waxing or is it waning? Explain how you can tell from the diagram.

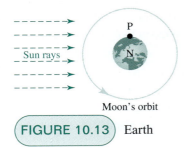

Sun rays

Moon's orbit

FIGURE 10.13 Earth

9. We can often see the moon during the day, even in the middle of the day. Mike said that he saw a full moon at 2 P.M. Draw a diagram (like one in Figure 10.3) to help you explain why Mike couldn't be right.

10. If you go outside shortly after sunrise and you see the moon up high in the sky, is the moon waxing or waning? Draw a diagram (like one in Figure 10.3) to help you explain your answer.

10.2 Angles

**Focal Points
Grades 3, 8**

Angles are used in two ways: to represent an amount of rotation (turning) about a fixed point, and to describe how two rays (or lines, line segments, or even planes) meet at a point. In this section we will consider basic facts about angles and applications of angles—some of which you might find surprising. We will study some key facts about angles, which are produced by configurations of lines in a plane, including the fact that the sum of the angles in a triangle is 180°.

Two Ways to Define Angles

The two ways of thinking about angles, as amounts of rotation and as rays meeting, are closely related. It is equally valid to define angles from either point of view. Since even very young children have experience spinning around, the "rotation" point of view is perhaps more primitive. So, our first definition is that an **angle** is an amount of rotation about a fixed point. An angle at a point P is said to be **congruent** to an angle at a point Q if both represent the same amount of rotation—even though this rotation takes place around different points. (See Figure 10.14.) Informally, we often simply say that congruent angles are "equal."

angle

congruent

Now we turn to the other point of view about angles. Suppose there are 2 rays in a plane and these rays have a common endpoint P, as illustrated in Figure 10.15. In this case, the 2 rays and the region between them is called the **angle** at P formed by the 2 rays. This definition of angle relates to the "rotation" definition by associating the 2 rays meeting at a point P to the smallest amount of counterclockwise rotation about P needed to rotate one of the rays to the position of the other ray. (See Figure 10.16.)

angle

When there are 2 line segments in a plane and these line segments have a common endpoint P, then we can define the angle between these line segments in the same way as for angles between rays. (Or we can simply extend the line segments to rays having P as an endpoint and use the definition for angles between rays.)

Measuring Angles and Special Angles

Angles are commonly measured in **degrees**. Degrees are usually indicated with a small circle: °. For example, 30 degrees is usually written 30°. If you stand at a fixed point and rotate counterclockwise in a full circle to return to your starting position, then you will have rotated 360°. If you stand at a point and rotate one-half of a full turn, then you will have rotated 180°; a quarter turn is 90°, and so on.

degree

FIGURE 10.14

The same amount of rotation around different points

FIGURE 10.15

An angle formed by two rays meeting at point P

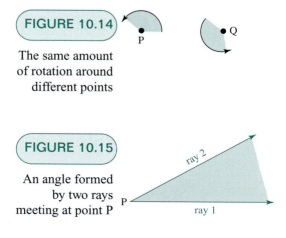

P   ray 1

ray 2

FIGURE 10.16

Amounts of
rotation
associated with
two rays

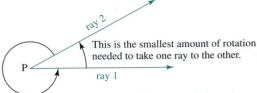

This is the smallest amount of rotation
needed to take one ray to the other.

This is another amount of rotation that takes one
ray to the other ray, but it is not the smallest possible
amount of rotation.

FIGURE 10.17

Two rays forming
a straight line

If 2 rays meet at a common endpoint Q to form a straight line, as illustrated in Figure 10.17, then the angle at Q is 180°. So, if you were standing at the point Q looking straight down one side of the line and rotated counterclockwise to look down the other side of the line, you would have rotated 180°.

A technical point: What about clockwise rotations? Clockwise rotations give rise to negative angles. So, for example, if you stand at a point and you rotate a quarter turn clockwise, then you have rotated −90°. You need not be concerned with negative angles because we will not be working with them.

Notice that a small angles formed by rays has a "pointier" look than a large angle formed by rays, as seen in Figure 10.18. Angles less than 90° are called **acute angles**, an angle of 90° is called a **right angle**, and angles greater than 90° are called **obtuse angles**. Right angles are often indicated by a small square, as shown in Figure 10.19.

right angle

protractor

Angles formed by rays can be measured with a simple device called a **protractor**. Figure 10.20 shows a protractor measuring an angle. Protractors usually have a small hole that should be placed directly over the point where the 2 rays meet. This hole lies on a horizontal line through the protractor. This horizontal line should be lined up with one of the rays that form the angle to be measured. The protractor in Figure 10.20 shows that the angle it is measuring is 65°.

FIGURE 10.18

Some angles

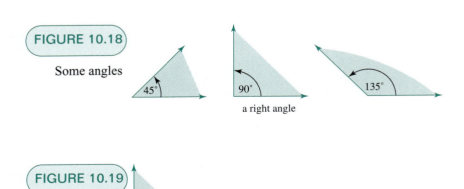

a right angle

FIGURE 10.19

Indicating a
right angle

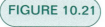

FIGURE 10.20

A protractor measuring an angle

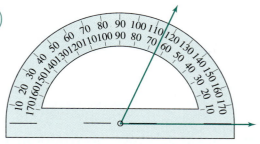

FIGURE 10.21

An "angle explorer"

Some teachers help their students learn about angles by making and using "angle explorers." An angle explorer is made by attaching 2 strips of cardboard with a brass fastener, as shown in Figure 10.21. By rotating the cardboard strips around, students get a feel for the relative sizes of angles.

Class Activity *Now Turn to Class Activities Manual*

10D Angle Explorers, p. 236

When 2 lines in a plane meet, they form 4 angles. Figure 10.22 shows some examples. When all 4 of the angles are 90°, we say that the 2 lines are **perpendicular**.

perpendicular

Two lines in a plane that *never* meet (even somewhere far off the page they are drawn on) are called **parallel**. Figure 10.23 shows examples of lines that are parallel and lines that are not parallel, even though you can't see where they meet.

parallel

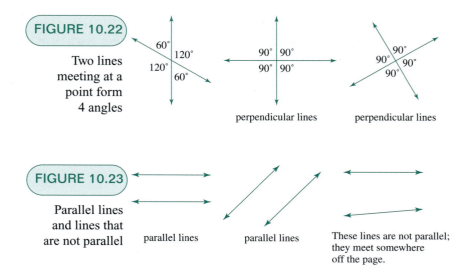

FIGURE 10.22

Two lines meeting at a point form 4 angles

FIGURE 10.23

Parallel lines and lines that are not parallel

Three configurations of lines in a plane

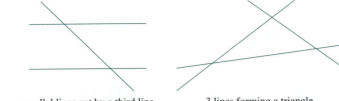

2 lines meeting in a plane   parallel lines cut by a third line   3 lines forming a triangle

Three Facts about Angles Produced by Configurations of Lines

Whenever there is a collection of lines in a plane, the lines produce angles where they meet. There are three key facts about angles that are produced when lines meet. The first fact is a theorem about the angles that are produced when 2 lines meet, as on the left in Figure 10.24. The second fact is the Parallel Postulate, which concerns parallel lines that are cut by another line, as in the middle of Figure 10.24. And the third fact is the famous theorem about the sum of the angles in a triangle. Why does this theorem fit here? We can think of triangles as arising from 3 lines in a plane that don't all meet at a single point, 1 line for each of the 3 sides of the triangle, as on the right in Figure 10.24.

postulate   Before we study these three key facts, what is a theorem and what is a postulate? A **postulate** (or axiom) is a mathematical statement that is considered foundational and is simply assumed to be true. In mathematics, we want to explain as much as possible why statements are true. But it is necessary to have some foundational assumptions, and these are called axioms or postulates. A **theorem** is a **theorem**   mathematical statement that has been proven to be true by the use of logical reasoning, based on previously proven theorems and on postulates.

Now study the first key fact (2 lines meeting in a plane) in the next Class Activity.

Class Activity   *Now Turn to Class Activities Manual*

10E Angles Formed by Two Lines, p. 237

Parallel   The second key fact is the Parallel Postulate. One version of the **Parallel Postulate** is this:
Postulate

If parallel lines are cut by another line, then the corresponding angles produced are equal. In other words, in Figure 10.25,

$$a = a', \quad b = b', \quad c = c', \quad d = d'$$

The Parallel Postulate certainly looks like it ought to be true. Why can't we prove it? In fact, mathematicians tried for many years to prove the Parallel Postulate from other, simpler postulates, but they were not able to. As it turns out, it is not possible to prove the Parallel Postulate from standard simpler

The Parallel Postulate

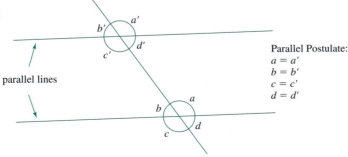

parallel lines

Parallel Postulate:
$a = a'$
$b = b'$
$c = c'$
$d = d'$

FIGURE 10.26

What is the third angle?

postulates, and there are versions of geometry in which the Parallel Postulate is actually not true. These were shocking discoveries for mathematicians. In fact, according to current theories in physics, the Parallel Postulate is not completely true in our universe. But in mathematics we often make simplifying assumptions and use models for the world around us that approximate the actual situation. So from here on, we will take the Parallel Postulate to be true.

Class Activity *Now Turn to Class Activities Manual*

10F Angles Formed When a Line Crosses Two Parallel Lines, p. 238

The third key fact is the theorem about the sum of the angles in a triangle. It states that the angles in a triangle add to 180°. Stop for a moment to think about this; it is really quite remarkable. No matter what the triangle, no matter how oddly shaped it may be—as long as it's a triangle—the angles *always* add to 180°. This theorem tells us about *all* triangles. Also, we can explain why this theorem is true *without checking every single triangle individually*. Many people are attracted to mathematics because the powerful reasoning of mathematics enables us to deduce general truths.

Class Activities 10G, 10H and 10J present several ways to think about why the angles in a triangle always add to 180°. Here is how we commonly use this fact: If we know 2 of the angles in a triangle, we can figure out what the third angle must be. For example, in Figure 10.26 the triangle was constructed to have an angle of 40° and an angle of 70°. These 2 angles add to 110°; thus, the third angle must be 70° so that all 3 angles will add to 180°.

The next Class Activity will help you see—informally—why the theorem about the sum of the angles in a triangle ought to be true.

Class Activity *Now Turn to Class Activities Manual*

10G Seeing That the Angles in a Triangle Add to 180°, p. 238

A proof of the theorem about the sum of the angles in a triangle requires the Parallel Postulate, as in the next Class Activity.

Class Activity *Now Turn to Class Activities Manual*

10H Using the Parallel Postulate to Prove That the Angles in a Triangle Add to 180°, p. 239

Yet another way to explain why the sum of the angles in a triangle add to 180° is related to describing routes in terms of distances and angles, which we consider next.

Angles and Navigation

As the next Class Activities show, angles are useful for describing routes. When we navigate a route it is the "angle as amount of turning" view of angles that applies. However, a map of the route shows angles in terms of lines meeting. Thus, in navigation we use both views of angles.

Class Activity   *Now Turn to Class Activities Manual*

10I Describing Routes, Using Distances and Angles, p. 240

We can explain why the sum of the angles in a triangle is 180° by considering what happens when we walk all the way around a triangle, as in the next Class Activity. When we explain the theorem about the sum of the angles in a triangle this way, it is not so easy to see how the Parallel Postulate is involved. (But it turns out that it is.)

Class Activity   *Now Turn to Class Activities Manual*

10J Explaining Why the Angles in a Triangle Add to 180° by Walking and Turning, p. 242

10K Angle Problems, p. 244

10L Students' Ideas and Questions about Angles, p. 245

Practice Exercises for Section 10.2

1. My son once told me that some skateboarders can do "ten-eighties." I said he must mean 180s, not 1080s. My son was right, some skateboarders can do 1080s! What is a 1080, and why is it called that? By contrast, what would a 180 be?

2. Use a protractor to measure the angles formed by the shape in Figure 10.27.

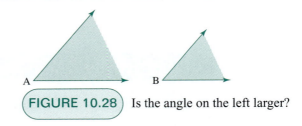

FIGURE 10.28   Is the angle on the left larger?

4. Suppose that 2 lines in a plane meet at a point, as in Figure 10.29. Use the fact that the angle formed by a straight line is 180° to explain why $a = c$ and $b = d$.

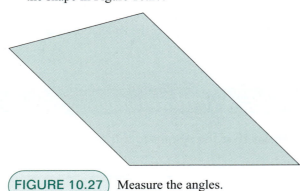

FIGURE 10.27   Measure the angles.

3. Explain why the angle at A in Figure 10.28 is not larger than the angle at B.

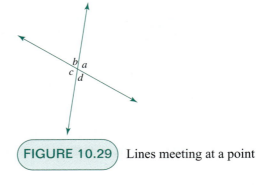

FIGURE 10.29   Lines meeting at a point

5. Given that the indicated lines in Figure 10.30 are parallel, determine the unknown angles without actually measuring them. Explain your reasoning briefly.

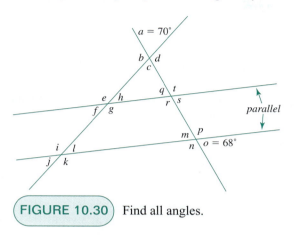

FIGURE 10.30   Find all angles.

6. Dorothy walks from point A to point F along the route indicated on the map in Figure 10.31.

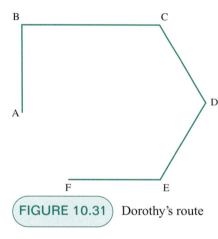

FIGURE 10.31   Dorothy's route

a. Show Dorothy's angles of turning along her route. Use a protractor to measure these angles.

b. Use your answers to part (a) to determine Dorothy's total amount of turning along her route.

c. Now determine Dorothy's total amount of turning along her route in a different way than in part (b).

7. For each of the triangles in Figure 10.32, determine the unknown angle without measuring it.

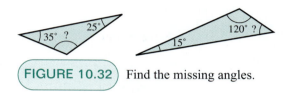

FIGURE 10.32   Find the missing angles.

8. Use the "walking and turning" method of Class Activity 10J to explain why the angles in a triangle add to 180°.

9. Figure 10.33 shows a square inscribed in a triangle. Since the square is *inside* the triangle, how can it be that the angles in the square add up to more degrees (360°) than the angles in the triangle (which only add to 180°)?

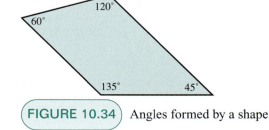

FIGURE 10.33   A square in a triangle

Answers to Practice Exercises for Section 10.2

1. A "ten-eighty" is 3 full rotations. This make sense because a full rotation is 360°, and

$$3 \times 360° = 1080°$$

A "180" would be half of a full rotation, which is not very impressive by comparison (although *I* certainly couldn't do it on a skateboard).

2. See Figure 10.34, which is a scaled version of the original shape.

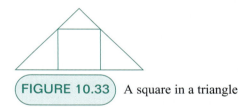

FIGURE 10.34   Angles formed by a shape

3. Even though the *line segments* forming the angle at A are longer, the angle at A is not larger than the angle at B. This is because the lower line segments of both angles would need to be rotated the same amount (about points A and B respectively) to get to the location of the upper line segments of the angles.

4. Since angles a and b together make up the angle formed by a straight line,

$$a + b = 180°$$

For the same reason,

$$b + c = 180°$$

So,

$$a = 180° - b$$

and

$$c = 180° - b$$

Since a and c are both equal to $180° - b$, they are equal to each other (i.e., $a = c$). The same argument (with the letters changed) explains why $b = d$.

5. Solving this problem is a bit like working out a puzzle! See Figure 10.35. Since $a = 70°$, the angle c must also be $70°$, since it is opposite a. Therefore, $b = 110°$, since $b + c = 180°$. Since d is opposite b, it follows that $d = 110°$, too. The very same reasoning allows us to deduce that $m = 68°$ and that $n = 112°$ and $p = 112°$. Because of the Parallel Postulate, $s = 68°$, $t = 112°$, $q = 68°$, and $r = 112°$. Since the angles in a triangle add to

180°, $h + 68° + 70° = 180°$, so $h = 42°$. As before, we deduce that $f = 42°$ and $e = 138°$, since $42° + 138° = 180°$. So $g = 138°$ as well. The Parallel Postulate now allows us to deduce that $i = 138°$, $j = 42°$, $k = 138°$, and $l = 42°$.

6. **a.** Figure 10.36 shows Dorothy's angles of turning. At each point where the path turns, the straight arrow shows the direction that Dorothy faces before she turns. The round arrow indicates Dorothy's angle of turning.

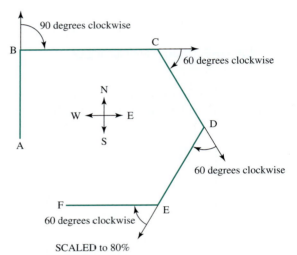

FIGURE 10.36 Dorothy's angles of turning along her route

b. Based on the results of part (a), Dorothy turns a total of

$$90° + 60° + 60° + 60° = 270°$$

along her route.

c. Another way to determine Dorothy's total amount of turning is by considering the directions she faces as she walks along her route. If we think of Dorothy as starting out facing north, then she ends up facing west, and in between she faces all the directions clockwise from north to west. That means Dorothy makes $\frac{3}{4}$ of a full 360° turn during her walk. Since $\frac{3}{4}$ of 360° is 270°, Dorothy turns a total of 270°.

7. Left triangle: 120°. Right triangle: 45°. In each case, use the fact that the angles in a triangle must add to 180° to determine the missing angle.

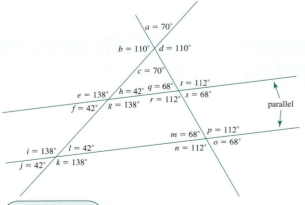

FIGURE 10.35 Angles formed by lines meeting

8. As seen in Class Activity 10J, if a person were to walk around a triangle, returning to her original position, she would have turned a total of 360°. This means that $d + e + f = 360°$, where d, e, and f are the exterior angles, as shown in Figure 10.37. But since

$$a + f = 180°,$$
$$b + d = 180°, \text{ and}$$
$$c + e = 180°;$$

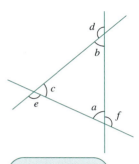

FIGURE 10.37 Exterior angles of a triangle add to 360°.

it follows that, on the one hand,

$$(a + f) + (b + d) + (c + e) = 180° + 180° + 180°$$
$$= 540°$$

while, on the other hand,

$$(a + f) + (b + d) + (c + e) = (a + b + c) + (d + e + f)$$
$$= (a + b + c) + 360°$$

Because $(a + f) + (b + d) + (c + e)$ is equal to both 540° and $(a + b + c) + 360°$, these last two expressions must be equal to each other. That is,

$$(a + b + c) + 360° = 540°$$

Therefore,

$$a + b + c = 540° - 360° = 180°$$

9. Unlike area, the angles in the square and the triangle don't depend on the sizes of either of these shapes. The fact that the square is inside the triangle has nothing to do with the sum of the angles in the square. We could even enlarge the square, so that the triangle fits inside the square, without changing the sum of the angles in the square.

Problems for Section 10.2

1. Given that the indicated lines in part A on the left of Figure 10.38 are parallel, determine the unknown angles without actually measuring them. Explain your reasoning briefly.

2. Given that the indicated lines in part B on the right of Figure 10.38 are parallel, determine the unknown angles without actually measuring them. Explain your reasoning briefly.

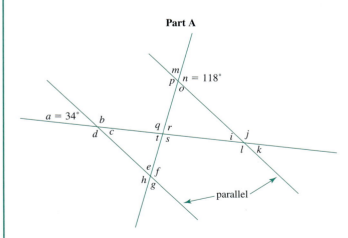

FIGURE 10.38 Determine the unknown angles

3. Amanda got in her car at point A and drove to point F along the route shown on the map in Figure 10.39.

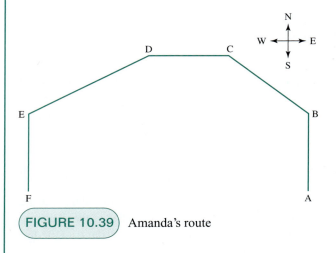

FIGURE 10.39 Amanda's route

a. Trace Amanda's route shown in Figure 10.39; show all of Amanda's angles of turning along her route. Use a protractor to measure these angles.

b. Determine Amanda's total amount of turning along her route by adding the angles you measured in part (a).

c. Now describe a way to determine Amanda's total amount of turning along her route *without* measuring the individual angles and adding them up. *Hint*: Consider the directions that Amanda faces as she travels along her route.

4. Ed the robot, is standing at point A and will walk to point D along the route shown on the map in Figure 10.40. On the map, 1 inch represents 10 of Ed's paces.

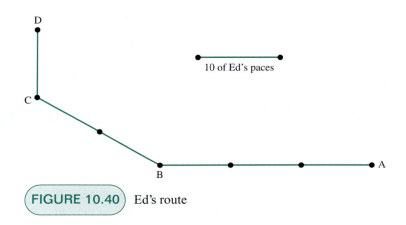

10 of Ed's paces

FIGURE 10.40 Ed's route

a. Trace the map in Figure 10.40. At the points where Ed will turn, indicate his angle of turning. Use a protractor to measure these angles, and mark them on your map.

b. Ed is a robot, so you must tell Ed exactly how to walk to point D. At points where Ed must turn, tell him how many degrees to turn, and which way.

c. Determine Ed's total amount of turning along his route by adding the angles you measured in part (a).

d. Now determine Ed's total amount of turning along his route *without* measuring the individual angles and adding them. *Hint*: Consider the directions that Ed faces as he travels along his route.

5. Draw a map that shows the following route leading to buried treasure:

Starting at the old tree, walk 10 paces heading straight for the tallest mountain in the distance. Turn clockwise, 90°. Walk 20 paces. Turn counterclockwise, 120°. Walk 40 paces. Turn clockwise, 60°. Walk 20 paces. This is the spot where the treasure is buried.

6. In your own words, explain clearly why the sum of the angles in a triangle must always add to 180°. Do not use the methods of Class Activity 10G.

7. Figure 10.41 shows a rectangle subdivided info four triangles. Jonathan says that because the angles in each of the four triangles add to 180°, the angles in the rectangle add to

$$180° + 180° + 180° + 180° = 720°$$

but Ben says that the angles in a rectangle add to

$$90° + 90° + 90° + 90° = 360°$$

Discuss the boys' ideas.

FIGURE 10.41 Rectangle subdivided into four triangles

8. Given that the lines in Figure 10.42 marked with arrows are parallel, determine the sum of the angles $a + b + c$ without measuring.

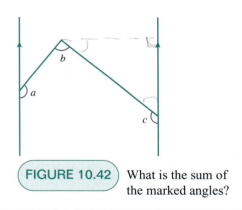

FIGURE 10.42 What is the sum of the marked angles?

9. Determine the sum $a + b + c + d + e + f$ of the angles in the 6-sided shape (hexagon) in Figure 10.43 without measuring. Explain your reasoning.

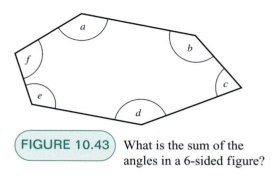

FIGURE 10.43 What is the sum of the angles in a 6-sided figure?

10. Determine the sum of the angles at the star points, $a + b + c + d + e$, in Figure 10.44 without measuring. Explain your reasoning.

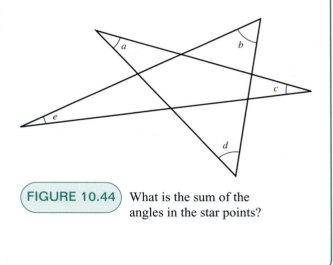

FIGURE 10.44 What is the sum of the angles in the star points?

10.3 Angles and Phenomena in the World

As we continue our study of angles, we now consider some ways that angles are relevant to phenomena we observe in the world around us. For example, as shown in the Class Activities, the seasons are not caused by variation in the distance from the earth to the sun, but rather by the angles that the sun's rays make with the ground. Another way that angles influence our daily experiences involves reflected light. The way you see your face in the mirror is more interesting and surprising than you might guess!

FIGURE 10.45

Some normal
lines to a surface
(cross-section
shown)

Angles and Sun Rays

How high or low is the sun in the sky? The angle that sun rays make with the horizontal ground is related to how high or low the sun appears in the sky. What causes the seasons? It is the tilt of the earth, namely, the angle that the earth's axis of rotation makes with the plane in which the earth orbits around the sun, that causes the seasons. The next Class Activities explore these ideas.

Class Activity   *Now Turn to Class Activities Manual*

10M Angles of Sun Rays, p. 246

10N How the Tilt of the Earth Causes Seasons, p. 248

Angles and Reflected Light

When a light ray strikes a smooth reflective surface, such as a mirror, the light ray reflects in a specific way that can be described with angles.

normal line   To describe how light reflection works, we need the concept of a **normal line** to a surface. The normal line at a point on a surface is the line that passes through that point and is perpendicular to the surface at that point. Think of "perpendicular to the surface at that point" as meaning "sticking straight out away from the surface at that point." Figure 10.45 shows a cross-section of a "wiggly" surface and its normal lines at various points.

Two fundamental physical laws govern the reflection of light from a surface. These laws also apply to the reflection of similar radiation, such as microwaves and radio waves:

1. The incoming and reflected light rays make the *same angle with the normal line* at the point where the incoming light ray hits the surface.

2. The reflected light ray lies in the same plane as the normal line and the incoming light ray. The reflected light ray is not in the same location as the incoming light ray unless the incoming light ray is in the same location as the normal line.

Figure 10.46 shows examples of how light rays reflect from surfaces. In each case, a cross-section of the surface is shown.

You can demonstrate the reflection laws by putting a mirror on a table, shining a penlight at the mirror, and seeing where the reflected light hits the wall. (Be careful to keep the light beam away from your and others' eyes.) If you raise and lower the light, while still pointing it at the mirror, the light beam traveling toward the mirror will form different angles with the mirror. By observing the reflected light beam on the wall, you can tell that the light beams going toward and from the mirror make the same angle with the normal line to the mirror. (If the mirror is on a horizontal table, then the normal lines to the mirror are vertical.)

The next Class Activities concern some interesting consequences of the laws of reflection.

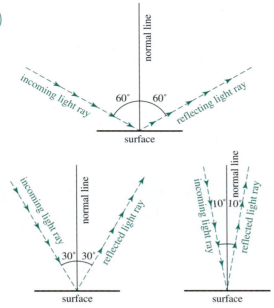

Light rays
reflecting off
surfaces

Class Activity   *Now Turn to Class Activities Manual*

10O How Big Is the Reflection of Your Face in a Mirror?, p. 249

10P Why Do Spoons Reflect Upside Down?, p. 251

10Q The Special Shape of Satellite Dishes, p. 251

Practice Exercises for Section 10.3

1. Draw pictures to show the relationship between the angle that the sun's rays make with horizontal ground and the length of the shadow of a telephone pole.

2. Figure 10.47 shows several diagrams (from the point of view of a fly looking down from the ceiling)

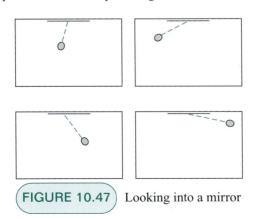

FIGURE 10.47 Looking into a mirror

of a person standing in a room, looking into a mirror on the wall. The direction of the person's gaze is indicated with a dashed line. What place in the room will the person see in the mirror?

3. Figure 10.48 shows a mirror seen from the top, and a light ray hitting the mirror. Draw a copy of this picture on a blank piece of paper. Use the following paper-folding method to show the location of the reflected light ray:

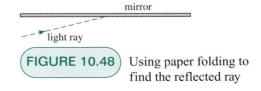

FIGURE 10.48 Using paper folding to find the reflected ray

- Fold and crease the paper so that the crease goes through the point where the light ray hits the mirror and so that the line labeled "mirror" folds onto itself. (This crease is perpendicular to the line labeled "mirror." You'll be asked to explain why shortly.)

- Keep the paper folded, and now fold and crease the paper again along the line labeled "light ray."

- Unfold the paper. The first crease is the normal line to the mirror. The second crease shows the light ray and the reflected light ray.

a. Explain why your first crease is perpendicular to the line labeled "mirror."

b. Explain why your second crease shows the reflected light ray.

Answers to Practice Exercises for Section 10.3

1. Figure 10.36 shows that when the sun's rays make a smaller angle with horizontal ground, a telephone pole makes a longer shadow than when the sun's rays make a larger angle with the ground.

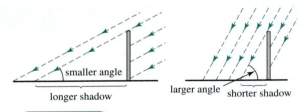

smaller angle

longer shadow

larger angle   shorter shadow

FIGURE 10.36 Sun rays hitting a telephone pole

2. Figure 10.49 shows that the person will see locations A, B, C, and D, respectively, which are on various walls. Location C is in a corner. As the pictures show, the incoming light rays and their reflections in the mirror make the same angle with the normal line to the mirror at the point of reflection.

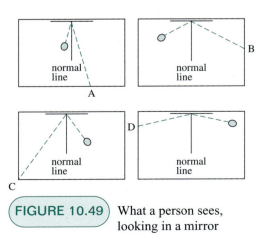

normal line   A

normal line   B

C   normal line   D   normal line

FIGURE 10.49 What a person sees, looking in a mirror

3. a. The first crease is made so that angles *a* and *b* shown in Figure 10.50 are folded on top of each other and completely aligned. Therefore, angles *a* and *b* are equal. But angle *a* and angle *b* must add up to 180° because the angle formed by a straight line is 180°. Hence, angle *a* and angle *b* must both be half of 180°, which is 90°. Therefore, this first crease is the normal line to the mirror at the point where the light ray hits the mirror.

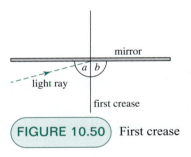

mirror

a *b*

light ray

first crease

FIGURE 10.50 First crease

b. The second crease is made so that angles *c* and *d*, shown in Figure 10.51 are folded on top of each other and completely aligned. Therefore, angle *c* is equal to angle *d*. Since the first crease is a normal line, by the reflection laws, the second crease shows the reflected light ray.

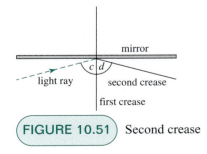

mirror

c *d*

light ray   second crease

first crease

FIGURE 10.51 Second crease

Problems for Section 10.3

1. Figure 10.52 depicts sun rays traveling toward 3 points on the earth's surface, labeled A, B, and C. Assume that these sun rays are parallel to the plane of the page.

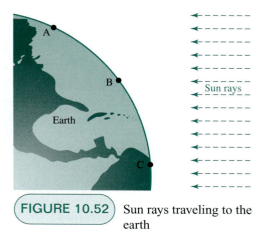

FIGURE 10.52   Sun rays traveling to the earth

a. Use a protractor to determine the angle that the sun's rays make with horizontal ground at points A, B, and C.

b. Suppose there are telephone poles of the same height at locations A, B, and C. Draw pictures like the ones in Figure 10.53 on which the vertical lines represent telephone poles, and show on your pictures the different lengths of the shadows of these telephone poles, using the information from part (a).

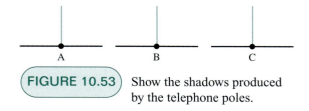

FIGURE 10.53   Show the shadows produced by the telephone poles.

2. Many people mistakenly believe that the seasons are caused by the earth's varying proximity to the sun. In fact, the distance from the earth to the sun varies only slightly during the year, and the seasons are caused by the tilt of the earth's axis. As the earth travels around the sun during the year, the tilt of the earth's axis causes the northern hemisphere to vary between being tilted toward the sun to being tilted away from the sun.

Figure 10.54 shows the earth as seen at a certain time of year from a point in outer space located in the plane in which the earth rotates about the sun. The diagram shows that the earth's axis is tilted 23.5° from the perpendicular to the plane in which the earth rotates around the sun. Throughout this problem, assume that the sun rays are parallel to the plane of the page.

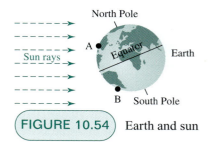

FIGURE 10.54   Earth and sun

a. Locations A and B in Figure 10.54 are shown at noon. Is the sun higher in the sky at noon at location A or at location B? Explain how you can tell.

b. During the day, locations A and B will rotate around the axis through the North and South Poles. Compare the amount of sunlight that locations A and B will receive throughout the day. Which location will receive more sunlight during the day?

c. Based on your answers to parts (a) and (b), what season is it in the northern hemisphere, and what season is it in the southern hemisphere in Figure 10.54? Explain.

d. At another time of year, the earth and sun are positioned as shown in Figure 10.55. At that time, what season is it in the northern and southern hemispheres? Why?

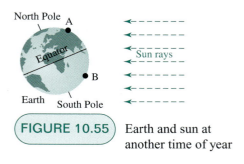

FIGURE 10.55   Earth and sun at another time of year

446   Chapter 10 • Geometry

e. At yet another time of year, the earth and sun are positioned as shown in Figure 10.56. At that time, what season is it in the northern and southern hemispheres? Why? Give an either-or-answer. (Notice that Figure 10.55 still shows the tilt of the earth's axis.)

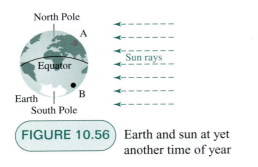

FIGURE 10.56   Earth and sun at yet another time of year

f. Refer to Figures 10.54, 10.55, and 10.56 and the results of the previous parts of this problem to answer the following: During which seasons are the sun's rays most intense at the equator? Look carefully before you answer—the answer may surprise you.

3. Refer to Figures 10.54, 10.55, 10.56, and 10.57 to help you answer the following: There are only certain locations on the earth where the sun can ever be seen *directly* overhead. Where are these locations? How are these locations related to the Tropic of Cancer and the Tropic of Capricorn? Explain.

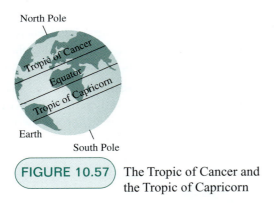

FIGURE 10.57   The Tropic of Cancer and the Tropic of Capricorn

4. Figure 10.58 shows a cross-section of Joey's toy periscope. What will Joey see when he looks in the periscope? Explain, using the laws of reflection (trace the periscope). What would be a better way to position the mirror in the telescope?

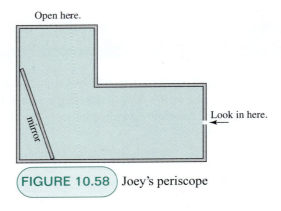

FIGURE 10.58   Joey's periscope

5. Figure 10.59 shows a sketch of a room that has a pair of perpendicular mirrors, drawn from the point of view of a fly on the ceiling. A person is standing in the room, looking into one of the mirrors at the indicated spot. On a separate piece of paper, draw a large sketch that looks (approximately) like Figure 10.59. Using your sketch, determine what the person will see in the mirror. (You may want to use paper folding, which is described in Practice Exercise 2.) Explain your answer.

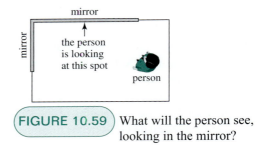

FIGURE 10.59   What will the person see, looking in the mirror?

6. Department store dressing rooms often have large mirrors that actually consist of three adjacent mirrors, put together as shown in Figure 10.60 (as seen looking down from the ceiling). Use the laws of reflection to show how you can stand in such a way as to see the reflection of your back. Draw a careful picture, using an enlarged version of Figure 10.60, that shows clearly how light reflected off your

FIGURE 10.60   Department store mirror

back can enter your eyes. Your picture should show where you are standing, the location of your back, and the direction of the gaze of your eyes. (You may wish to experiment with paper folding before you attempt a final drawing. See Practice Exercise 2.

7. A **concave** mirror is a mirror that curves in, like a bowl, so that the normal lines on the reflective side of the mirror point toward each other. Makeup mirrors are often concave. The left side of Figure 10.61 shows an eye looking into a concave makeup mirror. The right side of Figure 10.61 shows an eye looking into an ordinary flat mirror. Trace the two diagrams in Figure 10.61 and, in each case, show where a woman applying eye makeup sees her eye in the mirror. Use the laws of reflection to show approximately where the woman sees the top of her reflected eye and where she sees the bottom of her reflected eye. (Assume that the woman sees light that enters the center of her eye.) Notice that the figure shows the normal lines to the concave mirror. Based on your drawings, explain why a concave mirror makes a good mirror for applying makeup. (By the way, although a concave mirror is curved in the same direction as the bowl of a spoon, the smaller amount of curvature in a concave mirror prevents it from reflecting your image upside down, as a spoon does.)

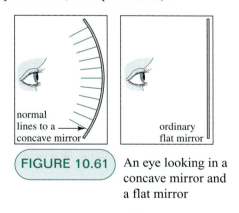

FIGURE 10.61 An eye looking in a concave mirror and a flat mirror

8. A **convex** mirror is a mirror that curves out, so that the normal lines on the reflective side of the mirror point away from each other, as shown at the top of

Figure 10.62. Convex mirrors are often used as side-view mirrors on cars and trucks. This problem will help you see why convex mirrors are useful for this purpose.

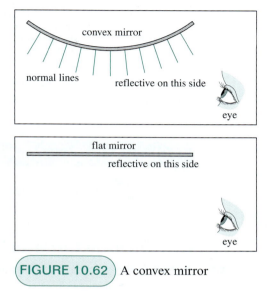

FIGURE 10.62 A convex mirror

Figure 10.62 shows a bird's-eye view of a cross-section of a convex mirror, a cross-section of a flat mirror, and eyes looking into these mirrors. Draw a copy of these mirrors and the eyes looking into them. Use the laws of reflection to help you compare how much of the surrounding environment each eye can see, looking into its mirror.

Now explain why convex mirrors are often used as side-view mirrors on cars and trucks.

9. A convex mirror is a mirror that curves out, so that the normal lines on the reflective side of the mirror point away from each other, as shown at the top of Figure 10.62. Convex mirrors are often used as side-view mirrors on cars and trucks, but these mirrors usually carry the warning sign that reads, "Objects are closer than they appear." Explain why objects reflected in a convex mirror appear to be farther away than they actually are. Use the fact that the eye interprets a smaller image as being farther away.

10.4 Circles and Spheres

What are circles and spheres? To the eye, circles and spheres are distinguished by their perfect roundness. Informally, we might say that a sphere is the surface of a ball. This describes circles and spheres from an informal or artistic point of view, but there is also a mathematical point of view. As we'll see, the

FIGURE 10.63

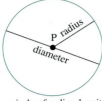

A circle and a noncircle

a circle of radius 1 unit centered at P

This is NOT a circle, but it could be a perspective drawing of a circle.

mathematical definitions of circles and spheres yield practical applications that cannot be anticipated by considering only the look of these shapes.

> **Class Activity**   *Now Turn to Class Activities Manual*
>
> **10R**   Points That Are a Fixed Distance from a Given Point, p. 253

Definitions of Circle and Sphere

circle   A **circle** is the collection of all the points in a plane that are a certain fixed distance away from a certain fixed point in the plane. This fixed point is called the **center** of the circle, and this distance is called the

radius   **radius** of the circle (plural: **radii**). So the radius is the distance from the center of the circle to any point

diameter   on the circle. The **diameter** of a circle is two times its radius. Informally, the diameter is the distance "all the way across" the circle.

For example, let's fix a point in a plane, and let's call this point P, as in Figure 10.63. All the points in the plane that are 1 unit away from the point P form a circle of radius 1 unit, centered at the point P, as shown in Figure 10.63. The diameter of this circle is 2 units.

sphere   A sphere is defined in almost the same way as a circle, but in space rather than in a plane. A **sphere** is the collection of all the points in space that are a certain fixed distance away from a certain fixed point in space. This fixed point is called the **center** of the sphere, and this distance is called the **radius** of the sphere. So the radius is the distance from the center of the sphere to any point on the sphere. The **diameter** of a sphere is two times its radius. Informally, the diameter is the distance "all the way across" the sphere.

For example, fix a point P in space. You might want to think of this point P as located in front of you, a few feet away. Then all the points in space that are 1 foot away from the point P form a sphere of radius 1 foot, centered at P. Try to visualize this. What does it look like? It looks like the surface of a very large ball, as illustrated in Figure 10.64. The diameter of this sphere is 2 feet.

compass   It's easy to draw almost perfect circles with the use of a common drawing tool called a **compass**, pictured in Figure 10.65. One side of a compass has a sharp point that you can stick into a point on a piece of paper. The other side of a compass has a pencil attached. To draw a circle centered at a point P and having a given radius, open the compass to the desired radius, stick the point of the compass in the point P, and spin the pencil side of the compass in a full revolution around P, all the while keeping the point of the compass at P.

FIGURE 10.64

A sphere and a nonsphere

a sphere of radius 1 unit, centered at P

This is NOT a sphere.

FIGURE 10.65

Drawing circles and spheres

drawing a circle with a compass   a simple drawing of a sphere

It is also possible to draw good circles with simple homemade tools. For example, you can use 2 pencils and a paper clip to draw a circle, as shown in Figure 10.66. Put a pencil inside one end of a paper clip and keep this pencil fixed at one point on a piece of paper. Put another pencil at the other end of the paper clip and use that pencil to draw a circle.

Because a sphere is an object in space, and not in a plane, it's harder to draw a picture of a sphere. Unless you are an exceptional artist, something like the simple drawing of a sphere in Figure 10.65 will do.

In the next Class Activity, you will apply the mathematical definition of circles.

Class Activity *Now Turn to Class Activities Manual*

10S Using Circles, p. 253

When Circles or Spheres Meet

What happens when 2 circles meet, or 2 spheres meet? Although this may seem like a topic of purely theoretical interest, it actually has practical applications.

As illustrated in Figure 10.67, there are only three possible arrangements of 2 distinct circles: The circles may not meet at all, the circles may meet at a single point, or the circles may meet at 2 distinct points.

Now what if you have 2 spheres—how do they meet? This is difficult, but try to visualize 2 spheres meeting. It might help to think about soap bubbles. When you blow soap bubbles you occasionally see a "double bubble" that is similar (although not identical) to 2 spheres meeting. As with circles, 2 distinct spheres might not meet at all, or they might barely touch, meeting at a single point. The only other possibility is that the 2 spheres meet along a circle, as shown in Figure 10.68. This is a circle that is common to each of the 2 spheres.

FIGURE 10.66

Drawing a circle with a paper clip

drawing a circle with a paper clip

FIGURE 10.67

Three ways that
two circles
can meet

These 2 circles
don't meet.

These 2 circles
meet at a single point.

These 2 circles
meet at 2 points.

FIGURE 10.68

Three ways that
two spheres
can meet

These 2 spheres
don't meet.

These 2 spheres meet
at a single point.

These 2 spheres meet
along a circle.

In the next Class Activity, you will see a practical application of spheres meeting.

Class Activity *Now Turn to Class Activities Manual*

10T The Global Positioning System (GPS), p. 255

10U Circle Designs, p. 257

If you did Class Activity 10S, you saw some applications of circles. We will use circles again in the next section to construct special triangles and 4-sided figures. The mathematical definition of circle will allow us to explain why our constructions work.

Practice Exercises for Section 10.4

1. Give the (mathematical) definitions of the terms *circle* and *sphere*.

2. Points P and Q are 4 centimeters apart. Point R is 2 cm from P and 3 cm from Q. Use a ruler and a compass to draw a precise picture of how P, Q, and R are located relative to each other.

3. A radio beacon indicates that a certain whale is less than 1 kilometer away from boat A and less than 1 kilometer away from boat B. Boat A and boat B are 1 kilometer apart. Assuming that the whale is swimming near the surface of the water, draw a map showing all the places where the whale could be located.

4. Suppose that during a flight an airplane pilot realizes that another airplane is 1000 feet away. Describe the shape formed by all possible locations of the other airplane at that moment.

5. An airplane is in radio contact with two control towers. The airplane is 20 miles from one control tower and 30 miles from another control tower. The control towers are 40 miles apart. Is this information enough to pinpoint the exact location of the airplane? Why or why not? What if you know the altitude of the airplane—do you have enough information to pinpoint the location of the airplane?

Answers to Practice Exercises for Section 10.4

1. See the text.

2. See Figure 10.69. Start by drawing points P and Q on a piece of paper, 4 centimeters apart. Now open a compass to 2 cm, stick its point at P, and draw a circle. Since point R is 2 cm from P, it must be located somewhere on that circle. Open a compass to 3 cm, stick its point at Q, and draw a circle. Since point R is 3 cm from Q, it must also be located somewhere on this circle. Thus, point R must be located at one of the two places where your two circles meet. Plot point R at either one of these two locations.

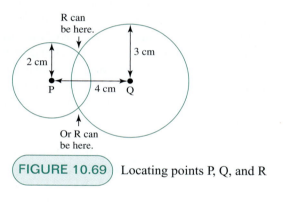

FIGURE 10.69 Locating points P, Q, and R

3. Since the whale is less than 1 km from boat A, it must be *inside* a circle of radius 1 km, centered at boat A.

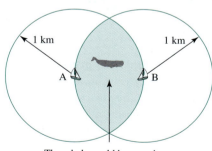

The whale could be anywhere in the shaded region.

FIGURE 10.70 Two boats and a whale

Similarly, the whale is *inside* a circle of radius 1 km, centered at boat B. The places where the insides of these two circles overlap are all the possible locations of the whale, as shown in Figure 10.70.

4. The other airplane could be at any of the points that are 1000 feet away from the airplane. These points form a sphere of radius 1000 feet.

5. The locations that are 20 miles from the first control tower form a sphere of radius 20 miles, centered at the control tower. (Some of these locations are underground, and therefore are not plausible locations for the airplane.) Similarly, the locations that are 30 miles from the second control tower form a sphere of radius 30 miles. The airplane must be located in a place that lies on *both* spheres (i.e., some place where the two spheres meet). The spheres must meet either at a single point (which is unlikely) or in a circle. So, most likely, the airplane could be anywhere on a circle. Therefore, we can't pinpoint the location of the airplane without more information.

Now suppose that the altitude of the airplane is known. Let's say it's 20,000 feet. The locations in the sky that are 20,000 feet from the ground form a very large sphere around the whole earth. This large sphere and the circle of locations where the airplane might be located either meet in a single point (unlikely) or in two points. So, even with this information, we still can't pinpoint the location of the airplane, but we can narrow it down to two locations.

Problems for Section 10.4

1. Smallville is 7 miles south of Gotham. Will is 8 miles from Gotham and 6 miles from Small-ville. Draw a map showing where Will could be. Be sure to show a scale for your map. Explain why you draw your map the way you do. Can you pinpoint Will to one location, or not?

2. A new mall is to be built to serve the towns of Sunnyvale and Gloomington, whose centers are 6 miles apart. The developers want to locate the mall not more than 3 miles from the center of Sunnyvale and also not more than 5 miles from the center of Gloomington. Draw a simple map showing Sunnyvale, Gloomington, and *all* potential locations for the new mall, based on the given information. Be sure to show the scale of your map. Explain how you determined the possible locations for the mall.

3. A radio beacon indicates that a certain dolphin is less than 1 mile from boat A and at least $1\frac{1}{2}$ miles from boat B. Boats A and B are 2 miles apart. Draw a simple map showing the locations of the boats and *all* the places where the dolphin might be located. Be sure to show the scale of your map. Explain how you determined all possible locations for the dolphin.

4. A new Giant Superstore is being planned somewhere in the vicinity of Kneebend and Anklescratch, towns that are 10 miles apart. The developers will say only that all the locations they are considering are more than 7 miles from Kneebend and more than 5 miles from Ankle-scratch. Draw a map showing Kneebend, Ankle-scratch, and *all* possible locations for the Giant Superstore. Be sure to show the scale of your map. Explain how you determined all possible locations for the Giant Superstore.

5. Part 2 of Class Activity 10U on circle designs shows a design made from circles. In part 3 of the same Class Activity, there is a beautiful circle design that was made by drawing a portion of the design of part 2. Make another interesting circle design by using a compass (or a paper clip) to draw a portion of the design of part 2 of Class Activity 10U. Leave construction marks to show how you made your drawing. You may wish to outline your final drawing in a different color to make it visible and distinct from the construction marks. (Many such designs can be found in church windows and in other art throughout the world.)

10.5 Triangles, Quadrilaterals, and Other Polygons

F·P Focal Points
Grade 3

In elementary school, children learn about the basic two-dimensional geometric shapes: circles, triangles, squares, rectangles, and other shapes. Not only do they learn to identify the basic shapes, but they also learn to describe the attributes of these shapes, and they learn to compare and classify shapes. In mathematics, shapes have succinct definitions that describe a few specific attributes that the shapes must have. But interestingly, shapes have additional attributes and characteristics, beyond those mentioned specifically in the definitions.

In this section, we will examine the definitions and additional characteristics of various basic shapes. We'll use these characteristics to organize and classify shapes and to determine how shapes are related to each other. Also, we'll see how to use Venn diagrams to organize shapes and to show relationships. If you do the Class Activities, you'll learn several ways to construct shapes. Throughout this section, we will see how the mathematical definition of a circle is essential in constructing shapes that have specific attributes.

Definitions of Triangles, Quadrilaterals, and Polygons

triangle   A **triangle** is a closed shape in a plane consisting of 3 line segments. The term **closed** means that every
closed   endpoint of one of the line segments meets exactly one endpoint of another line segment. Informally, we

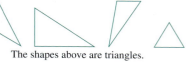

FIGURE 10.71

Triangles and shapes that are not triangles

The shapes above are triangles.

The shapes below are not triangles.

This is not made out of line segments.   This is not closed.

can say that a shape is closed if it has "no loose, dangling ends." Figure 10.71 shows examples of triangles and shapes that are not triangles.

Some kinds of triangles have special names. Figure 10.72 shows some examples of special kinds of triangles. A **right triangle** is a triangle that has a right angle (90°). In a right triangle, the side opposite the right angle is called the **hypotenuse**. A triangle that has 3 sides of the same length is called an **equilateral triangle**. A triangle that has as least 2 sides of the same length is called an **isosceles triangle**.

right triangle
hypotenuse

quadrilateral   A **quadrilateral** is a closed shape in a plane consisting of 4 line segments that do not cross each other. Figure 10.73 shows examples of quadrilaterals and shapes that are not quadrilaterals. The name *quadrilateral* makes sense because it means *four sided* (quad = four, lateral = side).

polygon   Triangles and quadrilaterals are kinds of polygons. A **polygon** is a closed, connected shape in a plane consisting of a finite number of line segments that do not cross each other. The line segments making the polygon are called its **sides** and the points where line segments meet are called the **vertices** (singular: vertex) of the polygon. Triangles are polygons with 3 sides, quadrilaterals are polygons with 4 sides, **pentagons** are polygons with 5 sides, **hexagons** are polygons with 6 sides, **octagons** are polygons with 8 sides. Figure 10.74 shows some examples. A polygon with *n* sides can be called an "*n*-gon." For example, a 13-sided polygon is a 13-gon.

side
vertex
pentagons
hexagons
octagons

The name *polygon* makes sense because it means *many angled* (poly = many, gon = angle). Similarly, for the names *pentagon, hexagon*, and so on, "penta" means 5, "hexa" means 6, and so on. It would be perfectly reasonable to call triangles *trigons* and quadrilaterals *quadrigons*, although these terms are not used conventionally.

regular   A polygon is called **regular** if all sides have the same length and all angles are equal. In Figure 10.74 the leftmost pentagon, hexagon, and octagon are all regular polygons, whereas the pentagon, hexagon, and octagon on the right are not regular polygons.

Some kinds of quadrilaterals are considered special; the most familiar are squares and rectangles. Following is a list of definitions of some special quadrilaterals (see Figure 10.75 for examples):

square—quadrilateral with 4 right angles whose sides all have the same length

rectangle—quadrilateral with 4 right angles

FIGURE 10.72

Special kinds of triangles

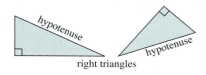

hypotenuse   hypotenuse

right triangles

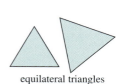

equilateral triangles   isosceles triangles

FIGURE 10.73

Quadrilaterals
and shapes that
are not
quadrilaterals

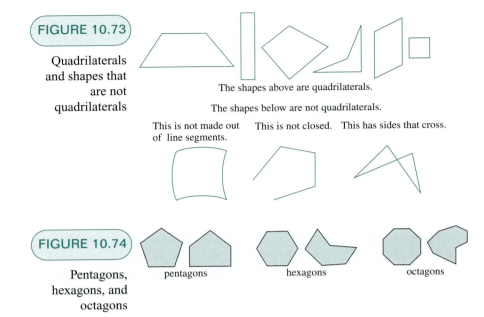

The shapes above are quadrilaterals.

The shapes below are not quadrilaterals.

This is not made out   This is not closed.   This has sides that cross.
of line segments.

FIGURE 10.74

Pentagons,
hexagons, and
octagons

pentagons   hexagons   octagons

rhombus—quadrilateral whose sides all have the same length. The name **diamond** is sometimes used instead of rhombus

parallelogram—quadrilateral for which opposite sides are parallel

trapezoid—quadrilateral for which at least one pair of opposite sides are parallel. (Some books define a trapezoid as a quadrilateral for which *exactly one* pair of opposite sides are parallel.)

Some of the Class Activities for this section explore special properties of various kinds of quadrilaterals. **diagonals** Several investigate diagonals of quadrilaterals. The **diagonals** of a quadrilateral are line segments connecting opposite corner points, as shown in Figure 10.76.

In the next Class Activity, the defining characteristic of circles is used to construct specific triangles and quadrilaterals.

Class Activity   *Now Turn to Class Activities Manual*

10V   Using a Compass to Draw Triangles and Quadrilaterals, p. 258

If you do the next Class Activity, you will find that the shapes you make have additional properties beyond the defining properties of the shapes. Some of these properties help in relating shapes. In Section 10.6, we will see that properties of rhombuses underlie basic constructions with ruler and compass.

FIGURE 10.75

Some special
quadrilaterals

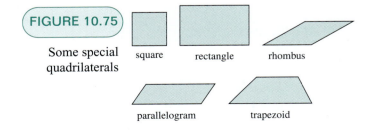

square   rectangle   rhombus

parallelogram   trapezoid

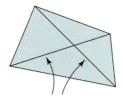

FIGURE 10.76

The diagonals of
a quadrilateral

the diagonals of a quadrilateral

Class Activity *Now Turn to Class Activities Manual*

10W Making Shapes by Folding Paper, p. 259

10X Constructing Quadrilaterals with Geometer's Sketchpad, p. 261

10Y Relating the Kinds of Quadrilaterals, p. 262

Showing Relationships with Venn Diagrams

Venn diagram

set

Venn diagrams provide a convenient way to use pictures to show relationships among collections of items. A **Venn diagram** is a picture that shows how certain sets are related. A **set** is a collection of objects.

Figure 10.77 shows a Venn diagram relating the set of mammals and the set of animals with 4 legs. The overlapping region represents mammals that have 4 legs because these animals fit in both categories: animals with 4 legs and mammals.

Venn diagrams are not just used in mathematics. Some teachers use Venn diagrams in language arts, for example, to compare the attributes of a child with the attributes of a story character, as in Figure 10.78.

A Venn diagram of two sets does not have to overlap the way the previous two examples did. Figure 10.79 shows two other kinds of Venn diagrams. The first shows a Venn diagram relating the set of boys and the set of girls. Of course, these sets do not overlap. The other Venn diagram in Figure 10.79 shows that the set of whole numbers is contained within the set of rational numbers. This is so because every whole number can also be expressed as a fraction (by "putting it over 1").

When representing three or more sets, Venn diagrams can become quite complex. In general, there can be double overlaps, triple overlaps, or more (if more than three sets are involved). For example, Figure 10.80 shows a Venn diagram relating the set of warm-blooded animals, the set of animals that lay eggs, and the set of carnivorous animals. There are three double overlaps and one triple overlap.

We can use Venn diagrams to show the relationships among the various kinds of special quadrilaterals.

Class Activity *Now Turn to Class Activities Manual*

10Z Venn Diagrams Relating Quadrilaterals, p. 263

FIGURE 10.77

A Venn diagram
showing the
relationship
between two sets

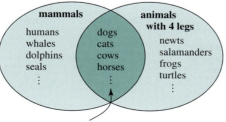

mammals that have 4 legs

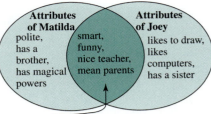

FIGURE 10.78

A language arts Venn diagram comparing Matilda's and Joey's attributes

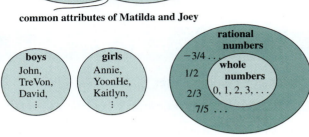

FIGURE 10.79

Other Venn diagrams showing the relationships between two sets

The following Class Activities will help you discover that some kinds of quadrilaterals have special properties that other quadrilaterals don't always have.

Class Activity   *Now Turn to Class Activities Manual*

10AA Investigating Diagonals of Quadrilaterals with Geometer's Sketchpad, p. 264

10BB Investigating Diagonals of Quadrilaterals (Alternate), p. 265

Definitions of Shapes versus Additional Properties the Shapes Have

If you did the Class Activities in this section, you discovered that shapes often have special properties that are not readily apparent from their definitions. For example, when you construct a rhombus and look at it carefully, you will notice that opposite sides of the rhombus appear to be parallel. (In fact, they are.) Therefore, every rhombus should also be a parallelogram. But now look back at the definition of rhombus: A rhombus is a quadrilateral whose 4 sides all have the same length. There is nothing in this *definition* that says that opposite sides of a rhombus should be parallel. We would somehow need to *deduce* that every rhombus is in fact also a parallelogram.

Here is another example: If you did activities investigating diagonals of quadrilaterals, then you may have found that the following seem to be true for every rhombus (in fact, they are true):

- The diagonals meet at a point that is halfway across each diagonal.

- The diagonals are perpendicular.

FIGURE 10.80

A Venn diagram showing the relationships among three sets

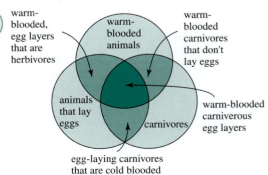

- The diagonals have different lengths unless the rhombus is a square.
- A diagonal cuts an angle of a rhombus in half.

You may also have found that the following seem to be true for every rectangle (in fact, they are true):

- The diagonals meet at a point that is halfway across each diagonal.
- The diagonals are only perpendicular when the rectangle is a square.
- The diagonals always have the same length.

Once again, these are examples of properties of shapes that we can't tell are true just from reading the definitions of these shapes—in fact, the definitions of rhombus and rectangle don't say anything at all about diagonals.

This observation leads to the heart of what theoretical mathematics is all about: The theoretical study of mathematics is about *starting with some assumptions and some definitions of objects and concepts, discovering additional properties that these objects or concepts must have, and then reasoning logically to deduce that the objects or concepts do indeed have those properties.* You probably already carried out such a logical deduction, for example, if you proved that the sum of the angles in a triangle is 180°. The *definition* of triangle does not tell us that the angles in a triangle will always add to 180°, but logical reasoning allows us to *deduce* this theorem. In the foregoing examples, we took the first step of *discovering* additional properties that rhombuses and rectangles must have, beyond those properties given in the definition. In these cases, we won't carry out the next step of deducing logically that rhombuses and rectangles really do have these additional properties, but it is possible to do so. In other words, it is possible to go beyond observing that rhombuses and rectangles always seem to have certain properties, to explain why these shapes must always have those properties. Such explanations are common in high school geometry courses.

Practice Exercises for Section 10.5

1. Define the following terms: quadrilateral, polygon, pentagon, hexagon, octagon, 13-gon, square, rectangle, rhombus, parallelogram, and trapezoid.

2. There are certain properties that all rectangles have but that are not described specifically in the definition of rectangle. Give examples of such properties.

3. There are certain properties that all rhombuses have but that are not described specifically in the definition of rhombus. Give examples of such properties.

4. Figure 10.81 shows a method for constructing an equilateral triangle. Explain why this method must always produce an equilateral triangle.

5. Use a ruler and a compass to construct a triangle that has one side of length 3 inches, one side of length 2 inches, and one side of length 1.5 inches. Describe your method, and explain why it must produce the desired triangle.

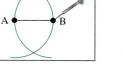

Step 1: Start with any line segment AB.

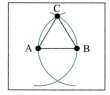

Step 2: Draw a circle centered at A, passing through B.

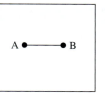

Step 3: Draw a circle centered at B, passing through A.

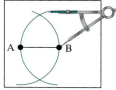

Step 4: Label one of the two points where the circles meet C. Connect A, B, and C with line segments.

FIGURE 10.81 Constructing an equilateral triangle

6. Use the definition of circles and rhombuses to explain why the quadrilateral ABDC produced by the method of Figure 10.82 must necessarily be a rhombus.

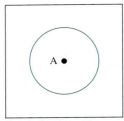

Step 1: Starting with any point A, draw a circle with center A.

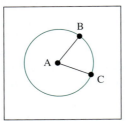

Step 2: Let B and C be any two points on the circle that are not opposite each other. Draw line segments AB and AC.

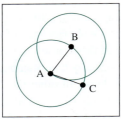

Step 3: Draw a circle centered at B and passing through A.

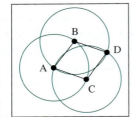

Step 4: Draw a circle centered at C and passing through A. Label the point other than A where these last two circles meet D.

FIGURE 10.82 Method for constructing rhombuses

7. Draw a Venn diagram showing the relationship between the set of rectangles and the set of rhombuses. Is there an overlap? If so, what does it consist of? Explain.

8. Draw a Venn diagram showing the relationship between the set of parallelograms and the set of trapezoids. Explain.

9. Some books define trapezoids as quadrilaterals that have *exactly one* pair of parallel sides. Draw a Venn diagram showing how the set of parallelograms and the set of trapezoids are related when this alternate definition of trapezoid is used. Explain.

10. Draw a Venn diagram showing the relationship between the set of rhombuses and the set of parallelograms. In doing this, can you rely only on the information given directly in the definition of rhombus and parallelogram, or would you need to deduce some additional information?

11. What special properties do diagonals of rhombuses have that diagonals of other quadrilaterals do not necessarily have?

12. What special properties do diagonals of rectangles have that diagonals of other quadrilaterals do not necessarily have?

Answers to Practice Exercises for Section 10.5

1. See text.

2. See text.

3. See text.

4. The method shown in Figure 10.81 produces an equilateral triangle because of the way it uses circles. Remember that a circle consists of all the points that are the same fixed distance away from the center point. The circle drawn in step 2 consists of all points that are the same distance from A as B is; since C is on this circle, the distance from C to A is the same as the distance from B to A. The circle drawn in step 3 consists of all points that are the same distance from B as A is; since C is also on this circle, the distance from C to B is the same as the distance from A to B. Therefore, the 3 line segments

AB, AC, and BC all have the same length, and the triangle ABC is an equilateral triangle.

5. Figure 10.83 shows the construction marks leading to the desired triangle. Starting with a line segment AB that is 3 inches long, draw (part of) a circle with center A and radius 2 inches. Then draw (part of) a circle with center B and radius 1.5 inches. (You can just as well draw the circle of radius 1.5 inches centered at A and the circle of radius 2 inches centered at B if you want to.) Let C be one of the two locations where the two circles meet. Connect A to C, and connect B to C to create the desired triangle.

The construction creates the desired triangle because it uses the defining property of circles. When you draw the circle with center A and radius

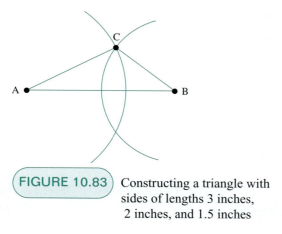

FIGURE 10.83 Constructing a triangle with sides of lengths 3 inches, 2 inches, and 1.5 inches

2 inches, the points on that circle are *all* 2 inches away from A. Similarly, when you draw the circle with center B and radius 1.5 inches, the points on that circle are *all* 1.5 inches away from B. So a point C where the two circles meet is *both* 2 inches from A and 1.5 inches from B. Hence, the side CA of the triangle ABC is 2 inches long, and the side CB is 1.5 inches long. The original side AB was 3 inches long, so the construction creates a triangle with sides of lengths 3 inches, 2 inches, and 1.5 inches.

6. A circle is the set of all points that are the same fixed distance away from the center point. Therefore, because points B and C are on a circle centered at A, the distance from B to A is equal to the distance from C to A. Because the point D is on a circle centered at B and passing through A, the distance from D to B is equal to the distance from B to A. Similarly, because D is on a circle centered at C and passing through A, the distance from D to C is equal to the distance from C to A. Therefore the 4 line segments AB, AC, BD, and CD have the same length. Rhombuses are quadrilaterals that have 4 sides of the same length. So the quadrilateral ABDC formed by the line segments AB, AC, BD, and CD is a rhombus.

7. See Figure 10.84. The shapes in the overlap are those shapes that have 4 right angles (as rectangles do) and

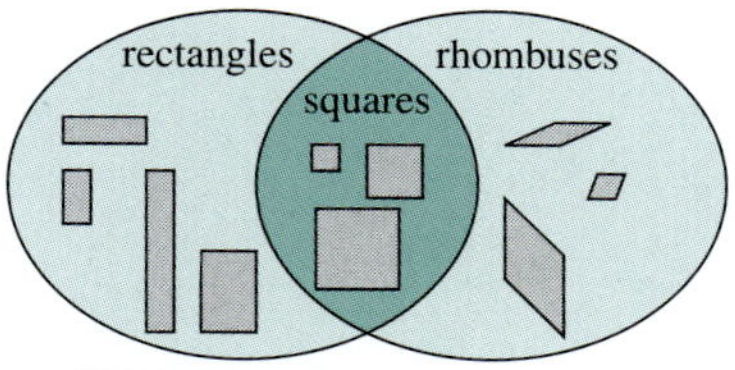

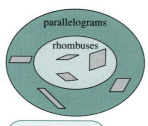

FIGURE 10.84 Venn diagram of rectangles and rhombuses

have 4 sides of the same length (as rhombuses do). According to the definition, those are exactly the shapes that are squares.

8. See Figure 10.85. According to the definition every parallelogram is also a trapezoid because parallelograms have two pairs of parallel sides, so they can be said to have at least one pair of parallel sides.

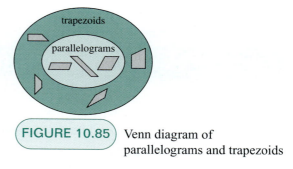

FIGURE 10.85 Venn diagram of parallelograms and trapezoids

9. See Figure 10.86. According to the alternative definition, no parallelogram is a trapezoid because parallelograms have two pairs of parallel sides, so they don't have exactly one pair of parallel sides. Therefore, with this alternative definition, the set of parallelograms and the set of trapezoids do not have any overlap.

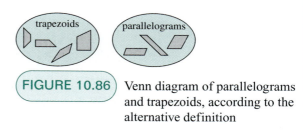

FIGURE 10.86 Venn diagram of parallelograms and trapezoids, according to the alternative definition

10. See Figure 10.87. Looking at rhombuses, we see that opposite sides appear parallel; therefore, it seems that every rhombus is also a parallelogram. Since the definition of rhombus does not say anything about opposite sides being parallel, we would need to deduce some additional information to explain why rhombuses really are parallelograms.

FIGURE 10.87 Venn diagram of rhombuses and parallelograms

11. The diagonals of a rhombus are perpendicular. The two diagonals of a rhombus meet at a point that is halfway across each diagonal. A diagonal of a rhombus cuts both angles it passes through in half.

12. The two diagonals of a rectangle have the same length and meet at a point that is halfway across each diagonal.

Problems for Section 10.5

1. a. Describe how four children could use 4 pieces of string to show a variety of different rhombuses.

b. Describe how you could show a variety of different rhombuses by threading straws onto string or by fastening sturdy strips of cardboard with brass fasteners.

2. Figure 10.88 shows a method for constructing isosceles triangles.

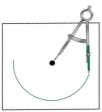
Step 1: Draw part of a circle.

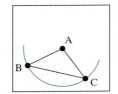
Step 2: Connect the center of the circle with two points on the circle.

FIGURE 10.88 A method for constructing isosceles triangles

a. Use the method of Figure 10.88 to draw two different isosceles triangles.

b. Use the definition of circles to explain why this method will always produce an isosceles triangle.

c. Use this method to draw an isosceles triangle that has two sides of length 6 inches and one side of length 4 inches.

3. a. Fold and cut paper to make an isosceles triangle that has one side that is 5 inches long and two sides that are 10 inches long. Describe your method, and explain why it must create an isosceles triangle.

b. Fold and cut paper to make an isosceles triangle that has one side that is 10 inches long and two sides that are 7 inches long. Describe your method.

4. Draw a line segment AB that is 4 inches long. Now modify the steps of the construction shown in Figure 10.81 to produce an isosceles triangle that has sides of lengths 7 inches, 7 inches, and 4 inches. Explain why you carry out your construction as you do.

5. a. Use a ruler and compass to help you draw a triangle that has one side of length 5 inches, one side of length 3.5 inches, and one side of length 4 inches.

b. Explain why your method of construction must produce a triangle with the required side lengths.

6. Is there a triangle that has one side of length 4 inches, one side of length 2 inches, and one side of length 1 inch? Explain.

7. Create two different rhombuses, both of which have 4 sides of length 4 inches. You may create your rhombuses by drawing, by folding and cutting paper, or by fastening objects together. In each case, explain how you know your shape really is a rhombus.

8. The line segments AB and AC in Figure 10.89 have been constructed so that they could be two sides of a rhombus.

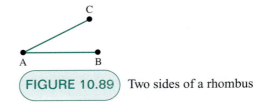

FIGURE 10.89 Two sides of a rhombus

a. Trace the angle BAC of Figure 10.89 on a piece of paper. Use a compass and straightedge to finish constructing a rhombus that has AB and AC as two of its sides.

b. By referring to the definition of rhombus, explain why your construction in part (a) must produce a rhombus.

9. In a paragraph, discuss how the definition of circle is used in constructing shapes (other than circles). Include at least two examples in your discussion.

10. a. Figure 10.90 shows 3 pairs of parallel line segments of the same length. In each case, describe how to get from point A to point B and from point C to point D by traveling along grid lines. Compare the instructions.

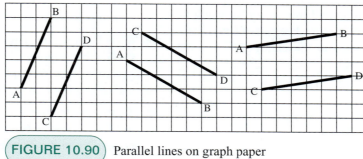

FIGURE 10.90 Parallel lines on graph paper

b. Figure 10.91 shows 3 pairs of perpendicular line segments of the same length. In each case, describe how to get from point A to point B and from point C to point D by traveling along grid lines. Compare the instructions and describe how they are related.

c. Copy the 3 line segments in Figure 10.92 onto graph paper (feel free to spread them out across the page) and use what you discovered in parts (a) and (b) to help you draw 3 squares that have those line segments as one side. Say briefly how to use the grid lines to draw the squares.

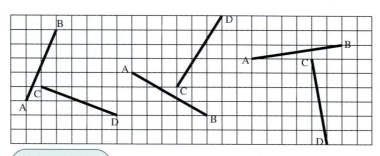

FIGURE 10.91 Perpendicular lines on graph paper

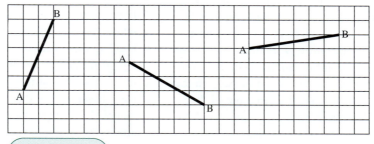

FIGURE 10.92 Draw squares on graph paper.

11. a. Given a parallelogram with angles a, b, c, and d, as in Figure 10.93, describe how angle a is related to the other 3 angles. Your answer should be general, in that it would hold for every parallelogram, not just for the particular parallelogram shown in the figure. To determine your answer, you might want to construct a parallelogram with Geometer's Sketchpad, measure the angles, and move the vertices in order to see many different parallelograms. Or, you could make several different parallelograms by cutting and folding paper, using the method described in the last part of Class Activity 10W, and measure the angles of the parallelograms you create.

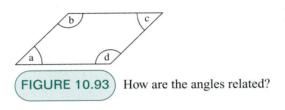

FIGURE 10.93 How are the angles related?

b. Use the definition of parallelogram and the Parallel Postulate to explain why the relationships you found in part (a) hold.

12. a. There are certain properties that all parallelograms have but that are not described specifically in the definition of parallelogram. Give at least two examples of such properties.

b. There are certain properties that all isosceles triangles have but that are not described specifically in the definition of isosceles triangle. Give at least one example of such a property.

13. a. Draw a Venn diagram showing the relationship between the set of rectangles and the set of parallelograms.

b. Discuss whether the relationship between the set of rectangles and the set of parallelograms that you found in part (a) can be determined immediately from the definitions of these shapes or whether additional information that derives from the definitions, but is not stated directly in the definitions, is necessary. If a blind person knew only the definitions of rectangles and parallelograms, would that person immediately be able to understand the relationship you found in part (a),

or did you rely on visual information about rectangles and parallelograms that is not stated in the definitions of these shapes?

c. Give a careful, thorough explanation for why the set of rectangles and the set of parallelograms are related the way they are. *Hint:* To show that opposite sides of a rectangle must be parallel, explain why these sides couldn't meet if they were extended to be lines. Show that if these two lines did meet somewhere, they would be part of a triangle whose angles would add to more than 180°.

14. Draw a Venn diagram that shows the relationships among the sets of quadrilaterals, squares, rectangles, parallelograms, rhombuses, and trapezoids. Explain briefly.

15. A problem that was given to fifth-graders is shown in Figure 10.94. Criticize the problem on *mathematical grounds*. Rewrite and redraw the problem to make it correct. Explain briefly.

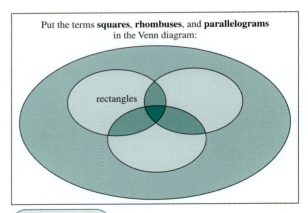

Put the terms **squares, rhombuses,** and **parallelograms** in the Venn diagram:

rectangles

FIGURE 10.94 A problem given to fifth-graders

16. Before builders lay the foundation for a rectangular house, they usually pound stakes into the ground where the 4 corners of the house will be. The builders then measure the diagonals of the quadrilateral formed by the 4 stakes. Suppose that these diagonals turn out to have different lengths. In this case, based on the discussion in this section about diagonals of quadrilaterals, what can you conclude about the angles at the corners? Explain. It is important for the corners of the foundation to be right angles, so that the walls above will be structurally stable. Should the builders be satisfied with the proposed foundation or should they make adjustments to it?

17. Given a right triangle that has angles *a, b,* and 90°, draw a line segment from the corner where the right angle is to the hypotenuse so that this line segment is perpendicular to the hypotenuse, as shown in Figure 10.95. The line segment divides the original right triangle into two smaller right triangles. Without measuring, determine the angles in these smaller right triangles. How are these angles related to the angles in the original large triangle? Give a general answer, one that does not depend on the specific values of *a* and *b*. Explain your reasoning.

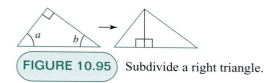

FIGURE 10.95 Subdivide a right triangle.

18. **a.** Use the reasoning of Class Activity 10J and the answer to Practice Exercise 8 on page 439 for triangles to determine the sum of the exterior angles of the quadrilateral in Figure 10.96. In other words, determine *e* + *f* + *g* + *h*. Measure with a protractor to check that your formula is correct for this quadrilateral.

   b. Will there be a similar formula for the sum of the exterior angles of pentagons, hexagons, 7-gons (also known as heptagons), octagons, and so on? Explain.

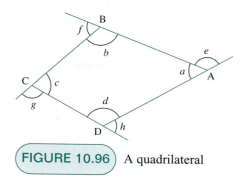

FIGURE 10.96 A quadrilateral

   c. Using your formula for the sum of the exterior angles of a quadrilateral, deduce the sum of the interior angles of the quadrilateral.

   In other words, find *a* + *b* + *c* + *d*, as pictured in Figure 10.96. Explain your reasoning. Measure with a protractor to verify that your formula is correct for this quadrilateral.

   d. Based on Class Activity 10J and your prior work, what formula would you expect to be true for the sum of the interior angles of a pentagon? What about for hexagons? What about for a polygon with *n* sides? Explain briefly.

19. **a.** Determine the angles in a regular pentagon. Explain your reasoning.

   b. Determine the angles in a regular hexagon. Explain your reasoning.

   c. Determine the angles in a regular *n*-gon. Explain your reasoning.

10.6 Constructions with Straightedge and Compass

straightedge   In this section, we will explore some easy constructions that can be done with a compass and a straightedge. These constructions work because of special properties of rhombuses. A **straightedge** is just a "straight edge," namely a ruler, except that it need not have any markings.

Constructions with straightedge and compass have their origins in the desire of those who wish to make precise drawings by using only simple, reliable tools. Think back to the times before there were

FIGURE 10.97

Constructing a line that is perpendicular to a line segment and divides it in half

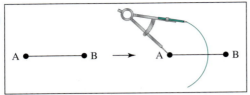

Step 1: Starting with a line segment AB, open a compass to at least half the length of the line segment, and draw (part of) a circle centered at A.

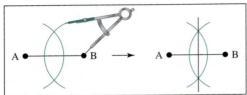

Step 2: Keep the compass opened to the same width, and draw (part of) a circle centered at B. Draw a line through the two points where the two circles meet. This line is perpendicular to AB and divides AB in half.

computers. If you were studying plane shapes, and if you wanted to discover additional properties these shapes had beyond those properties given in the definitions, then you would need some way to make precise drawings. A sloppy drawing could hide interesting features, or worse, could seem to show features that aren't really there. Nowadays computer technology can help us make accurate drawings easily, but we can still benefit from learning some of the old hands-on techniques.

Dividing a Line Segment in Half and Constructing a Perpendicular Line

Stop to think for a moment about how you could draw a perfect (or nearly perfect) right angle without using a previously made right angle. It's not so obvious.

Figure 10.97 shows how you can start with any line segment and construct a line that is perpendicular to the line segment *and* that passes through the point halfway across the line segment. The point halfway across a line segment is often called a **midpoint** of the line segment. So the procedure shown in Figure 10.97 actually does two things at once: It constructs a perpendicular line, and it finds a midpoint of a line segment. Instead of saying that the construction of Figure 10.97 divides the line segment *in half,* we can say that the construction **bisects** the line segment or that it constructs the *perpendicular bisector* to the line segment. Notice that the word *bisect* is similar to the word *dissect.* The former is used in geometry, whereas the latter in used in biology.

midpoint

bisect (line segment)

Dividing an Angle in Half

Figure 10.98 shows how you can start with any angle at a point P and divide the angle in half. In other words, given 2 rays that meet at a point P, the construction in Figure 10.98 produces another ray, also starting at P, that is halfway between the 2 original rays. Instead of saying that the construction of Figure 10.98 divides an angle in half, we can say that the construction **bisects** the angle.

bisect (angle)

The next Class Activity shows how the construction of a perpendicular bisector of a line segment and the bisector of an angle are related to properties of rhombuses.

FIGURE 10.98

Dividing an angle in half

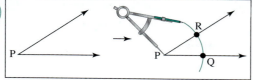

Step 1: Starting with two rays that meet at a point P, draw a circle centered at P. Let Q and R be the points where the circle meets the rays.

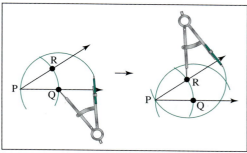

Step 2: Keep the compass opened to the same width. Draw a circle centered at Q and another circle centered at R.

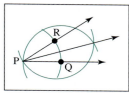

Step 3: Draw a line through point P and the other point where the last two circles drawn meet.

Class Activity   *Now Turn to Class Activities Manual*

10CC Relating the Constructions to Properties of Rhombuses, p. 266

10DD Constructing a Square and an Octagon with Straightedge and Compass, p. 267

Practice Exercises for Section 10.6

1. Draw a line segment. Use a straightedge and compass to construct a line that is perpendicular to your line segment, and divides your line segment in half. Repeat with several more line segments.

2. Draw an angle. Use a straightedge and compass to divide your angle in half. Repeat with several more angles.

3. **a.** Draw a rhombus that is naturally associated with the construction of a perpendicular bisector of a line segment shown in Figure 10.97.

   b. Explain why the quadrilateral that you identify as a rhombus in part (a) really must be a rhombus, according to the definition of rhombus and according to the way it was constructed.

   c. Which special properties of rhombuses are related to the construction of a perpendicular bisector of a line segment shown in Figure 10.97? Explain.

Answers to Practice Exercises for Section 10.6

1. See Figure 10.97.

2. See Figure 10.98.

3. **a.** Starting with the finished construction of a perpendicular bisector shown in Figure 10.99, let C and D be the points in the construction where the two circles meet. Form the 4 line segments AC, BC, AD, and BD, as shown in Figure 10.100.

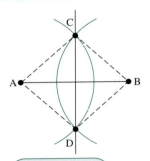

FIGURE 10.100   The construction for a perpendicular line forms a rhombus.

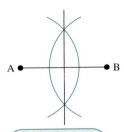

FIGURE 10.99   The result of using a straightedge and compass to construct a perpendicular bisector

b. Notice that AC and AD in Figure 10.100 are radii of the circle centered at A. Therefore, AC and AD have the same length. Similarly, BC and BD have the same length because they are radii of the circle centered at B. But the circles centered at A and at B have the same radius because they were constructed that way (without changing the width of the compass). Therefore, all 4 line segments AC, BC, AD, and BD have the same length. By definition, this means that the quadrilateral ACBD is a rhombus.

c. The original line segment AB is one of the diagonals of the rhombus ACBD of Figure 10.100. The perpendicular bisector CD is the other diagonal of the rhombus ACBD. According to the list of properties of rhombuses on pages 456 and 457, the diagonals of a rhombus meet at a point that is halfway across each diagonal, and the diagonals of a rhombus are perpendicular. These properties fit exactly with the properties of the line segments AB and CD because CD was constructed to be perpendicular to AB and to divide AB in half.

Problems for Section 10.6

1. **a.** Draw a ray with endpoint A. Use a straightedge and compass to carefully construct a 45° angle to your ray at A. Briefly describe your method, and explain why it makes sense.

   b. Draw a ray with endpoint B. Use a straightedge and compass to carefully construct a 22.5° angle to your ray at B. Briefly describe your method, and explain why it makes sense. (*Hint:* Notice that 22.5 is half of 45.)

2. **a.** On a blank piece of paper, draw a ray with endpoint A. Carefully fold your paper to create a fold line that makes a 90° angle with your ray at

A. Briefly describe your method, and explain why it makes sense.

   b. On a blank piece of paper, draw a ray with endpoint B. Carefully fold your paper to create a fold line that makes a 45° angle with your ray at B. Briefly describe your method, and explain why it makes sense.

   c. On a blank piece of paper, draw a ray with endpoint C. Carefully fold your paper to create a fold line that makes a 22.5° angle with your ray at C. Briefly describe your method, and explain why it makes sense.

3. Use a straightedge and compass to carefully construct a square. Leave the marks showing your construction.

4. **a.** Draw a rhombus that is naturally associated with the straightedge and compass construction of a ray dividing an angle in half, as shown in Figure 10.98.

   b. Explain why the quadrilateral that you identify as a rhombus in part (a) really must be a rhombus, according to the definition of rhombus and according to the way it was constructed.

   c. Which special properties of rhombuses described in Section 10.5 are closely related to the construction of a ray dividing an angle in half that is shown in Figure 10.98? Explain.

5. Describe how to use a compass to construct the pattern of 3 circles shown in Figure 10.101 so that the triangle shown inside the circles is an equilateral triangle. Then explain why the construction guarantees that the triangle really is an equilateral triangle.

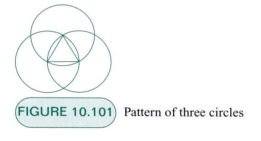

FIGURE 10.101 Pattern of three circles

6. **a.** Use a compass to draw a pattern of circles like the one in Figure 10.102. Briefly describe how to create this pattern.

   b. Use a straightedge and compass to carefully construct a regular hexagon. Construct your

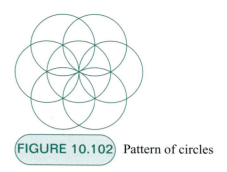

FIGURE 10.102 Pattern of circles

hexagon by first drawing another pattern of circles like the one in Figure 10.102, but this time draw only parts of the circles, so that you don't clutter your drawing.

7. Use a straightedge and compass to carefully construct a regular 12-gon. Briefly describe your method, and explain why it makes sense. It may help you to examine Figure 10.102.

8. Draw a large circle on a piece of paper. This circle will represent a pie. Use a straightedge and compass to carefully subdivide your pie into 6 equal pie pieces. Briefly describe your method, and explain why it makes sense. It may help you to examine Figure 10.102. (You may find this construction helpful in the future if you need to make pie pieces to help your students learn about fractions.)

9. Draw a large circle on a piece of paper. This circle will represent a pie. Use a straightedge and compass to carefully subdivide your pie into 12 equal pieces. Briefly describe your method, and explain why it makes sense. It may help you to examine Figure 10.102. (You may find this construction helpful in the future if you need to make pie pieces to help your students learn about fractions.)

Chapter Summary and Study Items

Section 10.1 Visualization

An important part of geometry is visualizing and mentally manipulating shapes.

Key skills and understandings:

- Visualize the earth, sun, and moon in space.
- Explain the phases of the moon and the direction of rotation of the earth.

Section 10.2 Angles

Angles can represent an amount of rotation or the region formed when two rays meet. There are three important facts about angles produced by configurations of lines. When two lines meet, the angles opposite each other are equal. When two parallel lines are cut by a third line, the corresponding angles that are produced are equal; this is called the Parallel Postulate. Three lines can meet so as to form a triangle. The angles in a triangle add to 180°.

Key skills and understandings:

- Use the fact that a straight line forms a 180° angle to explain why the angles opposite each other, which are formed when two lines meet, are equal.
- Use the Parallel Postulate to prove that the sum of the angles in a triangle is 180°.
- Think of walking around a triangle and explain why the sum of the angles in a triangle is 180°.
- Apply facts about angles produced by configurations of lines to find angles.

Section 10.3 Angles and Phenomena in the World

Practical situations in which angles are of interest involve angles formed by the sun's rays and angles concerning reflected light rays.

Key skills and understandings:

- Apply the laws of reflection to explain what a person looking into a mirror will see.

Section 10.4 Circles and Spheres

A circle is the collection of all the points in a plane that are a fixed distance away from a fixed point in the plane. A sphere is the collection of all the points in space that are a fixed distance away from a fixed point in space. Two circles in a plane can either not meet, meet at a single point, or meet at two points. Two spheres in space can either not meet, meet at a single point, or meet along a circle.

Key skills and understandings:

- Give the definition of circles and spheres.
- Given a situation that involves distances from one or more points, use circles or spheres to describe the relevant locations.

Section 10.5 Triangles, Quadrilaterals, and Other Polygons

A polygon is a closed, connected shape in a plane consisting of a finite number of line segments that do not cross each other. Triangles are 3-sided polygons. Quadrilaterals are 4-sided polygons. Squares, rectangles, parallelograms, rhombuses, and trapezoids are special kinds of quadrilaterals. These shapes have certain defining properties, but they also have other properties that are not stated in their definition (but which could be derived). Venn diagrams show relationships among the different quadrilaterals.

Key skills and understandings:

- Use a compass to construct triangles of specified side lengths (including equilateral and isosceles triangles), and use the definition of circle to explain why the construction must produce the required triangle.
- Use a compass to construct rhombuses, and use the definition of circle to explain why the construction must produce a rhombus.
- Fold and cut paper to produce triangles and quadrilaterals.
- Describe how various types of quadrilaterals are related to each other, and show the relationships with the aid of a Venn diagram. Explain why the relationships hold and whether they follow directly from the definitions of the shapes or whether they rely on other information about the shapes that is not stated directly in the definitions.

Section 10.6 Constructions with Straightedge and Compass

Using only a straightedge and a compass, we can construct the perpendicular bisector of a given line segment, and we can bisect a given angle. Both of these constructions produce an associated rhombus. The constructions give the desired results because of special properties of rhombuses: The diagonals in a rhombus are perpendicular and bisect each other, and the diagonals in a rhombus bisect the angles in the rhombus.

Key skills and understandings:

- Given a line segment, construct the perpendicular bisector by using a compass and straightedge. Describe the associated rhombus, and relate the construction to a special property of rhombuses: In a rhombus, the diagonals are perpendicular and bisect each other.

- Use a compass and straightedge to bisect a given angle. Describe the associated rhombus, and relate the construction to a special property of rhombuses: In a rhombus, the diagonals bisect the angles in the rhombus.

- Use the basic compass and straightedge constructions. For example, construct a square.

Measurement

Measurement is so familiar that we usually don't pay much attention to the underlying structure of measures or the requirements for taking measurements. We use measurements daily—buying food by the pound, gasoline by the gallon, and electricity by the kilowatt-hour. We also measure the passage of time, the amount of land a person owns, and the distance between cities. But what does it mean to measure a quantity, and what does it take to make the careful and accurate measurements we need for everyday tasks and for scientific applications?

In this chapter, we will first study fundamental ideas of measurement: the need to select a measurable attribute in order to measure an object or to describe its size; the concept that measurement involves comparison; the structure that most, but not all measures share; and the need for and role of units. We will study the common units of two systems of measurement: the U.S. customary system and the metric system. Then we will focus on length, area, and volume, and we will compare and contrast these measures. Finally, we will study some technical aspects of measurement: error and precision in measurement and conversion between different units. All measurements of actual objects are necessarily inexact. The way we report a measurement conveys the confidence we have in the measurement's precision.

The National Council of Teachers of Mathematics has made recommendations concerning the learning of measurement (see [54]): Instructional programs should enable all students from prekindergarten through grade 12 to

- understand measurable attributes of objects and the units, systems, and processes of measurement;
- apply appropriate techniques, tools, and formulas to determine measurements.

For the NCTM's specific measurement recommendations for grades PreK–2, grades 3–5, grades 6–8, and grades 9–12, go to www.pearsonhighered.com/beckmann and look for the grade band you are interested in.

11.1 Fundamentals of Measurement

Focal Points
PreK, K, Grade 2

Measurement is about determining the size of things. In this section, we focus on the ideas underlying the concept of measurement: the need to specify measurable attributes, the structure of measures, units of measurement. We examine systems of measurement and we also briefly discuss the process of measurement.

Selecting a Measurable Attribute

Measurement allows us to describe or determine the sizes of objects and to compare objects. But can we always say unambiguously that one object is larger than another? As we'll see, in some situations there is ambiguity about relative size because it is possible to measure an object with respect to several different kinds of attributes. Before you read on, please do the next Class Activity.

Class Activity   *Now Turn to Class Activities Manual*

11A   The Biggest Tree in the World, p. 269

If we compare the trees in the Class Activity by volume, then General Sherman is the largest. But if we compare the trees by height, then the Mendocino tree is largest. We could also compare the trees by the land area they cover, in which case the Banyan tree in Calcutta is probably largest. There is no single measurable attribute that we are required to use to determine which tree is largest. Before we can compare the trees, we must *select* a measurable attribute with which to compare them. Volume and height are natural choices, but we can also compare the circumference of the trees or the area of land they cover.

Similarly, we might say that one box of cereal is larger than another box of cereal, or that one bottle of juice is larger than another bottle of juice. In the case of the juice, we are probably comparing the sizes of the bottles by the volume of juice they contain. Maybe one bottle contains a liter of juice and the other bottle contains 750 milliliters of juice, which is less than a liter. But in the case of the cereal, we might be comparing the volume of cereal in each box, or we might be comparing the weight of the cereal in each box. Some cereals are more dense than others, so that a smaller volume of one cereal can weigh more than a larger volume of another. Therefore, the two ways of comparing the boxes of cereal—by volume or by weight—could lead to different conclusions about which box of cereal is larger.

To compare sizes of objects, we must specify, either implicitly or explicitly, a measurable attribute as a basis for the comparison. So be careful when comparing sizes of objects: If there is any chance of confusion, specify the measurable attribute with which you are comparing the objects.

A common activity for children in grades K–2 is to sort a collection of objects in different ways, such as by color or by shape. In this way children become aware that one object has various different attributes and that objects can be compared and contrasted in different ways.

The Order and Additive Structure of (Most) Measures

All measurement involves comparison, so we must be able to compare before we are able to measure. For example, before statements such as "This desk is 3 feet wide" or "This bottle holds 2 liters of juice" can make sense, more basic statements of comparison such as the following must make sense: "This desk is wider than that desk," "This bottle holds less juice than that bottle," "These 2 books weigh the same." So learning to compare objects with respect to a given attribute should precede learning to measure objects with respect to that attribute. Thus, *order*—namely, the concepts of *greater than, less than,* and *equal to*—fundamentally underlie the concept of measurement.

Another fundamental concept underlying most measurement is that of *additivity.* For example, before a statement such as "This rock weighs 2 pounds" can make sense, the more basic idea that the combined weight of 2 rocks is equal to the sum of the weights of the individual rocks must make sense, if only in an intuitive way. Why? When we say that a rock weighs 2 pounds, we mean that it weighs the same as two 1-pound objects together. Similarly, before a statement such as "This string is 3 meters long" can make sense, the more basic idea that when 3 strings are attached end-to-end, their combined length is the sum of the lengths of the 3 pieces must make sense. When we say that a string is 3 meters long, we mean that it is as long as three 1-meter-long objects placed end-to-end. Length, area, volume, weight, and time are familiar measures which are additive in the sense that the measure of a combined quantity is equal to the sum of the measures of the individual components. Interestingly, temperature is not additive. When substances of different temperatures are combined, the temperature of the combined substance is generally not equal to the sum of the temperatures of the component parts.

We will use the additivity of area and volume extensively in the next chapters when we determine areas and volumes of objects and when we develop area and volume formulas.

Units and the Meaning of Measurements

What does it mean to measure a quantity and what does it mean when we describe the size of an object with a specific measure such as 10 feet or 4 liters? We have discussed the need for specifying an attribute we wish to measure, and we have said that measurement involves comparison. But measurement is not just any kind of comparison; it is comparison with a fixed "reference amount" of a quantity. This reference amount is called a **unit**. For example, a mile is a unit of length, an acre is a unit of area, a gallon is a unit of capacity (or volume), and a kilowatt-hour is a unit of electrical power.

unit

measure   To **measure** a given quantity means to compare that quantity with a unit of the quantity. Usually, we must determine how many units of the quantity make up the given quantity. For example, the length of a rug in meters is the number of 1-meter sticks we would have to place end-to-end to traverse the entire length of the rug. The area of a rug in square feet is the number of 1-foot-by-1-foot squares we would need to cover the rug completely without gaps or overlaps. The weight of a piece of bread in grams is the number of 1-gram weights we would have to put on one side of a pan balance to balance the piece of bread on the other side. The volume of a compost pile in cubic yards is the number of 1-yard-by-1-yard-by-1-yard cubes we could fill, using the compost. Note that although we often write "1 cubic yard" as 1 yd^3, we should *read* 1 yd^3 as "1 cubic yard" (and similarly for other units). In the next Class Activity, you will confront some confusing aspects of working with units of area. Similar issues arise with units of volume.

Class Activity   *Now Turn to Class Activities Manual*

11B   What Does "6 Square Inches" Mean?, p. 271

What quantities can be used as units? *Any fixed* (positive) *amount* of a measurable quantity can be a unit, but we usually use commonly accepted standard units in order to communicate clearly. For example, to encourage a child to develop her concept of measurement, you might have her measure the length of a piece of ribbon by laying pens end-to-end. She might report that the ribbon is 5 pens long. In this case, the unit of length is "a pen," which is not a standard unit of length (see Figure 11.1).

On the other hand, if you want to buy ribbon from a spool at a store, you will probably use yards, feet, inches, or some combination of these to say how long a piece you want. That is, if the store is in the United States—if you were in France, you would use meters and centimeters to describe the length of ribbon.

FIGURE 11.1

Using a nonstandard unit to measure: a desk that is 5 pens wide

Systems of Measurement

A system of measurement is a collection of standard units. In the United States today, two systems of measurement are in common use: the U.S. customary system of measurement and the metric system (also known as the international system of units, or the SI system).

The U.S. Customary System Table 11.1 shows some standard units that are commonly used in the U.S. customary system of measurement.

TABLE 11.1 Common units of measurement in the U.S. customary system

Units in the U.S. Customary System		
Unit	Abbreviation	Some Relationships
Units of Length		
inch	in. or in	
foot	ft	1 ft = 12 in.
yard	yd	1 yd = 3 ft
mile	mi	1 mi = 1760 yd = 5280 ft
Units of Area		
square inch	in^2	
square foot	ft^2	$1\ ft^2 = 12^2\ in^2 = 144\ in^2$
square yard	yd^2	$1\ yd^2 = 3^2\ ft^2 = 9\ ft^2$
square mile	mi^2	
acre		$1\ acre = 43{,}560\ ft^2$
Units of Volume		
cubic inch	in^3	
cubic foot	ft^3	
cubic yard	yd^3	$1\ yd^3 = 3^3\ ft^3 = 27\ ft^3$
Units of Capacity (Volume)		
teaspoon	tsp	
tablespoon	T or tbs or tbsp	1 T = 3 tsp
fluid ounce (or liquid ounce)	fl oz	1 fl oz = 2 T
cup	c	1 c = 8 fl oz
pint	pt	1 pt = 2 c = 16 fl oz
quart	qt	1 qt = 2 pt = 32 fl oz
gallon	gal	1 gal = 4 qt = 128 fl oz
Units of Weight (Avoirdupois)		
ounce	oz	
pound	lb	1 pound = 16 ounces
ton	t	1 ton = 2000 pounds
Unit of Temperature		
degree Fahrenheit	°F	water freezes at 32°F water boils at 212°F

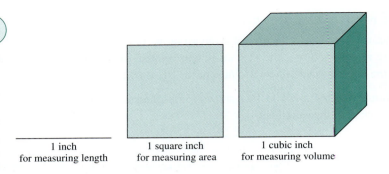

FIGURE 11.2

Examples of
units of length,
area, and volume
in the U.S.
customary
system. The unit
of length is the
inch, the unit of
area is the square
inch, and the unit
of volume is the
cubic inch.

1 inch
for measuring length

1 square inch
for measuring area

1 cubic inch
for measuring volume

For any unit of length, there are corresponding units of area and volume: a square unit, and a cubic unit, respectively. A square unit is the area of a square that is one unit wide and one unit long. A cubic unit is the volume of a cube that is one unit wide, one unit long, and one unit high. For example, a square inch is the area of a square that is one inch wide and one inch long. A cubic inch is the volume of a cube that is one inch wide, one inch long, and one inch high, as shown in Figure 11.2.

Capacity is essentially the same as volume, except that the term *capacity* can be used for the volume of a container, the volume of liquid in a container, or the volume of space taken up by a substance filling a container, such as flour or berries.

Notice the distinction between an *ounce*, which is a unit of weight, and a *fluid ounce*, which is a unit of capacity or volume.

Definitions of units have changed over time. As science and technology progress, we are able to measure more accurately, and units must be defined more precisely. For example, the inch was originally defined as the width of a thumb or as the length of 3 barleycorns. Since widths of thumbs and lengths of barleycorns can vary, these definitions are obviously not very precise. The inch is currently defined to be *exactly* 2.54 centimeters. (See [38].)

TABLE 11.2 The metric system uses prefixes to create larger and smaller units from base units

Some Metric System Prefixes		
Prefix	Meaning	
nano-	$10^{-9} = \frac{1}{1,000,000,000}$	billionth
micro-	$10^{-6} = \frac{1}{1,000,000}$	millionth
milli-	$10^{-3} = \frac{1}{1000}$	thousandth
centi-	$10^{-2} = \frac{1}{100}$	hundredth
deci-	$10^{-1} = \frac{1}{10}$	tenth
deka-	10	ten
hecto-	$10^2 = 100$	hundred
kilo-	$10^3 = 1000$	thousand
mega-	$10^6 = 1,000,000$	million
giga-	$10^9 = 1,000,000,000$	billion

The Metric System The metric system was first used in France around the time of the French Revolution (1790). At that time, which is often called the Age of Enlightenment, there was a focus on rational thought, and scientists sought to organize their subjects in a rational way. The metric system is a natural outcome of this desire, as it is an efficient, organized system that is designed to be compatible with the decimal system for writing numbers.

The metric system gives names to units in a uniform way. For each kind of quantity to be measured, there is a base unit (such as meter, gram, liter). Each related unit is labeled with a prefix that indicates the unit's relationship to the base unit. For example, the prefix *kilo* means *thousand*, so a *kilometer* is a thousand meters and a *kilogram* is a thousand grams. Many of the metric system prefixes are used only in scientific contexts, not in everyday situations. Some of the metric system prefixes are listed in Table 11.2.

Table 11.3 shows some of the most commonly used units in the metric system.

TABLE 11.3 Common units of measurement in the metric system

Units in the Metric System		
Unit	**Abbreviation**	**Some Relationships**
Units of Length		
millimeter	mm	$1 \text{ mm} = \frac{1}{1000} \text{ m}$
centimeter	cm	$1 \text{ cm} = \frac{1}{100} \text{ m} = 10 \text{ mm}$
meter	m	$1 \text{ m} = 100 \text{ cm}$
kilometer	km	$1 \text{ km} = 1000 \text{ m}$
Units of Area		
square millimeter	mm^2	
square centimeter	cm^2	$1 \text{ cm}^2 = 10^2 \text{ mm}^2 = 100 \text{ mm}^2$
square meter	m^2	$1 \text{ m}^2 = 100^2 \text{ cm}^2 = 10{,}000 \text{ cm}^2$
square kilometer	km^2	$1 \text{ km}^2 = 1000^2 \text{ m}^2 = 1{,}000{,}000 \text{ m}^2$
Units of Volume		
cubic millimeter	mm^3	
cubic centimeter	cm^3 or cc	$1 \text{ cm}^3 = 10^3 \text{ mm}^3 = 1000 \text{ mm}^3$
cubic meter	m^3	$1 \text{ m}^3 = 100^3 \text{ cm}^3 = 1{,}000{,}000 \text{ cm}^3$
cubic kilometer	km^3	$1 \text{ km}^3 = 1000^3 \text{ m}^3 = 1{,}000{,}000{,}000 \text{ m}^3$
Units of Capacity		
milliliter	mL or ml	$1 \text{ mL} = \frac{1}{1000} \text{ L}$
liter	L or l	$1 \text{ L} = 1000 \text{ mL}$
Units of Mass (Weight)		
milligram	mg	$1 \text{ mg} = \frac{1}{1000} \text{ g}$
gram	g	
kilogram	kg	$1 \text{ kg} = 1000 \text{ g}$
Unit of Temperature		
degree Celsius	°C	water freezes at 0°C water boils at 100°C

FIGURE 11.3

Comparing
millimeters,
centimeters,
and inches

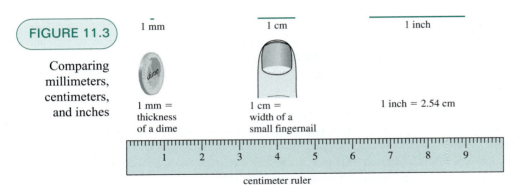

centimeter ruler
Small marks indicate millimeters.

FIGURE 11.4

Examples of units
of length, area, and
volume in the
metric system. The
unit of length is the
centimeter, the unit
of area is the square
centimeter, and the
unit of volume is the
cubic centimeter

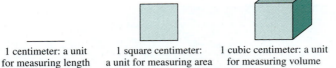

1 centimeter: a unit   1 square centimeter: 1 cubic centimeter: a unit
for measuring length   a unit for measuring area   for measuring volume

Figure 11.3 shows lines of length 1 millimeter, 1 centimeter, and 1 inch. The inch and the centimeter are related by 1 inch = 2.54 centimeters. A meter is about a yard and 3 inches. A kilometer is about 0.6 miles, so a bit more than half a mile.

Every unit of length has an associated unit of area and an associated unit of volume. Figure 11.4 shows a square centimeter and a cubic centimeter, which are the units of area and volume associated with the centimeter.

cc (cubic centimeter)

Soft drinks often come in 2-liter bottles. One liter is a little more than a quart. Doses of liquid medicines are often measured in milliliters. In this case, a milliliter is also often called a **cc,** for cubic centimeter.

metric ton (tonne)

Nutrition labels on food packages often refer to grams. Serving sizes are often given both in grams and in ounces. One ounce is about 28 grams. One kilogram is the weight of 1 liter of water, which is about 2.2 pounds. The weight of vitamins in foods is often given in terms of milligrams. Also, although 1000 kilograms should be called a megagram, it is usually called a **metric ton**, or **tonne.**

In the metric system, the units of length, capacity, and mass (weight) are related in a simple and logical way, as shown in Table 11.4.

TABLE 11.4 Relationships among units of capacity, mass, and length in the metric system

Relationships of Liter, Gram, and Meter
1 milliliter of water weighs 1 gram
1 milliliter of water (at 4°C) fills a cube that is 1 cm wide, 1 cm deep, and 1 cm high

TABLE 11.5 Relationships between U.S. customary and metric systems

U.S. Customary and Metric	
length	1 in. = 2.54 cm (exact)
capacity	1 gal = 3.79 L (good approximation)
weight (mass)	1 kg = 2.2 lb (good approximation)

Table 11.5 shows some basic relationships between the U.S. customary and metric systems of measurement.

Table 11.6 summarizes the basics of the metric system.

Even though the metric system was designed to be more logical and organized than the old customary systems, the definitions of units in the metric system have changed over time, too. For example, the meter was originally defined as one ten-millionth of the distance from the equator to the North Pole, whereas it is currently defined as the length of a path traveled by light in a vacuum in

$$\frac{1}{299,792,458}$$

of a second (see [38]).

Process of Measurement

How do we measure quantities? Often there are many ways to measure quantities, and we have special devices for measuring them. But the simplest and most direct way to measure a quantity is to count how many of the units are in the quantity to be measured. For example, to measure the volume of rice

TABLE 11.6 Summary of the metric system

Metric System					
		Prefix	Length	Capacity	Weight
0.001	$\frac{1}{1000}$	milli-	millimeter	milliliter	milligram
0.01	$\frac{1}{100}$	centi-	centimeter		centigram
0.1	$\frac{1}{10}$	deci-			
1		base unit	meter	liter	gram
10		deka-			
100		hecto-			
1000		kilo-	kilometer		kilogram
Fundamental Relationships					
1 milliliter of water weighs 1 gram and has a volume of 1 cubic centimeter.					

in a small bag, you can simply scoop the rice out, cup by cup, and count how many cups of rice you scooped. But to measure the volume of water in your bathtub, it would be tedious to scoop it out cup by cup or gallon by gallon. So you would probably want to use an indirect method for measuring this volume of water.

The simplest and most familiar measuring device is a ruler, which of course is for measuring length. A ruler whose unit is an inch displays a number of inch-long lengths. In order to use the ruler to measure how long an object is in terms of inches, we simply read off the number of inches on the ruler, rather than repeatedly laying down an inch-long object and counting how many it took to cover the length of the object. Similarly, other measuring devices such as scales, calipers, and speedometers allow us to measure quantities indirectly, rather than making direct counts of a number of units.

Class Activity *Now Turn to Class Activities Manual*

11C Using a Ruler, p. 272

Generally there is more than one unit that can be used to measure a given attribute. For example, within the metric system we can use millimeters, centimeters, meters, or kilometers (as well as other units) to measure lengths or distances. If a classroom is 10 meters long, then it is 1000 centimeters long and 10,000 millimeters long. We can also describe the classroom as 0.01 or $\frac{1}{100}$ kilometers long. Clearly, reporting the classroom as "10 meters long" is easier to grasp than any of the other options. When measuring or reporting measurements, use a unit that will best help in comprehending the answer.

Practice Exercises for Section 11.1

1. What does it mean to measure a quantity?

2. What is the simplest, most basic way to measure?

3. What is a unit?

4. What are the two main systems of measurement used in the United States today?

5. How many feet are in 1 mile? How many ounces are in 1 pound? How many pounds are in 1 ton?

6. What is the difference between an ounce and a fluid ounce?

7. What is special about the way units in the metric system are named?

8. How is 1 milliliter related to 1 gram and to 1 centimeter?

9. Use square centimeter tiles or centimeter graph paper to show or to draw three different shapes that have an area of 8 cm^2.

10. Use cubic-inch blocks (or cubic-centimeter blocks) to make a variety of solid shapes that have a volume of 24 in^3 (or 24 cm^3).

11. Which of the following have the same area or mean the same as 3 cm^2?

    • a 3-cm-by-3-cm square

- 3 square centimeters
- 3 cm × 3 cm

12. A construction site requires 40 cubic yards of concrete. What does "40 cubic yards of concrete" mean?

13. Table 11.7 displays information about two boxes of different types of breakfast cereals. Explain in several different ways why either of the two boxes of cereal can be considered the larger of the two.

TABLE 11.7 Comparing boxes of cereal

	Cereal Box 1	Cereal Box 2
Weight of cereal in box	16 oz	11 oz
Volume of cereal in box	4 cups	10 cups
Servings per box	8	10
Calories per box	1600	1100

Answers to Practice Exercises for Section 11.1

1. See page 476
2. See page 481
3. See page 476
4. See page 477
5. There are 5280 feet in a mile, 16 ounces in a pound, and 2000 pounds in a ton.

6. An ounce is a unit of weight, whereas a fluid ounce is a unit of capacity (or volume).

7. The metric system uses *base units* such as meter, liter, and gram, and it uses *prefixes* such as milli, centi, deci, deka, hecto, and kilo to create other units such as milliliter and kilogram.

8. One milliliter of water weighs 1 gram and occupies a volume of 1 cubic centimeter.

9. Arrange 8 square-centimeter tiles in any way to form a shape, or draw a shape that encloses 8 squares, as in the scale drawing in Figure 11.5.

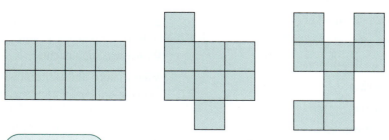

FIGURE 11.5   Three ways to show 8 cm^2

10. Arrange 24 cubic-inch blocks (or cubic-centimeter blocks) in any way to make a solid shape.

11. Only "3 square centimeters" means the same thing as 3 cm². A 3-cm-by-3-cm square has area 9 cm², that is, 9 square centimeters.

12. If you had a 1-yard-by-1-yard-by-1-yard cube that could be filled with concrete, then if you filled that cube up 40 times and poured the contents out into a pile, that big pile of concrete would have a volume of 40 cubic yards.

13. As Table 11.7 shows, if we compare the cereal either by weight or by the number of calories per box, then cereal box 1 is larger (16 ounces versus 11 ounces and 1600 calories versus 1100 calories). On the other hand, if we compare the cereal by volume or by the number of servings per box, then cereal box 2 is larger (4 cups versus 10 cups and 8 servings per box versus 10 servings per box).

Problems for Section 11.1

1. For each of the following metric units, give examples of two objects that you encounter in daily life whose sizes could appropriately be described using that unit:

   a. meter

   b. gram

   c. liter

   d. milliliter

   e. millimeter

   f. kilogram

   g. kilometer

2. For each of the following items, state which U.S. customary units and which metric units in common use would be most appropriate for describing the size of the item. In each case, say briefly why you chose the units.

   a. The volume of water in a full bathtub

   b. The length of a swimming pool

   c. The weight of a slice of bread

   d. The volume of a slice of bread

   e. The weight of a ship

   f. The length of an ant

3. What does it mean to say that a shape has an area of 8 square inches? Discuss your answer in as clear and direct a fashion as you can.

4. Discuss why it is easy to give an incorrect solution to the following area problem and what you must understand about measurement to solve the problem correctly:

   An area problem: Draw a shape that has an area of 2 square inches.

5. Discuss: Why is it not completely correct to describe volume as "length times width times height"? What is a better way to describe what volume means? Give some examples as part of your discussion, using different units of volume.

6. Describe how it could happen that three different animals could each be claimed—rightfully—to be the largest of the three. Discuss the implications of this kind of situation for teaching students about measurement.

7. Pick two ideas or concepts from your reading of this section that you found interesting or important for your future teaching. Write a paragraph about each one, describing what you would want to highlight or emphasize if you were teaching students these ideas or concepts.

8. Visit a store and write down at least 10 different items and their metric measurements that you find on their package labels. For each of the following units, find at least one item that is described using that unit: grams (g), kilograms (kg), milligrams (mg), liters (L), milliliters (mL).

11.2 Length, Area, Volume, and Dimension

All physical objects have several different measurable attributes. For example, what are some measurable attributes of a spool of wire at a hardware store? There is the length of wire that is wound onto the spool. There is the diameter of the wire (the cross-section of the wire is a small circle). There is the diameter of the spool and the length of the spool. There is the weight of the wire, and the weight of the wire and spool. The most common measurable attributes are length, area, volume, and weight. In this section, we will study length, area, volume, and the related concept of dimension. Although these concepts may seem advanced, the foundations for understanding them can be formed early in elementary school. When children make shapes by placing sticks end-to-end, or when they lay tiles or build with blocks, think of their work as exploring length, area, volume, and dimension.

The units for length, area, and volume are related, so it is easy to get confused about which unit to use when. Most real objects have one or more aspects or parts that should be measured by a unit of length, another aspect or part that should be measured by a unit of area, and yet another aspect or part that should be measured by a unit of volume.

We use a unit of *length*, such as inches, centimeters, miles, or kilometers, when we want to answer one of the following kinds of questions:

- How far?

- How long?

- How wide?

For example, "How much moulding (i.e., *how long* a piece of moulding) will I need to go around this ceiling?" The answer, in feet, is the number of foot-long segments that can be put end-to-end to go all around the ceiling. (See Figure 11.6).

length A **length** describes the size of something (or a part of something) that is one-dimensional; the length of that one-dimensional object is how many of a chosen unit of length (such as inches, centimeters, etc.) it takes to cover the object without gaps or overlaps, where it is understood that we may use parts of a unit,

one-dimensional too. Roughly speaking, an object is **one-dimensional** if at each location, there is only one independent direction along which to move within the object. An imaginary creature living in a one-dimensional world could only move forward and backward in that world. Living in a one-dimensional world would be like being stuck in a tunnel. Here are some examples of one-dimensional objects:

        a line segment

        a circle (only the outer part, not the inside)

        the four line segments making a square (not the inside of the square)

        a curved line

        the equator of the earth

        the edge where two walls in a room meet

        an imaginary line drawn from one end of a student's desk to the other end

FIGURE 11.6

A length problem

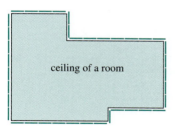

How many of these 1-foot lengths —— does it take to go around the ceiling of this room (i.e., around the outside)?

ceiling of a room

perimeter   In general, the "outer edge" around a (flat) shape is one-dimensional. The **perimeter** of a shape is the distance around a shape; it is the total length of the outer edge around the shape.

We use a unit of *area*, such as square feet or square kilometers, when we want to answer questions like the following:

- How much material (such as paper, fabric, or sheet metal) does it take to make this?
- How much material does it take to cover this?

For example, "How much carpet do I need to cover the floor of this room?" The answer, in square feet, is the number of 1-foot-by-1-foot squares it would take to cover the floor without gaps or overlaps. (See Figure 11.7.)

area
surface area

two-
dimensional

An **area** or a **surface area** describes the size of an object (or a part of an object) that is two-dimensional; the area of that two-dimensional object is how many of a chosen unit of area (such as square inches, square centimeters, etc.) it takes to cover the object without gaps or overlaps, where it is understood that we may use parts of a unit, too. The surface area of a solid shape is the total area of its outside surface. Roughly speaking, an object is **two-dimensional** if, at each location, there are two independent directions along which to move within the object. An imaginary creature living in a two-dimensional world could move only in the forward/backward and right/left directions, as well as in "in between" combinations of these directions (such as diagonally), but *not* up/down. What would it be like to live in a two-dimensional world? The book *Flatland* by A. Abbott [1] is just such an account. Here are some examples of two-dimensional objects:

  a plane

  the *inside* of a circle (this is also called a disk)

  the *inside* of a square

  a piece of paper (only if you think of it as having no thickness)

  the *surface* of a balloon (not counting the inside)

  the *surface* of a box

  some farmland (only counting the surface, not the soil below)

  the surface of the earth

Notice that most of the units that are used to measure area—such as cm^2, m^2, km^2, in^2, ft^2, mi^2—have a superscript "2" in the abbreviation. This 2 reminds us that we are measuring the size of a two-dimensional object.

We use a unit of *volume*, such as cubic yards or cubic centimeters, or even gallons or liters, when we want to answer questions like the following:

- How much of a substance (such as air, water, or wood) is in this object?
- How much of a substance does it take to fill this object?

For example, "How much air is in this room?" The answer, in cubic feet, is the number of 1-foot-by-1-foot-by-1-foot cubes that are needed to fill the room. (See Figure 11.8.)

FIGURE 11.7

An area problem

How many of these ▢ 1-foot-by-1-foot squares does it take to cover the floor of this room?

floor of a room

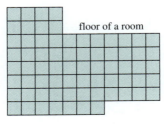

FIGURE 11.8

A volume problem

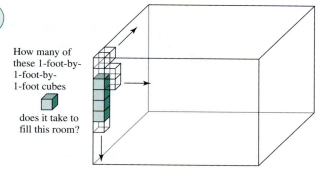

How many of these 1-foot-by-1-foot-by-1-foot cubes

does it take to fill this room?

volume

A **volume** describes the size of an object (or a part of an object) that is three-dimensional; the volume of that three-dimensional object is how many of a chosen unit of volume (such as cubic inches, cubic centimeters, etc.) it takes to fill the object without gaps or overlaps, where it is understood that we may use parts of a unit, too. Roughly speaking, an object is **three-dimensional** if at each location there are three independent directions along which to move within the object. Our world is three-dimensional because we can move in the three independent directions: forward/backward, right/left, and up/down, as well as all "in between" combinations of these directions. Here are some examples of three-dimensional objects:

three-dimensional

the air around us

the inside of a balloon

the inside of a box

the water in a cup

feathers filling a pillow

the compost in a wheelbarrow

the inside of the earth

Notice that many of the units that are used to measure volume—such as cm^3, m^3, in^3, ft^3—have a superscript "3" in their abbreviation. This 3 reminds us that we are measuring the size of a three-dimensional object.

One of the difficulties with understanding whether to measure a length, an area, or a volume is that many objects have parts of different dimensions. For example, consider a balloon. Its outside *surface* is two-dimensional, whereas the air filling it is three-dimensional. Then there is the circumference of the balloon, which is the length of an imaginary one-dimensional circle drawn around the surface of the balloon. The size of each of these parts should be described in a different way: The air in a balloon is described by its volume, the surface of the balloon is described by its area, and the circumference of the balloon is described by its length.

Another example in which the same object has parts of different dimensions is the earth. The *inside* of the earth is three-dimensional, its *surface* is two-dimensional, and the equator is a one-dimensional imaginary circle drawn around the surface of the earth. Once again, the sizes of these different parts of the earth should be described in different ways: by volume, area, and length, respectively.

Another example is farmland. The size of farmland is often measured in acres, which is a measure of area—this takes into account only how big the two-dimensional top surface of the farmland is. This is a practical way to talk about how big farmland is, even though it doesn't take into account the most important part of the land: the three-dimensional soil below the surface.

For a fascinating look at various dimensions, go to *www.pearsonhighered.com/beckmann*.

Class Activity   *Now Turn to Class Activities Manual*

11D Dimension and Size, p. 273

Practice Exercises for Section 11.2

1. **a.** Use centimeter or inch graph paper to make a pattern for a closed box (rectangular prism). The box should have 6 sides, and when you fold the pattern, there should be no overlapping pieces of paper.

   b. How much paper is your box made of? Be sure to use an appropriate unit in your answer.

   c. Describe one-dimensional, two-dimensional, and three-dimensional parts or aspects of your box. In each case, give the size of the part or aspect of the box, using an appropriate unit.

2. Describe one-dimensional, two-dimensional, and three-dimensional parts or aspects of a water tower.

In each case, name an appropriate U.S. customary unit and an appropriate metric unit for measuring or describing the size of that part or aspect of the water tower. What are practical reasons for wanting to know the sizes of these parts or aspects of the water tower?

3. Describe one-dimensional, two-dimensional, and three-dimensional parts or aspects of a store. In each case, name an appropriate U.S. customary unit and an appropriate metric unit for measuring or describing the size of that part or aspect of the store. What are practical reasons for wanting to know the sizes of these parts or aspects of the store?

Answers to Practice Exercises for Section 11.2

1. **a.** Figure 11.9 shows a scaled picture of a pattern for one possible box. Each small square represents a 1-inch-by-1-inch square.

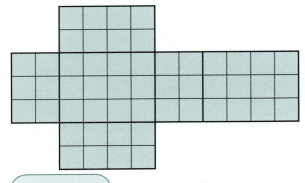

FIGURE 11.9 A pattern for a box

b. The box formed by the pattern in Figure 11.9 (in its actual size) is made of 52 square inches of paper because the pattern that was used to create the box consists of 52 1-inch-by-1-inch squares.

c. The edges of the box are one-dimensional parts of the box. Depending on how the box is oriented, the lengths of these edges are the height, width, and depth of the box. These measure 2 inches, 3 inches, and 4 inches. Another one-dimensional aspect of the box is its *girth* (i.e., the distance around the box). Depending on where you measure, the girth is either 10 inches, 12 inches, or 14 inches.

The full outer surface of the box is a two-dimensional part of the box. Its area is 52 square inches, as discussed in part (b).

The air inside the box is a three-dimensional aspect of the box. If you filled the box with 1-inch-by-1-inch-by-1-inch cubes, there would be 2 layers with 3×4 cubes in each layer. Therefore, according to the meaning of multiplication, there would be $2 \times 3 \times 4 = 24$ cubes in the box. Thus, the volume of the box is 24 cubic inches.

2. The height of the water tower (or an imaginary line running straight up through the middle of the water tower) is a one-dimensional aspect of the water tower. Feet and meters are appropriate units for describing the height of the water tower. Some towns do not want high structures; therefore, the town might want to know the height of a proposed water tower before it decides whether to build it.

The surface of the water tower is a two-dimensional part of the water tower. Square feet and square meters are appropriate units for describing the size of the surface of the water tower. We would need to know the approximate size of the surface of the water tower to know approximately how much paint it would take to cover it.

The water inside a water tower is a three-dimensional part of a water tower. Cubic feet, gallons, liters, and cubic meters are appropriate units for measuring the volume of water in the water tower. The owner of a water tower will certainly want to know how much water the tower can hold.

3. The edge where the store meets the sidewalk is a one-dimensional aspect of the store. Its length is appropriately measured in feet, yards, or meters. This length is the width of the store front, which is one way to measure how much exposure the store has to the public eye.

The surface of the store facing the sidewalk is a two-dimensional part of the store. Its area is appropriately described in square feet, square yards, or square meters. The area of the store front is another way to measure how much exposure the store has to the public eye. (A store front that is narrow but tall might attract as much attention as a wider, lower store front.)

Another important two-dimensional part of a store is its floor. The area of the floor is appropriately measured in square feet, square yards, or square meters, or possibly even in acres if the store is very large. The area of the floor indicates how much space there is on which to place items to sell to customers.

The air inside a store is a three-dimensional aspect of the store. Its volume is appropriately measured in cubic feet, cubic yards, cubic meters, or possibly even in liters or gallons. The builder of a store might want to know the volume of air in the store in order to figure what size air-conditioning units will be needed.

Problems for Section 11.2

1. Describe one-dimensional, two-dimensional, and three-dimensional parts or aspects of a soft drink bottle. In each case, specify an appropriate U.S. customary unit and an appropriate metric unit for measuring or describing the size of that part or aspect of the bottle. What are practical reasons for wanting to know the sizes of these parts or aspects of the soft drink bottle?

2. Describe one-dimensional, two-dimensional, and three-dimensional parts or aspects of a car. In each case, specify an appropriate U.S. customary unit and an appropriate metric unit for measuring or describing the size of that part or aspect of the car. What are practical reasons for wanting to know the sizes of these parts or aspects of the car?

3. Drawing on your reading from this section, describe how it could happen that 3 different cathedrals could each claim—rightfully—to be the largest of the 3. Discuss the implications of this kind of situation for teaching students about measurement.

4. Describe one-dimensional, two-dimensional, and three-dimensional parts or aspects of the blocks in Figure 11.10. In each case, compare the sizes of the 3 blocks, using an appropriate unit. Use this unit to show that each block can be considered biggest of all 3.

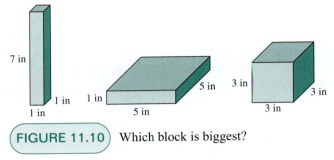

FIGURE 11.10   Which block is biggest?

5. The Lazy Daze Pool Club and the Slumber-N-Sunshine Pool Club have a friendly rivalry going. Each club claims to have the bigger swimming pool. Make up realistic sizes for the clubs' swimming pools so that each club has a legitimate basis for saying that it has the larger pool. Explain clearly why each club can say its pool is biggest. Be sure to use appropriate units. Considering the function of a pool, what is a good way to compare sizes of pools? Explain.

6. Suppose there are 2 rectangular pools: One is 30 feet wide, 60 feet long, and 5 feet deep throughout, and the other is 40 feet wide, 50 feet long, and 4 feet deep throughout. Show that each pool can be considered "biggest" by comparing the sizes of the pools in two meaningful ways other than by comparing one-dimensional aspects of the pools.

7. Explain why either of the 2 rectangles in Figure 11.11 can be considered the larger of the 2.

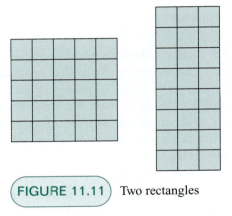

FIGURE 11.11   Two rectangles

11.3 Error and Precision in Measurements

Error and precision are natural parts of all actual measurements of physical quantities. It is important to distinguish between *theoretical* and *actual* measurements. When we say that a room that is 4 meters wide and 5 meters long has an area of 20 square meters, we are considering a theoretical room. On the other hand, if we are considering an actual room, we can never say that it is exactly 4 meters wide and 5 meters long. When measuring actual physical quantities, it is impossible to avoid a certain amount of error and uncertainty. In this section, we will see that the way we report a measurement indicates the accuracy of that measurement. Similarly, our interpretation of a measurement depends on the digits reported in the measurement.

Any reported measurement of an actual physical quantity is necessarily only approximate—you can never say that a measurement of a real object is exact. For example, if you use a ruler to measure the width of a piece of paper, you may report that the paper is $8\frac{1}{2}$ inches wide. But the paper isn't *exactly* $8\frac{1}{2}$ inches wide, it is just that the mark on your ruler that is closest to the edge of the paper is at $8\frac{1}{2}$ inches.

Interpreting Reported Measurements

Because measurements are never exact, we have conventions for reporting measurements. The way that a measurement is reported reflects its precision. For example, suppose that the distance between two

Interpreting a
reported distance
of 1200 miles

cities is reported as 1200 miles. Since there are zeros in the tens and ones places, we assume that this measurement is *rounded to the nearest hundred* (unless we are told otherwise). In other words, the actual distance between the two cities is between 1100 miles and 1300 miles, but closer to 1200 miles than to either 1100 miles or 1300 miles. Thus, the actual distance between the two cities could be anywhere between 1150 and 1250 miles, as indicated in Figure 11.12.

On the other hand, if the distance between two cities is reported as 1230 miles, then we assume (unless we are told otherwise) that this measurement is *rounded to the nearest ten*. In other words, the actual distance between the cities is between 1220 miles and 1240 miles, but closer to 1230 miles than to either 1220 miles or 1240 miles. In this case, the actual distance between the two cities could be anywhere between 1225 and 1235 miles, as indicated in Figure 11.13. Notice that in this case, we are getting a much smaller range of numbers that the actual distance could be than in the previous example. Here it's a range of 10 miles (1225 miles to 1235 miles), whereas in the previous case it's a range of 100 miles (1150 miles to 1250 miles).

Similarly, if your weight on a digital scale is reported as 130.4 pounds, then (assuming the scale's report is accurate) this is your actual weight, *rounded to the nearest tenth*. In other words, your actual weight is between 130.3 and 130.5 pounds, but closer to 130.4 pounds than to either 130.3 or 130.5 pounds. In this case, your actual weight is between 130.35 and 130.45 pounds.

Notice that there is a difference in reporting that the weight of an object is 130.0 pounds and reporting the weight as 130 pounds. When the weight is reported as 130.0 pounds, we assume that this weight is *rounded to the nearest tenth* of a pound. In this case, the actual weight is between 129.95 and 130.05 pounds. On the other hand, when the weight of an object is reported as 130 pounds, we assume this weight is *rounded to the nearest ten* pounds. In this case, we know only that the actual weight is between 125 and 135 pounds, which allows for a much wider range of actual weights than the range of 129.95 to 130.05 pounds (a 10-pound range versus a 0.1-pound range). So, when a weight is reported as 130.0 pounds, the weight is known with greater accuracy than when the weight is reported as 130 pounds.

What do we do when we know a distance more accurately than the way we write it would suggest? For example, suppose we know that a certain distance is 130,000 km, but that this is rounded to the nearest *hundred* kilometers, not to the nearest *ten-thousand* kilometers, as the way the number is written would suggest. In this case, we can report the distance as "130,000 km to 4 **significant digits.**"

**significant
digits**

Working with Measurements

In the next section and the next chapters we will calculate various lengths, areas, and volumes of objects by using lengths associated with these objects. In some cases, the lengths are taken to be *given* as

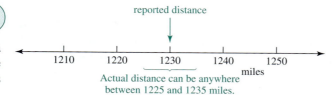

Interpreting a
reported distance
of 1230 miles

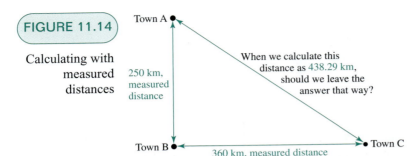

FIGURE 11.14

Calculating with measured distances

opposed to actually *measured*, and the area and volume calculations are then purely theoretical. For example, we will be able to calculate the area of a circle that has radius 6.25 inches. In this case, we are imagining a circle that has radius 6.25 inches, and we are calculating the exact area that such an imagined circle would have.

In other cases, we will work with actual or hypothetical measured distances to calculate other distances, or to calculate areas or volumes. In these cases, we should assume that the distances we are working with have been rounded, as described previously. Therefore, when we determine our final answer to a calculation involving measured distances, we should also round that answer, because the answer cannot be known more precisely than the initial distances with which we began the calculations. For example, we might calculate the distance from town A to town C, given that the distance from town A to town B is 250 km, and the distance from town B to town C is 360 km, as shown in Figure 11.14. By calculating, we determine that the distance from town A to town C is 438.29 km. But we should not leave our answer like that, because if we do, we are communicating that we know the distance from town A to town C rounded to the nearest *hundredth* of a kilometer. Because we know only the distances from town A to town B and town B to town C rounded to the nearest *ten* kilometers, we cannot possibly know the resulting distance from town A to town C with greater precision. Instead, we should report the distance from town A to town C as 440 km, which fits with the rounding of the distances we were given.

Keep the following point in mind when calculating with actual or hypothetical measurements: Even though you may be rounding your final answer, wait until you are done with all your calculations before you round. Waiting to the end to round will give you the most precise answer that fits with the measurements you started with.

Class Activity   *Now Turn to Class Activities Manual*

11E Reporting and Interpreting Measurements, p. 274

Practice Exercises for Section 11.3

1. What is the difference between reporting that an object weighs 2 pounds and reporting that it weighs 2.0 pounds?

2. If the distance between two cities is reported as 2500 miles, does that mean that the distance is exactly 2500 miles? If not, what can you say about the exact distance?

Answers to Practice Exercises for Section 11.3

1. If an object is reported as weighing 2 pounds, then we assume that the weight has been rounded to the nearest whole pound. Therefore, the actual weight is between 1 and 3 pounds, but closer to 2 pounds than to either 1 pound or 3 pounds. The actual weight must be between 1.5 pounds and 2.5 pounds. On the other hand, if an object is reported as weighing 2.0 pounds, then we assume that the weight has been rounded to the nearest *tenth* of a pound. Therefore, the actual weight is between 1.9 pounds and 2.1 pounds, but closer to 2.0 pounds than to either 1.9 pounds or 2.1 pounds. The actual weight must be between 1.95 pounds and 2.05 pounds. So, when the weight is reported as 2.0 pounds, the weight is known with greater accuracy than when it is reported as 2 pounds.

2. If the distance between two cities is reported as 2500 miles, then we assume that this number is the actual distance, rounded to the nearest hundred. Therefore, the actual distance is between 2400 miles and 2600 miles, but closer to 2500 miles than to either 2400 miles or 2600 miles. The actual distance between the two cities could be anywhere between 2450 miles and 2550 miles.

Problems for Section 11.3

1. One source says that the average distance from the earth to the moon is 384,467 kilometers. Another source says that the average distance from the earth to the moon is 384,000 kilometers. Can both of these descriptions be correct, or must at least one of them be wrong? Explain.

2. If an object is described as weighing 6.20 grams, then is this the exact weight of the object? If not, what can you say about the weight of the object? Explain your answer in detail.

3. Tyra is calculating the distance from town A to town C. Tyra is given that the distance from town A to town B is 120 miles, that the distance from town B to town C is 230 miles, that town B is due south of town A, and that town C is due east of town B. Tyra calculates that the distance from town A to town C is 259.4224 miles. Should Tyra leave her answer like that? Why or why not? If not, what answer should Tyra give? Explain. (You may assume that Tyra has calculated correctly.)

4. John has a paper square that he believes is 100 cm wide and 100 cm long, but in reality, the square is actually 1% longer and 1% wider than he believes. In other words, John's square is actually 101 cm wide and 101 cm long.

a. Since John believes the square is 100 cm by 100 cm, he calculates that the area of his square is 10,000 cm^2. What percent greater is the actual area of the square than John's calculated area of the square? Is the actual area of John's square 1% greater than John's calculated area, or is it larger by a different percent?

b. Draw a picture (which need not be to scale) to show why your answer in part (a) is not surprising.

c. Answer the following questions based on your answers to parts (a) and (b): In general, if you want to know the area of a square to within 1% of its actual area, will it be good enough to know the lengths of the sides of the square to within 1% of their actual lengths? If not, will you need to know the lengths more accurately or less accurately?

5. Sally has a Plexiglas cube that she believes is 100 cm wide, 100 cm long, and 100 cm tall, but in reality, the cube is actually 1% wider, 1% longer, and 1% taller than she believes. In other words, Sally's cube is actually 101 cm wide, 101 cm deep, and 101 cm tall.

a. Since Sally believes the cube is 100 cm by 100 cm by 100 cm, she calculates that the volume of her cube is 1,000,000 cm^3. What percent greater is the actual volume of the cube than Sally's calculation of the volume of the cube? Is the actual volume of Sally's cube 1% greater than her calculated volume, or is it larger by some different percent?

b. Answer the following questions based on your answer to part (a): In general, if you want to know the volume of a cube to within 1% of its actual volume, will it be good enough to know the lengths of the sides of the cube to within 1% of their actual lengths? If not, will you need to know the lengths more accurately or less accurately?

11.4 Converting from One Unit of Measurement to Another

As we've seen, we can use different units to measure the same quantity. For example, we can measure a length in miles, yards, feet, inches, kilometers, meters, centimeters, or millimeters. If we know a length in terms of one unit, how can we describe it in terms of another unit? This kind of problem is a conversion problem, which is the topic of this section.

To convert a measurement in one unit to another unit, we first need to know the relationship between the two units. For example, if we want to convert the weight of a bag of oranges given in kilograms into pounds, then we need to know the relationship between kilograms and pounds. Referring back to Table 11.5 in Section 11.1, we see that 1 kg = 2.2 1b.

Once we know how the units are related, we either multiply or divide to convert from one unit to another. How do we know which operation to use? Before you read on, please do the next Class Activity.

> ### Class Activity   *Now Turn to Class Activities Manual*
>
> **11F** Conversions: When Do We Multiply? When Do We Divide?, p. 274

How do we decide whether to multiply or divide to convert from one unit to another? A good way to decide is to think about the meanings of multiplication and division. For example, let's say a bag of oranges weighs 2.5 kg. What is the weight of the bag of oranges in pounds? If 1 kilogram is 2.2 pounds, then we can think of 2.5 kilograms as 2.5 groups with 2.2 pounds in each group, as illustrated in Figure 11.15. Thus, according to the meaning of multiplication, 2.5 kilograms is 2.5 × 2.2 = 5.5 pounds.

Here's another example: How many gallons are there in a 25-liter container? Referring back to Table 11.5, we see that 1 gal = 3.79 L. That is, every group of 3.79 liters is equal to 1 gallon. Therefore, to find how many gallons are in 25 liters, we need to find out how many groups of 3.79 liters are in 25 liters. This is a division problem, using the "how many groups?" interpretation of division. So there are 25 ÷ 3.79 = 6.6 gallons in 25 liters.

FIGURE 11.15

Converting 2.5 kilograms to pounds

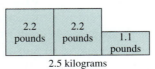

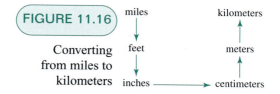

FIGURE 11.16

Converting from miles to kilometers

In some cases, the tables in Section 11.1 do not tell us directly how two units are related. For example, Table 11.5 does not tell us how kilometers relate to miles. To convert 3 miles to kilometers, we must work with the length relationships that we do know from the tables in Section 11.1. In this case, we use the relationship 1 in = 2.54 cm, the relationships among inches, feet, and miles, and the relationships among centimeters, meters, and kilometers to convert 3 miles to kilometers. We go through the following chain of conversions:

$$\text{miles} \to \text{feet} \to \text{inches} \to \text{centimeters} \to \text{meters} \to \text{kilometers}$$

Think of this as "ratcheting down," "going across," and then "ratcheting back up," as indicated in Figure 11.16.

Now let's convert 3 miles to kilometers. First, convert miles to feet: Since 1 mile is 5280 feet, 3 miles is 3 times as many feet (3 groups of 5280 feet), so

$$3 \text{ miles} = 3 \times 5280 \text{ feet} = 15{,}840 \text{ feet}$$

Next, convert feet to inches: Since 1 foot is 12 inches, 15,840 feet are 15,840 times as many inches (15,840 groups of 12), so

$$15{,}840 \text{ feet} = 15{,}840 \times 12 \text{ inches} = 190{,}080 \text{ inches}$$

Now convert inches to centimeters: Since 1 inch is 2.54 centimeters, 190,080 inches are 190,080 times as many centimeters (190,080 groups of 2.54 centimeters), so

$$190{,}080 \text{ inches} = 190{,}080 \times 2.54 \text{ cm} = 482{,}803.2 \text{ cm}$$

Next, convert centimeters to meters: Since 100 centimeters make 1 meter, we need to find how many groups of 100 centimeters are in 482,803.2 centimeters. According to the "how many groups?" interpretation of division, we can determine the number of meters by dividing:

$$482{,}803.2 \text{ cm} = (482{,}803.2 \div 100) \text{ m} = 4{,}828.032 \text{ m}$$

Finally, convert meters to kilometers: Since 1000 meters make 1 kilometer, we need to find how many groups of 1000 meters are in 4,828.032 meters. According to the "how many groups?" interpretation of division, we can determine the number of kilometers by dividing:

$$4{,}828.032 \text{ m} = (4{,}828.032 \div 1000) \text{ km} = 4.828032 \text{ km}$$

Thus, 3 miles is exactly 4.828032 kilometers, which is just a little less than 5 kilometers. If you ran 3 miles and your friend ran 5 kilometers, then your friend ran farther, but not by much.

It is easy to carry out the previous calculations for converting 3 miles to kilometers on a calculator. At each step, you either multiply or divide the number currently in the calculator's display. So after each step, be sure to leave the result of your calculation in your calculator, and use this number for the next calculation. For example, when you get 15,840 feet from multiplying 3 times 5280 feet, just leave the 15,840 in your calculator for the next calculation, which is multiplying by 12. That way you don't risk

Class Activity *Now Turn to Class Activities Manual*

11G Conversion Problems, p. 275

TABLE 11.8 A calculator flowchart for converting 3 miles to kilometers

enter 3,
times 5280 → 15840,
times 12 → 190080,
times 2.54 → 482803.2,
divided by 100 → 4828.032,
divided by 1000 → 4.828032.

making an error by punching the 15,840 back in. Table 11.8 shows a flowchart for solving the problem of converting 3 miles to kilometers by using a calculator. The arrows in the table point to results calculated by the calculator.

Dimensional Analysis

Class Activity   *Now Turn to Class Activities Manual*

11H Using Dimensional Analysis to Convert Measurements, p. 276

Many people use **dimensional analysis** to convert a measurement from one unit to another. When we use this method, we repeatedly multiply by the number 1 expressed as a fraction that relates two different units. The resulting calculations are identical to those done to convert units by using multiplication and division, as shown previously.

For example, suppose that a German car is 5.1 meters long. How long is the car in feet? To solve this problem, we will use the information provided in the tables of Section 11.1 to carry out the following chain of conversions:

$$\text{meters} \rightarrow \text{centimeters} \rightarrow \text{inches} \rightarrow \text{feet}$$

The following equation uses dimensional analysis to convert 5.1 meters to feet:

$$5.1 \text{ m} = 5.1 \text{ m} \times \frac{100 \text{ cm}}{1 \text{ m}} \times \frac{1 \text{ in.}}{2.54 \text{ cm}} \times \frac{1 \text{ ft}}{12 \text{ in.}} = 16.7 \text{ ft} \tag{11.1}$$

So, the car is 16.7 feet long.

How do we choose the fractions to use in dimensional analysis? Each fraction that we multiply by must be equal to 1. That way, when we multiply by the fraction, we do not change the original amount; we just express the amount in different units. We used the following fractions in Equation 11.1:

$$\frac{100 \text{ cm}}{1 \text{ m}} = 1, \quad \frac{1 \text{ in.}}{2.54 \text{ cm}} = 1, \quad \frac{1 \text{ ft}}{12 \text{ in.}} = 1$$

We also choose the fractions in dimensional analysis so that all the units will "cancel" except the unit in which we want our answer expressed. In Equation 11.1, the first fraction cancels the meters, but leaves centimeters; the second fraction cancels the centimeters, but leaves inches; and the third fraction cancels the inches, leaving feet, which is how we want our answer expressed. Thus,

$$5.1 \text{ m} = 5.1 \, \cancel{\text{m}} \times \frac{100 \, \cancel{\text{cm}}}{1 \, \cancel{\text{m}}} \times \frac{1 \, \cancel{\text{in.}}}{2.54 \, \cancel{\text{cm}}} \times \frac{1 \text{ ft}}{12 \, \cancel{\text{in.}}} = 16.7 \text{ ft}$$

Area and Volume Conversions

> **Class Activity**   *Now Turn to Class Activities Manual*
>
> **11I** Area and Volume Conversions, p. 277

Once you know how to convert from one unit of length to another, you can then convert between the related units of area and volume. However, to calculate area and volume conversions correctly, it's important to remember the meanings of the units for area and volume (such as square feet, cubic inches, square meters, and so on). Otherwise, it is easy to make the following common kind of mistake:

"Since 1 yard is 3 feet, 1 square yard is 3 square feet."

Figure 11.17 will help you see why this is a mistake. One square yard is the area of a square that is 1 yard wide and 1 yard long. Such a square is 3 feet wide and 3 feet long. Figure 11.17 shows that such a square has the area $3 \times 3 = 9$ square feet, *not* 3 square feet.

Let's say you are living in an apartment that has a floor area of 800 square feet, and you want to tell your Italian pen pal how big that is in square meters. There are several ways you might approach such an area conversion problem. One good way is this:

First determine what 1 foot (*linear*) is in terms of meters. Then use that information to determine what 1 *square* foot is in terms of *square* meters.

Here's how to find 1 foot in terms of meters:

$$1 \text{ ft} = 12 \text{ in.}$$
$$12 \text{ in.} = 12 \times 2.54 \text{ cm} = 30.48 \text{ cm}$$

This is because each inch is 2.54 cm and there are 12 inches.

$$30.48 \text{ cm} = (30.48 \div 100) \text{ m} = 0.3048 \text{ m}$$

This is because 1 m = 100 cm, so we must find how many 100s are in 30.48 to find how many meters are in 30.48 cm.

So,

$$1 \text{ ft} = 0.3048 \text{ m}$$

Remember that 1 square foot is the area of a square that is 1 foot wide and 1 foot long. According to the previous calculation, such a square is also 0.3048 meters wide and 0.3048 meters long. Therefore, by the "length times width" formula for areas of rectangles,

$$1 \text{ ft}^2 = (0.3048 \times 0.3048) \text{ m}^2 = 0.0929 \text{ m}^2$$

This means that 800 square feet is $800 \times 0.0929 = 74.32$ square meters, so the apartment is about 74 square meters.

FIGURE 11.17

One square yard is 9 square feet, not 3 square feet.

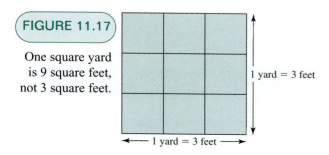

1 yard = 3 feet

1 yard = 3 feet

Volume conversions work similarly. Suppose we want to find 13 cubic feet in terms of cubic meters, for example. We already calculated that

$$1 \text{ ft} = 0.3048 \text{ m}$$

Now 1 cubic foot is the volume of a cube that is 1 foot high, 1 foot deep, and 1 foot wide. Such a cube is also 0.3048 meters high, 0.3048 meters deep, and 0.3048 meters wide. So, by the "height times width times length" formula, for volumes of rectangular prisms (boxes),

$$1 \text{ ft}^3 = 0.3048 \times 0.3048 \times 0.3048 \text{ m}^3 = 0.0283 \text{ m}^3$$

Therefore, 13 cubic feet is 13 times as much as 0.0283 m^3:

$$13 \text{ ft}^3 = 13 \times 0.0283 \text{ m}^3 = 0.37 \text{ m}^3$$

Class Activity   *Now Turn to Class Activities Manual*

11J Area and Volume Conversions: Which Are Correct and Which Are Not?, p. 278

Approximate Conversions and Checking Your Work

So far this discussion has focused on exact (or fairly exact) conversions. However, in many practical situations, you really don't need an exact answer—an estimate will do. Being able to make a quick estimate also gives you a way to check your work for an exact conversion: If your answer is far from your estimate, you've probably made a mistake. If it's not too far off, there's a good chance your work is correct.

The following approximate relationships will help you estimate conversions from one unit to another:

- An inch is about $2\frac{1}{2}$ centimeters, so 2 inches is about 5 centimeters.
- A meter is a little more than a yard.
- A kilometer is about $\frac{6}{10}$ of a mile (so a little over half a mile).
- A liter is a little more than a quart.
- A kilogram is a bit more than 2 pounds.

For example, you can mentally calculate the following quickly:

- A 5-kilometer run is about 3 miles.
- Gas that costs $1.00 per liter in Europe costs about $1.00 per quart, in other words, $4.00 per gallon.
- A 5-pound bag of oranges is about $2\frac{1}{2}$ kilograms (a little less).

Class Activity   *Now Turn to Class Activities Manual*

11K Problem Solving with Conversions, p. 279

Practice Exercises for Section 11.4

1. A class needs a 175-inch-long piece of rope for a project. How long is the rope in yards?

   a. Use multiplication or division or both to solve the rope problem. Explain your solution.

   b. Describe a number of different correct ways to write the answer to the rope problem. Explain briefly why these different ways of writing the answer mean the same thing.

2. The children in Mrs. Watson's class made chains of small paper dolls, as pictured in Figure 11.18. A chain of 5 dolls is 1 foot long. How long would the following chains of dolls be? In each case, give your answer in either feet or miles, depending on which answer is easiest to understand.

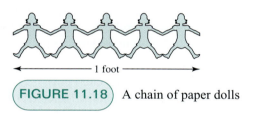

(FIGURE 11.18) A chain of paper dolls

   a. 100 dolls

   b. 1000 dolls

   c. 10,000 dolls

   d. 100,000 dolls

   e. one million dolls

   f. one billion dolls

3. Solve the following conversion problems, using the basic fact 1 inch = 2.54 cm in each case:

   a. A track is 100 meters long. How long is it in feet?

   b. If the speed limit is 70 miles per hour, what is it in kilometers per hour?

   c. Some farmland covers 2.4 square kilometers. How many square miles is it?

   d. Convert the volume of a compost pile, 1 cubic yard, to cubic meters.

   e. A man is 1.88 meters tall. How tall is he in feet?

   f. How many miles is a 10-kilometer race?

4. One mile is 1760 yards. Does this mean that 1 square mile is 1760 square yards? If not, how many square yards are in a square mile? Explain carefully.

5. How many cubic feet of mulch will you need to cover a garden that is 10 feet wide and 4 yards long with 2 inches of mulch? Can you find the amount of mulch you need by multiplying $2 \times (10 \times 4)$?

6. According to a research study, the average American consumes 63 doughnuts per year. Let's say that each doughnut is about an inch thick. If all the doughnuts consumed in the United States in one year were placed in one tall stack, how tall would this stack be? Would this stack reach to the moon? The moon is about 240 thousand miles away from the earth, and the population of the United States is about 300 million.

7. The capacity of a cube that is 10 cm wide, 10 cm deep, and 10 cm tall is 1 liter. One gallon is 0.134 cubic feet. One quart is one-quarter of a gallon. Which is more: one quart or one liter? Use the basic fact 1 inch = 2.54 cm to figure this out.

8. One acre is 43,560 square feet. The area of land is often measured in acres.

   a. What is a square mile in acres?

   b. What is a square kilometer in acres?

9. Suppose you want to cover a football field with artificial turf. You'll need to cover an area that's bigger than the actual field—let's say you'll cover a rectangle that's 60 yards wide and 130 yards long. How many square feet (not square yards) of artificial turf will it take? The artificial turf comes in a roll that is 12 feet wide. Let's say the turf costs $8 per linear foot cut from the roll. (So, if you cut off 10 feet from the roll, it would cost $80 and you'd have a rectangular piece of artificial turf that's 10 feet long and 12 feet wide.) How much will the turf you need cost? (It has been said that preparing the field costs even more than the artificial turf itself.)

Answers to Practice Exercises for Section 11.4

1. **a.** One method for determining the length of the rope in yards is to first figure out how many inches are in a yard. One foot is 12 inches. One yard is 3 feet. Since each foot is 12 inches, 3 feet is 3 groups of 12 inches, which is $3 \times 12 = 36$ inches. To find the length of the 175-inch-long piece of rope in yards, we must figure out how many groups of 36 inches are in 175 inches. This is a "how many groups?" division problem:

$$175 \div 36 = 4, \text{ remainder } 31$$

so there are 4 groups of 36 inches in 175 inches with 31 inches left over. Therefore, 175 inches is 4 yards and 31 inches.

b. We found that we can describe the length of the rope as 4 yards, 31 inches. The 31 inches can also be described as 2 feet, 7 inches, because there are 12 inches in each foot and

$$31 \div 12 = 2 \text{ remainder } 7$$

So we can also say that the rope is 4 yards, 2 feet, and 7 inches long. Instead of using a remainder in solving the division problem in part (a), we can give a mixed number or decimal answer to the division problem. So we can also say that the length of the rope is 4.86 yards or $4\frac{31}{36}$ yards, although these answers would probably not be very useful in practice. Similarly, we could give the answer to the division problem $31 \div 12$ as a mixed number or as a decimal and say that the rope is 4 yards and $2\frac{7}{12}$ feet long or 4 yards and 2.6 feet long. Although correct, these answers would probably not be useful in practice.

2. **a.** Because 5 dolls are 1 foot long, 10 dolls are 2 feet long. Thus, 100 dolls are 10 times as long, namely, 20 feet long.

b. 1000 dolls are 10 times as long as 100 dolls. Because 100 dolls are 20 feet long, 1000 dolls are 200 feet long.

c. 10,000 dolls are 10 times as long as 1000 dolls. Because 1000 dolls are 200 feet long, 10,000 dolls are 2000 feet long.

d. 100,000 dolls are 10 times as long as 10,000 dolls. Because 10,000 dolls are 2000 feet long, 100,000 dolls are 20,000 feet long. One mile is 5280 feet, so 20,000 feet is $20,000 \div 5280 = 3.8$ miles. So 100,000 dolls are nearly 4 miles long.

e. One million dolls are 10 times as long as 100,000 dolls. Because 100,000 dolls are 3.8 miles long, one million dolls are 38 miles long.

f. One billion dolls are 1000 times as long as one million dolls. Because one million dolls are 38 miles long, one billion dolls are 38,000 miles long. The circumference of the earth is about 24,000 miles, so one billion dolls would wrap around the earth more than $1\frac{1}{2}$ times!

3. **a.** $1 \text{ m} = 100 \text{ cm} = (100 \div 2.54) \text{ in.} = 39.37 \text{ in.} = (39.37 \div 12) \text{ ft} = 3.28 \text{ ft}$. So $100 \text{ m} = 328 \text{ ft}$.

b. $1 \text{ mile} = 5280 \text{ ft} = 63,360 \text{ in.} = 160934.4 \text{ cm} = 1609.344 \text{ m} = 1.609 \text{ km}$. So 70 miles per hour is 70×1.609 km per hour, which is 113 km per hour.

c. $1 \text{ km} = 1000 \text{ m} = 100,000 \text{ cm} = 39370.08 \text{ in.} = 3280.84 \text{ ft} = 0.621 \text{ miles}$. Therefore, 1 square kilometer is $0.621 \times 0.621 = 0.39$ square miles, and 2.4 square kilometers are 2.4×0.39 square miles, which is 0.9 square miles.

d. $1 \text{ yd} = 3 \text{ ft} = 36 \text{ in.} = 91.44 \text{ cm} = 0.9144 \text{ m}$. So, 1 cubic yard is $0.9144 \times 0.9144 \times 0.9144$ cubic meters $= 0.76$ cubic meters.

e. $1 \text{ m} = 100 \text{ cm} = 39.37 \text{ in.} = 3.28 \text{ ft}$. So $1.88 \text{ m} = 6.17 \text{ ft}$, which is 6 ft, 2 in.

Why is 0.17 ft equal to 2 in.? This is *not* explained by rounding the 0.17 to 2. Instead, since $1 \text{ ft} = 12 \text{ in.}$, $0.17 \text{ ft} = 17 \times 12 \text{ in.} = 2.04 \text{ in.}$, which rounds to 2 in.

f. $1 \text{ km} = 1000 \text{ m} = 100,000 \text{ cm} = (100,000 \div 2.54) \text{ in.} = 39,379 \text{ in.} = (39,379 \div 12) \text{ ft} = 3280.8 \text{ ft} = 3280.8 \div 5280 \text{ miles} = 0.62 \text{ miles}$.

4. One square mile is the area of a square that is 1 mile long and 1 mile wide. Such a square is 1760 yards long and 1760 yards wide, so we can mentally decompose this square into 1760 rows, each having 1760 squares that are 1 yard by 1 yard. Therefore, according to the meaning of multiplication, there are

$$1760 \times 1760 = 3,097,600$$

1-yard-by-1-yard squares in a square mile. This means that 1 square mile is 3,097,600 square yards and not 1760 square yards.

5. We know that 1 yard = 3 feet and 1 foot = 12 inches. Thus, 4 yards = 12 feet, and 2 inches = $\frac{1}{6}$ feet. So, you can think of the mulch as forming a box shape that is $\frac{1}{6}$ feet high, 10 feet deep (or wide), and 12 feet wide (or deep). This box has volume

$$\frac{1}{6} \times (10 \times 12) \text{ cubic feet} = 20 \text{ cubic feet}$$

So, you'll need 20 cubic feet of mulch.

You can't find the amount of mulch you need by multiplying $2 \times 10 \times 4$ because each number refers to a different unit of length.

6. The stack of doughnuts would be nearly 300,000 miles tall, which is more than the distance to the moon.

7. 1 liter = 1000 cubic centimeters. Convert 1 cm to feet: 1 cm = 3937 in. = 0.0328 ft. So 1 cubic centimeter is 0.000035314 cubic feet, and 1 liter = 0.035314 cubic feet. Since each gallon is 0.134 cubic feet, 1 liter = (0.035314 ÷ 0.134) gallons = 0.2635 gallons, which is a little more than $\frac{1}{4}$ gallon. So a liter is a little more than a quart.

8. a. 1 mile = 5280 feet, so 1 square mile = 5280 × 5280 square feet = 27,878,400 square feet. Since each acre is 43,560 square feet, we want to know how many 43,560 are in 27,878,400. This is a division problem. 27,878,400 ÷ 43,560 = 640. So there are 640 acres in a square mile.

   b. 1 km = 1000 m = 100,000 cm = (100,000 ÷ 2.54) in. = 39,370 in. = (39,370 ÷ 12) ft = 3280.8399 ft. So 1 km^2 = 3280.8399^2 ft^2 = 10,763,910.42 ft^2 = (10,763,910.42 ÷ 43,560) acres = 247 acres.

9. You'll need 60 × 130 = 7,800 square yards of turf. Each square yard is 3 × 3 = 9 square feet, so you'll need 7,800 × 9 = 70,200 square feet of turf. (Or, notice that 60 yards = 180 feet and 130 yards = 390 feet, so you'll need 180 × 390 = 70,200 square feet.)

$8 buys you 12 square feet of turf, so it will cost at least

$$8 \times (70,200 \div 12) = \$46,800$$

for the turf. But if you lay the turf in strips across the width of the field, then you will need 33 strips that are 60 yards = 180 feet long, which will cost $47,520. (The 33 strips come from the fact that each strip is 4 yards wide and 130 ÷ 4 = 32.5.)

Problems for Section 11.4

1. A recipe calls for 4 ounces of chocolate. If you make 15 batches of the recipe (for a large party), then how many pounds of chocolate will you need?

   a. Use multiplication or division or both to solve the chocolate problem. Explain why your method of solution makes sense in a way that children could understand.

   b. Describe a number of different correct ways to write the answer to the chocolate problem. Explain briefly why these different ways of writing the answer means the same thing.

2. A class needs 27 pieces of ribbon, each piece 2 feet, 3 inches long. How many yards of ribbon does the class need?

   a. Use multiplication or division or both to solve the ribbon problem. Explain why your method of solution makes sense in a way that children could understand.

   b. Describe a number of different correct ways to write the answer to the ribbon problem. Explain briefly why these different ways of writing the answer means the same thing.

3. To convert 24 yards to feet, should you multiply by 3 or divide by 3? Explain your answer in two different ways that a fifth-grader could understand. (Do not use dimensional analysis.)

4. To convert 2000 kilometers to meters, should you multiply by 1000 or divide by 1000? Explain your answer in two different ways that a fifth-grader could understand. (Do not use dimensional analysis.)

5. Shauntay used identical plastic bears to measure the length of a rope and found that the rope was 36 bears long. Next, Shauntay will measure the length of the rope using identical plastic giraffes. Shauntay found that 4 bears are as long as 3 giraffes. How many giraffes long will the rope be? Explain your reasoning clearly and in detail.

6. The following problem could appear in a mathematics textbook for use in elementary schools:

    A rectangle is 9 feet long and 6 feet wide. Which is greater, the perimeter of the rectangle or the area of the rectangle?

   a. Calculate the perimeter and the area of the rectangle in feet and square feet, respectively. Explain briefly.

   b. Calculate the perimeter and the area of the rectangle in yards and square yards, respectively. Explain briefly.

   c. Criticize the given textbook math problem on mathematical grounds. Why is the problem not a good problem?

7. a. A car is 16 feet, 3 inches long. How long is it in meters? Use the fact that 1 inch = 2.54 centimeters to determine your answer. Explain your reasoning briefly.

   b. A car is 415 centimeters long. How long is it in feet and inches? Use the fact that 1 inch = 2.54 centimeters to determine your answer. Explain your reasoning briefly.

8. The distance between two cities is described as 260 kilometers. What is this distance in miles? Make an estimate first, explaining your reasoning briefly. Then calculate your answer using the fact that 1 in. = 2.54 cm. Remember to round your answer appropriately: The way you write your answer should reflect its accuracy. Your answer should not be reported more accurately than your starting data.

9. In Germany, people often drive 130 kilometers per hour on the Autobahn (highway). How fast are they going in miles per hour? Make an estimate first, explaining your reasoning briefly. Then calculate your answer using the fact that 1 in. = 2.54 cm.

10. One foot is 12 inches. Does this mean that 1 square foot is 12 square inches? Draw a picture showing how many square inches are in a square foot. Use the meaning of multiplication to explain why you can calculate the area of 1 square foot in terms of square inches by multiplying.

11. A room has a floor area of 48 square yards. What is the area of the room in square feet? Solve this problem in two different ways, each time referring to the meaning of multiplication.

12. One kilometer is 1000 meters. Does this mean that 1 square kilometer is 1000 square meters? If not, what is 1 square kilometer in terms of square meters? Explain your answer in detail, referring to the meaning of multiplication.

13. One foot is 12 inches. Does this mean that one cubic foot is 12 cubic inches? Describe how to use the meaning of multiplication to determine what 1 cubic foot is in terms of cubic inches.

14. How much mulch will you need to cover a rectangular garden that is 20 feet by 30 feet with a 3-inch layer of mulch? Explain.

15. A classroom has a floor area of 600 square feet. What is the floor area of the classroom in square meters? Describe three different correct ways to solve this area problem. Each way of solving should use the fact that 1 in. = 2.54 cm. Explain your reasoning.

16. A house has a floor area of 800 square meters. What is the floor area of this house in square feet? Describe three different correct ways to solve this area problem. Each way of solving should use the fact that 1 in. = 2.54 cm. Explain your reasoning.

17. A house has a floor area of 250 square meters. Convert the house's floor area to square feet. Describe three different correct ways to solve this conversion problem. Each way of solving should use the fact that 1 in. = 2.54 cm. Explain your reasoning.

18. The Smiths will be carpeting a room in their house. In one store, they see a carpet they like that costs $35 per square yard. Another store has a similar carpet for $3.95 per square foot. Is this more or less expensive than the carpet at the first store? Explain your reasoning.

19. One acre is 43,560 square feet. If a square piece of land is 3 acres, then what is the length and width of this piece of land in feet? (Remember that the length and width of a square are the same.) Explain your reasoning.

20. A penny is $\frac{1}{16}$ of an inch thick.

   a. Suppose you had a thousand pennies. If you made a stack of these pennies, how tall would it be? Give your answer in feet and inches (e.g., 7 feet, 3 inches). Explain your reasoning.

b. Suppose you had a million pennies. If you made a stack of these pennies, how tall would it be? Give your answer in feet and inches (e.g., 7 feet, 3 inches). Would the stack be more than a mile tall or not? Explain your reasoning.

c. Suppose you had a billion pennies. If you made a stack of these pennies, how tall would it be? Give your answer in miles. Explain your reasoning.

d. Suppose you had a trillion pennies. If you made a stack of these pennies, how tall would it be? Give your answer in miles. Would the stack reach to the moon, which is about 240,000 miles away? Would the stack reach to the sun, which is about 93 million miles away? Explain your reasoning.

21. a. Write 100 zeros on a piece of paper and time how long it takes you. Based on the time it took you to write 100 zeros, approximately how long would it take you to write 1000 zeros? 10,000 zeros? 100,000 zeros? one million zeros? one billion zeros? one trillion zeros? In each case, give your answer either in minutes, hours, days, or years, depending on which answer is easiest to understand. Explain your answers.

b. Recall that a googol is the number whose decimal representation is a 1 followed by 100 zeros. Based on your answers in part (a), explain why it would not be possible for anyone to write a googol zeros.

c. Recall that a googolplex is the number whose decimal representation is a 1 followed by a googol zeros. Based on your answer in part (b), explain why it would not be possible for anyone to write a googolplex in its decimal representation. (Note, however, that it is possible to write a googolplex in scientific notation.)

22. For a certain type of rice, about 50 grains fill a 1-cm-by-1-cm-by-1-cm cube. Refer to this type of rice in all parts of this problem.

a. Describe the dimensions of a container that you could use to show your students 1 million grains of rice. Explain. Give a sense of approximately how big this container is by describing it as being about as big as a familiar object.

b. Describe the dimensions of a container that you could use to show your students 1 billion grains of rice. Explain. Give a sense of approximately how big this container is by describing it as being about as big as a familiar object.

c. Describe the dimensions of a container that you could use to show your students 1 trillion grains of rice. Explain. Give a sense of approximately how big this container is by describing it as being about as big as a familiar object.

23. A construction company has dump trucks that hold 10 cubic yards. If the company's workers dig a hole that is 8 feet deep, 20 feet wide, and 30 feet long, then how many dump truck loads will they need to haul away the dirt they dug out from the hole? Explain your reasoning.

24. Assuming that 1 gram of gold is worth $30, how much would 1 million dollars worth of gold weigh in pounds? Explain your reasoning.

25. Imagine that all the people on earth could stand side by side together. How much area would be needed? Would all the people on earth fit in Rhode Island? Explain your reasoning. Rhode Island, the smallest state in terms of area, has a land area of 1000 square miles. Assume that there are 6 billion people on earth, and assume that each person needs 4 square feet to stand in.

26. Joey has a toy car that is a 1:64 scale model of an actual car. (In other words, the length of the actual car is 64 times as long as the length of the toy car.) Joey's toy car can go 5 feet per second. What is the equivalent speed in miles per hour for the actual car? Explain.

Chapter Summary and Study Items

Section 11.1 Fundamentals of Measurement

In order to measure an object, or to describe its size, we must first select a measurable attribute. To measure a quantity is to compare the quantity with a unit amount. A unit is a fixed, reference amount. The most direct way to measure a quantity is to count how many of the unit amount make the quantity.

The U.S. customary system and the metric system are the two collections of units we use in the United States today. The metric system uses prefixes to name units in a uniform way. A fundamental relationship in the metric system is that 1 milliliter of water weighs 1 gram and has a volume of 1 cubic centimeter.

Key skills and understandings:

- Explain that, to measure a quantity, a measurable attribute must first be chosen and that measurement is comparison with a unit.

- Explain what it means for an object to have an area of 8 square centimeters, or a volume of 12 cubic inches, or a weight of 13 grams, or other similar measurements.

- Know and use proper notation for units of area and volume (e.g., square centimeters, cm^2).

- Explain how the metric system uses prefixes, and state the meaning of common metric prefixes: kilo, hecto, deka, deci, centi, and milli.

- Describe how capacity, weight (mass), and volume are linked in the metric system.

Section 11.2 Length, Area, Volume, and Dimension

A length describes the size of something (or part of something) that is one-dimensional. An area describes the size of something (or part of something) that is two-dimensional. A volume describes the size of something (or part of something) that is three-dimensional.

Key skills and understandings:

- Given an object, describe one-dimensional, two-dimensional, and three-dimensional parts or aspects of the object, and give appropriate U.S. customary and metric units for measuring or describing the size of those parts or aspects of the object.

- Compare objects with respect to one-dimensional, two-dimensional, and three-dimensional attributes.

- Recognize that in some cases, one object can be larger than another object with respect to one attribute, but smaller with respect to a different attribute.

Section 11.3 Error and Precision in Measurements

When it comes to measuring actual physical quantities, all measurements are only approximate. The way we write a measurement conveys how precisely the measurement is known. When calculating with actual measurements, the final result should be rounded to convey how precisely the result is known.

Key skills and understandings:

- Explain how a reported measurement conveys how precisely the measurement is known. For example, if the weight of an object is reported as 5 pounds, the actual weight of the object has been rounded to the nearest whole number and is therefore between 4.5 and 5.5 pounds. But if the weight of an object is reported as 5.0 pounds, the actual weight of the object has been rounded to the nearest tenth and is therefore between 4.95 and 5.05 pounds.

Section 11.4 Converting from One Unit of Measurement to Another

We can use multiplication and division to convert from one unit to another by considering how the units are related and by considering the meanings of multiplication and division. We can also use dimensional analysis to convert from one unit to another. Dimensional analysis relies on repeated multiplication by 1. Length measurements in the U.S. customary and metric systems are related by the fact that 1 inch = 2.54 cm.

Key skills and understandings:

- Use multiplication or division or both to convert a measurement from one unit to another. Explain why multiplication or division is the correct operation to use (without using dimensional analysis).

- Use dimensional analysis to convert from one measurement to another, explaining the method.

- Explain how to convert areas and volumes properly.

Area of Shapes

In this chapter, we continue our study of measurement by focusing on area of shapes. We examine how students in elementary school determine areas of shapes: progressing from primitive methods to more sophisticated methods and culminating in the use of area formulas. All methods for determining areas of shapes rely on what area means, and all use basic principles about area to decompose and recompose shapes to work with the pieces. These methods function in much the same way as calculation methods in arithmetic rely on what the operations mean and use the basic properties of arithmetic to decompose and recompose calculations. We review some familiar area formulas and discuss how these formulas are derived from the meaning of area and basic principles about area. We also examine the distinction between area and perimeter, which is a common source of confusion for students in elementary school. Finally, we consider how the Pythagorean theorem can be viewed as a fact about areas, and we see how the theorem can be derived from decomposing squares in different ways and equating areas.

12.1 Areas of Rectangles Revisited

 **Focal Points**
Grade 4

Area is a measure of how much two-dimensional space a shape takes up. We have discussed area and the area of rectangles in previous sections (Section 4.3 and Chapter 11). Here, we briefly revisit this important concept, but we examine progressively sophisticated thinking about areas of rectangles and attend more closely to the units of measurement involved.

Class Activity   *Now Turn to Class Activities Manual*

12A Units of Length and Area in the Area Formula for Rectangles, p. 280

Before students in elementary school learn the area formula for rectangles, they must learn what area means. For example, what does it mean to say that the area of a shape is 12 square centimeters? It means that the shape can be covered, without gaps or overlaps, with a total of twelve 1-cm-by-1-cm squares, allowing for squares to be cut apart and pieces to be moved if necessary. Given a 4-cm-by-3-cm rectangle, such as the one in Figure 12.1, a primitive way for students to determine its area is to cover the rectangle snugly with 1-cm-by-1-cm square tiles and count that it takes 12 tiles. Although this is a primitive method, it is important to understand it, because the method relies directly on the meaning of area and therefore emphasizes this meaning.

Covering rectangles with square tiles—though important—is a slow way to determine the area of rectangles. Fortunately, there is a quicker, more advanced way, which is based on the multiplicative structure that rectangles exhibit when they are broken into squares. These squares can be viewed as organized into equal rows (or columns). Because there are equal groups of squares, we can multiply to find the total number of squares. When the 4-cm-by-3-cm rectangle in Figure 12.1 is covered with 1-cm-by-1-cm squares there are 4 rows of squares with 3 squares in each row. Therefore, there are 4×3 squares covering the rectangle (without overlaps). Each square has area 1 cm², therefore, the total area of the rectangle is 4×3 square centimeters, which is 12 square centimeters.

The line of reasoning for the 4-cm-by-3-cm rectangle works generally to explain why a rectangle that is W units wide and L units long has an area of $L \times W$ square units. As Figure 12.1(b) indicates, if we cover such a rectangle with 1-unit-by-1-unit squares, then there will be L rows with W squares in each row. Therefore, there are $L \times W$ squares covering the rectangle. Each square has area 1 unit²; therefore, the total area of the rectangle is $L \times W$ square units.

Notice that in the $L \times W$ formula for the area of an L-unit-by-W-unit rectangle, the L and W refer to how many units long the sides of the rectangle are. In other words, in the area formula, the L and W refer to *one-dimensional* attributes of the rectangle. On the other hand, when we explain why we can multiply to find the area of the rectangle, we make equal groups of squares. We find that we can make L groups of squares and that there are W squares in each group, each of area of 1 square unit—a *two-dimensional* attribute.

The *length times width* formula,

$$L \times W$$

for areas of rectangles is valid not only when L and W are whole numbers, but also when L or W are fractions, mixed numbers, or decimals. The previous line of reasoning works in these cases as well. For

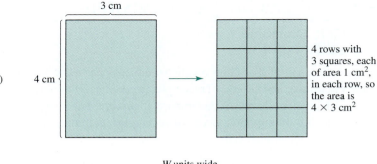

FIGURE 12.1

Why multiplying the side lengths of a rectangle gives the area of the rectangle

(a)   4 cm   3 cm

4 rows with 3 squares, each of area 1 cm², in each row, so the area is 4×3 cm²

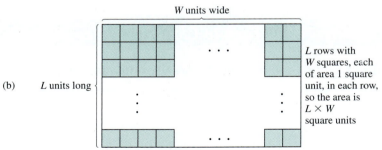

(b) L units long W units wide

L rows with W squares, each of area 1 square unit, in each row, so the area is $L \times W$ square units

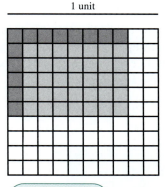

FIGURE 12.2

Why a $3\frac{1}{2}$-cm-by-$2\frac{1}{2}$-cm rectangle has area $3\frac{1}{2} \times 2\frac{1}{2}$ cm²

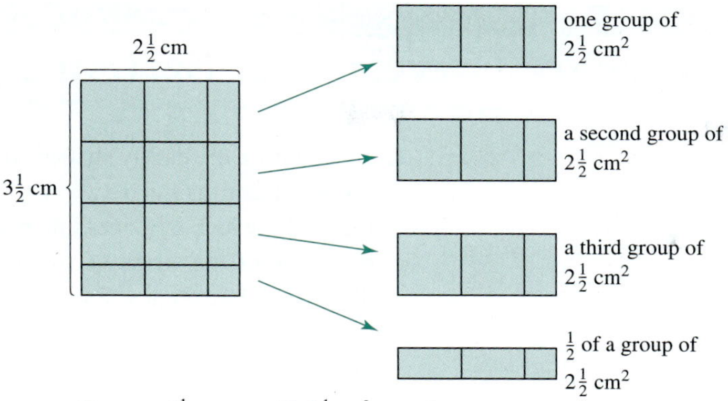

$2\frac{1}{2}$ cm

$3\frac{1}{2}$ cm

one group of $2\frac{1}{2}$ cm²

a second group of $2\frac{1}{2}$ cm²

a third group of $2\frac{1}{2}$ cm²

$\frac{1}{2}$ of a group of $2\frac{1}{2}$ cm²

There are $3\frac{1}{2}$ groups with $2\frac{1}{2}$ cm² in each group, for a total area of $3\frac{1}{2} \times 2\frac{1}{2}$ cm².

example, consider a rectangle that is $3\frac{1}{2}$ cm by $2\frac{1}{2}$ cm, as shown in Figure 12.2. If we cover such a rectangle with 1-cm-by-1-cm squares and parts of 1-cm-by-1-cm squares, then there will be $3\frac{1}{2}$ groups of squares with $2\frac{1}{2}$ squares in each group. So according to the meaning of multiplication, the area of the rectangle is

$$3\frac{1}{2} \times 2\frac{1}{2} \text{ cm}^2$$

Numerically, we can calculate that

$$3\frac{1}{2} \times 2\frac{1}{2} = \frac{7}{2} \times \frac{5}{2} = \frac{35}{4} = 8\frac{3}{4}$$

Hence, the area of the rectangle is $8\frac{3}{4}$ cm². If we count up whole squares and collect together partial squares in Figure 12.2, we also see that the rectangle's area is $8\frac{3}{4}$ square centimeters.

Practice Exercises for Section 12.1

1. What does it mean to say that a shape has an area of 12 square centimeters?

2. Given a 4-cm-by-3-cm rectangle:

   a. What is a primitive way to determine the area of the rectangle?

   b. What is a more advanced way to determine the area of the rectangle and why does it work?

3. Explain why it would be *incorrect* to use the darkly shaded squares in Figure 12.3 to describe the 0.6-unit and 0.8-unit lengths you use in the length × width formula for the area of the shaded rectangle.

1 unit

FIGURE 12.3   The darkly shaded squares should *not* be used to describe the length and width of the shaded rectangle.

Answers to Practice Exercises for Section 12.1

1. See text.

2. a. See text.

   b. You can multiply 4×3 to find the area of the shape in square centimeters. See the text for how to explain why.

3. By counting the darkly shaded squares, you set up a potential confusion between the side length of a rectangle, which is a one-dimensional attribute, and the areas of the strips of squares, which are two-dimensional attributes. Furthermore, notice that the side length of each small square is 0.1 units, whereas the area of each small square is 0.01 square units.

Problems for Section 12.1

1. You have a 5-foot-by-7-foot rectangular rug in your classroom. You also have a bunch of square foot tiles and some tape measures.

   a. What is the most primitive way to determine the area of the rug?

   b. What is a less primitive way to determine the area of the rug, and why does this method work?

2. Draw a 3-cm-by-7-cm rectangle. Then discuss the difference in the units we attach to the 3 and the 7 in these two situations:

   a. When applying the rectangle area formula to determine the area of the rectangle.

   b. When viewing the rectangle as decomposed into equal groups of 1-cm-by-1-cm squares in order to apply the definition of multiplication. Use drawings to support your discussion.

3. a. Explain how to see the area of the large rectangle in Figure 12.4 as consisting of $2\frac{1}{2}$ groups with $3\frac{1}{2}$ cm^2 in each group, thereby explaining why the area formula

$$2\frac{1}{2} \times 3\frac{1}{2}$$

to determine the area of the rectangle in square centimeters, fits with the meaning of multiplication as we have described it.

   b. Calculate $2\frac{1}{2} \times 3\frac{1}{2}$ without a calculator, showing your calculations. Then verify that this calculation has the same answer as when you determine the area of the rectangle in Figure 12.4 in square centimeters by counting full 1-cm-by-1-cm squares and combining partial squares.

4. a. Explain how to see the area of the large rectangle in Figure 12.5 as consisting of $4\frac{1}{2}$ groups with $5\frac{3}{4}$ cm^2 in each group, thereby explaining why the area formula

$$4\frac{1}{2} \times 5\frac{3}{4}$$

to determine the area of the rectangle in square centimeters, fits with the meaning of multiplication as we have described it.

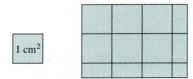

FIGURE 12.4 Explain why the area is $2\frac{1}{2} \times 3\frac{1}{2}$ cm^2.

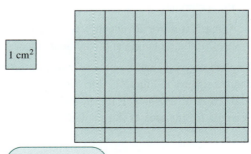

FIGURE 12.5 Explain why the area is $4\frac{1}{2} \times 5\frac{3}{4}$ cm^2.

b. Calculate $4\frac{1}{2} \times 5\frac{3}{4}$ without a calculator, showing your calculations. Then verify that this calculation has the same answer as when you determine the area of the rectangle in Figure 12.5 in square centimeters by counting full 1-cm-by-1-cm squares and combining partial squares.

5. a. Draw a (fairly long) line segment and designate it as being 1 unit long. Then draw a 0.6-unit-by-0.9-unit rectangle.

b. Apply the length × width formula for the area of the rectangle and verify that the formula gives you the correct area for your rectangle in part (a). Attend carefully to units of area.

c. When you applied the length × width formula to find the area of the rectangle in part (b), you used lengths of 0.6 and 0.9 units. Describe these lengths, show them in your drawing, and discuss an error that students might make in describing these lengths.

12.2 Moving and Additivity Principles about Area

Focal Points
Grade 4

How do you figure out the area of a shape? Your first thought might be to look for a formula to apply. But area formulas, as well as other ways of determining area by reasoning, are based on principles about area. In this section, we will study the two most fundamental principles that are used in determining the area of a shape. These principles agree entirely with common sense. In fact, you have probably used them without being consciously aware of it. Students in elementary school use these principles, if only informally or implicitly.

The moving and additivity principles about area are as follows:

- **Moving Principle:** If you move a shape rigidly without stretching it, then its area does not change.

- **Additivity Principle:** If you combine (a finite number of) shapes *without overlapping* them, then the area of the resulting shape is the sum of the areas of the individual shapes.

Just as the properties of arithmetic allow us to take numbers apart and calculate with the pieces, so too the moving and additivity properties allow us to take shapes apart and calculate areas of the pieces.

Class Activity   *Now Turn to Class Activities Manual*

12B Different Shapes with the Same Area, p. 282

A common way to use the additivity principle to find the area of some shape is as follows: Subdivide the shape into pieces whose areas are easy to determine; then add the areas of these pieces. The resulting sum is the area of the original shape, because we can think of the original shape as the combination of the pieces.

For example, what is the area of the surface of the swimming pool pictured in Figure 12.6(a) (a bird's-eye view)? To answer this question, imagine dividing the surface of the pool into two rectangular parts, as shown in Figure 12.6(b). According to the additivity principle, the area of the surface of the pool is the sum of the areas of the two rectangular pieces, which is

$$16 \times 28 + 30 \times 20 = 448 + 600 = 1048$$

square feet.

Another way to use the moving and additivity principles to determine the area of a shape is to subdivide the shape into pieces, then move and recombine those pieces, without overlapping, to make

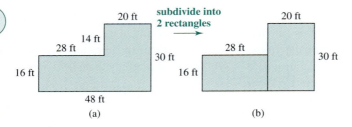

FIGURE 12.6

To determine the area of the surface of a swimming pool, subdivide it into two rectangles.

FIGURE 12.7

Finding the area of a patio by subdividing, moving, and recombining pieces

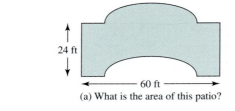

(a) What is the area of this patio?

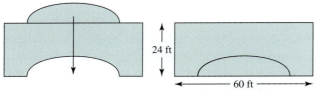

(b) Subdivide, move, and recombine the pieces to form a rectangle. The area of the rectangle is the same as the area of the original patio.

a new shape whose area is easy to determine. By the moving and additivity principles, the area of the original shape is equal to the area of the new shape.

For example, what is the area of the patio pictured in Figure 12.7(a)? To answer this, imagine slicing off the "bump" and sliding it down to fill in the corresponding "dent," as indicated in Figure 12.7(b). Note that the bump was moved rigidly without stretching, and the bump fills the dent perfectly, without any overlaps. Therefore, by the moving and additivity principles, the area of the resulting rectangle is the same as the area of the original patio. Since the rectangle has area

$$24 \times 60 = 1440$$

square feet, the patio also has an area of 1440 square feet.

By starting with a shape whose area we know and by subdividing the shape and recombining the pieces without overlapping to make a new shape, we can demonstrate that many different shapes can have the same area. (See Figure 12.8.)

FIGURE 12.8

Children show 35 square inches in several ways by subdividing a rectangle and recombining the parts without overlapping.

Class Activity   *Now Turn to Class Activities Manual*

12C   Using the Moving and Additivity Principles, p. 283

Practice Exercises for Section 12.2

1. Figure 12.9 shows the floor plan of a loft that will be getting a new wood floor. How many square feet of flooring will be needed?

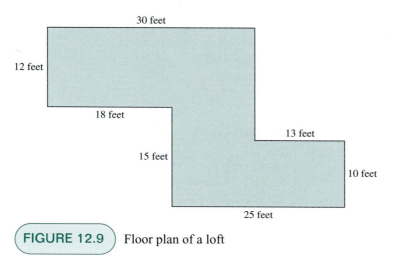

30 feet

12 feet

18 feet

13 feet

15 feet

10 feet

25 feet

FIGURE 12.9   Floor plan of a loft

2. What is the area of the shape in Figure 12.10? (This is not a perspective drawing; it is a flat shape.)

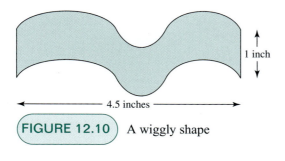

1 inch

4.5 inches

FIGURE 12.10   A wiggly shape

3. Use the fundamental principles about area to determine the area of the octagon in Figure 12.11.

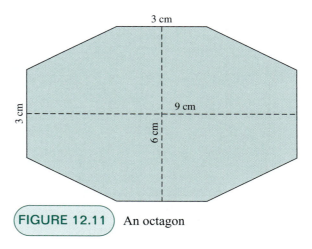

3 cm

9 cm

3 cm

6 cm

FIGURE 12.11   An octagon

Answers to Practice Exercises for Section 12.2

1. First, what is the length of the unlabeled side? To answer this question, compare the "vertical" distance across the floor on the left side of the room with the right side of the room. On the left side of the room, the vertical distance across the floor is 12 feet + 15 feet = 27 feet. On the right side of the room, the vertical distance across the room is *unlabeled length* + 10 feet. The two vertical distances across the floor are equal; therefore,

$$\text{unlabeled length} + 10 = 27$$

So the unlabeled length must be 17 feet.

Notice that we can subdivide the floor into three rectangular pieces: one 30 feet by 12 feet, one 12 feet by 5 feet and one 25 feet by 10 feet. We can think of the floor as the combination of these rectangular pieces. Therefore, according to the additivity principle about area, the number of square feet of flooring needed is $30 \times 12 + 12 \times 5 + 25 \times 10 = 670$.

2. We can subdivide the wiggly shape of Figure 12.10 into pieces and recombine the pieces as shown in Figure 12.12 to make a rectangle that is 1 inch wide and $4\frac{1}{2}$ inches long. (Slice off the two "bumps" on top and the bump on the bottom, and move them to fill the corresponding "dents.") The rectangle has area $4\frac{1}{2}$ square inches, so by the moving and additivity principles about area, the wiggly shape must also have area $4\frac{1}{2}$ square inches.

FIGURE 12.12   The wiggly shape becomes a rectangle after subdividing and recombining

3. The area of the octagon is 45 cm^2. One way to determine the area of the octagon is to think of the octagon as a 6-cm-by-9-cm rectangle with 4 triangles cut off, as indicated in Figure 12.13(a). Then, according to the additivity principle about areas,

$$\text{area of octagon} + \text{area of 4 triangles} = \text{area of rectangle}$$

But the 4 triangles can be combined, without overlapping, to form 2 small 1.5-cm-by-3-cm rectangles. So, by the previous equation, we have

$$\text{area of octagon} + 2 \times 1.5 \times 3 \text{ cm}^2 = 6 \times 9 \text{ cm}^2$$

Therefore,

$$\text{area of octagon} = 54 - 9 \text{ cm}^2 = 45 \text{ cm}^2$$

Another way to determine the area of the octagon is to subdivide it as shown in Figure 12.13(b). Once again, the 4 triangles can be combined to make 2 rectangles.

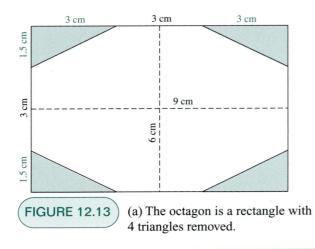

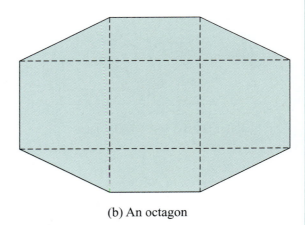

FIGURE 12.13   (a) The octagon is a rectangle with 4 triangles removed.

(b) An octagon

Problems for Section 12.2

1. Make a shape that has area 25 in.², but that has no (or almost no) straight edges. Explain how you know that your shape has area 25 in.²

2. Figure 12.14 shows the floor plan for a one-story house. Calculate the area of the floor of the house. Explain your method clearly.

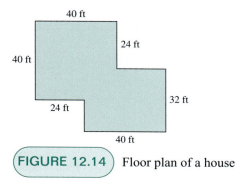

FIGURE 12.14   Floor plan of a house

3. An area problem: The Johnsons are planning to build a 5-foot-wide brick walkway around their rectangular garden, which is 20 feet wide and 30 feet long. What will the area of this walkway be? Before you solve the problem yourself, use Kaitlyn's idea:

   a. Kaitlyn's idea is to "take away the area of the garden." Explain how to solve the problem about the area of the walkway by using Kaitlyn's idea. Explain clearly how to apply one or both of the moving and additivity principles on area in this case.

   b. Now solve the problem about the area of the walkway in another way than you did in part (a). Explain your reasoning.

4. Figure 12.15 shows a design for a herb garden, with approximate measurements. Four identical plots of

land in the shape of right triangles (shown lightly shaded) are surrounded by paths (shown darkly shaded). Use the moving and additivity principles to determine the area of the paths.

5. Figure 12.16 shows the floor plan for a modern, one-story house. Bob calculates the area of the floor of the house this way:

$$36 \times 72 - 18 \times 18 = 2268 \text{ ft}^2$$

What must Bob have in mind? Explain why Bob's method is a legitimate way to calculate the floor area of the house, and explain clearly how one or both of the moving and additivity principles on area apply in this case.

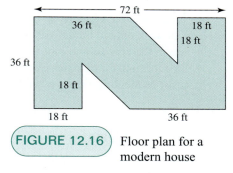

FIGURE 12.16   Floor plan for a modern house

6. Use the moving and additivity principles about area to determine the area, in square inches, of the shaded design in Figure 12.17. The shape is a 2-inch-by-2-inch square, with a square, placed diagonally inside, removed from the middle. In determining the area of the shape, use no formulas

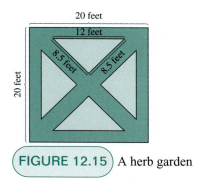

FIGURE 12.15   A herb garden

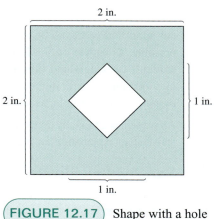

FIGURE 12.17   Shape with a hole

other than the one for areas of rectangles. Explain your method clearly.

7. Use the moving and additivity principles about area to determine the area, in square inches, of the shaded flower design in Figure 12.18. In determining the area of the shape, use no formulas other than the one for areas of rectangles. Explain your method clearly.

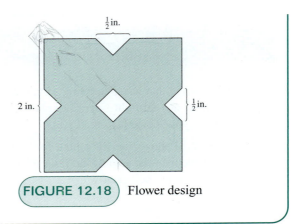

(FIGURE 12.18)   Flower design

12.3 Areas of Triangles

F·P **Focal Points**
Grade 5

In this section we focus on methods for determining areas of triangles, starting with primitive methods that work only in limited cases, progressing to more sophisticated and powerful methods, and culminating in the area formula for triangles. Elementary school students are expected to learn how to use the area formula for triangles *and* to understand the reasoning that underlies the formula. The reasoning that justifies the formula is a powerful problem-solving tool in its own right.

Using Moving and Additivity Principles to Determine Areas of Triangles

Before you read on, do the next Class Activity about determining areas of triangles using a variety of methods, from primitive to more sophisticated ones.

Class Activity *Now Turn to Class Activities Manual*

12D Determining Areas of Triangles in Progressively Sophisticated Ways, p. 284

Given a triangle, the most primitive method for determining its area is to count how many 1-unit-by-1-unit squares it takes to cover the shape without gaps or overlaps. This method requires squares to be cut apart and pieces to be moved and recombined with other pieces to make as many full squares as possible, as shown on the left in Figure 12.19. Although the method is primitive, it is important for students to understand this method because it highlights the meaning of area. More sophisticated methods for determining areas of triangles rely on relating the triangle to a rectangle and applying the rectangle area formula, as shown on the right in Figure 12.19.

As we will see next, by generalizing the more sophisticated methods for determining areas of triangles, we arrive at an explanation for the formula for the area of a triangle, that is, *one half the base times the height*.

Base and Height for Triangles

Before we discuss the *one half the base times the height* formula for the area of a triangle, we need to know **base** what base and height of a triangle mean. The **base** of a triangle can be any one of its three sides. In a formula, the word *base* or a letter, such as *b*, which represents the base, really means *length of the base*.

FIGURE 12.19

A progression of methods for determining the area of a right triangle.

Method 1

Method 2

Method 3

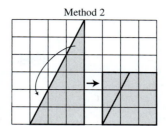

A primitive way to determine the area is to move small pieces and count the total number of squares.

More advanced methods relate the triangle to a rectangle, either by moving a big chunk or by embedding the triangle in a rectangle. These methods lead to the triangle area formula.

height Once a base has been chosen, the **height** is the line segment that is

- perpendicular to the base and

- connects the base, or an extension of the base, to the vertex of the triangle that is not on the base.

In a formula, the word *height* or a letter, such as *h*, which represents the height, actually means *length of the height*.

For example, Figure 12.20 shows two copies of a triangle ABC, and two of the three ways to choose the base *b* and the height *h*. (Notice that the base and height have different lengths for the different choices.) In the second case, the height is the (dashed) line segment CE. In this case, even though the height *h* doesn't meet *b* itself, it meets an *extension* of *b*.

Class Activity *Now Turn to Class Activities Manual*

12E Choosing the Base and Height of Triangles, p. 286

Formula for the Area of a Triangle

formula for the area of a triangle The familiar **formula for the area of a triangle** with base *b* and height *h* is

$$\text{area of triangle} = \frac{1}{2}(b \times h) \text{ square units}$$

If a triangle has a base that is 5 inches long, and if the corresponding height of the triangle is 3 inches long, then the area of the triangle is

$$\frac{1}{2}(5 \times 3) \text{ in.}^2 = 7.5 \text{ in.}^2$$

FIGURE 12.20

Two ways to select the base (*b*) and height (*h*) of triangle ABC

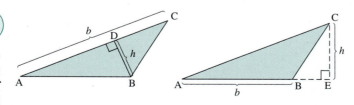

FIGURE 12.21

These triangles
are half of a
b-by-*h* rectangle.

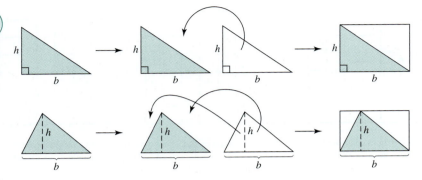

In the formula

$$\frac{1}{2}(b \times h)$$

the base *b* and the height *h* must be described with the *same unit*. For example, both the base and the height could be in feet, or both lengths could be in centimeters. But if one length is in feet and the other is in inches, for example, then you must convert both lengths to a common unit before calculating the area. The area of the triangle is then in *square units* of whatever common unit you used for the base and height. So if the base and height are both in centimeters, then the area resulting from the formula is in square centimeters (cm^2).

Why Is the Area Formula for Triangles Valid?

Class Activity   *Now Turn to Class Activities Manual*

12F   Explaining Why the Area Formula for Triangles Is Valid, p. 286

Suppose we have a triangle, and suppose we have chosen a base *b* and a height *h* for the triangle. Why is the area of the triangle equal to one half the base times the height? In some cases, such as those shown in Figure 12.21, two copies of the triangle can be subdivided (if necessary) and recombined, without overlapping, to form a *b*-by-*h* rectangle. In such cases, because of the moving and additivity principles about area,

$$2 \times \text{area of triangle} = \text{area of rectangle}$$

But since the rectangle has $b \times h$,

$$2 \times \text{area of triangle} = b \times h$$

Therefore,

$$\text{area of triangle} = \frac{1}{2}(b \times h)$$

But what about the triangle in Figure 12.22? For this triangle, it's not so clear how to turn it into half of a *b*-by-*h* rectangle. However, the formula for the area of the triangle is still valid. To explain why the $\frac{1}{2}(b \times h)$ formula is still valid for the triangle in Figure 12.22, enclose the triangle in a rectangle, as shown at the top

FIGURE 12.22

Why is the area
$\frac{1}{2}(b \times h)$?

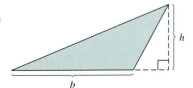

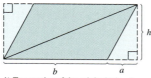

FIGURE 12.23

Enclose the original triangle in a rectangle of area $(b + a) \times h$ to explain why the area of the triangle is $\frac{1}{2}(b \times h)$

1) Two copies of the original triangle together with two copies of another triangle form a rectangle of area $(b + a) \times h = b \times h + a \times h$.

2) Removing the two additional triangles will remove an area of $a \times h$.

3) The remaining two copies of the original triangle must therefore have area $b \times h$.

4) The original triangle therefore has area $\frac{1}{2}(b \times h)$.

left in Figure 12.23. The rectangle consists of two copies of the original triangle (darkly shaded) and two copies of another triangle, which are lightly shaded. The rectangle has area $(b + a) \times h$, which is equal to $b \times h + a \times h$ by the distributive property. If we put the two lightly shaded triangles together, as at the top right of Figure 12.23, they form a rectangle of area $a \times h$. If we take this area away from the area of the large rectangle, the remaining area is the area of the two copies of the original triangle combined (by the moving and additivity principles). Therefore, the area of the two copies of the original triangle is

$$(b \times h + a \times h) - a \times h = b \times h$$

and so the original triangle has half this area, namely, area

$$\frac{1}{2}(b \times h)$$

So, given any triangle, and given any choice of base b and height h for the triangle, one of the arguments we have just given explains why the area of the triangle is $\frac{1}{2}(b \times h)$.

Class Activity *Now Turn to Class Activities Manual*

12G Determining Areas, p. 288

Practice Exercises for Section 12.3

1. Use the moving and additivity principles about area to determine the area of the triangle in Figure 12.24 in *two different ways*. In both cases, do not use a formula for areas of triangles.

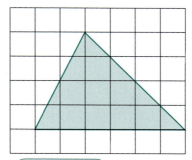

FIGURE 12.24 Determine the area of the triangle.

2. Show the heights of the triangles in Figure 12.25 that correspond to the bases that are labeled *b*. Then determine the areas of the triangles.

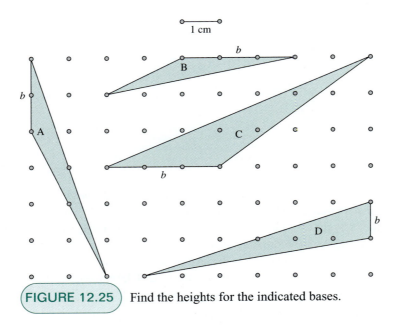

FIGURE 12.25 Find the heights for the indicated bases.

3. Determine the areas, in square units, of the shaded regions in Figure 12.26.

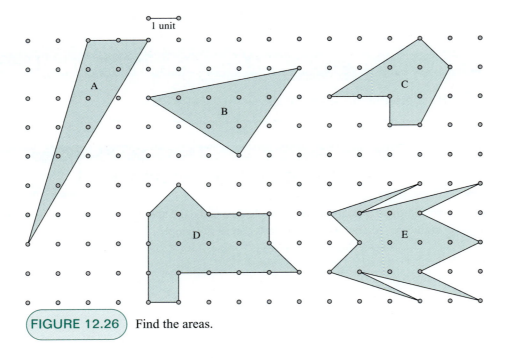

FIGURE 12.26 Find the areas.

4. In your own words, explain why the triangles in Figures 12.21 and 12.22 have area $\frac{1}{2}(b \times h)$ for the given choices of base *b* and height *h*.

Answers to Practice Exercises for Section 12.3

1. Method 1: Figure 12.27 (top, left) shows how to subdivide the triangle into three pieces. Imagine rotating two of those pieces down to form a 2-unit-by-6-unit rectangle (as indicated at the top right). Since the rectangle was formed by moving portions of the triangle and recombining them without overlapping, by the moving and additivity principles, the original triangle has the same area as the 2-unit-by-6-unit rectangle, namely 12 square units.

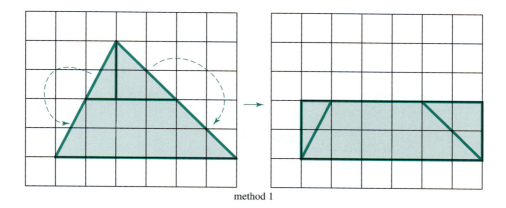

method 1

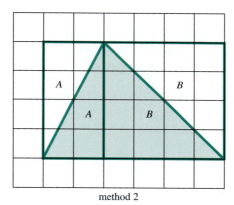

method 2

FIGURE 12.27   Two ways to determine the area of the triangle

Method 2: Figure 12.27 (bottom) shows how to subdivide the triangle into two smaller triangles that have areas A and B. Imagine making copies of the two triangles, rotating them, and attaching them to the original triangle to form a 4-unit-by-6-unit rectangle. Then, according to the moving and additivity principles,

$$2A + 2B = 24 \text{ square units}$$

By the distributive property,

$$2 \cdot (A + B) = 24 \text{ square units}$$

Therefore,

$$A + B = 12 \text{ square units}$$

Since $A + B$ is the area of the original triangle, the area of the original triangle is 12 square units.

2. See Figure 12.28. Notice that, for each of these triangles, we must extend the base in order to show the height meeting it at a right angle. Triangle A has height 2 cm and area $\frac{1}{2} \times 2 \times 2$ cm^2 = 2 cm^2. Triangle B has height 1 cm and area $\frac{1}{2} \times 3 \times 1$ cm^2 = $1\frac{1}{2}$ cm^2. Triangle C has height 3 cm and area $\frac{1}{2} \times 3 \times 3$ cm^2 = $4\frac{1}{2}$ cm^2. Triangle D has height 6 cm and area $\frac{1}{2} \times 1 \times 6$ cm^2 = 3 cm^2.

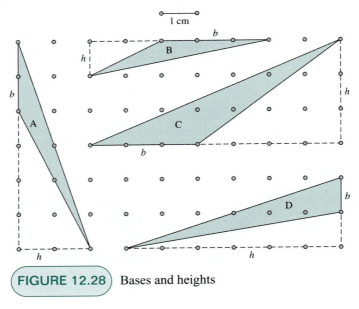

FIGURE 12.28 Bases and heights

3. Shape A is a triangle whose base can be chosen to be the top line segment of length 2 units. Then the height has length 7 units, so the area of triangle A is $\frac{1}{2} \times 2 \times 7 = 7$ square units.

As shown in Figure 12.29, triangle B and three additional triangles combine to make a 3-unit-by-5-unit rectangle. The rectangle has area 15 square units, and the three additional triangles have areas $\frac{1}{2}(1 \times 5) = 2\frac{1}{2}$ square units, $\frac{1}{2}(3 \times 2) = 3$ square units, and $\frac{1}{2}(2 \times 3) = 3$ square units. Therefore,

$$\text{(area of B)} + 2\frac{1}{2} + 3 + 3 = 15 \text{ square units}$$

So triangle B has area $15 - 8\frac{1}{2} = 6\frac{1}{2}$ square units.

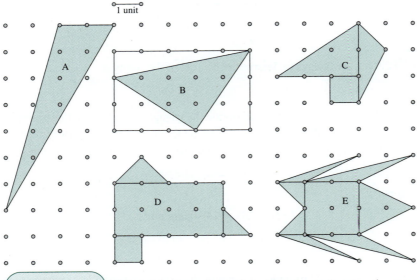

FIGURE 12.29 Subdivide these shapes into triangles and rectangles to calculate their areas.

The areas of the remaining shaded shapes can be calculated by subdividing them into rectangles and triangles as indicated in Figure 12.29. Shape C has area $5\frac{1}{2}$ square units. Shape D has area $10\frac{1}{2}$ square units. Shape E has area 10 square units.

4. See text.

Problems for Section 12.3

1. Use the moving and additivity principles to determine the area (in square units) of the triangle in Figure 12.30 in *two different ways*. Do not use a formula for areas of triangles. The grid lines in Figure 12.30 are 1 unit apart. Explain your reasoning.

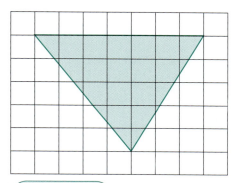

FIGURE 12.30 Determine the area of the triangle.

2. **a.** Use the moving and additivity principles to determine the areas (in square units) of the three lightly shaded triangles in Figure 12.31. (Do not use a formula for areas of triangles.) The grid lines are 1 unit apart. Explain your reasoning.

b. Use the moving and additivity principles and your results from part (a) to determine the area of the dark shaded triangle in the center of Figure 12.31 (in square units). Explain your reasoning.

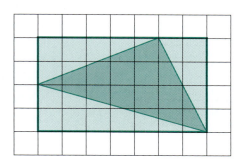

FIGURE 12.31 Four triangles forming a rectangle

3. Use the moving and additivity principles to determine the area (in square units) of the triangle in Figure 12.32. Do not use a formula for areas of triangles. Explain your method clearly. The grid lines in Figure 12.32 are 1 unit apart.

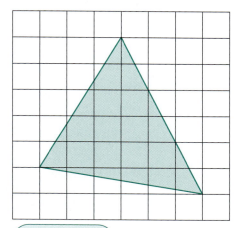

FIGURE 12.32 Determine the area of the triangle.

4. For each triangle in Figure 12.33, show the height of the triangle that corresponds to the indicated base b. Then use these bases and heights to determine the area of each triangle.

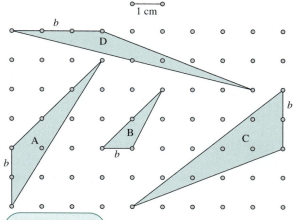

FIGURE 12.33 Determine the heights and areas of the triangles.

5. a. Use a ruler and compass to draw three identical triangles, having one side of length 4 inches, one side of length $2\frac{3}{4}$ inches, and one side of length $1\frac{3}{4}$ inches.

   b. For each of the three sides of the triangle you drew in part (a), let that side be the base of one triangle, and draw the corresponding height of the triangle.

   c. Measure each of the three heights and use each of these measurements to determine the area of the triangle. If your three answers for the area aren't exactly the same, explain why.

6. Becky was asked to divide a rectangle into 4 equal pieces and to shade one of those pieces. Figure 12.34 shows Becky's solution. Is Becky right or not? Explain your answer.

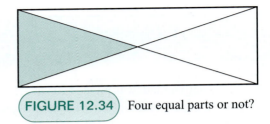

FIGURE 12.34   Four equal parts or not?

7. Explain clearly in your own words why the triangles in Figure 12.35 have area $\frac{1}{2}(b \times h)$ for the given choice of base b and height h.

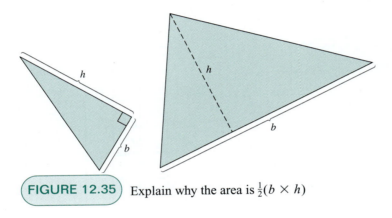

FIGURE 12.35   Explain why the area is $\frac{1}{2}(b \times h)$

8. Explain clearly in your own words why the triangle in Figure 12.36 has area $\frac{1}{2}(b \times h)$ for the given choice of base b and height h.

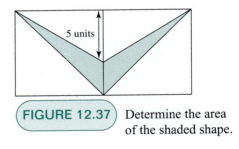

FIGURE 12.37   Determine the area of the shaded shape.

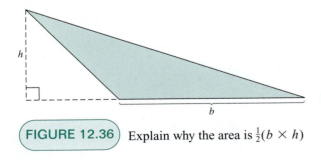

FIGURE 12.36   Explain why the area is $\frac{1}{2}(b \times h)$

10. Determine the area of the shaded triangle in Figure 12.38 in *two different ways*. Explain your reasoning in each case.

9. Determine the area of the shaded shape in Figure 12.37 in *two different ways*. The entire figure consists of two 8-unit-by-8-unit squares. Explain your reasoning in each case.

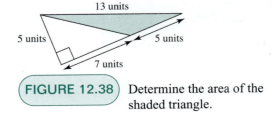

FIGURE 12.38   Determine the area of the shaded triangle.

11. Determine the area of the shaded shape in Figure 12.39. The entire figure consists of two 6-unit-by-6-unit squares with a 4-unit-by 4-unit square between them. Explain your reasoning.

FIGURE 12.39 Determine the area of the shaded shape.

12. Given that the rectangle ABCD in Figure 12.40 has area 108 square units, determine the area of the shaded triangle. Explain your reasoning.

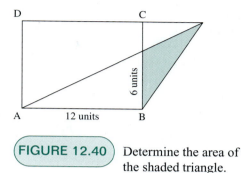

FIGURE 12.40 Determine the area of the shaded triangle.

12.4 Areas of Parallelograms and Other Polygons

F·P **Focal Points**
Grade 5

In this section, we'll study the area formula for parallelograms. Also, we'll continue to use the moving and additivity principles to determine areas of various polygons.

Since a rectangle is a special kind of parallelogram, and since the area of a rectangle is the width of the rectangle times the length of the rectangle, it would be natural to think that this same formula is true for parallelograms. But is it? Class Activity 12H examines this question.

Class Activity   *Now Turn to Class Activities Manual*

12H   Do Side Lengths Determine the Area of a Parallelogram?, p. 288

If you did Class Activity 12H, then you saw that, unlike the situation with rectangles, it is not possible to determine the area of a parallelogram from the lengths of its sides alone. However, as with triangles, there is a formula for the area of a parallelogram in terms of a base and a height.

base   As with triangles, the **base** of a parallelogram can be chosen to be any one of its four sides. In a formula, the word *base* or a letter, such as b, which represents the base, actually means *length of the base*.

height   Once a base has been chosen, the **height** of a parallelogram is a line segment that is

- perpendicular to the base and
- connects the base, or an extension of the base, to a vertex of the parallelogram that is not on the base.

In a formula, the word *height* or a letter, such as h, which represents the height, actually means *length of the height*. So the height is the distance between the base and the side opposite the base.

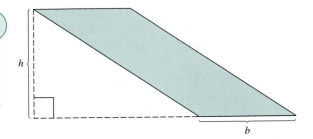

FIGURE 12.41

One way to choose the base and height of this parallelogram

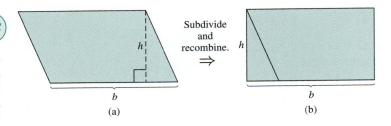

FIGURE 12.42

Subdividing and recombining a parallelogram (a) to make a rectangle (b)

Subdivide and recombine. ⇒

(a)   (b)

Figure 12.41 shows a parallelogram and one way to choose a base b and a height h.

formula for the area of a parallelogram

There is a very simple **formula for the area of a parallelogram.** If a parallelogram has a base that is b units long, and height that is h units long, then the area of the parallelogram is

$$b \times h \text{ square units}$$

In this formula, we assume that b and h are described with the same unit (e.g., both in centimeters, or both in feet). If b and h are in different units, then you must convert them to a common unit before using the formula. For example, if a parallelogram has a base that is 2 meters long and a height that is 5 centimeters long, then the area of the parallelogram is

$$200 \times 5 \text{ cm}^2 = 1000 \text{ cm}^2$$

because 2 meters = 200 cm.

> **Class Activity**   *Now Turn to Class Activities Manual*
>
> **12I** Explaining Why the Area Formula for Parallelograms Is Valid, p. 290

Why is the $b \times h$ formula for areas of parallelograms valid? In some cases, such as the parallelogram in Figure 12.42(a), we can explain why the area formula is valid by subdividing the parallelogram and recombining it to form a b by h rectangle, as shown in Figure 12.42(b). According to the moving and additivity principles about area, the area of the original parallelogram and the area of the newly formed rectangle are equal. Because the newly formed rectangle has area $b \times h$ square units, the original parallelogram also has area $b \times h$ square units.

Why is the $b \times h$ formula valid for the area of the parallelogram in Figure 12.43, for the given choices of b and h? In this case, we can enclose the parallelogram in a rectangle as shown on the top left in Figure 12.44. The rectangle consists of the parallelogram (darkly shaded) and two copies of a

FIGURE 12.43

Why is the area $b \times h$?

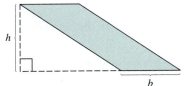

FIGURE 12.44

Enclose the parallelogram in a rectangle to explain why the area of the parallelogram is $b \times h$.

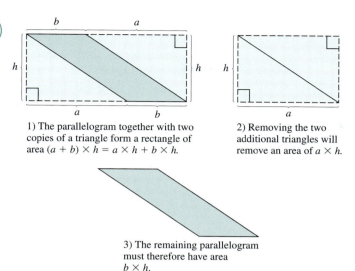

1) The parallelogram together with two copies of a triangle form a rectangle of area $(a + b) \times h = a \times h + b \times h$.

2) Removing the two additional triangles will remove an area of $a \times h$.

3) The remaining parallelogram must therefore have area $b \times h$.

right triangle. The rectangle has area $(a + b) \times h$ which is equal to $a \times h + b \times h$ by the distributive property. If we put the two triangles together, as on the top right in Figure 12.44, they form a rectangle of area $a \times h$. If we take this area away from the area of the large rectangle, the remaining area is the area of the parallelogram (by the moving and additivity principles). Therefore, the area of the parallelogram is

$$(b \times h + a \times h) - a \times h = b \times h$$

and so the parallelogram has area $b \times h$, which is what we wanted to show.

So, given any parallelogram, and given any choice of base b and height h for the triangle, one of the arguments we have just given explains why the area of the parallelogram is $b \times h$.

Practice Exercises for Section 12.4

1. Every rectangle is also a parallelogram. Viewing a rectangle as a parallelogram, we can choose a base and height for it, as for any parallelogram. In the case of a rectangle, what are other names for *base* and *height*?

2. How are the *base* × *height* formula for areas of parallelograms and the *length* × *width* formula for areas of rectangles related?

3. Why can there not be a parallelogram area formula that is expressed only in terms of the lengths of the sides of the parallelogram?

4. What is a formula for the area of a parallelogram, and why is this formula valid?

Answers to Practice Exercises for Section 12.4

1. In the case of a rectangle, the *base* and *height* are the *length* and *width* of the rectangle. The base can be either the length or the width.

2. The *base* × *height* formula for areas of parallelograms generalizes the *length* × *width* formula for areas of rectangles because these two formulas are the same in the case of rectangles. When a parallelogram is also a rectangle, the base can be chosen to be the length of the rectangle, and then the height is the width of the rectangle.

3. The three parallelograms in Figure 12.45 all have sides of the same length. (See also Class Activity 12H);

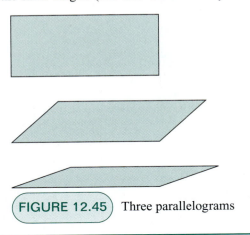

FIGURE 12.45   Three parallelograms

If there were a parallelogram area formula that was expressed only in terms of the lengths of the sides of the parallelogram, then all three of these parallelograms would have to have the same area. But it is clear that the three parallelograms have different areas. Therefore, there can be no parallelogram area formula that is expressed only in terms of the lengths of the sides of the parallelogram.

4. See text.

Problems for Section 12.4

1. Josie has two wooden beams that are 15 feet long and two wooden beams that are 10 feet long. Josie plans to use these four beams to form the entire border around a closed garden. Josie likes unusual designs. Without any other information, what is the most you can say about the area of Josie's garden? Explain.

2. Figure 12.46 shows a shaded parallelogram inside a rectangle. Use the moving and additivity principles to find an expression for the area of the shaded parallelogram in terms of all of the lengths x, y, and z. Explain your reasoning. Then simplify the expression you found.

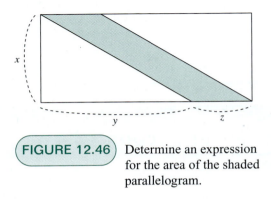

FIGURE 12.46   Determine an expression for the area of the shaded parallelogram.

3. Given that the shaded shape in Figure 12.47 is a parallelogram, find an equation relating the lengths a, b, c, and d. Explain why your equation must be true.

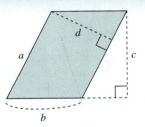

FIGURE 12.47   Find an equation relating the given lengths.

4. In the text, we saw a way to explain why the area of the parallelogram in Figure 12.48 is $b \times h$. Another way to explain the area formula for parallelograms is to use the area formula for triangles. Show how to subdivide the parallelogram in Figure 12.48 into two triangles. Then use the area formula for triangles to explain why the area of the parallelogram is $b \times h$.

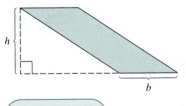

FIGURE 12.48   Why is the area $b \times h$?

5. 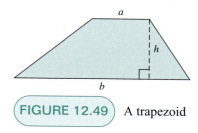 Figure 12.49 shows a trapezoid. This problem will help you find a formula for the area of the trapezoid and explain in several different ways why this formula is valid.

$$\boxed{\text{FIGURE 12.49}}\quad\text{A trapezoid}$$

a. By subdividing the trapezoid into two triangles, as shown in Figure 12.50, find a formula in

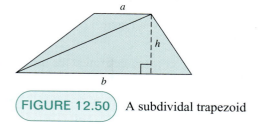

$$\boxed{\text{FIGURE 12.50}}\quad\text{A subdividal trapezoid}$$

terms of a, b, and h for the area of the trapezoid, and explain why your formula is valid.

b. Show how to combine two copies of the trapezoid in Figure 12.49 to make a parallelogram. Then use the formula for the area of a parallelogram to deduce the formula for the area of the trapezoid. Explain your reasoning.

c. Show how to cut off portions of the top half of the trapezoid and combine these portions with the bottom half of the trapezoid so as to make one or several rectangles, each of which has one side of length $\frac{1}{2}h$. Use this method to deduce the formula for the area of the trapezoid. Explain your reasoning.

d. Besides the methods of parts (a), (b) and (c) of this problem, there are still other methods for deducing the formula for the area of a trapezoid by using the moving and additivity principles. Find another method. Explain your reasoning.

6. A rug company weaves rugs that are made by repeating the design in Figure 12.51. Lengths of portions of the design are indicated in the figure. The yarn for the shaded portion of the design costs $5 per square unit, and the yarn for the unshaded portion of the design costs $3 per square unit. How much will the yarn for a 60-unit-by-84-unit rug cost? Explain your reasoning.

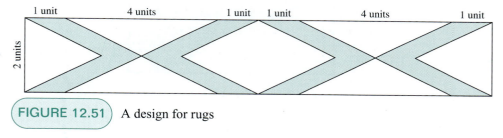

$$\boxed{\text{FIGURE 12.51}}\quad\text{A design for rugs}$$

7. Use the moving and additivity principles to determine the area (in square units) of the rhombus in Figure 12.52 *without* using the triangle or parallelogram area formulas. Adjacent dots are 1 unit apart. Explain your reasoning.

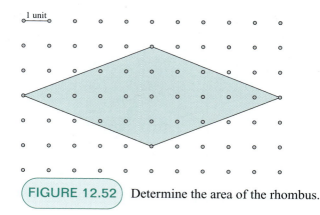

$$\boxed{\text{FIGURE 12.52}}\quad\text{Determine the area of the rhombus.}$$

8. Use the moving and additivity principles to find a formula for the area of the rhombus in Figure 12.53 in terms of the lengths of its diagonals. Do so *without* using the triangle or parallelogram area formulas. Let a and b be the lengths of the diagonals of the rhombus. Explain your reasoning.

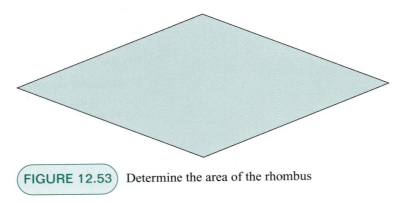

FIGURE 12.53   Determine the area of the rhombus

9. **a.** Determine the areas (in square units) of the 4 lightly shaded triangles in Figure 12.54. The grid lines are 1 unit apart. Explain your reasoning.

   b. Use the moving and additivity principles and your results from part (a) to determine the area of the dark shaded quadrilateral in Figure 12.54. Explain your reasoning.

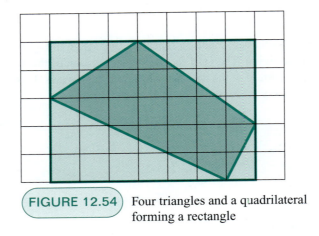

FIGURE 12.54   Four triangles and a quadrilateral forming a rectangle

10. Determine the area (in square units) of the quadrilateral in Figure 12.55 in *two different ways*. The grid lines are 1 unit apart. Explain your reasoning.

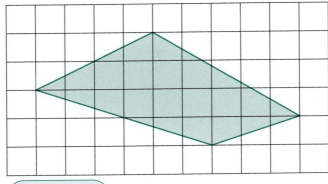

FIGURE 12.55   Determine the area of the quadrilateral.

11. Determine the area of the shaded shapes in Figure 12.56. Explain your reasoning.

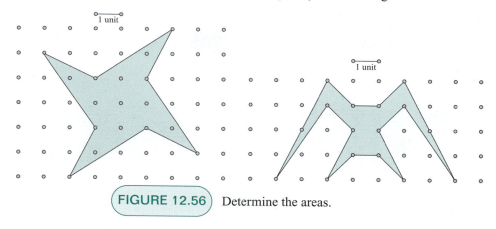

FIGURE 12.56   Determine the areas.

12. Figure 12.57 shows a map of some land. Determine the size of this land in acres. Recall that 1 acre is 43,560 square feet.

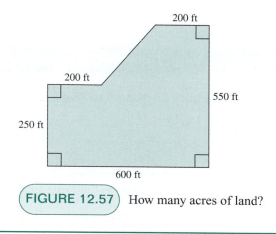

FIGURE 12.57   How many acres of land?

12.5 ⓒ Cavalieri's Principle about Shearing and Area

In addition to subdividing a shape and recombining its parts without overlapping them, there is another way, called *shearing*, to change a shape into a new shape that has the same area. Another way to explain the area formulas for triangles and parallelograms is with shearing.

To illustrate shearing, start with a polygon, pick one of its sides, and then imagine slicing the polygon into extremely thin (really, infinitesimally thin) strips that are parallel to the chosen side. Now imagine giving those thin strips a push from the side, so that the chosen side remains in place, but the thin strips slide over, remaining parallel to the chosen side and remaining the same distance from the chosen side throughout the sliding process. Then you will have a new polygon, as indicated in Figures 12.58 and 12.59. This process of "sliding infinitesimally thin strips" is called **shearing**.

shearing

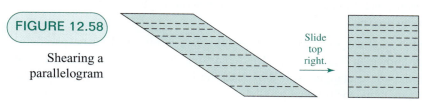

FIGURE 12.58

Shearing a
parallelogram

Slide
top
right.

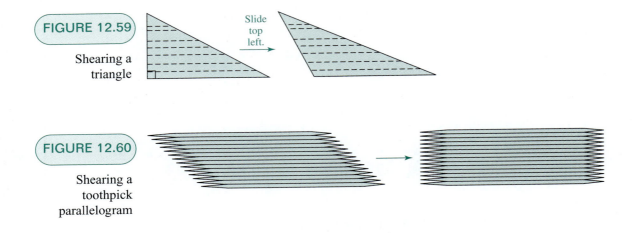

FIGURE 12.59

Shearing a triangle

FIGURE 12.60

Shearing a toothpick parallelogram

To simulate shearing, we replace the infinitesimally thin strips with toothpicks. If we then give the stack of toothpicks a push from the side, they will slide over, as in shearing. (See Figure 12.60.)

Class Activity   *Now Turn to Class Activities Manual*

12J Shearing a Toothpick Rectangle to Make a Parallelogram, p. 290

12K Is This Shearing?, p. 291

Cavalieri's principle

Cavalieri's principle for areas says that when a shape is sheared as just described, the areas of the original and sheared shapes are equal. Also note the following about shearing:

- During shearing, each point moves along a line that is *parallel* to the fixed side.

- During shearing, the thin strips *remain the same width and length*. The strips just slide over; they are not compressed either in width or in length.

- Shearing does not change the height of the "stack" of thin strips. In other words, if you think of shearing in terms of sliding toothpicks, the height of the stack of toothpicks doesn't change during shearing.

Class Activity   *Now Turn to Class Activities Manual*

12L Shearing Parallelograms, p. 292

12M Shearing Triangles, p. 292

We can use shearing to determine areas of shapes and to explain in another way why the area formulas for triangles and parallelograms are valid.

To explain why the area of the parallelogram in Figure 12.61(a) is $b \times h$, we can shear the parallelogram into a rectangle, as shown in Figure 12.61(b). According to Cavalieri's principle, the original parallelogram and the new rectangle formed by shearing have the same area. Because the rectangle has area $b \times h$ square units, the original parallelogram also has area $b \times h$ square units.

FIGURE 12.61

Shearing a
parallelogram
(a) into a
rectangle (b)

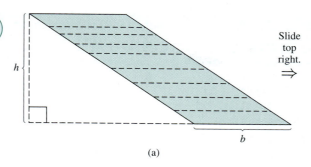
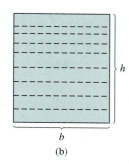

Slide
top
right.
$\Rightarrow$

(a)   (b)

FIGURE 12.62

Why is the area
$\frac{1}{2}(b \times h)$?

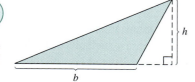

To explain why the area of the triangle in Figure 12.62 is $\frac{1}{2}(b \times h)$, shear the triangle into a right triangle, as shown in Figure 12.63. Recall that you can think of this shearing process as slicing the triangle into (infinitesimally) thin strips and giving those strips a push to move them over. Notice that, since the shearing is done parallel to the base of the triangle, the *sheared triangle still has the same base* b *and height* h *as the original triangle.*

The new right triangle has area $\frac{1}{2}(b \times h)$ square units, since two copies of this triangle can be combined without overlapping to form a *b*-by-*h* rectangle, as shown in Figure 12.64. Remember, this is the same *b* and *h* as for the original triangle. Now, according to Cavalieri's principle, the shearing process does not change the area. So the area of the original triangle is the same as the area of the new right triangle that was formed by shearing. Therefore, the area of the original triangle in Figure 12.62 is also $\frac{1}{2}(b \times h)$ square units, which is what we wanted to show. Notice that although we worked with the particular triangle in Figure 12.62, there wasn't anything special about this triangle. The same argument applies to show that the $\frac{1}{2}(b \times h)$ formula for the area of a triangle is valid for *any* triangle.

FIGURE 12.63

Shearing a
triangle

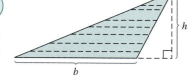
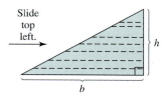

Slide
top
left.

FIGURE 12.64

The right
triangle is half
of a rectangle.

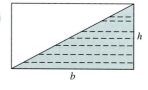

Practice Exercises for Section 12.5

1. Figure 12.65 shows a parallelogram on a peg-board. (Think of the parallelogram as made out of a rubber band, which is hooked around four pegs.) Show two ways to move points C and D of the parallelogram to other pegs, keeping points A and B fixed, in such a way that the area of the new parallelogram is the same as the area of the original parallelogram.

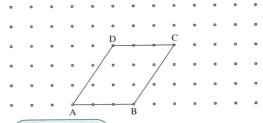

FIGURE 12.65 Parallelogram on a pegboard

2. Using two ordinary plastic drinking straws, cut two 4-inch pieces of straw and two 3-inch pieces of straw. Lace these pieces of straw onto a string in the following order: a 3-inch piece, a 4-inch piece, a 3-inch piece, a 4-inch piece. Tie a knot in the string so that the four pieces of straw form a quadrilateral, as pictured in Figure 12.66.

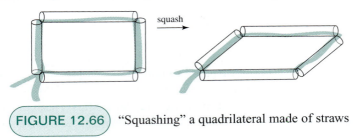

FIGURE 12.66 "Squashing" a quadrilateral made of straws

Put your straw quadrilateral in the shape of a rectangle. Gradually "squash" the quadrilateral so that it forms a parallelogram that is not a rectangle, as indicated in Figure 12.66. Is this "squashing" process for changing the rectangle into a parallelogram the same as the shearing process? Why or why not?

Answers to Practice Exercises for Section 12.5

1. See Figure 12.67. The parallelograms ABEF and ABGH are two examples of parallelograms that have the same area as ABCD. These parallelograms have the same area as ABCD by Cavalieri's principle, because they are just sheared versions of ABCD. In fact, if we move C and D anywhere along the line of pegs that goes through C and D, keeping the same distance between them, the resulting parallelogram will be a sheared version of ABCD and hence will have the same area as ABCD.

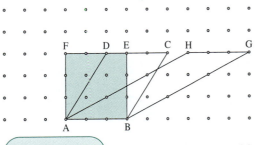

FIGURE 12.67 Three parallelograms with the same area

2. No, this "squashing" process is not the same as shearing. You can tell that it's not shearing because, if you made the rectangle out of very thin strips and you slid them over to make a parallelogram, they would have to become thinner to make the parallelogram that is formed from the straws (because the parallelogram is not as tall as the rectangle). But in the shearing process, the size of the strips does not change (either in length or in width). Therefore, "squashing" is not the same as shearing.

Problems for Section 12.5

1. Figure 12.68 shows a triangle on a pegboard. (Think of the triangle as made out of a rubber band, which is stretched around three pegs.) Describe or show pictures of at least two ways to move point C of the triangle to another peg (keeping points A and B fixed) in such a way that the area of the new triangle is the same as the area of the original triangle. Explain your reasoning.

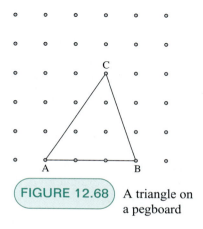

FIGURE 12.68 A triangle on a pegboard

2. **a.** Draw a picture showing the result of shearing the parallelogram in Figure 12.69 into a rectangle. Explain how you know you have sheared the parallelogram correctly.

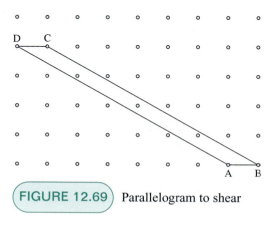

FIGURE 12.69 Parallelogram to shear

   b. Use part (a) to determine the area of the parallelogram in Figure 12.69. Explain why you can determine the area of the parallelogram this way.

3. **a.** Draw a picture showing the result of shearing the parallelogram in Figure 12.70 into a rectangle. Explain how you know you have sheared the parallelogram correctly. Note: Shearing does not have to be horizontal.

   b. Use part (a) to determine the area of the parallelogram in Figure 12.70. Explain why you can determine the area of the parallelogram this way.

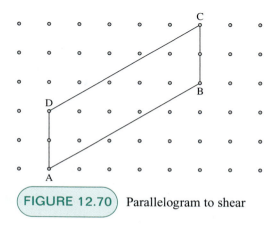

FIGURE 12.70 Parallelogram to shear

4. **a.** Draw a picture showing the result of shearing the triangle in Figure 12.71 into a right triangle. Explain how you know you have sheared the triangle correctly.

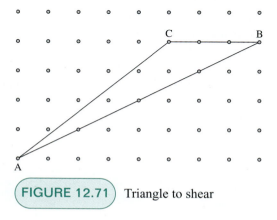

FIGURE 12.71 Triangle to shear

   b. Use the moving and additivity principles to help determine the area of your right triangle in part (a). Explain your reasoning. (Do not use a formula for areas of triangles.)

c. Use the results of parts (a) and (b) to determine the area of the triangle in Figure 12.71. Explain why you can determine the area of the triangle this way.

5. a. Draw a picture showing the result of shearing the triangle in Figure 12.72 into a right triangle. Explain how you know you have sheared the triangle correctly. Note: Shearing does not have to be horizontal.

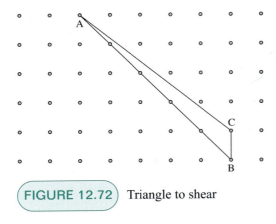

FIGURE 12.72   Triangle to shear

b. Use the moving and additivity principles to help determine the area of your right triangle in part (a). Explain your reasoning. (Do not use a formula for areas of triangles.)

c. Use the results of parts (a) and (b) to determine the area of the triangle in Figure 12.72. Explain why you can determine the area of the triangle this way.

6. The boundary between the Johnson and the Zhang properties is shown in Figure 12.73. The Johnsons and the Zhangs would like to change this boundary so that the new boundary is one straight line segment and so that each family still has the same amount of land area. Describe a precise way to redraw the boundary between the two properties. Explain your reasoning. *Hint*: Consider shearing the triangle ABC.

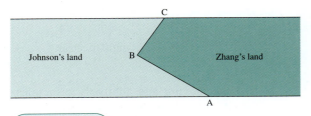

FIGURE 12.73   Boundary between two properties

7. Figure 12.74 shows a trapezoid. This problem will help you to find a formula for the area of the trapezoid, and to explain why this formula is valid.

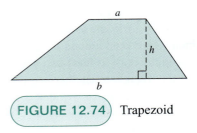

FIGURE 12.74   Trapezoid

a. Using Cavalieri's principle, explain why the trapezoid in Figure 12.74 has the same area as the new trapezoid in Figure 12.75.

b. Find a formula for the area of the *new trapezoid* in Figure 12.75 in terms of a, b, and h. Explain clearly why your formula gives the area of this new trapezoid.

c. Using parts (a) and (b) of this problem, give a formula in terms of a, b, and h for the area of the *original trapezoid* in Figure 12.74, and explain clearly why this formula gives the area of the original trapezoid.

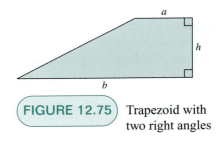

FIGURE 12.75   Trapezoid with two right angles

8. a. Suppose that in a trapezoid ABCD, as in Figure 12.76, AB and CD are parallel.

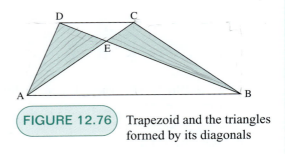

FIGURE 12.76   Trapezoid and the triangles formed by its diagonals

Let E be the point where the diagonals AC and BD meet. Explain why triangles AED and BEC must have the same area. (Do this without measuring actual lengths or areas.)

b.  What condition on the trapezoid ABCD will guarantee that AEB and CED have the same area? Explain.

9.  Given three points A, B, and C in a plane, as in Figure 12.77, describe geometrically all the places where you can put a point D so that if E is the point where the line segments BC and AD meet, the triangles ABE and CDE have the same area. Explain your reasoning.

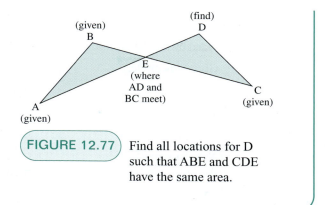

FIGURE 12.77 Find all locations for D such that ABE and CDE have the same area.

12.6 Areas of Circles and the Number Pi

F·P **Focal Points**
Grade 7

What is the area of a circle? You probably know the familiar formula

$$\pi r^2$$

for the area of a circle of radius r. The Greek letter π, pronounced "pie," stands for a mysterious number that is approximately equal to 3.14159. Since at least the time of the ancient Babylonians and Egyptians, nearly 4000 years ago, people have known about, and been fascinated by, the remarkable number π. In this section, we will discuss the number π, and then we will see why the area of a circle of radius r units is πr^2 square units.

circumference First, we need some terminology. The **circumference** of a circle is the distance around the circle. (See Figure 12.78.) Recall that the radius of a circle is the distance from the center of the circle to any point on the circle. Recall also that the diameter of a circle is the distance across the circle, going through the center; it is twice the radius.

pi (π) For any circle whatsoever—whether huge, tiny, or in between—the circumference divided by the diameter is always equal to the same number; this number is called **pi** and is written with the Greek letter π. That the circumference of a circle divided by its diameter always results in the same number can be

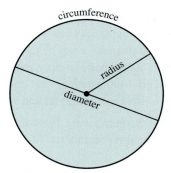

FIGURE 12.78 The circumference of a circle is the distance around the circle; the diameter is the distance across the circle through the center; the radius is the distance from the center to any point on the circle.

explained by establishing that every circle is a scaled version of one fixed circle. In scaling, both the circumference and the diameter are multiplied by the same number. Therefore, when the circumference is divided by the diameter, the same value results for all circles.

Given any circle,

$$\text{circumference} \div \text{diameter} = \pi$$

Thus,

$$\text{circumference} = \pi \times \text{diameter}$$

Now let r stand for the radius of a circle. Because the diameter of a circle is twice its radius, we obtain the following familiar expression $2\pi r$ for the circumference of a circle of radius r:

$$\text{circumference} = \pi \times 2r = 2\pi r$$

So, for example, if you plan to make a circular garden with a radius of 10 feet, and you want to enclose the garden with a fence, then you will need

$$2\pi \times 10 \text{ ft} = 63 \text{ ft, approximately}$$

of fence.

Class Activity   *Now Turn to Class Activities Manual*

12N How Big Is the Number π?, p. 294

12O A Hands-on Method to Approximate the Value of π, p. 294

You can do a simple experiment to determine the approximate value of π. Take a sturdy cup, can, or plate; measure the distance around the rim with a tape measure; and measure the distance across the cup, can, or plate at the widest part. When you divide these numbers you get an approximate value for π if the rim of the cup, can, or plate is a circle. But even with a big plate, an accurate tape measure, and careful measuring, you will probably be able to determine only that π is about 3.1. If you enter π on your calculator, however, you will see many more decimal places:

$$\pi = 3.14159265\ldots$$

Your calculator can show only a finite number of digits behind the decimal point, but it turns out that the decimal expansion of π goes on forever and does not have a repeating pattern. The mathematician Johann Lambert (1728–1777) first proved this in 1761.

Why is it that such a simple and perfect shape as a circle gives rise to such a mysterious and complicated number as π? Many people are attracted to mathematics because it provides a glimpse into the mysterious, the perfect, and the infinite. Although mathematics can be used to solve many practical problems, some people find the mystery and infinity in mathematics deeply appealing. It is not unlike the appeal of the greatest pieces of music, literature, and art. Mathematics is not only a practical subject, but it also informs us about what it means to be human, as music, literature, and art do.

How do we know the decimal expansion of the number π? There are many known formulas for π, and these formulas can be used to find the decimal representation of π. One elegant formula for π comes from the following equation:

$$\frac{\pi}{4} = 1 - \frac{1}{3} + \frac{1}{5} - \frac{1}{7} + \frac{1}{9} - \frac{1}{11} + \cdots$$

This equation was discovered by Indian mathematicians in the fifteenth century. The ellipsis in the expression to the right of the equal sign indicates that this expression goes on forever, continuing the pattern of adding and subtracting fractions whose denominators are the odd numbers in sequence. It is beyond the scope of this book to explain why this formula is true or where it comes from.

circle area formula   Not only is the circumference of a circle related to the number π, but the **area of a circle** is as well. A circle of radius r units has area

$$\pi r^2 \text{ square units}$$

For example, suppose we want to find the area of a circular patio of diameter 30 feet. If the diameter is 30 feet, then the radius is 15 feet, and the area of the patio is

$$\pi \times 15^2 \text{ ft}^2 = \pi \times 225 \text{ ft}^2 = 707 \text{ ft}^2$$

When you use the πr^2 area formula, be sure that you square only the value of r, and not the value of πr. For example, if you multiply π times r first, and then square that result, your answer will be π times too large.

Class Activity   *Now Turn to Class Activities Manual*

12P Overestimates and Underestimates for the Area of a Circle, p. 295

12Q Why the Area Formula for Circles Makes Sense, p. 295

12R Area Problems, p. 297

Practice Exercises for Section 12.6

1. Suppose we don't know the formula for the area of a circle, but we do know about areas of squares. What can we deduce about the area of a circle of radius r units from Figure 12.79?

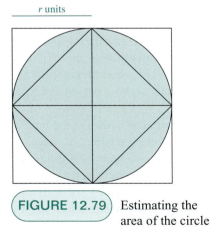

r units

FIGURE 12.79 Estimating the area of the circle

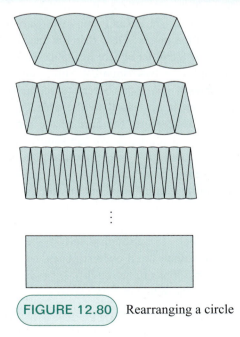

FIGURE 12.80 Rearranging a circle

2. Using Figure 12.80 (see also Class Activity 12Q), explain why it makes sense that a circle of radius r units has area πr^2 square units, assuming we already know that a circle of radius r has circumference $2\pi r$.

3. Some trees in an orchard need to have their trunks wrapped with a special tape to prevent an attack of pests. Each tree's trunk is about 1 foot in diameter and must be covered with tape from ground level up

to a height of 4 feet. The tape is 3 inches wide. Approximately how long a piece of tape will be needed for each tree?

4. The Browns plan to build a 5-foot-wide garden path around a circular garden of diameter 25 feet, as shown in Figure 12.81.

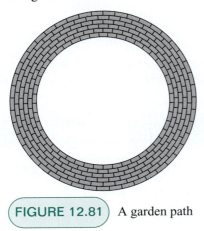

FIGURE 12.81   A garden path

What is the area of the garden path? Explain your answer.

5. What is the area of the 4-petal flower in Figure 12.82? The square is 6 cm by 6 cm, and the curves are half-circles.

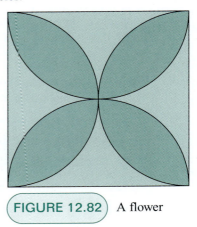

FIGURE 12.82   A flower

Answers to Practice Exercises for Section 12.6

1. We can conclude that the area of the circle is between $2r^2$ square units and $4r^2$ square units. The area of the circle is less than $4r^2$ because we can think of the outer square in Figure 12.79 as made up of four r-by-r squares. Since the circle lies completely inside this outer square, the area of the circle must be less than the area of the outer square. Therefore, the area of the circle is less than $4r^2$ square units. We can subdivide the inner square (diamond) into four triangles, which we can recombine to make two r-by-r squares. Since the circle completely contains the inner square, the area of the circle must be greater than $2r^2$ square units. Just by eyeballing, the area of the circle looks to be roughly halfway between these underestimates and overestimates, so the area of the circle looks to be roughly $3r^2$ (which is pretty close to the actual πr^2).

2. If you cut a circle into 8 pie pieces and rearrange them as at the top of Figure 12.80, you get a shape that looks something like a rectangle. If you cut the circle into 16 or 32 pie pieces and rearrange them as in the middle of Figure 12.80, you get shapes that look even more like rectangles. If you could keep

cutting the circle into more and more pie pieces, and keep rearranging them as before, you would get shapes that look more and more like the rectangle shown at the bottom of Figure 12.80.

The height of the rectangle in Figure 12.80 is the radius r of the circle. To determine the width of the rectangle (in the horizontal direction), notice that in the rearranged circles at the top and in the middle of Figure 12.80, half of the pie pieces point up and half point down. The circumference of the circle is divided equally between the top and bottom sides of the rectangle. The circumference of the circle is $2\pi r$; therefore, the width (in the horizontal direction) of the rectangle is half as much, which is πr. So the rectangle is r units by πr units, and therefore has area $\pi r \times r = \pi r^2$ square units. Since the rectangle is basically a cut-up and rearranged circle of radius r, the area of the rectangle ought to be equal to the area of the circle. Therefore, it makes sense that the area of the circle is also πr^2.

3. One "wind" of tape all the way around a tree trunk makes an approximate circle. Because the tree trunk has diameter 1 foot, each wind around the trunk uses about π feet of tape. The tape is 3 inches

wide, so it will take 4 winds for each foot of trunk height to be covered. Therefore, it will take 16 winds to cover the desired amount of trunk. This will use about $16 \times \pi$ ft, or about 50 ft of tape.

4. The diameter of the circular garden is 25 ft, so its radius is half as much, which is 12.5 feet. The garden together with the path form a larger circle of radius (12.5 + 5) feet = 17.5 feet. By the additivity principle about areas,

area of garden + area of path = area of larger circle

Therefore,

$$\begin{aligned} \text{area of path} &= \text{area of larger circle} \\ &\quad - \text{area of garden} \\ &= \pi 17.5^2 \text{ ft}^2 - \pi 12.5^2 \text{ ft}^2 \\ &= 471 \text{ ft}^2 \end{aligned}$$

5. You can make the design by covering the square with four half-circles of tissue paper; then the flower petals are exactly the places where *two* pieces of tissue paper overlap. The remaining parts of the square are covered with only *one* layer of tissue paper. Therefore,

area of four half-circles = area of flower
+ area of square

So that

$$\begin{aligned} \text{area of flower} &= \text{area of two circles} \\ &\quad - \text{area of square} \\ &= 18\pi - 36 \text{ cm}^2 \\ &= 20.5 \text{ cm}^2 \end{aligned}$$

Problems for Section 12.6

1. Tim works on the following exercise:

   For each radius r, find the area of a circle of that radius:

   $$r = 2 \text{ in.}, \quad r = 5 \text{ ft}, \quad r = 8.4 \text{ m}$$

   Tim gives the following answers:

   39.48, 246.74, 696.399

   Identify the errors that Tim has made. How did Tim likely calculate his answers? Discuss how to correct the errors; include a discussion on the proper use of a calculator in solving Tim's exercise. Be sure to discuss the appropriate way to write the answers to the exercise.

2. A large running track is constructed to have straight sections and two semicircular sections with dimensions given in Figure 12.83. Assume

that runners always run on the inside line of their track. A race consists of one full counterclockwise revolution around the track plus an extra portion of straight segment, to end up at the finish line shown. What should the distance x between the two starting blocks be in order to make a fair race? Explain your reasoning.

3.  Suppose you have a large spool used for winding rope (just like a spool of thread), such as the ones shown in Figures 12.84 and 12.85.

FIGURE 12.84 Large spools of wire

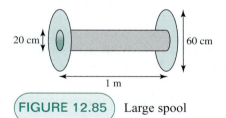

FIGURE 12.85 Large spool

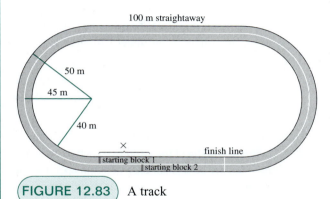

FIGURE 12.83 A track

Suppose that the spool is 1 m long and has an inner diameter of 20 cm and an outer diameter of 60 cm. Approximately how long a piece of 5-cm-thick rope can be wound onto this spool? (Assume that the rope is wound on neatly, in layers. Each layer will consist of a row of "winds," and each "wind" will be approximately a circle.) Explain your reasoning.

4. Jack has a truck that requires tires that are 26 inches in diameter. (Looking at a tire from the side of a car, a tire looks like a circle. The diameter of the tire is the diameter of this circle.) Jack puts tires on his truck that are 30 inches in diameter.

   a. A car's speedometer works by detecting how fast the car's tires are rotating. Speedometers do not detect how big a car's tires are. When Jack's speedometer reads 60 miles per hour is that accurate, or is Jack actually going slower or faster? Explain your reasoning. An exact determination of Jack's speed is not needed.

   b. Determine Jack's speed when his speedometer reads 60 mph. Explain your answer thoroughly.

5. Suppose that when pizza dough is rolled out it costs 25 cents per square foot, and that sauce and cheese, when spread out on a pizza, have a combined cost of 60 cents per square foot. Let's say sauce and cheese are always spread out to within 1 inch of the edge of the pizza. Compare the sizes and costs of a circular pizza of diameter 16 inches and a 10-inch-by-20-inch rectangular pizza.

6. Lauriann and Kinsey are in charge of the annual pizza party. In the past, they've always ordered 12-inch-diameter round pizzas, and each 12-inch pizza has always served 6 people. This year, the jumbo 16-inch-diameter round pizzas are on special, so Lauriann and Kinsey decide to get 16-inch pizzas instead. Lauriann and Kinsey think that since a 12-inch pizza serves 6 (which is half of 12), a 16-inch pizza should serve 8 (which is half of 16). But when Lauriann and Kinsey see a 16-inch pizza, they think it ought to serve even more than 8 people. Suddenly, Kinsey realizes the flaw in their reasoning that a 16-inch pizza should serve 8. Kinsey has an idea for determining how many people a 16-inch pizza will serve. What mathematical reasoning might Kinsey be thinking of, and how many people should a 16-inch-diameter pizza serve if a 12-inch-diameter pizza serves 6? (See Figure 12.86.) Explain your answers.

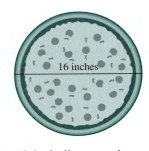

FIGURE 12.86 A 12-inch-diameter pizza and a 16-inch-diameter pizza

7. Penguins huddle together to stay warm in very cold weather. Suppose that a certain type of penguin has a circular cross-section approximately 14 inches in diameter (so that if you looked down on the penguin from above, the shape you would see would be a circle, 14 inches in diameter). Suppose that a group of this type of penguin is huddling in a large circular cluster, about 20 feet in diameter. (All the penguins are still standing upright on the ground; they are not piled on top of each other.)

FIGURE 12.87 Penguins in a huge cluster

   a. Assuming that the penguins are packed together tightly, estimate how many penguins are in this cluster. (You might use areas to do this.) Is this an overestimate or an underestimate? Explain.

   b. The coldest penguins in the cluster are the ones around the circumference. Approximately how many of these cold penguins are there at any given time? Explain.

   c. So that no penguin gets too cold, the penguins take turns being at the circumference. How many minutes per hour does each penguin spend at the circumference if each penguin spends the same amount of time at the circumference? Explain.

8. Let r units denote the radius of each circle in Figure 12.88. For each shaded circle portion in Figure 12.88, find a formula for its area in terms of r. Use the moving and additivity principles about areas to explain why your formulas are valid.

Angle ABC is 1°.   Angle DEF is 107°.

 FIGURE 12.88 Parts of circles

9.   The text gives the formula

$$\frac{\pi}{4} = 1 - \frac{1}{3} + \frac{1}{5} - \frac{1}{7} + \frac{1}{9} - \frac{1}{11} + \ldots$$

This formula can be used to give better and better approximations to π by using more and more terms from the expression to the right of the equal sign. Here's how this works. Read the symbol $\approx$ as *is approximately equal to*.

One term: $\frac{\pi}{4} \approx 1$, so $\pi \approx 4$

Two terms: $\frac{\pi}{4} \approx 1 - \frac{1}{3}$, so $\pi \approx 2.667$

Three terms: $\frac{\pi}{4} \approx 1 - \frac{1}{3} + \frac{1}{5}$, so $\pi \approx 3.467$

a. Use 4, 5, 6, 7, and 8 terms of the expression to the right of the equal sign to find five more approximations to π. Also find the 20th and the 21st approximations to π.

b. Looking at the three examples given and your results in part (a), describe a pattern to the approximations to π obtained by this method. (Look at the sizes of your answers.) How do these approximations compare with the actual value of π?

c. You can get better approximations of π by taking averages of successive approximations. For example, the average of the 1st and 2nd approximations to π is

$$\frac{4 + 2.667}{2} = 3.334$$

and the average of the 2nd and 3rd approximations to π is

$$\frac{2.667 + 3.467}{2} = 3.067$$

Find the average of the 3rd and 4th approximations to π, the 4th and 5th approximations, the 5th and 6th, the 6th and 7th, and the 7th and 8th approximations to π. Also find the average of the 20th and the 21st approximations. How close is this last average to the actual value of π?

12.7 Approximating Areas of Irregular Shapes

Moving and additivity principles can help us to easily find the areas of some shapes and to explain why area formulas are valid, but the principles can't always help us find a precise area for irregular shapes. For example, how could we determine the area of the bean-shaped region in Figure 12.89? There isn't any way to subdivide and recombine the bean-shaped region into regions whose areas are easy to determine. Usually, we cannot determine the area of an irregular region exactly; instead, we must be satisfied with determining the approximate area.

Class Activity *Now Turn to Class Activities Manual*

12S Determining the Area of an Irregular Shape, p. 297

FIGURE 12.89

A bean-shaped region

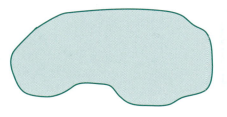

FIGURE 12.90

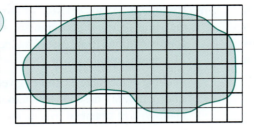

Determining
the area of a
bean-shaped
region

One hands-on way to approximate the area of an irregular region, such as the bean-shaped region in Figure 12.89, is to cut it out, weigh it, and compare its weight with the weight of a full piece of paper. The next step is to use proportional reasoning to find the area of the bean-shaped region. For example, if the irregular region weighs $\frac{3}{8}$ as much as a whole piece of paper, then its area is also $\frac{3}{8}$ as much as the area of the whole piece of paper.

We can also determine the area of an irregular region approximately by using graph paper, as indicated in Figure 12.90. The lines on this graph paper are spaced $\frac{1}{2}$ cm apart, with heavier lines spaced 1 cm apart (so that 4 small squares make 1 square centimeter).

By counting the number of 1-cm-by-1-cm squares (each consisting of 4 small squares) inside the bean-shaped region, and by mentally combining the remaining portions of the region that are near the boundary, we can determine that the bean-shaped region has an area of about 19 square centimeters.

We can also use graph paper to find underestimates and overestimates for the area of an irregular region. For example, the shaded squares in Figure 12.91 lie entirely within the bean-shaped region, so their combined area must be less than the area of the region. There are 58 small squares that lie inside the bean-shaped region. Each of these small squares is $\frac{1}{2}$ cm by $\frac{1}{2}$ cm and thus has an area of

$$\frac{1}{2} \cdot \frac{1}{2} \text{ cm}^2 = \frac{1}{4} \text{ cm}^2$$

Therefore, the 58 small squares have an area of

$$58 \cdot \frac{1}{4} \text{ cm}^2 = 14\frac{1}{2} \text{ cm}^2$$

So the area of the shaded region must be greater than $14\frac{1}{2}$ cm^2.

On the other hand, the dark outline in Figure 12.91 surrounds the squares that contain some portion of the bean-shaped region. The combined area of these squares must therefore be greater than the area of the region. There are 97 such small squares, each of which has area $\frac{1}{4}$ cm^2; thus, these 97 squares have a combined area of

$$97 \cdot \frac{1}{4} \text{ cm}^2 = 24\frac{1}{4} \text{ cm}^2$$

The area of the bean-shaped region must be less than $24\frac{1}{4}$ cm^2. So the area of the bean-shaped region must be between $14\frac{1}{2}$ cm^2 and $24\frac{1}{4}$ cm^2. If we used finer and finer graph paper, we would get narrower and narrower ranges between our underestimates and overestimates for the area of the region.

FIGURE 12.91

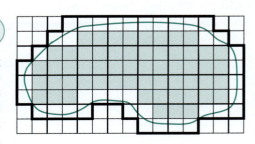

Determining
underestimates
and overestimates
for the area of the
bean-shaped
region

Practice Exercises for Section 12.7

1. a. In Figure 12.92, the small squares in the grid are 1 cm by 1 cm. Find an underestimate and an overestimate for the area of the shaded region by considering the squares that lie entirely within the region, and the squares that contain a portion of the region (as indicated in Figure 12.91).

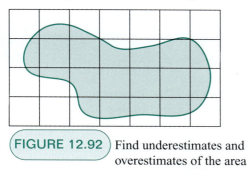

FIGURE 12.92 Find underestimates and overestimates of the area

b. Use the grid of Figure 12.92 to determine approximately the area of the shaded region in that figure.

c. Figure 12.93 shows the same shaded region as the region in parts (a) and (b), but this time with a finer grid. In this figure, the grid lines are $\frac{1}{2}$ cm apart. Use this finer grid to give better over-estimates and underestimates for the area of the shaded region than you found in part (a).

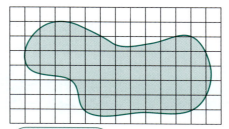

FIGURE 12.93 Find new underestimates and overestimates of the area

d. Use the finer grid of Figure 12.93 to determine approximately the area of the shaded region.

2. Suppose that you have a map on which 1 inch represents 50 miles. You trace a state on the map onto $\frac{1}{2}$-inch graph paper. (The grid lines are spaced $\frac{1}{2}$ inch apart.) You count that the state takes up about 91 squares of graph paper. Approximately what is the area of the state? Explain.

Answers to Practice Exercises for Section 12.7

1. a. There are 5 squares that lie entirely within the region; therefore, an underestimate for the area of the region is 5 square centimeters. There are 23 squares that contain some portion of the region. Therefore, 23 square centimeters is an overestimate for the area of the region.

b. Counting whole squares and combining partial squares, the area of the shaded region is about 13 square centimeters.

c. There are 41 small squares that are contained completely within the shaded region. Each small square is $\frac{1}{2}$ cm by $\frac{1}{2}$ cm; therefore, the area of each small square is $\frac{1}{2} \cdot \frac{1}{2}$ square centimeters, which is $\frac{1}{4}$ cm². Consequently, an underestimate for the area of the region is

$$41 \cdot \frac{1}{4} \text{ cm}^2 = 10\frac{1}{4} \text{ cm}^2$$

There are 77 small squares that contain some portion of the region. Therefore, an overestimate for the area of the region is

$$77 \cdot \frac{1}{4} \text{ cm}^2 = 19\frac{3}{4} \text{ cm}^2$$

Notice that this finer grid gives a narrower range of overestimates and underestimates for the area of the region.

d. Counting whole squares and combining partial squares, we find that the area is approximately the area of 55 squares. Since each square has an area of $\frac{1}{4}$ cm², the area of the region is approximately

$$55 \cdot \frac{1}{4} \text{ cm}^2 = 13\frac{3}{4} \text{ cm}^2$$

or about 14 square centimeters.

2. Each square on the graph paper is $\frac{1}{2}$ inch by $\frac{1}{2}$ inch; therefore, the area of each square of graph paper is $\frac{1}{2} \cdot \frac{1}{2}$ square inches, which is $\frac{1}{4}$ square inch. So, 91 squares of graph paper have a combined area of

$$91 \cdot \frac{1}{4} \text{ in.}^2 = 22\frac{3}{4} \text{ in.}^2$$

Since 1 inch on the map represents 50 miles, 1 square inch on the map represents

$$50 \cdot 50 \text{ mi}^2 = 2500 \text{ mi}^2$$

of actual land. Thus, $22\frac{3}{4}$ square inches on the map represents

$$22\frac{3}{4} \cdot 2500 \text{ mi}^2 = 56{,}875 \text{ mi}^2$$

or about 57,000 square miles.

Problems for Section 12.7

1. Suppose that you have a map on which 1 inch represents 100 miles. You trace a state on the map onto $\frac{1}{4}$-inch graph paper. (The grid lines are spaced $\frac{1}{4}$ inch apart.) You count that the state takes up about 80 squares of graph paper. Approximately what is the area of the state? Explain.

2. Suppose that you have a map on which 1 inch represents 25 miles. You cover a county on the map with a $\frac{1}{8}$-inch-thick layer of modeling dough. Then you re-form this piece of modeling dough into a $\frac{1}{8}$-inch-thick rectangle. The rectangle is

$1\frac{3}{4}$ inches by $2\frac{1}{4}$ inches. Approximately what is the area of the county? Explain.

3. Suppose that you have a map on which 1 inch represents 30 miles. You trace a state on the map, cut out your tracing, and draw this tracing onto card stock. Using a scale, you determine that a full $8\frac{1}{2}$-inch-by-11-inch sheet of card stock weighs 10 grams. Then you cut out the tracing of the state that is on card stock and weigh this card stock tracing. It weighs 5 grams. Approximately what is the area of the state? Explain.

12.8 Contrasting and Relating the Perimeter and Area of a Shape

If you know the distance around a shape, can you determine its area? If you know the area of a shape, can you determine the distance around the shape? What is the difference between perimeter and area? We will study these questions next.

Students sometimes get confused when deciding how to calculate perimeter and area. We *add* to calculate perimeter but we *multiply* to calculate areas of rectangles. When we calculate the perimeter of a rectangle, we add the lengths of all four sides (or we add the lengths of two adjacent sides and multiply by 2). On the other hand, when we calculate the area of a rectangle, we multiply only two of the sides' lengths. It's hard to keep all these different methods of calculation straight unless we have a clear idea of what perimeter and area mean and how they are different. But although they are different, there *is* a relationship between perimeter and area, as we will see in this section.

Calculating Perimeters of Polygons

Class Activity   *Now Turn to Class Activities Manual*

12T Calculating Perimeters, p. 299

Why do we add the lengths of the sides of a polygon to calculate its perimeter? The perimeter of a polygon (or other shape in the plane) is the total distance around the polygon. A hands-on way to think about

FIGURE 12.94

A shape of perimeter 26 cm and area 21 cm²

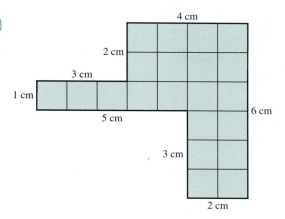

perimeter is as the length of a string that wraps snugly once around the shape. When we calculate the perimeter of a polygon by adding the lengths of the sides around the polygon, it is as if we had cut the string into pieces and are determining the total length of the string by adding the lengths of the pieces. We are calculating the total distance around the shape by adding up the lengths of pieces that together encircle the shape.

Class Activity *Now Turn to Class Activities Manual*

12U Perimeter Misconceptions, p. 300

Perimeter Is Different from Area

Note the difference between perimeter and area. The shape in Figure 12.94 has perimeter 26 cm, but it has area 21 cm². The perimeter of a shape is described by a unit of length, such as centimeters, whereas the area of a shape is described by a unit of area, such as square centimeters. The perimeter of the shape in Figure 12.94 is the number of 1-cm segments it takes to go all the way *around the shape*, whereas the area of the shape in Figure 12.94 is the number of 1-cm-by-1-cm squares it takes to *cover the shape* (without gaps or overlaps).

Class Activity *Now Turn to Class Activities Manual*

12V How Are Perimeter and Area Related?, p. 301

If you did Class Activity 12V, you discovered that for a given, fixed perimeter, there are many shapes that can have that perimeter, and these shapes can have different areas. By placing a loop of string on a flat surface, and by arranging the loop in different ways, you can show various shapes that have the same perimeters but different areas. (See Figure 12.95.) Therefore, perimeter does not determine area. But, for a given, fixed perimeter, which areas can occur?

Your intuition probably tells you that, of all shapes having a given, fixed perimeter, the circle is the one with the largest area. For example, of all shapes with perimeter 15 inches, a circle of circumference 15 inches is the shape that has the largest area. Since the circumference of this circle is 15 inches, its radius is

$$15 \div 2\pi \text{ in.} = 2.4 \text{ in. (approximately)}$$

FIGURE 12.95

Four strings of
equal length make
shapes of
different areas.

and therefore its area is

$$\pi \times 2.4^2 \text{ in.}^2 = 18 \text{ in.}^2 \text{ (approximately)}$$

By moving a 15-inch loop of string into various positions on a flat surface, you will probably find it plausible that *every* positive number less than 18 is the area, in square inches, of *some* shape of perimeter 15 in. So, for example, there is a shape that has perimeter 15 in. and area 17.35982 in.^{2}, and there is a shape that has perimeter 15 in. and area 2.7156 in.^{2} Notice that we can say this *without actually finding shapes* that have those areas and perimeters, which could be quite a challenge.

The following is true in general: Among all shapes of a given, fixed perimeter P, the circle of circumference P has the largest area, and every positive number that is less than the area of that circle is the area of some shape of perimeter P. So, although perimeter does not *determine* area, it does *constrain* which areas can occur. Explanations for why these facts are true are beyond the scope of this book.

What about areas of *rectangles* of a given, fixed perimeter? By stretching the loop of string between four thumbtacks pinned to cardboard, and by moving the thumbtacks, we can show various rectangles that all have the same perimeter, but have different areas. If you did Class Activity 12V, then you probably discovered that, of all rectangles of a given, fixed perimeter, the one with the largest area is a square. For example, among all rectangles of perimeter 24 inches, a square that has four sides of length 6 inches has the largest area, and this area is 6×6 in.$^2 = 36$ in.^{2} By moving a 24-inch loop of string to form various rectangles, you can probably tell that *every* positive number less than 36 is the area, in square inches, of *some* rectangle of perimeter 24 inches. So, for example, there is a rectangle that has perimeter 24 in. and area 35.723 in.^{2}, and there is a rectangle that has perimeter 24 in. and area 3.72 in.^{2}, even though it would take some work to find the exact lengths and widths of such rectangles.

The following is true in general: Among all rectangles of a given, fixed perimeter P, the square of perimeter P has the largest area, and every positive number that is less than the area of that square is the area of some rectangle of perimeter P. Although we will not do this here, these facts can be explained with algebra or calculus.

Class Activity   *Now Turn to Class Activities Manual*

12W Can We Determine Area by Measuring Perimeter?, p. 303

Practice Exercises for Section 12.8

1. The students in a class have been learning about the perimeter and area of rectangles. The teacher notices that one student consistently gives the wrong answer to problems on the area of rectangles: The student's answer is almost always 2 times the correct answer. For example, if the correct answer is 12 square feet, the student gives the answer 24 square feet. What might be the source of this error?

2. Describe a common error that students make when finding the perimeter of a shape such as the shaded shape in Figure 12.96. Describe the misunderstanding that is at the root of this error.

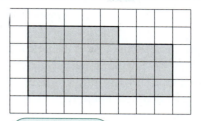

FIGURE 12.96 What error do students often make concerning perimeter?

3. A piece of property is described as having a perimeter of 4.7 miles. Without any additional information about the property, what is the most informative answer you can give about the area of the property? If you assume that the property is shaped like a rectangle, then what is the most informative answer you can give about its area?

Answers to Practice Exercises for Section 12.8

1. If the student had been calculating perimeters of rectangles by adding the lengths of 2 adjacent sides of the rectangle and multiplying this result by 2 (to account for the other 2 sides), then the student might be attempting to calculate area in an analogous way. Instead of just multiplying the length times width of the rectangle, the student may think that the result must be multiplied by 2, as it was for the perimeter.

2. Students sometimes count the number of squares along the outer border of the shape, as in Class Activity 12U. When they do so, they are not treating perimeter as a one-dimensional attribute. When students count squares, they are determining the area of the region formed by the border squares, which is a two-dimensional attribute.

3. Among all shapes that have perimeter 4.7 miles, the circle with circumference 4.7 miles has the largest area. This circle has radius

$$4.7 \div 2\pi \text{ miles} = 0.75 \text{ miles (approximately)}$$

and therefore it has area

$$(\pi \times 0.75^2) \text{ miles}^2 = 1.8 \text{ miles}^2 \text{ (approximately)}$$

Any area that is less than the area of this circle is a possible area for the property. So, without any additional information, the best we can say about the property is that its area is at most 1.8 square miles and the actual area could be anywhere between 0 and 1.8 square miles.

Now suppose that the property is shaped like a rectangle. Among all rectangles of perimeter 4.7 miles, the square of perimeter 4.7 miles has the largest area. This square has 4 sides of length $4.7 \div 4$ miles = 1.175 miles, and therefore this square has area 1.175×1.175 miles^2 = 1.4 miles^2. Any area that is less than the area of this square is a possible area for the property. So, if we know that the property is in the shape of a rectangle, then the most informative answer we can give about the area of the property is that its area is at most 1.4 square miles, and the actual area could be anywhere between 0 and 1.4 square miles.

Problems for Section 12.8

1. Suppose that a student in your class wants to know why we multiply only 2 of the lengths of the sides of a rectangle to determine the rectangle's area. After all, when we calculate the perimeter of a rectangle, we add the lengths of the *4 sides* of the rectangle, so why don't we multiply the lengths of the *4 sides* to find the area?

   a. Explain to the student what perimeter and area mean, and explain why we carry out the perimeter and area calculations for a rectangle the way we do.

   b. Describe some problems or activities that might help the student understand the calculations.

2. Sarah is confused about the difference between the perimeter and the area of a polygon. Explain the two concepts and the distinction between them.

3. Anya wants to draw many different rectangles that have a perimeter of 16 centimeters. Anya draws a few rectangles, but then she stops drawing and starts looking for pairs of numbers that add to 8.

   a. Why does it make sense for Anya to look for pairs of numbers that add to 8? How is Anya likely to use these pairs of numbers in drawing additional rectangles?

   b. How could you adapt Anya's idea if you were going to draw many different rectangles that have a perimeter of 14 centimeters?

   c. Use Anya's idea to help you draw 3 different rectangles that have a perimeter of 5 inches. Label your rectangles with their lengths and widths.

4. a. On graph paper, draw 4 different rectangles that have perimeter $6\frac{1}{2}$ inches.

   b. Without using a calculator, determine the areas of the rectangles you drew in part (a). Show your calculations, or explain briefly how you determined the areas of the rectangles.

5. Which of the lengths that follow could be the length of 1 side of a rectangle that has perimeter 7 inches? In each case, if the length is a possible side length of a rectangle of perimeter 7 inches, then determine the lengths of the other 3 sides without using a calculator. If the length is not a

possible side length of such a rectangle, then explain why not. Show your calculations.

   a. $2\frac{3}{4}$ inches

   b. $5\frac{1}{2}$ inches

   c. $1\frac{7}{8}$ inches

   d. $3\frac{3}{8}$ inches

   e. $3\frac{5}{8}$ inches

6. a. Without using a calculator, find the lengths and widths of 5 different rectangles that have perimeter $4\frac{1}{2}$ inches. Show your calculations and explain them briefly.

   b. Without using a calculator, find the areas of the 5 rectangles you found in part (a). Show your calculations.

7. a. Draw 4 different rectangles, all of which have a perimeter of 8 inches. At least 2 of your rectangles should have side lengths that are not whole numbers (in inches). Label your rectangles with their lengths and widths.

   b. Determine the areas of each of your 4 rectangles in part (a) without using a calculator. Show your calculations, or explain briefly how you determined the areas. Then label your rectangles A, B, C, and D in decreasing order of their areas, so that A has the largest area and D has the smallest area among your rectangles.

   c. Qualitatively, how do the larger-area rectangles you drew in part (a) look different from the smaller-area rectangles? Describe how the shapes of the rectangles change as you go from the rectangle of largest area to the rectangle of smallest area.

8. a. Draw 4 different rectangles, all of which have area 4 square inches. Label your rectangles with their lengths and widths.

   b. Determine the perimeters of each of your rectangles in part (a). Then label your rectangles A, B, C, and D in increasing order of their perimeters, so that A has the smallest perimeter and D has the largest perimeter among your rectangles.

   c. Qualitatively, how do the smaller-perimeter rectangles you drew in part (a) look different from the larger-perimeter rectangles? Describe how the shapes of the rectangles change as you go from the rectangle of smallest perimeter to the rectangle of largest perimeter.

9. A forest has a perimeter of 210 miles, but no information is given about the shape of the forest. Justify your answers to the following (in all parts of this problem, the perimeter is still 210 miles):

   a. Is it possible that the area of the forest is 3000 square miles? Explain.

   b. Is it possible that the area of the forest is 3600 square miles? Explain.

   c. If the forest is shaped like a rectangle, then is it possible that the area of the forest is 3000 square miles? Explain.

   d. If the forest is shaped like a rectangle, then is it possible that the area of the forest is 2500 miles? Explain.

10. Bob wants to find the area of an irregular shape. He cuts a piece of string to the length of the perimeter of the shape. He measures to see that the string is about 60 cm long. Bob then forms his string into a square on top of centimeter graph paper. Using the graph paper, he determines that the area of his string square is about 225 cm². Bob says that therefore the area of the irregular shape is

also 225 cm². Is Bob's method for determining the area of the irregular shape valid or not? Explain. If the method is not valid, what can you determine about the area of the irregular shape from the information that Bob has? Explain.

11. Write several paragraphs discussing whether Nick's idea for estimating the area of an irregular shape given in Class Activity 12W provides a valid way to estimate the area of the irregular shape. If Nick's method is not valid, is there a way that he could use the string to get some sort of information about the area of the shape?

12. a. Describe a concrete way to demonstrate that many different shapes can have the same perimeter.

   b. Describe a concrete way to demonstrate that many different shapes can have the same area.

13. Consider all rectangles whose *area* is 4 square inches, including rectangles that have sides whose lengths are not whole numbers. What are the possible perimeters of these rectangles? Is there a smallest perimeter? Is there a largest?

   Answer this question either by pure thought or by actual examination of rectangles. Then write a paragraph describing your answer (including pictures, if relevant) and stating your conclusions clearly.

12.9 Using Moving and Additivity Principles to Prove the Pythagorean Theorem

F·P **Focal Points**
Grade 8

In this section, we will apply the moving and additivity principles about area to prove the Pythagorean theorem. The Pythagorean theorem is named for the Greek mathematician Pythagoras, who lived around 500 B.C. There is evidence, though, that this theorem was known long before that time, perhaps even by the ancient Babylonians in around 2000 B.C. Pythagoras founded a secretive school, whose motto was "All is number." The members of this school, the Pythagoreans, tried to explain all of science, philosophy, and religion in terms of numbers.

Pythagorean theorem The Pythagorean theorem (or Pythagoras's theorem) is a theorem about right triangles. Recall that, in a right triangle, the side opposite the right angle is called the hypotenuse. The **Pythagorean theorem** says,

> In a right triangle, the square of the length of the hypotenuse is equal to the sum of the squares of the lengths of the other two sides. In other words, if c is the length of the hypotenuse of a right triangle, and if a and b are the lengths of the other two sides, then
> $$a^2 + b^2 = c^2$$ (12.1)

FIGURE 12.97

Right triangles

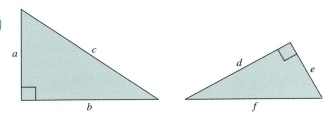

For example, in Figure 12.97, the right triangle on the left has sides of lengths a, b, and c, and the hypotenuse has length c, so, according to the Pythagorean theorem,

$$a^2 + b^2 = c^2$$

The right triangle on the right of Figure 12.97 has sides of lengths d, e, and f, and the hypotenuse has length f, so, in this case, according to the Pythagorean theorem,

$$d^2 + e^2 = f^2$$

In Equation 12.1 it is understood that *all sides of the triangle are expressed in the same units.* For example, all sides can be given in centimeters or all sides can be given in feet. If you have a right triangle where one side is measured in feet and the other in inches, for example, then you need to convert both sides to inches or to feet (or to some other unit) in order to use Equation 12.1.

Class Activity   *Now Turn to Class Activities Manual*

12X Using the Pythagorean Theorem, p. 304

12Y Can We Prove the Pythagorean Theorem by Checking Examples?, p. 305

If you have a particular right triangle in front of you, such as those in Figure 12.97, can measure the lengths of its sides and check that the Pythagorean theorem really is true in that case. For example, in the triangle on the left of Figure 12.97,

$$a = 3 \text{ cm}, \quad b = 4 \text{ cm}, \quad \text{and} \quad c = 5 \text{ cm}$$

so

$$a^2 + b^2 = 25, \quad \text{and} \quad c^2 = 25$$

and $a^2 + b^2$ really is equal to c^2. For the other triangle in Figure 12.97, $d = 4$ cm, $e = 2$ cm, and f is approximately 4.5 cm, so $d^2 + e^2$ is 20 and f^2 is approximately 20, also.

We could continue to check many right triangles to see if the Pythagorean theorem really is true in those cases. If we checked many triangles, it would be compelling evidence that the Pythagorean theorem is always true, but we could never check all the right triangles because they are infinite in number.

One of the cornerstones of mathematics, an idea that dates back to the time of the mathematicians of ancient Greece, is that a lot of evidence for a statement is not enough to know that the statement is **proof**   true. *Proof* is required to know for sure that a statement really is true. A **proof** is a thorough, precise, logical explanation for why a statement is true, based on assumptions or facts that we already know or assume to be true. So, a proof is what establishes that a theorem is true. Proofs are one of the important aspects of this book, too, even if we don't usually call our explanations proofs.

Professional and amateur mathematicians have developed literally hundreds of proofs of the Pythagorean theorem. If you do Class Activity 12Z, you will work through and discover one of these proofs. Class Activity 14BB in Chapter 14 and Problem 8 on page 584 guide you through two more proofs.

Class Activity *Now Turn to Class Activities Manual*

12Z A Proof of the Pythagorean Theorem, p. 305

Practice Exercises for Section 12.9

1. A garden gate that is 3 feet wide and 4 feet tall needs a diagonal brace to make it stable. How long a piece of wood will be needed for this diagonal brace? See Figure 12.98.

FIGURE 12.98 Garden gate with diagonal brace

2. A boat's anchor is on a line that is 75 feet long. If the anchor is dropped in water that is 50 feet deep, then how far away will the boat be able to drift from the spot on the water's surface that is directly above the anchor? Explain.

3. An elevator car is 8 feet long, 6 feet wide, and 9 feet tall. What is the longest pole you could fit in the elevator? Explain.

4. Imagine a right pyramid with a square base (like an Egyptian pyramid). Suppose that each side of the square base is 200 yards and that the distance from one corner of the base to the very top of the pyramid (along an outer edge) is 245 yards. How tall is the pyramid? Explain.

5. The front and back of a greenhouse have the shape and dimensions shown in Figure 12.99. The greenhouse is 40 feet long from front to back, and the angle at the top of the roof is 90°. The entire roof of the greenhouse will be covered with screening to block some of the light entering the greenhouse. How many square feet of screening will be needed?

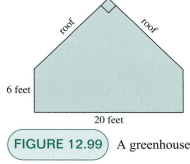

FIGURE 12.99 A greenhouse

6. Given a right triangle with short sides of lengths *a* and *b* and hypotenuse of length *c*, use Figure 12.100 to explain why

$$a^2 + b^2 = c^2$$

(You may assume that all shapes that look like squares really are squares and that, in each picture, all four triangles with side lengths *a, b, c* are identical right triangles.)

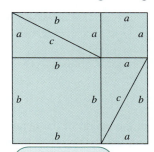
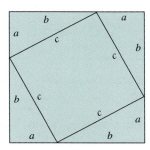

FIGURE 12.100 Shapes forming identical large squares

7. In the previous practice exercise, you used Figure 12.100 to provide a proof of the Pythagorean theorem. Even though it is somewhat hidden, discuss how this proof actually relies on the fact that the angles in a triangle add to 180°.

8. When Kyle was asked to state the Pythagorean theorem, he responded by writing the equation

$$a^2 + b^2 = c^2$$

Is this a correct statement of the Pythagorean theorem? Why or why not?

Answers to Practice Exercises for Section 12.9

1. If we let x be the length in feet of the diagonal brace, then, according to the Pythagorean theorem,

$$3^2 + 4^2 = x^2$$

Therefore, $x^2 = 25$, so $x = 5$, which means that the diagonal brace is 5 feet long.

2. If we let x be the distance that the boat can drift away from the spot on the water's surface that is directly over the anchor, then, according to Figure 12.101 and the Pythagorean theorem,

$$x^2 + 50^2 = 75^2$$

Therefore,

$$x^2 = 5625 - 2500 = 3125$$

so

$$x = \sqrt{3125} = 55.9$$

This tells us that the boat can drift 56 feet horizontally.

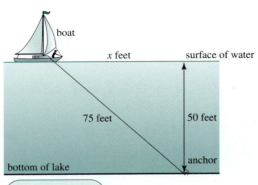

FIGURE 12.101 How far can an anchored boat drift?

3. The length of the longest pole that will fit is the distance between point A and point C in Figure 12.102, which is about 13 feet. To determine the distance from A to C, we will use the Pythagorean theorem twice: first with the right triangle on the floor of the elevator, to determine the length of AB, and then with triangle ABC to determine the length of AC. By applying the Pythagorean theorem to the triangle on the floor of the elevator, which has short sides of lengths 8 feet and 6 feet, we conclude that $8^2 + 6^2 = AB^2$. So $AB = \sqrt{100} = 10$ feet. Notice that ABC is also a right triangle, with the right angle at B. Therefore, by the Pythagorean theorem, $AB^2 + BC^2 = AC^2$, so $AC = \sqrt{181} = 13.5$ feet. Thus, the longest pole that can fit in the elevator is about 13 feet.

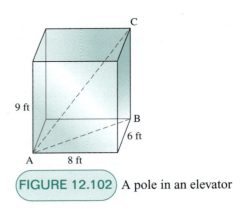

FIGURE 12.102 A pole in an elevator

4. The pyramid is about 200 yards tall. Here's why. Notice that the height of the pyramid is the length of BC in Figure 12.103. We will determine the length of BC by using the Pythagorean theorem twice, first with the triangle ABD, to determine the length of AB, and then with the triangle ABC, to determine the length of BC. By the Pythagorean theorem, $AD^2 + BD^2 = AB^2$, so since AD and BD are each 100 yards, $AB = \sqrt{20000} = 141.4$ yards.

(Actually, as you'll see in the next step, we really need only AB², not AB, so we don't even have to calculate the square root.) The triangle ABC is a right triangle, with right angle at B. So, by the Pythagorean theorem, $141.4^2 +$ $BC^2 = 245^2$, and it follows that $BC = \sqrt{40025} = 200$ yards. So the pyramid is about 200 yards tall.

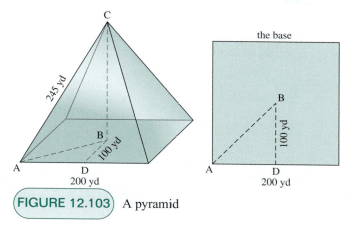

FIGURE 12.103 A pyramid

5. We need 1131 square feet of screening. The roof is made out of two rectangular pieces, each of which is 40 feet long and A feet wide, where A is shown in Figure 12.104. Because the angle at the top of the roof is 90°, we can use the Pythagorean theorem to determine A:

$$A^2 + A^2 = 20^2$$

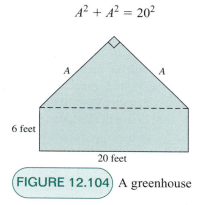

FIGURE 12.104 A greenhouse

Therefore, $2A^2 = 400$, so $A^2 = 200$, which means that $A = 14.14$ ft. Hence, each rectangular piece of roof needs 40×14.14 ft² = 565.69 ft² of screening. The two pieces of roof require twice as much. Rounding our answer, we see that we need about 1131 ft² of screening.

6. One way to prove that $a^2 + b^2 = c^2$ is to imagine taking away the 4 triangles in each large square of Figure 12.100. Both of the large squares in Figure 12.100 have sides of length $a + b$, so both large squares have the same area. Hence, according to the moving and additivity principles, if we remove the 4 copies of the right triangle from each large square, the remaining areas will still be equal. From the

square on the left, 2 smaller squares remain, one with sides of length a and one with sides of length b. Thus, the remaining area on the left is $a^2 + b^2$. From the square on the right, a single square with sides of length c remains. Therefore, the remaining area on the right is c^2. The remaining area on the left is equal to the remaining area on the right, therefore $a^2 + b^2 = c^2$.

7. When we used Figure 12.100 to prove the Pythagorean theorem, we assumed that the 2 large squares (each of which is made of 4 copies of the right triangle and either 1 or 2 more squares) really are squares of side length $a + b$ in order to know that they have the same area. Now refer to Figure 12.105. Focusing on point P on the square on the right, we see that the edge there should be a straight line in order for the square really to be a square. If we let A be the angle in the triangle opposite the side of length a and we let B be the angle opposite the side of length b (as shown on the triangle on the left in Figure 12.105), then

$$A + B + 90 = 180$$

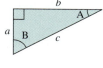

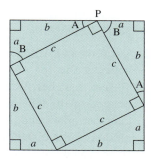

FIGURE 12.105 Explaining why the edges of the square really are straight

because the angles in the right triangle add to 180°. Therefore at point P on the right in Figure 12.105, the angles add to 180°, so we can conclude that, in fact, we do have a straight line there. The situation will be similar at the other points along the edge of this square, as well as the points inside and on the edges of the square on the left in Figure 12.100.

8. Kyle's statement of the Pythagorean theorem is not a correct statement because it is not complete. Kyle should say that the Pythagorean theorem is about the lengths of the sides of right triangles. If a right triangle has short sides of lengths a and b and hypotenuse of length c, then $a^2 + b^2 = c^2$.

Problems for Section 12.9

1. Jessica says she doesn't understand the Pythagorean theorem. She put numbers in for a, b, and c in the equation $a^2 + b^2 = c^2$, but the equation isn't always true. For example, when she put in $a = 2$, $b = 3$, and $c = 4$, she got $4 + 9 = 16$, which isn't correct. Jessica wants to know why the Pythagorean theorem isn't working. Write an informative paragraph discussing Jessica's conundrum. Why does the Pythagorean theorem appear not to be correct? What misunderstanding does Jessica have?

2. Town B is 380 km due south of town A. Town C is 460 km due east of town B. What is the distance from town A to town C? Explain your reasoning. Be sure to round your answer appropriately.

3. How long of a piece of ribbon will you need to stretch from the top of a 25-foot pole to a spot on the ground that is 10 feet from the bottom of the pole? Explain.

4. Rover, the dog, is on a 30-foot leash. One end of the leash is tied to Rover, who is 2 feet tall. The other end of the leash is tied to the top of a 6-foot pole. How far can Rover roam from the pole? Explain.

5. Carmina and Antone measure that the distance between the spots where they are standing is 10 feet, 7 inches. When measuring, Antone held his end of the tape measure up 4 inches higher than Carmina's end. If Carmina and Antone had measured the distance between them along the floor, would the distance be appreciably different than what they measured? Explain.

6. Use the Pythagorean theorem to help you determine the area of an equilateral triangle with sides of length 1 unit. Explain your reasoning.

7. Assuming that the earth is a perfectly round, smooth ball of radius 4000 miles and that 1 mile = 5280 feet, how far away does the horizon appear to be to a 5-foot-tall person on a clear day? Explain your reasoning. To solve this problem, start by drawing a picture that shows the cross-section of the earth, a person standing on the surface of the earth, and the straight line of the person's gaze reaching to the horizon. (Obviously, you won't want to draw this to scale.) You will need to use the following geometric fact: If a line is *tangent* to a circle at a point P (meaning it just "grazes" the circle at the point P; it meets the circle only at that one point), then that line is *perpendicular* to the line connecting P and the center of the circle, as illustrated in Figure 12.106.

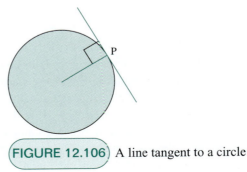

FIGURE 12.106 A line tangent to a circle

Chapter Summary and Study Items

Section 12.1 Areas of Rectangles Revisited

The area of a shape, in square units, is the number of 1-unit-by-1-unit squares it takes to cover the shape exactly without gaps or overlaps, allowing for squares to be cut apart and pieces to be moved if necessary. The most primitive way to determine the area of a shape is to count how many squares it takes to cover it exactly. A quicker and more advanced way to determine the area of a rectangle is to multiply its width and length. We can explain this formula by viewing the rectangle as decomposed into rows of squares.

Key skills and understandings:

- Know what area is and know the most primitive way to determine the area of a shape.

- Explain why it is valid to multiply the length and width of a rectangle to determine its area.

- Attend carefully to units of length and area when discussing and explaining the area formula for rectangles.

- Explain and use the area formula for rectangles in the case of fractions, mixed numbers, and decimals.

Section 12.2 Moving and Additivity Principles about Area

Two powerful principles are often used—almost subconsciously—to determine areas. These are the moving and additivity principles. The moving principle states that if you move a shape rigidly without stretching it, then its area does not change. The additivity principle states that if you combine shapes without overlapping them, then the area of the resulting combined shape is the sum of the areas of the individual shapes.

Key skills and understandings:

- Use the moving and additivity principles to determine areas of shapes, including cases where ultimately subtraction is used to determine the area of a shape of interest.

Section 12.3 Areas of Triangles

Any of the three sides of a triangle can be the base of the triangle. The height corresponding to a given base is perpendicular to the base and connects the base (or an extension of the base) to the vertex of the triangle that is not on the base. The area of the triangle is $\frac{1}{2}$ (base $\times$ height). The moving and additivity principles explain why this area formula is valid.

Key skills and understandings:

- Determine the area of a triangle in various ways, including using less and more sophisticated methods.

- Use the moving and additivity principles to explain why the area formula for triangles is valid.

- Use the area formula for triangles to determine areas and to solve problems.

Section 12.4 Areas of Parallelograms and Other Polygons

The lengths of the sides of a parallelogram do not determine its area. Instead, the area of a parallelogram is base $\times$ height. As with triangles, the base can be any of the four sides of the parallelogram. The height is the distance between the base and the side opposite the base. The moving and additivity principles explain why this area formula is valid.

Key skills and understandings:

- Use the moving and additivity principles to explain why the area formula for parallelograms is valid.

- Use the area formula for parallelograms to determine areas and to solve problems.

- Determine areas of various polygons.

Section 12.5 Cavalieri's Principle about Shearing and Area

Shearing is a process of changing a shape by sliding infinitesimally thin strips of the shape. Cavalieri's principle says shearing does not change areas. When we shear triangles and parallelograms parallel to a base, the height does not change and the area does not change. We can use a combination of Cavalieri's principle and the moving and additivity principles to explain in another way why the formulas for the areas of triangles and parallelograms are valid.

Key skills and understandings:

- Show how to shear a shape.

- Know that when you shear a shape, the original shape and the sheared shape have the same area (Cavalieri's principle).

- Use a combination of shearing and the moving and additivity principles to determine areas of shapes.

Section 12.6 Areas of Circles and the Number Pi

For any circle, its circumference divided by its diameter is always the same number; this number is π (pi) and it is equal to $3.14159265\ldots$. So the circumference of a circle of radius r is $2\pi r$. The area of a circle of radius r is πr^2. We can see why this area formula is plausible by subdividing a circle into wedges and rearranging the wedges to form an approximate rectangle (or parallelogram) of dimensions r by πr.

Key skills and understandings:

- Know how the number π is defined, and use the definition and the fact that the diameter of a circle is twice its radius to explain the $2\pi r$ formula for the circumference of a circle of radius r.

- Explain why the πr^2 formula for the area of a circle of radius r is plausible by subdividing a circle and rearranging pieces.

- Use the formulas for circumference and area of a circle to determine lengths and areas and to solve problems.

Section 12.7 Approximating Areas of Irregular Shapes

When it comes to irregular shapes, we usually have to make do with estimating their areas, because we often cannot determine their areas exactly. We can determine the approximate area of a shape by overlaying it with a grid and determining approximately how many grid squares the shape takes up.

Key skills and understandings:

- Use grids to determine approximate areas of shapes.

- Use other informal methods, such as working with modeling dough and weighing card stock, to determine approximate areas of shapes.

Section 12.8 Contrasting and Relating the Perimeter and Area of a Shape

The perimeter of a polygon (or other shape) is the total distance around the polygon. When we calculate the perimeter of a polygon by adding the lengths of the sides, it is as if we had wrapped a string snugly around the polygon and cut it into pieces corresponding to each side.

For a given, fixed perimeter, there are many shapes that have the same perimeter but have different areas. Thus, perimeter does not determine area. Among all rectangles of a given, fixed perimeter, the square with that perimeter has the largest area. Given a fixed perimeter, every area less than or equal to the area of the square of that perimeter is the area of some rectangle of that perimeter. Among all closed shapes of a given, fixed perimeter, the circle of that circumference has the largest area. Given a fixed perimeter, every area less than or equal to the area of the circle of that circumference is the area of some shape of that perimeter.

Key skills and understandings:

- Explain why we calculate perimeters of polygons the way we do. Discuss misconceptions with perimeter calculations.

- Recognize that perimeter does not determine area.

- Given a fixed perimeter, determine the areas of all rectangles of that perimeter and determine the areas of all shapes of that perimeter.

Section 12.9 Using Moving and Additivity Principles to Prove the Pythagorean Theorem

The Pythagorean theorem states that in any right triangle, the square of the length of the hypotenuse is equal to the sum of the squares of the lengths of the other two sides. We can prove the Pythagorean theorem by applying the moving and additivity principles to two congruent squares subdivided into four copies of the right triangle and additional squares.

Key skills and understandings:

- Prove the Pythagorean theorem.

- Use the Pythagorean theorem to determine lengths and distances.

Solid Shapes and Their Volume and Surface Area

*E*ven the very youngest schoolchildren learn about the common solid shapes of geometry when they play with a set of building blocks and observe the different characteristics and qualities of the shapes. Later in elementary school, students learn how to make patterns for shapes as they distinguish the outer surface of a shape and its surface area from the interior of a shape and its volume. And later still, students learn formulas for volumes and surface areas of shapes. Why do we study the common solid shapes of geometry? Most objects in the world around us can be thought of as approximated by a combination of the perfect shapes of geometry. For example, a tree trunk is approximately a cylinder, a pile of sand could be roughly in the shape of a cone, and a soft drink bottle is roughly like the combination of a cylinder and a cone. When children build buildings or other structures from a set of blocks, they are discovering how to compose complex new shapes from simple ones.

13.1 Polyhedra and Other Solid Shapes

 **Focal Points**
K, Grade 5

In this section we study some of the most familiar solid shapes of geometry, most of which are common in sets of building blocks but some of which are less common yet surprisingly beautiful. Some solid shapes have curved outer surfaces, whereas others have only flat surfaces consisting of triangles, squares, rectangles, pentagons, or other polygons. These latter kinds of shapes are called *polyhedra*.

Polyhedra

polyhedron (polyhedra)

A closed shape in space that is made out of polygons such as triangles, squares, or pentagons is called a **polyhedron.** The plural of polyhedron is polyhedra. Figure 13.1 shows some polyhedra and a shape that is not a polyhedron. The

FIGURE 13.1

Polyhedra and a shape that is not a polyhedron

a polyhedron with 6 faces, 12 edges, and 8 vertices

a polyhedron with 5 faces, 8 edges, and 5 vertices

A sphere is *not* a polyhedron.

faces
edge
vertex (vertices)

polygons that make up the surface of the polyhedron are called the **faces** of the polyhedron. The place where two faces come together is called an **edge** of the polyhedron. A corner point where several faces come together is called a **vertex** or corner of the polyhedron. The plural of vertex is vertices.

The name *polyhedron* comes from the Greek; it makes sense because *poly* means *many* and *hedron* means *base,* so that *polyhedron* means *many bases.*

Given a polyhedron, students can study its characteristics and qualities by noticing its faces, edges, and vertices. What kinds of shapes are the faces? How many faces are there? How many vertices and how many edges does the shape have? Two common types of polyhedra are prisms and pyramids. Related to these shapes are cylinders and cones, but they have curved surfaces and so are *not* polyhedra.

Prisms, Cylinders, Pyramids, and Cones

One special type of polyhedron is a prism. Some people hang glass or crystal prisms in their windows to bend the incoming light and cast rainbow patterns around the room.

right prism

From a mathematical point of view, a *right prism* is a polyhedron that can be thought of as "going straight up over a polygon," as shown in Figure 13.2. Think of **right prisms** as formed in the following way: Take two paper copies of any polygon and lay both flat on a table, one on top of the other so that they match up. Move the top polygon *straight up* above the bottom one. If vertical rectangular faces are now placed so as to connect corresponding sides of the two polygons, then the shape formed this way is a right prism. The two polygons that you started with are called the **bases** of the right prism.

bases

prism

oblique prism

By modifying the previous description, we get a description of *all* prisms, not just right prisms. As before, start with two paper copies of any polygon and lay both flat on a table, one on top of the other so that they match up. Move the top polygon up without twisting, away from the bottom polygon, keeping the two polygons parallel. This time, the top polygon does not need to go straight up over the bottom polygon. Instead, it can be positioned to one side, as long as it is not twisted and it remains parallel to the bottom polygon. Once again, if faces are now placed so as to connect corresponding sides of the two polygons, then the shape formed this way is a **prism**. (This time the faces will be parallelograms.) Every right prism is a prism, but Figure 13.3 shows prisms that are not right prisms. A prism that is not a right prism can be called an **oblique prism**. As before, the two polygons that you started with are called the bases of the prism.

FIGURE 13.2

Some right prisms

FIGURE 13.3

Oblique prisms

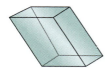

FIGURE 13.4

A prism turned sideways is still a prism.

If a prism is moved to a different orientation in space, it is still a prism. So, for example, the polyhedra shown in Figure 13.4 are prisms, although you may need to rotate them mentally to be convinced that they really are prisms.

Prisms are often named according to the kind of polygons that make the bases of the prism. So, for example, a prism with a triangle base can be called a **triangular prism**, and a prism with a rectangle base can be called a **rectangular prism**.

triangular, rectangular prism

A cylinder is a kind of shape that is related to prisms. Roughly speaking, a cylinder is a tube-shaped object, as shown in Figure 13.5. The tube inside a roll of paper towels is an example of a cylinder. **Cylinders** can be described in the following way: Draw a closed curve on paper (such as a circle or an oval), cut it out, make a copy of it, and lay the two copies on a table, one on top of the other, so that they match. Next, take the top copy and move it up without twisting, away from the bottom copy, keeping the two copies parallel. Now imagine paper or some other kind of material connecting the two curves in such a way that every line between corresponding points on the curves lies on this paper or material. The shape formed by this paper or other material is a cylinder. The regions formed by the two starting curves can again be called the bases of the cylinder. In some cases, you want to consider the two bases as part of the cylinder; in other cases, you do not. Both kinds of shapes can be called cylinders.

cylinder

As with prisms, a cylinder can be a **right cylinder** or an **oblique cylinder**, according to whether one base is or is not "straight up over" the other. The outer two cylinders in Figure 13.5 are right cylinders, whereas the middle cylinder is an oblique cylinder.

right cylinder

oblique cylinder

In addition to prisms, another special type of polyhedron is a *pyramid*. You have probably seen pictures of the famous pyramids in Egypt. Mathematical pyramids include shapes like the Egyptian pyramids as well as variations on this kind of shape, as shown in Figure 13.6. **Pyramids** can be described as follows: Start with any polygon and a separate single point that does not lie in the plane of the polygon. Now use additional polygons to connect the point to the original polygon. This should be done in such a way that all the lines that connect the point to the polygon lie on the sides of these additional polygons. These new polygons, together with the original polygon, form a pyramid. As usual, the original polygon is called the base of the pyramid. Figure 13.7 shows a roof that is a pyramid with an octagon base.

pyramid

As with prisms and cylinders, certain kinds of pyramids are called right pyramids. A **right pyramid** is a pyramid for which the point lies "straight up over" the center of the base. A pyramid that is not a right

right pyramid

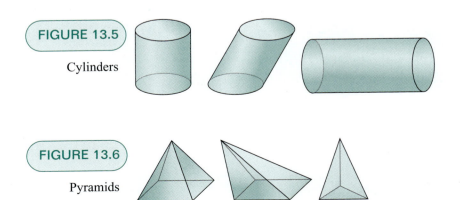

FIGURE 13.5

Cylinders

FIGURE 13.6

Pyramids

FIGURE 13.7

Pyramid roof with an octagon base

FIGURE 13.8 Oblique pyramid on a building

oblique pyramid pyramid can be called an **oblique pyramid**. In Figure 13.6, the outer two pyramids are right pyramids, whereas the middle pyramid is an oblique pyramid. Figure 13.8 shows an oblique pyramid decorating the top of a building.

cones Just as cylinders are shapes that are related to prisms, *cones* are shapes that are related to pyramids. Roughly speaking, cones are objects like ice cream cones or cone-shaped paper cups, as well as related objects, as shown in Figure 13.9. As with pyramids, **cones** can be described by starting with a closed curve in a plane and a separate point that does not lie in that plane. Now imagine paper or some other kind of material that joins the point to the curve in such a way that all the lines that connect the point to the curve lie on that paper (or other material). That paper (or other material), together with original curve, form a cone. As usual, the starting curve and the region inside it is called the base of the cone. Sometimes the base of a cone is considered a part of the cone, and sometimes it isn't. Either shape (with or without the base) can be called a cone.

right cone As with prisms, cylinders, and pyramids, certain kinds of cones are called right cones. A **right cone** is a cone whose point lies directly over the center of its base. A cone that is not a right cone can be called an

oblique cone **oblique cone**. In Figure 13.9, the outer two cones are right cones, whereas the cone in the middle is an oblique cone.

Class Activity *Now Turn to Class Activities Manual*

13A Making Prisms and Pyramids, p. 308

13B Analyzing Prisms and Pyramids, p. 309

13C What's Inside the Magic 8 Ball?, p. 310

FIGURE 13.9

Cones

FIGURE 13.10

The Platonic solids

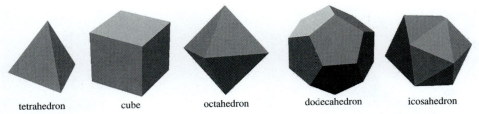

tetrahedron   cube   octahedron   dodecahedron   icosahedron

The Platonic Solids

Five special polyhedra that are very regular and uniform all around are called the **Platonic solids**, in honor of Plato, who thought of these solid shapes as associated with earth, fire, water, air, and the whole universe. Examples are pictured in Figure 13.10, and the five Platonic solids are described next:

- **Tetrahedron** has 4 equilateral triangle faces, with 3 triangles coming together at each vertex.
- **Cube** has 6 square faces, with 3 squares coming together at each vertex.
- **Octahedron** has 8 equilateral triangle faces, with 4 triangles coming together at each vertex.
- **Dodecahedron** has 12 regular pentagon faces, with 3 pentagons coming together at each vertex.
- **Icosahedron** has 20 equilateral triangle faces, with 5 triangles coming together at each vertex.

The names of the Platonic solids make sense because *hedron* comes from the Greek for "base" or "seat," while *tetra, octa, dodeca,* and *icosa* means 4, 8, 12, and 20, respectively. So, for example, *icosahedron* means *20 bases*, which makes sense because an icosahedron is made of 20 triangular "bases."

The Platonic solids are special because each one is made of only one kind of regular polygon, and the same number of polygons come together at each vertex. In fact, the Platonic solids are the only con-

convex   vex polyhedra having these two properties. A shape in the plane or in space is **convex** if any line segment connecting two points on the shape lie entirely within the shape. So, convex polyhedra do not have any protrusions or indentations.

Although cubes are commonly found in daily life (e.g., boxes and ice cubes), the other Platonic solids are not commonly seen. However, these shapes do occur in nature occasionally. For example, the mineral pyrite can form a crystal in the shape of a dodecahedron. (This crystal is often called a pyritohedron.) The mineral fluorite can form a crystal in the shape of an octahedron, and although rare, gold can, too. Some viruses are shaped like an icosahedron. Figure 13.11 shows a picture of a dodecahedron sculpture in front of a school in Jesup, Georgia. See *www.pearsonhighered.com/beckmann* for additional examples of Platonic solids.

The real attraction of the Platonic solids is their beautiful perfection and symmetry. You can't really appreciate the Platonic solids unless you make models of these wonderful shapes, so I recommend that you construct some models—see Problem 6 on page 561 and Class Activity 13D.

FIGURE 13.11

A dodecahedron sculpture in Jesup, Georgia

Class Activity   *Now Turn to Class Activities Manual*

13D Making Platonic Solids with Toothpicks and Gumdrops, p. 310

13E Why Are There No Other Platonic Solids?, p. 311

The following Class Activity applies to all polyhedra, not just the Platonic solids.

Class Activity   *Now Turn to Class Activities Manual*

13F Relating the Numbers of Faces, Edges, and Vertices of Polyhedra, p. 313

Practice Exercises for Section 13.1

1. Describe each of the following shapes: prisms, cylinders, pyramids, and cones.

2. Try to visualize a (right) prism that has hexagonal bases. How many faces, edges, and vertices does such a prism have?

3. Try to visualize a pyramid that has an octagonal base. How many faces, edges, and vertices does such a pyramid have?

4. What other name can you call a tetrahedron? (See Figure 13.12.)

5. What happens if you try to make a convex polyhedron whose faces are all equilateral triangles

FIGURE 13.12   A tetrahedron

and for which 6 triangles come together at every vertex?

6. Is it possible to make a convex polyhedron so that 7 or more equilateral triangles come together at every point? Explain.

Answers to Practice Exercises for Section 13.1

1. See text.

2. The prism has 8 faces: 6 rectangular faces and 2 hexagonal faces. The prism has 18 edges and 12 vertices.

3. The pyramid has 9 faces: 8 triangular faces and 1 octagonal face. The pyramid has 16 edges and 9 vertices.

4. A tetrahedron is special kind of pyramid with a triangle base.

5. If you put 6 equilateral triangles together so that they all meet at one point, you will find that they make a flat hexagon, as seen in Figure 13.13. It makes sense that 6 triangles put together at a point make a flat shape because the angle at a vertex of an equilateral triangle is 60°, so 6 of these angles side by side will make a full 360°. If you tried to make a

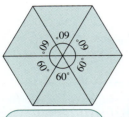

(FIGURE 13.13)   Six triangles put together at a point

convex polyhedron in such a way that 6 equilateral triangles came together at every vertex, you could never get the polyhedron to "close up."

6. If you put 7 or more equilateral triangles together so that they all meet at one point, you will find that some triangles will have to slant up and others will have to slant down. Thus, you will have to create indentations or protrusions when you put triangles together in this way, therefore, the polyhedron will not be convex. It makes sense that 7 triangles meeting at a point will do this because the angle at a vertex of an equilateral triangle is 60°, so 7 or more of these angles side by side will make more than 360°. This forces some triangles to slant up and others to slant down in order to put so many triangles together at a point.

Problems for Section 13.1

1. For each of the following solid shapes, find at least two examples of a real-world object that has that shape:

   a. prism

   b. cylinder

   c. pyramid

   d. cone

2. Answer the following questions without using a model:

   a. How many faces (including the bases) does a prism with a parallelogram base have? What shapes are the faces? Explain briefly.

   b. How many edges does a prism with a parallelogram base have? Explain.

   c. How many vertices does a prism with a parallelogram base have? Explain.

3. Answer the following questions without using a model:

   a. How many faces (including the base) does a pyramid with a rhombus base have? What shapes are the faces? Explain briefly.

   b. How many edges does a pyramid with a rhombus base have? Explain.

   c. How many vertices does a pyramid with a rhombus base have? Explain.

4. Recall that an *n*-gon is a polygon with *n* sides. For example, a polygon with 17 sides can be called a 17-gon. A triangle could be called a 3-gon. Find formulas (in terms of *n*) for the following, and explain why your formulas are valid:

   a. the number of faces (including the bases) on a prism that has *n*-gon bases

   b. the number of edges on a prism that has *n*-gon bases

   c. the number of vertices on a prism that has *n*-gon bases

   d. the number of faces (including the base) on a pyramid that has an *n*-gon base

   e. the number of edges on a pyramid that has an *n*-gon base

   f. the number of vertices on a pyramid that has an *n*-gon base

5. This problem goes with Class Activity 13C on the Magic 8 Ball. Make a closed, three-dimensional shape out of some or all of the triangles in Figure 13.14. Make any shape you like—be creative. Answer the following questions:

   a. Do you think your shape could be the one that is actually inside the Magic 8 Ball? Why or why not?

   b. If you were going to make your own advice ball, but with a (possibly) different shape inside, what would be the advantages or disadvantages of using your shape?

6. Referring to the descriptions of the Platonic solids on page 558, construct the five Platonic solids by cutting out the triangles, squares, and rectangles in the appendix, and taping these shapes together. You may wish to copy the pages onto card stock to make sturdier models that are easier to work with.

7. A cube is polyhedron that has 3 square faces meeting at each vertex. What would happen if you tried to make a polyhedron that has 4 square faces meeting at each vertex? Explain.

8. The Platonic solids are convex polyhedra with faces that are either equilateral triangles, squares, or regular pentagons. If you try to make a convex polyhedron whose faces are all regular hexagons, what will happen? Explain.

9.  Two gorgeous polyhedra can be created by *stellating* an icosahedron and a dodecahedron. *Stellating* means *making starlike*. Imagine turning each face of an icosahedron into a "star point," namely, a pyramid whose base is a triangular face of the icosahedron. Likewise, imagine turning each face of a dodecahedron into a star point, namely, a pyramid whose base is a pentagonal face of the dodecahedron. A stellated icosahedron will then have 20 star points, while a stellated dodecahedron will have 12 star points.

a. Make 20 copies of Figure 13.14 on card stock. Cut, fold, and tape them to make 20 triangle-based pyramid star points. (The pattern makes star-point pyramids that don't have bases.) Tape the star points together as though you were making an icosahedron out of their (open) bases.

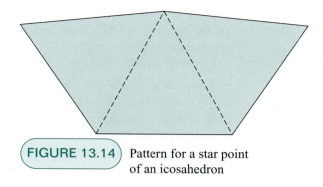

FIGURE 13.14 Pattern for a star point of an icosahedron

b. Make 12 copies of Figure 13.15 on card stock. Cut, fold, and tape them to make 12 pentagon-based pyramid star points. (The pattern makes star point pyramids that don't have bases.) Tape the star points together as though you were making a dodecahedron out of their (open) bases.

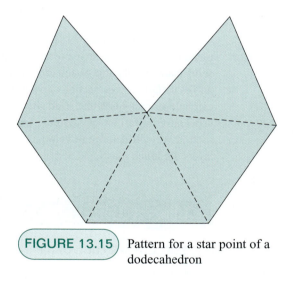

FIGURE 13.15 Pattern for a star point of a dodecahedron

13.2 Patterns and Surface Area

F·P **Focal Points**
Grades 5, 7

In the previous section we saw that important characteristics of polyhedra are their faces, edges, and vertices. So three-dimensional shapes have essential characteristics that are two-dimensional, one-dimensional, and even zero-dimensional (points). In this brief section, we focus on a two-dimensional aspect of solid shapes: their outer surface. We also consider other two-dimensional aspects of solid shapes: cross-sections and shadows.

One way to make a solid shape is to make a pattern on paper (or light cardboard) for its outer surface, cut out the pattern, fold it as needed, and join the cut edges. Making a shape in this way helps to concentrate on the characteristics of the outer surface of the shape, and the process provides an opportunity to practice visualization.

Patterns are useful not only for making models of shapes but also for determining and finding formulas for surface areas of solid shapes. Recall that the surface area of a solid shape is the total area of the outer surface of the shape. To determine the surface area of a shape, we can make or think about how to make a pattern for the shape. We can then apply the additivity principle for area and add the areas of the component parts of the pattern to determine the surface area of the shape.

Class Activity   *Now Turn to Class Activities Manual*

13G What Shapes Do These Patterns Make?, p. 314

13H Patterns for Prisms and Pyramids, p. 315

13I Patterns for Cylinders, p. 316

13J Patterns for Cones, p. 317

---

cross-section   Given a solid shape, a cross-section of it is formed by slicing the shape with a plane. The places where the plane meets the solid shape form a shape in the plane. This shape in the plane is a **cross-section** of the solid shape. Both patterns and cross-sections provide two-dimensional information about a solid shape, but unlike patterns, cross-sections give information about the *inside* of a shape. Thinking about cross-sections can help us recognize that solid shapes have an interior in addition to an outer surface. Cross-sections are especially important in calculus, where they are used to determine the volume of a shape. In the next section we will determine volumes of prisms by thinking about them as decomposed into layers. These layers are much like thickened cross-sections.

Class Activity   *Now Turn to Class Activities Manual*

13K Cross-Section of a Pyramid, p. 318

13L Cross-Sections of a Long Rectangular Prism, p. 318

The outer surface of a solid shape and its cross-sections are two-dimensional aspects of the shape. Additional two-dimensional aspects of a solid shape are the shadows the shape casts when light shines on it. Students who are learning about light and shadows are often asked to predict the shadow that a solid shape will make. Depending on how a shape is oriented with respect to the light source, a single shape can have several very different shadows. Experiment by holding solid shapes under a light and seeing what shadows you can get!

Class Activity   *Now Turn to Class Activities Manual*

13M Shadows of Solid Shapes, p. 319

Practice Exercises for Section 13.2

1. Figure 13.16 shows small versions of several patterns for shapes. Larger versions of these patterns are shown in Figure A.1 on page 713. Before you cut, fold, and tape the larger patterns in Figure A.1, visualize the folding process in order to help you visualize the final shapes. Then cut out the patterns on page 713 along the solid lines, fold down along the dotted lines, and tape sides with matching labels together. Does the shape match your visualized shape?

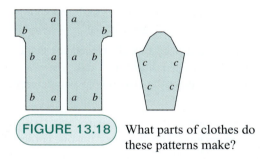

FIGURE 13.17 Small version of patterns for shapes

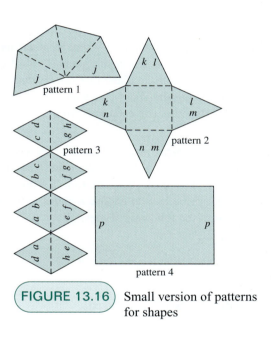

FIGURE 13.16 Small version of patterns for shapes

2. Figure 13.17 shows small versions of 6 patterns for shapes. Larger versions of these patterns are shown in Figure A.2 on page 715. Before you cut and tape the larger patterns in Figure A.2, try to visualize the final shapes. Predict how the shapes will be alike and how they will be different. Then cut out the patterns on page 715 along the solid lines, and tape sides with matching labels together. (Don't do any folding!)

3. What familiar parts of clothing are made from the patterns in Figure 13.18 when sides with matching labels are sewn together? Visualize!

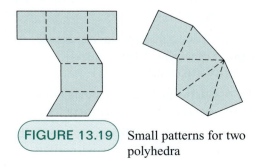

FIGURE 13.18 What parts of clothes do these patterns make?

4. Examine the small patterns in Figure 13.19. Try to visualize the polyhedra that would result if you were to cut these patterns out on the heavy lines, fold them on the dotted lines, and tape various sides together.

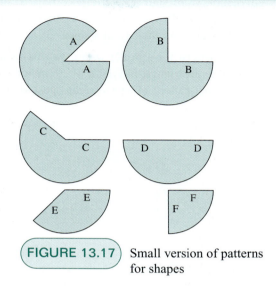

FIGURE 13.19 Small patterns for two polyhedra

Now cut out the large versions of these patterns on page 719 and 721 along the heavy lines. Fold the shapes, and tape the sides together to make polyhedra. Were your predictions correct? If not, undo your polyhedra and try to visualize again how they turn into their final shapes.

5. Make a pattern for a prism whose two bases are identical to the triangle in Figure 13.20. Include the bases in your pattern. (Use a straightedge and compass to make a copy of the triangle in Figure 13.20.) Label all sides that have length a, b, and c on your pattern.

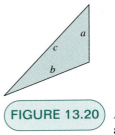

FIGURE 13.20   A base for a prism and a pyramid

6. Make a pattern for a pyramid whose base is identical to the triangle in Figure 13.20. Include the base in your pattern. (Use a straightedge and compass to make a copy of the triangle in Figure 13.20.) Label all sides that have length a, b, and c on your pattern. You may use graph paper.

7. What is the surface area of a closed box that is 4 feet wide, 3 feet deep, and 5 feet tall? Draw a scaled-down pattern for the box to help you determine its surface area.

8. A right pyramid has a square base with sides 60 meters long. The distance from one corner of the base to the very top of the pyramid (along an outer edge) is 50 meters. Determine the surface area of the pyramid (not including the base).

9. Suppose you take a rectangular piece of paper, roll it up, and tape two ends together, without overlapping them, to make a tube. If the tube is 12 inches long and has a diameter of $2\frac{1}{2}$ inches, then what were the length and width of the original rectangular piece of paper?

10. Find a formula for the surface area of a cylinder of radius r units and height h units. Include the top and bottom of the cylinder.

11. A cone is to be made from a circle of radius 3 cm (for the base) and a quarter-circle (for the portion without the base). Determine the radius of the quarter-circle.

12. ↻ Make a paper model of a cone (like an ice cream cone). Now visualize a plane slicing through the cone. The places where the plane meets the cone form a shape in the plane. Describe all possible shapes in the plane that can be made this way, by slicing the cone. Use your model to help you, but also visualize each case without the use of your model.

13. ↻ Make a paper model of a cube. Now visualize a plane slicing through the cube. The places where the plane meets the cube form a shape in the plane. Describe how to choose a plane so that the slice forms the following shapes:

• a square

• a rectangle that is not a square

• a triangle

• a rhombus that is not a square

• a hexagon

Answers to Practice Exercises for Section 13.2

1. Patterns 1 and 2 make pyramids, without and with a base, respectively. Pattern 3 makes an octahedron. Pattern 4 makes a cylinder.

2. All 6 patterns make cones. Some are shorter and wider; some are taller and narrower.

3. The 2 patterns on the left form a pant leg when they are sewn together. The curved portion of side b makes the crotch. The pattern on the right makes a sleeve. The curved portion at the top makes the arm hole. The seam at edge c runs straight down the arm, from the armpit to the wrist.

4. The pattern on the left of Figure 13.19 makes an oblique square prism. The pattern on the right makes an oblique pyramid with a square base.

5. & 6. See Figure 13.21.

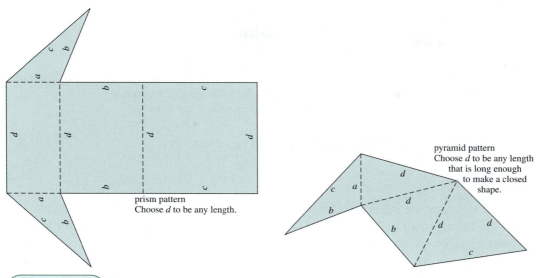

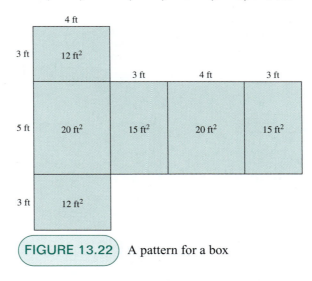

FIGURE 13.21   Patterns for a prism and a pyramid with triangle bases

7. Figure 13.22 shows a scaled-down pattern for the desired box. By subdividing the pattern into rectangles and using the additivity principle about areas, we see that the surface area of the box is

$$2 \times (4 \times 3) + 2 \times (5 \times 4) + 2 \times (3 \times 5) = 94 \text{ ft}^2$$

FIGURE 13.22   A pattern for a box

8. The surface of the pyramid (not including the base) consists of 4 isosceles triangles, each of which has a base of 60 meters and 2 sides of length 50 meters, as shown in Figure 13.23. To find the area of one of these triangles, we must determine the triangle's height. If h stands for the height of the triangle, then, by the Pythagorean theorem,

$$30^2 + h^2 = 50^2$$

Therefore,

$$h^2 = 50^2 - 30^2 = 2500 - 900 = 1600$$

and $h = 40$, so the height of the triangle is 40 meters. The area of each of the 4 triangles making the surface of the pyramid is therefore

$$\frac{1}{2}(60 \cdot 40) = 1200$$

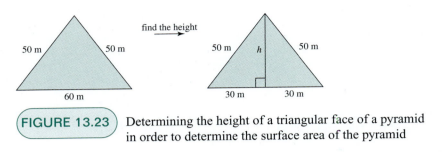

FIGURE 13.23   Determining the height of a triangular face of a pyramid in order to determine the surface area of the pyramid

square meters. So the surface area of the pyramid (not including the base) is

$$4 \cdot 1200 = 4800$$

square meters.

9. The two edges of paper that are rolled up make circles of diameter 2.5 inches. Therefore, the lengths of these edges are $\pi \times 2.5$ inches, which is about 8 inches. The other two edges of the paper run along the length of the cylinder, so they are 12 inches long. Thus, the original piece of paper was about 8 inches by 12 inches.

10. The surface of the cylinder consists of two circles of radius r units (the top and the bottom) and a tube. Imagine slitting the tube open along its length and unrolling it, as indicated in Figure 13.24. The tube then becomes a rectangle. The height, h, of the tube becomes the length of two sides of the rectangle. The circumference of the tube, $2\pi r$, becomes the length of two other sides of the rectangle. Therefore, the rectangle has area $2\pi rh$. According to the moving and additivity principles about area, the surface area of the cylinder is equal to sum of the areas of the two circles (from the top and bottom) and the area of the rectangle (from the tube), which is

$$(2\pi r^2 + 2\pi rh) \text{ units}^2$$

FIGURE 13.24

Taking a cylinder apart

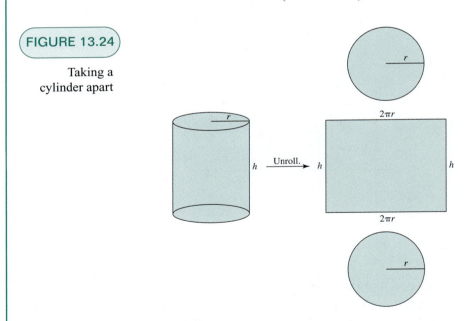

11. Think about how the portion of the cone without the base would be attached to the base if you made the cone out of paper. The $\frac{1}{4}$ portion of the circle making the cone must wrap completely around the base. Since the base has radius 3 cm, the circumference of the base is $2\pi \cdot 3$ cm. So if r is the radius of the larger circle, $\frac{1}{4}$ of $2\pi r$ must be equal to $2\pi \cdot 3$. In other words,

$$\frac{1}{4} \cdot 2\pi r = 2\pi \cdot 3$$

Therefore, $r = 12$, and so the radius of the quarter-circle making the cone is 12 cm.

12. With the cone positioned as shown in Figure 13.25, horizontal planes slice the cone either in a single point or in a circle. (You get a single point if the plane goes through the very bottom, small circles near the bottom, larger circles as you go up.) A vertical plane slices the cone in either a V-shape or a curve called a **hyperbola**. A slanted plane slices the cone in either an **ellipse** (oval shape) or in a curve called a **parabola**. This is why circles, ellipses, hyperbolas, and parabolas are collectively referred to as **conic sections**, because they come from slicing a cone.

FIGURE 13.25

Slicing a cone with a plane

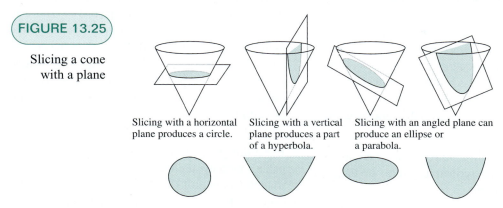

Slicing with a horizontal plane produces a circle.

Slicing with a vertical plane produces a part of a hyperbola.

Slicing with an angled plane can produce an ellipse or a parabola.

13. See Figure 13.26. Horizontal planes slice the cube in squares. A vertical plane slices the cube in either a square or a rectangle that is not a square. You can get a triangle, a rhombus, or a hexagon by slicing with angled planes. To see how to get a hexagon, cut out the two patterns in Figure A.3 on page 717 along the solid lines, fold them down along the dotted lines, and tape them to create two solid shapes. The two shapes can be put together at the hexagon to make a cube; therefore, a plane can slice a cube to create a hexagon at the slice.

FIGURE 13.26

Slicing a cube with a plane

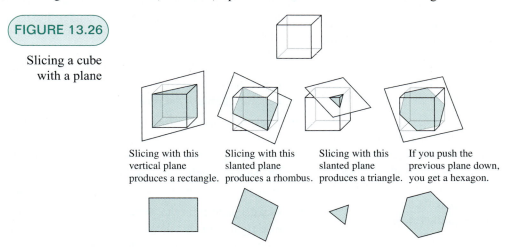

Slicing with this vertical plane produces a rectangle.

Slicing with this slanted plane produces a rhombus.

Slicing with this slanted plane produces a triangle.

If you push the previous plane down, you get a hexagon.

Problems for Section 13.2

1. Find all the different patterns for a closed three-dimensional shape that are made out of 4 copies of the triangle in Figure 13.27 in such a way that each triangle is joined to another triangle along a whole edge, not just at a corner. (This kind of pattern is sometimes called a *net*.) For two patterns to be considered different, you should not be

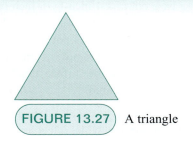

FIGURE 13.27   A triangle

able to match the patterns up when they are cut out. If you cut out, fold, and tape your patterns, the shape that you make is called a *tetrahedron* or a pyramid with a triangle base.

2. **a.** Make three different patterns that could be cut out, folded, and taped to make an open-top 1-inch-by-1-inch-by-1-inch cube. In your patterns, each square should be joined to another square along a whole edge of the square, not just at a corner. (This kind of pattern is sometimes called a *net*.) For two patterns to be considered different, you should not be able to match up the patterns when they are cut out.

   b. How many different patterns for an open-top cube as described in part (a) are there? Find all such patterns.

3. **a.** Make three different patterns that could be cut out, folded, and taped to make a closed 1-inch-by-1-inch-by-1-inch cube. In your patterns, each square should be joined to another square along a whole edge of the square, not just at a corner. (This kind of pattern is sometimes called a *net*.) For two patterns to be considered different, you should not be able to match the patterns up when they are cut out.

   b. How many different patterns for a closed cube as described in part (a) are there? Find all such patterns.

4. **a.** Describe or show how to make a cylinder without bases out of a 4-inch-by-5-inch rectangular piece of paper in such a way that all the paper is used and the paper does not overlap. Then describe or show how to make a different cylinder without bases out of another 4-inch-by-5-inch rectangular piece of paper. As before, all the paper should be used and there should be no overlaps. Determine the surface area (without the bases) of each cylinder and explain your reasoning.

   b. Describe or show how to make a pyramid without a base out of a 4-inch-by-5-inch rectangular piece of paper in such a way that all the paper is used and the paper does not overlap. (You may form faces of the pyramid by joining pieces of paper.) Determine the surface area (without the base) of your pyramid.

5. If a cardboard box with a top, a bottom, and 4 sides is W feet wide, L feet long, and H feet

tall, then what is the surface area of this box? Find a formula for the surface area in terms of W, L, and H. Explain clearly why your formula is valid.

6. Use a straightedge and compass to help you make a pattern for a prism whose two bases are identical to the triangle in Figure 13.20 on page 564. You may wish to draw your pattern on graph paper. Make your pattern significantly different from the one for a prism shown in Figure 13.21 on page 565. Include the bases in your pattern. (Use a straightedge and compass to make a copy of the triangle in Figure (13.20.) Label all sides that have length a, b, and c on your pattern.

7. Use a straightedge and compass to help you make a pattern for a pyramid whose base is identical to the triangle in Figure 13.20. Make your pattern significantly different from the one for a pyramid shown in Figure 13.21. Include the base in your pattern. (Use a straightedge and compass to make a copy of the triangle in Figure 13.20.) Label all sides that have length a, b, and c on your pattern.

8. Use a ruler and compass to help you make a pattern for a prism whose two bases are triangles that have one side of length 2.5 inches, one side of length 2 inches, and one side of length 1.5 inches. It may help you to draw your pattern on $\frac{1}{4}$-inch graph paper. Indicate which sides of your pattern would be joined to make the prism.

9. Use a ruler and compass to help you make a pattern for a pyramid with a triangular base that has one side of length 2 inches, one side of length 3 inches, and one side of length 4 inches. Indicate which sides of your pattern would be joined to make the pyramid.

10. Make a pattern for the "bottom portion" **(frustum)** of a right cone, as pictured in Figure 13.28. What article of clothing is often shaped like this?

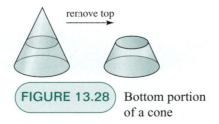

FIGURE 13.28 Bottom portion of a cone

11. **a.** Make a pattern for an oblique cylinder with a circular base. You may leave the bases off your pattern. (*Advice:* Be willing to experiment first. You might start by making a pattern for a right cylinder and modifying this pattern.)

b. The sleeves of most shirts and blouses are more or less in the shape of an oblique cylinder. If your pattern in part (a) were to make a sleeve, what part of your pattern would be at the shoulder? What part of your pattern would be at the armpit?

12. Make a pattern for an oblique cone with a circular base. You may leave the base off your pattern. (*Advice*: Be willing to experiment first. You might start by making a pattern for a right cone and modifying this pattern.)

13. Tim needs a sturdy cardboard box that is 3 feet tall by 2 feet long by 1 foot wide. Tim wants to make the box out of a large piece of cardboard that he will cut, fold, and tape. The box must close up completely, so it needs a top and a bottom. Show Tim how to make such a box out of one rectangular piece of cardboard: Tell Tim what size cardboard he'll need to get (how wide, how long) and explain or show how he should cut, fold, and tape the cardboard to make the box. Be sure to specify exact lengths of any cuts Tim will need to make. Include pictures where appropriate. Your instructions and box should be practical; Tim should be able to actually make and use the box. You might want to make a scale model for your box out of paper. *Note*: Many boxes have flaps around all 4 sides that one can fold down and interlock to make the top and bottom of the box. You can make this kind of top and bottom, or something else if you prefer, but make sure it will make a sturdy box that closes completely.

14. A company will manufacture a tent that will have a square, 15-foot × 15-foot base, 4 vertical walls, each 10 feet high, and a pyramid-shaped top made out of 4 triangular pieces of cloth, as shown in Figure 13.29. At its tallest point in the center of the tent, the tent will reach a height of 15 feet.

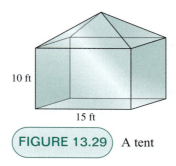

10 ft

15 ft

FIGURE 13.29 A tent

Draw a pattern for 1 of the 4 triangular pieces of cloth that will make the pyramid-shaped top of the tent. Label your pattern with enough information so

that someone making the tent would know how to measure and cut the material for the top of the tent.

15. Carla wants to make the triangular faces for a right pyramid with a $2\frac{3}{4}$-inch-by-$2\frac{3}{4}$-inch square base. The tallest point in the center of Carla's pyramid must be $2\frac{1}{2}$ inches above the base. Using a ruler that has tick marks every $\frac{1}{16}$ of an inch, carefully measure and draw one of the triangular faces for Carla's pyramid, labeling the lengths you measured. Show all calculations, explaining your reasoning. Be sure to explain how to measure decimal lengths, using a ruler that has tick marks every $\frac{1}{16}$ of an inch.

16. Imagine an Egyptian right pyramid with a square base. Suppose the sides of the square base are 80 meters long and the tallest point in the center of the pyramid is 80 meters above ground. Determine the surface area of the pyramid (not including the base).

17. Imagine a pyramid with a square base (like an Egyptian pyramid). Suppose that the sides of the square base are all 200 yards long and that the distance from one corner of the base to the top of the pyramid (along an outer edge) is 245 yards. Determine the surface area of the pyramid (not including the base). Explain your reasoning.

18. A popular brand of soup comes in cans that are $2\frac{5}{8}$ inches in diameter and $3\frac{3}{4}$ inches tall. Each such can has a paper label that covers the entire side of the can (but not the top or the bottom).

   a. If you remove a label from a soup can, you'll see that it's made from a rectangular piece of paper. How wide and how long is this rectangle (ignoring the small overlap where two ends are glued together)? Use mathematics to solve this problem, even if you have a can to measure. Explain your reasoning.

   b. Ignoring the small folds and overlaps where the can is joined, determine how much metal sheeting is needed to make the entire can, including the top and bottom. Use the moving and additivity principles about area to explain why your answer is correct. Be sure to use an appropriate unit to describe the amount of metal sheeting.

19. The portion of a cone without the base is made from $\frac{1}{4}$ of a circle of radius 8 inches (by joining radii). By calculating, determine the radius of the circle that will make a base for the cone, and determine the total surface area of the cone (with the base). Explain your reasoning. You may wish to make the cone and check that the base you propose really does work.

20.  **a.** Make a pattern for a cone such that the portion of the cone without the base is made from a portion of a circle of radius 4 inches (by joining radii) and such that the base of the cone is a circle of radius 3 inches. Show any relevant calculations, explaining your reasoning.

   b. Determine the total surface area (including the base) of your cone in part (a). Explain your reasoning.

21. **a.** Make a pattern for a cone such that the portion of the cone without the base is made from $\frac{2}{3}$ of a circle (by joining radii) and such that the base of the cone is a circle of radius 6 cm. Show any relevant calculations, explaining your reasoning.

   b. Determine the total surface area (including the base) of your cone in part (a). Explain your reasoning.

22. Describe in detail or draw precise patterns for two different cones that each has a surface area of approximately 30 cm², not including the base. Explain why your patterns produce cones of the desired surface area.

23. A cone with a circular base of radius 6 cm is to be made so that the distance from the vertex (point) of the cone to the center of its base is 8 cm.

   a. Determine the total surface area of the cone (including the base). Explain your reasoning.

   b. Carefully draw a pattern for the portion of the cone without the base, labeling the pattern with relevant lengths and angles. Explain all relevant calculations.

13.3 Volumes of Solid Shapes

F·P **Focal Points**
Grades 5, 7

The volume of a solid shape is a measure of how much three-dimensional space the shape takes up. As with area, there is a progression of increasingly sophisticated methods for determining volumes of solid shapes, culminating in the development of volume formulas for various shapes. Basic principles underlie all these methods.

What Volume Is

To understand how to determine volumes and to understand volume formulas, students must first know what volume means. For example, what does it mean to say that the volume of a solid shape is 30 cubic centimeters? It means that the solid shape could be made (without leaving any gaps) with a total of 30 1-cm-by-1-cm-by-1-cm cubes, allowing cubes to be cut apart and pieces to be moved if necessary. So the most primitive, basic way to determine the volume of a solid shape is to make the shape out of unit cubes (filling the inside completely) and to count how many cubes it took. Although primitive, this method is important, because it relies directly on the definition of volume and therefore emphasizes the meaning of volume. Even though students should know this approach to determining volume, they need to move beyond this primitive method to more efficient ways of determining volumes that can be used to solve problems. The first step in understanding how to determine volumes in other ways is to use the moving and additivity principles for volume, if only implicitly and subconsciously.

The Moving and Additivity Principles about Volumes

As with shapes in a plane, fundamental principles determine how volumes behave when solid shapes are moved or combined:

1. If we move a solid shape rigidly without stretching or shrinking it, then its volume does not change.

2. If we combine (a finite number of) solid shapes *without overlapping* them, then the volume of the resulting solid shape is the sum of the volumes of the individual solid shapes.

We have already used these principles implicitly, in explaining why we can determine the volume of a box by multiplying its height times its width times its length (see Chapter 4). We thought of the box as

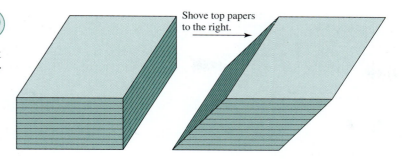

FIGURE 13.30

Shearing a stack
of paper

Shove top papers
to the right.

subdivided into layers, and each layer as made up of 1-unit-by-1-unit-by-1-unit cubes. Each small cube has volume 1 cubic unit, and the volume of the whole box (in cubic units) is the sum of the volumes of the cubes, which is just the number of cubes.

If you have a solid lump of clay, you can mold it into various different shapes. Each of these different shapes is made of the same volume of clay. From the point of view of the moving and additivity principles, it is as if the clay had been subdivided into many tiny pieces, and then these tiny pieces were recombined in a different way to form a new shape. Therefore, the new shape is made of the same volume of clay as the old shape. Similarly, if you have water in a container and you pour the water into another container, the volume of water stays the same, even though its shape changes.

Cavalieri's Principle about Shearing and Volumes

As with a shape in a plane, we can shear a solid shape and obtain a new solid shape that has the same volume.

Let's start with a polyhedron, pick one of its faces, and then imagine slicing the polyhedron into extremely thin (really, infinitesimally thin) slices that are parallel to the chosen face—this is rather like slicing a salami with a meat slicer. Now imagine giving those thin slices a push from the side, so that the chosen slice remains in place but so that the other thin slices slide over, remaining parallel to the chosen face and remaining the same distance from the chosen slice throughout the sliding process. The result is a new solid shape, as indicated in Figure 13.30. This process of "sliding infinitesimally thin slices" is called **shearing**.

shearing

You can show shearing nicely with a stack of paper. Give the stack of paper a push from the side, so that the sheets of paper slide over as shown in Figure 13.30. To understand shearing of solid shapes, think of the thin slices as made of paper. Note that in the shearing process, each thin slice remains unchanged: each slice is just slid over and is not compressed or stretched.

Cavalieri's principle

Cavalieri's principle for volume says that when you shear a solid shape as described, the volume of the original and sheared solid shapes are equal.

The Volume Formula for Prisms and Cylinders

Before introducing the simple volume formula for prisms and cylinders, recall that prisms and cylinders can be thought of as formed by joining two parallel, congruent bases. The **height of a prism or cylinder** is the distance between the planes containing the two bases of the prism or cylinder, measured in the direction perpendicular to the bases, as indicated in Figure 13.31. *The height is measured in the direction perpendicular to the bases, not on the slant.*

height (of prism or cylinder)

prism and cylinder volume formula

There is a very simple **formula for volumes of prisms and cylinders**:

$$(\text{height}) \times (\text{area of base})$$

In the volume formula it is understood that if the height is measured in some unit, then the area of the base is measured in square units of the same unit. The volume of the prism or cylinder resulting from the formula is then in cubic units of the same basic unit. For example, what is the volume of a 4-inch-tall can that has a circular base of radius 1.5 inches? The area of the base is $\pi(1.5)^2$ in.^{2} = 7.07 in.^{2}; therefore, the volume of the can is 4×7.07 in.^{3} = 28 in.^{3} (approximately).

FIGURE 13.31

Volume of prism
or cylinder =
(height) ×
(area of base).

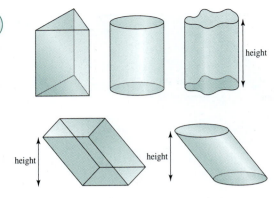

The next Class Activity will help you explain why the (height) × (area of the base) volume formula for prisms and cylinders makes sense.

Class Activity   *Now Turn to Class Activities Manual*

13N Why the Volume Formula for Prisms and Cylinders Makes Sense, p. 320

13O Filling Boxes and Jars, p. 321

Why do we multiply the height by the area of the base to calculate the volume of a prism or cylinder? Consider a rectangular prism that is 4 units high and has a base of area 6 square units, as pictured in Figure 13.32. If we fill this prism with 1-unit-by-1-unit-by-1-unit cubes, then the prism will be made of 4 layers. Each layer has 6 cubes in it—1 cube for each square unit of area in the base. So the whole prism is made of 4 groups (layers), with 6 cubes in each group, and therefore there are

$$4 \times 6$$

cubes in the prism. Each cube has a volume of 1 cubic unit; therefore, the volume of the prism is 4 × 6 cubic units. The same reasoning explains why the volume of a right prism or cylinder is

$$(\text{height}) \times (\text{area of base})$$

cubic units.

The volume formula for prisms and cylinders is valid not only for right prisms and cylinders but also for oblique ones. Why? Because an oblique prism or cylinder can be sheared into a right one. During shearing, neither the height, nor the area of the base, nor the volume changes, so the same formula applies to the oblique prism or cylinder as to the right prism or cylinder.

FIGURE 13.32

Why a 4-unit-
high prism with a
base of area
6 square units has
volume 4 × 6
cubic units

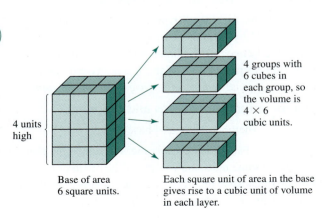

The volume formula for prisms and cylinders is also valid when the height or the area of the base is a fraction or a decimal. In fact, the reasoning we used previously still applies to the case of fractions and decimals, except that we must consider layers that contain partial cubes, such as layers containing $7\frac{1}{2}$ cubes. And we must consider partial layers. For example, if a prism is $4\frac{1}{2}$ units high instead of 4 units high, then we would have $4\frac{1}{2}$ groups of cubes instead of 4 groups of cubes. So we would have 4 full groups of cubes and another half-group of cubes, meaning another group of cubes that has only $\frac{1}{2}$ as much volume as the other groups.

Volume Formulas for Pyramids and Cones

height (of pyramid or cone)

Recall that pyramids and cones can be thought of as formed by joining a base with a point. The **height of a pyramid or cone** is the perpendicular distance between the point of the pyramid or cone and the plane containing the base. As indicated in Figure 13.33, *the height is measured in the direction perpendicular to the base, not on the slant.*

pyramid and cone volume formula

The **formula for volumes of pyramids and cones** is

$$\frac{1}{3}(\text{height}) \times (\text{area of base})$$

In the volume formula it is understood that if the height is measured in some unit, then the area of the base is measured in square units of the same unit. The volume of the pyramid or cone resulting from the formula is then in cubic units of the same basic unit. For example, what is the volume of sand in a cone-shaped pile that is 15 feet high and has a radius at the base of 7 feet? According to the volume formula, the volume of sand is

$$\frac{1}{3} \times 15 \times \pi(7)^2 \text{ ft}^3$$

which is about 770 cubic feet of sand.

Where does the $\frac{1}{3}$ in the volume formula for pyramids and cones come from? Class Activities 13P and 13Q consider this question.

Class Activity   *Now Turn to Class Activities Manual*

13P Comparing the Volume of a Pyramid with the Volume of a Rectangular Prism, p. 322

13Q The $\frac{1}{3}$ in the Volume Formula for Pyramids and Cones, p. 322

13R Volume Problem Solving, p. 323

13S The Volume of a Rhombic Dodecahedron, p. 324

FIGURE 13.33

Volume of pyramid or cone = $\frac{1}{3}$ (height) × (area of base)

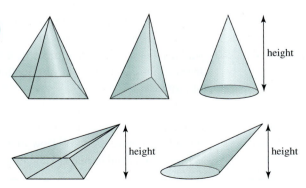

Volume versus Surface Area

A common source of confusion is the distinction between volume and surface area of a solid shape. Informally, we can think of surface area as a measure of how much paper, cloth, or other thin substance it would take to cover the outer surface of the shape. On the other hand, the volume of a shape is a measure of how much stuff it takes to fill the shape. You will think more about the distinction between surface area and volume in the next Class Activity, as well as in the exercises and problems that follow.

Class Activity   *Now Turn to Class Activities Manual*

13T Volume versus Surface Area and Height, p. 325

Practice Exercises for Section 13.3

1. What does it mean to say that a solid shape has a volume of 30 cubic centimeters?

2. Your students have an open-top box that has a 2-inch-by-4-inch rectangular base and is 3 inches high. They also have a bunch of cubic-inch blocks and some rulers.

   a. What is the most primitive way for your students to determine the volume of the box?

   b. What is a more advanced way for your students to determine the volume of the box, and why does this method work?

3. Explain why the (height) × (area of the base) volume formula is valid for right prisms using the example of a prism that is 4 units high and that has a base of area 6 square units.

4. Discuss why the $\frac{1}{3}$ in the volume formula for pyramids (and cones) is plausible.

5. What is the difference between the surface area of a solid object and its volume?

6. The water in a full bathtub is roughly in the shape of a rectangular prism that is 54 inches long, 22 inches wide, and 9 inches high. Use the fact that 1 gallon = 0.134 cubic feet to determine how many gallons of water are in the bathtub.

7. A concrete patio will be made in the shape of a 15-foot-by-15-foot square with half-circles attached at two opposite ends, as pictured in Figure 13.34. If the concrete will be 3 inches thick, how many cubic feet of concrete will be needed?

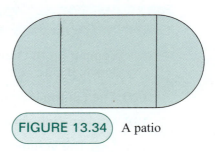

FIGURE 13.34 A patio

8. Suppose that a tube of toothpaste contains 15 cubic inches of toothpaste and that the circular opening where the toothpaste comes out has a diameter of $\frac{5}{16}$ inch. If every time you brush your teeth you squeeze out a $\frac{1}{2}$-inch-long piece of toothpaste, how many times can you brush your teeth with this tube of toothpaste?

9. A typical ice cream cone is $4\frac{1}{2}$ inches tall and has a diameter of 2 inches. How many cubic inches does it hold (just up to the top)? How many fluid ounces is this? (One fluid ounce is 1.8 cubic inches.)

Answers to Practice Exercises for Section 13.3

1. See text.

2. The most primitive way for your students to determine the volume of the box is to count how many cubes it takes to fill the box. That number of cubes is the volume of the box in cubic inches. A more advanced way to determine the volume of the box is to multiply the height (3 inches) by the area of the base (8 square inches) to obtain the volume, 24 cubic inches. See text for how to explain why this method works.

3. See text.

4. As the experiment in Class Activity 13P showed, you must pour the contents of a pyramid into a prism of the same height and base three times to fill the prism. Class Activity 13Q showed that 3 oblique pyramids that have square bases fit together to make a cube.

5. See the text in this section as well as the previous section (and Section 11.2) for a discussion of surface area. See this section for a discussion of volume.

6. There are about 46 gallons of water in the bathtub. Because the water in the tub is $(54 \div 12)$ ft = 4.5 ft long, $(22 \div 12)$ ft = 1.83 ft wide, and $(9 \div 12)$ ft = 0.75 ft tall, the volume of water in the tub is $4.5 \times 1.83 \times 0.75$ ft^3 = 6.17625 ft^3. Since each gallon of water is 0.134 ft^3, the number of gallons of water in the tub is the number of 0.134 ft^3 in 6.1875 ft^3, which is $6.1875 \div 0.134$ gallons. This is about 46 gallons.

7. We need 100 cubic feet of concrete for the patio. To explain why, notice that the patio is a kind of cylinder, so its volume can be calculated with the (height) $\times$ (area of base) formula. The base, which is the surface of the patio, consists of a square and two half-circles. Therefore, by the moving and additivity principles for areas, the area of the base is the area of the square plus the area of the circle that is created by putting the two half-circles together. All together, the area of the patio (the base) is $15^2 + \pi(7.5)^2 = 401.7$ square feet. The height of the concrete patio is $\frac{1}{4}$ of a foot—notice that we must convert 3 inches to feet to have consistent units. Therefore, according to the (height) $\times$ (area of base) formula, the volume of concrete needed for the patio is $\frac{1}{4} \times 401.7$ cubic feet, which is about 100 cubic feet of concrete.

8. You will be able to brush your teeth 391 times. Each time you brush, the amount of toothpaste you use is the volume of a cylinder that is $\frac{1}{2}$ inch high and has a radius of $\frac{5}{32}$ inch. According to the volume formula for cylinders, this volume is $\frac{1}{2} \times \pi(\frac{5}{32})^2$ in.^{3} = 0.0383 in.^{3}. The number of times you can brush is the number of 0.0383 in.^{3} in 15 in.^{3}, which is $15 \div 0.0383 = 391$ times.

9. The cone holds 4.7 cubic inches, which is 2.6 fluid ounces. The diameter of an ice cream cone is 2 in., therefore its radius is 1 in., and so the area of the base of an ice cream cone (the circular hole that holds the ice cream) is $\pi \times 1^2$ in.^{2} = 3.14 in.^{2}. According to the volume formula for cones, the volume of an ice cream cone is $\frac{1}{3} \times 4.5 \times 3.14$ in.^{3} = 4.7 in.^{3}. Since each fluid ounce is 1.8 in.^{3}, the number of fluid ounces the cone holds is the number of 1.8 in.^{3} in 4.7 in.^{3}, which is $4.7 \div 1.8 = 2.6$ fluid ounces.

Problems for Section 13.3

1. Suppose that a student in your class wants to know why we multiply only three of the lengths of the edges of a box in order to calculate the volume of the box. Why don't we have to multiply *all* the lengths of the edges?

  a. Explain to this student why it makes sense to calculate the volume of a box as we do.

  b. Describe some problems or activities that might help the student understand the calculation.

2. 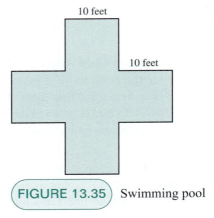 **a.** Justin is not clear about the difference between the *surface area* and the *volume* of a box. Explain the two concepts to Justin in a way that could help him learn to distinguish between the two concepts.

   b. Determine the surface area and the volume of a closed box that is 5 inches wide, 4 inches deep, and 6 inches tall. Explain in detail why you calculate as you do.

3. Young children sometimes think that tall containers necessarily hold more than shorter containers. Use graph paper to make two patterns for open-top boxes so that one box is taller than the other, but so that the shorter box has the greater volume. Explain briefly why your boxes meet the required conditions.

4. Students often confuse the surface area and the volume of a solid shape.

   a. Describe what surface area and volume are and discuss how they are different.

   b. Describe a hands-on activity that could help students understand the distinction between surface area and volume.

5. In your own words, explain why the (height) × (area of the base) volume formula is valid for right prisms using the example of a prism that is 5 cm high and that has a base of area 8 cm².

6. Discuss how to use blocks to explain why the (height) × (area of base) formula for the volume of a prism is true (for the case of right prisms where the height and area of the base are whole numbers). In your explanation, be sure to attend to the meaning of multiplication, the meaning of volume, and the units you are using.

7. A cylindrical container has a base that is a circle of diameter 8 centimeters. When the container is filled with 250 milliliters of liquid, the container becomes $\frac{2}{5}$ full. How tall is the container? Explain your reasoning.

8. One liter of water is in a cylindrical container. The water is poured into a second cylindrical container whose radius is $\frac{3}{4}$ the radius of the first container. Describe how the water level in the second container compares to the water level in the first container. Be specific and explain your reasoning.

9. **a.** Measure how fast water comes out of some faucet of your choice. Give your answer in gallons per minute, and explain how you arrived at your answer. (You do not have to fill up a gallon container to do this.)

   b. Figure 13.35 shows a bird's-eye view of a swimming pool in the shape of a cross (four 10-foot by 10-foot squares surrounding a 10-foot by 10-foot square). The pool is 4 feet deep, but doesn't have any water in it. There is a small faucet on one side with which to fill the pool. How long would it take to fill up this pool, assuming that the water runs out of this faucet at the same rate as water from your faucet? Use the fact that 1 gallon = 0.134 cubic feet. Explain your answer.

   10 feet

   10 feet

   FIGURE 13.35   Swimming pool

10. Find a gallon container, a half-gallon container, a quart container, a pint container, or a one-cup measure. Measure the lengths of various parts of your chosen container and use these measurements to determine the volume of your container in cubic inches. Use your answer to estimate how many gallons are in a cubic foot. Explain your answer. (There are about $7\frac{1}{2}$ gallons in a cubic foot.)

11. One gallon is 3.79 liters, and 1 cm³ holds 1 milliliter of liquid. Use these facts to determine the number of gallons in a cubic foot. Explain.

12. 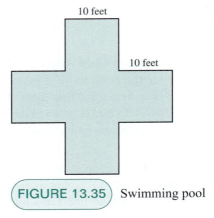 A cake recipe will make a round cake that is 6 inches in diameter and 2 inches high.

   a. If you use the same recipe, but pour the batter into a round cake pan that is 8 inches in diameter, how tall will the cake be? Explain.

   b. Suppose you want to use the same cake recipe to make a rectangular cake. If you use a rectangular pan that is 8 inches wide and 10 inches long, and if you want the cake to be about 2 inches tall, then how much of the recipe should you make? (For example, should you make twice as much of the recipe, half as much, three-quarters as much, or some other amount?) Give an approximate, but *practical*, answer. Explain.

13. A recipe for gingerbread makes a 9-in. × 9-in. × 2-in. pan full of gingerbread. Suppose that you want to use this recipe to make a gingerbread house. You decide that you can either use a 10-in. × 15-in. pan or a 11-in. × 17-in. pan, and that you can either make a whole recipe or half of the recipe of gingerbread.

   a. Which pan should you use, and should you make the whole recipe or just half a recipe, if you want the gingerbread to be between $\frac{1}{4}$ inch and $\frac{1}{2}$ inch thick? Explain.

   b. Draw a careful diagram showing how you would cut the gingerbread into parts to assemble into a gingerbread house. Indicate the different parts of the house (front, back, roof, etc.). When assembled, the gingerbread house should look like a real three-dimensional house—but be as creative as you like in how you design it. You do not have to use every bit of the gingerbread to make the house, but try to use as much as possible.

   c. If you want to put a solid, 2-inch-tall fence around your gingerbread house so that the fence is 6 inches away from the house, what will be the perimeter of this fence? Explain.

   d. To make the fence in part (c), how many batches of gingerbread recipe will you need,

and what pan (or pans) will you use? Indicate how you will cut the fence out of the pan (or pans). Explain.

14. The front (and back) of a greenhouse have the shape and dimensions shown in Figure 13.36. The greenhouse is 40 feet long, and the angle at the top of the roof is 90°. A fungus has begun to grow in the greenhouse, so a fungicide will need to be sprayed. The fungicide is simply sprayed into the air. To be effective, one tablespoon of fungicide is needed for every cubic yard of volume in the greenhouse. How much fungicide should be used? Give your answer in terms of units that are practical. (For example, it would not be practical to have to measure 100 tablespoons, nor would it be practical to have to measure 3.4 quarts. But it would be practical to measure 1 quart and 3 fluid ounces.) Explain.

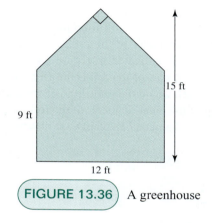

FIGURE 13.36 A greenhouse

15. A volume problem:

The front and back of a storage shed are shaped like isosceles right triangles with two sides of length 10 feet, as shown on the left of Figure 13.37. The storage shed is 40 feet long. Determine the volume of the storage shed.

Qing solves the volume problem as follows: First, he uses the Pythagorean theorem to determine that the length of the unknown side of the triangle shown at the left is $\sqrt{200}$ feet long. Then, Qing uses the Pythagorean theorem again to calculate the height of the triangle h if the base is the side of length $\sqrt{200}$. Qing carries this out as follows:

$$\left(\frac{\sqrt{200}}{2}\right)^2 + h^2 = 10^2$$

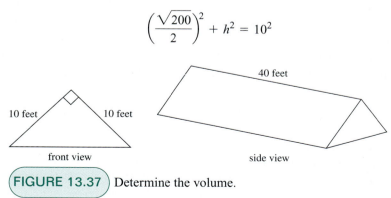

FIGURE 13.37 Determine the volume.

$$h^2 = 100 - \frac{200}{4} = 100 - 50 = 50$$

$$h = \sqrt{50}$$

Next, Qing determines that the floor area of the storage shed is

$$40 \cdot \sqrt{200}$$

square feet. Finally, Qing calculates that the volume of the storage shed is given by the floor area of the shed times the height of the triangle:

$$40 \cdot \sqrt{200} \cdot \sqrt{50}$$

a. Is Qing's method of calculation correct? Discuss Qing's work. Which parts (if any) are correct; which parts (if any) are incorrect?

b. Solve the volume problem in a different way than Qing did, explaining your reasoning.

16. Fifty pounds of wrapping paper are wound onto a roll. Looking at the roll from the side, we see that the outer diameter of the roll is 3 times the inner diameter, as indicated in Figure 13.38. The inner (unshaded) part of the roll is not filled with paper. How much would the same kind of wrapping paper weigh if the outer diameter of the roll were 2 times the inner diameter? Explain your reasoning.

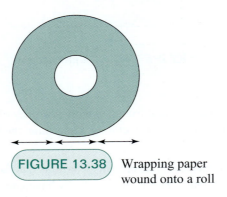

FIGURE 13.38 Wrapping paper wound onto a roll

17. A conveyor belt dumps 2500 cubic yards of gravel to form a cone-shaped pile. How high could this pile of gravel be, and what could the circumference of the pile of gravel be at ground level? Give *two different realistically possible pairs of answers* for the height and circumference of the pile of gravel (both piles of volume 2500 cubic yards), and compare how the piles of gravel would look in the two cases. Explain.

18. A construction company wants to know how much sand is in a cone-shaped pile. The company measures that the distance from the edge of the pile at ground level to the very top of the pile is 55 feet. The company also measures that the distance around the pile at ground level is 220 feet.

a. How much sand is in the pile? (Be sure to state the units in which you are measuring this.) Explain.

b. The construction company has trucks that carry 10 cubic yards in each load. How many loads will it take to move the pile of sand? Explain.

19. One of the Hawaiian volcanoes is 30,000 feet high (measured from the bottom of the ocean) and has volume 10,000 cubic miles. Assuming that the volcano is shaped like a cone with a circular base, find the distance around the base of the volcano (at the bottom of the ocean). In other words, if a submarine were to travel all the way around the base of the volcano, at the bottom of the ocean, how far would the submarine go? Explain.

20. Make a pattern for a cone-shaped cup that will hold about 120 mL when filled to the rim. Use a pattern like the one for a cone-shaped cup in Figure A.11 on page 733. Indicate measurements on your cup's pattern, and explain why the pattern will produce a cup that holds 120 mL. A centimeter tape measure may be helpful—you may wish to cut out the one on page 731.

21. Cut out the pattern in Figure A.11 on page 733 and tape the two straight edges together to make a small cone-shaped cup. You might want to leave a small "tab" of paper on one of the edges, especially if you use glue instead of tape. In any case, make sure that the two straight edges are joined.

a. Determine approximately how many fluid ounces the cone-shaped cup holds by filling it with a dry, pourable substance (such as rice, sugar, flour, or even sand), and pouring the substance into a measuring cup.

b. The pattern for the cone-shaped cup indicates that the two straight edges you taped together are 10 cm long, and the curved edge that makes the rim of the cup is 25 cm long. Use these measurements to determine the volume of the cone-cup in milliliters. Explain your method.

c. If you look on a can of soda, you'll see that 12 fluid ounces is 355 ml. Use this relationship, and your answer to part (b), to determine how many fluid ounces the cone-cup holds. Compare your answer to your estimate in part (a).

22. 🏺 A cone without a base is made from a quarter-circle. The base of the cone is a circle of radius 3 cm. What is the volume of the cone? Explain your reasoning.

23. a. Determine the volume of a cone that has a circular base of radius 6 cm and height 8 cm.

    b. Make a pattern for the cone (without its base) in part (a). Show all relevant calculations, explaining your reasoning.

24. A cone without a base is made from $\frac{3}{4}$ of a circle of radius 20 cm. Determine the volume of the cone. Explain your reasoning.

25. a. A cone-shaped cup has a circular opening at the top of diameter 10 cm. When the cup is filled with 240 ml of liquid, it is $\frac{4}{5}$ full in terms of volume (so 240 ml is $\frac{4}{5}$ of the volume of the cup). What is the height of the cup? Explain your reasoning.

    b. A cone-shaped cup has a circular opening at the top of diameter 10 cm. When the cup is filled with 240 ml of liquid, it is filled to $\frac{4}{5}$ of its height and the diameter of the surface of the liquid is 8 cm. What is the height of the cup? Explain your reasoning.

13.4 Volume of Submersed Objects versus Weight of Floating Objects

In this brief section we consider a hands-on way to determine volumes: submersing in water. When we submerse an object in water, we can obtain information about the volume of the object. On the other hand, when an object *floats* in water, we can get information about the *weight* of the object.

> ## Class Activity   *Now Turn to Class Activities Manual*
>
> **13U** Determining Volumes by Submersing in Water, p. 326
>
> **13V** 🏺 Underwater Volume Problems, p. 326
>
> **13W** Floating versus Sinking: Archimedes's Principle, p. 327

A hands-on way to determine the volume of an object is to submerse the object in a known volume of water and measure how much the water level goes up. For example, if we have a measuring cup filled with 300 mL of water and if the water level goes up to 380 mL when we put a plastic toy that sinks in the water, then the volume of the toy is 80 mL. Recall that 1 cubic centimeter holds 1 mL of liquid, so we can also say that the volume of the toy is 80 cm³. Why does it make sense that the volume of the toy is 80 mL or 80 cm³? When we place the toy in the water, the toy takes up space where water had been. The water that had been where the toy is now, is the "extra" water at the top of the measuring cup. So the volume of this "extra" water at the top of the measuring cup is the volume of the toy.

But what if an object is not *submersed* in water but *floats* on the water instead? For example, suppose we have a measuring cup filled with 300 mL of water and suppose that when we put a toy in the water, the toy floats, and the water level goes up to 380 mL. In this case, the amount of water that the toy displaced does not have the same *volume* as the toy; instead, it has the same *weight* as the toy. This fundamental physical fact, that *an object that floats displaces an amount of water that weighs as much as the object,*

Archimedes's principle   is called **Archimedes's principle**, in honor of Archimedes, the great mathematician and physicist who lived in ancient Greece, 287–212 B.C. So in the case of this floating toy, the toy weighs 80 grams because 80 mL of water weighs 80 grams.

Archimedes's principle indicates why some objects float. Think about gradually lowering an object into water. Will the object float or not? As you lower the object into the water, it displaces more and more water. At some point, the object may have displaced a volume of water that weighs as much as the object. If so, then the object will float at that point. But if you lower an object into water, and if the amount of water it displaces never weighs as much as the object, then the object will sink. Notice that floating is not just a matter of how *light* the object is: It has to do with the *shape* of the submersed part of the object, and whether this shape displaces enough water. Otherwise, heavy ships made of steel would never be able to float.

Practice Exercises for Section 13.4

1. A fish tank in the shape of a rectangular prism is 40 cm tall, 60 cm long, and 25 cm wide. A rock is placed on the bottom of the tank. Then 25 liters of water are poured into the tank. At that point, the tank is $\frac{2}{3}$ full. What is the volume of the rock in cubic meters?

2. A measuring cup contains 300 mL of water. When a ball is put into the water in the measuring cup, the ball sinks to the bottom and the water level rises to 400 mL.

   a. What, if anything, can you deduce about the volume of the ball?

   b. What, if anything, can you deduce about the weight of the ball?

3. Suppose you have a paper cup floating in a measuring cup that contains water. When the paper cup is empty, the water level in the measuring cup is at 250 mL. When you put some flour into the measuring cup, the cup is still floating and the water level in the measuring cup goes up to 350 mL. What information about the flour in the measuring cup can you deduce from this experiment? Explain.

Answers to Practice Exercises for Section 13.4

1. The volume of the rock is 0.015 cubic meters, which we can see as follows: The total volume of the tank is $40 \times 60 \times 25 = 60{,}000$ cm^3. Since 1 liter fills a 10-cm-by-10-cm-by-10-cm cube, 1 liter is 1000 cm^3. Therefore, 60,000 cm^3 is 60 liters. So when the tank is $\frac{2}{3}$ full, it is filled with $\frac{2}{3} \cdot 60 = 40$ liters. Since 25 liters of water were poured in, the rock must take up $40 - 25 = 15$ liters, which is 15,000 cm^3. But we want the volume in cubic meters. Since 1 m = 100 cm,

$$1 \text{ m}^3 = 100 \cdot 100 \cdot 100 \text{ cm}^3 = 1{,}000{,}000 \text{ cm}^3$$

Therefore, the volume of the rock is

$$\frac{15{,}000}{1{,}000{,}000} = \frac{15}{1000} = 0.015$$

cubic meters.

2. a. Since the water level rose 100 mL, the ball displaced 100 mL of water. One hundred milliliters has a volume of 100 cubic centimeters because 1 mL has a volume of 1 cm^3. So the ball has a volume of 100 cm^3.

   b. If the ball were *floating*, then we would be able to say that the ball weighs 100 grams, because according to Archimedes's principle, a floating body displaces an amount of water that weighs as much as the body. But because the ball is not floating, we cannot determine the exact weight of the ball from this experiment. However, we can say that the ball must weigh more than 100 grams. Here's why. Think of gradually lowering the ball into the water. When we lower the ball into the water, there can never be a time when the amount of water that the ball displaces weighs as much as the ball—otherwise, the ball would float, according to Archimedes's principle. So the amount of water that the ball displaces as it is lowered into the water must always weigh less than the ball. Since the ball displaces 100 mL of water, and since 100 mL of water weighs 100 g, the ball must weigh more than 100 g.

3. Since the water level rose from 250 mL to 350 mL, the flour floating in the cup displaced 100 mL of water. Because the cup with the flour in it is floating, the weight of the displaced water is equal to the weight of the flour, according to Archimedes's principle. Because 1 mL of water weighs 1 gram, 100 mL of water weighs 100 g, so the flour weighs 100 g.

Problems for Section 13.4

1. A tank in the shape of a rectangular prism has a base that is 20 cm wide and 30 cm long. The tank is partly filled with water. When a rock is put in the tank and sinks to the bottom, the water level in the tank goes up 2 cm. What is the volume of the rock? Explain your reasoning.

2. A container holds 5 liters. Initially, the container is $\frac{1}{2}$ full of water. When an object is placed in the container, the object sinks to the bottom and the container becomes $\frac{7}{8}$ full. What is the volume of the object in cubic centimeters? Explain your reasoning.

3. A fish tank in the shape of a rectangular prism is 40 cm wide and 60 cm tall. At first, the tank is empty. Then, some stones with a total volume of 15,000 cm^3 are put in the tank. Finally, 120 liters of water are poured into the tank. At that point, the tank is $\frac{3}{4}$ full. How wide is the tank? Explain your reasoning.

4. Suppose that you have a recipe that calls for 200 grams of flour, but you don't have a scale. Explain in detail how to apply Archimedes's principle to measure 200 grams of flour by using a measuring cup that has metric markings.

5. Suppose that you have a recipe that calls for $\frac{1}{2}$ pound of flour, but you don't have a scale. Explain in detail how to apply Archimedes's principle and the fact that 1 kilogram = 2.2 pounds to measure $\frac{1}{2}$ pound of flour using a measuring cup that has metric markings. As a point of interest, we often say that "a pint is a pound"; therefore, you might think that 1 cup of flour (which is $\frac{1}{2}$ of a pint) weighs $\frac{1}{2}$ pound. However, the "pint is a pound" rule applies to water and to similar liquids like milk and clear juices. But ordinary flour is less dense than water, so 1 cup of flour actually weighs less than $\frac{1}{2}$ pound.

Chapter Summary and Study Items

Section 13.1 Polyhedra and Other Solid Shapes

Some of the basic solid shapes are prisms, cylinders. pyramids, and cones. There are five Platonic solids: the tetrahedron, the cube, the octahedron, the dodecahedron, and the icosahedron.

Key skills and understandings:

* Describe prisms, cylinders, pyramids, and cones.

* Determine the number of vertices, edges, and faces of a given type of prism or pyramid, and explain why the numbers are correct.

Section 13.2 Patterns and Surface Area

An essential two-dimensional aspect of a solid shape is its outer surface. Patterns can be made to form the outer surface of a solid shape. The surface area of a solid shape is determined by adding the areas of all the component parts of the outer surface of the shape. Additional two-dimensional aspects of solid shapes are cross-sections and shadows.

Key skills and understandings:

- Visualize what shape a pattern will make.
- Make patterns for prisms, cylinders, pyramids, and cones.
- Determine the surface area of prisms, cylinders, pyramids, and cones.

Section 13.3 Volumes of Solid Shapes

The volume of a solid shape, in cubic units, is the number of 1-unit-by-1-unit-by-1-unit cubes it takes to make the shape (without leaving any gaps), allowing for cubes to be cut apart and pieces to be moved if necessary. The most primitive way to determine the volume of a shape is to count how many cubes it takes to make it.

The moving and additivity principles about volumes are basic principles that we use (consciously or not) when we determine volumes. The moving principle states that if we move a shape rigidly without stretching it, its volume does not change. The additivity principle states that if we combine shapes without overlapping them, the volume of the resulting combined shape is the sum of the volumes of the individual shapes. Cavalier's principle also applies to volumes. It tells us that if we shear a solid shape, the volume of the shape does not change.

The volume of a prism or a cylinder is (height) $\times$ (area of base). The volume of a pyramid or a cone is $\frac{1}{3}$(height $\times$ area of base). In all cases, the height is measured perpendicular to the base, not on the slant. We can explain the volume formula for prisms and cylinders by viewing a prism or cylinder as cut into slices parallel to the base. We can explain the $\frac{1}{3}$ in the volume formula for pyramids and cones by putting 3 oblique pyramids together to form a cube and by shearing the oblique pyramids. We can also see that this $\frac{1}{3}$ is plausible because the contents of a prism will fill a pyramid of the same base and height 3 times.

The volume of a solid shape is distinct from its surface area and its height, although these distinctions are a source of confusion for some students.

Key skills and understandings:

- Know what volume is and know the most primitive way to determine the volume of a solid shape.
- Know and use the moving and additivity principles for volume.
- Explain why the volume formula for prisms and cylinders is valid.
- Explain why the $\frac{1}{3}$ in the volume formula for pyramids and cones is plausible.
- Use the volume formulas for prisms, cylinders, pyramids, and cones to determine volumes and to solve problems.
- Discuss the distinction between the volume, the surface area, and the height of a solid shape.

Section 13.4 Volume of Submersed Objects versus Weight of Floating Objects

A hands-on way to determine the volume of an object is to *submerse* the object in water and to determine how much water the object displaced. If an object *floats*, it does not displace its volume, but rather the amount of water that *weighs* as much as it does (Archimedes's principle).

Key skills and understandings:

- Determine the volume of an object that sinks in liquid from how much liquid the object displaces.
- Determine the weight of an object that floats from how much liquid the floating object displaces.

Geometry of Motion and Change

*I*n this chapter we extend our study of shapes by allowing them to move and to change size. The movements of shapes leads to the subject of symmetry: By copying and moving shapes in different ways, we can create designs with different kinds of symmetry. We will also consider the extent to which a given shape can be changed. Certain shapes are rigid and inflexible, whereas others are "floppy" and movable. This difference in flexibility and rigidity is related to construction practices. Finally, we will ask what happens when objects change size but otherwise retain the same shape. The theory of size change has widely used practical applications, such as the measurement of distances in land surveying.

14.1 Reflections, Translations, and Rotations

 **Focal Points**
Grade 3

One especially attractive property of some shapes and designs is symmetry. In order to study symmetry, we will first examine certain transformations of planes—namely, reflections, translations, rotations, and glide-reflections. In the next section, we will use reflections, translations, rotations, and glide-reflections to define what we mean by symmetry, as well as to create symmetrical designs. From this point of view, reflections, translations, rotations, and glide-reflections are the building blocks of symmetry.

Definitions of Special Transformations: Reflections, Translations, Rotations, and Glide-Reflections

Roughly speaking, a **transformation** of a plane is just what the name implies: an action that changes or transforms a plane. We are interested in transformations that start with a plane and change it back into the same plane. Even though we will start

FIGURE 14.1

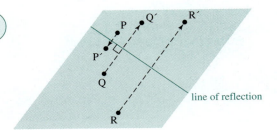

and end with the same plane, the transformation will usually have caused most or all of the individual points on the plane to change locations. The three kinds of transformations that we will study are reflections, translations and rotations. We will also briefly consider a fourth kind of transformation: glide-reflections.

reflection
line of
reflection

A **reflection** (or **flip**) of a plane across a chosen line, called the **line of reflection**, is the following kind of transformation: For any point Q in the plane, imagine a line that passes through Q and is perpendicular to the line of reflection, as shown in Figure 14.1. Now imagine moving the point Q to the other point on the line that is the same distance from the line of reflection. (If the point Q had been on the line of reflection, then it would stay in its place.) If we move each point in the plane as just described, then the end result is a reflection of the plane.

One way to think about reflections is that, under a reflection, each point in the plane goes to its "mirror image" on the other side of the line of reflection (hence the name "reflection").

An informal way to see the effect of a reflection is to draw a point on paper in wet ink or paint and quickly fold the paper along the desired line of reflection: The location where the wet ink rubs off onto the paper shows the final position of the point after reflecting. Another way to see the effect of a reflection is by drawing a point on a semitransparent piece of paper and flipping the paper upside down by twirling it around the desired line of reflection: You will see the location of the reflected point on the other side of the paper.

translation

A **translation** (or **slide**) of a plane by a given distance in a given direction is the end result of moving each point in the plane the given distance in the given direction. To illustrate a translation, put a flat piece of paper on a table top and slide the piece of paper in some direction (without rotating it). Now imagine doing this sliding process with a whole plane instead of a piece of paper: This would produce a translation of the plane. Figure 14.2 indicates initial and final positions of various points under a translation. Notice that the direction and distance of a translation can be specified by an arrow, as in Figure 14.2.

rotation

A **rotation** (or **turn**) about a point through a given angle is a transformation of a plane that is the end result of rotating all points in the plane about a fixed point, through a fixed angle. To

FIGURE 14.2

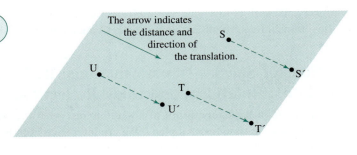

Effect of a
translation on
points S, T, and
U. The
solid arrow
represents the
distance and
direction of the
translation.

FIGURE 14.3

Effect of a rotation about point A on points V, W, and X

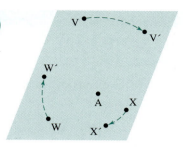

illustrate a rotation about a point, put a flat piece of paper on a table top, stick a pin through the point of rotation so as to hold that point fixed on the table, and rotate the piece of paper about that point. Now imagine rotating an infinite plane instead of a piece of paper: This movement would produce a rotation of the plane about the fixed point. Figure 14.3 shows the initial and final positions of some points in a plane under a rotation about the point A. Notice that points farther from the point A move a greater *distance* than points closer to A, even though all points rotate through the same *angle*.

glide-reflection A **glide-reflection** is the end result of combining a reflection and then a translation in the direction of the line of reflection.

Class Activity   *Now Turn to Class Activities Manual*

What Is Special about Reflections, Translations, Rotations, and Glide-Reflections?

Rotations, reflections, translations, and glide-reflections are special because these transformations *do not change distances*. For example, suppose you pick two points P and Q in the plane, and suppose that you rotate the plane. Then, even though the position and orientation of the pair of points P and Q may change after rotation, *the distance between the two points* is still the same, even after they have been rotated. Similarly, translations, reflections, and glide-reflections preserve the distances between pairs of points. Furthermore, mathematicians have proven that every transformation of the plane that preserves distances between all pairs of points must be either a rotation, a translation, a reflection, or a glide-reflection.

Practice Exercises for Section 14.1

1. Match the specified transformations in A, B, C, and D of Figure 14.4 to the effects shown in 1, 2, 3, and 4.

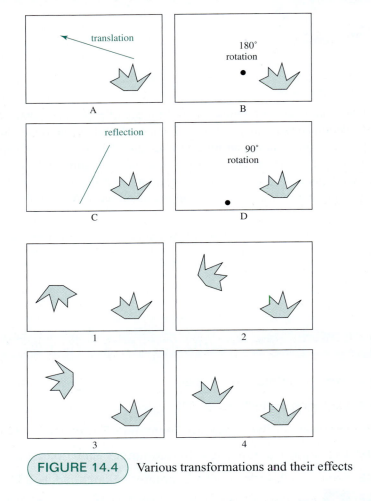

FIGURE 14.4 Various transformations and their effects

2. For each picture in Figure 14.5, determine what *single* transformation—reflection, translation, or rotation—will take the initial shape to the final shape.

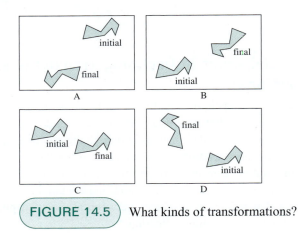

FIGURE 14.5 What kinds of transformations?

3. Draw the result of rotating the shaded shapes in Figure 14.6 by 90° counterclockwise around the origin. Explain how you know where to draw your rotated shapes.

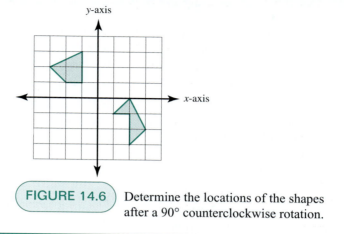

FIGURE 14.6   Determine the locations of the shapes after a 90° counterclockwise rotation.

Answers to Practice Exercises for Section 14.1

1. A–4, B–1, C–2, D–3.

2. See Figure 14.7.

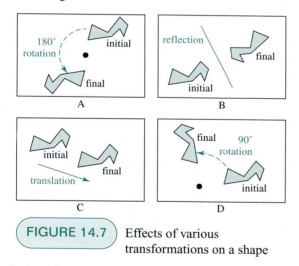

FIGURE 14.7   Effects of various transformations on a shape

3. See Figure 14.8. To determine where to draw the rotated shapes, notice that the *x*-axis rotates to the *y*-axis and the *y*-axis rotates to the *x*-axis. So the rotation takes the point A, which is 2 units to the right of the origin, to the point A′ which is 2 units up from the origin. The point B, which is 3 units below point A, must therefore rotate to the point B′, which is 3 units to the right of point A′. Point C, which is 3 units up and 1 unit to the left of the origin, will

rotate to the point C′, which is 3 units to the left and 1 unit down from the origin. Similarly, by considering the location of each corner point in the shaded shapes relative to the *x*- and *y*-axes, we can determine where these points go when they are rotated, thereby determining the location of the rotated shapes.

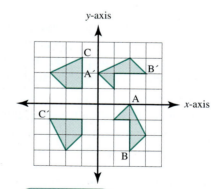

FIGURE 14.8   Shapes and their locations after a 90° counterclockwise rotation

Problems for Section 14.1

1. 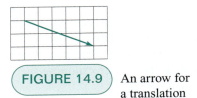 On graph paper, draw x- and y-axes, and draw two shapes that are not symmetrical. Draw the result of reflecting the shapes across the x-axis. Explain how you know where to draw your reflected shapes.

2. On graph paper, draw x- and y-axes, and draw two shapes that are not symmetrical. Draw the result of reflecting the shapes across the y-axis. Explain how you know where to draw your reflected shapes.

3. a. On graph paper, draw x- and y-axes, and draw two shapes that are not symmetrical. Draw the result of translating the shapes 5 units to the right.

   b. On graph paper, draw x- and y-axes, and draw two shapes that are not symmetrical. Draw the result of translating the shapes in the direction given by the arrow in Figure 14.9. Explain how you know where to draw your translated shapes.

FIGURE 14.9 An arrow for a translation

4. On graph paper, draw x- and y-axes, and draw two shapes that are not symmetrical. Draw the result of rotating the shapes 180° around the origin. Explain how you know where to draw your rotated shapes.

5. On graph paper, draw x- and y-axes, and draw two shapes that are not symmetrical. Draw the result of rotating the shapes 90° counterclockwise around the origin. Explain how you know where to draw your rotated shapes.

6. On graph paper, draw x- and y-axes, and draw two shapes that are not symmetrical. Draw the result of rotating the shapes 90° clockwise around the origin. Explain how you know where to draw your rotated shapes.

7. a. On graph paper, draw x- and y-axes, and plot the following points, labeling them A, B, C, and D:

   $$(5, 2), \quad (-3, 4), \quad (-4, -2), \quad (4, -5)$$

   Plot the locations of these points after they have been translated 3 units to the right, labeling the translated points A′, B′, C′, and D′.

   b. If (a, b) is a point in a coordinate plane, what will its location be after it has been translated 3 units to the right? Explain your answer.

8. a. On graph paper, draw x- and y-axes and plot the following points, labeling them A, B, C, and D:

   $$(3, 1), \quad (-4, 2), \quad (-5, -2), \quad (3, -2)$$

   Plot the locations of these points after they have been translated according to the arrow in Figure 14.9, labeling the translated points A′, B′, C′, and D′.

   b. If (a, b) is a point in a coordinate plane, what will its location be after it has been translated according to the arrow in Figure 14.9. Explain your answer.

9. a. On graph paper, draw x- and y-axes and plot the following points, labeling them A, B, C, and D:

   $$(4, 5), \quad (-4, 3), \quad (-3, -5), \quad (2, -1)$$

   Plot the locations of these points after they have been reflected across the x-axis, labeling the reflected points A′, B′, C′, and D′.

   b. If (a, b) is a point in a coordinate plane, what will its location be after it has been reflected across the x-axis? Explain your answer.

10. a. On graph paper, draw x- and y-axes and plot the following points, labeling them A, B, C, and D:

    $$(3, 4), \quad (-5, 2), \quad (-4, -5), \quad (3, -4)$$

    Plot the locations of these points after they have been reflected across the y-axis, labeling the reflected points A′, B′, C′, and D′.

    b. If (a, b) is a point in a coordinate plane, what will its location be after it has been reflected across the y-axis? Explain your answer.

11. a. On graph paper, draw x- and y-axes and plot the following points, labeling them A, B, C, and D:

    $$(3, 4), \quad (-5, 2), \quad (-4, -5), \quad (3, -4)$$

    Plot the locations of these points after they have been rotated 180° about the origin, labeling the rotated points A′, B′, C′, and D′.

b. If (a, b) is a point in a coordinate plane, what will its location be after it has been rotated 180° about the origin? Explain your answer.

12. a. On graph paper, draw x- and y-axes and plot the following points, labeling them A, B, C, and D:

$$(5, 1), \quad (-3, 4), \quad (-2, -5), \quad (4, -1)$$

Plot the locations of these points after they have been rotated 90° counterclockwise about the origin, labeling the rotated points A′, B′, C′, and D′.

b. If (a, b) is a point in a coordinate plane, what will its location be after it has been rotated 90° counterclockwise about the origin? Explain your answer.

13. a. On graph paper, draw x- and y-axes and plot the following points, labeling them A, B, C, and D:

$$(3, 4), \quad (-1, 4), \quad (-3, -5), \quad (2, -3)$$

Plot the locations of these points after they have been rotated 90° clockwise about the origin, labeling the rotated points A′, B′, C′, and D′.

b. If (a, b) is a point in a coordinate plane, what will its location be after it has been rotated 90° clockwise about the origin? Explain your answer.

14. Investigate the following questions, either with Geometer's Sketchpad (preferable) or by making drawings on (graph) paper. Start by drawing a shape (or design) that is not symmetrical. What is the *net effect* if you rotate your shape about some point by 180° and then rotate the resulting shape by 180° *about some other point?* What *single* transformation (reflection, translation, or rotation) will take your initial shape to your final shape?

15. Investigate the following questions, either with Geometer's Sketchpad (preferable) or by making drawings on (graph) paper. Start by drawing a shape (or design) that is not symmetrical. What is the *net effect* if you reflect your shape across a line and then reflect the resulting shape across another line that is perpendicular to the first line? What *single* transformation (reflection, translation, or rotation) will take your initial shape to your final shape?

16. Investigate the questions that follow, either with Geometer's Sketchpad or by making drawings on (graph) paper. Draw a shape (or design) that is not symmetrical, draw a separate line, and draw a point on the line.

a. If you reflect the shape across the line and then rotate the reflected shape 180° about the point, will the final position of the shape be the same as if you had *first* rotated the shape 180° and *then* reflected the rotated shape across the line?

b. If you reflect the shape across the line and then rotate the reflected shape 90° counterclockwise about the point, will the final position of the shape be the same as if you had *first* rotated the shape 90° counterclockwise and *then* reflected the rotated shape across the original, unrotated line?

14.2 Symmetry

Focal Points
Grade 3

Symmetry is an area shared by mathematics, the natural world, and art, so it offers opportunities for cross-disciplinary study. Why is symmetry deeply appealing to most people? Maybe it is because objects with symmetry seem more perfect than objects that don't have symmetry. Or maybe it is because objects with symmetry involve repetition. Is there something about human nature that causes us to enjoy repetition? For example, almost all music involves repetition of themes—different themes are repeated throughout a piece of music in a certain pattern. Young children love to hear the same story over and over again. Maybe we like repetition because it helps us learn and understand, and maybe our enjoyment of repetition makes symmetrical designs appealing.

When we look at natural objects in the world around us, we find a mix of symmetry and asymmetry. Most creatures are mostly symmetrical. Plants typically have symmetrical parts, even if they are not symmetrical over all: Leaves and flowerheads are usually symmetrical. On the other hand, geological features are rarely symmetrical. It would be surprising to see a mountain, a lake, a river, or a rock that

FIGURE 14.10

A shape with reflection symmetry

a shape   a shape and its
                        line of symmetry

we would describe as symmetrical, even though these objects are usually made up of smaller, repeated parts. Other natural objects such as some volcanoes, snowflakes, crystals, beaches, river stones, and waves are symmetrical, or nearly so.

Just as the concept of *circle* has an informal or artistic interpretation as well as a mathematical definition, the notion of *symmetry* has both an informal interpretation and a specific mathematical definition that applies to shapes in a plane or in space. Now we will consider the mathematical definition of symmetry, which requires that we draw upon the transformations just studied: rotations, reflections, and translations. First, we will use rotations, reflections, and translations to *define* symmetry. In other words, we will use these transformations to say what it means for a shape or a design to have specific kinds of symmetry. Next, we will use rotations, reflections, and translations to *create* shapes and designs that have symmetry.

Defining Symmetry

There are four kinds of symmetry that a shape or design in a plane can have: reflection symmetry, translation symmetry, rotation symmetry, and glide-reflection symmetry. We will use reflections, translations, and rotations to say what it means for a shape or design to have symmetry of one of these types.

reflection (mirror) symmetry   A shape or design in a plane has **reflection symmetry** or **mirror symmetry** if there is a line in the plane such that the shape or design as a whole occupies the same place in the plane both before and after reflecting across the line. This line is called a **line of symmetry**. For example, Figure 14.10 shows a design and a line of symmetry of the design. Notice that reflecting across the line of symmetry causes most points on the design to swap locations with another point on the design, but the design *as a whole* occupies the same place in the plane. Shapes or designs can have more than one line of symmetry, as shown in Figure 14.11.

To determine whether a design or shape drawn on a semitransparent piece of paper has reflection symmetry with respect to a certain line on the paper, fold the paper along that line. If you can see that the parts of the design on the two parts of the paper match one another, then the design has reflection symmetry, and the line you folded along is a line of symmetry. Another way to see whether a design or shape has reflection symmetry is to use a mirror that has a straight edge. Place the straight edge of the mirror along the line that you think might be a line of symmetry, and hold the mirror so that it is perpendicular to the design. When you look in the mirror, is the design the same as it was without the mirror in place? If so, then the design has reflection symmetry. These methods may help you get a better feel for reflection symmetry when you are first learning about it, but you should also be able to use *visualization* to determine whether a design has reflection symmetry.

FIGURE 14.11

A design with four lines of symmetry

a design   the design and its
                      four lines of symmetry

FIGURE 14.12

A wallpaper
pattern with
translation
symmetry

The pattern continues forever in all directions.

**translation
symmetry**

A design or pattern in a plane has **translation symmetry** if there is a translation of the plane such that the design or pattern *as a whole* occupies the same place in the plane both before and after the translation. True translation symmetry occurs only in designs or patterns that take up an infinite amount of space. So when we draw a picture of a design or pattern that has translation symmetry, we can only show a small portion of it; we must imagine the pattern continuing on indefinitely. Figure 14.12 shows a pattern with translation symmetry. In fact, notice that this pattern has two independent translations that take the pattern as a whole to itself: One is "shift right," another is "shift up." Wallpaper patterns generally have translation symmetry with respect to translations in two independent directions.

Frieze patterns—often seen on narrow strips of wallpaper positioned around the top of the walls of a room—provide additional examples of designs with translation symmetry. Figure 14.13 shows a frieze pattern that has translation symmetry. Unlike a wallpaper pattern, a frieze pattern will only have tranlation symmetry in one direction (and its "reverse") instead of in two independent directions.

**rotation
symmetry**

A shape or design in a plane has **rotation symmetry** if there is a rotation of the plane, of more than 0° but less than 360°, such that the shape or design *as a whole* occupies the same points in the plane both before and after rotation. For example, see Figure 14.14. Rotating 72° about the center of the design takes the design as a whole to the same location in the plane, even though each individual curlicue in the design moved to the position of another curlicue.

The design of Figure 14.14 is said to have 5-fold rotation symmetry because, by applying the $72° = 360° \div 5$ rotation about the center of the design 5 times, all points on the design return to their starting positions. Generally, shapes or designs can have 2-fold, 3-fold, 4-fold, and so on, rotation symmetry. A design has ***n*-fold rotation symmetry**, provided that a rotation of $360 \div n$ degrees takes the design as a whole to the same location. Applying this rotation n times returns every point on the design to its initial position. Figure 14.15 shows some examples of designs with various rotation symmetries.

***n*-fold
rotation
symmetry**

You can determine whether a design or shape drawn on a piece of paper has rotation symmetry: Make a copy of the design on semitransparent paper. Place the semitransparent paper over the original design so

FIGURE 14.13

A frieze pattern
with translation
symmetry

The pattern continues forever to the right and to the left.

FIGURE 14.14

A design with
5-fold rotation
symmetry

FIGURE 14.15

Designs with rotation symmetry

2-fold rotation symmetry   3-fold rotation symmetry   4-fold rotation symmetry   6-fold rotation symmetry

FIGURE 14.16

A frieze pattern with glide-reflection symmetry

The pattern continues forever to the right and left.

that the two designs match one another. If you can rotate the semitransparent paper less than a full turn so that the two designs match one another again, then the design has rotation symmetry. If you can keep rotating by the same amount for a total of 2, 3, 4, 5, and so on, times until the design on the semitransparent paper is back to its initial position, then the design has 2-fold, 3-fold, 4-fold, 5-fold, and so on, rotation symmetry, respectively. When you are first learning about rotation symmetry, it may help you to physically rotate designs; however, you should also use *visualization* to determine if a design has rotation symmetry.

glide-reflection symmetry

A design or pattern in a plane has **glide-reflection symmetry** if there are a reflection and a translation such that, after applying the reflection followed by the translation, the *design as a whole* occupies the same location in the plane. Glide-reflection symmetry is often seen on frieze patterns. For example, Figure 14.16 shows an example of a frieze pattern with glide-reflection symmetry. If this design is reflected across a horizontal line through its middle and then translated to the right (or left), the design as a whole will occupy the same location in the plane as it did originally. (See Figure 14.17.)

Notice that the frieze pattern in Figure 14.16 also has translation symmetry. Many designs have more than one type of symmetry. For example, the design shown in Figure 14.18 has both 2-fold rotation symmetry as well as reflection symmetry with respect to two lines: one horizontal and one vertical.

Class Activity *Now Turn to Class Activities Manual*

14G 🏛 Checking for Symmetry, p. 335

14H Frieze Patterns, p. 336

FIGURE 14.17

Understanding glide-reflection symmetry

1. Reflect across the horizontal.

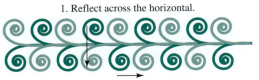

2. Translate right.

FIGURE 14.18

A design on
Egyptian fabric

Creating Symmetry

One way to create symmetrical designs is to start with any design and then reflect it, repeatedly rotate it, or repeatedly translate the design, keeping the original and all copies of the design produced in the process. The new design, consisting of the original and all copies, will have reflection symmetry, rotation symmetry, or translation symmetry, respectively. Figure 14.19 indicates the process of creating symmetrical designs this way.

FIGURE 14.19

Creating
symmetrical
designs

Reflect a design. The new design has
reflection symmetry.

Repeatedly rotate a design. The new design has
rotation symmetry.

Repeatedly translate a design.
If you could keep going forever, to the right and the
left, the new design would have translation symmetry.

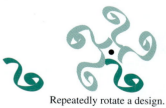

To create a design with 2-fold rotation symmetry, start with any design and rotate it $\frac{360°}{2} = 180°$ about any point in the plane. The original design and the rotated design together form a new design that has 2-fold rotation symmetry. To create a design with 6-fold rotation symmetry, start with any design and rotate it $\frac{360°}{6} = 60°$ five times, keeping the original design and all the rotated designs. All six designs together form a design with 6-fold rotation symmetry. In general, to create a design with n-fold rotation symmetry, start with any design and rotate it $\frac{360°}{n}$ repeatedly until the original design gets back to its initial position. Thereafter, if you rotate the whole design, the constituent designs will cycle around, but the design as a whole occupies the same location in the plane. Since n rotations of $\frac{360°}{n}$ produce a total of 360°, which is a full rotation, the design produced this way has n-fold rotation symmetry.

Class Activity *Now Turn to Class Activities Manual*

14I Traditional Quilt Designs, p. 337

14J Creating Symmetrical Designs with Geometer's Sketchpad, p. 338

14K Creating Symmetrical Designs (Alternate), p. 339

The artist M. C. Escher (1898–1972) created many interesting symmetrical designs involving congruent, interlocking shapes—something like Figure 14.20. You can see some of Escher's artwork at *www.pearsonhighered.com/beckmann*.

FIGURE 14.20

One kind of Escher-type design

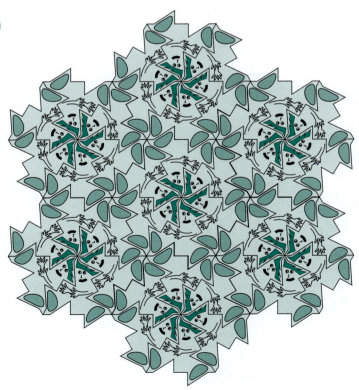

Software called TesselMania! enables even young children to create Escher-type designs. However, if you do the next Class Activity, you will see that you can also use Geometer's Sketchpad to create and modify Escher-type designs. Try Geometer's Sketchpad and enjoy seeing what makes these fascinating designs work.

Class Activity *Now Turn to Class Activities Manual*

14L Creating Escher-Type Designs with Geometer's Sketchpad (for Fun), p. 339

Practice Exercises For Section 14.2

1. Symmetrical designs are found throughout the world in all cultures. Figure 14.21 shows a small sample of such designs. Determine the kinds of symmetry these designs have.

Amish design found in
central Pennsylvania

Norwegian knitting design

Native American design

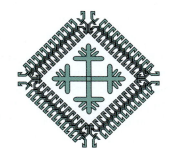

design from a Persian rug

FIGURE 14.21 Symmetrical designs from around the world

2. Practice Class Activities 14J or 14K.

3. Draw a design that is made out of copies of the shaded shape in Figure 14.22 and has 4-fold rotation symmetry, but no reflection symmetry.

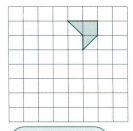

FIGURE 14.22 Create a symmetrical design.

4. Draw a design that is made out of copies of the shaded shape in Figure 14.22 and has 2 lines of symmetry.

5. What is the difference between a reflection and reflection symmetry?

Answers to Practice Exercises for Section 14.2

1. The Amish design has 6-fold rotation symmetry in addition to reflection symmetry with respect to 6 different lines: 3 lines that pass through the middle of opposite flower petals and 3 lines that pass through the middle of opposite hearts. (Notice that because the design has 6-fold rotation symmetry, it also has 3-fold and 2-fold rotation symmetry.)

The Norwegian knitting design has 2-fold rotation symmetry in addition to reflection symmetry with respect to 2 different lines: one horizontal line and one vertical line. (Notice that the individual "snowflake" designs within the design have 4-fold rotation symmetry and have reflection symmetry with respect to horizontal, vertical, and 2 diagonal lines, but the entire Norwegian knitting design has less symmetry.)

The Native American design has the same symmetries as the Norwegian knitting design, namely 2-fold rotation symmetry and reflection symmetry with respect to a horizontal and a vertical line.

The design from a Persian rug has 2-fold rotation symmetry in addition to reflection symmetry with respect to 2 different lines: one vertical and one horizontal. (The design at first appears to have 4-fold rotation symmetry, but it does not because of the way the "teeth" on the border are oriented.)

3. See Figure 14.23. The center of rotation is marked.

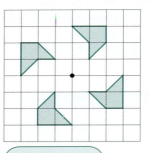

FIGURE 14.23   A design with 4-fold rotation symmetry

4. See Figure 14.24. The 2 lines of symmetry are marked. Notice that the design also has 2-fold rotation symmetry.

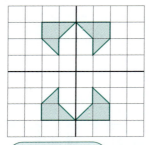

FIGURE 14.24   A design with 2-lines of symmetry

5. A reflection is a *transformation* of the plane. Reflection symmetry is a *property* that certain shapes or designs have.

Problems for Section 14.2

1. Find examples of symmetrical designs from a modern-day culture or a historical one. Make copies of the designs (either by photocopying or by drawing them) and determine what kinds of symmetry the designs have.

2. Determine all the symmetries of Design 1 and Design 2 in Figure 14.25. Consider each design as a whole. For each design, describe all lines of symmetry (if the design has reflection symmetry), and determine whether the design has 2-fold, 3-fold, 4-fold, or other rotation symmetry. Explain your answers.

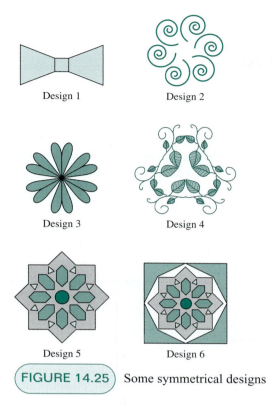

Design 1   Design 2

Design 3   Design 4

Design 5   Design 6

FIGURE 14.25   Some symmetrical designs

and determine whether the design has 2-fold, 3-fold, 4-fold, or other rotation symmetry. Don't forget to look for translation symmetry and glide-reflection symmetry, too. Explain your answers.

Design 7, pattern continues forever to the right and left.

Design 8, pattern continues forever to the right and left.

Design 9, pattern continues forever in all directions.

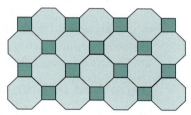

Design 10, pattern continues forever in all directions.

FIGURE 14.26   Some patterns

3. Determine all the symmetries of Design 3 and Design 4 in Figure 14.25. Consider each design as a whole. For each design, describe all lines of symmetry (if the design has reflection symmetry), and determine whether the design has 2-fold, 3-fold, 4-fold, or other rotation symmetry. Explain your answers.

4. Determine all the symmetries of Design 5 and Design 6 in Figure 14.25. Consider each design as a whole. For each design, describe all lines of symmetry (if the design has reflection symmetry), and determine whether the design has 2-fold, 3-fold, 4-fold, or other rotation symmetry. Explain your answers.

5. Determine all the symmetries of Design 7 and of Design 8 in Figure 14.26. Consider each design as a whole. For each design, describe all lines of symmetry (if the design has reflection symmetry), and determine whether the design has 2-fold, 3-fold, 4-fold, or other rotation symmetry. Don't forget to look for translation symmetry and glide-reflection symmetry, too. Explain your answers.

6. Determine all the symmetries of Design 9 and Design 10 in Figure 14.26. Consider each design as a whole. For each design, describe all lines of symmetry (if the design has reflection symmetry),

7. a. What types of triangles have reflection symmetry? Explain.

   b. What types of triangles have rotation symmetry? Explain.

8. a. Determine all the symmetries of a square. Explain.

   b. Determine all the symmetries of a rectangle that is not a square. Explain.

   c. Determine all the symmetries of a parallelogram that is not a rectangle or a rhombus. Explain.

9. Compare translation and translation symmetry. Explain how these two concepts are related.

10. Compare rotation and rotation symmetry. Explain how these two concepts are related.

11. Compare (mathematical) reflection and reflection symmetry. Explain how these two concepts are related.

12. Draw a simple asymmetrical shape or design. Then copy your shape or design so as to create a single new design that has both 2-fold rotation symmetry and translation symmetry *simultaneously*. (You may wish to use graph paper or software.)

13. Draw a simple asymmetrical shape or design. Then copy your shape or design so as to create a single new design that has both 4-fold rotation symmetry and translation symmetry *simultaneously*. (You may wish to use graph paper or software.)

14. Draw a simple asymmetrical shape or design. Then copy your shape or design so as to create a single new design that has both 2-fold rotation symmetry and reflection symmetry *simultaneously*. (You may wish to use graph paper.)

15. Draw a simple asymmetrical shape or design. Then copy your shape or design so as to create a single new design that has both 4-fold rotation symmetry and reflection symmetry *simultaneously*. (You may wish to use graph paper or software.)

16. Draw a simple asymmetrical shape or design. Then copy your shape or design so as to create a single new design that has rotation, reflection, and translation symmetry *simultaneously*. (You may wish to use graph paper or software.)

14.3 Congruence

In this section we focus on *congruence*, which is the mathematical term for saying two geometrical objects are the same. We can define congruence in terms of rotations, translations, and reflections, which we studied in Section 14.1. What information about two shapes will guarantee that they are congruent? Surprisingly, the answer for triangles differs from the answer for other shapes. Triangles, unlike other polygons, are structurally rigid, and a criterion for the congruence of triangles is related to this property. This distinction between triangles and other polygons explains some standard practices in building construction. So even though the study of congruence may seem purely abstract and theoretical, congruence has many practical applications.

Definition of Congruence

Informally, two shapes (either in a plane or solid shapes in space) that are the same size and shape are called congruent. But if we have two shapes or designs drawn on two semitransparent pieces of paper, how would we determine whether the shapes are congruent? We would probably slide one shape over the other and rotate the shapes to see if they match up. If not, we might flip one piece of paper over and try again. So to check if two shapes are congruent, we would use translations, rotations, and reflections. Therefore, we define congruence formally in the following way: Two shapes or designs in a plane are

congruent   **congruent** if there is a rotation, a reflection, a translation, or a combination of these transformations, that takes one shape or design to the other shape or design. For example, Figure 14.27 shows two shaded shapes that are congruent. They are congruent because the shaded shape on the right was created by first translating the shaded shape on the left horizontally to the right, and then rotating the translated shape 90° about the point shown.

FIGURE 14.27

Congruent shapes

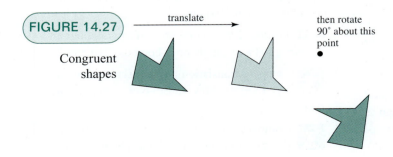

translate →

then rotate
90° about this
point
●

FIGURE 14.28

A triangle and a quadrilateral made of straws

In the next Class Activity, you will see a structural difference between triangles and quadrilaterals. The structural aspect of triangles that you will see is a physical manifestation of a condition for triangle congruence.

Class Activity   *Now Turn to Class Activities Manual*

14M Triangles and Quadrilaterals of Specified Side Lengths, p. 340

The Side-Side-Side (SSS) Congruence Criterion

If you form a triangle by threading a 3-inch, a 4-inch, and a 5-inch piece of straw together and a quadrilateral by threading a 3-inch, a 4-inch, a 3-inch, and a 4-inch piece of a straw together in that order (see Figure 14.28 and the instructions in Class Activity 14M), you will find that the triangle is rigid, whereas the quadrilateral is "floppy." That is, the triangle's sides cannot be moved independently, whereas the quadrilateral's sides can be moved independently, forming many different quadrilaterals. (See Figure 14.29.) We can interpret this difference about triangles and quadrilaterals in terms of congruence. *All triangles with sides of length 3 inches, 4 inches, and 5 inches are congruent,* whereas there are quadrilaterals with sides of length 3 inches, 4 inches, 3 inches, 4 inches (in that order) that are *not* congruent.

What if you have a triangle with sides of lengths other than 3 inches, 4 inches, and 5 inches? Imagine making this different triangle out of straws as well. Do you think it would also be rigid like the 3 inch, 4 inch, 5 inch one? In fact, any triangle, no matter what the lengths of its sides are, is structurally rigid. This fact about triangles can be stated in terms of congruence.

> Given a triangle that has sides of length *a, b,* and *c* units, it is congruent to all other triangles that have sides of length *a, b,* and *c* units.

FIGURE 14.29

Straws forming a rectangle and the same straws forming a parallelogram

FIGURE 14.30

Extra pieces
of wood create
triangles for
stability

FIGURE 14.31

A crane is made
out of many
triangles

SSS congruence

This last statement is the **side-side-side congruence** criterion, which is often simply called **SSS congruences** for triangles; it is the mathematical way to say that triangles are structurally rigid and determined by their side lengths.

The structural rigidity of triangles makes them common in building construction. For example, builders temporarily brace the frame of a house under construction with additional pieces of wood to keep the walls from falling over. (See Figure 14.30.) These additional pieces of wood create triangles; as triangles are rigid, the walls are held securely in place.

Triangles are also used in structures that need to be sturdy, but relatively lightweight. For example, the crane in Figure 14.31 consists of many triangles. Because the triangles are rigid, they make the crane strong without using solid metal, which would be very heavy.

In contrast, the structural flexibility of quadrilaterals makes them useful in objects that must move and change shape. For example, the folding laundry rack in Figure 14.32 is able to collapse because it is made from hinged rhombuses. The triangle at the top keeps the rack from collapsing when it is in use. When the rack is ready to be stowed, the triangle at the top is disconnected and the rhombuses collapse.

FIGURE 14.32

A folding laundry
rack uses hinged
rhombuses in its
construction. The
triangle at the top
keeps the rack
from collapsing
when in use.

FIGURE 14.33

Constructing a
triangle with
3 given side
lengths

We've seen from a physical point of view that side-side-side congruence for triangles should be true, but we can also explain this mathematically. Suppose we have two triangles that have the same side lengths. Then, by translating and rotating one of the triangles, we can arrange for one side of each triangle to be matched up. Now think about creating the two triangles by drawing circles whose radii are the two other side lengths of the triangles, as in Figure 14.33. The circles meet in two points, so each of our two triangles must be one of the two triangles shown in Figure 14.33. Since we can reflect one triangle to the other, our two triangles must be congruent.

In the next Class Activity, you may find ways to describe a triangle exactly other than specifying its side lengths.

Class Activity *Now Turn to Class Activities Manual*

14N Describing a Triangle, p. 341

14O Triangles with an Angle, a Side, and an Angle Specified, p. 342

The Angle-Side-Angle (ASA) Congruence Criterion

In addition to the side-side-side congruence criterion, there are other cases where all triangles with specific properties are congruent. One such case is when we specify the length of one side of a triangle and we also specify the two angles at either end of this side. For example, given a line segment AB, which triangles have a 40° angle at A and a 60° angle at B? Figure 14.34 shows that there are two such triangles: one triangle is "above" AB and the other is "below" AB. However, these two triangles are congruent, as you can see by reflecting one triangle across AB.

So, all triangles that have AB as one side and that have a 40° angle at A and a 60° angle at B are congruent. If a triangle has a different side of the same length as AB—say, CD—and if the triangle has a 40° angle at C and a 60° degree angle at D, then we can translate and rotate the triangle so that CD and AB

FIGURE 14.34

The two triangles
that have AB as
one side and
have specified
angles at A
and B

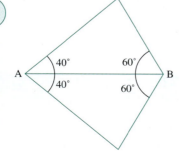

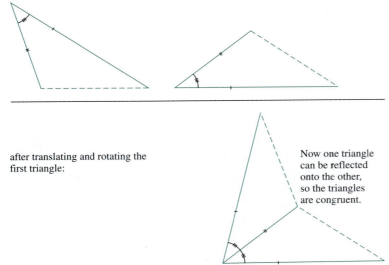

FIGURE 14.35

Showing why SAS congruence holds

The lengths of two sides (marked with dash and double dash) and the angle between the two sides (marked with triple dash) are given:

after translating and rotating the first triangle:

Now one triangle can be reflected onto the other, so the triangles are congruent.

match up (C landing at A and D landing at B). Once again, either the two triangles already match up, or one triangle can be reflected so as to match the other, as in Figure 14.34. The same will be true for other side lengths and other angles as well—as long as the two specified angles add to less than 180°. (Otherwise, a triangle cannot be formed, since all three positive angles in a triangle must add to 180°.) Thus, in general,

> if we specify the length of a side of a triangle, and if we specify two angles that add to less than 180° at the two ends of a line segment, then all triangles formed with those specifications are congruent.

ASA congruence   This last statement is the **angle-side-angle congruence** criterion, often simply called **ASA congruence**, for triangles.

The Side-Angle-Side (SAS) Congruence Criterion

There is yet another case in which all triangles with specific properties are congruent:

> If we specify the lengths of two sides of a triangle, and if we specify the angle between these two sides, then all triangles formed with those specifications are congruent.

SAS congruence   This last statement is the **side-angle-side congruence** criterion, often simply called **SAS congruence**, for triangles. Why is this criterion for triangle congruence valid? Think about creating two triangles from two sides of given lengths and with a given angle between them, as at the top of Figure 14.35. By translating and rotating one of the triangles, we can arrange for one side of the same length on each triangle to be matched up, as at the bottom of Figure 14.35. If the triangles aren't already matched up, then after reflecting one of them, they must match.

Class Activity *Now Turn to Class Activities Manual*

14P   Using Triangle Congruence Criteria, p. 343

Practice Exercises for Section 14.3

1. What is the SSS congruence criterion?

2. What is the ASA congruence criterion?

3. What is the SAS congruence criterion?

4. Suppose you make a triangle by threading 3 pieces of straw onto a string and tying the ends of the string together to make a loop. Similarly, suppose you make a quadrilateral by threading 4 pieces of straw onto a string and tying a loop. Describe the structural difference between the triangle and the quadrilateral (*other than* the fact that the triangle is made out of 3 pieces of straw and the quadrilateral is made out of 4), and explain how this structural difference is related to the concept of congruence.

5. Suppose someone tells you that he has a garden with 4 sides, 2 of which are 10 feet long and opposite each other, and the other 2 of which are 15 feet long and opposite each other. With only this information, can you determine the exact shape of the garden? Take into account that the person might like gardens in unusual shapes.

6. Visually, it appears that the "base angles" of an isosceles triangle are the same size (congruent). Use a triangle congruence criterion to prove that the base angles of an isosceles triangle are always congruent in the following way: Let A, B, and C be the vertices of an isosceles triangle, where AB and AC have the same length. Let D be the midpoint between B and C. Use a triangle congruence criterion to explain why triangles ADB and ADC are congruent. Thereby conclude that the angles at B and C (the base angles) have the same size.

7. Some barbecue tongs are made by fastening a pair of identical metal pieces halfway across their lengths, as shown in Figure 14.36. Thus, the tongs are fastened halfway between points A and D and halfway between points B and C. Visually, it appears that the distance between A and B is equal to the distance between C and D, and that this will be the case no matter what angle the tongs are spread apart by. Use a triangle congruence criterion, and previous facts we have studied about angles, to explain clearly why the distance between A and B must always be equal to the distance between C and D, no matter what angle the tongs are spread apart by. Be sure to state which triangle congruence criterion you use and why it applies.

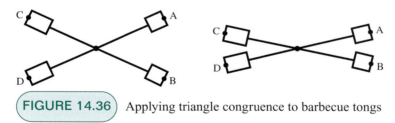

(FIGURE 14.36) Applying triangle congruence to barbecue tongs

Answers to Practice Exercises for Section 14.3

1. See text.

2. See text.

3. See text.

4. See text.

5. No, this information alone does not allow you to determine the shape of the garden. Although most people would probably make the garden rectangular, someone with an artistic flair might make the garden in the shape of a parallelogram that is not a rectangle. You can simulate this with 4 strung-together straws, as in Class Activity 14M and Figure 14.29.

6. See Figure 14.37. Since D is the midpoint between B and C, BD and CD have the same length. We are given that AB and AC have the same length. Therefore, the triangles ADB and ADC have the same side lengths. According to the SSS congruence criterion, the triangles ADB and ADC must therefore

be congruent. Since these triangles are congruent, all their angles must match, too. Therefore, the base angles at B and C must have the same size.

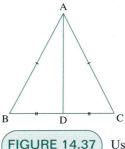

FIGURE 14.37 Using SSS to explain why the base angles of an isosceles triangle must be equal

7. Let E be the point where the metal pieces are fastened to make the tongs, as shown in Figure 14.38. In Chapter 10, we saw that when two lines meet at a point, the opposite angles formed that way are congruent (of the same size). Therefore, the angles

AEB and CED have the same size. Since the metal pieces are identical and are fastened halfway across their lengths, the lengths of AE, BE, CE, and DE are equal. Now we can apply the SAS congruence criterion to the triangles AEB and CED. Since AE and CE have the same length, and BE and DE have the same length, and since the angle between AE and BE is congruent to the angle between CE and DE, the SAS criterion tells us that the triangles AEB and CED are congruent. Since these triangles are congruent, all their side lengths must match, too. Therefore, the distance between A and B must be equal to the distance between C and D.

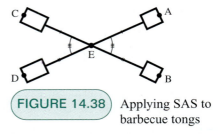

FIGURE 14.38 Applying SAS to barbecue tongs

Problems for Section 14.3

1. Suppose that Ada, Bada, and Cada are three cities and that Bada is 20 miles from Ada, Cada is 30 miles from Bada, and Ada is 40 miles from Cada. There are straight-line roads between Ada and Bada, Ada and Cada, and Bada and Cada.

   a. Draw a careful and precise map showing Ada, Bada, and Cada and the roads between them, using a scale of 10 miles to 1 inch. Describe how to use a compass to make a precise drawing.

   b. If you were to draw another map, or if you were to compare your map to a classmate's, how would they compare? In what ways might the maps differ? In what ways would they be the same? Which criterion for triangle congruence is most relevant to these questions?

2. Write a paragraph in which you discuss, in your own words, what you learned about triangles and quadrilaterals from Class Activity 14M, and how the concept of congruence is related to what you learned.

3. Is there a side-side-side-side congruence criterion? Discuss.

4. Is there an angle-angle-angle congruence criterion? Discuss.

5. This problem continues the investigation of Class Activity 10O in Section 10.3 on the size of your reflected face in a mirror. Figure 14.39 shows a side view of a person looking into a mirror. The mirror is parallel to AE. Lines BF and DG are normal lines to the mirror.

   a. What is the significance of the points F and G in Figure 14.39? Explain why, referring to the laws of reflection.

   b. Using the theory of congruent triangles discussed in this section, explain why triangles ABF and CBF are congruent and explain why triangles CDG and EDG are congruent.

   c. Use your answer to part (b) to explain why the length of BD is half the length of AE. Therefore, explain why the reflection of your face in a mirror is half as long as your face's actual length. (Notice that BD and FG have the same length.)

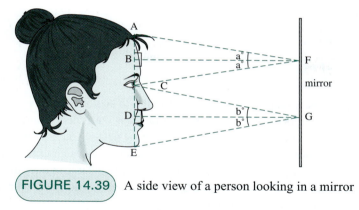

FIGURE 14.39   A side view of a person looking in a mirror

6. Suppose you fasten two rods of different lengths at their midpoints, as shown in Figure 14.40. Let the endpoints of the rods be A, B, C, and D. Visually, it appears that the distance between A and B is equal to the distance between C and D and that the distance between A and C is equal to the distance between B and D, no matter what angle the rods are spread apart by. Use a triangle congruence criterion, and previous facts we have studied about angles, to explain why the previous statements about distances must always be true, no matter what angle the rods are spread apart by. Be sure to state which triangle congruence criterion you used and why it applies.

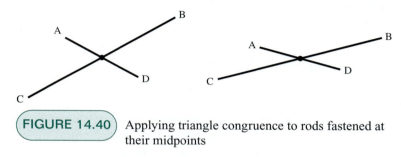

FIGURE 14.40   Applying triangle congruence to rods fastened at their midpoints

7. Ann and Kelly are standing on a river bank, wondering how wide the river is. Ann is wearing a basball cap, so she comes up with the following idea: She lowers her cap until she sees the tip of the visor just at the opposite bank of the river. She then turns around 180°, to face away from the river, being careful not to tilt her head or cap, and has Kelly walk to the spot where she can just see Kelly's shoes. By pacing off the distance between them, Ann and Kelly figure that Kelly was 50 feet away from Ann. If the ground around the river is level, what, if anything, can Ann and Kelly conclude about how wide the river is? Relate this to triangle congruence.

14.4 Similarity

 **Focal Points**
Grade 7

Two shapes are congruent if they are the same shape and size. This description is another way to say that two geometric objects are "equal." But how do we describe two objects if they are identical *except* for their size? Shapes, such as those in Figure 14.41, are called *similar*. The notion of similarity is used in solving many problems in geometry; it also has a wide range of practical applications, such as in surveying and mapmaking and representational drawing.

similar (informal)   Informally, we say that two objects that have the same shape, but not necessarily the same size, are **similar**. Another way to say this is as follows: Two objects or shapes are similar if one object represents a scaled version of the other (scaled up or down). For example, a scale model of a train is similar (at least on the outside) to the actual train on which it is modeled. The network of streets on a street map is similar to the network of real streets it represents—at least if the streets are on flat ground.

FIGURE 14.41

Similar shapes

One way to create a shape or design that is similar to another shape or design is by using grids of lines. If you enjoy doing crafts, then you may be familiar with this technique. Draw a network of equally spaced parallel and perpendicular grid lines over the design you want to scale (or use an overlay of such grid lines), as shown in Figure 14.42. If you want the new design to be, say, twice as wide and twice as long, then make a new network of parallel and perpendicular grid lines that are spaced twice as wide as the original grid lines. Now copy the design onto the new grid lines, square by square. The new design will be similar to the original design.

Class Activity   *Now Turn to Class Activities Manual*

14Q Mathematical Similarity versus Similarity in Everyday Language, p. 343

14R A First Look at Solving Scaling Problems, p. 343

Similarity and Scale Factors

Examining the craft project method for creating similar designs that is illustrated in Figure 14.42 will help us define more precisely what we mean by similarity. The grid for the large bunny was made so that the width and height of the large bunny would be twice the width and height of the small bunny. But notice that *all* lengths on the large bunny are twice as long as the corresponding lengths on the small bunny. This scaling of all lengths is a key aspect of similarity.

In contrast, the two bunnies on the right in Figure 14.43 are *not* similar to the original bunny on the left, because in each case the width has a different relationship to the original than the height does. The middle bunny is just as tall as the original bunny, but it is only half as wide. The bunny on the right is just as wide as the original bunny, but it is only half as tall.

similar   Formally, we say two shapes or objects (in a plane or in space) are **similar** if every point on one object corresponds to a point on the other object and there is a positive number, k, such that the distance between any two points on the second object is k times as long as the distance between the corresponding points on the

scale factor   first object. This number k is called the **scale factor** from the first object to the second object.

For example, the box of a toy model car might indicate that the toy car is a 24 : 1 scale model of an actual car. This means that the toy car and the actual car are similar and the scale factor from the toy car to the actual car is 24, or equivalently, that the scale factor from the actual car to the toy car is $\frac{1}{24}$.

FIGURE 14.42

Using a grid to make similar pictures

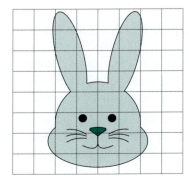
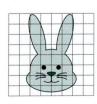

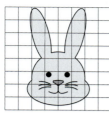

FIGURE 14.43

Examples of pictures that are *not* similar

The original bunny   These bunnies are *not* similar to the original bunny.

In particular, the length, width, and height of the actual car are 24 times the length, width, and height, respectively, of the toy car. Similarly, for other distances, such as the width of the windshield or the length of the hood, each distance on the actual car is 24 times the corresponding distance on the toy car.

Warning: Scale factors apply only to lengths *not* to areas or volumes. In Section 14.5 we will see how areas and volumes behave under scaling.

Three Methods for Solving Problems about Similar Objects or Shapes

In many practical situations, two shapes or objects are given to be similar. If we know various lengths on one object, then we can determine all the corresponding lengths on the other object, as long as we know at least one of the corresponding lengths. There are three common ways to do this, and all use the scale factor, even if indirectly. Now let's study these three ways by using a specific example.

Problem: Suppose that an artist creates a scale model of a sculpture. The scale model is 10 inches wide and 20 inches tall. If the actual sculpture is to be 60 inches wide, then how tall will the actual sculpture be?

Solutions:

scale factor method

1. **Scale Factor Method** (See Figure 14.44.) As the scale model and the actual sculpture are to be similar, there is a scale factor, k, from the model to the actual sculpture such that every length on the actual sculpture is k times as long as the corresponding length on the scale model. Therefore, the width of the actual sculpture is $k \times 10$ inches and the height of the actual sculpture is $k \times 20$ inches. Since the width of the sculpture is given as 60 inches,

$$k \times 10 \text{ inches} = 60 \text{ inches}$$

so

$$k = 60 \div 10 = 6$$

FIGURE 14.44

Using the scale factor method

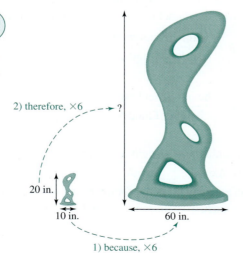

2) therefore, ×6 - - -→ ?

20 in.

10 in.   60 in.

1) because, ×6

FIGURE 14.45

Using the
internal factor
method

►20 in.

10 in.
1) because, ×2

60 in.
2) therefore, ×2

Thus, the height of the sculpture is

$$k \times 20 \text{ inches} = 6 \times 20 \text{ inches} = 120 \text{ inches}$$

Notice that the scale factor method is the method of *comparing like quantities* discussed in Section 7.2, where it was described as an "external" comparison method.

**internal
factor
method**

2. **Internal Factor Method** (See Figure 14.45.) As the scale model is 10 inches wide and 20 inches tall, it is $\frac{20}{10} = 2$ times as tall as it is wide. The actual sculpture should, therefore, also be 2 times as tall as it is wide. Since the sculpture is to be 60 inches wide, it should be

$$2 \times 60 \text{ inches} = 120 \text{ inches}$$

tall. Why should the actual sculpture also be 2 times as tall as it is wide? There is a scale factor, k, such that the width and height of the sculpture are $k \times 10$ and $k \times 20$, respectively. Therefore,

$$\frac{\text{sculpture height}}{\text{sculpture width}} = \frac{k \times 20}{k \times 10} = \frac{20}{10} = 2$$

Notice that the internal factor method is the method of *comparing unlike quantities* discussed in Section 7.2, where it was described as an "internal" comparison method. Internal factors are studied in trigonometry, where they are viewed as sines, cosines, and other trigonometric functions. Internal factors are also used to describe slopes of lines.

**proportion
method**

3. **Set up a Proportion Method** As before, since the scale model and the actual sculpture are to be similar, there is a scale factor, k, from the model to the actual sculpture. k can be calculated as

$$k = \frac{\text{width of sculpture}}{\text{width of model}}$$

and

$$k = \frac{\text{height of sculpture}}{\text{height of model}}$$

Therefore,

$$\frac{\text{width of sculpture}}{\text{width of model}} = \frac{\text{height of sculpture}}{\text{height of model}}$$

All the quantities in this last equation are known, except for the height of the sculpture. Let's let h stand for the height of the sculpture in inches. Then, substituting the known quantities, we have

$$\frac{60 \text{ in.}}{10 \text{ in.}} = \frac{h \text{ in.}}{20 \text{ in.}} \tag{14.1}$$

FIGURE 14.46

Applying the scale factor method and the internal factor method to a proportion

Since,   *therefore,*

$$\times 6 \left(\frac{60 \text{ in.}}{10 \text{ in.}} = \frac{h \text{ in.}}{20 \text{ in.}} \right) \times 6 \qquad \text{so } h = 6 \times 20 \text{ in.} = 120 \text{ in.}$$

therefore, $\times 2$

$$\frac{60 \text{ in.}}{10 \text{ in.}} = \frac{h \text{ in.}}{20 \text{ in.}} \qquad \text{so } h = 2 \times 60 \text{ in.} = 120 \text{ in.}$$

Since, $\times 2$

But we know that we can determine whether two fractions are equal by cross-multiplying. So, because the previous equation is true,

$$60 \times 20 = 10 \times h$$

Dividing both sides of this equation by 10, we see that

$$h = \frac{60 \times 20}{10} = 120$$

so the sculpture is 120 inches tall.

Although the third method is more common, the logic behind it is a little more subtle than that of the first two methods.

Notice that we can link the scale factor method and internal factor method to the proportion in Equation 14.1, as shown in Figure 14.46.

Class Activity   *Now Turn to Class Activities Manual*

14S Using the Scale Factor, Internal Factor and Set up a Proportion Method, p. 344

14T A Common Misconception about Scaling, p. 345

14U Using Scaling to Understand Astronomical Distances, p. 345

14V More Scaling Problems, p. 346

When Are Two Shapes Similar?

In the situations we have encountered thus far, objects have been *given* as being similar. In other words, the very nature of the situation tells us that the objects are similar. However, there can be cases where it is not entirely obvious or clear that two shapes in question are similar. In those cases, how can we tell whether the shapes are similar? It is tempting to think that one can always tell "by eye," but this method is not reliable. Figure 14.47 shows two triangles that appear to be similar but are not.

triangle similarity criterion

Here is a **criterion for similarity of triangles**: Two triangles are similar exactly when the two triangles have the same-size angles. To clarify, two triangles are similar exactly when it is possible to

FIGURE 14.47

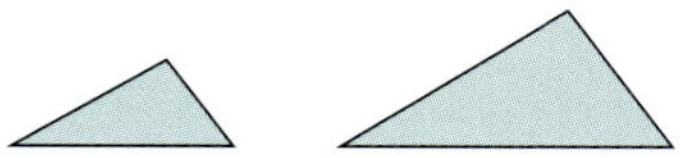

These triangles may look similar but they are not.

FIGURE 14.48

Similar triangles

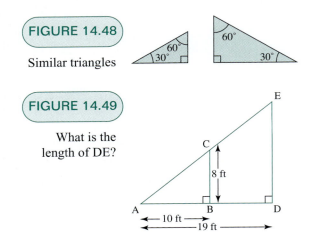

FIGURE 14.49

What is the
length of DE?

match each angle of the first triangle with an angle of the second triangle in such a way that matched
angles have equal measures. So, for example, we can determine that the triangles in Figure 14.48 are
similar because they have the same-size angles.

Actually, notice that to determine whether two triangles are similar, we really only need to check that *two*
of their angles are the same size, because the sum of the angles in a triangle is always 180°.
If two angles are known, the third one is determined as the angle that makes the three add to 180°. Thus, if
two angles in one triangle are the same size as two respective angles in another triangle, then the third
angles will automatically also be equal in size.

Figure 14.49 provides an example in which we can apply the criterion for triangle similarity. In Figure 14.49, what is the length of side DE? First, notice that the figure shows that triangles ABC and ADE
have two angles of the same size: the angle at A and a right angle. Therefore, these triangles are similar. Now we can apply any of our three methods to determine the length of side DE. Using the scale factor method, the scale factor is $\frac{19}{10}$, so the length of DE is

$$\frac{19}{10} \times 8 \text{ ft} = 15.2 \text{ ft}$$

In general, triangles created by parallel lines, as in Figure 14.50, are similar. Why? A line that crosses two
parallel lines makes the same angle with both parallel lines because of the parallel postulate (which we discussed in Section 10.2. In Figure 14.50, the two triangles ABC and ADE share an angle at A, and BC and
DE are parallel, so the angles at B and D are the same size, and the triangles ABC and ADE are similar.

Another common situation in which similar triangles are created is illustrated in Figure 14.51. In this case,
two lines cross at a point A. If the lines BC and DE are parallel, then the triangles ABC and ADE are similar. This is because the angle at A in ABC is the same size as the angle at A in ADE, as we discussed in
Section 10.2. Also, the angles at B and at D are equal in size because these are the points where the parallel line segments BC and DE meet the line segment BD. Therefore, two out of three angles in ABC and
ADE have the same size, so all three must have the same size, and the triangles are similar.

Why does the "three equal angles" criterion for triangle similarity work? Basically, it is because the
process of scaling doesn't change angles. If two triangles are similar, then one triangle can be scaled to
become congruent to the other, and since this scaling doesn't change angles, the triangles must have the

FIGURE 14.50

Parallel lines
create similar
triangles.

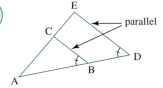

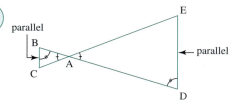

FIGURE 14.51

Parallel lines on opposite sides of the point where two lines meet create similar triangles.

same angles. Conversely, suppose two triangles have the same angles, say, *a*, *b*, and *c*. Then one of the triangles can be scaled so that the sides common to angles *a* and *b* have the same length in both triangles. In the scaling process, the angles don't change, so if you match the sides that have the same length, they both have angles *a* and *b* at either end, and these triangles are congruent (by ASA congruence). Therefore, the original triangles must have been similar.

Using Similar Triangles to Determine Distances

Similar triangles can be used in practical situations to find an unknown distance or length when several other related distances and lengths are known. Often, the tricky part in applying the theory of similar triangles is locating the similar triangles and sketching them. You will need to apply your visualization skills to help you sketch the situation, showing the relevant components. In the following examples, think carefully about why the pictures were drawn the way they were. If they had been drawn from a different perspective, would you see the relevant similar triangles? (In some cases, yes, but in many cases, no.)

Class Activity   *Now Turn to Class Activities Manual*

14W   Measuring Distances by "Sighting", p. 346

Finding Distances by "Sighting"

How far away is an object in the distance? When we look at a distant object, it appears smaller than it actually is. Can we quantify the object's apparent size? Similar triangles can help us answer both of these questions.

sighting To describe how big an object appears to be, you can compare it with the size of your thumb by stretching your arm out straight in front of you, closing one eye, and "**sighting**" from your thumb to the object you are considering. This situation creates a pair of similar triangles, as shown in Figure 14.52. The similar triangles are ABC (eye, base of thumb, top of thumb) and ADE (eye, bottom of picture, top of picture). These triangles are similar because the angles at A are equal and the angles at B and D are equal if the thumb is held parallel to the picture. (See Figure 14.50.)

FIGURE 14.52

"Thumb sighting" a picture on a wall

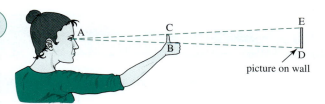

FIGURE 14.53

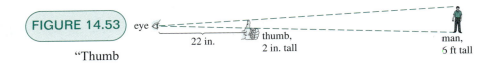

FIGURE 14.53

"Thumb sighting" a man

FIGURE 14.54

Surveying

You can also use this "thumb sighting" to find a distance. Let's say that I see a man standing in the distance, and at this distance he appears to be "1 thumb tall." If the man is actually 6 feet tall, then approximately how far away is he? As just explained, this thumb sighting creates similar triangles. (See Figure 14.53, which is *not* drawn to scale.) The distance from my sighting eye to the base of my thumb on my outstretched arm is 22 inches. My thumb is 2 inches tall. So, the distance from my eye to my thumb is $\frac{22}{2}$ times as long as the length of my thumb. Therefore, according to the internal factor method, the man is

$$\frac{22}{2} \times 6 \text{ feet} = 66 \text{ feet}$$

away. You might object that this tells us the distance from my eye to the man's feet, but, because the man is fairly far away, this distance is almost identical to the distance along the ground from my feet to his feet. Plus, we are only determining *approximately* how far away the man is, because all the measurements used are very rough. So this extra bit of inaccuracy is insignificant.

How could you modify the method of thumb sighting to make it more useful and more accurate? You could use objects besides your thumb for sighting—a ruler held vertically would be much more accurate, for example. And what if you didn't just hold the ruler in your hand but had it attached to a fixed length of pole—that would also create greater accuracy. These sorts of improvements have led to some of the surveying equipment used today. People at construction sights who are looking through a device on a tripod at a pole in the distance (such as in Figure 14.54), are surveying.

Some kinds of surveying equipment measure distances based on the theory of similar triangles using a more elaborate method than primitive thumb sighting. For example, to measure the distance between the surveying equipment and a pole of known height in the distance, the person surveying looks through the equipment and "sights" the pole. A kind of ruler inside the surveying equipment replaces the thumb.

Finding Heights by Using Sun Rays and Shadows

Is there a way we can find the height of a tree, pole, or other tall object without climbing up it to measure how tall it is? In fact, yes. We can use the shadow that an object casts, together with the theory of similar triangles, to determine how tall the object is.

Class Activity   *Now Turn to Class Activities Manual*

14X Using a Shadow or a Mirror to Determine the Height of a Tree, p. 348

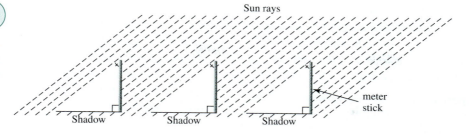

FIGURE 14.55

Sun rays make the same angles with vertical meter sticks.

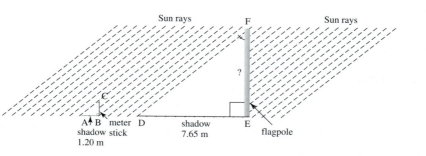

FIGURE 14.56

Similar triangles are formed by objects and their shadows.

Suppose Ms. Ovrick takes her class outside on a sunny day to determine the height of the flagpole. The students hold meter sticks perpendicular to ground and measure the length of the shadows that the meter sticks cast. Although the numbers vary, all the shadows are about 120 cm long. Why should all the shadows be about the same length? Since the sun is very far away, when its light rays reach the earth, the rays are virtually parallel. So, as we see in Figure 14.55, the angles that the sun rays make with the meter sticks should all be the same, and so the shadows cast by the meter sticks should all be the same too.

Next, the students measure the length of the flagpole's shadow and find that it is 7 meters and 65 centimeters long. How tall is the flagpole? Let's use the theory of similar triangles to determine the height of the flagpole. Think about how shadows of the meter sticks and the flagpole are created by the sun's rays. Because the sun rays are virtually parallel when they get to the earth, the rays make the same angle with the top of the flagpole as they do with the tops of the meter sticks, as we see in Figure 14.56. The meter sticks and the flagpole are perpendicular to the ground, so the triangle formed by a meter stick and its shadow (ABC) and the triangle formed by the flagpole and its shadow (DEF) have the same angles and are therefore similar. Since the length of the shadow of the meter stick is 1.2 times its height, the length of the shadow of the flagpole must also be 1.2 times its height. Therefore, the flagpole is

$$7.65 \div 1.2 = 6.375$$

meters tall, or about 6.4 meters tall.

Practice Exercises for Section 14.4

1. Amber has drawn a simple design on a rectangular piece of paper that is 4 inches wide and 12 inches long. Amber wants to make a scaled-up version of her design on a rectangular piece of paper that is 10 inches wide. She must figure out how long to make the larger rectangle. Amber says that since the larger rectangle is 6 inches wider than the smaller rectangle (because $10 - 4 = 6$), she should make her larger rectangle 6 inches longer than the smaller rectangle. She figures that she should make the larger rectangle $12 + 6 = 18$ inches long. Will Amber be able to make a correctly proportioned scaled-up version of her design on a rectangle that is 10 inches wide and 18 inches long? If not, how long should she make her rectangle, and why?

2. A postcard problem: A 24-inch-by-48-inch rectangular picture will be scaled down to fit on a postcard. On the postcard, the short side of the picture will be 4 inches long. What will the long side of the picture be on the postcard?

Solve the postcard problem in two ways: with the scale factor method and with the internal factor method. In each case:

- Briefly explain the idea and the reasoning of the method.

- Link the method to a proportion that more advanced students might set up, in which they set two fractions equal to each other.

3. Hannah wants to make a scale drawing of herself standing straight with her arms down. Hannah is 4 feet 6 inches tall. Her arm is 22 inches long. If Hannah wants to make the drawing of herself 10 inches tall, then how long should she draw her arm?

Solve this problem in two different ways: one using the scale factor method and one using the internal factor method, explaining your reasoning both times.

4. Ms. Bullock's class is making a display of the sun, the planets, and Pluto. Each planet and Pluto will be depicted as a circle. The students in Ms. Bullock's class want to show the earth as a circle of diameter

TABLE 14.1 Diameters of heavenly bodies

Heavenly Body	Approximate Diameter
Sun	1,392,000 km
Mercury	4,900 km
Venus	12,100 km
Earth	12,700 km
Mars	6,800 km
Jupiter	138,000 km
Saturn	115,000 km
Uranus	52,000 km
Neptune	49,500 km
Pluto	2,300 km

10 cm, and they want to show the correct relative sizes of the planets, Pluto, and the sun. Given the information in Table 14.1, what should the diameters of these scale drawings be? Since the sun is so big, it wouldn't be practical to make a full model of the sun; how could the students make a sliver of the sun of the correct size?

5. Istabrag is 4 feet 9 inches tall, and her shadow is 3 feet 6 inches long. At the same time, the shadow of a tree (measured from the base of the tree to the tip of the shadow) is 24 feet 6 inches long. How tall is the tree?

a. Draw a picture of the similar triangles that are relevant to solving this problem and explain why the triangles are similar.

b. Determine the height of the tree and explain your reasoning.

6. Suppose that you go outside on a clear night when much of the moon is visible and use a ruler to "sight" the moon (as described in Class Activity 14W). Use the information that follows to determine how big the moon will appear to you on your ruler. (You will also need to measure the distance from your eye to a ruler that you are holding with your arm stretched out.) On a night when the moon is visible, try this and verify it.

The distance from the earth to the moon is approximately 384,000 km. The diameter of the moon is approximately 3,500 km.

7. A sculptor makes a scale model for a sculpture she plans to carve out of marble that is 15 inches long, 8 inches wide, and 27 inches high. She finds a block of marble that is 5 feet long, 4 feet wide, and 10 feet high. What will the finished dimensions of the sculpture be if she makes the sculpture as large as possible?

8. Go to the Web site *www.pearsonhighered.com/ beckmann* to learn how to make a **camera obscura** from a Pringles® potato chip can. A camera obscura is a fun device that projects images onto a screen through a small hole. It illustrates a fundamental idea of photography: projecting an image through a small hole. You can make a camera obscura from any tube by cutting off a piece of the tube (about 2 inches from one end), placing semitransparent paper or plastic over the place where you cut, and taping the tube back together, with the semitransparent paper now in the middle of the tube. Cover one end of the tube with aluminum foil, and use a

pin to poke a small hole in the middle of the foil. Cover the side of the tube with aluminum foil to keep light out. Now look through the open end of the tube at a well-lit scene. You should see an upside-down projected image of the scene on the semitransparent paper.

Suppose you make a camera obscura out of a tube of diameter 3 inches, and suppose you put the semi-transparent paper or plastic 2 inches from the hole in the aluminum foil. How far away would you have to stand from a 6-foot-tall man in order to see the entire man on the camera obscura's screen?

Answers to Practice Exercises for Section 14.4

1. Amber will not be able to make a scaled-up version of her design on paper that is 10 inches wide and 18 inches long. As we see in Figure 14.57, if Amber makes the larger rectangle 10 inches by 18 inches, it will not be proportioned in the same way that the original rectangle is. Adding the same amount to the length and width generally does not preserve the ratio of length to width in a rectangle. Instead, Amber could reason that since the original rectangle is 3 times as long as it is wide, the larger rectangle should also be 3 times as long as it is wide. So if the larger rectangle is 10 inches wide, it should be $3 \times 10 = 30$ inches wide.

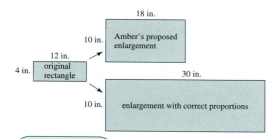

FIGURE 14.57 How should Amber scale up her 4-inch-by-12-inch rectangle?

2. *Scale factor method*: The short side of the postcard is $\frac{1}{6}$ as long as the short side of the picture, so the scale factor from the picture to the postcard is $\frac{1}{6}$. The long side of the postcard must also be $\frac{1}{6}$ as long as the long side of the picture because all side lengths must be scaled down by the same factor. Therefore, the long side of the postcard must be $\frac{1}{6} \cdot 48 = 8$ inches. This solution method is linked to a proportion students could set up to solve the problem, as shown next:

Internal factor method: The long side of the picture is 2 times as long as the short side of the picture. The long side of the postcard must therefore also be 2 times as long as the short side of the postcard in order to be proportioned in same way as the picture. Therefore, the long side of the postcard must be $2 \cdot 4 = 8$ inches. This solution method is linked to a proportion students could set up to solve the problem, as shown next:

$$\times 2 \left(\frac{48}{24} = \frac{x}{4} \right) \times 2$$

3. First, let's determine Hannah's height in inches. Each foot is 12 inches, so 4 feet is $4 \times 12 = 48$ inches. Therefore, 4 feet 6 inches is $48 + 6 = 54$ inches, so Hannah is 54 inches tall.

Internal Factor Method: Hannah's arm is 22 inches long, which is $\frac{22}{54}$ of her height. Therefore, in her drawing, Hannah's arm should also be $\frac{22}{54}$ of her height in the drawing, namely,

$$\frac{22}{54} \times 10 \text{ inches} = 4 \text{ inches}$$

approximately. So Hannah should draw her arm 4 inches long.

Scale Factor Method: Because Hannah is 54 inches tall and the scale drawing of herself is to be 10 inches tall, the height of Hannah's drawing is $\frac{10}{54}$ of Hannah's actual height. In other words, the scale factor from Hannah to the drawing of herself is $\frac{10}{54}$. The length of Hannah's arm in her drawing should also be $\frac{10}{54}$ of the length of her actual arm. Therefore, the drawing of Hannah's arm should be

$$\frac{10}{54} \times 22 \text{ inches} = 4 \text{ inches}$$

long approximately.

4. From the problem statement, we infer that we want the collection of circles representing the planets, Pluto, and the sun to be similar to cross-sections of the actual planets—-all with the same scale factor. If we use the internal factor method, we don't have to worry about converting kilometers to centimeters (or vice versa), and we won't have to work with huge numbers.

Because the earth's diameter is 12,700 km and Mercury's diameter is 4,900 km, Mercury's diameter is $\frac{4,900}{12,700}$ times as long as the earth's diameter. The same relationship should hold for the models of Mercury and the earth. Since the model earth is to have a diameter of 10 cm, the model Mercury should have a diameter of

$$\frac{4,900}{12,700} \times 10 \text{ cm} = 3.9 \text{ cm}$$

Table 14.2 shows the calculations for the other heavenly bodies.

TABLE 14.2 Diameters of scale of heavenly bodies.

Heavenly Body	Approximate Diameter of Circle Representing It
Sun	$\frac{1,392,000}{12,700} \times 10 \text{ cm} = 1096 \text{ cm} = 10.96 \text{ m}$
Mercury	$\frac{4,900}{12,700} \times 10 \text{ cm} = 3.9 \text{ cm}$
Venus	$\frac{12,100}{12,700} \times 10 \text{ cm} = 9.5 \text{ cm}$
Earth	given as 10 cm
Mars	$\frac{6,800}{12,700} \times 10 \text{ cm} = 5.4 \text{ cm}$
Jupiter	$\frac{138,000}{12,700} \times 10 \text{ cm} = 108.7 \text{ cm}$
Saturn	$\frac{115,000}{12,700} \times 10 \text{ cm} = 90.6 \text{ cm}$
Uranus	$\frac{52,000}{12,700} \times 10 \text{ cm} = 40.9 \text{ cm}$
Neptune	$\frac{49,500}{12,700} \times 10 \text{ cm} = 39 \text{ cm}$
Pluto	$\frac{2,300}{12,700} \times 10 \text{ cm} = 1.8 \text{ cm}$

The circle representing the sun should have a diameter of about 11 meters and therefore a radius of about 5.5 meters. The students could measure a piece of string $5\frac{1}{2}$ meters long. One student could hold one end and stay in a fixed spot, while another student could attach a pencil to the other end and use it to draw a piece of a 5.5-meter-radius circle representing the sun.

5. As explained with the meter stick and flagpole in Figure 14.56, the triangle formed by Istabrag and her shadow is similar to the triangle formed by the tree and its shadow. Istabrag is 4 feet 9 inches, or 4.75 feet tall. (The 0.75 is because 9 inches is $\frac{9}{12} = \frac{3}{4} = 0.75$ of a foot.) Her shadow is 3 feet 6 inches, or 3.5 feet long. So, Istabrag is

$$\frac{4.75}{3.5} = 1.357\ldots$$

times as tall as her shadow is long. The same relationship must hold for the tree. So, the tree must be 1.357 times as tall as its shadow is long. Therefore, the tree is

$$1.357 \times 24.5 = 33.25$$

feet tall, or 33 feet 3 inches tall. (The 3 inches is because $0.25 = \frac{1}{4}$ and $\frac{1}{4}$ of a foot is 3 inches.)

6. When you sight the moon with a ruler, holding the ruler parallel to the moon's diameter, you create similar triangles ABC and ADE, as shown in Figure 14.58 (not to scale). These triangles are similar as explained on page 610 with Figure 14.50. Since the diameter of the moon is 3,500 km and the distance to the moon is 384,000 km, the moon's diameter (DE) is $\frac{3,500}{384,000}$ times as long as the distance to the moon (AD). The same relationship must hold for the apparent size of the moon on the ruler (BC) and the distance from your eye to the ruler (AB): The apparent size of the moon on the ruler must be $\frac{3,500}{384,000}$ times as long as the distance from your eye to the ruler. (This is the internal factor method.) Notice that $\frac{3,500}{384,000}$ is about $\frac{1}{100}$. So, if the distance from your eye to the ruler is about 22 inches, then the moon's diameter will appear to be about 0.2 inches on your ruler, or about $\frac{1}{5}$ of an inch.

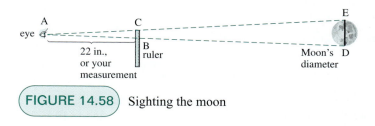

FIGURE 14.58 Sighting the moon

7. One way to solve this problem is to think about the scale factors we might use. Since each foot is 12 inches, 5 feet is $5 \times 12 = 60$ inches, 4 feet is $4 \times 12 = 48$ inches, and 10 feet is $10 \times 12 = 120$ inches. Thinking only about the length, we see that the scale factor from the model to the sculpture, k, should be such that

$$k \times 15 = 60$$

so

$$k = \frac{60}{15} = 4$$

Similarly, thinking only about the width and the height, we find that the scale factors would be

$$k = \frac{48}{8} = 6$$

and

$$k = \frac{120}{27} = 4.4$$

respectively. The artist will need to use the smallest of these scale factors; otherwise her sculpture would require more marble than she has. So the artist should use $k = 4$, in which case the dimensions of the sculpture are as follows:

$$\text{width} = 4 \times 15 \text{ in.} = 60 \text{ in.}$$
$$\text{depth} = 4 \times 8 \text{ in.} = 32 \text{ in.}$$
$$\text{height} = 4 \times 27 \text{ in.} = 108 \text{ in.}$$

8. Figure 14.59 shows how light from an object DE enters the pinhole (at point A) and projects onto the screen inside a camera obscura. If the object being looked at and the screen BC of the camera obscura are both vertical, then BC and DE are both vertical and are therefore parallel. Therefore, as described in the text, ABC and ADE are similar. Since these two triangles are similar, all corresponding distances on the triangle scale by the same scale factor. Since the distance from the screen of the camera obscura to the pinhole is $\frac{2}{3}$ times the length of the screen, by the internal factor method, the distance from the pinhole to the man should be $\frac{2}{3}$ times the length of the man, which is $\frac{2}{3} \times 6$ feet = 4 feet. So you need to stand at least 4 feet from the man to see the full length of him on the camera obscura.

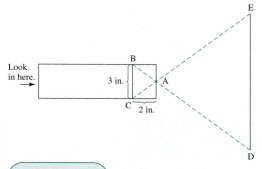

FIGURE 14.59 Side view of a camera obscura

Problems for Section 14.4

1. Frank's dog, Fido, is 16 inches tall and 30 inches long. Frank wants to draw Fido 4 inches tall. Frank figures that because Fido's height in the drawing will be 12 inches less than Fido's actual height (because $16 - 12 = 4$), Fido's length in the drawing should also be 12 inches less than Fido's actual length. Therefore, Frank figures that he should draw Fido $30 - 12 = 18$ inches long. Will the drawing of Fido be correctly proportioned if Frank makes it 18 inches long? If not, how long should Frank draw Fido? Explain.

2. **a.** If two shapes are congruent, are they also similar? Explain.

   b. If two shapes are similar, are they necessarily also congruent? Explain.

3. Tyler's flag problem: Tyler has designed his own flag on a rectangle that is 3 inches tall and 6 inches wide. Now Tyler wants to draw a larger version of his flag on a rectangle that is 9 inches tall. How wide should Tyler make his larger flag?

   Solve this problem in two ways: with the scale factor method and with the internal factor method. In each case:

   - Explain clearly the idea and the reasoning of that method as if you were explaining it to students who know about multiplication and division but not about setting up proportions.

   - Link the method to a proportion that more advanced students might set up, in which they set two fractions equal to each other.

4. Jasmine's flag problem: Jasmine has designed her own flag on a rectangle that is 8 inches tall and 12 inches wide. Now Jasmine wants to draw a smaller version of her flag on a rectangle that is

6 inches tall. How wide should Jasmine make her smaller flag?

Solve this problem in two ways: with the scale factor method and with the internal factor method. In each case:

- Explain clearly the idea and the reasoning of that method as if you were explaining it to students who know about multiplication and division but not about setting up proportions.

- Link the method to a proportion that more advanced students might set up, in which they set two fractions equal to each other.

5. Kelsey wants to make a scale drawing of herself standing straight and with her arms down. Kelsey is 4 feet 4 inches tall. Her leg is 24 inches long. If Kelsey wants to make the drawing of herself 10 inches tall, then approximately how long should she draw her leg?

Solve this problem in two ways: with the scale factor method and with the internal factor method. In each case:

- Explain clearly the idea and the reasoning of that method as if you were explaining it to students who know about multiplication and division but not about setting up proportions.

- Link the method to a proportion that more advanced students might set up, in which they set two fractions equal to each other.

6. A painting that is 4 feet 3 inches by 6 feet 4 inches will be reproduced on a card. The side that is 4 feet 3 inches long will become 3 inches long on the card. Determine how long the 6 feet 4 inches side will become on the card. Explain why your method of solution is valid.

7. Ms. Winstead's class went outside on a sunny day and measured the lengths of some of their classmates' shadows. The class also measured the length of a shadow of a tree. Inside, the students made a table like the one shown in Table 14.3.

TABLE 14.3 Heights of students and lengths of shadows

	Tyler	Jessica	Sunjae	Lameisha	Tree
shadow length	33 in.	34 in.	32 in.	34 in.	22 feet
height	53 in.	57 in.	52 in.	58 in.	?

In your own words, explain clearly how the theory of similar triangles applies in this situation. Include a picture as part of your explanation. Use the theory of similar triangles to determine the approximate height of the tree.

8. A thumb sighting problem: Suppose you are looking down a road and you see a person ahead of you. You hold out your arm and "sight" the person with your thumb, finding that the person appears to be as tall as your thumb is long. Assume that your thumb is 2 inches long and that the distance from your sighting eye to your thumb is 22 inches. If the person is 5 feet 4 inches tall, then how far away are you from the person?

a. Draw a picture showing that the thumb sighting problem involves similar triangles. Explain why the triangles are similar.

b. Solve the thumb sighting problem in two different ways. In both cases, explain the logic behind the method you use.

9. Explain and draw a picture to show how you could use the theory of similar triangles to determine the height of a flagpole by looking into a mirror on the ground. Indicate the similar triangles that would be involved and explain why these triangles really must be similar. Say which measurements you will need to make, and show or describe briefly how you will use these measurements to determine the height of the pole. (You may either make up numbers for these measurements or use letters to stand for these numbers.)

10. Mr. Jackson's class wants to make a display of the solar system showing the correct relative distances of the planets and Pluto from the sun (in other words, showing the distances of the planets and Pluto from the sun to scale). Table 14.4 shows the approximate actual distances of the planets and Pluto from the sun.

TABLE 14.4 Distances of planets and Pluto from the sun

Heavenly Body	Approximate Distance from Sun
Mercury	36,000,000 miles
Venus	67,000,000 miles
Earth	93,000,000 miles
Mars	141,000,000 miles
Jupiter	484,000,000 miles
Saturn	887,000,000 miles
Uranus	1,783,000,000 miles
Neptune	2,794,000,000 miles
Pluto	3,666,000,000 miles

If the distance from the sun to Mercury is to be represented in the display as 1 inch, then how should the distances from the sun to the other planets and Pluto be represented? Explain your reasoning in detail for one of the planets in such a way that fifth graders who understand multiplication and division, but who do not know about setting up proportions, might be able to understand. Calculate the answers for the other planets without explanation.

11. Suppose you have a TV whose screen is 36 inches wide. You decide that you want to simulate the experience of watching a movie on a 30-foot-wide screen as closely as possible. How far away should you sit from the TV so that the TV screen will appear as wide in your field of vision as a 30-foot-wide screen would from 40 feet away? Explain how the theory of similar triangles applies to help you answer this question. Include a sketch of similar triangles in your explanation, and explain in detail why these triangles really are similar.

12. Which TV screen appears bigger: a 50-inch screen viewed from 8 feet away or a 25-inch screen viewed from 3 feet away? Explain carefully. (TV screens are typically measured on the diagonal, so to say that a TV has a "50-inch screen" means that the diagonal of the screen is 50 inches long.)

13. An art museum owns a painting that it would like to reproduce in reduced size onto a 24-inch-by-36-inch poster. The painting is 85 inches by 140 inches. Give your recommendation for the size of the reproduced painting on the poster——how wide and long do you suggest that it be? Draw a scale picture showing how you would position the reproduced painting on the poster. Explain your reasoning.

14. A city has a large cone-shaped Christmas tree that stands 20 feet tall and has a diameter of 15 feet at the bottom. The lights will be wound around the tree in a spiral, so that each "row" of lights is about 2 feet higher than the previous row of lights. Approximately how long a strand of lights (in feet) will the city need? To answer this, it might be helpful to think of each "row" (one "wind") of lights around the tree as approximated by a circle. Explain your reasoning.

15. Most ordinary cameras produce a negative, which is a small picture of the scene that was photographed (with reversed colors). A photograph is produced from a negative by printing onto photographic paper. When a photograph is printed from a negative, one of two things happens: Either the printed picture shows the full picture that was captured in the negative, or the printed picture is cropped, showing only a portion of the full picture on the negative. In either case, the rectangle forming the printed picture is similar to the rectangular portion *of the negative* that it comes from.

An ordinary 35-mm camera produces a negative in the shape of a $1\frac{7}{16}$-inch-by-$\frac{15}{16}$-inch rectangle. Some of the most popular sizes for printed pictures are $3\frac{1}{2}$ in. × 5 in., 4 in. × 6 in. and 8 in. × 10 in. Can any of these size photographs be produced without either cropping the picture or leaving blank space around the picture? Why, or why not? Explain your reasoning clearly.

16. The text discussed the angle-angle-angle criterion for two triangles to be similar. Is there an angle-angle-angle-angle criterion for quadrilaterals to be similar? Discuss.

17. Let's say that you are standing on top of a mountain looking down at the valley below, where you can see cars driving on a road. You stretch out your arm, use your thumb to "sight" a car, and find that the car appears to be as long as your thumb is wide. Estimate how far away you are from the car. Explain your method clearly. (You will need to make an assumption to solve this problem. Make a realistic assumption, and make it clear what your assumption is.)

18. a. During a total solar eclipse, the moon moves in front of the sun, obscuring the view of the sun from the earth (at some locations on the earth). Surprisingly, the moon seems to be superimposed on the sun during a solar eclipse, appearing to be almost identical in size as seen from the earth. How does this situation give rise to similar triangles? Draw a sketch. (It does not have to be to scale.)

b. Use the data that follow and either the scale factor method or the internal factor method to determine the approximate distance of the earth to the sun. Explain the reasoning behind the method you use.

- The distance from the earth to the moon is approximately 384,000 km.
- The diameter of the moon is approximately 3500 km.
- The diameter of the sun is approximately 1,392,000 km.

19. On pages 611 and 612 there is a description of how the theory of similar triangles can be used to measure distances in land surveying. This problem will help you understand the theory for how the altitude above sea level of a location can be determined by surveying as well.

Suppose a person stands on the side of a hill at a point A. Using surveying equipment, the person measures the distance AB to a point B up the hill as 38 feet. Surveying equipment can also be used to measure angles. Suppose the surveying equipment reports that the line connecting A and B makes an angle of 15 degrees with a horizontal line, AC, as shown in Figure 14.60. Suppose that, by previous surveying, the point A is already known to be 523 feet above sea level.

a. The triangle in Figure 14.60 is drawn to scale. Use a ruler to measure parts of this triangle. Then use these measurements, together with mathematical reasoning and the information about the given points A and B, to find the altitude of the point B above sea level.

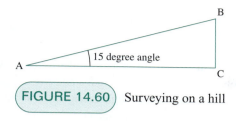

FIGURE 14.60 Surveying on a hill

b. Even if you measured carefully, your results won't be really accurate, so in practice you need a better method than measuring a scale drawing. A more accurate method uses trigonometry. Here is what trigonometry tells us in this situation: If we declare the distance from A to B to be 1 unit, then the distance from B to C is $\sin(15°)$ units = 0.258819 units and the distance from A to C is $\cos(15°)$ units = 0.965926 units.

Use the more accurate information from the previous paragraph to find the altitude of point B above sea level *to the nearest inch*. Assume that point A is actually 523 feet and 3 inches above sea level, and that the distance from point A to point B is actually 38 feet and 2 inches. Give your answer in feet and inches (e.g., 682 feet and 10 inches).

20. A pinhole camera is a very simple camera made from a closed box with a small hole on one side. Film is put inside the camera, on the side opposite the small hole. The hole is kept covered until the photographer wants to take a picture, when light is allowed to enter the small hole, producing an image on the film. Unlike an ordinary camera, a pinhole camera does not have a view finder, so with a pinhole camera, you can't see what the picture you are taking will look like.

a. Suppose that you have a pinhole camera for which the distance from the pinhole to the opposite side (where the film is) is 3 inches. Suppose that the piece of film to be exposed (opposite the pinhole) is about 1 inch tall and $1\frac{1}{2}$ inches wide. Let's say you want to take a picture of a bowl of fruit that is about 15 inches wide and piled 6 inches high with fruit and that you want the bowl of fruit to fill up most of the picture. Approximately how far away from the fruit bowl should you locate the camera? Explain, using similar triangles.

b. Explain why the image produced on film by a pinhole camera is upside down.

21. If the map in Figure 14.61 has a scale such that 1 inch represents 15 miles, then what is the scale of the map in Figure 14.62, which shows the same region? Explain your reasoning.

1 inch represents 15 miles

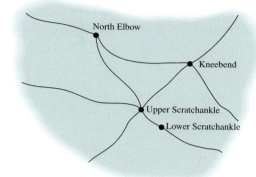

FIGURE 14.61 Map with a scale of 1 inch = 15 miles

1 inch represents ? miles

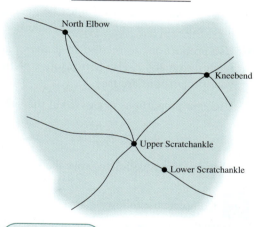

FIGURE 14.62 What is the scale of this map?

22. Imagine this scenario: The moon remains the same size as it is now, but it orbits the earth at *half* the distance from the earth as it currently does. If you went outside at night and "sighted" the moon with a ruler (as described in Class Activity 14W and Practice Exercise 6), how would its new sighted diameter compare with its current sighted diameter? Use pictures (which do not need to be to scale) to help explain your answer. (The distance from the earth to the moon is approximately 384,000 km, and the diameter of the moon is approximately 3500 km.)

23. Sue has a rectangular garden. If Sue makes her garden twice as wide and twice as long as it is now, will the area of her garden be twice as big as its current area? Examine this problem *carefully* by working out some examples and drawing pictures. Explain your conclusion. If the garden is not twice as big (in terms of area), how big is it compared with the original?

24. This problem will help you prove the Pythagorean theorem with the aid of similar triangles instead of with the moving and additivity principles. Given any right triangle with short sides of length a and b and hypotenuse of length c, as on the left in Figure 14.63, subdivide the triangle into two smaller right triangles, as shown on the right in Figure 14.63. Let s and t be the lengths of sides of these smaller triangles, as shown in the figure. To prove the Pythagorean theorem, you must prove that $a^2 + b^2 = c^2$, which you will do by completing parts (a), (b), and (c).

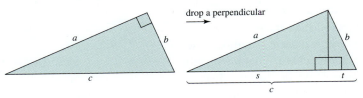

FIGURE 14.63   Right triangle subdivided into two right triangles

a. Use angles to explain why the two smaller right triangles on the right in Figure 14.63 are similar to the original triangle on the left in Figure 14.63. (Do not use any actual measurements of angles because the proof must be general—it must work for *any* initial right triangle.)

b. Use proportions to relate $\frac{s}{a}$ to a and c, and to relate $\frac{t}{b}$ to b and c. Use part (a) to explain why your proportions are true.

c. Use your proportions in part (b) to show that

$$c = \frac{a^2}{c} + \frac{b^2}{c}$$

Then use this equation to prove the Pythagorean theorem.

25. Complete the following area puzzle.

a. Trace the 8-unit-by-8-unit square in Figure 14.64.

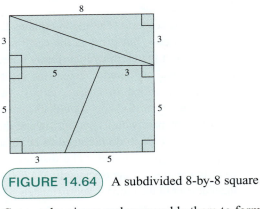

FIGURE 14.64   A subdivided 8-by-8 square

b. Cut out the pieces and reassemble them to form the 5-unit-by-13-unit rectangle in Figure 14.65.

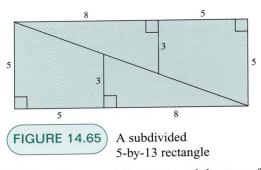

FIGURE 14.65   A subdivided 5-by-13 rectangle

c. Calculate the area of the square and the area of the rectangle. Is there a problem? Can the pieces from your 8-unit-by-8-unit square really fit together perfectly, without overlaps or gaps, to form a 5-unit-by-13-unit rectangle?

d. Explain the problem that you discovered in part (c). *Hints:* Can the two pieces shown in Figure 14.66, which you cut out of the square, really fit together as shown in the rectangle to form an actual triangle? If so, wouldn't there be some similar triangles? Look at the lengths of corresponding sides of the supposedly similar triangles. Do these numbers work the way they should for similar triangles?

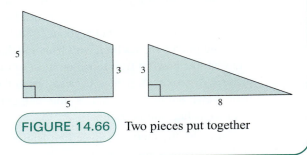

FIGURE 14.66   Two pieces put together

14.5 Areas, Volumes, and Scaling

How are the surface areas and volumes of similar objects related?

When two objects are similar, as in Figure 14.67, there is a scale factor such that corresponding *lengths* on the objects are related by multiplying by the scale factor. Does this mean that the *surface areas* of the objects are related by multiplying by the scale factor? Does this mean that the *volumes* of the objects are related by multiplying by the scale factor? We will examine these questions in this section.

Class Activity   *Now Turn to Class Activities Manual*

14Y   Areas and Volumes of Similar Boxes, p. 349

14Z   Areas and Volumes of Similar Cylinders, p. 350

If you did Class Activities 14Y and 14Z, you probably discovered that if an object is scaled with a scaled factor k, then even though the *lengths* of various parts of the object scale by the factor k, the *surface area* of the object scales by the factor k^2, and the *volume* of the object scales by the factor k^3. In other words, if one object (object 1) is the same as another object (object 2), except that it is k times as wide, k times as long, and k times as tall as object 2, then *the surface area of object 1 is k^2 times as big as the surface* area of object 2, and *the volume of object 1 is k^3 times as big* as the volume of object 2.

For example, if you make a scale model of a building, and if the scale factor from the model to the building is 150 (so that all lengths on the building are 150 times as long as the corresponding length on the model), then the surface area of the building is $150^2 = 22,500$ times as large as the surface area of the model, and the volume of the building is $150^3 = 3,375,000$ times as large as the volume of the model.

Notice how nicely these scale factors for surface area and volume fit with their dimensions and with the units used to measure surface area and volume: The surface of an object is two-dimensional. It is measured in units of in.^{2}, cm^2, . . . , and when an object is scaled with scale factor k, the surface area scales by k^2. Similarly, the size of a full, three-dimensional object can be measured by volume, which is measured in units of in.^{3}, cm^3, . . . , and when an object is scaled with scale factor k, the volume scales by k^3.

Class Activity   *Now Turn to Class Activities Manual*

14AA   Determining Areas and Volumes of Scaled Objects, p. 351

14BB   A Scaling Proof of the Pythagorean Theorem, p. 352

14CC   Area and Volume Problem Solving, p. 354

FIGURE 14.67

Two similar cups

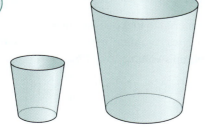

Practice Exercises for Section 14.5

1. If there were a new Goodyear blimp that was $2\frac{1}{2}$ times as long, $2\frac{1}{2}$ times as wide, and $2\frac{1}{2}$ times as tall as the current one, then how much material would be needed to make the new blimp, compared with the current one? How much gas would be needed to fill the new blimp (at the same pressure), compared with the current one? (See Figure 14.68.)

FIGURE 14.68   A blimp and the blimp scaled with scale factor $2\frac{1}{2}$

2. Suppose you make a scale model for a pyramid by using a scale of 1:100 (so that the scale factor from the model to the actual pyramid is 100). If your model pyramid is made of 4 cardboard triangles, using a total of 60 square feet of cardboard, then what is the surface area of the actual pyramid?

3. If a giant were 12 feet tall, but proportioned like a typical 6-foot-tall man, about how much would you expect the giant to weigh? (See Figure 14.69.)

FIGURE 14.69   A 6-foot-tall man and a 12-foot-tall giant

4. Suppose that a larger box is 3 times as wide, 3 times as long, and 3 times as high as a smaller box. Use a formula for the surface area of a box to explain why the larger box's surface area is 9 times the smaller box's surface area. Use a formula for the volume of a box to explain why the larger box's volume is 27 times the smaller box's volume.

5. Suppose a large cylinder has twice the radius and twice the height of a small cylinder. Use a formula for the surface area of a cylinder to explain why the surface area of the large cylinder is 4 times the surface area of the small cylinder. Use the formula for the volume of a cylinder to explain why the volume of the large cylinder is 8 times the volume of the small cylinder.

Answers to Practice Exercises for Section 14.5

1. According to the way surface area behaves under scaling, the new blimp's surface area would be

$$\left(2\frac{1}{2}\right)^2 = \frac{25}{4} = 6\frac{1}{4}$$

times as large as the current blimp's surface area. So the new blimp would require $6\frac{1}{4}$ times as much material to make as the current blimp. According to the way volume behaves under scaling, the new blimp's volume would be

$$\left(2\frac{1}{2}\right)^3 = \frac{125}{8} = 15\frac{5}{8}$$

times as large as the current blimp's volume. So the new blimp would require $15\frac{5}{8}$ times as much gas to fill it as the current blimp.

2. According to the way surface areas behave under scaling, the actual pyramid's surface area is 100^2, or 10,000 times as large as the surface area of the pyramid. The model is made of 60 square feet of cardboard, so this is its surface area. Therefore, the surface area of the actual pyramid is $10,000 \times 60$ square feet, which is 600,000 square feet.

3. Since the giant is twice as tall as a typical 6-foot-tall man, and since the giant is proportioned like a 6-foot-tall man, the giant should be a scaled version of a 6-foot-tall man, with scale factor 2. Therefore, the volume of the giant should be $2^3 = 8$ times the volume of the 6-foot-tall man. Assuming that weight is proportional to volume, the giant's weight should be 8 times the weight of a typical 6-foot-tall man. If a typical 6-foot-tall man weighs between 150 and 180 pounds, then the giant should weigh 8 times as much, or between 1200 and 1440 pounds.

4. A box that is w units wide, l units long, and h units high has a volume of

$$wlh$$

cubic units. A box that is 3 times as wide, 3 times as long, and 3 times as high is $3w$ units wide, $3l$ units deep, and $3h$ units high, so it has volume

$$(3w)(3l)(3h) = 27wlh$$

cubic units. Therefore, a box that is 3 times as wide, 3 times as long, and 3 times as high as another smaller box has a volume that is 27 times the volume of the smaller box.

A box that is w units wide, l units long, and h units high has a surface area of

$$2wl + 2wh + 2lh$$

square units. This is because the box has two faces that are w units by l units, two faces that are w units by h units, and two faces that are l units by h units, as you can see by looking at the pattern for a box in Figure 14.70. If another box is 3 times as wide, 3 times as long, and 3 times as high, then this bigger box will have surface area

$$2(3w)(3l) + 2(3w)(3h) + 2(3l)(3h)$$

square units because its width, length, and height are $3w$, $3l$, and $3h$ units, respectively. But

$$2(3w)(3l) + 2(3w)(3h) + 2(3l)(3h)$$

$$= 9(2wl + 2wh + 2lh)$$

Therefore, a box that is $3w$ units wide, $3l$ units long, and $3h$ units high has 9 times the surface area of a box that is w units wide, l units long, and h units high.

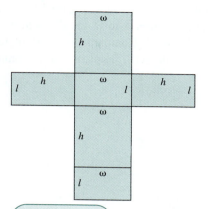

FIGURE 14.70 Pattern for a box of width w, height h, and length l

5. Let's call the radius of the small cylinder r units and the height of the small cylinder h units. Then the volume of the small cylinder is

$$h\pi r^2$$

cubic units, according to the *(height) $\times$ (area of base)* volume formula. The larger cylinder has twice the radius and twice the height of the smaller cylinder; therefore, the larger cylinder has radius $2r$ units and height $2h$ units. So the larger cylinder has volume

$$(2h)\pi(2r)^2 = 8(h\pi r^2)$$

cubic units. Because $8h\pi r^2$ is 8 times $h\pi r^2$, the volume of the large cylinder is 8 times the volume of the small cylinder.

The surface area of the small cylinder of radius r and height h is

$$2\pi r^2 + 2\pi rh$$

square units. (See the answer to Practice Exercise 10 of Section 13.2.) Using the same formula again, but now with $2r$ substituted for r and $2h$ substituted for h, we obtain the surface area of the big cylinder as

$$2\pi(2r)^2 + 2\pi(2r)(2h) = 8\pi r^2 + 8\pi rh$$

$$= 4(2\pi r^2 + 2\pi rh)$$

square units. Because $4(2\pi r^2 + 2\pi rh)$ is 4 times $2\pi r^2 h + 2\pi rh$, the surface area of the large cylinder is 4 times the surface area of the small cylinder.

Problems for Section 14.5

1. In the triangle ADE pictured in Figure 14.71, the point B is halfway from A to D and the point C is halfway from A to E. Compare the areas of the triangle ABC and the ADE. Explain your reasoning clearly.

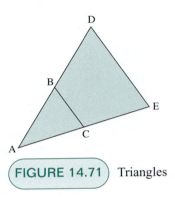

FIGURE 14.71   Triangles

2. A scale model is constructed for a domed baseball stadium with a scale of 1 foot to 100 feet. The model's dome is made of 40 square feet of cardboard. The model contains 50 cubic feet of air. How many square feet of material will the actual stadium's dome be made of? How many cubic feet of air will the actual stadium contain?

3. Og, a giant mentioned in the Bible, might have been 13 feet tall. If Og was proportioned like a 6-foot-tall man weighing 200 pounds, then how much would Og have weighed? Explain your reasoning.

4. An artist plans to make a large sculpture of a person out of solid marble. She first makes a small scale model out of clay, using a scale of 1 inch for 2 feet. The scale model weighs 1.3 pounds. Assuming that a cubic foot of clay weighs 150 pounds and a cubic foot of marble weighs 175 pounds, how much will the large marble sculpture weigh? Explain your reasoning.

5. According to one description, King Kong was 19 feet, 8 inches tall and weighed 38 tons. Typical male gorillas are about 5 feet, 6 inches tall and weigh between 300 and 500 pounds. Assuming that King Kong was proportioned like a typical male gorilla, does his given weight of 38 tons agree with what you would expect? Explain.

6. If you know the volume of an object in cubic inches, can you find its volume in cubic feet by dividing by 12? Discuss.

7. Suppose that a gasoline-powered engine has a gas tank in the shape of an inverted cone, with radius 10 inches and height 20 inches, as shown in Figure 14.72. The gas flows out through the tip of the cone, which points down. The shaded region represents gas in the tank. The height of this "cone of gas" is 10 inches, which is half the height of the gas tank.

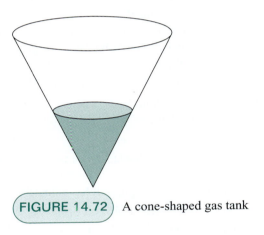

FIGURE 14.72   A cone-shaped gas tank

a. What do you think? In terms of volume, is the tank half full, less than half full, or more than half full? Draw a picture of the gas tank, and mark approximately where you think the gas would be when the tank is $\frac{3}{4}$ full, $\frac{1}{2}$ full, and $\frac{1}{4}$ full in terms of volume.

b. Find the volume of the gas in the tank. Is this half, less than half, or more than half the volume of the tank?

d. Now find the volume of gas if the height of the gas was $\frac{3}{4}$ of the height of the tank. Is this more or less than half the volume of the tank? Is this surprising?

8. A cup has a circular opening and a circular base. A cross-section of the cup and the dimensions of the cup are shown in Figure 14.73.

a. Determine the volume of the cup. Explain your reasoning.

b. If the cup is filled to $\frac{1}{2}$ of its height, what percent of the volume of the cup is filled? Explain your reasoning.

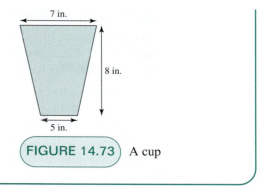

(FIGURE 14.73) A cup

Chapter Summary and Study Items

Section 14.1 Reflection, Translations, and Rotations

Reflections, translations, and rotations are the fundamental distance-preserving transformations of the plane. A glide-reflection is a combination of a translation and a reflection.

Key skills and understandings:

- Determine the location of a shape after a translation, a reflection, a rotation of 90° clockwise or counterclockwise, or a 180° rotation is applied.

Section 14.2 Symmetry

A shape or design has reflection symmetry, or mirror symmetry, if there is a line in the plane such that the shape or design as a whole occupies the same place in the plane both before and after reflecting across the line. A shape or design has *n*-fold rotation symmetry if there is a point such that, after rotating by $\frac{360°}{n}$ around the point, the shape or design as a whole occupies the same place in the plane both before and after rotation. A shape or design has translation symmetry if there is a translation such that the shape or design as a whole occupies the same place in the plane both before and after translation.

Key skills and understandings:

- Given a shape or design, determine its symmetries.
- Create designs that have specified symmetries.

Section 14.3 Congruence

Informally, we say two shapes are congruent if they are the same size and shape. More formally, we describe two shapes as congruent if there is a rotation, a reflection, a translation, or a combination of these transformations that takes one shape to the other. There are several standard criteria for two triangles to be congruent. If two triangles have the same side lengths (so that corresponding sides have the same lengths), then they are congruent (SSS congruence). If two triangles have a side of the same length, and the corresponding angles at each end of the sides have the same measure in both triangles, then the triangles are congruent (ASA congruence). If two triangles have two corresponding sides of the same length and the same angle between the two sides, then they are congruent (SAS congruence). The structural stability of triangles is related to SSS congruence.

Key skills and understandings:

- Describe and use the SSS, ASA, and SAS criteria for congruence.

Section 14.4 Similarity

Informally, we call two shapes similar if they are the same shape, but not necessarily the same size. If two shapes or objects are similar, then all distances between corresponding parts of the objects scale by the same factor. This factor is called a scale factor.

Three methods allow us to solve problems about similar shapes or objects: the scale factor method, the internal factor method, and setting up and solving a proportion. To use the scale factor method, we use corresponding known lengths in the two similar objects in order to determine the scale factor. We then apply the scale factor to find an unknown length. To use the internal factor method, we find the factor by which two known lengths within one of the objects are related. The corresponding lengths in the other, similar object will be related by the same factor. We then apply this factor to find an unknown length in the other object. We can also solve problems about similar objects or figures by setting up and solving a proportion.

Two triangles are similar exactly when the two triangles have corresponding angles of the same size. Similar triangles can arise when parallel lines cross a pair of lines that cross. We can explain why triangles formed this way are similar by explaining why the corresponding angles have the same size (using the Parallel Postulate). We can use similar triangles to determine distances and heights.

Key skills and understandings:

- Use the scale factor, internal factor, and set up a proportion method to solve problems about similar shapes, and explain the rationale for each method.

- Explain why triangles are similar by showing that corresponding angles must have the same size. When the Parallel Postulate applies, use it to show that corresponding angles are produced by a line crossing two parallel lines and are therefore of the same size.

- Use similar triangles to determine heights and distances.

- Use similar shapes or objects to solve problems.

Section 14.5 Areas, Volumes, and Scaling

If the lengths in one object are k times the corresponding lengths in a second, similar object, then the surface area of the first object is k^2 times the surface area of the second object and the volume of the first object is k^3 times the volume of the second object. We can see why these facts about how areas and volumes scale are plausible by extrapolating from patterns and models for prisms and cylinders and from formulas for their surface areas and volumes.

Key skills and understandings:

- Use patterns, models, and formulas to explain how areas and volumes of similar objects are related.

- Apply the way that areas and volumes scale in similar objects to solve problems.

Statistics

The field of statistics provides tools for studying questions that can be answered with data. A seemingly endless variety of data is available for populations, health, financial and business activities, and the environment. It is virtually impossible to think of a social activity or a physical phenomenon for which we cannot collect data. Statistical concepts help to interpret these data and to recognize trends.

According to the Curriculum Framework for PreK–12 Statistics Education adopted by the American Statistical Association, statistical problem solving is an investigative process that involves four components:

- Formulate Questions
  —Clarify the problem at hand.
  —Formulate one or more questions that can be answered with data.
- Collect Data
  —Design a plan to collect appropriate data.
  —Implement the plan to collect the data.
- Analyze Data
  —Select appropriate graphical or numerical methods.
  —Use these methods to analyze the data.
- Interpret Results
  —Interpret the analysis.
  —Relate the interpretation to the original question.
  (See [3]).

The National Council of Teachers of Mathematics has made the following recommendations (see [64]) concerning the learning of statistics and probability: Instructional programs from prekindergarten through grade 12 should enable all students to

- formulate questions that can be addressed with data and collect, organize, and display relevant data, to answer them;
- select and use appropriate statistical methods to analyze data;

- *develop and evaluate inferences and predictions that are based on data;*
- *understand and apply basic concepts of probability.*

For NCTM's specific statistics and probability recommendations for grades PreK–2, grades 3–5, grades 6–8, and grades 9–12, go to www.pearsonhighered.com/beckmann.

15.1 Formulating Questions, Designing Investigations, and Gathering Data

Focal Points
Grade 8

We humans are naturally curious about each other and about the world. We want to know what others around us think about the political questions of the day; the latest movies, TV shows, and music; and similar topics of interest. Local, state, and national governments want to know facts about population, employment, income, and education in order to design appropriate programs. Scientists want to know how to cure and prevent diseases, as well as to know about the world around us. Therefore, news organizations, government bodies, and scientists regularly design investigations and gather data to help answer specific questions.

According to guidelines for assessment and instruction in statistics education endorsed by the American Statistical Association (see [3]), statistical problem solving involves four components:

1. Formulate questions
2. Collect data
3. Analyze data
4. Interpret results

In this section we'll start by examining the first two of these components, formulating questions and collecting data. Then we'll turn our attention to analyzing data and interpreting results in the case when a sample is used to predict characteristics of a full population. The rest of this chapter will be devoted to tools we can use to analyze data and interpret results.

Formulating Questions

The first step in a statistical investigation is to formulate a question that can be answered by the collection and analysis of data. For example, if a class is planning a party, the students might like to know what the favorite snacks are among the students in the class. If everyone in a class received a small packet of candies as a treat, the students might be curious whether all the bags contain the same number of candies. If the candies come in different colors, the students might wonder whether there is a color which generally appears the most, and if so, which color it is. Students might be curious about how heavy their backpacks are. Are girls' names typically longer than boys' names? Do bean seeds sprout better in the dark or in light? How does the length of a pendulum affect the time it takes to swing back and forth? All of these questions can be answered (at least in part) by the collection and analysis of data. And all of these questions anticipate variability in the data that will be collected. For example, we don't expect all backpacks to weigh the same, nor are all girls' names the same length. So statistical investigations must anticipate and take into account the variability that naturally exists in most data.

Observational Studies and Experiments

There are two basic types of statistical studies: observational studies and experiments.

Observational Studies Observational studies observe characteristics and quantities but do not attempt to influence these characteristics or quantities. For example, in an observational study, researchers might count the number of turtle eggs in a certain area, measure the height of a plant every day, or collect participants' responses to a questionnaire in a survey.

> ### Class Activity *Now Turn to Class Activities Manual*
>
> **15A** Challenges in Formulating Survey Questions, p. 355

It may seem like a simple matter to write survey questions to determine facts about people or to determine people's opinions, but if you did Class Activity 15A, then you probably began to realize that it is not. Different people may interpret the same question in different ways. For example, even a simple question such as "How many children are in your family?" can be interpreted in different ways. Does it mean how many people under the age of 18 are in your family? Does it mean how many siblings you have, plus yourself? Should stepchildren be included? A good survey question must be clear.

Experiments In contrast to observational studies, in experiments researchers don't simply observe characteristics or quantities; instead, they try to determine if certain factors will influence the characteristics or quantities. For example, in an experiment to determine if a new drug is safe and effective, a group of volunteers will be divided into two (or more) groups. One group receives the new drug, and the other group receives a placebo—an inert substance that is made to look like a real drug. The volunteers are observed to see if their symptoms improve and if there are any side effects in the group receiving the new drug.

Populations and Samples

population

Regardless of the nature of the investigation, a statistical study focuses on a certain population. The **population** of a statistical study is the full set of people or things that the study is designed to investigate. For example, before a presidential election, a study might investigate how registered voters plan to vote. In this case, the population of the study is all registered voters in the country. Another study might investigate properties of ears of corn in a cornfield. In this case, the population of the study is all ears of corn in the cornfield. In most cases, it is not possible to study every single

sample

member of a population; instead, a sample is chosen for study. A **sample** of a population of a study consists of some collection of members of the full population. A pollster will select a sample of voters to survey about how they will vote in an upcoming election instead of trying to ask every voter. A scientist studying ears of corn in a cornfield will select a sample of ears of corn to study rather than try to study every single ear of corn in the cornfield.

> ### Class Activity *Now Turn to Class Activities Manual*
>
> **15B** Choosing a Sample, p. 356

Using Random Samples to Predict Characteristics of a Full Population

If you did Class Activity 15B, then you probably realize that some samples may not accurately reflect the full population. The best way to pick a sample so that it will be likely to reflect the full population accurately is to pick a *random sample*.

representative sample

Random Samples For a sample of a population under study to be useful, the characteristics of the sample must reflect the characteristics of the full population. In other words, the sample should be **representative** of the full population. But unless you choose a sample carefully, the sample's characteristics might not be representative of the full population's characteristics. For example, if a group studying child nutrition in a county took their sample of children from a school in an affluent neighborhood, their findings could differ

random sample

significantly from the actual status of child nutrition in the county. The best way to choose a representative sample of a population is to pick a random sample. To choose a **random sample** means to choose the members of the sample in such a way that every member of the population has an equal chance of becoming a member of the sample.

Sometimes students may interpret *random sample* incorrectly as a sample that is chosen in an unconventional, haphazard, or out-of-the-blue way. For example, say there are 100 students in the 6th grade and a random sample of 10 students is to be chosen. Although Johnny may have an unusual method in mind for picking the 10 students, the method won't produce a random sample of 10 students unless each student was equally likely to be chosen for the sample. A simple way for students to pick a random sample is to cut a list of the students' names into identical slips of paper, put the slips of paper in a bag, mix them well, and then pick out exactly 10 of the slips without looking. The slips of paper could instead be numbered from 1 to 100, with each number corresponding to a student in a class list. The 10 names or numbers that are picked correspond to the 10 students that are to be in the random sample.

In settings where the population is too large for it to be practical to label a slip of paper for each individual in the population, random samples are often chosen with the aid of a list of random numbers, which can be generated by computers and calculators.

Predicting the Characteristics of a Full Population If a random sample of a population is chosen, then the characteristics of the sample are likely to be very close to the characteristics of the full population if the sample is large enough. In other words, sufficiently large random samples are likely to be representative of the full population. For example, if 54% of a (large enough) random sample of voters said they were likely to vote for candidate X, then it is likely that close to 54% of the full population would also vote for candidate X.

Class Activity   *Now Turn to Class Activities Manual*

15C Using Random Samples, p. 356

15D Using Random Samples to Estimate Population Size by Marking (Capture–Recapture), p. 358

Here is an example of how we can use a random sample to predict the characteristics of a full population. Suppose that at a factory, a batch of 450 transistors have just been produced. A random sample of 75 transistors are pulled out for testing. Of the 75 transistors, 3 are found to be defective. How many defective transistors should we expect to find among the full batch of 450?

As the sample of 75 was chosen randomly, it is likely to be representative of the full batch (approximately). There are a number of ways we can determine approximately how many of the 450 transistors are likely to be defective. If the sample is representative of the full batch, then the fraction of defective transistors in the sample and in the full batch will be equivalent. Thus, we should solve the proportion

$$\frac{3}{75} = \frac{x}{450} \tag{15.1}$$

for *x*, where *x* stands for the number of defective transistors in the full batch. We can solve this proportion by cross-multiplying and solving the resulting equation:

$$3 \cdot 450 = 75x$$

We find that *x* = 18, so there should be about 18 defective transistors in the full batch of 450.

Another way to determine approximately how many defective transistors there are in the full batch is to think about repeatedly picking out random samples of 75 until all 450 transistors have been picked out.

If each sample is just like the first one, then each sample of 75 would have 3 defective transistors. The next two tables can help us think about repeatedly picking samples of 75 and finding 3 defective transistors in each sample:

$$
\begin{array}{ccc}
75 \rightarrow 3 & \qquad & 75 \rightarrow 3 \\
75 \rightarrow 3 & & 150 \rightarrow 6 \\
75 \rightarrow 3 & & 225 \rightarrow 9 \\
75 \rightarrow 3 & & 300 \rightarrow 12 \\
75 \rightarrow 3 & & 375 \rightarrow 15 \\
75 \rightarrow 3 & & 450 \rightarrow 18
\end{array}
$$

since $450 \div 75 = 6$, we can pull out 6 samples of 75. If each of the 6 samples of 75 yields 3 defective transistors, then there would be a total of $6 \cdot 3 = 18$ defective transistors in the full batch of 450. Of course, if we were to pick out samples of 75, it is unlikely that each and every batch would contain exactly 3 defective transistors, and most likely the full batch wouldn't contain exactly 18 defective transistors either. However, this method gives us a good way to determine the approximate number of defective transistors we should expect.

We could also notice that the sample consists of $\frac{1}{6}$ of the full batch, since $6 \cdot 75 = 450$. Since the sample should be representative of the full batch, it should contain $\frac{1}{6}$ of the defective transistors. Therefore, the full batch should contain 6 times as many defective transistors, namely, $6 \cdot 3 = 18$ transistors. Observe that this line of reasoning fits with solving the proportion in Equation 15.1 as on the left in Figure 15.1.

Yet another way to sovle the problem is to notice that $\frac{1}{25}$ of the sample was defective, since $\frac{3}{75} = \frac{1}{25}$. Since the sample should be representative of the full batch, $\frac{1}{25}$ of the full batch should be defective, which is $450 \div 25 = 18$ transistors. Observe that this line of reasoning fits with solving the proportion in Equation 15.1 as on the right in Figure 15.1.

Importance of Randomness

Consider an experiment to determine if a new drug is safe and effective. A group of volunteers is divided into two (or more) groups. One group receives the new drug; the other group receives a placebo. Such an experiment is useful only if it accurately predicts the outcome of the use of the drug in the population at large. Therefore, it is crucial that the two groups, the one receiving the drug and the one receiving the placebo, are as alike as possible—in all ways except in whether or not they receive the drug. The two groups should also be as much as possible like the general population who would take the drug. Therefore, the two groups are usually chosen randomly (such as by flipping a coin—heads goes to group 1, tails to group 2). In addition, such experiments are usually double-blind; that is, neither the volunteer nor the treating physician knows who is in which group.

In controlled drug experiments, the randomness of the assignment of patients to the two groups and the double-blindness ensure that the study will accurately predict the outcome of the use of the new drug in the whole population. If the patients were not assigned randomly to the two groups, then the person assigning the patients to a group might subconsciously pick patients with certain attributes for one or the other group, thereby making the group that is treated by the new drug slightly different from the group receiving the placebo. This difference could result in inaccurate predictions about the safety and effectiveness of the drug in the population at large. Similarly, if the patient knows whether or not he has received the new drug, his resulting mental state may affect his recovery. If the physician knows whether or not the new drug has been administered, her treatment of the patient may be different, and the patient may again be affected differently.

FIGURE 15.1

Determining how many defective transistors there are

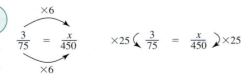

Practice Exercises for Section 15.1

1. Kaitlyn, a fifth-grader, asked 5 of her friends in class which book is their favorite. All 5 said *Harry Potter and the Sorcerer's Stone*. Can Kaitlyn conclude that most of the children at her school would say *Harry Potter and the Sorcerer's Stone* is their favorite book? Why or why not?

2. A large bin is filled with 200 table-tennis balls. Some of the balls are white and some are orange. Tyler reaches into the bin and randomly pulls out 10 table-tennis balls. Three of the balls are orange and 7 are white. Based on Tyler's sample, what is

the best estimate we can give for the number of orange table-tennis balls in the bin? Explain your reasoning.

3. There is a large bin filled with table-tennis balls, but we don't know how many. There are 40 orange table-tennis balls in the bin; the rest are white. Amalia reaches into the bin and randomly picks out 20 table-tennis balls. Of the 20 Amalia picked, 6 are orange. Based on Amalia's sample, what is the best estimate we can give for the number of table-tennis balls in the bin? Explain your reasoning.

Answers to Practice Problems for Section 15.1

1. No, Kaitlyn can't conclude that most of the children at her school would say *Harry Potter and the Sorcerer's Stone* is their favorite book. Kaitlyn's friends may all have a similar taste in books, and their taste in books may not be representative of the tastes of the whole school. Also, younger children at lower reading levels might prefer different books because *Harry Potter and the Sorcerer's Stone* is too advanced for them to read.

2. Tyler picked a random sample, so the characteristics of his sample should reflect the characteristics of the full population of the balls in the bin (more or less). Three of the 10 balls that Tyler picked are orange, so 30% of the balls in Tyler's sample are orange. The full population of balls in the bin should also be about 30% orange. Since 30% of 200 is 60, the best estimate is that 60 balls are orange and 140 are white.

A more elementary way to solve this problem is to think about repeatedly picking out 10 table-tennis balls until all 200 balls have been picked out. If each sample of 10 is just like the first one, then each sample will contain 3 orange balls and 7 white ones. Since $200 \div 10 = 20$, we can pick 20 sets of 10 balls. This would yield a total of $20 \cdot 3 = 60$ orange balls and $20 \cdot 7 = 140$ white balls. Of course, since not every set of 10 balls will contain exactly 3 orange balls, this is just an estimate, and it is not likely to be exactly correct.

3. Amalia picked a random sample, so the characteristics of her sample should reflect the characteristics of the full population of the balls in the bin (more or less). Since 6 of 20 balls that Amalia picked are orange, 30% of the balls Amalia picked are orange. Therefore, approximately 30% of the balls in the bin should be orange. Since there are 40 orange table-tennis balls in the bin, and since these 40 balls should be about 30% of all the balls in the bin, we have

$$30\% \cdot (\text{balls in the bin}) = 40$$

So

$$\text{balls in the bin} = 40 \div 0.30 = 133$$

Therefore, there should be approximately 133 balls in the bin.

A more elementary way to solve this problem is to think about repeatedly picking out 20 table-tennis balls until all the orange balls have been picked out. If each sample of 20 is just like the first one, then each sample will contain 6 orange balls. The next table shows what would happen:

$$20 \rightarrow 6$$
$$40 \rightarrow 12$$
$$60 \rightarrow 18$$
$$80 \rightarrow 24$$
$$100 \rightarrow 30$$
$$120 \rightarrow 36$$

Once we get up to 36 orange balls, there are only 4 orange balls left, which is $\frac{4}{6} = \frac{2}{3}$ as much as 6. So if we pick another $\frac{2}{3}$ of 20 balls, which is about another 13 balls, we should get the remaining orange balls.

Thus, in all, we would pick $120 + 13 = 133$ balls in order to get the full 40 orange balls. So there should be approximately 133 table-tennis balls in the bin.

Problems for Section 15.1

1. A class has a collection of 100 bottle caps and wants to pick a random sample of 20 of them. Kyle has the idea that he will pick every third one he looks at in order to pick these 20. Discuss whether Kyle's method is appropriate for picking a random sample. If not, how else could the class pick a random sample of 20 bottle caps?

2. Neil, a third-grader, asked 10 of his classmates whether they prefer to wear shoes with laces or without laces. Most of the classmates Neil asked prefer to wear shoes without laces. Discuss whether it would be reasonable to assume that most of Neil's class or most of the students at Neil's school would also prefer to wear shoes without laces.

3. An announcer of a TV program invited viewers to vote in an Internet poll, indicating whether or not they are better off economically this year than last year. Most of the people who participated in the poll indicated they are worse off this year than last year. Based on this information, can we conclude that most people are worse off this year than last year? If so, explain why. If not, explain why not.

4. At a factory that produces doorknobs, a batch of 1500 doorknobs has just been produced. To check the quality of the doorknobs, a random sample of 100 doorknobs is selected to test for defects. Of these 100 doorknobs, 2 were found to be defective. Based on these results, what is the best estimate you can give for the number of defective doorknobs in the batch of 1500? Solve this problem in three different ways, explaining your reasoning in each case.

5. At a factory that produces switches, a batch of 3000 switches has just been produced. To check the quality of the switches, a random sample of 75 switches is selected to test for defects. Of these 75 switches, 2 were found to be defective. Based on these results, what is the best estimate you can give for the number of defective switches in the batch of 3000? Solve this problem in three different ways, explaining your reasoning in each case.

6. At a light bulb factory, 1728 light bulbs are ready to be packaged. A worker randomly pulls 60 light bulbs out and tests each one. Two light bulbs are found to be defective.

    a. Based on the information given, what is your best estimate for the number of defective light bulbs in the full batch of 1728 light bulbs? Explain your reasoning.

    b. If the light bulbs are put in boxes with 144 light bulbs in each box (in 12 packages of 12), then approximately how many defective light bulbs would you expect to find in a box? Explain your reasoning.

7. Carter has a large collection of marbles. Carter knows that he has exactly 20 blue marbles in his collection, but he does not know how many marbles he has in all. Carter mixes up his marbles and randomly picks out 40 marbles. Of the 40 marbles he picked, 4 are blue. Based on these results, what is the best estimate you can give for the total number of marbles in Carter's collection? Explain your reasoning. Solve this problem in three different ways, explaining your reasoning in each case.

8. The following problem is an example of the *capture–recapture* method of population estimation. A researcher wants to estimate the number of rabbits in a region. The researcher sets some traps and catches 30 rabbits. After the rabbits are tagged, they are released unharmed. A few days later the researcher sets some traps again and this time catches 35 rabbits. Of the 35 rabbits trapped, 5 are tagged, which indicates that they had been trapped a few days earlier. Based on these results, what is the best estimate you can give for the number of rabbits in the region? Solve this problem in three different ways, explaining your reasoning in each case.

9. A group studying violence wants to determine the attitudes toward violence of all the fifth-graders in a county. The group plans to conduct a survey, but the group does not have the time or resources to survey every fifth-grader. The group can afford to survey 120 fifth-graders. For each of the methods in (a) through (c) of choosing 120 fifth-graders to survey, discuss advantages and disadvantages of that method. Based on your answer, which of the three methods will be best for determining the attitudes toward violence of all the fifth-graders in the county? (There are 6 elementary schools in the county, and each school has 4 fifth grades.)

   a. Have each elementary school principal in the county select 20 fifth-graders.

   b. Have each fifth-grade teacher in the county select 5 fifth-graders.

   c. Obtain a list of all fifth-graders in the county. Obtain a list of 120 random numbers from 1 to the number of fifth-graders in the county. Use the list of random numbers to select 120 random fifth-graders in the county.

15.2 Displaying Data and Interpreting Data Displays

 **Focal Points**
Grade 8

Every day we see the results of surveys in newspapers, on television newscasts, or on the Internet. Sometimes, results are reported with numbers (such as a number of people or as a percentage), but other times results are shown in a table, or in a chart or graph. Charts and graphs help us get a sense of what data mean, because a chart or graph can show us the "big picture" about data in a way that tables and numbers often can't.

In this section we study common data displays as well as different types of questions that we can ask about displays of data. Interpreting data can range from reading information directly off a graph, to comparing and combining this information, to thinking about implications of this information. Producing a display of data is not a major mathematical goal in its own right, but data displays can provide a context for applying mathematical analysis. Also in this section, we will examine some common errors and misleading practices that sometimes occur with data displays.

Some data are numerical, such as the weights of apples picked at an orchard, the heights of students in a class, or the volume of juice poured into bottles at a bottling plant. But other data are categorical, such as the colors of the plastic dinosaurs in a bucket, the yes or no vote of each person in a class on some issue, the favorite piece of playground equipment of each student at school. Different types of displays are appropriate for these different kinds of data.

Displays of Categorical Data

Categorical data can be displayed in "real graphs," pictographs, or pie graphs.

Real Graphs and Pictographs The most basic kinds of graphs are real graphs and pictographs. Both of these are suitable for use with young children and can act as a springboard for understanding other kinds of graphs. A **real graph**, or *object graph*, displays actual objects in a graph form; a **pictograph** is like a real graph, except that it uses pictures of objects instead of actual objects. In some pictographs, a single picture may represent more than one object. We can use real graphs and pictographs to show how a collection of related objects is sorted into different groups. The real graph or pictograph allows us to see at a glance which groups have more, and which groups have fewer, objects in them.

real graph
pictograph

Suppose we have a tub filled with small beads that are alike, except that they have different colors: yellow, pink, purple, and green. We can mix up the beads and scoop some out. How many beads of each color do we have? Which color beads do we have the most of? Which do we have the least of? To answer these questions at a glance, we can line the beads up, as indicated in Figure 15.2. This creates a real graph.

We can turn a real graph into a pictograph by replacing the objects with pictures of the objects, as on the left in Figure 15.3.

FIGURE 15.2

A real graph: beads of different colors

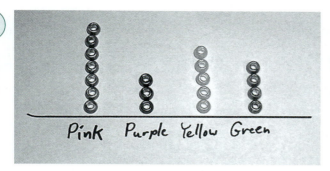

When we want to represent a large number of objects in a pictograph, it makes sense to allow each picture to stand for more than one object. Each picture could stand for 2, 5, 10, 100, 1000, or some other convenient number of objects. If each picture stands for more than 1 object, then a key near the graph should show clearly how many objects the picture stands for.

Suppose a fifth-grade class of students in Wyoming has pen pals in Idaho, Oregon, Washington state, and Georgia. The class can look up the populations of these states on the Internet by going to the U.S. Census Bureau homepage (or see *www.pearsonhighered.com/beckmann*) and get a list as follows:

Wyoming	532,668
Idaho	1,523,816
Oregon	3,790,060
Washington	6,549,224
Georgia	9,685,744

Although this list shows the (estimated) populations quite precisely, to get a good feel for the relative populations of the states, the class can first round the populations to the nearest half-million and then draw a pictograph as in Figure 15.4. In this pictograph, the picture of a person represents 1 million people, so half a million people can be represented by half a person. In the pictograph we can see at a glance how much bigger in population some of these states are than others.

Pictographs can be vertical as in Figure 15.3 or horizontal as in Figure 15.4.

bar graph   **Bar Graphs** A **bar graph,** or *bar chart,* is essentially just a streamlined or fused version of a pictograph. A bar graph shows at a glance the relative sizes of different categories. The pictographs on the left of Figures 15.3 and 15.4 have been turned into bar graphs on the right.

double bar graph   A **double bar graph** is a bar graph in which each category has been subdivided into two subcategories. For example, if the bar graph is about people, then the two subcategories might be male and female. Figure 15.5 shows median weekly earnings in 2009 of full-time workers, 25 years and older. The different colored bars for men and women show the different earnings of these two

FIGURE 15.3

Colors of beads displayed in a pictograph and bar graph

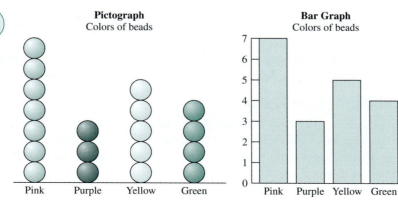

FIGURE 15.4

Populations of some states

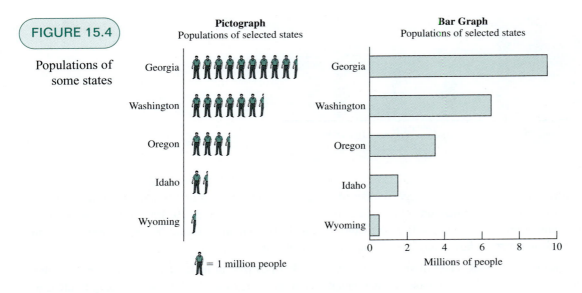

Pictograph
Populations of selected states

Bar Graph
Populations of selected states

= 1 million people

Millions of people

FIGURE 15.5

Double bar graph: median weekly earnings by education and gender

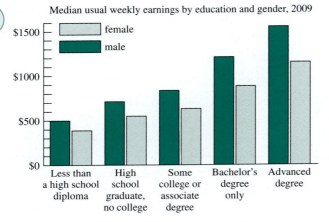

Median usual weekly earnings by education and gender, 2009

groups in each of the different educational levels. The data used in constructing the double bar graph were gathered by the Bureau of Labor Statistics and are available on the Internet. (Go to *www.pearsonhighered.com/beckmann.*)

pie graph   **Pie Graphs** A **pie graph**, *pie chart*, or *circle graph* uses a subdivided circle to show how data partition into categories. Figure 15.6 shows in a pie graph how U.S. firms in 2006 were divided according to how many employees they had. We can see at a glance that most firms had fewer than 20 employees. The data for this pie graph were taken from the Bureau of Labor Statistics. (Go to *www.pearsonhighered.com/beckmann.*)

By their nature, pie graphs involve percentages or fractions. In fact, simple activities with pie graphs can be used to introduce fractions and percents to children. For example, let's say we have a small bag filled with 6 black snap cubes, 8 purple snap cubes, 6 green snap cubes, and 4 white snap cubes. We can arrange these snap cubes evenly around a paper plate, keeping like colors together, as in Figure 15.7. We can then make "pie wedges" showing $\frac{1}{3}$, $\frac{1}{4}$, $\frac{1}{4}$ and $\frac{1}{6}$ of the plate. By filling the plate with pie wedges, we can see that $\frac{1}{3}$ of the snap cubes are purple, $\frac{1}{4}$ are black, $\frac{1}{4}$ are green, and $\frac{1}{6}$ are white.

Displays of Numerical Data

Numerical data can be displayed in dot plots, histograms, and stem-and-leaf plots, which show how frequently certain numbers or intervals of numbers occur in a set of data; line graphs, which are usually used to show how data changes over time; and scatterplots, which are used to investigate relationships between two kinds of data.

FIGURE 15.6

Pie graph: percent of U.S. firms with various numbers of employees (2006)

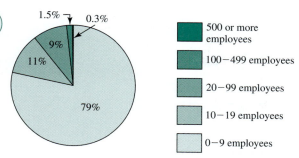

- 500 or more employees
- 100−499 employees
- 20−99 employees
- 10−19 employees
- 0−9 employees

dot plot **Dot Plots** A **dot plot** is a kind of pictograph in which the categories are numbers and the pictures are dots. The dots function as tally marks.

Let's say we have a bag of black-eyed peas and a measuring teaspoon. How many peas are in a teaspoon? We can scoop out a teaspoon of peas and count the number of peas. If we do it again and again, we may get different numbers of peas in a teaspoon. The following numbers could be the number of peas in a teaspoon that we get from many different trials:

$$18, \quad 19, \quad 19, \quad 17, \quad 19, \quad 18, \quad 20, \quad 18, \quad 19, \quad 19, \quad 20, \quad 20, \quad 19$$
$$20, \quad 18, \quad 19, \quad 18, \quad 17, \quad 18, \quad 20, \quad 18, \quad 18, \quad 16, \quad 18, \quad 19$$

We can organize these data in a dot plot, as shown in Figure 15.8. Each dot in the dot plot represents a number in the previous list. We can see right away from the dot plot that a teaspoon of black-eyed peas usually consists of 18 or 19 peas. By counting that there are 5 dots above 20, we see that there were 5 times when the teaspoon held 20 peas.

Dot plots are easy to draw, and you can even draw a dot plot as you collect data—you don't necessarily have to write your data down first before you draw a dot plot. For this reason, dot plots can be a quick and handy way to organize numerical data.

We can also use dot plots to record how data falls into intervals. Consider the average January temperatures in degrees Fahrenheit of 25 U.S. cities from this list:

$$22, \quad 43 \quad 30, \quad 18, \quad 33, \quad 26, \quad 44, \quad 30, \quad 7, \quad 20, \quad 73, \quad 40, \quad 45,$$
$$38, \quad 57, \quad 40, \quad 50, \quad 31, \quad 36, \quad 54, \quad 29, \quad 58, \quad 14, \quad 61, \quad 63$$

We can group these data by the intervals 0–9, 10–19, 20–29, and so on, up to 70–79 and make the dot plot on the left of Figure 15.9. By counting that there are 6 dots over the interval 30 to 39, we learn that there were 6 cities (among the cities in the list) that have average January temperatures between 30 and 39 degrees Fahrenheit.

histogram **Histograms** A **histogram** is a bar graph for which the categories are individual numbers or equal-length intervals of numbers. Just as we can think of fusing the pictures in a pictograph to make a bar graph, we can think of fusing the dots in a dot plot to make a histogram. Figure 15.9 shows a dot plot and corresponding histogram of average January temperatures in some U.S. cities. But histograms allow for

FIGURE 15.7

A pie chart in a paper plate

FIGURE 15.8

Dot plot: the number of black-eyed peas in a teaspoon

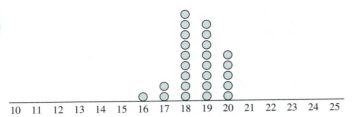

FIGURE 15.9

A dot plot and histogram with intervals of temperatures as categories

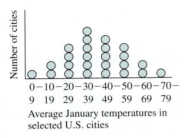

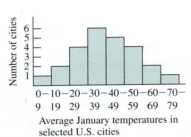

a bit more flexibility than dot plots: In histograms, the height of a bar can refer either to the *number* of pieces of data in that category, as in Figure 15.9, or to the *percentage* of the data in that category, as in the histogram on household income in Figure 15.10. Note that the implied intervals in the household income histogram are $0–$9,999, $10,000–$19,999, $20,000–$29,999, and so on. From this histogram, we can see that about 12% of U.S. households had income between $10,000 and $19,999 and about 3% of households had income between $100,000 and $109,999.

stem-and-leaf plot

Stem-and-Leaf Plots A **stem-and-leaf plot** or *stem plot,* can be used to quickly organize the numerical data you are collecting; the plot looks like a horizontal dot plot. Stem-and-leaf plots are most useful for organizing two-digit data, such as the average January temperatures discussed previously, which are displayed in Figure 15.9. Table 15.1 shows these temperatures organized into a stem-and-leaf plot. The *stem* in a stem-and-leaf plot consists of the numbers to the left of the vertical line, and the *leaves* are the numbers to the right. For example, the third row in Table 15.1 shows the stem 2 and the leaves 2, 6, 0, and 9, which stand for the data 22, 26, 20, and 29. Although you can put the leaves in order, it is also acceptable simply to put them in as you encounter them. In general, the stem should consist of the digits in the places above the ones place (the tens and hundreds places and perhaps even the thousands place or higher), and the leaves should consist of ones digits in the data.

line graph

Line Graphs A **line graph** is a graph in which adjacent data points are connected by a line. Often, a line graph is just a graph of a function, such as the graph of a population of a region over a period of time or the graph of the temperature of something over a period of time.

TABLE 15.1 Stem-and-leaf plot of average January temperatures in 25 U.S. cities

0	7
1	84
2	2609
3	030816
4	34050
5	7048 key: 2	6 = 26
6	13
7	3
8	
9	

FIGURE 15.10

Histogram: household income in the United States in 2007

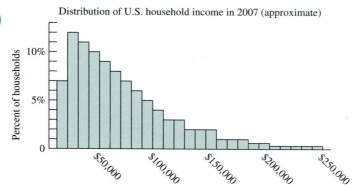

Distribution of U.S. household income in 2007 (approximate)

Line graphs are appropriate for displaying "continuously varying" data, for example, data that vary over a period of time. Figure 15.11 shows that the average math scores of 9-year-old children have been slowly increasing, as measured by NAEP, the National Assessment of Educational Progress. NAEP is often called "The Nation's Report Card." It is a national assessment of what children know and can do in various subjects. Notice that the points that are plotted in Figure 15.11 indicate that the NAEP math test was given in 1978, 1982, 1986, 1990, 1992, 1994, 1996, 1999, 2004, and 2008. It makes sense to connect these points because if the test were given at times in between, and if the children's average math scores were plotted as points, these points would probably be close to the lines drawn in the graph of Figure 15.11.

Line graphs are not appropriate for displaying data in categories that don't vary continuously. For example, it would not be appropriate to use a line graph to display the data on state populations that is displayed in the pictographs and bar graphs of Figure 15.4. Figure 15.12 shows an *inappropriate* line graph of this sort. It doesn't make sense to use a line graph in this case because there is no logical reason to connect the state populations—there is no continuous variation between the population in Georgia and the population in Washington, for example.

scatterplot **Scatterplots** A **scatterplot** consists of a collection of data points that are plotted in a plane. We use scatterplots to see how two kinds of data are related. For example, for each of the 50 states, how is the percent of students in 8th grade who are proficient in reading on the NAEP exam related to the percent of students who are proficient in mathematics on the NAEP exam? Figure 15.13 shows a scatterplot that consists of 51 points, one for each of the 50 states and one for the District of Columbia. Each point consists of a pair of numbers, namely,

(percent proficient in reading, percent proficient in math)

For example, the District of Columbia contributes the point (12, 7), because 12% of its 8th-grade students are proficient in reading and 7% are proficient in math. Massachusetts contributes the point (44, 43), because 44% of its 8th-grade students are proficient in reading and 43% are proficient in math.

We can see from the scatterplot in Figure 15.13 that states with a low percentage of students proficient in reading also have a low percentage of students proficient in math, and states with a higher percentage of students proficient in reading also have a higher percentage of students proficient in math. So proficiency in math seems to be correlated with proficiency in reading.

FIGURE 15.11

Line graph: math NAEP scores of 9-year-old children

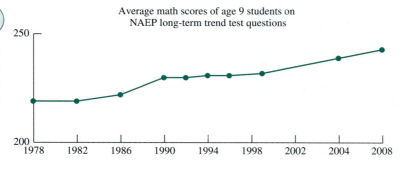

Average math scores of age 9 students on NAEP long-term trend test questions

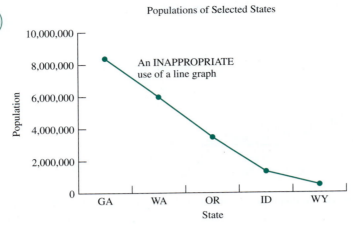

In contrast, the scatterplot in Figure 15.14 shows a different kind of situation. This scatterplot shows populations and areas of countries in the world. We see that there are a few countries with very large populations and large areas and that there are a few countries with very large areas, but that most countries have much smaller populations and areas. Land and population do not seem to be correlated.

Creating and Interpreting Data Displays

Data displays are created to communicate and interpret data. Therefore, in creating a data display, choose a display that will communicate the point you want to make most clearly. Give your display a title so that the reader will know what it is about. Label your display clearly.

Read data displays carefully before interpreting them. For example, the vertical axis of the line graph in Figure 15.11 starts at 200, not at 0, which can make changes appear greater than they are. For example, you might initially think that the math scores in 1999 were almost 50% higher than they were in 1978. In fact, the math scores in 1999 were only about 6% higher than they were in 1978.

Be careful when drawing conclusions about cause and effect from a data display. For example, in Figure 15.5, we see that women earn less money than men in all categories of education. We might be tempted to conclude that this is due to discrimination against women on the basis of gender. Although this could be the case, there could be other reasons for the difference in earnings. For example, many of the women might have taken time off from work to raise children. This break in employment could cause many women to have less work experience and therefore earn less. In any case, further study would be needed to determine the cause of the difference in earnings.

Although we can read specific numerical information from data displays, the main value of displays is in showing qualitative information about the data. So when you read a data display, look for qualitative information that it conveys and think about the implications of this information. The pie chart in

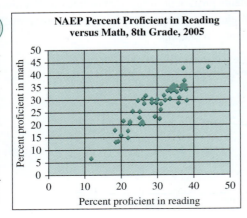

FIGURE 15.13

Scatterplot of percent of 8th-grade students proficient in reading versus mathematics for the 50 states and District of Columbia, 2005

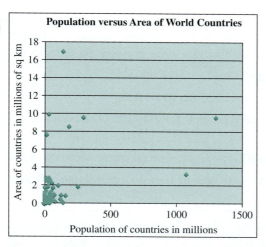

FIGURE 15.14

Scatterplot of
population versus
area
of world
countries, 2005

Figure 15.6, for example, shows that only a tiny fraction of U.S. firms have 100 or more employees. Therefore, a law that would apply only to firms with 100 or more employees would not affect most businesses. On the other hand, from the display in Figure 15.6, we do not know what percent of all *employees* work in firms with 100 or more employees. Further data would need to be gathered to determine this.

Class Activity *Now Turn to Class Activities Manual*

15E What Is Wrong with These Displays? p. 361
15F What Is Wrong with the Interpretation of These Displays? p. 362

Reading Graphs at Different Levels

Data displays are produced to organize data so that we can make sense of the data and analyze the data to answer questions. Merely producing data displays is not an important mathematical goal. To make data displays valuable in mathematics instruction, we must apply mathematical ideas to analyze the displays and answer questions.

In the book *Developing Data-Graph Comprehension in Grades K–8* [19], author Frances Curcio describes three different levels of graph comprehension: reading the data, reading between the data, and reading beyond the data. The following descriptions of these three levels are adapted from Curcio's book [19, p. 7]:

1. *Reading the data.* This level of comprehension requires a literal reading of the graph. The reader simply "lifts" the facts explicitly stated in the graph, or the information found in the graph title and axes labels, directly from the graph. No interpretation occurs at this level.
2. *Reading between the data.* This level of comprehension includes the interpretation and integration of the data in the graph. It requires the ability to compare quantities (e.g., greater than, tallest, smallest) and the use of other mathematical concepts and skills (e.g., addition, subtraction, multiplication, division) that allow the reader to combine and integrate data and identify the mathematical relationships expressed in the graph.
3. *Reading beyond the data.* This level of comprehension requires the reader to predict or infer from the data by tapping existing knowledge and knowledge developed from "reading the data" and "reading between the data" for information that is neither explicitly nor implicitly stated in the graph.

The book gives a number of graphing activities for grades K–8. Each graphing activity includes a list of questions for discussion. Each question is labeled according to the level of graph-reading comprehension the question is designed to foster in the students. For example, one activity is to create a real graph and pictograph showing the different ways that the children in a class get to school in the morning. (The real graph uses toy cars and buses, and it uses toy shoes to represent walking to school; the pictograph uses pictures of these objects.) The following is an excerpt of the questions for discussion (16, p. 37):

- What is a good title for this graph? (Read between the data)
- How many children ride the bus to school? (Read the data)
- Which is the most popular way for children to travel to school? How do you know? How can you be sure? (Read between the data)
- How do you think this graph would change if it were raining or snowing? (Read beyond the data)
- About how many children live close to school? How do we know this? (Read beyond the data)

Class Activity   *Now Turn to Class Activities Manual*

15G Three Levels of Questions about Graphs, p. 364

15H Display These Data about Pets, p. 367

15I Investigating Small Bags of Candies, p. 368

15J The Length of a Pendulum and the Time It Takes to Swing, p. 368

15K Balancing a Mobile, p. 370

Practice Exercises for Section 15.2

1. A class plays a fishing game in which there is a large tub filled with plastic fish that are identical, except that some are red and some are white. A child is blindfolded and pulls 10 fish out of the tub. The child removes the blindfold, writes down how many of each color fish she got, and then puts the fish back in the tub. Each child takes a turn. The results are shown in Table 15.2.

   a. Display these data in a dot plot.

   b. Write and answer (to the extent possible) at least three questions about the data display in part (a); include at least one question at each of the three graph-reading levels discussed in this section.

2. When is it appropriate to use a line graph?

3. If you have data consisting of percentages, is it always possible to display these data in a single pie graph?

TABLE 15.2 Numbers of red and white fish picked out of a tub by children

Name	Fish	Name	Fish	Name	Fish
Michelle	9 red, 1 white	Peter	6 red, 4 white	Sarah	8 red, 2 white
Tyler	7 red, 3 white	Brandon	9 red, 1 white	Adam	7 red, 3 white
Antrice	7 red, 3 white	Brittany	6 red, 4 white	Lauren	6 red, 4 white
Yoon-He	6 red, 4 white	Orlando	4 red, 6 white	Letitia	9 red, 1 white
Anne	6 red, 4 white	Chelsey	7 red, 3 white	Jarvis	7 red, 3 white

Answers to Practice Exercises for Section 15.2

1. a. Each dot in Figure 15.15 represents one child. The dot is plotted over the number of red fish that were picked (of course the dot plot could also be organized according to the number of white fish that were picked).

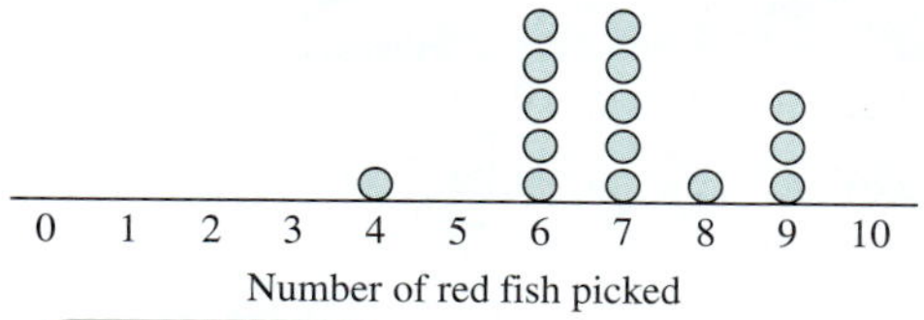

FIGURE 15.15 Dot plot showing how many children picked each number of red fish

b. How many people picked 9 red fish? (Level: Read the data) Answer: 3 people picked 9 red fish, which we can see from the 3 dots above 9.

How many people picked either 6 or 7 red fish? (Level: Read between the data) Answer: 10 people picked either 6 or 7 fish because there are 5 dots over 6 and another 5 dots over 7.

Are there more red fish or white fish in the tub? (Level: Read beyond the data) Answer: there are probably more red fish than white fish because most people picked more red fish than white fish. These random samples are probably representative of the fish in the tub.

2. See text. Line graphs are appropriate for displaying only data that vary continuously.

3. It is not always possible to display data consisting of percentages in a single pie graph. If the percentages aren't *separate parts of the same whole,* then you can't display them in a pie graph. For example, the table on children's eating, shown next, is based on a table found on the Internet. Go to *www.pearsonhighered.com/beckmann.*

Food	Percent of 4- to 6-Year-Olds Meeting the Dietary Recommendation for the Food
Grains	27%
Vegetables	16%
Fruits	29%
Saturated fat	28%

Even though the percentages add to 100%, they probably do not represent *separate* groups of children. For example, many of the 16% of the children who meet the recommendations on vegetables probably also meet the recommendations on fruits. Therefore, it would not be appropriate to display these data in a pie graph.

Problems for Section 15.2

1. Three third-grade classes are having a contest to see which class can read more pages during the month of February. The classes want to create a display, to be posted in the hall, that will show their progress. What do you recommend? Be specific, and make sure your recommendation can be carried out realistically. Draw a picture of what the display might look like at some point during February. Explain briefly why you think the display might look like that.

2. Find 3 coins of any type (as long as all 3 have a head side and a tail side).

a. Take 2 of the coins and flip the pair 30 times. While you flip the coins, make a dot plot to show how many times there were 0 heads, 1 head, and 2 heads.

b. Now take all 3 coins and flip the triple 30 times. While you flip the coins, make a dot plot to show how many times there were 0 heads, 1 head, 2 heads, and 3 heads.

c. Write at least six questions about your dot plots in parts (a) and (b); include at least two at each of the three levels of graph reading discussed in this section. Label each question with its approximate level. Your questions may be suitable for each dot plot separately or for the two dot plots together. Answer each question (to the extent possible).

3. Table 15.3 shows women's 400-meter freestyle Olympic winning times in recent history.

TABLE 15.3 Women's 400-meter freestyle Olympic winning times

Year	Winning Time minutes:seconds	Year	Winning Time minutes:seconds
1924	6:02.2	1972	4:19.44
1928	5:42.8	1976	4:09.89
1932	5:28.5	1980	4:08.76
1936	5:26.4	1984	4:07.10
1948	5:17.8	1988	4:03.85
1952	5:12.1	1992	4:07.18
1956	4:54.6	1996	4:07.25
1960	4:50.6	2000	4:05.80
1964	4:43.3	2004	4:05.34
1968	4:31.8	2008	4:03.22

a. Make a graphical display of the data in Table 15.3.

b. Write at least four questions about your graph in part (a); include at least one at each of the three graph-reading levels. Label each question with its level. Answer your questions (to the extent that it is possible) and explain your answers briefly.

4. Table 15.4 shows the percent of the U.S. population in various age ranges in various years.

a. Make at least two graphical displays of some (or all) of the data in Table 15.4.

b. For each of two of your graphical displays in part (a), write at least four questions about the graph; include at least one at each of the three graph-reading levels. Label each question with its level. Answer your questions (to the extent that it is possible) and explain your answers briefly.

TABLE 15.4 Percent of U.S. population by age and year

Year	Under 5	5–19	20–44	45–64	65 and over
1860	15%	36%	36%	10%	3%
1880	14%	34%	36%	13%	3%
1900	12%	32%	38%	14%	4%
1920	11%	30%	38%	16%	5%
1940	8%	26%	39%	20%	7%
1960	11%	27%	32%	20%	9%
1980	7%	25%	37%	20%	11%
2000	7%	22%	37%	22%	12%

5. Table 15.5 shows data about eighth-graders' mathematics performance on the 2005 National Assessment of Educational Progress (NAEP).

TABLE 15.5 Percentages of students at each achievement level (below basic, basic, proficient, advanced) for mathematics, grade 8, by jurisdiction, 2005

Jurisdiction	Below Basic	At Basic	At Proficient	At Advanced
Midwest	28%	40%	26%	6%
Northeast	27%	39%	26%	7%
South	34%	40%	21%	5%
West	38%	37%	20%	5%

a. Display the data in Table 15.5 in several pie graphs.

b. Display the data in Table 15.5 in a bar graph.

c. Compare and contrast your graphical displays in parts (a) and (b). What are the pie graphs most useful for? What is the bar graph most useful for?

6. Table 15.6 shows data about fourth-graders' mathematics performance on the 2005 National Assessment of Educational Progress (NAEP).

a. Use some of the data in Table 15.6 to make a graphical display useful for comparing the percentage of fourth-graders performing at or above the proficient level in mathematics in the states in the table.

b. Use some of the data in Table 15.6 to make a graphical display useful for comparing the percentage of fourth-graders performing at or below the basic level in mathematics in the states in the table.

c. Use some of the data in Table 15.6 to make a graphical display useful for seeing how the percentage of fourth-graders performing below the basic level in mathematics is related to the percentage of fourth-graders performing at the proficient level.

TABLE 15.6 Percentages of students at each achievement level (below basic, basic, proficient, advanced) for mathematics, grade 4, by selected states, 2005

Selected States	Below Basic	At Basic	At Proficient	At Advanced
Alabama	34%	46%	19%	2%
California	29%	43%	24%	4%
Georgia	24%	47%	26%	4%
Massachusetts	9%	42%	41%	8%
South Carolina	19%	46%	31%	5%
Texas	13%	47%	35%	5%

7. Tables 15.7–15.12 show data about eighth-graders' reading performance on the 2005 National Assessment of Educational Progress (NAEP). (NAEP reading scores range from 0 to 500.) Make graphical displays of the data in at least two of the tables, and write a paragraph about what you learn from your displays.

TABLE 15.7 Average NAEP scores for reading, grade 8, by national school lunch program eligibility, 2005

National School Lunch Eligibility	Average Score
Eligible	247
Not eligible	270

TABLE 15.8 Average NAEP scores for reading, grade 8, by school location, 2005

School Location	Average Score
Large city	250
Mid-size city	257
Fringe/large city	264
Fringe/mid-size city	264
Large town	261
Small town	260
Rural (metropolitan suburban area)	262
Rural (not a metropolitan suburban area)	265

TABLE 15.9 Average NAEP scores for reading, grade 8, by reported importance of the reading test, 2005

Importance of Success On This Reading Test	Average Score
Not very important	258
Somewhat important	265
Important	263
Very Important	257

TABLE 15.10 Average NAEP scores for reading, grade 8, by how much reading is favored as an activity, 2005

Reading Is a Favorite Activity	Average Score
Strongly disagree	251
Disagree	259
Agree	269
Strongly agree	280

TABLE 15.11 Average NAEP scores for reading, grade 8, by number of books in home, 2005

Books in Home	Average Score
0–10 books	238
11–25 books	247
26–100 books	263
More than 100 books	276

TABLE 15.12 Average NAEP scores for reading, grade 8, by parental education level, 2005

Parental Education Level	Average Score
Did not finish H.S.	244
Graduated H.S.	252
Some education after H.S.	265
Graduated college	270

8. Information about the National Assessment of Education Progress (NAEP) is available on the Internet. Using the Internet, find the achievement level results from recent NAEP assessments at grades 4 and 8. Go to *www.pearsonhighered.com/beckmann*. Click on the "advanced" button for access to the most interesting data. Browse through these data. Select some data that interest you, write a paragraph about the data, and make a display of the data to help convey your point.

9. Information about the National Assessment of Education Progress (NAEP) is available on the Internet. Using the Internet, find the NAEP 2004 Long-Term Trend Summary Data Tables. Go to *www.pearsonhighered.com/beckmann*. Browse through the long-term trend data in mathematics and reading. Notice that the "contextual variables" tables include a variety of information, such as data on watching television, time spent on homework, reading practices in and out of school, and the use of scientific instruments. Select some data that interest you, write a paragraph about the data, and make a display of the data to help convey your point.

10. **a.** Describe *in detail* an activity suitable for use with elementary school children in which the children pose a question, gather data, and create an appropriate display for the data to help answer the question.

   b. Write at least four questions that you could ask the children about the proposed graph in part (a), once the graph is completed. Include at least one question at each of the three graph-reading levels. Label each question with its level.

15.3 The Center of Data: Mean, Median, and Mode

 Focal Points
Grade 8

When we have a collection of numerical data, such as the collection of scores on a test or the heights of a group of 6-year-old children, our first questions about the data are usually, "What is typical or representative of these data?" and "What is the center of these data?" When a data set is large, it is especially helpful to know the answers to these questions. For example, if the data consist of the test scores of all students taking the SAT in a given year, no one person would want to look at the entire data set (even if he or she could); it would be overwhelming and impossible to comprehend. Instead, we want ways of understanding the nature of the data. In particular, it is helpful to have a single number that summarizes a set of data.

To say what is representative, in the center, or typical of a list of numbers, we commonly use the terms *mean,* the *median,* or the *mode.* The mean, median, and mode each provide a single-number summary of a set of numerical data and are often called **measures of center.**

The Mean

mean
average

To calculate the **mean**, also called the *arithmetic mean,* or the **average**, of a list of numbers, add all the numbers and divide this sum by the number of numbers in the list. For example, the mean of

$$7, \ 10, \ 11, \ 8, \ 10$$

is

$$(7 + 10 + 11 + 8 + 10) \div 5 = 46 \div 5 = 9.2$$

We divide by 5 because there are 5 numbers in the list: 7, 10, 11, 8, 10.

Class Activity   *Now Turn to Class Activities Manual*

15L The Mean as "Making Even" or "Leveling Out," p. 371

The Mean as "Leveling Out" Although we defined the mean in terms of adding and dividing, we can think of the mean of a list of numbers as obtained by "leveling out" the numbers so that all numbers become the same. For example, consider the list

$$4, \ 3, \ 6, \ 5, \ 2$$

FIGURE 15.16

The mean as "leveling out"

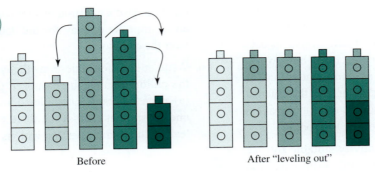

Before   After "leveling out"

as represented by towers of blocks: one 4 blocks tall, one 3 blocks tall, one 6 blocks tall, one 5 blocks tall, and one 2 blocks tall, as shown in Figure 15.16. If we rearrange the blocks so that all 5 towers become the same height, the common height is the mean. Why? When we calculate the mean numerically, we first add the numbers

$$4 + 3 + 6 + 5 + 2$$

In terms of the block towers, adding the numbers tells us the total number of blocks in all 5 towers. When we then divide the sum of the numbers by 5,

$$(4 + 3 + 6 + 5 + 2) \div 5$$

we are determining the number of blocks in each tower when the full collection of blocks is divided equally among 5 towers. In other words, we are determining the number of blocks in each tower when the 5 towers are made even. As you will see in the next Class Activity, some problems about the mean will be easy to solve by thinking of the mean in terms of leveling out.

Class Activity   *Now Turn to Class Activities Manual*

15M Solving Problems about the Mean, p. 372

The Mean as Balance Point Another way to think about the mean is as a "balance point." This way of thinking about the mean is especially appropriate when data are represented in a dot plot or a histogram. Think of a dot plot or histogram as being like a seesaw that will balance on a fulcrum when the fulcrum is placed at an appropriate spot. The location for the fulcrum—the balance point—turns out to be the mean of the set of data. For example, consider the data set

$$4, \quad 4, \quad 4, \quad 12$$

which is represented in the dot plot in Figure 15.17. You can probably tell that the dot plot will balance if a fulcrum is placed under the number 6. And in fact, the mean of the data is also 6.

Why is the mean at the location of a fulcrum? We will not discuss this subject in detail here, but two factors are involved: (1) knowing how levers work and (2) knowing that for data to the right of the mean, the sum of all the distances from the mean of these data is equal to the sum of all the distances from the mean of the data that lie to the left of the mean.

FIGURE 15.17

The mean as "balance point"

By thinking about the mean as a balance point, we can quickly estimate the mean of data that are displayed in a dot plot or a histogram.

> **Class Activity** *Now Turn to Class Activities Manual*
>
> **15N** The Mean as "Balance Point," p. 373

The Median

median To calculate the **median** of a list of numbers, organize the list from smallest to largest (or vice versa). The number in the middle of the list is the median. If there is no number in the middle, then the median is halfway between the two middle numbers. For example, to find the median of

$$7, \quad 10, \quad 11, \quad 8, \quad 10$$

rearrange the numbers from smallest to largest as follows:

$$7, \quad 8, \quad \boxed{10}, \quad 10, \quad 11$$

The number 10 is exactly in the middle, so it is the median. There are 6 numbers in the list 4, 6, 5, 3, 5, 3. When we put them in order, we see there is no single number in the middle of the list:

$$3, \quad 3, \quad \boxed{4, \quad 5}, \quad 5, \quad 6$$

The median is halfway between the two middle numbers 4 and 5; therefore, the median is 4.5.

> **Class Activity** *Now Turn to Class Activities Manual*
>
> **15O** Same Median, Different Mean, p. 374

The Median versus the Mean

The mean and the median are both single-number summaries of a set of data, and both give us a sense of what is typical or representative of the data. But the median and mean need not be the same, as you will see in the next Class Activity.

The median is usually not affected much by a few pieces of data that are very far from the majority of the data. On the other hand, the mean can be significantly affected by data that are far from the majority of the data, as Figure 15.18 shows.

Because of the mean's sensitivity to extreme values, the median sometimes gives a better sense of what is typical or representative for a set of data. For example, when discussing household income of families in the United States, the median is generally preferred over the mean, because a small percentage of very

FIGURE 15.18

Data that are far from the majority of the data affect the mean more than the median.

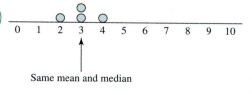

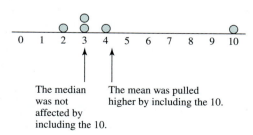

wealthy households pulls the mean up. The vast majority of households have a much more modest income. Thus, the median is more typical, or representative, of the household income of most families in the United States than is the mean.

On the other hand, when a total combined amount is to be calculated from a representative amount, the mean is a better choice for the representative amount than is the median. For example, to calculate the total income of all U. S. households, we should multiply the *mean* household income by the number of households. If we multiplied the median household income by the number of households we would get a value that is lower than the total income of all households. Some problems at the end of this section further explore the mean versus the median.

> ### Class Activity   *Now Turn to Class Activities Manual*
>
> **15P** Can More Than Half Be Above Average?, p. 376

To help remember which is the median and which is the mean, think of this example: On a road, the median line is the line down the middle of the road. Similarly, the median is the middle of all the data values when they are put in order.

Errors with the Mean and the Median

> ### Class Activity   *Now Turn to Class Activities Manual*
>
> **15Q** Errors with the Mean and the Median, p. 377

Students who are beginning to learn statistical concepts sometimes mistakenly try to apply the median in cases where it does not apply (see [47]). The median (and the mean) apply only to numerical data, not to categorical data. For example, if students in a class have each chosen a favorite sport and constructed a bar graph showing how many students chose football, basketball, baseball, soccer, or another sport as favorite, there is no "median favorite sport" because there is no natural or preferred way to order the sports. Similarly, a "mean favorite sport" would not make sense either. Expressed more generally, it would not make sense to consider a balance point on a bar graph of categorical data as the mean of the data.

When data are displayed in a dot plot or a histogram, students sometimes calculate means and medians incorrectly by finding the mean or median of the frequency counts instead of the data. For example, a student might attempt to find the mean of the data shown in the dot plot in Figure 15.19 incorrectly by averaging the frequencies 3, 5, 4 instead of by averaging the data values

$$6, \ 6, \ 6, \ 7, \ 7, \ 7, \ 7, \ 7, \ 8, \ 8, \ 8, \ 8,$$

to determine the mean correctly.

FIGURE 15.19

What is the mean of the data in the dot plot?

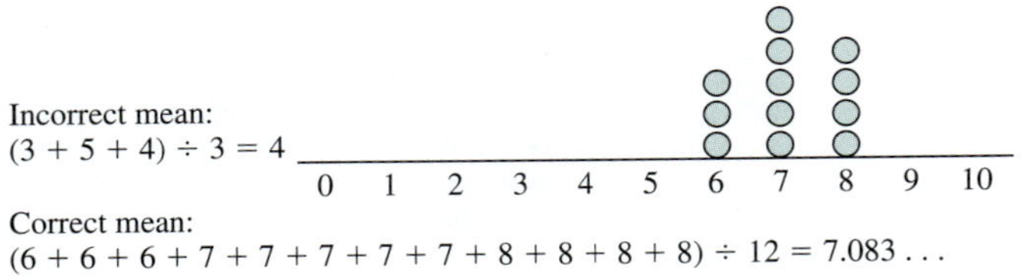

Incorrect mean:
$(3 + 5 + 4) \div 3 = 4$

Correct mean:
$(6 + 6 + 6 + 7 + 7 + 7 + 7 + 7 + 8 + 8 + 8 + 8) \div 12 = 7.083\ldots$

The Mode

mode   To calculate the **mode** of a list of numbers, reorganize the list so as to group equal numbers in the list together. The mode is the number or numbers that occur most frequently in the list. For example, to find the mode of

$$11, \quad 9, \quad 10, \quad 8, \quad 11, \quad 10, \quad 11, \quad 9, \quad 12$$

first reorganize the list as follows:

$$8, \quad 9, \quad 9, \quad 10, \quad 10, \quad 11, \quad 11, \quad 11, \quad 12$$

The number 11 occurs 3 times, which is more than any other number in the list. Therefore 11 is the mode of this list of numbers. Statisticians do not use the mode as much as the mean and the median, because relatively small changes in data can produce large changes in the mode. Consequently, we will not work with the mode as a measure of center in the Class Activities, Practice Exercises, or Problems.

Unlike the mean or the median, the mode can be used with categorical data to identify a **modal category,** which is a category (or categories) that contains the largest number of entries. For example, suppose 20 students in a class vote on which of baseball, basketball, football, or soccer is their favorite sport to play. Let's say that 9 students choose basketball and each of the other sports receives fewer than 9 votes. Then basketball is the modal category in that case.

We can use the modal category to predict which category a new piece of data is most likely to fall into. Return to the example of voting for a favorite sport to play. If a new student were to join the class and we had to guess about what sport this student would like to play, our best guess is that it would be the modal category, basketball.

Practice Exercises for Section 15.3

1. Explain how to use physical objects to describe the mean of a data set as "leveling out" the data set.

2. We can determine the mean of a list of whole numbers such as 4, 3, 6, 5, and 2 by building 5 block towers made of 4 blocks, 3 blocks, 6 blocks, 5 blocks, and 2 blocks and by redistributing the blocks among the 5 towers until they are all the same height; this common height is the mean. We can also determine the mean of a list of numbers by adding the numbers and then dividing the sum by the number of numbers in the list. Interpret the numerical procedure for calculating mean in terms of block towers. Explain why the mean you determine by physically "leveling out" block towers and the mean you determine by calculating numerically must always come out the same.

3. Describe how to show the mean of a data set as the "balance point" of the data set.

4. When numerical data are displayed in a dot plot, what is an especially appropriate way to think about the mean?

5. What is a common error that students make when they compute the mean of numerical data that are displayed in a dot plot?

6. Describe some common errors students make with the median.

7. Ten students take a 10-point test. The mean score is 8. Could the median be 10? Could the median be 5? In each case, explain why or why not.

8. Juanita read an average of 3 books a day for 4 days. How many books will Juanita need to read on the fifth day so that she will have read an average of 5 books a day over 5 days? Solve this problem in several ways, and explain your solutions.

9. George's average score on his math tests in the first quarter is 60. George's average score on his math tests in the second quarter is 80. George's semester score in math is the average of all the tests George took in the first and second quarters. Can George necessarily calculate his semester score by averaging 60 and 80? If so, explain why. If not, explain

why not. Explain what other information you would need to calculate George's semester average, and show how to calculate this average.

10. Suppose that all fourth-graders in a state take a writing competency test that is scored on a 5-point

scale. Is it possible (in theory) for 80% of the fourth-graders to score below average? If so, show how that could occur. If not, explain why not.

Answers to Practice Exercises for Section 15.3

1. See text.

2. See text.

3. See text.

4. As a balance point. See text.

5. Students sometimes calculate the mean of the frequencies instead of using the data values. See the text and Class Activities.

6. See the text and the Class Activities. Students sometimes forget to put the data in order before determining the middle value. They sometimes try to apply the median to categorical data, where it does not apply. As with the mean, students may mistakenly work with frequencies rather than with data values when calculating the median.

7. Since the mean score is 8, the total number of points scored by all 10 students is $10 \times 8 = 80$. For the median to be 10, at least 6 students would have to score a 10. These 6 students contribute a total of 60 points. The remaining 20 points could then be distributed in many ways among the remaining 4 students, for example, say all 4 remaining students scored a 5. So yes, it's possible for the median to be 10.

For the median to be 5, at least 6 students would have to score 5 or less. So say there are 6 students who each score a 5 or less; therefore, they contribute a total of at most 30 points. For the mean to be 8, the total number of points scored by all students is 80, so the remaining 4 students would have to contribute the remaining 50 points, which is not possible since it's a 10-point test. So no, it is not possible for the median to be 5.

8. *Method 1*: Imagine putting the books that Juanita has read into 4 stacks, one for each day. Because she has read an average of 3 books per day, the books can be redistributed so that there are 3 books in each

of the 4 stacks. This uses the "average as leveling out" point of view. Now picture a fifth stack with an unknown number of books in it—the number of books Juanita must read to make the average 5 books per day. If the books in this fifth stack are distributed among all 5 stacks so that all 5 stacks have the same number of books, then this common number of books is the average number of books Juanita will have read over 5 days. Therefore, this fifth stack must have 5 books plus another $4 \times 2 = 8$ books, in order to distribute 2 books to each of the other 4 stacks. So Juanita must read $5 + 8 = 13$ books on the fifth day.

Method 2: To read an average of 5 books a day over 5 days, Juanita must read a total of 25 books. This is because

$$(\text{total books}) \div 5 = \text{average}$$

Therefore,

$$(\text{total books}) \div 5 = 5$$

so

$$\text{total books} = 5 \times 5 = 25$$

Similarly, because Juanita has already read an average of 3 books per day for 4 days, she has read a total of $4 \times 3 = 12$ books. Therefore, Juanita must read $25 - 12 = 13$ books on the fifth day.

9. George cannot necessarily calculate his semester score by averaging 60 and 80. We don't know if George took the same number of tests in the first and second quarters. To calculate George's average score over the whole semester, we need to know how many tests he took in the first and second quarters. Suppose George took 3 tests in the first quarter and 5 tests in the second quarter. Then the sum of his first 3 test scores is $3 \times 60 = 180$, and the sum of his next 5 test scores is $5 \times 80 = 400$, so the sum

of all 8 of his test scores is 180 + 400 = 580. Therefore, George's average score on all 8 tests is 580 ÷ 8 = 72.5. Notice that this is not the same as the average of 60 and 80, which is only 70.

10. It is theoretically possible for 80% of the fourth-graders to score below average. For example, suppose there are 200,000 fourth-graders and that

    10,000 score 1 point,

    150,000 score 2 points,

    10,000 score 3 points,

    25,000 score 4 points, and

    5000 score 5 points,

as shown in the bar graph of Figure 15.20. You can see from the bar graph that the "balance point" is between 2 and 3, and 160,000 children, namely, 80% of the 200,000 children, score below this average.

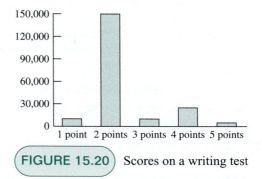

FIGURE 15.20 Scores on a writing test

Problems for Section 15.3

1. Explain why the mean of a list of numbers must always be in between the smallest and largest numbers in the list.

2. Explain why the mean of two numbers is exactly halfway between the two numbers.

3. Shante caught 17 ladybugs every day for 4 days. How many ladybugs does Shante need to catch on the fifth day so that she will have caught an average of 20 ladybugs per day over the 5 days? Solve this problem in two different ways, and explain your solutions.

4. John's average annual income over a 4-year period was $25,000. What would John's average annual income have to be for the next 3 years so that his average annual income over the 7-year period would be $50,000? Solve this problem in two different ways, and explain your solutions.

5. Tracy's times swimming 200 yards were as follows:

    2:45,   2:47,   2:44

How fast will Tracy have to swim her next 200 yards so that her mean time for the 4 trials is 2:45? Explain how you can solve this problem without adding the times.

6. Explain how you can quickly calculate the mean of the following list of test scores without adding the numbers:

    83,   79,   81,   76,   81

7. Explain how you can quickly calculate the mean of the following list of test scores without adding the numbers:

    84,   79,   81,   78,   83

8. Julia's average on her first 3 math tests was 80. Her average on her next 2 math tests was 95. What is Julia's average on all 5 math tests? Solve this problem in two different ways, and explain your solutions.

9. A teacher gives a 10-point test to a class of 10 children.

    a. Is it possible for 9 of the 10 children to score above average on the test? If so, give an example to show how. If not, explain why not.

    b. Is it possible for all 10 of the children to score above average on the test? If so, give an example to show how. If not, explain why not.

10. In your own words, describe how to view the mean of a set of numerical data in two different ways: in terms of leveling out and as a balance point. In each case, give an example to illustrate.

11. Discuss Jessica's reasoning about calculating the mean of the data displayed in the dot plot in Figure 15.21.

Jessica's reasoning: I took 2 dots from above the 8 and moved one to 6 and one to 7. Then all the towers were leveled out and all were 4 tall, so the mean is 4.

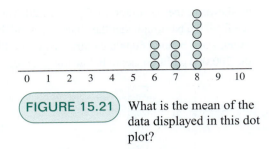

FIGURE 15.21 What is the mean of the data displayed in this dot plot?

Is Jessica's reasoning valid? If not, discuss how her idea could be used to make a correct statement about the mean.

12. The dot plot in Figure 15.22 represents the number of small candies found in several packets. Michael found the mean number of candies in a packet by calculating this way:

$$\frac{21 + 22 + 23}{3} = 22$$

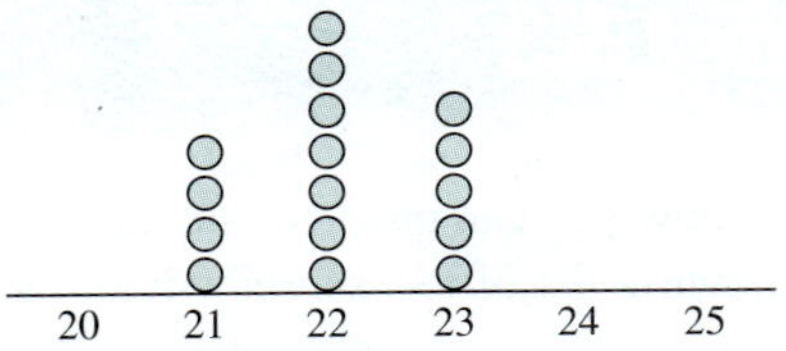

FIGURE 15.22 Dot plot of the number of candies in packets

Anne found the mean number of candies in a packet by calculating this way:

$$\frac{4 + 7 + 5}{3} = 5\frac{1}{3}$$

Is either of these methods correct? If not, explain what is wrong, and explain how to calculate the mean number of candies in the packets correctly.

13. For each of the following situations, decide whether the mean or the median would likely give a better sense of what is typical or representative of the data for the purpose that is described. Discuss your reasoning in each case.

   a. The starting salaries of a class of college graduates in which one of the graduates is drafted by the National Football League. Prospective students would like a sense of the starting salary they might expect when they get a job after they graduate.

   b. The net weights of one type and size of boxes of laundry detergent produced at a factory over the course of a year. Government regulators would like to know if consumers who use the laundry

detergent on a regular basis are getting the amount of detergent they have paid for in the long run.

   c. The number of turtle eggs in nests along a beach. Researchers estimate that 80% of all the turtle eggs will hatch and they want to estimate how many baby turtles there will be.

14. The students in Mrs. Marshall's class are estimating the number of blades of grass in the school's lawn. Each student places a piece of cardboard with a 1-inch-by-1-inch square cut out of the middle over a patch of grass, cuts off the grass that pokes through the hole, and counts the number of blades of grass that were cut. The students determine that the entire lawn consists of about 21,600 1-inch-by-1-inch squares. The students also determine that the mean number of blades of grass in their samples is 37 and the median number is 34. The students want to use their samples to estimate the number of blades of grass in the lawn, but they can't decide if they should use the mean or the median number of blades of grass in their samples. In other words, should they multiply 37 by 21,600 or should they multiply 34 by 21,600 in order to estimate the number of blades of grass in the lawn? Parts a, b, c will help you determine whether the mean or the median will be better for the purpose of estimating the number of blades of grass.

   a. Suppose Mrs. Marshall's students use their method to determine the number of blades of grass in a 5-inch-by-5-inch square of grass. In this case, they can subdivide the patch of grass into 5 × 5 = 25 1-inch-by-1-inch squares and determine the number of blades of grass in each such square. The students can then determine the mean number of blades of grass in the squares and multiply this number by 25. Explain why this method, which uses the mean, will always produce the correct number of blades of grass in the 5-inch-by-5-inch patch of grass.

   b. As in part (a), suppose Mrs. Marshall's students use their method to determine the number of blades of grass in a 5-inch-by-5-inch square of grass. But this time, suppose the students use the *median* number of blades of grass in the 25 1-inch-by-1-inch squares instead of the mean. Give an example to show that multiplying the median number of blades of grass in the 1-inch-by-1-inch squares by

25 may not produce the correct number of blades of grass in the 5-inch-by-5-inch patch. Explain your example.

   c. Returning to the original problem about the grass in the school lawn, how should Mrs. Marshall's students estimate the number of blades of grass in the lawn? Explain.

15. A teacher gives a 10-point test to a class of 9 children. All scores are whole numbers. If the median score is 8, then what are the highest and lowest possible mean scores? Explain your answers.

16. A teacher gives a 10-point test to a class of 9 children. All scores are whole numbers. If the mean score is 8, then what are the highest and lowest possible median scores? Explain your answers.

17. In Ritzy County, the average annual household income is $100,000. In neighboring Normal County, the average annual household income is $30,000.

Does it follow that in the two-county area (consisting of Ritzy County and Normal County), the average annual household income is the average of $100,000 and $30,000? If so, explain why; if not, explain why not. What other information would you need to calculate the average annual household income in the two-county area? Show how to calculate this average if you had this information.

18. In county A, the average score on the grade 5 Iowa Test of Basic Skills (ITBS) was 50. In neighboring county B, the average score on the grade 5 ITBS was 71. Can you conclude that the average score on the grade 5 ITBS in the two-county region (consisting of counties A and B) can be calculated by averaging 50 and 71? If so, explain why; if not, explain why not. What other information would you need to know to calculate the average grade 5 ITBS score in the two-county region? Show how to calculate this average if you had this information.

19. a. The histogram at the top of Figure 15.23 shows the distribution of annual household incomes in Swank County. Without calculating the median or average annual household income in Swank County, make an educated guess about which one is greater. Describe your thinking.

   b. Determine the approximate values of the median and average annual household incomes in Swank County from the histogram at the top of Figure 15.23. Which is greater, the average or the median? Compare with part (a).

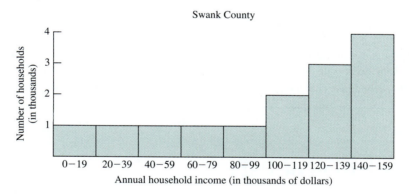

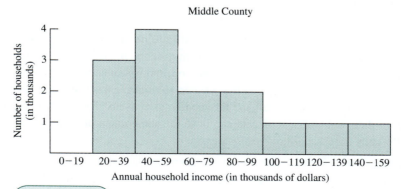

FIGURE 15.23 Annual household incomes in Swank and middle Counties

c. The histogram at the bottom of Figure 15.23 shows the distribution of annual household incomes in Middle County. Without calculating the median or average annual household income in Middle County, make an educated guess about which one is greater. Describe your thinking.

d. Determine the approximate values of the median and average annual household income in Middle County from the histogram at the bottom of Figure 15.23. Which is greater, the average or the median?

20. Ms. Smith needs to figure her students' homework grades. There have been 5 homework assignments: the first with 30 points, the second with 40 points, the third with 30 points, the fourth with 50 points, and the fifth with 25 points. Ms. Smith is debating between two different methods of calculating the homework grades.

Method 1: Add up the points scored on all 5 of the homework assignments, and divide by the total number of points possible. Then multiply by 100 to get the total score out of 100.

Method 2: Find the percentage correct on each homework assignment, and find the average of these five percentages.

a. Jodi's homework grades are $\frac{27}{30}, \frac{25}{40}, \frac{26}{30}, \frac{30}{50}, \frac{24}{25}$, where, for example, $\frac{27}{30}$ stands for 27 points out of 30 possible points in a homework assignment. Compare how Jodi does with each of the two methods. Which method gives her a higher score?

b. Make up scores (on the same assignments) for Robert so that Robert scores higher with the other method than the one that gave Jodi the higher score.

21. The *average speed* of a moving object during a period of time is the distance the object traveled divided by the length of the time period. For example, if you left on a trip at 1:00 P.M., arrived at 3:15 P.M., and drove 85 miles, then the average speed for your trip would be

$$85 \div 2.25 = 37.8$$

miles per hour.

a. Jane drives 100 miles from Philadelphia to New York with an average speed of 60 miles per hour. When she gets to New York, Jane turns around immediately and heads back to Philadelphia along the same route. Using this information, find several different possible average speeds for Jane's *entire trip* from Philadelphia back to Philadelphia.

b. Ignoring practical issues, such as speed limits and how fast cars can go, would it be theoretically possible for Jane to average 100 miles per hour for the whole trip from Philadelphia to New York and back that is described in part (a)? If so, explain how; if not, explain why not.

c. Theoretically, what are the largest and smallest possible average speeds for Jane's entire trip from Philadelphia back to Philadelphia that is described in part (a)? Explain your answers clearly.

15.4 The Distribution of Data

F·P   **Focal Points**
           Grade 8

When we collect data, we don't expect all the data to be the same; rather, we expect to see variability in the data. If a country gives a test to all fourth-graders, the test scores will surely vary from student to student. If a company measures the amount of mercury in cans of tuna fish, the amount will likely vary from can to can. We have seen that we can summarize a set of numerical data with the mean or the median of the data. The mean and the median give us an idea of what is typical or representative for a set of data, but two sets of data can have the same mean or median and yet may be distributed very differently across a dot plot or histogram.

In this section we will examine some of the common shapes that data distributions take. We also will study some statistical tools for summarizing how data are distributed within a data set. Some of these tools involve the use of percentiles. Although you may not teach students about percentiles, you will probably receive reports about your students' performance on standardized tests in terms of percentiles. You may also receive reports about your students in terms of normal distributions; therefore, we will study normal distributions briefly at the end of the section.

Data Distributions of Different Shapes

A large set of numerical data often has a particular type of shape when it is plotted in a histogram or a dot plot. The overall shape of the data distribution can give insight into the nature of the data. Before you read on, do the next Class Activity to get a feel for what different shapes of data distributions tell about the data.

> ### Class Activity   *Now Turn to Class Activities Manual*
>
> **15R** What Does the Shape of a Data Distribution Tell about the Data?, p. 378

Some data distributions have a long tail that extends far to the right or far to the left of the majority of the data, as shown in Figure 15.24. These distributions are called **skewed to the right** and **skewed to the left** respectively. For example, in a mature forest, the majority of the trees may be approximately the same height—the height of the tree canopy. There may be a far smaller number of shorter trees, perhaps because there is not enough sunlight below the canopy for many new trees to start growing or to survive. A histogram of tree heights in this forest would be skewed to the left. On the other hand, if most of the fish in a lake are small and there are far fewer large fish, then a histogram of the weights of fish in this lake would be skewed to the right.

Some data distributions have two (or more) peaks, as shown in Figure 15.25. A distribution that has two peaks is called **bimodal**. Some forests have one bunch of tall trees that form the tree canopy and another bunch of shorter trees that form an understory. A histogram of tree heights in such a forest is bimodal.

FIGURE 15.24

Distributions skewed to the right and left

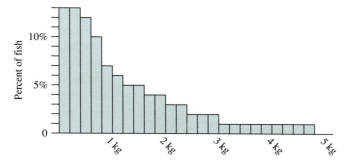

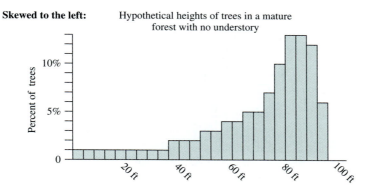

FIGURE 15.25

A bimodal distribution

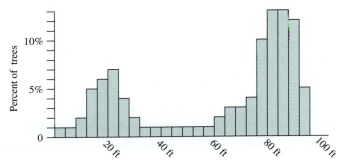

Bimodal distribution:

Hypothetical heights of trees in a forest that has an understory

Data distributions that are symmetrical, or nearly so, such as the one in Figure 15.26 are called **symmetric**. Data sets such as the weights of children of a fixed age, the lengths of a certain species of fish in a lake, or the scores of students on a large, standardized test, tend to be symmetric. One especially important type of symmetric distribution arises from random samples of a population.

Symmetric Distributions Arising from Random Samples of a Population

Class Activity *Now Turn to Class Activities Manual*

15S Distributions of Random Samples, p. 380

Suppose a population is divided into two categories. The population could be light bulbs produced at a factory and the categories could be "defective" and "not defective," or the population could be the deer in a region and the categories could be "infected" and "not infected" with a certain disease. For concreteness and simplicity, let's think of the population as voters, and let's say the categories are "yes" and "no," as if the voters will vote for or against a certain referendum. Now suppose we take a bunch of reasonably large random samples of the same size. In each case, we determine the percent of the sample that voted yes and we plot these data in a histogram or a dot plot, as in Figure 15.27. An amazing and very useful fact is displayed: These random samples tend to form a distribution that is symmetric about the percent of yes votes in the full population. Furthermore, the larger the sample size, the more tightly clustered the distribution tends to be about its center of symmetry. These facts about random samples allow statisticians to use samples to make predictions with a certain degree of confidence (e.g., when predicting the outcome of an election based on a poll). The details about making predictions based on samples is beyond the scope of this book, but they can be found in any book on statistics in a chapter on statistical inference.

FIGURE 15.26

A symmetric distribution

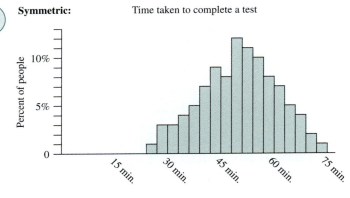

Symmetric:

Time taken to complete a test

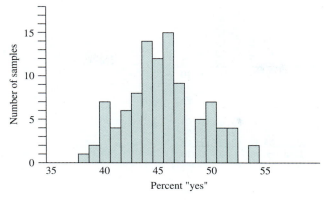

FIGURE 15.27

Random samples of voters

Samples of **200** voters from a population of 100,000 voters, of which 45% vote yes (100 such samples were taken.)

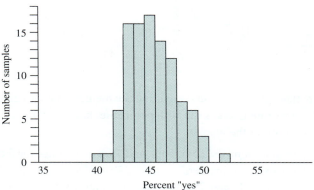

Samples of **400** voters from a population of 100,000 voters, of which 45% vote yes (100 such samples were taken.)

Notice that both of the distributions in Figure 15.27 are symmetrical, both have approximately the same mean, and both have approximately the same median. So although the shape, mean, and median are useful summaries of each data set, these pieces of information are not enough to distinguish the two distributions. The two distributions are distinguished by how dispersed they are. The second distribution is less dispersed and more tightly clustered than the data in the first distribution. One way to summarize this difference in dispersion is through the use of percentiles and box plots, which we examine next.

Summarizing Distributions with Percentiles and Box Plots

> **Class Activity**   *Now Turn to Class Activities Manual*
>
> **15T** Comparing Distributions: Mercury in Fish, p. 383

If you did the previous Class Activity, you probably noticed that the means or medians of two data distributions may not provide enough information to compare the data adequately. It can be helpful to have additional information, such as the high and low values of the data, as well as information about how spread out the data are. Although a histogram can show such data, information that summarizes a histogram is also helpful. One way to summarize data sets is with the aid of several percentiles.

Percentiles   Consider the three dot plots in Figure 15.28. The mean and median of the data in all three dot plots is 7, so neither the mean nor the median help distinguish one data set from another. But information about percentiles can help distinguish these data sets.

percentile   A **percentile** is a generalization of the median. The 90th percentile of a set of numerical data is the number such that 90% of the data is less than or equal to that number and 10% is greater than or equal to that number. (If there is no single number that satisfies this criterion, then use the midpoint of all

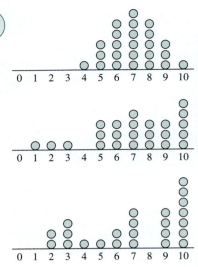

FIGURE 15.28

Dot plots of three different hypothetical test results, each with median 7 and average 7

numbers that satisfy it, as in computing the median.) Other percentiles are defined similarly. The median of a list of numbers is the 50th percentile, because 50% of the list is at or below the median, and 50% is at or above the median.

How can we determine the 25th, 50th, and 75th percentiles for the three data sets in Figure 15.28? Each data set consists of 24 numbers, so since 25% of 24 is 6, we should break the data after every 6th value to find the 25th, 50th, and 75th percentiles, as indicated in Figure 15.29. Thus we get the values of the 25th, 50th, and 75th percentiles shown in Table 15.13. Notice that in the second data set, the lowest 25% of the data ends at 5 and the rest of the data begin at 6, so the 25th percentile is 5.5, because 5.5 is halfway between 5 and 6. Similarly, in the third data set the 25th percentile is 4.5, which is halfway between 4 (where the lowest 25% of the data ends) and 5 (where the rest of the data begin).

Results of standardized tests are often reported as percentiles. So if a student scores at the 75th percentile on a test, then that student has scored the same as or better than 75% of the students taking the test.

Class Activity   *Now Turn to Class Activities Manual*

15U Determining Percentiles, p. 384

FIGURE 15.29

Finding the 25th, 50th, and 75th percentiles of three sets of data

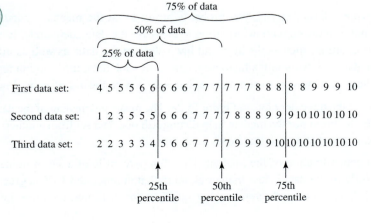

TABLE 15.13 The 25th, 50th, and 75th percentiles for three data sets

Data Set in Figure 15.28	Percentile		
	25th	50th	75th
First data set	6	7	8
Second data set	5.5	7	9
Third data set	4.5	7	10

Refer again to the three data sets displayed in the dot plots in Figure 15.28. Even though we cannot distinguish the data sets by their medians or means because all of these numbers are 7, we can see a difference among the data sets when we examine several percentiles. Table 15.13 lists the 25th, 50th, and 75th percentile for each of the three data sets. From the table, we can see that the first data set is more tightly clustered around the median than the other two data sets, because the 25th and 75th percentiles are closer to the 50th percentile in the first data set than in the other two. In the other two data sets, the 25th and 75th percentiles are farther from the 50th percentile, which indicates that these data sets are more spread out. So if we didn't have the dot plots for the three data sets, but we had the values of the 25th, 50th, and 75th percentiles, we could get a sense of how the data are distributed in these data sets.

When it comes to large data sets, such as data on household income or data on student performance on a test, we are often presented with tables of percentiles. The purpose of such tables is to summarize the data and to give us a sense of how the data are distributed.

Box Plots Instead of listing percentiles in a table, another way to summarize how data are distributed is with a **box plot**, or box-and-whiskers plot. In simple cases, a box plot shows the location of the lowest data value, the 25th percentile, the 50th percentile (median), the 75th percentile, and the highest data value. A box is drawn from the 25th percentile to the 75th percentile, and "whiskers" are drawn from the lowest data value to the 25th percentile and from the 75th percentile to the highest data value, as in Figure 15.30, which shows box plots for the three data sets in Figure 15.28.

box plot

Although we will not work with examples of this type here, in general, the whiskers in a box plot extend to the lowest and highest data values *that are not outliers* instead of to the lowest and highest values. Informally, an **outlier** is a data value that is much higher or lower than most of the data. An outlier is indicated in a box plot by placing a star at the location of its value.

Notice that even if we didn't have the dot plots in Figure 15.28, we could still obtain information about how the data are distributed from the box plots in Figure 15.30. We can see that the first data set is clustered more closely around 7 and is spread over a narrower range than either of the other two data sets.

A box plot shows at a glance where the middle 50% of the data lie. This is so because the data that lie between the 25th and 75th percentiles form the middle 50% of the data (75% − 25% = 50%). Box plots highlight the median and allow us to determine quickly whether the middle 50% of the data are symmetrically distributed about the median.

Class Activity *Now Turn to Class Activities Manual*

15V Box Plots, p. 385

Box plots for the
three data sets in
Figure 15.28

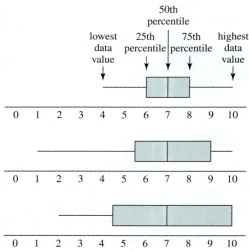

Percentiles versus Percent Correct

> **Class Activity**   *Now Turn to Class Activities Manual*
>
> **15W** Percentiles versus Percent Correct, p. 386

A common source of confusion is the distinction between a student's percentile on a test and what percent of the test the student got correct. The percentile and the percent correct report very different things about a student's performance on a test. When a student's performance on a test is reported as a percentile, that information tells us how the student did in comparison with other students who took the test. If a student scored at the 75th percentile, then that means that 75% of the students who took the test had the same or a lower score than this student. But this percentile information only tells us how the student did relative to other students who took the test; it doesn't tell us what percent of the test the student did correctly. For example, it could be that the student got 90% of the test correct and that 75% of the students who took the test scored 90% or less too. Or it could be that the student got 50% of the test correct and that 75% of the students who took the test scored 50% or less too.

In interpreting students' scores on standardized tests, be sure to distinguish between percentiles and percent correct.

Normal Curves and Standardized Test Results

In this section, we focus briefly on normal distributions. Normal distributions form the shape of the familiar symmetrical bell-shaped curve, such as the one in Figure 15.31. Many data sets have a normal distribution (or roughly so). For example, data sets such as the weights of children of a fixed age, the length of adult fish of a certain species, or the scores of students on a large, standardized test, generally have a normal distribution. When we discussed random samples previously, we noted that under certain circumstances, they tend to form symmetric distributions. In fact, these distributions tend to approximate a normal distribution. This fact about random samples allows statisticians to use samples to make predictions with a certain degree of confidence.

criterion referenced tests When you teach, your students may take different kinds of large-scale standardized tests. They may take criterion referenced tests as well as norm referenced tests. **Criterion referenced tests** are used to determine whether students have learned a specific body of material, such as subject material

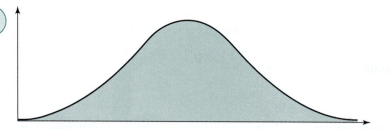

FIGURE 15.31

A normal distribution

norm referenced tests specified in state standards for their grade level. In contrast, **norm referenced tests** are used to compare students. The NAEP and SAT exams are norm referenced tests. When your students take a norm referenced test, some of the reports you receive about their performance may be based on information about normal distributions. In order to discuss these ways of reporting test results, we must first consider the standard deviation.

A **standard deviation** is a measure of how "spread out" a numerical set of data is. (You can read the exact definition of standard deviation in any book on statistics; we do not need it here.) For data that are in a normal distribution, approximately 68% of the data lies within 1 standard deviation of the mean of the data, approximately 95% of the data lies within 2 standard deviations of the mean, and approximately 99.7% of the data lies within 3 standard deviations of the mean.

A student's score on a norm referenced test will probably be reported as a percentile, but it may also be reported as a *z-score* (or *normal score*), as a *normal curve equivalent*, or as a *stanine*. Z-scores, normal curve equivalents, and stanines are obtained by using the standard deviation to rescale test scores.

A student's **z-score** tells how many standard deviations away from the mean the student's test score is. So a z-score of 1.5 is 1.5 standard deviations above the mean, and a z-score of -0.6 is 0.6 standard deviations below the mean. If a test had an average of 250 points and a standard deviation of 50 points, then a student with a z-score of 1.5 scored

$$250 + 1.5 \cdot 50 = 325$$

points on the test, and a student with a z-score of -0.6 scored

$$250 - 0.6 \cdot 50 = 220$$

points on the test. See Figure 15.32 for the locations of the z-scores 0, 1, 2, -1, and -2.

A student's **normal curve equivalent** score on a test is a score between 1 and 99. Normal curve equivalent scores are like z-scores, except that they are scaled so that a student who scores the average score on the test receives a 50 and a student who scores at the 99th percentile receives a 99. To do this, it turns out that 1 standard deviation needs to be 21.06 points. So for a student who has a z-score of 1.5, the normal curve equivalent score is

$$50 + 1.5 \cdot 21.06 = 81.59$$

or 82 points. See Figure 15.32 for locations of some normal curve equivalent scores.

FIGURE 15.32

Z-scores, normal curve equivalents, and stanines for a normal distribution

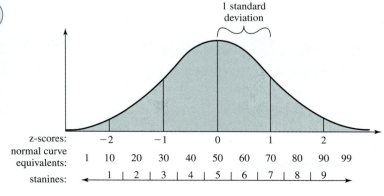

A student's **stanine** score on a test is a whole number score between 1 and 9. Except for the 1st and 9th stanine, each stanine covers $\frac{1}{2}$ of a standard deviation worth of scores.

Students who score within $\frac{1}{4}$ of a standard deviation from the mean receive a stanine score of 5. See Figure 15.32 for the locations of other stanines.

Practice Exercises for Section 15.4

1. For each of the following situations, describe what shape you would expect a histogram or dot plot of the data to take, and briefly discuss why you would expect that shape.

   a. Student arrival times at school

   b. Number of sit-ups third-graders can do in a minute.

   c. Amount of money students on a college campus carry with them

2. At a math center in a class, there is a bag filled with 35 black blocks and 15 white blocks. Each student in a class of 30 will do the following activity at the math center: randomly pick 10 blocks out of the bag without looking and write the number of black blocks picked on a sticky note.

   Make a hypothetical dot plot that could arise from this situation. Make your dot plot so that it has characteristics you would expect to see in a dot plot arising from actual data. Briefly describe these characteristics of the dot plot.

3. **a.** Determine the 25th, 50th, and 75th percentiles for the hypothetical test scores shown in the dot plots of Figure 15.33.

   b. Suppose you only have the percentiles from part (a) and you don't have the dot plots in

Figure 15.33. Discuss what you can tell about the data sets. Can you tell which data set is more tightly clustered and which is more dispersed?

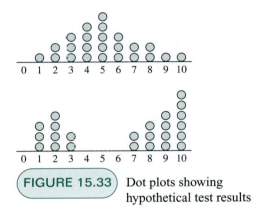

FIGURE 15.33 Dot plots showing hypothetical test results

4. Discuss the difference between getting 75% correct on a test and being at the 75th percentile on a test.

5. A test is given to students across the country. The test has a distribution that is approximately normal. How can we characterize a student's performance on this test relative to other students if the student's score is 1 standard deviation above the mean? What if the score is 1 standard deviation below the mean? What if the score is 2 standard deviations above the mean?

Answers to Practice Exercises for Section 15.4

1. **a.** A few students arrive early, but most students probably arrive just before school starts. So these data would probably be skewed to the left, with a long tail to the left consisting of those students who arrive early.

   b. These data will probably be symmetrical around the mean number of sit-ups that third-graders can do.

c. Most students probably don't carry a lot of money with them, so the data will probably be skewed to the right, with a long tail to the right consisting of those students who do have a lot of money on them.

2. See Figure 15.34. Since 70% of the blocks in the bag are black and since the samples are random samples, we expect that in most of the samples, the percent of black blocks will also be close to 70%. The dot plot should be roughly symmetrical around 7.

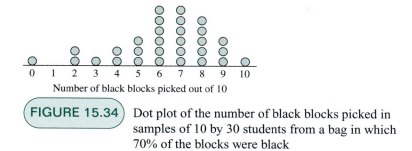

Number of black blocks picked out of 10

FIGURE 15.34 Dot plot of the number of black blocks picked in samples of 10 by 30 students from a bag in which 70% of the blocks were black

3. **a.** For both of the hypothetical tests, there are 24 test scores. Since 25% of 24 is 6, the 25th percentile is the score between the score of the 6th lowest and 7th lowest test scores. Therefore, the 25th percentile is 3.5 in the first case and 2 in the second case. The 50th percentile is the score between the 12th and 13th lowest scores, which is 5 in the first case and 8 in the second case. The 75th percentile is between the 18th and 19th lowest scores, which is 6.5 in the first case and 9.5 in the second case. The following table summarizes these results:

	25th Percentile	50th Percentile	75th Percentile
Test 1	3.5	5	6.5
Test 2	2	8	9.5

b. The first data set is more tightly clustered and less dispersed than the second data set. We know this because the 25th and 75th percentiles are closer together on the first test than on the second test. So the middle 50% of the data, which lies between the 25th and 75th percentiles, is less dispersed in the first data set than in the second data set. We can also see that many of the test scores on the second test are quite high, and much of the data is clumped toward the right of the distribution. But we can't tell from the percentiles alone whether the test scores on the second test are distributed in a long tail or form a second clump (which they do).

4. See text.

5. In a normal distribution, approximately 68% of the data lie within 1 standard deviation of the mean. Since the data are symmetrical about the mean, half of the 68%, namely 34% of the scores, lie above the mean. Another 50% of the scores lie below the mean. So a student who scored 1 standard deviation above the mean scored higher than about 50% + 34% = 84% of the students who took the test. In comparison with the other students, this student did quite well.

Similarly, a student who scored 1 standard deviation below the mean scored lower than about 84% of the students who took the test. In comparison with the other students, this student did poorly.

In a normal distribution, approximately 95% of the data lie within 2 standard deviations of the mean. Since the data are symmetrical about the mean, half of the 95%, namely about 48% of the scores, lie above the mean. Another 50% of the scores lie below the mean. So a student who scored 2 standard deviations above the mean scored higher than about 50% + 48% = 98% of the students who took the test. In comparison with the other students, this student did exceptionally well.

Problems for Section 15.4

1. What is the difference between scoring in the 90th percentile on a test and scoring 90% correct on a test? Discuss this question carefully, giving examples to illustrate.

2. What is the purpose of reporting a student's percentile on a state or national standardized test? How is this purpose different from reporting the student's percent correct on a test?

3. The three histograms in Figure 15.35 show the hypothetical performance of students in three different school districts on the same test. A score below 40 on the test is considered failing. A score of 80 or above is considered excellent.

   a. Estimate the mean score on the test for each school district by viewing the mean as a balance point, as discussed in Section 15.3.

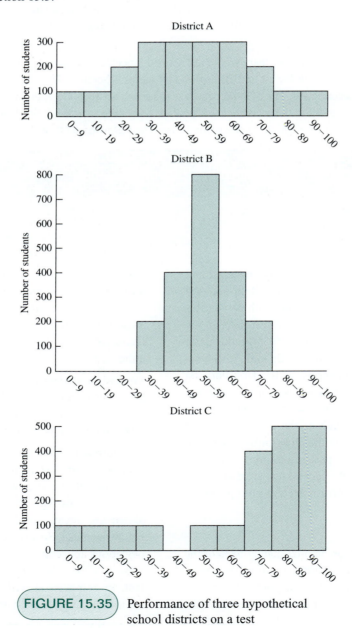

FIGURE 15.35 Performance of three hypothetical school districts on a test

b. Discuss what information you can glean from the histograms that wouldn't be apparent just from knowing the mean or median scores on the test.

c. Discuss how each school district could argue that it did better than at least one other school district.

4. At a math center in a class, there is a bag filled with 40 red blocks and 10 blue blocks. Each child in the class of 25 will do the following activity at the math center: Randomly pick 10 blocks out of the bag without looking, and write on one sticky note the number of red blocks and the number of blue blocks picked.

a. Describe a good way to display the whole class's data. Your proposed display should be a realistic and practical way to show every child's piece of data (in a class of 25).

b. Sketch a graph that could be the graph you proposed in part (a). Briefly describe the characteristics you expect the graph to have.

c. Write at least four questions about your hypothetical graph in part (b) that the teacher could ask the children. Include at least one question at each of the three levels discussed in Section 15.2 (Read the data, Read between the data, and Read beyond the data). Label each question with its level. Answer your questions.

5. Refer to Figure 15.27 on page 661.

a. Refer to the first histogram: samples of 200. What percent of those samples would predict correctly that 45% of the voters would vote yes? Explain briefly.

b. Refer to the second histogram: samples of 400. What percent of those samples would predict correctly that 45% of the voters would vote yes? Explain briefly.

c. Refer to the first histogram: samples of 200. What percent of those samples have percentages of yes votes between 43% and 47%? Explain briefly.

d. Refer to the second histogram: samples of 400. What percent of those samples have percentages of yes votes between 43% and 47%? Explain briefly.

6. Refer to Figure 15.27 on page 661.

a. Refer to the first histogram: samples of 200. What percent of those samples would predict (incorrectly) that more than 50% of the voters would vote yes? Explain briefly.

b. Refer to the second histogram: samples of 400. What percent of those samples would predict (incorrectly) that more than 50% of the voters would vote yes? Explain briefly.

7. Refer to Figure 15.27 on page 661.

a. Write at least three questions that can be asked about either the first or the second histogram. Include at least one question for each of the three graph-reading levels discussed in Section 15.2. Answer each question for the first histogram and for the second histogram (as best you can).

b. Write a "read beyond the data" question about the two histograms together. Answer your question as best you can.

8. Use the NAEP long-term trend data about mathematics performance of 9-year-olds in Table 15.14 to make box plots for the years 1978, 1992, 2004, and 2008. You may assume that in each year, scores ranged from 0 to 400. Use your box plots to compare the mathematics performance of 9-year-olds in these years.

TABLE 15.14 NAEP long-term trend mathematics scores for 9-year-olds at selected percentiles in selected years

Percentile	1978	1992	2004	2008
25th	195	208	220	222
50th	220	231	243	246
75th	244	253	264	266

9. a. Determine the 25th, 50th, and 75th percentiles for the amounts of vitamin C found in two different hypothetical brands of vitamin pills shown in Figure 15.36. For each brand, 40 samples were tested.

b. Make a box plot for the data in Figure 15.36.

c. Suppose you only have the box plot from part (b) and you don't have the dot plots in Figure 15.36. Discuss what you can tell about the data sets.

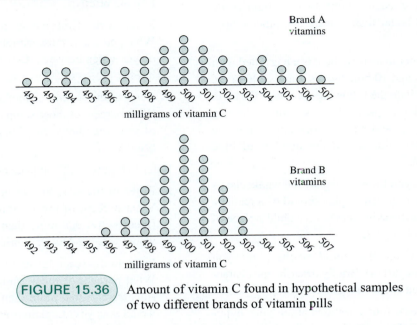

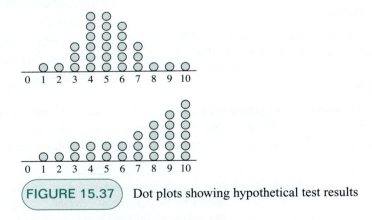

(FIGURE 15.36) Amount of vitamin C found in hypothetical samples of two different brands of vitamin pills

10. a. Determine the 25th, 50th, and 75th percentiles for the hypothetical test scores shown in the dot plots of Figure 15.37. Each dot plot has 28 pieces of data.

b. Make box plots for the dot plots in Figure 15.37.

c. Suppose you only have the box plot from part (b) and you don't have the dot plots in Figure 15.37. Discuss what you can tell about the data sets.

(FIGURE 15.37) Dot plots showing hypothetical test results

11. A fifth-grade class is wondering if girls' names tend to be longer than boys' names. Table 15.15 shows the names of all the fifth-grade girls and boys at the school.

a. Display the lengths of the fifth-grade girls' and boys' names in dot plots. Compare the dot plots.

b. Write and answer at least four questions about your dot plots from part (a), including at least one question at each of the three graph-reading levels discussed in Section 15.2 (Read the data, Read between the data, and Read beyond the data).

c. Display the lengths of the fifth-grade girls' and boys' names in box plots.

d. Compare the box plots from part (c). What do they tell you about the fifth-grade girls' and boys' names?

TABLE 15.14 Fifth-grader names

Girls' Names				
Felecia	Laquiera	Kayla	Sandra	Brittany
Shanique	Kelly	Darica	Crystal	Sasha
Shana	Gloria	Autumn	Katlyn	Shalamar
Shelli	Lauren	Alexis	Keshaundra	Brittney
Brandi	Kendra	Olivia	Sierra	Beth
Jessica	Ebonique	Christiana	Talia	Morgan
Ashley	Kimsey	Jade	Brianna	Amanda
Adebola	Kierra	Candace		

Boys' Names				
Kody	Michael	Tyler	Shane	Jamal
Adam	Dwight	McKinley	Michael	Marcus
Troy	Charles	Ryan	Shalon	Matthew
Mario	Bobby	Brandon	Nicholas	Chandler
Ali	Austin	Samuel	Jeron	Kwame
Everett	Fredrick	Michael	Christopher	Joseph
Tyler	Nikos	Jon	James	Bradley
Orlando	Mark	Russell		

12. A 400-point test is given to a group of students. The mean score is 240. A student scores a 300 on the test.

a. Is it possible that the student's score falls within the middle 50% of the scores on the test? Explain your answer and use an example to illustrate.

b. Is is possible that the student scored higher than 90% of the students taking the test? Explain your answer and use an example to illustrate.

Chapter Summary and Study Items

Section 15.1 Formulating Questions, Designing Investigations, and Gathering Data

A statistical study begins with the formulation of a question that can be answered (at least partially) by the collection and analysis of data. In statistics, there are observational studies and experiments; both involve the consideration of a population. Often it is not practical to study the entire population of interest, so a sample is studied instead. Large enough random samples are generally representative of the full population. Therefore, proportions can be used to make reasonable predictions about a full population on the basis of a random sample.

Key skills and understandings:

- Recognize that some samples may not be representative of a whole population, but that large enough random samples generally are.

- Use random samples to make predictions about a full population with the aid of proportions.

Section 15.2 Displaying Data and Interpreting Data Displays

Common displays of categorical data include real graphs, pictographs (such as bar graphs), and pie graphs. Displays of numerical data include dot plots, histograms, stem-and-leaf plots, line graphs, and scatterplots. All such graphical displays can assist in the interpretation of data. To help your students think about data displays, you can pose three levels of graph-reading questions: read the data, read between the data, and read beyond the data.

Key skills and understandings:

* Make data displays to help convey information about data.

* For a given data display, formulate and answer questions at the three levels of graph reading.

* Recognize erroneous or misleading data displays.

Section 15.3 The Center of Data: Mean, Median, and Mode

The mean (average) and median are good one-number summaries of a collection of numerical data. We can think of the mean in terms of "leveling out" the data. This way of thinking about the mean agrees with the way we calculate the mean. It also provides a way of solving problems about the mean. Another way to think about the mean is as a "balance point." The mean and median can be the same or different for a given set of data.

Key skills and understandings:

* Describe how to view the mean as leveling out, and explain why this way of viewing the mean agrees with the way we calculate the mean (by adding and then dividing).

* Use the leveling out view of the mean to solve problems about the mean. Also use the standard way of calculating the mean to solve problems about the mean.

* View the mean as a balance point and understand that this point of view is especially appropriate for data displayed in histograms and dot plots.

* Create data sets with different means and medians.

* Discuss errors that students commonly make with the mean and the median.

Section 15.4 The Distribution of Data

Data can be distributed in different ways, so two sets of data with the same means and medians can be quite different. Percentiles can help to discern differences in data sets.

Percentiles are generalizations of the median. A student's percentile on a test is not the same as the percent the student got correct on the test. Box plots show several percentiles graphically. Reports about student performance on standardized tests often include percentiles and may also include other measures that are based on rescaling normal distributions.

Key skills and understandings:

* Indicate the shape a data set is likely to take, especially in the case of random samples.

* Given a data set, determine various percentiles.

* Given the value of various percentiles, describe roughly how data are distributed, especially in comparison with other data sets.

* Make box plots. Given box plots, describe roughly how data are distributed, especially in comparison with other data sets.

* Discuss the difference between percentiles and percent correct.

Probability

The study of probability arises naturally in a variety of situations. Games in which we flip coins, roll dice, spin spinners, or pick cards all have an element of chance. We don't know what the spinner will land on, and we don't know how the dice will roll. This element of chance adds excitement to the game and makes it easy to design class activities and problems that present probability in a way that can be fun for students to learn.

If probability were used only in analyzing games of chance, it would not be an important subject. But probability has far wider applications. In business and finance, probability can be used in determining how best to allocate assets. Farmers can use data on the probability of various amounts of rain to determine how best to plant their fields. In medicine, probability can be used to determine how likely it is that a person actually has a certain disease, given the outcomes of test results. Some diagnostic techniques, particularly genetic tests, can be used to determine the likelihood that a person will contract a disease.

The National Council of Teachers of Mathematics has made the following recommendations concerning the learning of statistics and probability (see [64]). Instructional programs should enable all students from prekindergarten through grade 12 to

- formulate questions that can be addressed with data and collect, organize, and display relevant data, to answer them;

- select and use appropriate statistical methods to analyze data;

- develop and evaluate inferences and predictions that are based on data;

- understand and apply basic concepts of probability.

For NCTM's specific statistics and probability recommendations for grades PreK–2, grades 3–5, grades 6–8, and grades 9–12, go to www.pearsonhighered.com/beckmann.

16.1 Basic Principles of Probability

In this section, we focus on some of the basic principles of probability. We begin with some simple contexts in which to introduce basic ideas of probability. We then discuss how to use basic principles of probability to calculate probabilities in simple situations. Finally, we examine the relationship and distinction between theoretical and experimental probability.

In probability, we study experiments or situations where we know which outcomes are possible, but we do not know precisely which outcome will occur at a given time. Examples include flipping a coin and seeing if it lands with a head or a tail facing up, spinning a spinner and seeing which color the arrow lands on, and randomly picking a person out of a group and determining if the person has a certain disease. Given an experiment or a situation in which various outcomes are possible, the **probability**   theoretical **probability** of a given outcome is a number quantifying how likely that outcome is. The probability of a given outcome is the fraction or percentage of times that outcome should occur in the ideal. For example, the probability that a spinner lands on green is the fraction of times that the spinner should land on green in the ideal.

Probabilities are always between 0 and 1, or equivalently, when they are given as percentages, between 0% and 100%. A probability of 0% means that outcome will not occur; a probability of 100% means that outcome is certain to occur; a probability of 50% means the outcome is as likely to occur as not to occur. For example, when we flip a coin, the probability of getting *either* a head *or* a tail is 100% because one or the other is certain to occur. The probability of getting a head is 50%, or $\frac{1}{2}$, because heads and tails are equally likely to occur.

If we have a coin in hand, how do we know for sure that a coin is *fair* (i. e., that heads and tails are equally likely to occur when we flip the coin)? In fact, we never know for sure that the coin is fair; however, unless the coin has been weighted lopsidedly, there is no reason to believe that one of heads or tails should be more likely to occur than the other. If it were important for the coin to be fair—for example, if the coin were to be used in a state lottery—then the coin would be flipped many times, perhaps thousands of times, to determine if the coin is likely to be fair. Statistical methods exist for determining the likelihood that a coin is fair, but we will not discuss these methods.

Principles for Determining Probability

principles of probability   Several key **principles of probability** can be used to calculate probabilities, and we will discuss and apply these principles in the rest of this section. Here are the principles:

1. If two outcomes of an experiment or situation are equally likely, then their probabilities are equal.

2. If there are several outcomes of an experiment or situation that cannot occur simultaneously, then the probability that one of those outcomes will occur is the sum of the probabilities of each of the individual outcomes.

3. If an experiment is performed many times, then the fraction of times that a given outcome occurs is likely to be close to the probability of that outcome occurring. The greater the number of times the experiment is performed, the more likely it is that the fraction of times a given outcome occurs is close to the probability of that outcome occurring.

Probabilities of Equally Likely Outcomes

If an experiment or situation has N distinct possible outcomes, none of which can occur simultaneously, and if all of these outcomes are equally likely, then for each possible outcome, the probability of that outcome is $\frac{1}{N}$. This statement follows from the first two principles of probability, as the following simple case shows.

Suppose you have an ordinary number cube (die) that has dots on each of its six faces. Each face has a different number of dots on it, ranging from 1 dot to 6 dots. You can roll the number cube and count the number of dots on the side that lands face up. Let's assume that the number cube is not weighted, so that each side is equally likely to land face up. Then by principle 1, the probability of rolling a 1 is equal to the probability of rolling a 2, which is equal to the probability of rolling a 3, and so on, up to 6. In other words,

prob. of 1 = prob. of 2 = prob. of 3 = prob. of 4 = prob. of 5 = prob. of 6

Now when you roll a number cube, the probability of rolling either a 1 or 2 or 3 or 4 or 5 or 6 is 1, because *some* face has to land up. And you can't *simultaneously* roll a 2 and a 3 on one number cube, or any other pair of distinct numbers. Therefore, by principle 2,

prob. of 1 + prob. of 2 + prob. of 3 + prob. of 4 + prob. of 5 + prob. of 6 = 1

Because all these probabilities are equal and add up to 1, each probability must be $\frac{1}{6}$. That is,

$$\text{prob. of } 1 = \frac{1}{6}$$

$$\text{prob. of } 2 = \frac{1}{6}$$

$$\text{prob. of } 3 = \frac{1}{6}$$

$$\text{prob. of } 4 = \frac{1}{6}$$

$$\text{prob. of } 5 = \frac{1}{6}$$

$$\text{prob. of } 6 = \frac{1}{6}$$

The Probability That One of Several Possible Outcomes Occurs

What is the probability of rolling an odd number with a number cube? The only way to roll an odd number with a number cube is to roll a 1, 3, or 5. Since you can't roll any two of those numbers simultaneously on one number cube, principle 2 applies and tells you that the probability of rolling a 1, 3, or 5 is the sum of the probabilities of rolling a 1, of rolling a 3, and of rolling a 5. Therefore, the probability of rolling an odd number is

$$\frac{1}{6} + \frac{1}{6} + \frac{1}{6} = \frac{3}{6} = \frac{1}{2}$$

Class Activity   *Now Turn to Class Activities Manual*

16A Probabilities with Spinners, p. 387

16B Some Probability is Misconceptions, p. 388

Experimental Probability

Although the theoretical probability that a given outcome of an experiment will occur is the fraction of times that that outcome occurs in the ideal, when you perform an experiment a number of times, the *actual* fraction of times that the given outcome occurs will usually *not* be equal to the probability of that outcome. When you perform an experiment a number of times, the fraction or percentage of **experimental** times that a given outcome occurs is called the **experimental probability** of that outcome. For **probability** example, if you flipped a coin 100 times and you got 54 heads, then the experimental probability of getting heads was 54%.

Class Activity   *Now Turn to Class Activities Manual*

16C Using Experimental Probability to Make Predictions, p. 389

16D Experimental versus Theoretical Probability: Picking Cubes from a Bag, p. 390

16E If You Flip 10 Pennies, Should Half Come Up Heads?, p. 391

Probability principle 3 says that the experimental probability is likely to be close to the theoretical probability of an outcome of an experiment if the experiment was performed many times. In other words, if you perform an experiment many times, the actual fraction that a given outcome occurs is usually close to the probability of that outcome occurring. This statement is similar to the statement that a random sample of a population is likely to be representative of the full population if the sample is large enough. (See Sections 15.1 and 15.4.)

Because of principle 3, we can use experimental probabilities to estimate theoretical probabilities. For example, suppose there are 10 blocks in a bag. Some of the blocks are red and some are blue, but we don't know how many of each color block are in the bag. Let's say we reach into the bag 50 times, each time removing one block, recording its color, and putting it back into the bag. If we picked 16 red blocks, then the experimental probability of picking a red block was 32%. This 32% is likely to be close to the theoretical probability of picking a red block, which is

$$\frac{\text{(number of red blocks)}}{10}$$

because in the ideal, out of every 10 blocks we pick, (number of red blocks) of the blocks should be red. If there are 3 red blocks in the bag, the probability of picking red is 30%, which is close to 32%. Our best guess for the number of red blocks in the bag is therefore 3. Even though we based our guess on the available evidence, our guess could turn out to be wrong because of the variability that is inherent in random choices.

When we perform an experiment a number of times to determine an experimental probability, the greater the number of times we perform the experiment, the more likely it is that the experimental probability is close to the theoretical probability. For example, suppose that the probability of winning a certain game is 25%. Imagine that you could play the game 100 times; imagine that you could play the game 1000 times. If you play the game 100 times, which is a fairly large number of times, you should expect to win close to 25% of the games. But if you play the game 1000 times, it's even more likely that the percent of games you win will be close to 25%. We can see these statements reflected in the dot plots in Figure 16.1, which were obtained by repeatedly playing the game 100 times and then repeatedly playing the game 1000 times. Each dot at the top of Figure 16.1 represents 100 games; the dot is plotted at the percentage of 100 games that were won. Each dot at the bottom of Figure 16.1 represents 1000 games; the dot is plotted at the percentage of the 1000 games that were won (rounded to the nearest whole number). Notice that in both dot plots in Figure 16.1, the dots cluster around 25%, but that more of the dots in the lower dot plot are close to 25% than in the top dot plot. Therefore, when 1000 games are played, the percentage of games won is more likely to be close to 25% than when 100 games are played.

FIGURE 16.1

Percent of games
won in 100 sets
of 100 games
and in 100 sets
of 1000 games,
where the
probability of
winning each
game was 25%

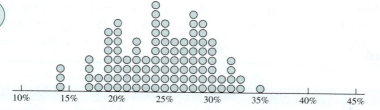

Each dot represents 100 games, each of which had probability 25% of being won.
The dot is plotted at the percentage of the 100 games that were won.

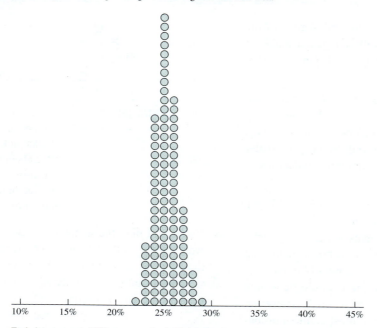

Each dot represents 1000 games, each of which had probability 25% of being won.
The dot is plotted at the percentage of the 1000 games that were won.

Practice Exercises for Section 16.1

1. Determine the probability of spinning either a red or a yellow on the spinner shown in Figure 16.2. Explain briefly.

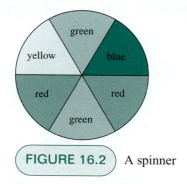

FIGURE 16.2   A spinner

2. A family math night at school features the following game. There are two opaque bags, each containing red blocks and yellow blocks. Bag 1 contains 3 red blocks and 5 yellow blocks. Bag 2 contains 4 red blocks and 9 yellow blocks. To play the game, you pick a bag and then you pick a block out of the bag without looking. You win a prize if you pick a red block. Tom says he is more likely to pick a red block out of bag 2 than bag 1 because bag 2 contains 1 more red block than bag 1. Is this correct?

3. At a math center in a class, there is a bag filled with 30 red blocks and 20 blue blocks. Each child in the class of 25 will complete the following activity at the math center: Pick a block out of the bag without looking, record the block's color, and put the block back into the bag. Each child will do this 10 times in a row. Then the child will write the number of blue blocks picked on a sticky note. Describe a

good way to display the data of the whole class. Show roughly what you expect the display you suggest to look like, and say why.

4. The probability of winning a game is $\frac{17}{100}$. Does this mean that if you play the game 100 times you will win 17 times? If not, what does it mean?

Answers to Practice Exercises for Section 16.1

1. The spinner is equally likely to land in each of the 6 sections. Since red and yellow together make up 3 of the 6 sections, the probability of spinning red or yellow is $\frac{3}{6} = \frac{1}{2}$.

2. No, the probability of picking a red block out of bag 1 is $\frac{3}{8}$ because 3 of 8 blocks are red, whereas the probability of picking a red block out of bag 2 is $\frac{4}{13}$ because 4 of 13 blocks are red. Since $\frac{3}{8}$ is greater than $\frac{4}{13}$, the probability of picking a red block is higher for bag 1 than for bag 2.

3. A dot plot like the one in Figure 16.3 would be a good way to display these data. The class could make this dot plot on the chalkboard, using sticky notes instead of dots. Since $\frac{2}{5}$ of the blocks in the bag are blue, the probability of picking a blue block is $\frac{2}{5} = 40\%$. So, in the ideal, there would be 4 blue blocks chosen in 10 picks of a block. However, since the choices are random, most children probably won't get exactly 4 blue blocks, but most will probably get somewhere around 4 blocks. A few children may even get 8, 9, or 10 blue blocks in their 10 picks. So the dot plot will probably look something like the one in Figure 16.3, where most dots cluster around 4, but some dots are not close to 4.

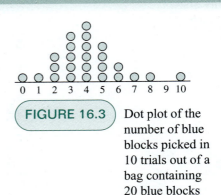

FIGURE 16.3 Dot plot of the number of blue blocks picked in 10 trials out of a bag containing 20 blue blocks and 30 red blocks

4. No, if you play the game 100 times you probably won't win exactly 17 times. The fraction of times that you do win is the experimental probability, which is usually not exactly equal to the theoretical probability. The theoretical probability of winning, $\frac{17}{100}$, is the number of times you would win "in the ideal" if you played 100 times. If you play the game many times, the fraction of times you win will usually be close to this theoretical probability and will usually get closer and closer to this theoretical probability the more games you play.

Problems for Section 16.1

1. Some games have spinners. When the arrow in a spinner is spun, it can land in any one of several different colored regions. Determine the probability of spinning each of the following on a spinner like the one in Figure 16.4:

   a. Red

   b. Either red or green

   c. Either red or yellow

   Explain your answers, referring to ideas discussed in this section.

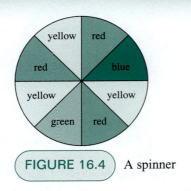

FIGURE 16.4 A spinner

2. a. Draw a spinner such that the probability of landing on red is $\frac{1}{3}$, the probability of landing on green is $\frac{1}{4}$, and the only other color that the spinner could land on is yellow.

b. For your spinner in part (a), what is the probability of landing on yellow? Explain.

c. Now draw a different spinner that has the same probabilities as in part (a). Explain briefly.

3. a. Draw a 4-color spinner (red, green, yellow, blue) such that

- landing on green is two times as likely as landing on red;

- landing on yellow is two times as likely as landing on green;

- landing on blue is more likely than landing on red.

b. Determine the probabilities of landing on each of the colors on your spinner in part (a). Briefly explain your reasoning.

c. Draw another spinner that satisfies the conditions in part (a) and is different from your spinner in part (a).

4. Write a paragraph discussing the following:

a. Michael says that if you flip a coin 100 times, it will have to come up heads 50 times out of the 100. Is Michael correct?

b. If the probability that a certain outcome of an experiment will occur is 30%, does that mean that when you carry out the experiment 100 times, the outcome will occur 30 of those times?

5. A family math night at school features the following game. There are two opaque bags, each containing red blocks and yellow blocks. Bag 1 contains 3 red blocks and 5 yellow blocks. Bag 2 contains 5 red blocks and 15 yellow blocks. To play the game, you pick a bag and then you pick a block out of the bag without looking. You win a prize if you pick a red block. Kate thinks she should pick from bag 2 because it contains more red blocks. Is Kate more likely to pick a red block if she picks from bag 2 rather than bag 1? Explain why or why not.

6. There are 50 small balls in a tub. Some balls are white and some are orange. Without being able to see into the tub, each student in a class of 25 is allowed to pick a ball out of the tub at random. The color of the ball is recorded and the ball is put back into the tub. At the end, 7 orange balls and 18 white balls were picked. What is the best estimate you can give for the number of orange balls and the number of white balls in the tub? Describe how to calculate this best estimate, and explain why your method of calculation makes sense in a way that a fifth-grader might understand. Is your best estimate necessarily accurate? Why or why not?

7. In a classroom, there are 100 plastic fish in a tub. The tub is hidden from the students' view. Some fish are green and some are yellow. The students know that there are 100 fish, but they don't know how many of each color there are. The students go fishing, each time picking a random fish from the tub, recording its color, and throwing the fish back in the tub. At the end of the day, 65 fish have been chosen, 12 green and 53 yellow. What is the best estimate you can give for the number of green fish and the number of yellow fish in the tub? Describe how to calculate this best estimate, and explain why your method of calculation makes sense in a way that a fifth-grader might understand. Is your best estimate necessarily accurate? Why or why not?

8. There is a bag filled with 4 red blocks and 16 yellow blocks. Each child in a class of 25 will pick a block out of the bag without looking, record the block's color, and put the block back into the bag. Each child will repeat this for a total of 10 times. Then the child will write on a sticky note the number of red blocks picked. Describe a good way to display the whole class's data. Show roughly what you expect the display you suggest to look like, and say why.

9. Write several paragraphs in which you describe and discuss some of the misconceptions that students often have about probability and that you learned about from the Class Activities and Practice Exercises in this section.

16.2 Counting the Number of Outcomes

In the previous section, we saw that if an experiment has equally likely outcomes, then we can determine the probability that one or several of those outcomes occurs. In simple cases, probabilities are easy to determine; think of a spinner that is equally likely to land in its sections or a number cube for which each face is equally likely to land up. But what if we want to determine the probability that several experiments in a row each have a certain outcome? For example, what is the probability that a certain spinner lands on red, and then a number cube lands with 6 up? To calculate such a probability, we need to first determine how many different, equally likely outcomes there are.

In this section, we study some methods for calculating numbers of outcomes. Our main tools will be multiplication and the ways of displaying multiplicative structure that we studied in Section 4.1. We can use these ways of thinking about multiplication and showing multiplicative structure to calculate numbers of outcomes in a variety of situations, including situations commonly encountered in the study of probability. Next we will apply our techniques for determining numbers of outcomes to calculate probabilities.

multi-stage experiments
Note that we'll be working with two-stage and, more generally, with multi-stage experiments. **Multi-stage experiments** consist of performing several experiments in a row, such as spinning a spinner and then rolling a number cube (a two-stage experiment) or flipping a coin three times in a row (a three-stage experiment).

Two-Stage Experiments and Ordered Pair Problems

Section 4.1 described how ordered pair problems are one type of problem that can be solved by multiplication. For example, if there are 3 pants and 4 shirts, how many outfits consisting of pants and a shirt can be made? Each outfit can be viewed as an ordered pair (pants, shirt). Each pants can be paired with 4 shirts, so each pants contributes a group of 4 outfits. There are 3 pants, so there are 3 groups of 4 outfits that can be made. According to the meaning of multiplication, this is 3×4 outfits, which is 12 outfits.

Probability problems about two-stage experiments usually involve ordered pair problems. For example, suppose we first spin a spinner that is equally likely to land on 1 of 4 colors (red, blue, yellow, or green) and then we roll a number cube for which each of the 6 faces labeled with 1 through 6 dots is equally likely to land face up. The problem of determining how many different possible outcomes there are to this two-stage experiment is an ordered pair problem. Each outcome consists of a pair (color, number of dots). For each color that can be spun, there are 6 possible outcomes for the number cube, so each color contributes 6 ordered pairs. Since there are 4 colors, there are 4 groups of 6 ordered pairs altogether, which is 4×6 ordered pairs according to the meaning of multiplication. Therefore, there are 24 possible outcomes for the two-stage experiment of spinning the spinner and rolling the number cube. We can use a tree diagram or an organized list of the ordered pairs to display the outcomes visually and to show their multiplicative structure, as in Figure 16.5.

Multi-Stage Experiments and Related Counting Problems

In some situations we want to determine the number of possible outcomes when multiple stages are involved. Try the next Class Activity before you read on.

Class Activity *Now Turn to Class Activities Manual*

16F How Many Keys Are There?, p. 391

FIGURE 16.5

Outcomes of a
two-stage
experiment

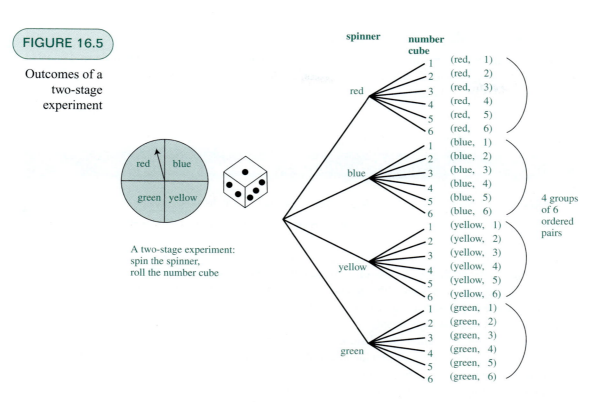

A two-stage experiment:
spin the spinner,
roll the number cube

In the same way that we can use multiplication to determine a total number of ordered pairs, we can use multiplication to determine a total number of ordered triples or ordered 4-tuples, and so on, which is useful for determining the total number of outcomes in multi-stage experiments.

For example, suppose you have 3 coins: a penny, a nickel, and a dime. How many possible outcomes are there if you flip all 3 coins? Think of each outcome as an ordered triple, consisting of an outcome on the penny, an outcome on the nickel, and an outcome on the dime. As with ordered pairs, ordered triples can be displayed in organized lists and tree diagrams (as in Figure 16.6), and they have a multiplicative structure.

For each of the two outcomes for the penny, the nickel has two possible outcomes (because the outcome of flipping the nickel does not depend on the outcome of flipping the penny), so there are

$$2 \times 2$$

possible outcomes for the penny and nickel. For each of the 2×2 possible outcomes for the penny and nickel, there are two possible outcomes for the dime (because the outcome of flipping the dime does not depend on the outcome of flipping the penny and nickel). So all together, there are

$$2 \times 2 \times 2 = 8$$

total possible outcomes when we flip the 3 coins.

FIGURE 16.6

A tree diagram
showing all and
organized list
possible
outcomes from
flipping 3 coins

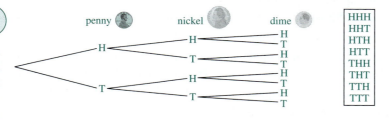

Counting Problems That Involve Dependent Outcomes

In some counting problems and multi-stage experiments, what happens at one stage may depend on what happened at previous stages—and this will influence the number of possible outcomes. Try the next Class Activity before you read on.

Class Activity *Now Turn to Class Activities Manual*

16G Counting Outcomes: Independent versus Dependent, p. 392

Suppose that every student in a class of 25 writes his or her name on a slip of paper and puts the slip in a bag. The teacher mixes the slips up and picks a name out of the bag. This student will be class president for the day. The teacher reaches into the bag again and picks another name out of the bag. This student will be class vice president for the day. How many possible outcomes are there? There are 25 possible outcomes for class president. But once the president has been chosen, only 24 additional possibilities remain for the vice-president. So altogether, each of the 25 possible presidents can be paired with 24 possible vice presidents. This activity involves a situation of 25 groups of 24 pairs, each pair consisting of a president and vice president. Therefore, there are $25 \times 24 = 600$ possible different outcomes when a class president and vice president are picked as described. Even though there are 25 different students who could become vice-president, we multiply 25×24 rather than 25×25 because the choice of vice-president depends on the choice of president.

Practice Exercises for Section 16.2

1. How many different 3-digit numbers can you write using only the digits 1, 2, and 3 if you do not repeat any digits (so that 121 and 332 are not counted)? Show how to solve this problem with an organized list and with a tree diagram. Explain why this problem can be solved by multiplying.

2. How many different 3-digit numbers can be made by using only the digits 1, 2, and 3, where repeated digits are allowed (so that 121 and 332 are counted)? Show how to solve this problem with an organized list and with a tree diagram. Explain why this problem can be solved by multiplying.

3. How many different keys can be made if there are 10 places along the key that will be notched and if each notch will be 1 of 8 depths?

4. Annette buys a wardrobe of 3 skirts, 3 pants, 5 shirts, and 3 sweaters, all of which are coordinated so that she can mix and match them any way she likes. How many different outfits can Annette create from this wardrobe? (Every day Annette wears either a skirt or pants, a shirt, and a sweater.)

5. A delicatessen offers 4 types of bread, 20 types of meats, 15 types of cheese, a choice of mustard, mayonnaise, both or neither, and a choice of lettuce, tomato, both or neither for their sandwiches. How many different types of sandwiches can the deli make with 1 meat and 1 cheese? Should the deli's advertisement read "hundreds of sandwiches to choose from," or would it be better to substitute thousands or even millions for hundreds?

6. A pizza parlor offers 10 toppings to choose from. How many different large pizzas are there with exactly 2 different toppings? (For example, you could order pepperoni and mushroom, but not double pepperoni.)

Answers to Practice Exercises for Section 16.2

1. Figure 16.7 shows a tree diagram and an organized list displaying all such 3-digit numbers. Both show a structure of 3 big groups, 1 group for each possibility for the first digit. Once a first digit has been chosen, there are 2 choices for the second digit (since repeats aren't allowed). This means there are 3 groups of 2 possibilities for the first 2 digits. So by the meaning of multiplication, there are $3 \times 2 = 6$ possibilities for the first 2 digits. Once the first 2 digits are chosen, there is only 1 choice for the third digit—it must be the remaining unused digit. So there are $3 \times 2 = 6$ three-digit numbers using only the digits 1, 2, 3, with no digit repeated.

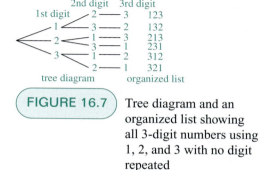

FIGURE 16.7 Tree diagram and an organized list showing all 3-digit numbers using 1, 2, and 3 with no digit repeated

2. The tree diagram and the organized list in Figure 16.8 both show 3 big groups. Each of those 3 big groups has 3 groups of 3. Therefore, according to the meaning of multiplication, there are

$$3 \times (3 \times 3) = 27$$

3-digit numbers that use only the digits 1, 2, and 3.

3. Figure 16.9 shows part of a tree diagram for all such keys. There are 8 primary branches corresponding to the 8 choices for the first notch. For each of those 8 choices on the first notch, there are 8 choices for the second notch. Therefore, there are 8 groups of 8 choices for the first 2 notches, and so, by the meaning of multiplication, there are 8×8 choices for the first 2 notches. For each of the 8×8 choices for the first 2 notches, there are 8 choices for the third notch. Therefore, there are $8 \times 8 \times 8$ choices for the first 3 notches. And so on. When all 10 notches are taken into account, there are $8 \times 8 \times 8 \times 8 \times 8 \times 8 \times 8 \times 8 \times 8 \times 8 = 8^{10} = 1,073,741,824$ that can be made, which is more than a billion keys!

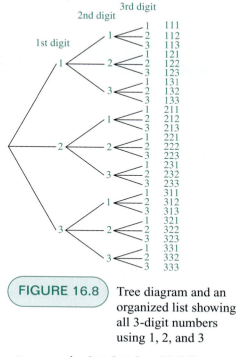

FIGURE 16.8 Tree diagram and an organized list showing all 3-digit numbers using 1, 2, and 3

4. Annette can make $6 \times 5 \times 3 = 90$ different outfits.

5. There are $4 \times 20 \times 15 \times 4 \times 4 = 19,200$ different sandwiches that can be made with a meat and a cheese. The deli should substitute thousands for hundreds.

6. For each of the 10 choices for the first topping, there are 9 choices for the second topping (there are only 9 because repeats are not allowed). This forms 10 groups of 9, so there are 10×9 choices

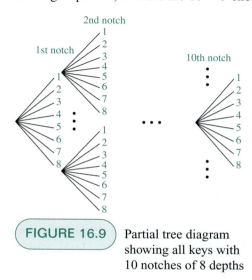

FIGURE 16.9 Partial tree diagram showing all keys with 10 notches of 8 depths

of double toppings. However, when they are counted this way, pepperoni and mushrooms counts as different from mushrooms and pepperoni.

Therefore, we must divide by 2 so as not to double-count pizzas. So there are actually only $\frac{10 \times 9}{2} = 45$ different large pizzas with 2 different toppings.

Problems for Section 16.2

1. A bakery makes 4 different kinds of cake. Each cake can have 3 different kinds of frosting. Each frosted cake can be decorated in 2 different ways. How many ways are there of ordering a decorated, frosted cake?
   Show how to solve the problem by using an organized list and a tree diagram. Explain why you can solve the problem by multiplying.

2. Allie and Betty want to know how many 3-letter combinations, such as BMW or DDT, are possible. (Letters are allowed to repeat, as in DDT or BOB.) Allie thinks there can be $26 + 26 + 26$ three-letter combinations, whereas Betty thinks the number is $26 \times 26 \times 26$. Which girl, if either, is right, and why? Explain your answers clearly and thoroughly, drawing on the meaning of multiplication.

3. Explain your answers to the following:

   a. How many 9-digit numbers are there that use only the digits 1, 2, 3, . . . , 8, 9? (Repetitions are allowed, so, for example, 123211114 is allowed.)

   b. How many 9-digit numbers are there that use each of the digits 1, 2, 3, . . . , 8, 9 exactly once?

4. In all 3 parts in this problem, explain your solution clearly.

   a. How many whole numbers are there that have exactly 10 digits and that can be written by using only the digits 8 and 9?

   b. How many whole numbers are there that have at most 10 digits and that can be written by using only the digits 0 and 1? (Is this situation related to part (a)?)

   c. How many whole numbers are there that have exactly 10 digits and that can be written by using only the digits 0 and 1?

5. Most Georgia car license plates currently use the format of 3 numbers followed by 3 letters (such as 123 ABC). How many different license plates can be made this way? Explain your solution clearly.

6. a. A 40-member club will elect a president and then elect a vice-president. How many possible outcomes are there?

   b. A 40-member club will elect a pair of copresidents. How many possible outcomes are there?

   c. Are the answers to parts (a) and (b) the same or different? Explain why they are the same or why they are different.

7. A dance club has 10 women and 10 men. In each of the following parts, give a clear explanation.

   a. How many ways are there to choose one woman and one man to demonstrate a dance step?

   b. How many ways are there to pair each woman with a man (so that all 10 women and all 10 men have a dance partner at the same time)?

8. *A pizza parlor problem.* How many different large pizzas with no double toppings can be made? (For example, mushroom, pepperoni, and sausage is one possibility, and so is mushroom and sausage, but not double mushroom and sausage. Your count should also include the case of no toppings.) You can't answer the question yet because you don't know how many toppings the pizza parlor offers.

   a. Determine the number of different pizzas when there are exactly 3 toppings to choose from (e.g., mushroom, pepperoni, and sausage).

   b. Determine the number of different pizzas when there are exactly 4 toppings to choose from.

   c. Determine the number of different pizzas when there are exactly 5 toppings to choose from.

   d. Look for a pattern in your answers in parts (a), (b), and (c). Based on the pattern you see, predict the number of different pizzas when there are 10 toppings to choose from.

   e. Now find a different way to determine the number of different pizzas when there are 10 toppings to choose from. This time, think about the situation in the following way: Pepperoni can be either on or off, mushrooms can be either on or off, sausage can be either on or off, and so on, for all 10 toppings. Explain clearly how to use this idea to answer the question and why this method is valid.

9. ↻ A pizza parlor offers 10 different toppings to choose from. How many different large pizzas can be made if double toppings, but not triple toppings or more, are allowed? For example, double pepperoni and (single) mushroom is one possibility, as is double pepperoni and double mushroom, but triple pepperoni is not allowed. Notice that you can choose each topping in one of three ways: off, as a single topping, or as a double topping.

16.3 Calculating Probabilities in Multi-Stage Experiments

Now we will apply the counting techniques we studied in the previous section to solve probability problems involving multi-stage experiments. Before we do so though, we consider the distinction between independent and dependent outcomes in multi-stage experiments. We conclude this section with a discussion of expected value—a useful application of probability.

Independent and Dependent Outcomes

An important factor in calculating probabilities for multi-stage experiments is the distinction between independent and dependent outcomes at a stage. If you flipped a coin 10 times in a row and all 10 flips came up heads, would you think that your next flip is more likely to be a tail because a tail is "due"? Would you think that you were on a "hot streak" and so are more likely to get another head on your next coin flip? In fact, on your next flip you are just as likely to get a tail as a head. There is nothing in the flipping history **independent** of the coin that can influence the next flip. The outcome of one coin flip is **independent** of the outcome of any other coin flip. In other words, the outcome of one coin flip has no bearing on the outcome of any other coin flip.

Suppose you have a bag filled with 9 blue marbles and 1 green marble. If you randomly pick a marble out of the bag, record its color, and put it back in the bag, then your next random marble pick is independent of your first marble pick. On the other hand, if you leave the first marble out of the bag after selecting it, then you *do* influence the next marble pick. If the first marble you selected was the green one, then your second marble must be blue. If the first marble you selected was blue, then on your second marble pick, the probability of picking the green marble is $\frac{1}{9}$ instead of $\frac{1}{10}$ because now there are only 9 marbles in the bag. When the first marble is left out of the bag, the second marble pick is **dependent** **dependent** on the first.

Calculating Probabilities

Let's consider the case of flipping 3 coins: a penny, a nickel, and a dime. What is the probability that all 3 will land heads up when flipped? In Section 16.2 we saw that there are

$$2 \times 2 \times 2 = 8$$

possible outcomes when 3 coins are flipped, and all these outcomes are equally likely. In only 1 of those outcomes do all 3 coins land heads up. Therefore, the probability of all 3 coins landing heads up is $\frac{1}{8}$, or 12.5%.

We can use the organized list and tree diagram of Figure 16.6 to calculate other probabilities as well. For example, what is the probability of getting 2 heads and 1 tail when we toss the 3 coins? Two heads and one tail occur in 3 of the 8 possible outcomes, namely, HHT, HTH, and THH. So the probability that either 1 of these 3 outcomes will occur is

$$\frac{1}{8} + \frac{1}{8} + \frac{1}{8} = \frac{3}{8}$$

or 37.5%.

FIGURE 16.10

A spinner for a
children's game

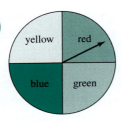

In some games that use spinners, such as the one in Figure 16.10, the spinner is equally likely to land in red, yellow, blue, or green. What is the probability that in 2 spins, the spinner will land first on red and then on blue? We could make an organized list or a tree diagram to show all possible outcomes. We can also make an array to show all possible outcomes, as in Figure 16.11. Because the outcome of the second spin is independent of the outcome of the first spin, the array shows 4 rows of 4 possible outcomes, so there are $4 \times 4 = 16$ possible outcomes for the 2 spins. Each of the 16 spins is equally likely, so the probability of spinning a red and then a blue is $\frac{1}{16} = 6.25\%$.

Class Activity   *Now Turn to Class Activities Manual*

16H   Number Cube Rolling Game, p. 393

16I   Picking Two Marbles from a Bag of 1 Black and 3 Red Marbles, p. 394

Be careful about using an organized list, a tree diagram, or an array when working with outcomes that are not equally likely. For example, suppose there are 3 marbles in a bag, 1 red and 2 green. If we randomly pick a marble from the bag, record its color, put it back in the bag, and then randomly pick another marble, can we use the tree diagram in Figure 16.12 to calculate the probability of picking a green marble followed by the red marble? No, we cannot use this tree diagram, because even though there are only 2 possible outcomes when picking a marble—picking red and picking green—these 2 outcomes are not equally likely. We are more likely to pick a green marble than the red one. If we think of the 2 green marbles as labeled "green 1" and "green 2," then we can use the tree diagram in Figure 16.13. The 9 outcomes in Figure 16.13 are equally likely. Two of these outcomes are a green marble followed by the red marble. Therefore, the probability of picking green followed by red is $\frac{2}{9} = 22\%$.

We can use organized lists, tree diagrams, and arrays even when working with dependent outcomes. For example, suppose there are 3 marbles in a bag, 1 red and 2 green. If we randomly pick a marble from the bag, record its color, and *without* returning the marble to the bag, randomly pick another marble, what is the probability of picking a green marble followed by a red marble? There are 6 equally likely outcomes, as shown in Figure 16.14. Two of these outcomes are a green marble followed by the red marble. Therefore, the probability of picking green followed by red is $\frac{2}{6} = \frac{1}{3} = 33\%$.

FIGURE 16.11

Array showing
all possible
outcomes on
2 spins of a
spinner

Spin 2

		R	G	B	Y
	R	RR	RG	RB	RY
Spin 1	G	GR	GG	GB	GY
	B	BR	BG	BB	BY
	Y	YR	YG	YB	YY

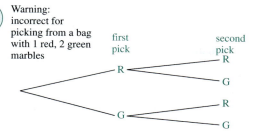

FIGURE 16.12

An incorrect tree diagram for picking 2 marbles from a bag with 1 red and 2 green marbles (if the first marble is replaced after picking)

Warning: incorrect for picking from a bag with 1 red, 2 green marbles

---

Class Activity *Now Turn to Class Activities Manual*

16J More Probability Misconceptions, p. 396

Expected Value

Expected value is an important application of probability because it allows a variety of businesses to make educated decisions about how to allocate money. For an experiment that has numerical outcomes, the **expected value** of the experiment is the average outcome of the experiment in the ideal, over the long term.

expected value

Class Activity *Now Turn to Class Activities Manual*

16K Expected Earnings from the Fall Festival, p. 396

Suppose a farmer is considering planting a crop. Let's say that the probability the crop is successful is 75%. If the crop is successful, the crop will earn the farmer a net amount of $100,000, but if the crop is not successful, then the farmer will lose the $20,000 he invested in seed and supplies. The two outcomes of this experiment are $100,000 and −$20,000. Although one or the other of these outcomes will occur, if we think about the farmer planting the crop *repeatedly over the long term*, we can ask what amount the farmer should expect to earn on average, in the ideal. This amount is the expected value.

Think about the farmer planting the crop over and over, many times. In the ideal (according to the given probabilities), the crop will be successful 75% of those times and will fail in the remaining 25% of

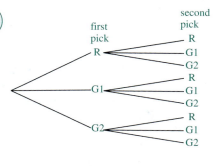

FIGURE 16.13

A correct tree diagram for picking 2 marbles from a bag with 1 red and 2 green marbles (if the first marble is replaced after picking)

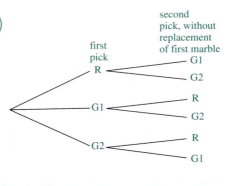

FIGURE 16.14

A tree diagram for picking 2 marbles from a bag with 1 red and 2 green marbles (if the first marble is *not* replaced after picking)

the times. Let's say that the farmer could plant the crop 100 times. Then of those 100 times, in 75 of them, the farmer would earn $100,000, but in 25 of them the farmer would loose $20,000. Altogether, over those 100 times, the farmer would earn a total of

$$75 \cdot \$100,000 - 25 \cdot \$20,000$$

If we divide by 100, we will have the average amount the farmer should expect to earn per crop in the long run, namely

$$\frac{75}{100} \cdot \$100,000 - \frac{25}{100} \cdot \$20,000 = \$70,000$$

So on average, over the long run, the farmer should expect to earn $70,000 from planting the crop. Notice that we found this average amount by multiplying the probability of each outcome by the outcome and then adding over all outcomes. This is how to calculate expected value in general.

Practice Exercises for Section 16.3

1. Consider the experiment of rolling a number cube 2 times.

   a. Determine the probability of getting a 1 on both rolls of the number cube. Explain your answer.

   b. Determine the probability of getting a 1 on either the first or the second roll of the number cube. Explain your answer.

   c. Determine the probability of getting either a 1 or a 2 on both rolls of the number cube.

2. What is the probability of getting 4 heads in a row on 4 tosses of a coin?

3. Suppose you have 3 marbles in a bag, 1 red and 2 green. If you reach into the bag without looking and randomly pick out 2 marbles at once, what is the probability that both of the marbles you pick will be green?

4. A school's fall festival includes the following game. There are 3 enclosed tubs, each containing red, yellow, and blue plastic bears. To play the game, a contestant reaches once into each tub and randomly picks out a bear. The contestant must reach through a sleeve attached to the tub, so that the color of the bear being chosen cannot be seen. If all 3 bears are red, the contestant wins a prize. The first tub contains 5 red bears, 1 yellow bear, and 1 blue bear. The second tub contains 3 red bears, 2 yellow bears, and 1 blue bear. The third tube contains 1 red bear, 2 yellow bears, and 2 blue bears. What is the probability of winning the prize?

5. Continue Practice Exercise 4: The school is expecting 350 people to play this game. Each contestant pays 50 cents to play. Each prize costs the school 75 cents. How much money (net) should the school expect to earn from this game? Explain.

Answers to Practice Exercises for Section 16.3

1. **a.** Since the 2 rolls of the number cube are independent, the following array shows all possible outcomes:

(1,1) (1,2) (1,3) (1,4) (1,5) (1,6)
(2,1) (2,2) (2,3) (2,4) (2,5) (2,6)
(3,1) (3,2) (3,3) (3,4) (3,5) (3,6)
(4,1) (4,2) (4,3) (4,4) (4,5) (4,6)
(5,1) (5,2) (5,3) (5,4) (5,5) (5,6)
(6,1) (6,2) (6,3) (6,4) (6,5) (6,6)

Here, (4,3) stands for a 4 on the first roll and a 3 on the second roll. There are $6 \times 6 = 36$ possible outcomes, all of which are equally likely. So, by principles 1 and 2, the probability of getting a 1 on both rolls is $\frac{1}{36}$, or 2.8%.

b. Referring to the array in part (a), there are 11 ways to get a 1 on either the first roll or the second roll, or both (the first row together with the first column). Therefore, the probability of getting at least 1 on 2 rolls of a number cube is $\frac{11}{36}$, or 30.6% (so you'd expect to get at least 1 a little less than one-third of the time).

c. Referring to the array in part (a), there are 4 ways to get either a 1 or a 2 on both rolls, namely, (1,1), (1,2), (2,1), and (2,2). Therefore, the probability of getting either a 1 or a 2 each time on 2 rolls of a number cube is $\frac{4}{36} = \frac{1}{9}$, or about 11%.

2. The tree diagram in Figure 16.15 shows all possible outcomes of the 4 coin tosses, since the coin tosses are independent. There are $2 \times 2 \times 2 \times 2 = 16$ possible outcomes, all of which are equally likely. Only one of those outcomes produces 4 heads in a row. So the probability of getting 4 heads in a row on 4 tosses of a coin is $\frac{1}{16}$, or 6.25%.

3. We can use the tree diagram in Figure 16.14 to solve this problem by thinking of one of the marbles as the first one chosen and the other as the second one chosen (where the first marble is not replaced before the second one is chosen). The tree diagram shows that there are 6 equally likely outcomes. Two of those outcomes correspond to both green marbles being chosen. So the probability of picking the 2 green marbles is $\frac{2}{6} = \frac{1}{3} = 33\%$.

Another way to solve the problem is to create the following organized list, which shows all possible ways of picking a pair of marbles from the bag:

(G1, G2)   (G1, R)
(G2, R)

(Notice that (G2, G1) would be the same as (G1, G2). Results are similar for other cases.) These 3 possible ways of picking 2 marbles are all equally likely. Out of the 3 ways of picking 2 marbles, there is 1 way to pick 2 green marbles. Therefore, the probability of picking 2 green marbles is $\frac{1}{3}$.

4. From the (partial) tree diagram in Figure 16.16, we see that for each of the 7 bears that could be picked from the first tub, there are 6 bears that could be

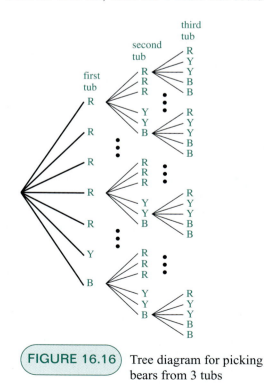

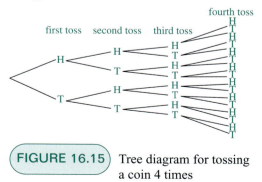

FIGURE 16.15 Tree diagram for tossing a coin 4 times

FIGURE 16.16 Tree diagram for picking bears from 3 tubs

picked from the second tub, and for each of these $7 \cdot 6$ possibilities, there are 5 bears that could be chosen from the third tub. Therefore, there are $7 \cdot 6 \cdot 5$ equally likely possible outcomes for the game. How many of these outcomes consist of picking a winning combination of 3 red bears? The tree diagram shows that for each of the 5 red bears that could be chosen from the first tub, there are 3 red bears that could be chosen from the second tub, and for each of these $5 \cdot 3$ possibilities, there is only 1 red bear that could be chosen from the third tub. So there are $5 \cdot 3 \cdot 1$ outcomes in which all three bears are red. Since there are $7 \cdot 6 \cdot 5$ equally likely outcomes and since $5 \cdot 3 \cdot 1$ of those outcomes result in winning the game, the probability of winning the game is

$$\frac{5 \cdot 3 \cdot 1}{7 \cdot 6 \cdot 5} = \frac{1}{14}$$

5. If 350 people play the game, then in the ideal, $\frac{1}{14}$ of the people should win the game. So in the ideal,

$$\frac{1}{14} \cdot 350 = 25$$

people should win. The 25 prizes the school would give out to the winners would cost

$$25 \cdot \$0.75 = \$18.75$$

But the school will collect

$$350 \cdot \$0.50 = \$175$$

if 350 people play the game. So net, the school can expect to earn

$$\$175 - \$18.75 = \$156.25$$

from this game in the ideal case. Of course, in actuality, the school might earn more or less than this amount since more or fewer people might win the game and since more or fewer than 350 people might play the game.

Problems for Section 16.3

1. A children's game has a spinner that is equally likely to land on any 1 of 4 colors: red, blue, yellow, or green. Determine the probability of spinning a red both times on 2 spins. Explain your reasoning.

2. A children's game has a spinner that is equally likely to land on any 1 of 4 colors: red, blue, yellow, or green. Determine the probability of spinning a red at least once on 2 spins. Explain your reasoning.

3. A children's game has a spinner that is equally likely to land on any 1 of 4 colors: red, blue, yellow, or green. Determine the probability of spinning a red 3 times in a row on 3 spins. Explain your reasoning.

4. ![icon] A children's game has a spinner that is equally likely to land on any 1 of 4 colors: red, blue, yellow, or green. Determine the probability of spinning either a red followed by a yellow or a yellow followed by a red in 2 spins. Explain your reasoning.

5. A children's game has a spinner that is equally likely to land on any 1 of 4 colors: red, blue, yellow, or green. Determine the probability of *not* spinning a red on either of 2 spins. Explain your reasoning.

6. Determine the probability of spinning a blue followed by a green in 2 spins on the spinner in Figure 16.17. Explain your reasoning.

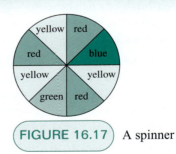

FIGURE 16.17   A spinner

7. Determine the probability of spinning a blue followed by a red in 2 spins on the spinner in Figure 16.17. Explain your reasoning.

8. Determine the probability of spinning a red followed by a yellow in 2 spins on the spinner in Figure 16.17. Explain your reasoning.

9. Suppose you have a penny, a nickel, a dime, and a quarter in a bag. You shake the bag and dump out the coins. What is the probability that exactly 2 coins will land heads up and 2 coins will land heads down? Is the probability 50% or not? Solve this problem by drawing a tree diagram that shows all the possible outcomes when you dump the coins out. Explain.

10. You have a bag containing 2 yellow and 3 blue blocks.

a. You reach in and randomly pick a block, record its color, and put the block back in the bag. You then reach in again and randomly pick another block. What is the probability of picking 2 yellow blocks? Explain your answer.

b. You reach in and randomly pick a block, record its color, and without replacing the first block, you reach in again and randomly pick another block. What is the probability of picking 2 yellow blocks? Explain your answer.

11. There are 3 plastic bears in a bag. The teacher tells Bob that there are 3 bears in the bag, but she doesn't tell Bob what color the bears are. Bob picked a bear out of the bag 3 times and each time he got a red bear. (He puts the bear he picked back in the bag before picking the next bear).

   a. Bob says that the 3 bears in the bag must all be red. Is Bob necessarily correct?

   b. Now suppose that of the 3 bears in the bag, 2 are red and 1 is blue. Calculate the probability of picking 3 red bears in 3 picks.

12. There are 4 black marbles and 5 red marbles in a bag. If you reach in and randomly select 2 marbles, what is the probability that both are red? Explain your reasoning clearly.

13. Suppose you have 100 light bulbs and one of them is defective. If you pick out 2 light bulbs at random (either both at the same time, or first one, then another, without replacing the first light bulb), what is the probability that one of your chosen light bulbs is defective? Explain your answer.

14. *A game at a fund-raiser*: There are 20 rubber ducks floating in a pool. One of the ducks has a mark on the bottom, indicating that the contestant wins a prize. Each contestant pays 25 cents to play. A contestant randomly picks 2 of the ducks. If the contestant picks the duck with the mark on the bottom, the contestant wins a prize that costs $1.

   a. What is the probability that a contestant will win a prize? Explain.

   b. If 200 people play the game, about how many people would you expect to win? Why?

   c. Based on your answer to part (b), how much money should the duck game be expected to earn for the fund-raiser (net) if 200 people play it? Explain.

15. You are making up a game for a fund-raiser. You take 5 table-tennis balls, number them from 1 to 5, and put all 5 in a brown bag. Contestants will pick the 5 balls out of the bag one at a time, without looking, and line the balls up in the order they were picked. If the contestant picks 1, 2, 3, 4, 5, in that order, then the contestant wins a prize of $2. Otherwise, the contestant wins nothing. Each contestant pays $1 to play.

   a. What is the probability that a contestant will win the prize? Explain your reasoning.

   b. If 240 people play your game, then approximately how many people would you expect to win the prize? Why?

   c. Based on your answer to part (b), about how much money would you expect your game to earn (net) for the fund-raiser if 240 people play? Explain.

16. a. A waitress is serving 5 people at a table. She has the 5 dishes they ordered (all 5 are different), but she can't remember who gets what. How many different possible ways are there for her to give the 5 dishes to the 5 people? Explain your answer.

   b. Based on part (a), if the waitress just hands the dishes out randomly, what is the probability that she will hand the dishes out correctly?

17. Social Security numbers have 9 digits and are presented in the form

   $$xxx - xx - xxxx$$

   where each x can be any digit from 0 to 9.

   a. How many different Social Security numbers can be made this way? Explain.

   b. If the population of the United States is 300 million people, and if everybody living in the United States has a Social Security number, then what fraction of the possible Social Security numbers are in current use? Explain your answer.

18. Standard dice are shaped like cubes. They have 6 faces, each of which is a square. Because there are 6 faces on a standard die and because each face is equally likely to land face up, the probability that a given face will land up is $\frac{1}{6}$. Could you make other three-dimensional "dice" so that for each face, the probability of that face landing up is $\frac{1}{4}$? $\frac{1}{8}$? $\frac{1}{10}$? $\frac{1}{12}$? Other fractions? Explain your answers. *Hint*: see Section 13.1.

16.4 Using Fraction Arithmetic to Calculate Probabilities

The probability of an outcome of an experiment can be calculated by determining the ideal fraction of times the given outcome should occur if the experiment were performed many times. When we apply this point of view, we will see how to use fraction arithmetic to calculate probabilities.

Using Fraction Multiplication to Calculate Probabilities

Class Activity *Now Turn to Class Activities Manual*

16L   Using the Meaning of Fraction Multiplication to Calculate a Probability, p. 397

We can use basic principles of probability and the meaning of fraction multiplication to calculate certain probabilities. In order to do so, we will rely on the following way of thinking about the outcome of an experiment: Consider what would happen ideally if the experiment were repeated a very large number of times. Calculate directly the ideal fraction of times the outcome you are interested in occurs. This fraction is the probability of that outcome.

For example, if we roll a number cube (die) 2 times, what is the probability that we will get two 1s in a row? Think about doing the experiment of rolling a number cube 2 times in a row many times over. The numbers 1, 2, 3, 4, 5, and 6 are all equally likely to come up on one roll of a number cube; therefore, in the ideal, the first roll will be a 1 in $\frac{1}{6}$ of the rolls. Now consider those $\frac{1}{6}$ of the experiments when the first roll is a 1. Of those double rolls in which the first roll is a 1, in the ideal, the second roll will be a 1 in $\frac{1}{6}$ of the time, too, because the outcomes of the first and second rolls are independent. Therefore, in the ideal, in $\frac{1}{6}$ of $\frac{1}{6}$ of all the experiments, both rolls will be a 1. According to the meaning of multiplication, "$\frac{1}{6}$ of $\frac{1}{6}$ of all the experiments" is

$$\frac{1}{6} \cdot \frac{1}{6} = \frac{1}{36}$$

of all the experiments. Therefore, in the ideal, in $\frac{1}{36}$ of all the experiments, you will roll two 1s, so the probability of rolling two 1s in a row on 2 rolls of a number cube is $\frac{1}{36}$.

We can also see the "$\frac{1}{6}$ of $\frac{1}{6}$ of all the experiments" pictorially in Figure 16.18. The large rectangle represents "all the experiments." The lightly shaded region represents $\frac{1}{6}$ of the experiments. The darkly shaded region represents $\frac{1}{6}$ of the lightly shaded region, or $\frac{1}{6}$ of $\frac{1}{6}$ of all the experiments. In this case, this probability can also be calculated with an array or a tree diagram.

FIGURE 16.18

$\frac{1}{6}$ of $\frac{1}{6}$ of the experiments is $\frac{1}{6} \cdot \frac{1}{6} = \frac{1}{36}$ of the experiments

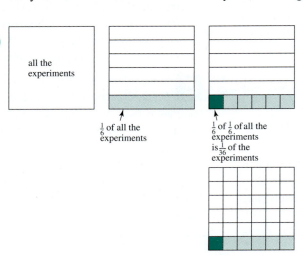

Using Fraction Multiplication and Addition to Calculate Probabilities

> **Class Activity** *Now Turn to Class Activities Manual*
>
> **16M** Using Fraction Multiplication and Addition to Calculate a Probability, p. 398

Besides using fraction multiplication, we can use a combination of multiplication and addition to calculate probabilities. Again, think about performing an experiment a large number of times. View the probability of a given outcome occurring as the fraction of times the outcome should occur in the ideal.

For example, consider the game that consists of choosing a rubber duck from among 5 rubber ducks floating in a duck pond and then choosing a rubber frog from among 4 frogs in a frog pond. One of the ducks and 3 of the frogs are marked with Xs on the bottom, as indicated in Figure 16.19. The other 4 ducks and the other frog are marked with Os on the bottom. To win the game, both the duck and the frog that are chosen should have the same marking on the bottom (either both X or both O). The markings cannot be seen when the ducks and frogs are chosen; all the ducks look alike and all the frogs look alike. What is the probability of winning this game?

Think about playing this game of picking a duck and a frog many times over. Sometimes both animals chosen will be marked with Xs; sometimes both will be Os; sometimes one will be marked with an X and the other with an O. Each of these outcomes occurs a certain fraction of the time, and none of these different outcomes can occur at the same time. So the fraction of times that the game will be won is the fraction of times that both animals have an X plus the fraction of times both animals have an O.

So let's first find the fraction of times when both the duck and the frog that are chosen should have Xs, in the ideal. Since 1 out of the 5 ducks is marked with an X, the duck with an X will be chosen $\frac{1}{5}$ of the time, in the ideal. Of those times when the duck with an X is chosen, the fraction of times that the frog chosen has an X is $\frac{3}{4}$, in the ideal, since 3 out of the 4 frogs have Xs. So in the ideal, in $\frac{3}{4}$ of $\frac{1}{5}$ of the times, both the duck and the frog that are chosen will have an X. Since "$\frac{3}{4}$ of $\frac{1}{5}$" is

$$\frac{3}{4} \cdot \frac{1}{5}$$

we can find the probability that both the duck and the frog that are chosen have an X by multiplying $\frac{3}{4}$ and $\frac{1}{5}$. We can also see this probability represented pictorially on the left in Figure 16.20.

What is the probability that both animals chosen will be marked with an O? In the ideal, the duck that is chosen will be marked with an O $\frac{4}{5}$ of the time, since 4 out of the 5 ducks are marked with an O. Of those times when a duck with an O is chosen, the frog that is chosen should be marked with an O $\frac{1}{4}$ of the time, in the ideal, since 1 out of the 4 frogs has an O. So in the ideal, in $\frac{1}{4}$ of $\frac{4}{5}$ of the times, both the duck and the frog will have an O. Since "$\frac{1}{4}$ of $\frac{4}{5}$" is

$$\frac{1}{4} \cdot \frac{4}{5}$$

this product is the probability that both the duck and the frog that are chosen have an O. We can also see this probability represented pictorially on the right in Figure 16.20.

FIGURE 16.19

A game: pick a duck and a frog

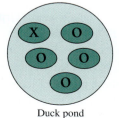

Duck pond

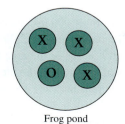

Frog pond

FIGURE 16.20

Probability that both the duck and the frog chosen are marked the same

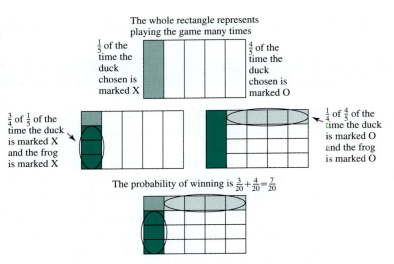

Finally, the probability of picking both a duck and a frog that are marked the same is the sum of the 2 probabilities just calculated, namely,

$$\frac{3}{4} \cdot \frac{1}{5} + \frac{1}{4} \cdot \frac{4}{5} = \frac{7}{20}$$

Therefore, the probability of winning the game is $\frac{7}{20}$, or 35%.

The Case of Spot, the Drug-Sniffing Dog

Now let's look at a more advanced probability calculation. As in the previous example, we first consider the outcome of a large number of experiments in the ideal. We then divide the number of times the outcome we are interested in occurs by the total number of experiments to calculate the probability that the outcome we are interested in occurs.

Spot is a dog who has been trained to sniff luggage to detect drugs. But Spot isn't perfect. His trainers conducted careful experiments and found that in luggage that *doesn't* contain any drugs, Spot will nevertheless bark to indicate the presence of drugs for 1% of such luggage. In luggage that *does* contain drugs, Spot will bark to indicate he thinks drugs are present for about 97% of this luggage. (So he fails to recognize the presence of drugs in about 3% of luggage containing drugs.)

Think of the consequences of putting Spot in a busy airport where police estimate that 1 in 10,000 pieces of luggage (or 0.01%) passing through that airport contain drugs. Here's a question to answer before placing Spot in a busy airport:

> If Spot barks to indicate that a piece of luggage contains drugs, what is the probability that the luggage actually does contain drugs? (Remember, Spot is not perfect; sometimes he barks when there are no drugs.)

Before you read on, guess what this probability is. You might be surprised at the actual answer, which we will now calculate. It is not 97%.

Imagine that we made Spot sniff a very large number of pieces of luggage—let's say, 1,000,000 pieces of luggage. For these 1,000,000 pieces of luggage, we will first determine in the ideal how many times Spot will bark, indicating that he thinks they contain drugs. Then we will determine how many of these pieces of luggage actually do contain drugs.

According to the police estimates, 0.01% of the 1,000,000 pieces of luggage at this airport, or about 100 pieces of luggage, will contain drugs. The remaining 999,900 will not contain drugs. According to the information we have about Spot, of the 100 pieces of luggage that contain drugs, Spot will bark recognizing the presence of drugs in 97 pieces of luggage, in the ideal. In the ideal, for the 999,900 pieces of luggage that don't contain drugs, Spot will still bark for 1% of these pieces of luggage, which is 9999 pieces of luggage. So, all together, Spot will bark indicating that he thinks drugs are present for

$$9999 + 97 = 10,096$$

pieces of luggage, in the ideal. But of those 10,096 pieces of luggage, only 97 actually contain drugs (because the other 9999 of the 10,096 were ones where Spot barked, but they didn't actually contain any drugs). So, of the 10,096 times that Spot barks, indicating he thinks there are drugs, only 97 of those times are drugs *actually present*. Now,

$$\frac{97}{10,096} = 0.0096\ldots = 0.96\%$$

so that, of the times when Spot barks, indicating he think drugs are present, only 0.96% of the time are drugs actually present. So if Spot barks to indicate he *thinks* drugs are present, the probability that drugs actually *are* present is 0.96%, which is *less than one percent*. Even though Spot's statistics sounded pretty good at the start, we might think twice about putting him on the job, since many innocent people could be delayed at the airport.

Practice Exercises for Section 16.4

1. What is the probability of spinning a yellow followed by a blue in 2 spins on the spinner in Figure 16.21? Explain how to solve this problem with fraction multiplication, and explain why this method makes sense.

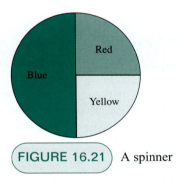

FIGURE 16.21   A spinner

2. At a doughnut factory, 3 of the 200 boxes of doughnuts are bad. Each bad box contains only 10 doughnuts (instead of the usual 12), and in each bad box, 3 of the 10 doughnuts are tainted. (No doughnuts in any other boxes are tainted.) If a person buys a random box of these doughnuts and eats a random doughnut from the box, what is the probability that the person will eat a tainted

doughnut? Explain how to solve this problem with fraction multiplication, and explain why the method makes sense.

3. You are on a game show where you can pick 1 of 3 doors. A prize is placed randomly behind 1 of the 3 doors; the other doors do not have a prize behind them. You pick a door. The host opens 1 of the other doors and it does not have the prize behind it. The host then offers you the opportunity to switch to the other unopened door before revealing where the prize is.

   a. What is probability that you will win the prize if you stick with your door?

   b. What is the probability that you will win the prize if you switch to the other unopened door? Solve this problem by imagining that you could play the game a large number of times; determine what fraction of the time you would win if you switched to the other unopened door each time you played.

   c. If you want to win the prize, what strategy is better: sticking with the first door you chose or switching to the other unopened door? Explain.

Answers to Practice Exercises for Section 16.4

1. Imagine spinning the spinner twice in a row many times. In the ideal, in $\frac{1}{4}$ of the times, the first spin will land on yellow. In the ideal, in $\frac{1}{2}$ of those times when the first spin is yellow, the second spin will be blue, since the second spin does not depend on the

first spin. Therefore, in the ideal, in $\frac{1}{2}$ of $\frac{1}{4}$ of the times, the spinner will land on yellow followed by blue. Since $\frac{1}{2}$ of $\frac{1}{4}$ is

$$\frac{1}{2} \cdot \frac{1}{4}$$

the probability of spinning yellow followed by blue is

$$\frac{1}{2} \cdot \frac{1}{4} = \frac{1}{8}$$

2. Imagine that the person could buy a random box of doughnuts and eat a random doughnut from the box many times. Then in the ideal, in $\frac{3}{200}$ of the times, the person will get a bad box. Of the $\frac{3}{200}$ of the times the person gets a bad box, the person will pick a tainted doughnut $\frac{3}{10}$ of the time, in the ideal. Therefore, in $\frac{3}{10}$ of $\frac{3}{200}$ of the times, the person will eat a tainted doughnut from a bad box. Since $\frac{3}{10}$ of $\frac{3}{200}$ is

$$\frac{3}{10} \cdot \frac{3}{200}$$

the probability of eating a tainted doughnut is

$$\frac{3}{10} \cdot \frac{3}{200} = \frac{3 \cdot 3}{10 \cdot 200} = \frac{9}{2000} = 0.45\%$$

which is less than half a percent.

3. **a.** Since the prize is placed randomly behind 1 of the 3 doors, your probability of winning the prize is $\frac{1}{3}$ if you stick with the door you picked originally.

b. If you played the game many times, then in the ideal, in $\frac{1}{3}$ of those times, the prize would be behind the first door you pick. So in these cases, when you switch, you will not get the prize. But in the $\frac{2}{3}$ of the times when the door you picked does not have the prize behind it, the other unopened door must have the prize behind it. So when you switch doors, you will get the prize. Therefore, when you switch doors, you will win the prize $\frac{2}{3}$ of the times, in the ideal.

c. The switching strategy is better because when you switch, your probability of winning is $\frac{2}{3}$, whereas when you don't switch, your probability of winning is only $\frac{1}{3}$.

Problems for Section 16.4

1. A children's game has a spinner that is equally likely to land on any 1 of 4 colors: red, blue, yellow, or green. What is the probability of spinning a red followed by a green in 2 spins? Explain how to solve this problem with fraction multiplication, and explain why this method makes sense.

2. Suppose you flip a coin and roll a number cube (die). What is the probability of getting a head and rolling a 6? (The numbers 1 through 6 are all equally likely to occur.) Explain how to solve this problem with fraction multiplication, and explain why this method makes sense.

3. Use fraction arithmetic to solve problem 1 on page 690 of Section 16.3. Explain why you can use fraction arithmetic to solve this problem.

4. Use fraction arithmetic to solve problem 3 on page 690 of Section 16.3. Explain why you can use fraction arithmetic to solve this problem.

5. Use fraction arithmetic to solve problem 6 on page 690 of Section 16.3. Explain why you can use fraction arithmetic to solve this problem.

6. Use fraction arithmetic to solve problem 7 on page 690 of Section 16.3. Explain why you can use fraction arithmetic to solve this problem.

7. Use fraction arithmetic to solve problem 8 on page 690 of Section 16.3. Explain why you can use fraction arithmetic to solve this problem.

8. There are 3 boxes. One of the 3 boxes contains 2 envelopes, and the other 2 boxes contain 3 envelopes each. In the box with the 2 envelopes, 1 of the 2 envelopes contains a prize. No other envelope contains a prize. If you pick a random box and a random envelope from that box, what is the probability that you will win the prize? Explain how to solve this problem with fraction multiplication, and explain why this method makes sense.

9. A game consists of spinning a spinner and then rolling a number cube. The spinner is equally likely to land on any 1 of the 4 colors red, yellow, green, or blue. The number cube is equally likely to land with any of the 6 sides labeled 1 through 6 up. To win the game, a contestant must either spin red and roll any number or spin a color other than red and roll a 6. In other words, the contestant must

either spin red or roll a 6 to win. What is the probability of winning this game? Explain why you can use fraction arithmetic to solve this problem.

10. A game consists of spinning a spinner and then rolling a number cube. The spinner is equally likely to land on any 1 of the 4 colors red, yellow, green, or blue. The number cube is equally likely to land with any of the 6 sides labeled 1 through 6 up. To win the game, a contestant must either spin red and roll a 1 or spin green and roll a 6. What is the probability of winning this game? Explain why you can use fraction arithmetic to solve this problem.

11. A game consists of spinning a spinner and then rolling a number cube. The spinner is equally likely to land on any 1 of the 4 colors red, yellow, green, or blue. The number cube is equally likely to land with any of the 6 sides labeled 1 through 6 up. To win the game, a contestant must either spin a red or a yellow and roll a 1 or a 2 or spin a green or blue and roll a 5 or a 6. What is the probability of winning this game? Explain why you can use fraction arithmetic to solve this problem.

12. Use fraction arithmetic to solve problem 2 on page 690 of Section 16.3. Explain why you can use fraction arithmetic to solve this problem.

13. Use fraction arithmetic to solve problem 4 on page 690 of Section 16.3. Explain why you can use fraction arithmetic to solve this problem.

14. Use fraction arithmetic to solve problem 5 on page 690 of Section 16.3. Explain why you can use fraction arithmetic to solve this problem.

15. Suppose you have 2 boxes, 50 black pearls and 50 white pearls. You can mix the pearls up any way you like and put them all back into the 2 boxes. You don't have to put the same number of pearls in each box, but each box must have at least 1 pearl in it. You will then be blindfolded, and you will get to open a random box and pick a random pearl out of it.

   a. Suppose you put 25 black pearls in box 1 and 25 black pearls and 50 white pearls in box 2. Calculate the probability of picking a black pearl by imagining that you were going to pick a random box and a random pearl in the box a large number of times.

   In the ideal, what fraction of the time would you pick box 1? What fraction of those times that you picked box 1 would you pick a black pearl from box 1, in the ideal?

On the other hand, in the ideal, what fraction of the time would you pick box 2? What fraction of those times that you picked box 2 would you pick a black pearl from box 2, in the ideal?

Overall, what fraction of the time would you pick a black pearl, in the ideal?

   b. Describe at least 2 other ways to distribute the pearls than what is described in part (a). Find the probability of picking a black pearl in each case.

   c. Try to find a way to arrange the pearls so that the probability of picking a black pearl is as large as possible.

16. Due to its high population, China has a stringent policy on having children. In rural China, couples are allowed to have either 1 or 2 children according to the following rule: If their first child is a boy, they are not allowed to have any more children. If their first child is a girl, then they are allowed to have a second child. Let's assume that all couples follow this policy and have as many children as the policy allows.

   a. Do you think this policy will result in more boys, more girls, or about the same number of boys as girls being born? (Answer without performing any calculations—just make a guess.)

   b. Now consider a random group of 100,000 rural Chinese couples who will have children. Assume that any time a couple has a child the probability of having a boy is $\frac{1}{2}$.

   i. In the ideal, how many couples will have a boy as their first child, and how many will have a girl as their first child?

   ii. In the ideal, of those couples who have a girl first, how many will have a boy as their second child and how many will have a girl as their second child?

   iii. Therefore, overall, what fraction of the children born in rural China under the given policy should be boys and what fraction should be girls?

17. The Pretty Flower Company starts plants from seed and sells the seedlings to nurseries. They know from experience that about 60% of the calla lily seeds they plant will sprout and become a seedling. Each calla lily seed costs 20 cents, and a pot containing at least one sprouted calla lily seedling can be sold for $2.00.

Pots that don't contain a sprouted seedling must be thrown out. The company figures that costs for a pot, potting soil, water, fertilizer, fungicide and labor are $0.30 per pot (whether or not a seed in the pot sprouts). The Pretty Flower Company is debating between planting 1 or 2 seeds per pot. Help them figure out which choice will be more profitable by working through the following problems:

a. Suppose the Pretty Flower Company plants 1 calla lily seed in each of 100 pots. Using the previous information, approximately how much profit should the Pretty Flower Company expect to make on these 100 pots? Profit is income minus expenses.

b. Now suppose that the Pretty Flower Company plants 2 calla lily seeds (1 on the left, 1 on the right) in each of 100 pots. Assume that whether or not the left seed sprouts has no influence on whether or not the right seed sprouts. So, the right seed will still sprout in about 60% of the pots in which the left seed does not sprout. Explain why the Pretty Flower Company should expect about 84 of the 100 pots to sprout at least 1 seed.

c. Using part (b), determine how much profit the Pretty Flower Company should expect to make on 100 pots if 2 calla lily seeds are planted per pot. Compare your answer with part (a). Which is expected to be more profitable: 1 seed or 2 seeds per pot?

d. What if calla lily seeds cost 50 cents each instead of 20 cents each (but everything else stays the same)? Now which is expected to be more profitable: 1 seed or 2 seeds per pot?

18. Suppose that in a survey of a large, random group of people, 17% were found to be smokers and 83% were not smokers. Suppose further that 0.08% of the smokers and 0.01% of the nonsmokers from the group died of lung cancer. (Be careful: Notice the decimal points in these percentages—they are not 8% and 1%.)

a. What percent of the group died of lung cancer?

b. Of the people in the group who died of lung cancer, what percent were smokers? (It may help you to make up a number of people for the large group and to calculate the number of people who died of lung cancer and the number

of people who were smokers and also died of lung cancer.)

19. Suppose that 1% of the population has a certain disease. Also suppose that there is a test for the disease, but it is not completely accurate: It has a 2% rate of false positives and a 1% rate of false negatives. This means that the test *reports* that 2% of the people who don't have the disease do have it, and the test *reports* that 1% of the people who do have the disease don't have it. This problem is about the following question:

If a person tests positive for the disease, what is the probability that he or she actually has the disease?

Before you start answering the next set of questions, go back and read the beginning of the problem again. Notice that there is difference between *actually having* the disease and *testing positive* for the disease, and there is difference between *not having* the disease and *testing negative* for it.

a. What do you think the answer to the previous question is? (Answer without performing any calculations—just make a guess.)

b. For parts (b) through (f), suppose there is a random group of 10,000 people and all of them are tested for the disease.

Of the 10,000 people, in the ideal, how many would you expect to actually have the disease and how many would you expect not to have the disease?

c. Continuing part (b), of the people from the group of 10,000 who do not have the disease, how many would you expect to test positive, in the ideal? (Give a number, not a percentage.)

d. Continuing part (b), of the people from the group of 10,000 who have the disease, how many would you expect to test positive? (Give a number, not a percentage.)

e. Using your work in parts (c) and (d), of the people from the group of 10,000, how many in total would you expect to test positive, in the ideal?

f. Using your previous work, what percent of the people who test positive for the disease should actually have the disease, in the ideal? How does this compare to your guess in part (a)? Are you surprised at the actual answer?

Chapter Summary and Study Items

Section 16.1 Basic Principles of Probability

The probability that a given outcome of an experiment will occur is the fraction of times that outcome should occur "in the ideal." Several principles of probability are key: (1) if two outcomes of an experiment are equally likely, then their probabilities are equal; (2) if there are several outcomes of an experiment that cannot occur simultaneously, then the probability that one of those outcomes will occur is the sum of the probabilities of each individual outcome; and (3) if an experiment is performed many times, the fraction of times that a given outcome occurs is likely to be close to the probability of that outcome occurring. Using these principles, we can determine probabilities of many simple experiments. When we perform an experiment a number of times, the fraction of times that a given outcome occurred is the experimental probability of that outcome occurring.

Key skills and understandings:

- Use principles of probability to determine probabilities in simple cases.

- Recognize that an experimental probability is likely to be close to the theoretical probability when the experiment has been performed many times.

- Apply experimental probability to make estimates.

Section 16.2 Counting the Number of Outcomes

We can often use multiplication to calculate the number of outcomes in two-stage and multi-stage experiments and other situations.

Key skills and understandings:

- Apply multiplication to count the total number of outcomes in various situations, including those of multi-stage experiments and cases where the outcome at one stage depends on the outcome at previous stages.

Section 16.3 Calculating Probabilities in Multi-Stage Experiments

We can apply the counting techniques of Section 16.2 to calculate probabilities for multi-stage experiments. A useful application of probability is the calculation of expected value.

Key skills and understandings:

- Apply the counting techniques of Section 16.2 to calculate probabilities for multi-stage experiments, including cases where the outcome at one stage depends on the outcome at previous stages.
- Calculate an expected amount of earnings when probabilities are involved.

Section 16.4 Using Fraction Arithmetic to Calculate Probabilities

One way to determine the probability of an outcome of an experiment is to think about performing the experiment many times. The probability of that outcome occurring is the fraction of times that the outcome should occur in the ideal. By applying this point of view, we can see how to calculate probabilities with fraction arithmetic.

Key skills and understandings:

- Explain why certain probabilities can be calculated by the multiplication of fractions.
- Use fraction arithmetic to determine certain probabilities, and explain why the method of calculation makes sense.

Photo Credits

Bibliography

[1] Edwin Abbott. *Flatland*. Princeton University Press, 1991.

[2] Alan Agresti and Christine Franklin. *Statistics, the Art and Science of Learning from Data*. Pearson Prentice Hall, 2007.

[3] American Statistical Association. A Curriculum Framework for PreK–12 Statistics Education. Available at *http://www.amstat.org/education/gaise/* and *http://www.amstat.org/education/gaise/ GAISEPreK-12.htm*

[4] W. S. Anglin. *Mathematics: A Concise History and Philosophy*. Springer-Verlag, 1994.

[5] Deborah Loewenberg Ball. Prospective elementary and secondary teachers' understanding of division. *Journal for Research in Mathematics Education*, 21(2):132–144, 1990.

[6] P. Barnes-Svarney, *New York Public Library Science Desk Reference*. Stonesong Press, 1995.

[7] Tom Bassarear. *Mathematics for Elementary School Teachers*. Houghton Mifflin, 1997.

[8] Peter Beckmann. *A History of Pi*. St. Martin's Press, 1971.

[9] Sybilla Beckmann, *Focus in Grade 5: Teaching with Curriculum Focal Points*. National Council of Teachers of Mathemtics, 2009.

[10] E. T. Bell. *The Development of Mathematics*. Dover, 1992. Originally published in 1945.

[11] George W. Bright. Helping elementary- and middle-grades preservice teachers understand and develop mathematical reasoning. In *Developing Mathematical Reasoning in Grades K–12*, pages 256–269. National Council of Teachers of Mathematics, 1999.

[12] Bureau of Labor Statistics, U.S. Department of Labor. Core subjects and your career. *Occupational Outlook Quarterly*, pages 26–40, Summer 1999. Available online at *http://stats.bls.gov/opub/ooq/ooqhome.htm*

[13] Bureau of Labor Statistics, U.S. Department of Labor. More education: Higher earnings, lower unemployment. *Occupational Outlook Quarterly*, page 40, Fall 1999. Available online at *http://www.bls.gov/ opub/ooq/ooqhome.htm*

[14] California State Board of Education. *Mathematics Framework for California Public Schools*, 1999.

[15] The Carnegie Library of Pittsburgh Science and Technology Department. *Science and Technology Desk Reference*, 1993.

[16] T. P. Carpenter, E. Fennema, M. L. Franke, S. B. Empson, and L. W. Levi. *Children's Mathematics: Cognitively Guided Instruction*. Heinemann, Portsmouth, NH, 1999.

[17] Suzanne H. Chapin, Catherine O'Connor, and Nancy Canavan Anderson. *Classroom Discussions, Using Math Talk to Help Students Learn*. Math Solutions Publications, 2003.

[18] Conference Board of the Mathematical Sciences (CBMS). *The Mathematical Education of Teachers*-Volume 11 of *CBMS Issues in Mathematics Education*. The American Mathematical Society and the Mathematical Association of America, 2001. See also *http://www.cbmsweb.org/MET_Document/ index.htm*

[19] Frances Curcio. *Developing Data-Graph Comprehension in Grades K–8*. National Council of Teachers of Mathematics, 2nd ed., 2001.

[20] Stanislas Dehaene. *The Number Sense*. Oxford University Press, 1997.

[21] Demi. *One Grain of Rice*. Scholastic, 1997.

[22] Carol S. Dweck. *Mindset, The New Psychology of Success*. Random House, Inc., 2006.

[23] Tatiana Ehrenfest-Afanassjewa. *Uebungensammlung zu einer Geometrischen Propaedeuse*. Martinus Nijhoff, 1931.

[24] Encyclopaedia Britannica *Encyclopaedia Britannica*. Encyclopaedia Britannica, Inc., 1994. Available at *www.britannica.com*

[25] Euclid. *The Thirteen Books of the Elements, Translated with Commentary by Sir Thomas Heath*. Dover, 1956.

[26] David Freedman, Robert Pisani, and Roger Purves. *Statistics*. W. W. Norton and Company, 1978.

[27] K. C. Fuson. Developing mathematical power in whole number operations. In J. Kilpatrick, W. G. Martin, and D. Schifter, eds. *A Research Companion to Principles and Standards for School Mathematics*. National Council of Teachers of Mathematics, 2003.

[28] K. C. Fuson. *Math Expressions*. Houghton Mifflin Company, 2006.

[29] K. C. Fuson, S. T. Smith, and A. M. Lo Cicero. Supporting first graders' ten-structured thinking in urban classrooms. *Journal for Research in Mathematics Education*, vol. 28, Issue 6, pages 738–766, 1997.

[30] Georgia Department of Education. Georgia Performance Standards, 2005. Available at *http://www. georgiastandards.org*

[31] Georgia Department of Education. *Georgia Quality Core Curriculum*.

[32] Gersten, R., Beckmann, S., Clarke, B., Foegen, A., Marsh, L., Star, J. R., & Witzel, B. *Assisting students struggling with mathematics: Response to Intervention (RtI) for elementary and middle schools* (NCEE 2009-4060), 2009.

[33] Alexander Givental. The Pythagorean theorem: What is it about? *The American Mathematical Monthly*, vol. 113, number 3, March 2006.

[34] Sir Thomas Heath. *Aristarchus of Samos*. Oxford University Press, 1959.

[35] N. Herscovics and L. Linchevski. A cognitive gap between arithmetic and algebra. *Educational Studies in Mathematics*, 27:59–78, 1994.

[36] Heisuke Hironaka and Yoshishige Sugiyama, ed., *Mathematics for Elementary School*. Tokyo Shoseki Co., Ltd. 2006 Available at *www.globaledresources.com*, 2006.

[37] Infoplease. *http://www.infoplease.com*

[38] H. G. Jerrard and D. B. McNeill. *Dictionary of Scientific Units*. Chapman and Hall, 1963.

[39] Graham Jones and Carol Thornton. *Data, Chance, and Probability, Grades 1–3 Activity Book*. Learning Resources, 1992.

[40] Constance Kamii, Barbara A. Lewis, and Sally Jones Livingston. Primary arithmetic: Children inventing their own procedures. *Arithmetic Teacher*, 41:200–203, 1993.

[41] Felix Klein. *Elementary Mathematics from an Advanced Standpoint*. Dover, 1945. Originally published in 1908.

[42] Morris Kline. *Mathematics and the Physical World*. Dover, 1981.

[43] Kunihiko Kodaira, ed. *Japanese Grade 7 Mathematics*. UCSMPTextbook Translations. The University of Chicago School Mathematics Project, 1992.

[44] Kunihiko Kodaira, ed. *Japanese Grade 8 Mathematics*. UCSMPTextbook Translations. The University of Chicago School Mathematics Project, 1992.

[45] Susan J. Lamon. *Teaching Fractions and Ratios for Understanding*. Lawrence Erlbaum Associates, 1999.

[46] Glenda Lappan, James T. Fey, William M. Fitzgerald, Susan N. Friel, and Elizabeth Difanis Phillips. *Connected Mathematics 2*. Pearson Prentice Hall, 2006.

[47] Aisling M. Leavy, Susan N. Friel, and James D. Marner. *It's a Fird! Can You Compute a Median of Categorical Data?* Vol. 14, 6, February 2009.

[48] Liora Linchevski and Drora Livneh. Structure sense: The relationship between algebraic and numerical contexts. *Educational Studies in Mathematics*, 40(2):173–196, 1999.

[49] Liping Ma. *Knowing and Teaching Elementary Mathematics*. Lawrence Erlbaum Associates, 1999.

[50] Edward Manfre, James Moser, Joanne Lobato, and Lorna Morrow. *Heath Mathematics Connections*. D. C. Heath and Company, 1994.

[51] Mathematical Association of America (MAA) Committee on the Mathematical Education of Teachers. *A Call For Change: Recommendations for the Mathematical Preparation of Teachers of Mathematics*. The Mathematical Association of America, 1991.

[52] Francis H. Moffitt and John D. Bossler. *Surveying*. Addison-Wesley, 1998.

[53] Joan Moss and Robbie Case. Developing children's understanding of the rational numbers: A new model and an experimental curriculum. *Journal for Research in Mathematics Education*, 30(2): 122–147, 1999.

[54] Gary Musser and William F. Burger. *Mathematics for Elementary Teachers*, 4th ed., Prentice Hall, 1997.

[55] National Center for Education Evaluation and Regional Assistance, Institute of Education Sciences, U.S. Department of Education. Retrieved from *http://ies.ed.gov/ncee/wwc/publications/practiceguides/*, Washington, DC.

[56] National Center for Education Statistics. Findings from Education and the Economy: An indicators Report. Technical Report NCES 97-939, National Center for Education Statistics, 1997.

[57] National Center for Education Statistics. *Highlights from the Third International Mathematics and Science Study*. Technical Report NCES 2001-027, National Center for Education Statistics, 2000. Available online at *http://nces.ed.gov/timss/*

[58] National Council of Teachers of Mathematics. *Curriculum and Evaluation Standards for School Mathematics*. National Council of Teachers of Mathematics, 1989.

[59] National Council of Teachers of Mathematics. *Curriculum Focal Points for PreKindergarten through Grade 8 Mathematics: A Quest for Coherence*. National Council of Teachers of Mathematics, 2006.

[60] National Council of Teachers of Mathematics. *Navigating through Algebra in Grades 3–5*. National Council of Teachers of Mathematics, 2001.

[61] National Council of Teachers of Mathematics. *Navigating through Algebra in Prekindergarten–Grade 2*. National Council of Teachers of Mathematics, 2001.

[62] National Council of Teachers of Mathematics. *Navigating through Geometry in Grades 3–5*. National Council of Teachers of Mathematics, 2001.

[63] National Council of Teachers of Mathematics. *Navigating through Geometry in Prekindergarten–Grade 2*. National Council of Teachers of Mathematics, 2001.

[64] National Council of Teachers of Mathematics. *Principles and Standards for School Mathematics*. National Council of Teachers of Mathematics, 2000. See the Web site at *www.nctm.org*

[65] National Mathematics Advisory Panel. *Foundations for Success*. U.S. Department of Education, 2008.

[66] National Research Council. *Adding It Up: Helping Children Learn Mathematics*. J. Kilpatrick, J. Swafford, and B. Findell, eds. Mathematics Learning Study Committee, Center for Education, Division of Behavioral and Social Sciences and Education. National Academy Press, 2001.

[67] National Research Council. *How People Learn*. National Academy Press, 1999.

[68] National Research Council. *Mathematics Learning in Early Childhood: Paths Toward Excellence and Equity*. Committee on Early Childhood Mathematics, Christopher T. Cross, Taniesha A. Woods, and Heidi Schweingruber, eds. Center for Education, Division of Behavioral and Social Sciences and Education. The National Academies Press, 2009.

[69] Phares G. O'Daffer, Randall Charles, Thomas Cooney, John Dossey, and Jane Schielack. *Mathematics for Elementary School Teachers*. Addison-Wesley, 2004.

[70] A. M. O'Reilley. Understanding teaching/teaching for understanding. In Deborah Schifter, ed., *What's Happening in Math Class?*, vol. 2: Reconstructing Professional Identities, pages 65–73. Teachers College Press, 1996.

[71] Dav Pilkey. *Captain Underpants and the Attack of the Talking Toilets*. Scholastic, 1999.

[72] George Polya. *How To Solve It; A New Aspect of Mathematical Method*. Princeton University Press, 1988. Reissue.

[73] Susan Jo Russell and Karen Economopoulos. *Investigations in Number, Data, and Space*. Pearson Scott Foresman, 2008.

[74] Deborah Schifter. Reasoning about operations, early algebraic thinking in grades K–6. In *Developing Mathematical Reasoning in Grades K–12*, 1999 Yearbook. National Council of Teachers of Mathematics, 1999.

[75] Steven Schwartzman. *The Words of Mathematics*. The Mathematical Association of America, 1994.

[76] The Secretary's Commission on Achieving Necessary Skills. *What Work Requires of Schools, A SCANS Report for America 2000*. U.S. Department of Labor, 1991.

[77] Singapore Ministry of Education. EPB Pan Pacific, Panpac Education Private Limited. *The Singapore Model Method for Learning Mathematics*. Singapore: 2009.

[78] Singapore Curriculum Planning and Development Division, Ministry of Education. *Primary Mathematics*, vol. 1A–6B. Times Media Private Limited, Singapore, 3d ed., 2000. Available at *http://www.singaporemath.com*

[79] Singapore Curriculum Planning and Development Division, Ministry of Education. *Primary Mathematics Workbook*, vol. 1A–6B. Times Media Private Limited, Singapore, 3d ed., 2000. Available at *http://www.singaporemath.com*

[80] K. Stacey. Traveling the road to expertise: a longitudinal study of learning. In H. L. Chick and J. L. Vincent, eds. *Proceedings of the 29th Conference of the International Group for the Psychology of Mathematics Education*, vol. 1, pp. 19–36. PME, 2005.

[81] K. Stacey, S. Helme, S. Archer, and C. Condon. The effect of epistemic fidelity and accessibility on teaching with physical materials: A comparison of two models for teaching decimal numeration. *Educational Studies in Mathematics*, 47, pp. 199–221, 2001.

[82] K. Stacey, S. Helme, V. Steinle, A. Baturo, K. Irwin, and J. Bana. Preservice teachers' knowledge of difficulties in decimal numeration. *Journal of Mathematics Teacher Education*, 4(3), 205–225, 2001.

[83] K. Stacey, S. Helme, and V. Steinle. Confusions between decimals, fractions and negative numbers: A consequence of the mirror as a conceptual metaphor in three different ways. In M. van den Heuvel-Panhuizen, ed., *Proceedings of the 25th Conference of the International Group for the Psychology of Mathematics Education*, vol. 4, 217–224. PME, 2001.

[84] V. Steinle, K. Stacey, and D. Chambers. *Teaching and Learning about Decimals* [CD-ROM]: Department of Science and Mathematics Education, The University of Melbourne, 2002. Online sample at *http://extranet.edfac.unimelb.edu.au/DSME/decimals/*

[85] John Stillwell. *Mathematics and Its History*. Springer-Verlag, 1989.

[86] Dina Tirosh. Enhancing prospective teachers' knowledge of children's conceptions: The case of division of fractions. *Journal for Research in Mathematics Education*, 31(1):5–25, 2000.

[87] Ron Tzur. An integrated study of children's construction of improper fractions and the teacher's role in promoting learning. *Journal for Research in Mathematics Education*, 30(4):390–416, 1999.

[88] The University of Chicago School Mathematics Project. *Everyday Mathematics*. McGraw Hill, 2006.

[89] U.S. Department of Education. *Mathematics Equals Opportunity*. Technical report, U.S. Department of Education, 1997. Available online at *http://www.ed.gov/pubs/math/index.html*

Index

FIGURE A.1

Patterns for
shapes for
Exercise 1 on
page 563

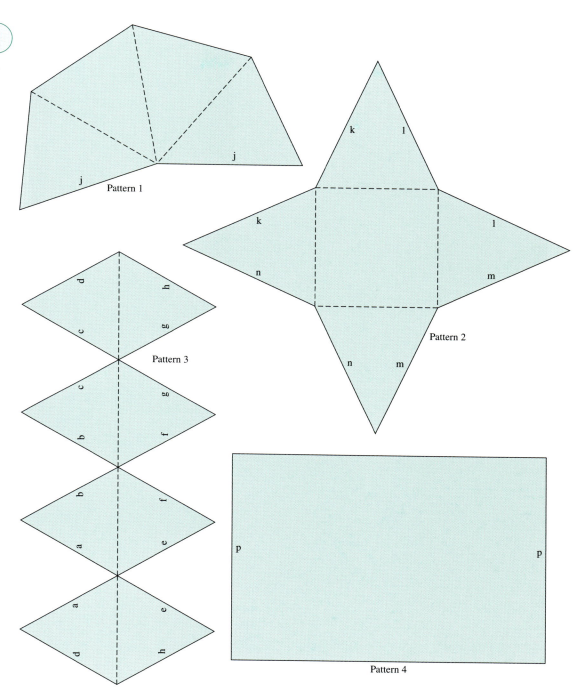

Pattern 1

Pattern 2

Pattern 3

Pattern 4

FIGURE A.2

Patterns for shapes for Exercise 2 on page 563

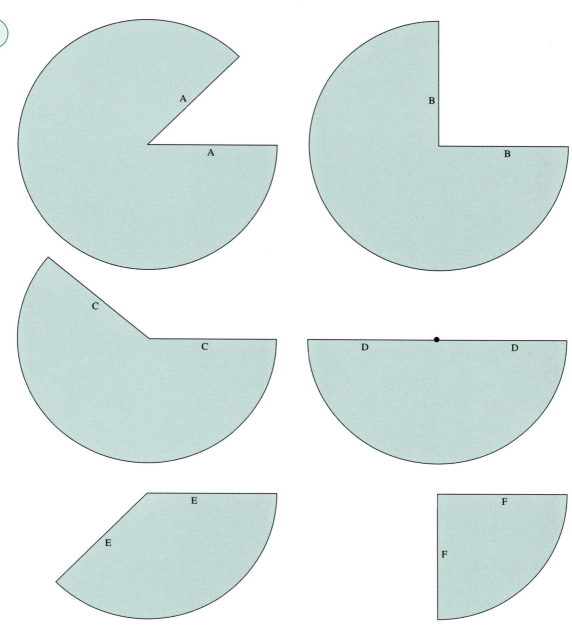

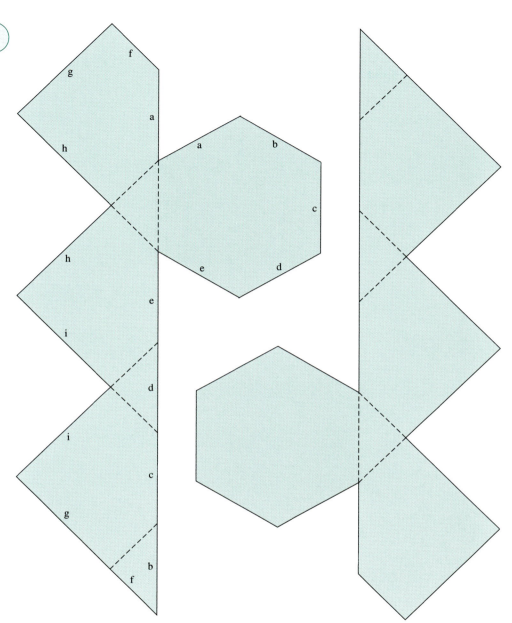

FIGURE A.3

Patterns for a sliced cube for Exercise 13 on page 567

FIGURE A.4

For Exercise 4
on page 563

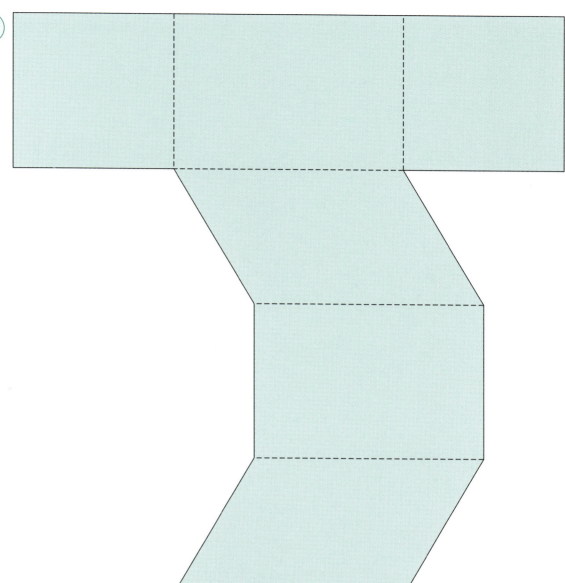

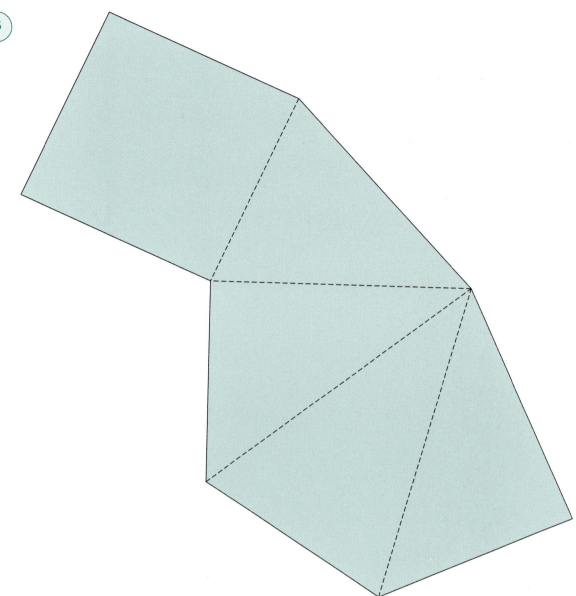

FIGURE A.5

For Exercise 4
on page 563

FIGURE A.6

For Problem 6
on page 561

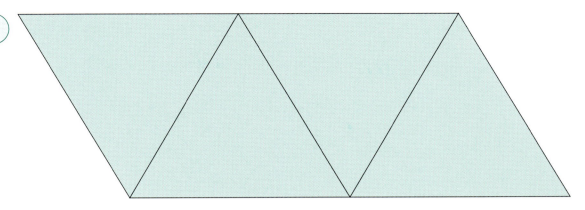

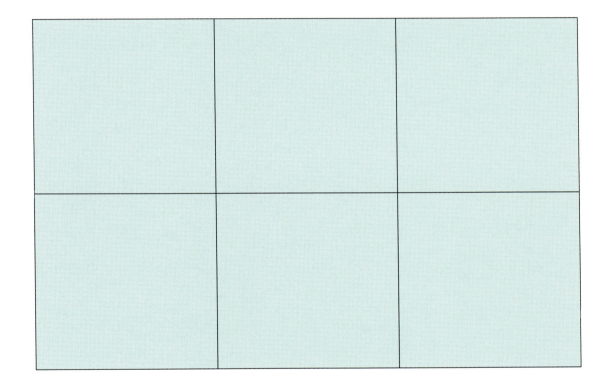

FIGURE A.7

For Problem 4
on page 560

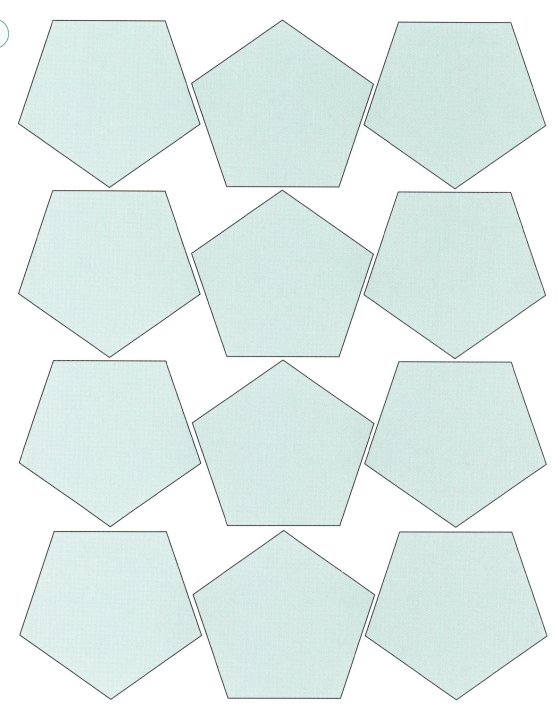

FIGURE A.8

For Problem 6
on page 561

FIGURE A.9

Triangles for
Problem 5 on
page 560

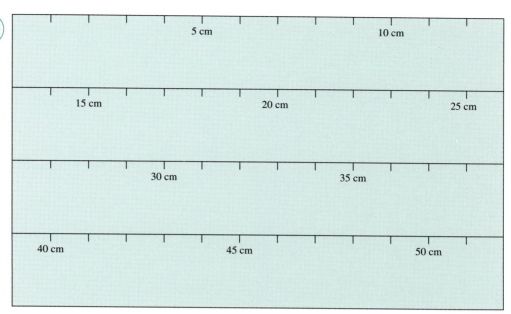

FIGURE A.10

A centimeter tape measure for Problem 20 on page 578

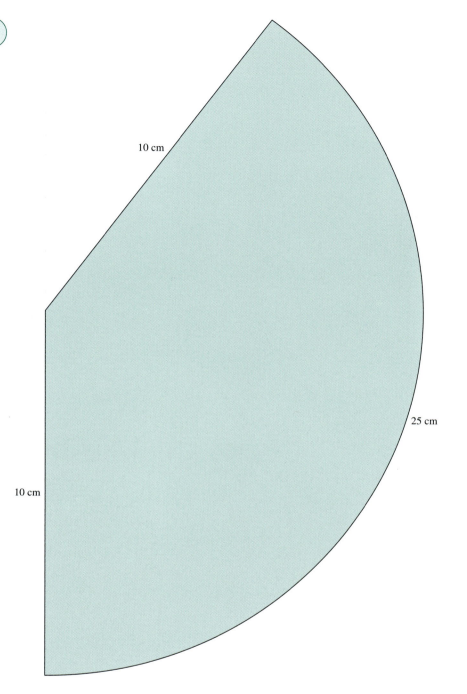

FIGURE A.11

A pattern for a
cup for
Problem 20 on
page 578

10 cm

25 cm

10 cm